# FRESHWATER FISH CULTURE

*— Volume 1 —*

**S.K. Sarkar**
Department of Zoology
Netaji Nagar College (Day), Kolkata

**2016**
**Daya Publishing House®**
*A Division of*
**Astral International Pvt. Ltd.**
**New Delhi – 110 002**

First Impression, 2011
Reprinted, 2016

ISBN 978-81-7035-262-4 (Hardbound)
ISBN 978-93-5124-062-4 (International Edition)



*Published by* : **Daya Publishing House®**
A Division of
**Astral International Pvt. Ltd.**
– ISO 9001:2008 Certified Company –
4760-61/23, Ansari Road, Darya Ganj
New Delhi-110 002
Ph. 011-43549197, 23278134
E-mail: info@astralint.com
Website: www.astralint.com

*Laser Typesetting* : **Classic Computer Services**, Delhi - 110 035

*Digitally Printed at* : **Replika Press Pvt. Ltd.**

PRINTED IN INDIA

*To*
*Samadrita (Rimi)*
*Priyangana (Pulash)*
*Tirna (Manik)*

*Devoted to my students*
*who are interested in the field*
*of fisheries science*

# Foreword

For the last twenty years, significant development in the field of freshwater fisheries and aquaculture has taken place in many countries of the world. Inspite of this, the inland capture fisheries are limited in their production potential. Inland water resources face a lot of developmental problems such as pollution of water and soil caused by excessive application of chemical fertilizers and pesticides, under-utilization of the existing waters, lack of appropriate knowledge of culture methods, lack of scientific management in fisheries sector and in fish culture systems, excessive growth of aquatic plants, and lack of adequate fund for rural upliftment.

For good health and vigor of human society, appropriate nourishment is essential. Fish is a highly nutritious food because fish contains a lot of nutrient substances in significant quantities. For this reason, fish is considered as a source of world's nutrition.

Fish production in Europe, North America, Near East and North Africa are moderate while in Sub-Saharan Africa and South Pacific regions, fish production is comparatively less than other parts of the world. Obviously, attention should be focused on the regions where fish production is moderate or less and where water resources are not being effectively utilized.

The need for the book on fishery science to cater the current information in a new style incorporating modern trends and recent thinking on the subject material has long been felt. This need has become more pronounced in the light of proposed work.

Published literature on freshwater fish culture and other related topics in the form of research articles, bulletins, and reports is scattered. In the book *Freshwater Fish Culture* the author has provided some selected references which ranges from simple bulletins through books and review articles to key scientific papers reporting original findings. These will serve as a bridge between the very extensive literature and the proposed book which every student of fishery science will want to explore. The author has tried to outline the essentials of fish culture and some other related topics, with as little elaboration as possible, so that the students and progressive fish farmers may have an opportunity to see the entire subject material in perspective.

I hope this book will be well received by the teachers, fishery students, fish culturists, and farm managers. I congratulate Dr. S. K. Sarkar for his work in the field of freshwater fish culture.

**Dr. G. N. Sarkar**

Principal
Netaji Nagar College (Day)
Calcutta - 700 092

# Preface

Freshwater fish culture is an important issue in many countries of the world, replenishing the gap between the demand and supply for fish and fishery products. It also helps to improve nutrition of human health, creating additional employment, and contributing to the economy for the benefit of the rural people.

Scientific and technological approaches are necessary to increase production of fish through aquaculture and also to reduce or make up the expected shortfall in food fish supplies. These approaches are, however, greatest in those countries and regions where population growth is rapidly expanding the gap between all potential sources of supply and demand for fish. Priority should be given to the advancement of aquatic production techniques, which are viable from a socio-economic and environmental points of view, to increase production in aquatic ecosystem which are not considered for culture. Of special importance is the management of culture-based fisheries and in open-water system.

At present, man has confronted with a lot of problems such as pollution of soil and water, population explosion etc. resulting in increasing food deficit and malnutrition, competition, proverty, unemployment, and depletion of natural resources. Both aquaculture and agriculture have come forward to solve these problems for human interest to large extent.

Water and soil are the chief essential ingredients in a fish culture ecosystem for high production. Effective management of these ingredients to provide the necessary background for better appreciation of fish culture, impact of pollutants on fish and aquatic ecosystems, recycling of organic wastes and wastewaters for fish culture, role of nutrient carriers in fish culture, management principles in fisheries sector, role of micro-organisms in the productivity of ecosystem etc. are becoming increasingly important for sustainable development of fish culture ecosystems and productivity generally. This book addresses some of these aspects and approaches pertaining to fish culture and fisheries which are necessary to understand how fish production may be improved which will paved the way for supplying proteins to rural people.

It should be remembered that it is also necessary to stimulate ideas whereby fish culture technology can be applied for increasing fish production. Moreover, fish production through fisheries and aquaculture is extremely variable in different countries of the world because production depends on the geographical and climatic conditions of any given region. Therefore, it is difficult to make a justifiable generalization on fish production potential.

Due to gradual decline in fish catches in major traditional fishing grounds in a number of developing countries in the world, fin-fish and shell-fish culture in confinement was seem as the best option to catching fish to feed the growing people and to provide them with alternative livelihood opportunities for their socio-economic upliftment.

Fish is essential in improving the diet in region where a great number of people are engaged in labour-intensive work but depend on low-protein diets. A cereal-based diet in combination with a small amount of fish protein would undoubtedly improve the quality of the diet and increase the nutritional quality of the cereal protein.

Development of fish culture through application of latest technologies (such as bio-technology, integrated farming system, computer application etc.) is the most visible face of fish production. Although India is blessed with a wealth of aquatic resources, awareness about their importance is still lacking; of course, some significant advancements have been made on this aspect. However, it is expected that bright days will come ahead for fish culture-related industries in support of production of fish and fishery products and the country will be ready for the next century – a year in advance.

The question over the future availability of productive aquatic resources places heavy responsibility in the Government and Non-Government organizations. With the adoption of experience gained over the years, these organizations could bring to bear their technological capabilities in increasing the yield efficiency and quality of products from aquatic resources for mass consumption. Development of aquaculture through successive stages has contributed enormously to projecting new strategies to make the nation into runaway successes.

The importance of micro-organisms in fish culture should not be overlooked. The ability of micro-organisms to recycle natural resources, to degrade the waste materials for fish culture, to decompose toxic substances such as ammonia, nitrate, pesticides, and herbicides to establish bio-geochemical cycles; to establish food chain and food web – these and other importance of micro-organisms are highly significant to fish production.

The subject material is presented in twenty two chapters. Each chapter has been divided into a number of sections. Each chapter of this book is introduced by a summary, followed by a series of main entries. All entries made are important and it is set apart for a particular purpose that it has been presented in a convincing and an attractive form. At the end of each chapter, conclusion, references and a series of self-study questions of several types have been incorporated.

The main aim of writing this book is to cater some informations on freshwater fish culture and other related topics for fishery students. The subject material in the book has been presented in a simple and clear manner, brief and to the point to enable the students to have a general idea on the subject to large extent. Elaborate discussions have, however, been avoided. It is expected that the teachers, students, and the progressive fish farmers can find current informations on the subject material.

The writing of a book on a specific subject as fish culture require considerable help from several colleagues. Among these, the author wishes to acknowledge the following persons : R. N. Das, WBCADC, Calcutta; A. Pramanik, RBC College, Naihati; P. Basu Chowdhury, Netaji Nagar Day College, Calcutta; Arvind Kumar, S.K. University, Dumka; and U. K. De, Bongoan College. The most significant has been the contribution of Sankhajit Jana, Artist, who spend a lot of energy and time to draw most of the diagrams in the book.

As expansion of fish culture is necessary for food requirement to ever-increasing populations and productive employment, it became apparent that the availability of current informations and the wish for embracing reasonable well important facets of the treatise "Freshwater Fish Culture" made increase into two separate volumes necessary. The first volume covers the general considerations of freshwater fish culture ecosystems, importance of water and soil for fish culture, use of nutrient carriers, natural food and artificial feed in fish culture, role of nutrients and their dynamics, inland fisheries activities, recycling of wastes and wastewaters, carp and composite fish culture, management strategies in fish culture, pollution, and world's fish supply as food. The second general organizations of fish, history and development of fish culture, culture of carnivorous and exotic fishes, fish culture in rice fields, fisheries in mountain regions and in reservoirs, riverine fisheries, transport of fish and fish seed, fish diseases and health management, recirculatory systems and genetics in fish culture, toxicity and ecotoxicology, role of nutrients in fish production, and fish culture in relation to public health.

The author is grateful to the Editor and other technical staffs at Daya Publishing House, Delhi, for their cooperation and assistance in the task of preparing and publishing the book. Suggestions to improve the book will be gratefully accepted.

**S. K. Sarkar**

# Contents

# 1

# Water and Soil Around Fish

Fish is entirely dependent upon healthy and unpolluted aquatic environment for his continued survival, growth and reproduction. Fish culture ecosystem is basically composed of water and soil which are considered as dynamic natural bodies and composed of organic materials, minerals, and living forms. These natural bodies are very important for development of aquatic ecosystem that connotes the space within which production of fish and other aquatic forms of commercially important as well as their activities take place. Water and soil are, therefore, considered as the most encompassing substances on this planet. Though a few quantities of freshwater areas in the world are suitable for commerical fish cultivation, these waters produce fish and fish products for mankind.

Fish and other aquatic animals have their specific surroundings, medium or environment to which they interact ceaselessly on and remain fully adapted. The environment is a collective term that encompasses all the conditions in which aquatic and terrestrial animals live. Therefore, the environment of fish can be divided into abiotic and biotic ones. Abiotic environment includes two principal media which also act like factors such as lithosphere (soil) and hydrosphere (water). The lithosphere is the solid portion of the crust and part of the upper mantle having a thickness of 8-30 and 65-80 kilometres, respectively. On the contrary, hydrosphere is the water of the earth as distinguished from the rocks, living things, and the air. It includes the waters of the snow, ice, and glaciers above the soil surface and water below the soil surface.

People always favour good soil which are the natural bodies of the earth. This is due to the fact that good and fertile soils helped to construct flourishing civilizations for humankind. Soils have several meanings for humankind. Excellent soils are considered as the starting point for successful fish culture. Water and soil are also used to consume wastes from sewage, industrial, municipal, and animal sources. Water and soil are as important to fish farmers as to city dwellers. Since good and fertile soils are generally considered as the most productive, fish farmers always try to cultivate fish in such ponds. Therefore, it is necessary for a fish farmer to gather a brief knowledge about the soil where a potential fish farm has to be constructed.

It is worth remembering that water is ubiquitous and is a basic need of plants and animals. Water is often regarded as a blend, inert liquid, a space filler in living organisms and the earth. Without water there could be no life of any kind on the earth. It is very clear to establish the degree of importance of water to fish and other aquatic forms. Some tend to measure the progress of aquaculture by the amount and the quality of water it requires. Aquaculture progress will be jeopardized if water is not available in adequate quantity and quality.

## 1.1 Types of Water to be Used for Fish Culture

Suppose one is asked what types of water he would like to use for fish culture. He will say that he would prefer to use good quality of water. But question may arise from fish farmers whether the water is pure or not. It should always be mentioned that good quality of water does not mean that it is pure. For optimum health and survival of fish in ponds, it must contain certain dissolved substances; of course, within the tolerance limits. Besides dissolved substances, water must contain microscopic plankton organisms in suspended conditions.

## 1.2 Reasons for Fish Culture

A fish may be defined as a vertebrate adapted for a purely aquatic life, propelling and balancing itself by means of fins, and obtaining oxygen from the water for breathing purposes by means of gills. Fishes, thus defined, were formerly regarded as representing a single class of the great sub-kingdom of vertebrates; but a more thorough knowledge of their anatomy and evolutionary history has led to a different conclusion. A history of fishes, however, depicts that the practice of fish farming is older than agriculture. Although agricultural activities have been dramatically expanded throughout the world, great development of fish farming systems through adoption of biotechnology has drawn attention of the humankind for the following reasons :

1. Unlike major agricultural crops, fishes do not consume water but utilize only dissolved oxygen from it.
2. In contrast to poultry birds, fishes have the highest fecundity.
3. Catching of fish from cultured waters can be ascertained than from vast natural waters.
4. Because less energy is required to metabolize proteins, fish can synthesize more protein per calory of energy consumed than livestock and poultry.
5. Food conversion ratio of fish is much less (1.6 : 1). It means that 1.6 kg of food is required to yield 1 kg of fish. In case of poultry, it is 2.2 : 1.
6. Wastewater and other organic wastes can be recycled through aquaculture. This is a subject to considerable importance in recent years and has been discussed in chapter 16.
7. Cost of fish production per unit area is much less than that of poultry.

## 1.3 Quality of Water

Quality of water is essential to understand whether a particular water body is suitable for fish culture operation or not. Since survival of different species of fish in different geographical areas of the world depends on different qualities of water which is extremely variable under different climatic conditions, every country has prescribed its own water quality standards for fish culture. Although the standards of water quality laid down by the Environmental Protection Agency (EPA), USA, are usually considered to be most rigorous, all countries may not accept such stringent limits. This is due to the fact that some species of fish which live in temperate regions may not live in tropical ones due to variability in water qualities. For example, in some temperate countries, the range of alkalinity of water in fish culture ponds has been recommended as 20-50 mg/l whereas in tropical countries, the range varied between 70 and 150 mg/l. Consequently, all countries may not accept such stringent limits as suggested by the EPA.

Physical standards of water in fish ponds are characterized by the turbidity, color, taste, and odour. Turbidity of water is mainly due to the presence of suspended matter such as clay, silt, finely divided inorganic and organic matter, and plankton. Taste and odour of water indicate by the existance of fish and should be nothing.

Chemical standards are mainly due to the presence of several types of elements at desired levels such as calcium, hardness, iron, magnesium, zinc, chloride, nitrates, selenium, arsenic, etc. Permissible levels of different elements in pond water are shown in Table 1.1.

**Table 1.1 : Permissible Levels of Various Types of Elements in a Pond Water. Except pH, all Values are in Terms of mg/l**

| Elements | Permissible Levels | Elements | Permissible Levels |
|---|---|---|---|
| pH | 7.5-8.5 | Sulphate | 170 |
| Total solids | 500 | Fluoride | 0.2 |
| Total hardness (as $CaCO_3$) | 150-200 | Nitrate | 0.07-0.8 |
| Calcium | 70 | Ammonia | 0.3-0.8 |
| Magnesium | 40 | Arsenic | 0.02 |
| Iron | 0.2 | Chromium | 0.01 |
| Manganese | 0.1 | Lead | 0.01 |
| Copper | 0.7 | Selenium | 0.02 |
| Zinc | 3 | Phosphate | 0.09-0.8 |
| Chloride | 150 | | |

Compiled by the author from different published literature

## 1.4 Forms of Natural Waters

Water is the first thing to be considered for the construction of a fish farm. If a farm is constructed, the place would have been selected according to its demand on water supply. In natural environment, however, various types of water are found such as acid and alkaline, cold and warm, polluted and unpolluted, soft and hard, super-saturated and sub-saturated, running (slow and rapid) and stagnant, surface and groundwater, sea water, mineral water, and rain water.

1. *Sea Water :* It is a hard water and contains large amounts of dissolve impurities (about 2.5 per cent of sodium chloride and 1 per cent other soluble salts).
2. *River Water :* Rivers receive water from rain or melting of ice on mountain peaks. During its course, river water collects impurities such as clay and soluble salts of sodium, potassium, magnesium, calcium, and iron. The presence of last three salts makes the water hard. It is fit for drinking, irrigation, and fisheries provided that it is not polluted.
3. *Well and Spring Waters :* These waters are free from suspended impurities owing to their filtration through the porous strata of the earth and also contain high amounts of dissolved salts than those present in river waters. The water is hard and used for drinking, irrigation, and fisheries.
4. *Rain and Tap Waters :* Rain water contains some dissolved gases of the atmosphere such as nitrogen, oxygen and carbon dioxide. It also contains traces of nitric and ammonium nitrate which are produced in the atmosphere during thunderstorm as well as traces of sulphurus and sulphuric acids. The rain water is very soft.

Tap water is, in many cases, hard and the total alkalinity varies between 160 and 250 mg/l as calcium carbonate. Since this water comes from underground level, a number of soluble salts such as calcium, magnesium, iron, potassium, chlorine etc. are present in appreciable quantities.

For fish culture in ponds and lakes, tap or rain water may be used. The rain water (though it may be acidic to a lesser extent) when introduced in fish culture ponds, should be treated with agricultural limestone. Tap water from city waterworks is used in ponds both for fish culture and hatchery with considerable caution. Water coming from city waterworks, contain some undesirable substances which needs treatment with chlorine at the 0.3 mg/l rate. In rural communities, the tap water is pure enough, as it comes from artesian wells and hence treatment is not necessary.

Although carp and goldfish are very tolerance to chlorine at the level of 0.3 mg/l, other species of fish are susceptible. Continual presence of the chlorine in flowing waters makes it unfit for fish and other aquatic life. In properly balanced stagnant water bodies, however, there is no harmful effect because the chlorine will rapidly disappear.

In ponds and tanks where tap water is frequently used, it is advisable to remove chlorine with sodium bisulphite or sodium nitrate at the 1 mg/l rate. Of course, sodium bisulphite is by far the best because it is cheapest and the most efficient.

5. *Acid and Alkaline Waters :* All physico-chemical reactions in any aquatic environment dramatically affect the aquatic life and are sensitive to all variations of pH. The pH value extends from zero to 7 is considered as acidic while 7 to 14 is considered as alkaline. For better survival and growth of fish, the pH value 7.0-8.5 is essential. In acid water (pH 6 and less than 6) fish are succumbed to various diseases.

6. *Polluted and Unpolluted Waters :* In nature, water is present in three forms such as atmospheric moisture, precipitation, and soil water. The precipitation or rainfall is the main source of soil moisture. Generally oceans, lakes, streams, rivers, springs, and ponds are the chief sources of water. In nature, however, most of the water bodies become polluted by adding different toxic substances to natural waters through human interference that makes the water less or not suitable for fish culture. Spring water is mostly unpolluted and may be used for fish culture in altitudes. Water is polluted by sewage and industrial wastes. Humans have always dumped sewage in different freshwater bodies where it would be carried away and gradually disappear. Industrial plants release their organic and inorganic wastes and contaminate the aquatic environment to greater or lesser extent and poison of aquatic flora and fauna. Wastes and effluents of various industries such as fertilizers, sugar, textiles, paper and pulp, rubber, distillaries, chemical works, soda and chlorine cause pollution of natural waters. The wastes of coal mining produce acids in water due to the presence of sulfur that sometimes lower the pH of streams to as low as one. Pesticides and herbicides also percolate through soil and get dissolved in soil water. Consequently, groundwater table becomes highly polluted.

## 1.5 Distribution of Water on Earth

Water is not uniformly distributed throughout the earth (Table 1.2). About 95 per cent of the total water on earth is bound with rocks and does not cycle. Of the remaining 5 per cent, about 97 per cent is in the ocean, 2 per cent exists as ice and glaciers and about 1 per cent is freshwater present in groundwater, soil water, and inland surface water.

**Table 1.2 : Distribution of Water on the Earth. Values are in Terms of Kilogram × $10^{17}$**

| | |
|---|---|
| **A. Chemically bound water with rocks (Does not cycle)** | |
| 1. Sedimentary rocks | 2,100 |
| 2. Crystalline rocks | 250,000 |
| **B. Free water (Undergo hydrolic cycle)** | |
| 1. Oceans | 13,200 |
| 2. Ice cap and glaciers | 292 |
| 3. Ground water to a depth of 4,000 metres | 83.5 |
| 4. Freshwater lakes/ponds | 125 |
| 5. Inland seas and saline lakes | 1.04 |
| 6. Soil moisture | 0.67 |
| 7. Water vapour | 0.13 |
| 8. Rivers | 0.013 |

*Source* : Clapham, Jr. (1973).

## 1.6 Concept of Water

For the welfare of mankind, it is necessary to know the concepts as to what waters are. For example, an agricultural farmer will consider the water as a main source for survival and growth of plants. A fish farmer always try to maintain good quality and productive water for survival and growth of commercially important species of fish (Fig. 1.1). While a homeowner uses water for washing, bathing, and cocking purposes, water is a nuisance substance to a mining engineer and always try to remove from the coal mine.

The main reasons for studying waters is to obain a general concept as to how they should be used. Such a concept is very important to understand how water can serve the agriculturist, the aquaculturist, and the mining engineer.

## 1.7 Three Approaches – Eutrophic, Oligotrophic, and Dystrophic

Quality of water suggests that on the basis of chemical quality, three concepts of fish culture ecosystem may be considered. In general, water may be treated as (1) a natural entity – a chemically synthesized product of nature, (2) a natural habitat for aquatic life including fish, and (3) a natural wealth for bumper fish crop. Natural habitat and bumper fish crop in any aquatic ecosystem illustrate the three approaches that is generally used in considering the productivity of water in ponds and lakes – that of the *eutrophic, oligotrophic,* and *dystrophic.*

Eutrophic waters are productive, usually broad and shallow, whose inputs comes from base-rich soils. They have a high content of dissolved ions, essential elements, and organic contents and hence these water characteristics are very favourable for fish survival and growth.

**Fig. 1.1 : Productive water and soil mean bumper fish. Progressive fish farmers are always interested to produce healthy fish and are aware of the importance of productive aquatic ecosystem.**

In contrast, oligotrophic waters are derived from hard, base-poor rocks, contains only low levels of nutrients and hence unproductive, often deep, very transparent as there is little microscopic fish food organisms or suspended matter. While production in oligotrophic water is low, the number, growth and abundance of plants and animals are also low.

Dystrophic waters are characteristic of acid soils such as heathlands or peat moors and are distinguished from other water bodies by the color. The color of the water is brown owing to the presence of humic acids and undecomposed plant materials. Many ions and nutrient elements are lacking, and conditions for fish life are poor.

In addition to these widely represented types, there are other types of water areas with more extreme conditions.

In general, the classification, description and properties of water in natural environment are considered in aquatic study. It does not focus on the practical use of water. But in aquaculture study, water in relation to production of plants and animals of economically important species are considered. To produce more food from aquatic sources, it is necessary to determine the reasons for variations in the productivity of waters and find means of improving and managing productivity.

## 1.8 Practical View of Water

Light determine the productivity of aquatic environment. The upper 1-2 metres of the water column is very productive. Microscopic fish-food organisms concentrated in the upper surface of water than that of middle or lower surfaces where light penetration is not intense and hence unproductive. This condition does not create problems during fish culture operation in case of shallow waters where light is penetrated even at the botom of water. Productivity of water is the result of synthetic force which acts on aquatic environment due to interactions of several abiotic factors of soil, air, and water. Synthetic forces are the combination of light, carbon dioxide, and water that together triggers the production of food organisms owing to the photosynthetic activity.

Characteristics of water vary from region to region. For example, the water in tropical and sub-tropical zones are highly productive than that of temperate and arctic ones. Aquaculturists have examined the variations of water quality located in different geographical areas that recognize a number of different types of water bodies, having distinguishing characteristic features. However, the water is a collective term for different types of water as vegetation is used to designate all plants.

## 1.9 Water-Soil Relationship

Several concepts emphasize the importance of water-soil relationship in relation to ecosystem productivity.

1. A large number of inorganic and organic elements are present in water and soil in dissolved condition.
2. Different types of bacteria present in soils help to decompose dead and decaying matter and thereby releasing inorganic nutrients into water – essential for fish growth.
3. When the concentrations of essential nutrients in water decreased, nutrients are released from soil slowly to replenish those removed from water for fish and fish food organisms. Therefore, the water-soil relationship occurs as a dynamic equilibrium state.
4. Fertilization programs in fish ponds also add nutrients to soil which are preserved for future utilization. Such exchanges of elements are dependent on soil particles and water and interactions of abiotic as well as biotic factors of soil and water.

### *Water Solution*

As noted earlier, good quality of water should be recommended for successful fish cultivation. Good quality of water means that it will not contain any undesirable and toxic substances in dissolved state. Furthermore, water has high solvent power. No other liquid except water dissolve so many different kinds of solutes. It is a general solvent of all essential nutrient elements necessary for aquatic life. Besides elements, most of the organic compounds (carbohydrates and proteins) are more soluble in water. Water solution usually refers to a liquid phase consisting of small but significant quantities of all essential nutrient elements and organic compounds released either from the surface of pond soil or from soluble materials or from artificial feeds.

### *Soil Solution*

The term soil solution refers to as the aqueous liquid phase of the soil and its solutes consisting of ions dissociated from the surfaces of the soil particles and of other soluble materials. It contains small but significant quantities of soluble inorganic and organic compounds. Fine organic and inorganic particles and soil solids release these elements to the soil solution from which they are released to the overlying water.

### *Acidity and Alkalinity*

The most important characteristic feature of water and soil is its acidity or alkalinity. As regards the nature of water and soil, some are neutral, some are acidic, and some basic. In an aquatic ecosystem, many chemical and biological reactions take place which are dependent on the levels of hydrogen ion ($H^+$) and hydroxyl ion ($OH^-$) in the soil and water. These ions influence the activity of micro-organisms and, in turn, the availability of several nutrients to fish food organisms.

A solution, whether it is soil or water, containing unequal number of $H^+$ and $OH^-$ is called either *acidic* or *alkaline solution*, respectively. But pure water contains equal number of $H^+$ and $OH^-$, is called *neutral solution* and can be expressed as follows :

$$[H^+] = [OH^-] = 10^{-7} \text{ gram equivalent/litre}$$

If there is more $H^+$ ion concentration than $OH^-$ ion concentration in water, it is said to be *acidic* and if there is more $OH^-$ ion concentration than $H^+$ ion concentration in water, it is said to *alkaline*. For example, if $[H^+]$ of water is $10^{-4}$, then $[OH^-]$ is equal to $10^{-4}$ since $kw = 10^{-4}$ (kw is the ionic product for water). Similarly, if $[H^+]$ of water is $10^{-8}$, then $[OH^-]$ is equal to $10^{-6}$ and the water is alkaline. Therefore, water having $[H^+]$ greater than $10^{-7}$ gm equivalent per litre are acid and that of water having $[H^+]$ lesser than $10^{-7}$ gm equivalent per litre are alkaline.

The concentration (activity) of $H^+$ and $OH^-$ ions in soil and water is ascertained by determining its pH. Technically the pH is the negative logarithm of the concentration of $H^+$ ($-\log H^+$) in the medium. Thus, each unit change in pH represents a tenfold change in the concentration of $H^+$ and $OH^-$ ions. A simple relationship between the concentration of $H^+$ and $OH^-$ ions and pH is shown in Fig. 1.2. During fish culture operation, the pH of water should be studied carefully along with Fig. 1.3, which shows the ranges in pH generally considered in different waters. The pH of the water should be considered as highly significant when aquaculture program is undertaken on scientific basis.

**Acidic pH = 6.0** **Neutral pH = 7.0** **Alkaline pH = 8.0**

**Fig. 1.2 : Diagramatic representation of acidity, alkalinity, and neutrality. At pH 6.0, the $H^+$ ions are dominant, being 10 times greater whereas the $OH^-$ ions have decreased proportionately. The sample is, therefore, acid at pH 6.0, these being 100 times more $H^+$ ions than $OH^-$ ions present. At pH 8.0, the $OH^-$ ions are 100 times more than $H^+$ ions. Therefore, the pH 8.0 is alkaline. At pH 7.0, the $H^+$ and $OH^-$ ions of a samples are balanced, and their numbers are same.**

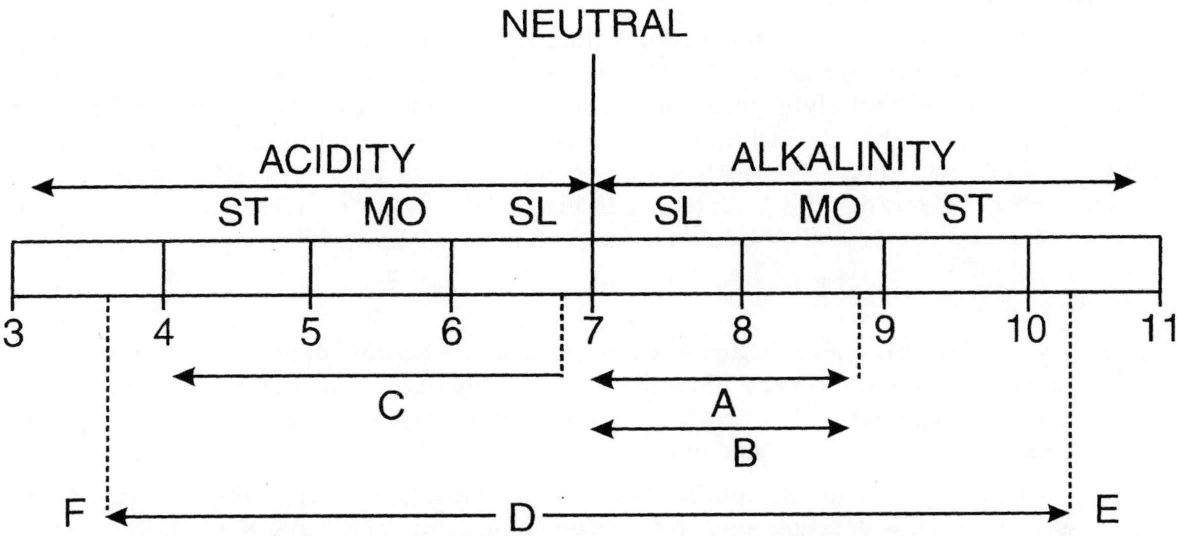


**Fig. 1.3 : Extreme range in pH of water and soil for most fish culture ecosystems. Slight to moderate alkalinity in pond water is considered for fish culture. ST, Strong; MO, Moderate; SL, Slight; (A) pH range ideal for fish culture; (B) Range in pH common for arid region mineral soils; (C) pH range not ideal for fish culture; (D) Extreme range in pH for most mineral soils; (E) Attained only by alkali mineral soils, (F) Extreme pH for acid soils.**

The pH value of water controls the concentration and availability of other minerals. For example, certain plankton species require considerable amounts of calcium and phosphorus and thus grow well on alkaline waters. At low pH, copper, iron, and zinc are generally toxic to aquatic life. Highly acidic or alkaline waters are detrimental to micro-organisms, fish, and fish food organisms. Neutral or slightly alkaline are, however, the best for their growth.

## 1.10 Relationship Between Water and Fish

The degree of absorption of essential nutrient elements usually depends on the concentration of elements present in water. As soon as fish food organisms (plankton and bottom fauna) are developed in a pond ecosystem through manuring and fertilization, nutrients are absorbed by bacteria. These bacteria are consumed by fish food organisms. Ultimately, fish consumes these food organisms.

Generally phytoplankton absorb nutrient elements in an active way. Certain carriers in phytoplankton can actively transport nutrients from water to the body. Micro-organisms coupled with toxicants stimulate or inhibit nutrient element uptake. Therefore, a balanced relationship among micro-organisms, fish food organisms, and physico-chemical properties of water and soil undoubtedly ensure the effective use of nutrient elements for sustaining fish production.

Relationship between fish growth and concentration of natural food organisms in water is very significant and such relationship is directly controlled by the availability of nutrients in water. As fish consumes relatively sufficient quantities of food organisms from water, the level of nutrients gradually become less than is needed for fish yield. As a result, nutrient elements in water and soil must be constantly replenished from either soil organic and inorganic matters or from manures and fertilizers.

## 1.11 Importance of Water

Water is an important constituent of the aquatic ecosystem because it is considered as a medium for various chemical reactions. Its solvent properties, capacity for heat regulation, hydrogen ion concentration and electrolytic properties clearly indicate the significance of water. Different physico-chemical properties of water make it as the best solvent adapted to aquatic life. It is a highly reactive substance with unusual properties, and it is very different, both physically and chemically, from most other common liquids. Water takes part in a variety of reaction in any aquatic ecosystem. It reacts with metals, non-metals, oxides of metals and non-metals as well as organic compounds.

### *Properties of Water*

1. *Solvent :* Water dissolves or disperse many substances because of its dipolar nature. Many crystalline salts and other ionic compounds readily dissolve in water. Their solubility is due to strong electrolytic attraction between water dipoles and the ions of compounds to form the very stable hydrated ions.

   Non-ionic but polar compounds such as sugars, urea etc. which is due to the propensity of water molecules to hydrogen bond with polar functional groups (such as OH groups of sugars). Solvent properties are very important because different chemical reactions in aquatic ecosystem take place in ceaseless motion.

2. *Absorption of Water :* Sulphuric acid, phosphorus pentaoxide, fused calcium chloride, calcium oxide, anhydrous sodium sulphate, dehydrated silica gel etc. has the capacity to absorb certain amount of water. They are known as desiccating agents. Although such absorption of water by desiccating agents is often a physical process in certain cases, this is attended with chemical changes. Thus, phosphorus pentaoxide and calcium oxide absorb water and produce phosphoric acid and calcium hydroxide, respectively.

$$P_2O_5 + 3H_2O \longrightarrow 2H_2PO_4$$

$$CaO + H_2O \longrightarrow Ca(OH)_2$$

3. *Electrolytic Properties :* Water is dissociated and form two types of ions ($OH^-$ and $H^+$ ions) and therefore it is, to some extent, an amphoteric electrolyte.

4. *Dielectric Constant :* Dielectric constant is high. The effect of solvent on the electrolytic dissociation of the solute is referred to as *dielectric constant* of that solvent. Water not only dissociate itself but also it has a specially high dissociating effect on its solute.

5. *Heat Capacity :* Water has high heat capacity. More heat is needed to raise the temperature of a given mass of water. On the contrary, during cooling water gives out more heat per unit of temperature fall. This property of water acts as *buffer* against sudden and extreme changes of temperature that act as protecting device on aquatic animals. The most general way in which the heat capacity of water is important to life is the tendency of seas, lakes, and ponds to prevent any great change in temperature.

6. *Latent Heat of Water :* Seas and oceans absorb huge amount of heat than that of lakes or ponds without any change in temperature. This is partly due to high heat capacity of water but largely due to high latent heat of evaporation of water. Water evaporates at all temperatures and it always absorbs heat which is released again when water condences into ice. Seas and oceans are thus act as heat distributing agents.

7. *Polarity* : The property of polarity helps to explain how water molecules interact with each other. The hydrogen (positive) end of one molecule attracts oxygen (negative) end of another, resulting in a chain-like grouping. The polarity of water accounts for several reactions so important in aquatic science. For example, polarity of water molecules encourages the dissolution of salts in water since components have greater attraction for water molecules than for each other.
8. *Cohesion Versus Adhesion* : Hydrogen bonding (hydrogen atoms act as links between water molecules) accounts for two basic forces responsible for water retention in an aquatic ecosystem : the attraction of water molecules for each other (cohesion) and the attraction of water molecules for soil surfaces (adhesion). By adhesion, some water molecules are held at the soil surfaces. In turn, these bound water molecules further removed from the soil surfaces. The forces of adhesion and cohesion make it possible for soil to retain water.
9. *Orientation of Water Molecules* : The molecules of water are polar and the charges of water are evenly distributed (i.e. $H_2O^+$ and $H_2O^-$). Molecules are attracted either to positive or negative ions on other charged surfaces and tend to orient on themselves. This can modify the behavior of ion on charged surfaces in water.
10. *Aggregation of Water Molecules* : Water molecules aggregate with each other. Previously it was thought that water contains two molecules of water (either dihydrol or trihydrol). But recent view emphasize the probability of a large aggregation of molecules although molecular aggregation of water depends on various environmental conditions such as pH, alkalinity, and oxygen.
11. *Ion Product of Water – The pH Scale* : The dissociation of water is an equilibrium process :

$$H_2O \rightleftharpoons H^+ + OH^-$$

for which we can write the equilibrium constant

$$K_{eq} = \frac{[H^+]\ [OH^-]}{[H_2O]}$$

where the brackets indicate concentrations in moles per litre. The magnitude of this equilibrium constant at any given temperature can be calculated from conductivity measurements on pure water. Since, the concentration of pure water is high (it is equal to the number of grams of water in a litre divided by the gram mol. wt., or 1,000/18=55.5M) and the concentration of $H^+$ and $OH^-$ ions very low ($1 \times 10^{-7}$M at 25°C), the molar concentration of water is not changed by its very slight ionization. The equilibrium constant expression may thus be simplified to

$$55.5 \times K_{eq} = [H^+]\ [OH^-]$$

and the term $55.5 \times K_{eq}$ can then be replaced by a constant Kw, called the *ion product* of water,

$$Kw = [H^+]\ [OH^-]$$

The value of Kw at 25°C is $1.0 \times 10^{-4}$

Kw is the basis for the pH scale, a means of designating the actual concentration of $H^+$ (and thus of $OH^-$) ions in any aqueous solution in the acidity range between

1.0 M $H^+$ and 1.0 M $OH^-$. The pH is defined as the negative logarithm of hydrogen ion concentration.

$$pH = \log_{10} \frac{1}{[H^+]} = -\log_{10} [H^+]$$

In a precisely neutral solution at 25°C

$$[H^+] = [OH^-] = 1.0 \times 10^{-7} M$$

The pH of such a solution is

$$pH = \log_{10} \frac{1}{1 \times 10^{-7}} = 7.0$$

### *Chief Constituents*

It is needless to mention that fish culturists should know about the potentialities of water. Oxygen and carbon dioxide are the chief gaseous constituents of water present in dissovled state. Besides these two gases, organic elements such as ammonia, chlorides and nitrates are also present. Oxygen and carbon dioxide in water are obtained from aquatic plants and animals through their respiration. The concentrations of oxygen and carbon dioxide in water increased through photosynthetic process of plants and respiration in animals, respectively. Oxygen and carbon dioxide gases are very dissimilar. Carbon dioxide is 30 per cent heavier, 15 per cent less diffuse, and 40 per cent more soluble than oxygen.

There are two schools of thought among different research works on water problems. According to one school, oxygen should be considered as the most important element in water for aquatic life, while the other thought suggest that oxygen is a secondary consideration and carbon dioxide is considered as the index to the degrees of profit and suitability of water.

## 1.12 The Fitness of the Aquatic Environment for Living Organisms

Living organisms have effectively adapted to their aqueous environment and have been evolved means of exploiting the unusual properties of water. The high specific heat of water is useful to the cell because cell water act as a "heat buffer" allowing the temperature of the cell to remain relatively constain as the temperature of the environment fluctuates. Moreover, the high heat of evaporation of water is exploited as an effective means for vertebrates to lose heat by evaporation of sweat. The high degree of internal cohesion of liquid water, due to hydrogen bonding, is exploited by higher plants to transport dissolved nutrients from the roots to the leaves. Ice has a lower density than liquid water and therefore floats has important biological consequences in the ecology of aquatic organisms.

## 1.13 Water – As Profit and Recreation

In ancient times when man had established intimate relationship with fish in their daily life, they began to utilize fish as food and to cultivate for recreation. Subsequently, fish cultivation was considered as a profitable business. So far as recreational aspect is considered, it can be said that it includes the stocking of a pond or lake to provide fish as food or to catch fish with an angler or to beautify ones garden with a pool and tank containing fish, lotus or lily plants, and a fountain.

The present era is a golden one for the culture of mullet, goldfish, eel, carp, shrimp, lobster, and oyster in freshwater and brackish water ponds. Using sophisticated techniques, it has paved the way for the farmers to produce healthy fish in ponds and tanks, sale the fish to the market for higher returns and again restocked the ponds to maintain a continuous yield and supply to the market throughout the year.

The very latest branch for aquaculture is the carp, tilapia, shrimp, and prawn. This branch of fish culture is undoubtedly a most profitable business although environmental degradation particularly in coastal and fish processing industrial regions are not uncommon. However, these delicious food items are being introduced into freshwater and brackish water farming systems in many countries of the world where suitable conditions for their culture prevail. The successful results mean a boom in cultivation of carp and shellfish. Due to its lucrative taste and high protein content, there will be no marketing problem. In this book, the main viewpoint will be that of the profit aspect, although profit as well as recreational aspects are closely related in many cases and therefore, these two aspects cannot be fully separated and should be monitored from a common platform.

## 1.14 Water – A Rich Source of Food

The world has vast and varied types of water bodies such as rivers, reservoirs, lakes, lagoons, ponds, swamps, estuaries, backwaters, brackish waters, water-logged areas, oceans and seas. These offer immense potentialities of fisheries and aquaculture development. Since the evolution of mankind, some aquatic animals (shrimp, fish, mussel, and oyster) and plants (sea weeds and phytoplankton) form an important part of human diet. Different methods have been developed from time to time to collect all sorts of vast aquatic food. The world fish production is estimated at 97.2 million tonnes contributed by over 50 countries. However, the production potential of aquatic food items is tremendous and there is still vast opportunity to harvest food items through use of sophisticated technology with utmost care. For examples, indiscriminate use of technology by unskilled personnel without considering the detrimental impact resulted in collapse of shrimp farming industries in Indonesia, Taiwan, Phillippines, and Bangladesh. In contrast, lack of skilled personnel and adequate infrastructure has delayed to adapt the latest techniques in several under developed countries. However, the production potential of aquatic resources in India can be estimated as shown in Table 1.3. Table shows that even if about 50 per cent of these areas are possible to bring under cultivation, prawn and fish production will be increased at a gallopin rate.

**Table 1.3 : Aquatic Resources of India and Their Potential Fish Production**

| Resources | Area* (in Million/ha) | Estimated Production potential per unit (Tonne/ha) | Total estimated production potential (in Million Tonne) |
|---|---|---|---|
| Reservoirs | 2.09 | 0.5 | 1.0 |
| Ponds and tanks | 2.25 | 5.0 | 11.2 |
| Brackish water | 1.25 | 3.0 | 3.6 |
| Water-logged areas | 1.5 | 5.0 | 7.5 |
| Inland and Coastal saline soils | 3.0 | 2.0 | 6.0 |

* Data from the Handbook on Fisheries Statistics (1993).

*Source* : Dwivedi and Ninawe (1998).

An explanation will provide a vivid picture about the efficiency of production potential in productive water bodies. Freshwater and marine capture fisheries though less capital intensive, are based on natural production processes and are, therefore, highly dependent on the environmental conditions affecting the prouductive water bodies. Therefore, aquaculture has become an important sector to exploit artificially productive water bodies. The adoption of advanced technologies in about one million hectares of brackish water could yield about 4 million metric tonnes of prawn. It has also been estimated that the total yield is expected to bring foreign exchange earnings of the value of about 2,800 million US dollars to the country. Moreover, estimated production of 2.5 million tonnes from 2.1 million hectares of reservoirs and 19 million tonnes from about 4.4 million hectares of ponds, lakes, and water-logged areas are expected to earn an additional 600 million US dollars. Therefore, it is evident from the fact that all these resources (except inland and marine capture and culture) possess tremendous possibilities to boost the economic condition of the nation.

## 1.15 The Soil of an Aquatic Ecosystem

The soil is one of the most important ecological factor and feature of aquatic environment. It is the loose, unconsolidated top layer of earth's crust and is the mixture of organic detritus and weathered rock materials. These materials are formed through physical, chemical, and biological processes that occurred for a long period of time on the earth's surface. Therefore, soil is considered as the main source of nutrients as it enhances the productivity of ponds. The soil is also considered as the seat of detritus food chain. Nutrients are released in detritus are decomposed by soil microbes such as algae, bacteria, fungi, and protozoa and bound in soil particles. Pond mud is also the main source of nutrients for all sorts of aquatic plants. In addition, pond mud is also the means of support for bottom-dwelling animals such as bottom fish, shellfish, and bottom organisms (earthworms, some molluscs, algae, insects etc.) which are used as food by fishes.

### *Concept of Soil*

The science which deals with the study of origin, development, formation, and properties of soil is called the *soil science*. From the study of soil science, two concepts of soil have developed. One concept considers soil as a natural entity, a biochemically weathered and synthesized product of nature while the other concept considers soil as a natural habitat of plants and animals and on this basis, the study of soil can be justified. These conceptions are considered in studying soils of different regions.

The classification, its origin, and its description are examined in *pedology*. It is the study of the soil as a natural body and does not focus on the practical use of soils. Pedological studies are very important for construction of a fish farm.

*Edaphology* is the study of various properties of soils in relation to the productivity of ecosystem. Edaphologists are practical and have high production of aquatic food as the ultimate target. To achieve that target, they always try to determine the reasons for variation in the productivity of different types of ecosystems.

In this book some edaphological aspects in relation to the production of fish and fish food organisms have been considered. Pedological aspects, though important for fish farm construction, have also been described briefly in a general way which, to some or large extent, will cater some conceptions about soils to progressive fish farmers. Since the studies of the chemical, physical, and biological features of soils contribute equally to pedology and edaphology, these conceptions are closely related to each other and cannot be separated.

## 1.16 Formation of Soil

Soil is a stratified mixture of organic and inorganic materials. These materials are produced through decomposition of dead remains of animals and plants by activities of living organisms. The mineral constituents are derived from some parent material, the soil-forming rocks, and by weathering.

There are three types of soil-forming rocks such as *igneous rocks, sedimentary rocks,* and *metamorphic rocks.* However, the soil-forming rocks are the chemical mixture of various types of minerals. The chemical nature of most common and abundant minerals of soil-forming rocks are given in Table 1.4.

**Table 1.4 : Characteristic Features and Chemical Composition of Some Common Soil Minerals**

| Minerals | Characteristic features | Chemical Constituents |
|---|---|---|
| **I. Silt and Sand Minerals** | | |
| 1. *Silica* | Occurs as crystalline quartz and amorphous opal; dominant in sand and diatomite; combined in silicates as an essential constituent of many minerals | $SiO_2$ |
| 2. *Feldspars* | A group of abundant rock-forming minerals of the general formula, MAl $(Al.Si)_2O_8$, where M = Na, K, Ca or Fe. It constitutes 60 per cent of the earth's crust. They are white and gray to pink | |
| (a) *Plaqioclase* | | (Na.Ca) Al (Si, Al) $Si_2O_8$ |
| (b) *Orthoclase* | | $K_2Al\ Si_3O_8$ |
| 3. *Micas* | Prominent rock-forming constituents of metamorphic rock and a group of monoclinic minerals of the general formula (K, Na, Ca) $(Mg, Fe, Li, Al)_{2-3}$ $(Al, Si)_4O_{10}(OH,F)_2$ | |
| (a) *Biotite* | It is black in hand specimen and brown or green in thin section | $K(Mg, Fe^{+2})_3$ $(Al, Fe^{+3})$ $Si_3O_{10}(OH)_2$ |
| (b) *Muscovite* | It is colorless to pale brown and a very common mineral in many sedimentary rocks speci-sandstones | $KAl_2(AlSi_3)O_{10}(OH)_2$ |
| 4. *Pyroxene* | It is characterized by short and stout crystals. A group of common rock-forming minerals | $(Mg,Fe^{+2}$ or Na) $Si_2O_6$ or (Mg, $Fe^{+2}$ or Al) $Si_2O_6$ |
| 5. *Amphibole* | | Mg, Fe, Ca, or Na $(Si, Al)_8\ O_{22}\ (OH)_2$ or Mg, $Fe^{+2}$, $Fe^{+3}$, or Al $(Si, Al)_8O_{22}(OH)_2$ |
| 6. *Olivine* | A green or brown mineral and a common rock-forming mineral of basic and low-silicate igneous rocks. It weathers readily at the earth's surface | $(Mg.Fe)_2\ SiO_4$ |
| 7. *Serpentines* | They have a greasy or silky lusters, a slightly, soap feel, and a tough fracture. They are usually compact, are green or greenish gray and always secondary minerals | $(Mg.Fe)_3Si_2O_5(OH)_4$ |
| 8. *Calcite* | White or gray and reacts with cold dilute hydrochloric acid. A chief constituent of limestone and most marble | $CaCO_3$ |

***Contd...***

**Table 1.4 – *Contd...***

| Minerals | Characteristic features | Chemical Constituents |
|---|---|---|
| 9. *Magnesite* | A white to gray mineral and generally found as earthly masses. | $MgCO_3$ |
| 10. *Dolomite* | A rock containing more than 90 per cent mineral dolomite and less than 10 per cent calcite. It is white to light-colored | $Ca.Mg(CO_3)_2$ |
| 11. *Iron oxide* | | |
| (a) *Limonite* | A group of brown amorphous hydrous ferric oxides, a common secondary mineral – formed by oxidation of iron-bearing minerals. It occurs as precipitate in lakes, ponds or bogs | $FeO(OH)$, $X\ H_2O$ |
| (b) *Magnetite* | A black, isometric, and opaque mineral. It often contains titanium oxide and occurs in rocks of all kinds and in sands | $(Fe.Mg)Fe_2O_4$ |
| (c) *Hematite* | A common iron mineral, occurs in rhomboidal crystals, in reniform masses or in deep-red earthy forms. It is found in sedimentary and metamorphic rocks | $Fe_2O_3$ |
| **II. Clay Minerals** | | |
| 1. *Kaolin* | A group of clay mineral with a two-layer crystal minerals in which each silicon-oxygen sheet is alternate linned with one aluminium hydroxyl sheet. It is derived from alkali feldspar and micas. A soft, white and non-plastic clay | $Al_2Si_2O_5\ (OH)_4$ |
| 2. *Montmorillonite* | A dioctahedral clay mineral and represents a high-alumina end-member that has some slight replacement of $Al^{+3}$ by $Mg^{+2}$ and substantially no replacement of $Si^{+4}$ by $Al^{+3}$ | $Na_{0.33}\ Al_{1.6},\ Mg_{0.33}\ Si_4O_{10}(OH)_2\ .\ n\ H_2O$ |

Besides the above-mentioned minerals, some other minerals are also found in very low amounts in soil-forming rocks. These are *Rutile* (titanium oxides), *Zircon* (zirconium oxides), *Toarnalive* (boron silicates or aluminium with alkali metals and iron and magnesium), and *Gauconite* (hydrated silicon of iron and potassium).

## *Process of Soil Formation*

1. *Weathering of Soil-forming Rock :* It is the destructive process by which rocks are changed on exposure to atmospheric agents (such as physical, chemical and biological) at or near the earth's surface, with little or no transport of the altered materials. The fragmental and unconsolidated rock material, whether residual or transported, that forms the surface of the land and overlies the bedrock and includes rock debris of all kinds such as glacial drift, alluvium, volcanic ash, and vegetal accumulations, is called *regolith.*

   Regoliths are the basic materials which under the influence of other pedogenic processes finally develop into manure soil.

(a) *Physical weathering :* The physical weathering agents are climatic in character such as water, temperature, ice, gravety, and wind where it does not cause any chemical transformation of rock minerals. Temperature causes breakdown of those rocks which have heterogenous structure. Rock minerals expand in high temperature and contract when the temperature falls markedly. The differential expansion and contraction of different minerals of rocks develop internal tensions and produce cracks in the rocks and consequently, the rock weather into finer particles.

Natural water (rain water) or rapidly flowing water (torrent water) or wave action also cause mechanical weathering of rocks. These types of waters dislocate solid particles of varying diameters from lands to the bottom of rivers and seas.

During land slides and rock slippages, the rock is fragmented by abrasion. Stormy wind causes abrasion of exposed rock. It acts as like as dunes and thus fine suspended particles of rocks are transported to long distances either through water currents or storm wind.

(b) *Chemical weathering* : When physical disintegration of rock occurs, a greater surface area of rock is exposed and entered into the chemical weathering which occurs simultaneously with the physical weathering. During chemical weathering, decomposition of parent mineral materials into new mineral complexes or secondary products take place. For example, parent minerals that contain silicon and aluminium (feldspar), are converted to secondary mineral such as clay. Chemical weathering includes number of reactions such as solution (water – helps remove water-soluble minerals of weathered rocks), hydrolysis (exchange of constituent parts between water and minerals and forming hydroxides of iron, magnesium, calcium, aluminium etc.), oxidation (such as ferrous oxide to ferric oxide), reduction (ferric oxide to ferrous oxide), carbonation (formation of carbonic acid through the combination of water and carbon dioxide and reduction of carbonic acid with hydroxides of rock minerals such as calcium, magnesium etc), and hydration (attraction of water molecules with rock mineral – causes increase in volume of parent minerals and become soft and rapidly weathered).

(c) *Biological weathering* : Although exposed rocks are not suitable for many forms of life, a number of micro-organisms, lichens, and mosses grow successfully. The activities of these forms of life alter the mineral composition and physical structure of the rock. Micro-organisms absorb mineral nutrients from the rock. After death, decay, and decomposition of micro-organisms, nutrient elements are added to the developing soil.

The soil which is formed by weathering of soil-forming rock is called *embryonic* or *primary soil.* The embryonic soil is transformed into various types of soils such as (i) *Sedimentary* or *Residual Soil* – the mature soil lying immediately over the parent rock, (ii) *Immature* or *Skeletal Soil* – the partly weathered material in which maturation of soil has not occurred and (iii)*Transported* or *Secondary Soil* – the weathered parent materials are shifted to different places by streams and rivers (alluvial soil), wind (aeolian soil), sand storms (sand dunes), standing water and wave action (lacustrine soil), and the oceanic waves (marine soil).

2. *Addition of Organic Matter, Humification and Mineralization* : Addition of organic matter to the embryonic soil through deciduous portions of the plants takes place ceaselessly on. Similarly, animals add their excreta regularly to the soil. Plants and animals contribute their own bodies to the soil surface when they die.

(a) *Litter* : Various types of organic plant debris are dropped to the soil is called *litter.* It is chiefly composed of twigs, dead leaves, woods, dead roots, and various plant products. In terrestrial ecosystem, the main sources of litter are forests and grasslands but in case of aquatic ecosystem, aquatic vegetation significantly contributes to the addition of litter. In these two types of ecosystems, however, there is much accumulation of thousands of tonnes of litter on the soil surface. Litter consists of carbohydrates in the form of cellulose, hemicellulose, lignin, resins, waxes, and proteins. The inorganic constituents of litter are potassium, calcium, iron, manganese, copper, aluminium, magnesium, phosphorus, and nitrogen.

(b) *Humification :* The litter is decomposed by combined efforts of microbes and consequently various types of organic and inorganic nutrients are produced. These nutrients are incorporated into the mineral particles of soil and becomes dark in color. The residual and partly decomposed black colored organic matter which is added to the mineral matter of soil is called *humus* and the process of formation of humus is called *humification.* In an aquatic ecosystem, humus includes two types of organic matter : (i) excreta of aquatic animals and bottom-dwelling organisms, and (ii) partly decomposed organic matter – derived from aquatic plants. However, the rate of humification and humus accumulation are primarily determined by the organisms involved in the decaying process and the temperature, nitrogen supply, and pH of the soil. For example, the greater the supply of nitrogen and rise in temperature, the higher is the rate of formation of humus.

Humus is the chief component of soil and contains a number of organic compounds such as purines, pyrimidines, aromatic compounds, hexose sugars, fats, oils, waxes, amino acids, resins, proteins, tannins, and lignins. Humus is odorless, black colored, and homogeneous complex substance.

(c) *Mineralization :* It is a process by which humus is completely decomposed or reduced into simple compounds such as carbon dioxide, water, minerals, and salts. In this process the organic components are replaced by inorganic materials.

(d) *Nutrient cycle :* The mineral potential of aquatic ecosystem operates in a dynamic state through a series of inputs and outputs for essential elements. Autotrophic organisms and the soil serve as storage chamber. Fluxes of nutrients keep on aquatic plants to soil pathway and soil to plant pathway. These pathways are supported by water-soil channel. The output of nutrient from the system takes place through drainage and fish harvest. Therefore, the nutrient cycle is the pathway of nutrient materials from their occurrence in the abiotic environment to their incorporation in living organisms and their return to the abiotic environment through the metabolic pathway, death, and decay of organisms.

3. *Formation of Organo-Mineral Complexes :* At the final stage of pedogenesis, colloidal particles which are formed due to weathering, humification and mineralization, accumulate and form *concretions.* Some colloidal humus particles may become associated with mineral particles to form organo-mineral complexes. These complexes are formed by two mechanisms such as *cementing* and *electo-chemical bonding*. Cementing mechanism of concretion formation involves the action of substances that are absorbed on the surface of the soil particles. In the second mechanism, aggregation of negatively charged colloidal clay and humus particles of water molecules and calcium takes place. However, during formation of organo-mineral complexes, a characteristic profile is developed under the influence of vegetation. climate, temperature, parent material and the activities of the soil communities. Thus, the mature soil forms a complex system of non-living and living substances.

## 1.17 The Soil Profile

Examination of a vertical section of a soil reveals the presence of more or less distinct horizontal layers (Figure 1.4). Such a vertical section of a soil that displays all its horizons is called the *soil profile* and the individual layers are called *horizons*. Every undisturbed and well-developed soil has its own specific profile characteristics. Each horizon varies in thickness, color, texture, structure, consistency, composition, and acidity. The soil profile is very important for classifying and surveying soils and also in determining how the soils can effectively be used for various purposes.

The upppermost horizons (A horizon) of a soil profile are darker in color than the lower horizons. This difference is due to the accumulation of organic matter that results from the death and decay of organisms incorporated into the A horizons. Weathering also tends to be more intense in the upper horizons than in the lower horizons.

The underlying layers (B horizons) contain comparatively less organic matter and are characterized by the accumulation of silicate clays, iron and aluminium oxides, gypsum, and calcium carbonate in varying amounts. These materials may have been (1) washed down from upper layers or (2) formed through the process of weathering.

The A and B horizons comprise the *solum* (from the Latin world *solum,* which means soil), which shows the evidence of the genesis of a natural body and is distinct from the parent materials below. These horizons have resulted from the soil-building processes of biochemical breakdown, weathering, and synthesis.

The different layers of a soil profile are not always vivid and well defined. Often the transition from one layer to the other is very slow. Consequently, it becomes very difficult to obtain a clear picture about the soil profile.

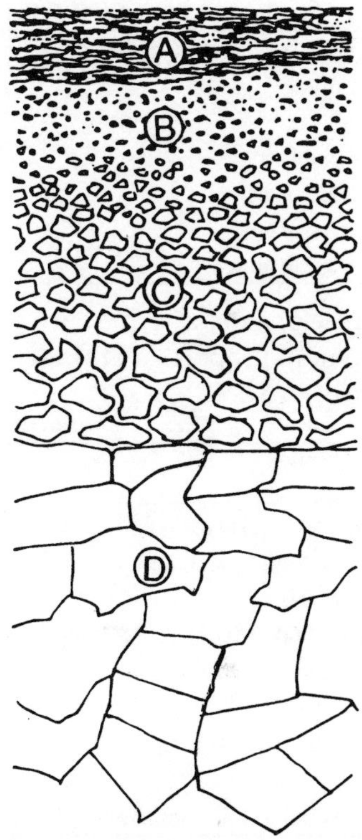


**Fig. 1.4 : Digramatic representation showing the profiles of soil. (A) Horizon A (Organic debris in different stages of decomposition and minerals); (B) Horizon B (Fine particles and minerals); (C) Horizon C (Weathered mineral materials); (D) Horizon D (Unweathered rock and bedrock).**

### *Top Soil*

When water of a fish pond is drained or dried up during summer season, it is advisable to a fish farmer to plough the soil. At the time of ploughing, the natural state of the upper 5-8 inches is altered. This altered part of the soil is referred to as the *top soil.* Top soil contains many nutrients available to water. Incorporation of organic residue, fertilizers and agricultural limestone enhance the fertility of soil and productivity of water at level compatible with the economic production of aquatic species of commercial value.

## 1.18 Soil Texture

The inorganic portion of soils vary in size and composition. It is composed of small minerals and rock fragments of various kinds. The rock fragments are composed of aggregates of minerals and are the small remaining quantities of massive rocks from which the regolith and the soil have been formed by weathering.

The mineral particles present in soils are extremely variable in size and the range of particle size in terms of millimetre (mm) varied between less than 0.002 and above 5 (Table 1.5). The clay particles have colloidal properties and can be seen only with the aid of electron microscope. Sand particles can be seen easily with the naked eye, do not stick together and feel gritty when rubbed between the fingers. Silt particles are much smaller than the sand particles, which are powdery when dry and are not sticky. Gravel pieces are as large as the smaller rock fragments.

**Table 1.5 : Size of Different Mineral Particles in Soil**

| Types of the Mineral Particles | Diametre (mm) |
|---|---|
| Clay | Less than 0.002 |
| Silt | 0.002-0.020 |
| Fine sand | 0.020-0.200 |
| Coarse sand | 0.200-2.000 |
| Fine gravel | 2.000-5.000 |
| Coarse gravel | Above 5.000 |

The ability of a given soil to supply chemical nutrients and physical properties are determined by the proportions of different sized particles, a property called *soil texture*. The term is applied to the smaller features, as seen on a smooth surface of a rock. Terms such as silty clay, sandy loam, and clay are used to identify the soil texture. The main properties of mineral particles are show in Table 1.6.

**Table 1.6 : Properties of Three Major Inorganic Soil Particles**

| Properties | Silt (0.002-0.020 mm) | Clay (< 0.002 mm) | Sand (0.200-2.000 mm) |
|---|---|---|---|
| Types of observation | Microscopic | Electron microscopic | Naked eye |
| Attraction of particles for each other | Medium | High | Low |
| Attraction of particles for water | Medium | High | Low |
| Consistency Properties : | | | |
| 1. Wet Condition | Smooth | Sticky | Loose |
| 2. Dry Condition | Powdery | Hard | Very loose |
| Dominant minerals | Primary and Secondary | Secondary | Primary |
| Capacity to hold nutrients | Low | High | Very low |

### *Primary and Secondary Minerals*

Minerals that have formed at the same time as the rock enclosing it, by igneous (a rock or mineral that solidified from molten material or lava such as mica, quartz, and feldspars) or hydrothermal (mineral deposit precipitated from a hot aqueous solution) processes and that remain its original composition and form, are known as *primary minerals*. They are very prominent in the sand and silt fractions. Other minerals such as iron oxide and silicate clays, that have formed later than the rock enclosing it, at the expense of an earlier-formed primary mineral, as a result of weathering, metamorphism, or solution, are called *secondary minerals*. These minerals are dominant in silt and clay fractions. Inorganic fraction of the soil is the principal source of most of the mineral elements and have found to be essential for productivity of any ecosystem.

## 1.19 Soil Structure

The arrangement of the silt, sand, and clay particles within the soil are termed as *soil structure*. It is as important as the relative amount of these particles present (soil texture). These particles are commonly found associated together in aggregates, but remain independent of each other. Structural forms of soil are important in classifying soils.

## 1.20 Classification of Soil

Classification is the systematic arrangement of soils into groups or categories on the basis of their characters. For agricultural and aquacultural purposes, soils are considered as the most important criteria that determines the fertility of soil. Therefore, it is necessary to classify different types of soils which helps to solve the problems related to the construction of a fish farm, productivity of ponds and lakes, and water-holding capacity.

Soil surveys have been prepared for most countries of the world for myriad purposes. The soil taxonomy system includes : orders, sub-orders, groups, sub-groups, families, and series. Since the family includes soils with common purposes that affect the soil's response to management, the family name must include terms that describes these properties. This is accomplished by using the following classes :

### *Particle-size Classes*

The distribution of particle-size is for the whole soil and several classes are used for most soils such as *sandy* (fine earth is sand or loamy sand), *loamy* (fine earth is loamy, very fine or finer sand), *fragmental* (gravel, very coarse sand), *clayey* (fine earth contains 35-40 per cent clay), *sandy skeletal* (fine earth is loamy sand or sand), *loamy skeletal* (loamy, fine or finer sand, but less than 35 per cent clay), and *clayey skeletal* (fine earth is clay).

### *Calcareous and Reaction Classes*

These classes are characterized by the presence or absence of carbonates to the soil. Different terms such as *calcareous, non-calcareous,* acid etc. are used.

### *Shape and Slope Classes*

These classes are used to indicate shape and slope.

### *Classes of Coating on Sand*

These classes are characterized by the coating of sand particles with clay and silt.

### *Permanent Crack Classes*

These classes are used to recognize soils with permanent cracks.

### *Mineralogical Classes*

These classes refer to the mineral or minerals that are most prominent in the soil. Common mineralogical classes of soils include :

1. *Ferritic* : Soils contain oxides of iron.
2. *Carbonatic* : Soils contain some gypsum and carbonates.
3. *Gypsic* : Soils contain gypsum.
4. *Oxidic* : Soils contain high quantities of aluminium and iron oxides.
5. *Gibbsitic* : Soils contain oxides of aluminium.
6. *Serpentinic* : A group of common rock-forming minerals. They have a greasy or silky luster, a slightly soapy feel, compact, and are commonly green or greenish gray.
7. *Micaeous* : Soils contain micas.
8. *Siliceous* : Soils contain abundant silica especially as free silica rather than silicates.
9. *Chloritic* : Soils contain chlorite.
10. *Mixed* : Mixture of clays.
11. *Kaolinitic* : Soils contain common clay mineral of the kaolin group. The kaolin minerals are generally derived from alkali feldspars (aluminosilicates) and micas. It does not appreciably expand under varying water content and does not exchange iron or magnesium.
12. *Vermiculitic* : Soils contain a group of micaeous clay minerals closely related to chlorite and montomorillionite.

The soils of different parts of the world have been classified under several main heads such as alluvial soil, black soil, lateritic soil, red soil, forest soil, desert soil, saline and alkaline soils, peat, podzol soil, grey-brown podzol soil, chernozem soil, and brown soil. From fishery standpoint of view, these types of soils are important to greater or lesser extent and hence will be briefly surveyed.

### *Alluvial Soil*

It is a soil developing from recently deposited alluvium and exhibiting essentially no horizon development or modification of the recently deposited minerals. This soil is formed by silt deposition of a number of river systems such as the Amazan, the Mississippi, the Yangtse-Kiang etc. These soils differ in consistency from sand to loam and silt to clay. The fertility status of alluvial soil in different states of India is shown in Table 1.7.

### *Black Soil*

This soil, also termed as *regur*, is particularly found in different states of India such as Tamil Nadu, Karnataka, Andhra Pradesh, Madhya Pradesh, Maharashtra, and Uttar Pradesh. In general, black soil is highly argillaceous (soil substances composed of clay minerals and has a peculiar "earthy" odor which they emit when breathed upon), composed of fine grains, dark in color with a high proportion of magnesium and calcium carbonates, iron, potassium and aluminium but poor in nitrogen, phosphorus, and organic matter. Black soil possesses a high quantity of montomorillionite and beidellitic group of minerals. This soil is, however, highly impervious to water and when wet, it becomes sticky.

**Table 1.7 : Nutrient Status of Alluvial Soil in Different States of India**

| State | Nutrient status |
|---|---|
| Assam | High phosphorus and potash but moderate quantity of nitrogen and organic matter; fine-textured soil |
| Bihar | Availability of phosphorus and potash is high; lime content is low; soil is acidic to neutral in reaction |
| West Bengal | Soil is more fertile and productive |
| Uttar Pradesh | Soil contains varying amounts of calcium carbonate and soluble salts; neutral to alkaline in reaction |
| Gujarat | Poor in organic matter and nitrogen but rich in phosphorus and potash |
| Orissa | Potash content is high but low in phosphorus |
| Andhra Pradesh | Fertile soil; rich in calcium, phosphorus, and potash |
| Tamil Nadu | Soil is poor in nutrients |
| Kerala | Water-retention capacity is low; soil is poor in nutrients |

*Source :* Selected data from Roychowdhuri (1966)

## *Lateritic Soil*

This type of soil is more compact and composed of vesicular rocks. It is a subsoil layer and mixture of iron and aluminium oxides with small amounts of titanium and manganese oxides.

Lateritic soil is found in Southeast Asian countries, Mayanmar, Laos, Cambodia, in humid regions of South America and Africa, and Northern parts of Australia. This type of soil is predominant in highly weathered humid tropical soils that, when exposed and allowed to dry, becomes very hard and will not soften when rewetted. When erosion removes the overlying layers, the laterite is exposed and a virtual pavement results. In general, lateritic soils are poor in nitrogen, phosphorus, potash, and calcium; but rich in organic matter (10-20 per cent), iron, and aluminium with low base-exchange capacity and pH value (4.5-5.5).

## *Red Soil*

This soil is developed as a result of biochemically weathering of ancient and metamorphic rocks and is found in different states of India (Table 1.8), Argentina, Paraguay, South Africa, Eastern parts of North America, Southern parts of China, and in the Mediterranean regions. In general, soils are poor in potash, nitrogen, calcium and phosphorus, iron oxide, and humus with the mineral Kaolinite. The retention capacity of potash, calcium, and magnesium by exchange is very limited. Phosphorus is fixed in red soil and its capacity is increased with decrease in soil pH.

**Table 1.8 : Nutrient Status of Red Soil in Different States of India**

| State | Nutrient status |
|---|---|
| Bihar | High percentage of ferric oxide; rich in potash but low in phosphorus; soil is acidic (pH 5.0-6.8) |
| Uttar Pradesh | Brownish-grey in color; loam to sandy or clay loam; soils are not fertile |
| Andhra Pradesh | Reddish-brown or pale-brown in color; loamy sand to very coarse texture; neutral in reaction; organic matter and soluble salts are very low |
| Mysore | Loamy soil; iron, aluminium, potash, and phosphorus contents are high but low in nitrogen |
| Tamil Nadu | Soil is acidic to alkaline (pH 6.6-8.0); base-exchange capacity, organic matter, nitrogen, phosphorus, and potash contents are very low |

*Source :* Selected data from Roychowdhuri (1966).

### *Forest Soil*

This type of soil is found in arid or semi-arid areas such as India (Than regions of Punjab and Rajasthan), Africa (Sahara region), and South America (Atakana desert) and it is considered as a group of zonal soils having a light-colored surface horizon overlying calcareous material. It is developed under conditions of aridity, warm to cool climate, and scant scrub vegetation. Since, the soil contains various types of minerals, the color is extremely variable such as grey, yellow, and brown depending on the color of mineral particles. Some soils in these regions contain high amounts of soluble salts and low in organic matter.

### *Brown (or Chestnut) Soil*

Brown soil is very common in Central Asia, Spain, Argentina, and Mongolia. It is an intermediate stage of Chernozen and Desert soils, and are formed in dry grassland areas. Organic matter and mineral salt contents are moderate. Since dry weather in these regions does not encourage decomposition, the color of soil becomes brown. The soil is moderately fertile.

### *Saline and Alkaline Soils*

Saline is a term that connotes a natural deposit of halite or of any soluble salts such as an evaporite. Saline soils are generally nonsodic (soil that contains no sodium and does not interface with the growth of most crop plants) and containing sufficient soluble salts to impair its productivity. The exchangeable sodium adsorption ratio is less than about 15, and the pH is less than 8. Sodium is found in the form of sodium chloride and sodium sulfate.

Alkaline soils are those that has pH more than 7. It is usually applied to surface layer but may be used to characterize any horizon. Due to absence of vegetation, bacterial activity and organic matter are absent. These soils have an alkaline action and it is the sodium carbonate that renders the soil more sterile.

In some parts of the tropical and sub-tropical regions where soils have no adequate surface drainage, salts are gradually accumulated from the surrounding regions. During the dry season, soluble salts are sucked up in solution to the surface and are deposited in the form of white efflorescent nitrogen.

### *Peat Soil*

It is an unconsolidated organic soil containing more than 50 per cent organic matter and used to refer to the stage of decomposition of the organic matter in stagnant water with small amounts of oxygen. "Peat" referring to the slightly decomposed or undecomposed deposit of semi-carbonized plant remains in a water-saturated environment such as a bog, of persistently high moisture content (at least 75 per cent) and "Muck" to the highly decomposed materials. In India, peat soil is found in some parts of Kerala, Orissa, West Bengal, Uttar Pradesh, and Bihar. However, the water-logged areas are formed by dried basins of different river systems and lakes in coastal and alluvial areas. Soils of these regions are blue or blue-black in color due to the presence of ferrous iron.

### *Podzol Soil*

It is found in coniferous forest of the United States and Canada. Plant materials fall on the topsoil, are decomposed and therefore, the soil is rich in organic matter. The soil is acidic in reaction.

### *Grey-Brown Soil*

This type of soil covers the temperate mixed forest in the North-eastern parts of the USA, North-western Europe, and Northern parts of China. Organic matter content is much lower than that of Podzol soil, but contains high amounts of calcium and potassium.

### *Chernozen Soil*

This soil is characterized by a deep, highly fertile soil and by an extensive covering of grassland in the temperate latitudes of the interior of North America, especially in the Mississippi valley region (Prairie soil) and also in the semi-arid mid-latitudes of South-eastern Europe and Asia (Steppe soil). The calcium and organic matter contents of the top-soil are very high and generally black in color.

## 1.21 Soil Fertility

The fertility of soil refers to the quality of a soil that enables it to provide essential chemical elements in quantities and proportions for the growth of plants (aquatic and terrestrial), fish and other aquatic animals for providing a wide array of benefits to humankind. From the standpoint of fish production, it can be said that the soil fertility is related to several stages such as (i) nutrients are directly required for the growth and development of natural food organisms (plankton and bottom fauna), (ii) certain micro-organisms in fish culture ecosystems are developed which permit biogeochemical cycles to sustain productivity, (iii) the methods permitting to restore or maintain soil fertility (such as fertilization, manuring and liming), and (iv) soil fertility is directly related to fish production because it helps to establish a food chain in water and soil for fishes. Therefore, fish production is considered as an integral part of the soil fertility.

In general, most of the pond soils are not as fertile as required for fish growth and still remains unproductive. A holistic approach towards improvement of pond soil fertility is necessary. However, it has now become imperative to implement some appropriate strategies to enhance soil fertility for the generations to come.

## 1.22 Nature of Soil Separate

Soil separates are the individual-sized groups of mineral soil particles such as sand, clay and silt. The various types of mineral soil particles and their characteristic features will be briefly discussed.

### *Sand*

Sand grains may be irregular or round depending on the amount of abrasion they have undergone. Sand particles are generally not sticky and hence are not plastic. The water-holding capacity of sandy soil ponds is low and because of the large spaces between the particles, water along with nutrients pass through rapidly.

### *Clay*

The size of the individual particles is small and therefore, the surface area per unit mass of clay is very high (about 10,000 times as much surface area as the same weight of medium-sized sand). The size of clay particles varies from round to plate-like and the water-holding capacity of clay soils is very high. The adsorption capacity of water and nutrients, plasticity, and cohesion are all surface phenomena. High specific surface of clay is significant in determining soil properties. The presence of clay in a fish pond soil provides a fine texture and water as well as nutrient losses significantly reduced.

### *Silt*

Soil particles are intermediate in properties and size between clay and sand particles. They are irregularly fragmental and diverse in shape. The dominant mineral is quartz. A thin film of clay is formed over silt separates and consequently, they possess some plasticity, stickiness and adsorptive capacity, but much less than the clay separate. Silt may cause the soil to be compact unless it is mixed with organic matter.

## 1.23 Chemical Composition of Soil Separates

Silt and sand are made up of silicon dioxide and other primary minerals and known for their resistance to physico-chemical changes. Consequently, these two separates have low chemical activity. The primary minerals contain nutrient elements are so insoluble as to make their nutrient-supplying ability insignificant.

Chemcially, silicate clays vary widely. Some silicate clays contain varying quantities of aluminium, iron, potassium, magnesium, calcium, and other elements. Clay soils hold small but significant quantities of $Ca^{2+}$, $K^{+}$, $Al^{3+}$, $NH_4^{+}$, and $H^{+}$. These cations are exchangeable and can be released for absorption by algae and macrophytes.

In the hot and humid tropics where soils are very much susceptible to physico-chemical changes, iron and aluminium oxides are more prominent. Hence, such changes can have a profound effect on the chemical composition of soil separates.

Different types of soil separates differ significantly in chemical composition and consequently, they vary in their content of mineral nutrients. For example, the clay separate would be expected to be highest and the sand separate to be lowest in nutrients (Table 1.9). This generalization is, however, true for most soils, although some exceptions may exist.

**Table 1.9 : Common Names of Soils and Their Texture in Relation to the Soil Textural Class Names**

| Common Names | Texture | Soil textural call names |
|---|---|---|
| Sandy soils | Coarse | Sand and loamy sand |
| Loamy soils | | |
| | Coarse | Sandy loam and fine sandy loam |
| | Medium | Loam, silt loam, silt |
| | Fine | Sandy-clay loam, silty-clay loam, clay loam |
| Clayey soils | Fine | Sandy clay, silty clay, clay |

## 1.24 Textural Classes of Soil

It is a grouping of textural units based on the relative properties of the various size groups of mineral soil particles (sand, clay and silt). These textural classes, listed from the coarsest to the finest in texture, are given in Table 1.10. Textural classes of soil have been broadly divided into three main groups such as *sand, loam* and *clay.* Within each group specific textural class names have been given. These textural class names are, however, primarily determined by particle-size distribution.

**Table 1.10 : Calcium, Nitrogen, and Phosphorus Contents of Clay, Silt and Sand Separates Found in Tropical Pond Soils**

| Soil separates | Nutrient content (per cent) | | |
|---|---|---|---|
| | Calcium | Nitrogen | Phosphorus |
| Clay | 2.8-4.8 | 0.15-0.30 | 0.20-0.40 |
| Silt | 2.6-3.7 | 0.05-0.15 | 0.05-0.20 |
| Sand | 1.0-2.5 | 0.02-0.07 | 0.01-0.06 |

## 1.25 Aggregation in Soils

Generally two factors are responsible for dealing with soil aggregation. The first factor is responsible for aggregate formation and the second one gives the aggregate stability once they are formed. Since these two factors act simultaneously, it is hard to distinguish their relative effects on aggregate formation in soil.

Several specific factors such as role of organic matter, physical processes, role of adsorption cations, and soil tillage influence crumbs and granules formation in soil.

### *Influence of Organic Matter*

Organic matter stimulates the formation and stabilization of aggregates. After decomposition, organic matter generates viscous microbial substances. These substances alongwith associated fungi and bacteria initiate crumb formation. Polysaccharides of organic exudates react with the silicate clays and aluminium and iron oxides. Polysaccharides orient the clays in such a way that they form bridges between soil particles. Consequently, water-stable aggregates are formed. The aggregates of soils high in organic matter are more stable than those low in organic matter.

### *Physical Processes*

Any action that will force contract between soil particles stimulate aggregate formation. In an aquatic ecosystem, however, the mixing action of soil micro-organisms and other bottom fauna encourage contracts and stimulate aggregation.

### *Role of Adsorbed Cations*

Aggregate formation is influenced by the type of the cations adsorbed by soil colloids. For example, the adsorption of some cations (such as $Ca^{2+}$, $Al^{3+}$ or $Mg^{2+}$) play an important role in aggregate through a process termed as *flocculation*. These cations help soil colloids to come together in small aggregates.

### *Soil Tillage*

In many tropical countries, most of the fish culture ponds are dried up. This condition generally prevails in seasonal water bodies. Before the monsoon season, the pond soil is ploughed for the maintenance of stable soil structure and to incorporate organic residues into the soil and create a more favourable condition.

Tillage involves the mechanical manipulation of the pond soil. At the same time, the physical condition of the pond soil is also improved. Tillage operation has other advantages. For example, mixing and stirring the soil generally hastens the toxic gases. Thus, tillage is considered necessary in the management of pond soil.

Although the classification of different types of soils, their profiles, and their relationships between soil fertility and agriculture are extremely significant to crop production, the complexities of soils and their properties are beyond the purview of this treatise. At the same time, aquacultural implications must be recognized along with those relating to fish production and fish farm management generally.

## 1.26 Conclusion

Water is very critical to aquatic plants and animals. Water is the natural body which is necessary for myriad purposes. Quality of water is very important to understand whether a particular water body is suitable for fish culture or not.

Water is the most abundant compound in living organisms. Its relatively heat of vaporization and surface tension are the result of intermolecular attraction in the form of hydrogen bonding between water molecules. Liquid water has short-range order and consists of "flickering clusters" of very short half-life.

The polarity and hydrogen bonding properties of the water molecules make it a potent solvent for many ionic compounds and neutral molecules. Water also disperses amphipathic molecules, such as soaps, to form micelles, clusters of molecules in which the hydrophobic groups are hidden from exposure to water and the polar groups are located on the external surfaces.

Though water and soil are considered as the bridge between the organic, inorganic, and living world, most of the people studied freshwater bodies simply through their naked eyes, who is interested with the physical standards, nature, origin, and types of water. Some people use water bodies as recreational purposes. Some are studied water bodies as a habitat for plants and animals. But those who are actively engaged in cultivating aquatic animals of commercial importance, are interested with the quality and classification of waters regardless of how they are used. Water-soil relationship in suitable fish culture ecosystems under favourable conditions is very significant for fish growth. Generally nutrients are released from soil to the overlying water where natural but microscopic food organisms grow profusely for optimum fish production. Variations of essential nutrients, organic matter, abiotic-biotic factor of pond/lake environments determine most of the potential usefulness of waters. The interactions among these components are of great significance so far as the fish production on commerical basis is considered.

Soil is the main source of organic and inorganic nutrients for productivity of water bodies. In addition, good and fertile soil should be considered when construction of a suitable fish farm is contemplated. The soil and water are excellent example of the two ecosystems, containing huge number of different types of animals and plants that form an inter-related biological complex.

## *References*

Clampham, Jr. W.B. 1973. *Natural Ecosystem*. Mac-Millan Company, New York.

Dwivedi, S.N. and A.S. Ninawe. 1998. Freshwater resource utilization for production, employment and social benefits. *In : Advances in Fisheries and Fish Production* (*Ed*. S.H. Ahmad), Hindustan Publishing Corporation, New Delhi, India, pp. 22-33.

*Handbook on Fisheries Statistics*. 1993. Ministry of Agriculture, Government of India.

Roychowdhuri, S.P. 1966. *India : Land and Soil*. National Book Trust of India, New Delhi.

## Questions

1. Briefly describe the importance of water and soil in fish culture. What are the reasons of fish culture?
2. What types of water a fish farmers will use for fish culture? Mention the quality of water required for fish.
3. How many types of natural waters are found? What types of water will you select for freshwater fish culture?
4. Define the following terms : (a) eutrophic, (b) oligotrophic, (c) dystrophic, (d) acidity and alkalinity, (e) regolith, (f) chemical weathering, (g) biological weathering, (h) humification, (i) soil structure and texture.
5. Explain the importance of water-soil relationship in relation to ecosystem productivity.
6. Why water is so important in fish culture? What are the chief constituents of water?
7. Water is regarded as profit, recreation, and a source of food -- explain.
8. Why soil is considered as one of the most important ecological factor for construction of a fish farm?
9. Describe soil horizons and explain the importance of their role in determining the properties of soil.
10. Distinguish between primary and secondary minerals.
11. Briefly explain how the major processes involved in soil formation.
12. Why soils are classified for the construction of a fish farm?
13. Which of the classes of soil classification are found in the soils of the following regions and why?

    (a) South-east Asian Countries, (b) Madhya Pradesh, (c) Uttar Pradesh, (d) Southern parts of China, (d) Semi-arid areas of India, (e) Orissa and West Bengal of India.
14. How soil aggregates are formed and which factors are rsponsible for their stability?
15. How soil fertility is related to fish production?
16. A fish culture expert advised a fish farmer to incorporate organic manure to improve the texture of his pond soil. Do you follow his advise?

17. How do the properties of water and soil influence the use of water and soil for fish culture?
18. Which entry of this chapter is most important for a fish culturist?
19. A progressive fish farmer wants to construct a farm. What criteria he will select for the purpose?

# 2

# Nutrient Carriers Used in Fish Culture

For optimum growth of fish and overall productivity of fish culture ecosystem, a variety of nutrient carriers probably have been subjected to most study, and for many good reasons receives much attention. The amount of different nutrients present in any natural aquatic environment is generally low and indicates unproductive, while the quantity required for the growth of fish and fish food organisms (phytoplankton, zooplankton, and bottom fauna/flora) is comparatively large. When too much nutrient carriers (organic and inorganic) are used for fish culture, substantial amounts of nutrients are lost through several ways and may become pollutants. Therefore, large applications of nutrient carriers may cause hazards to fish. Nutrients are utilized by the fish through food chains. Although inorganic fertilizers and organic manures contain various essential elements, all of them are not necessary for fish growth. However, the most important elements essential in fish culture include nitrogen, phosphorus, and calcium. These nutrients must be conserved and carefully managed.

While the use of organic manures in fish culture was practised long before, chemical fertilizers have been extensively used as early as the late thirties. In fish culture practices, commercial grade fertilizers are widely used. Any organic or inorganic material of synthetic or natural origin applied to a fish culture ecosystem to supply certain elements essential to the growth of fish and the productivity of ecosystem, is considered as a *fertilizer* or *nutrient carrier.*

## 2.1 Fertilizer Elements

At least fourteen essential elements have been identified. Of them, six elements are considered as *macronutrients* while other eight elements are *micronutrients.* Magnesium and calcium are applied as lime as and when required. Although lime is considered as fertilizer, it does exert nutritive effect. Except sulphur, three other macronutrients such as nitrogen, phosphorus, and potash are generally used as commercial fertilizers (For excellent discussion on the subject, see Collings 1962). The consumption of NPK fertilizers in India markedly increased from 1970 to 1991, but thereafter, the consumption of fertilizers declined because of the announcement of decontrol of fertilizers by the Union Government from August 25, 1992. Of course, the consumption of fertilizers further increased significantly since 1996.

Because fertilizers and manures are very important for fish culture development, it is necessary to understand about their types, behavior, and significance and therefore, different types of nutrient carriers have been discussed in a general way. The chemical fertilizers will be considered first.

## 2.2 Nitrogen Fertilizers

Different types of nitrogen fertilizers are used to supply nitrogen in fish ponds as shown in Table 2.1. Table indicates that there is a wide range of nitrogen contents of nitrogen carriers – from

11 to 46 per cent in different forms of carriers. Although nitrogen carriers are produced either from ammonia (Figure 2.1) or from atmospheric nitrogen which are easily available in nature, the costs of nitrogen fertilizers are high because the energy required to produce is very high and exceeds that required to produce other fertilizers and as a result, it places constraints on the production and use of nitrogen carriers. Therefore, efficient use of these nitrogen carriers is necessary.

**Table 2.1 : Nitrogen Carriers**

| Fertilizer | Chemical form | Nitrogen (%) |
|---|---|---|
| Calcium ammonium nitrate | $Ca(NO_3)_2$, $(NH_4)_2NO_3$ | 25 |
| Ammonium sulfate nitrate | $(NH_4)_2SO_4$, $(NH_4)_2NO_3$ | 20 |
| Ammonium sulfate | $(NH_4)_2SO_4$ | 21 |
| Ammonium nitrate | $(NH_4)_2NO_3$ | 33 |
| Calcium nitrate | $Ca(NO_3)_2$ | 15 |
| Urea | $CO(NH_2)_2$ | 44-46 |
| Calcium cyanamid | $CaCN_2$ | 22 |
| Anhydrous ammonia | Liquid ammonia | 82 |
| Diammonium phosphate | $(NH_4)_2HPO_4$ | 21 |
| Ammonium polyphosphate | $(NH_4)_3HP_2O_7$;$NH_4H_2PO_4$, $(NH_4)_3H_2P_2O_{10}$ | 12-15 |
| Sodium nitrate | $NaNO_3$ | 16 |
| Potassium nitrate | $KNO_3$ | 13 |

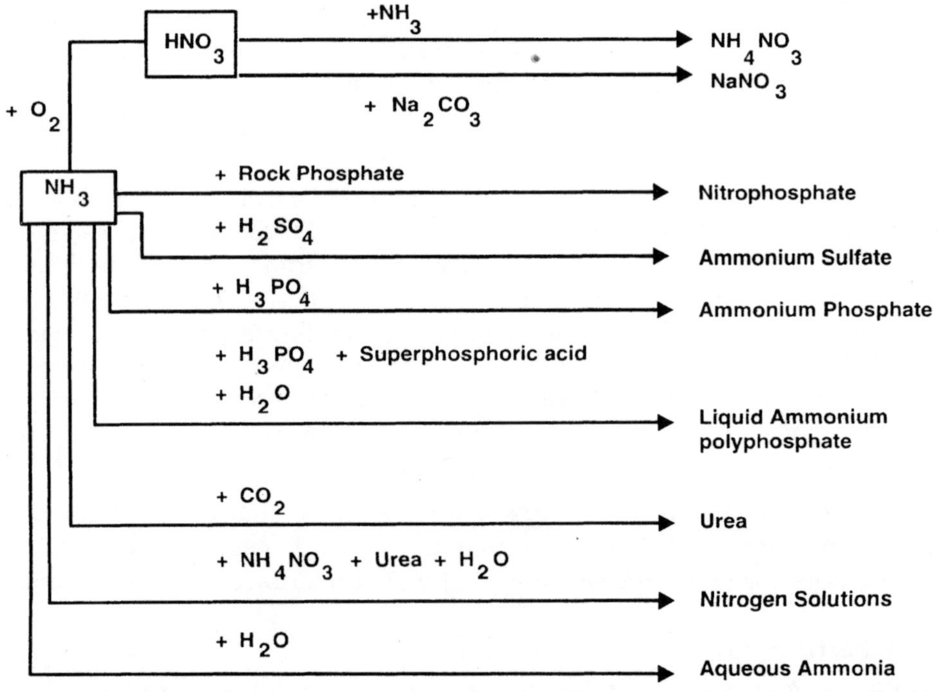


**Fig. 2.1 : Synthesis of various nitrogenous fertilizer materials from ammonia. Ammonia is obtained in large quantities either as a by-product of coke manufacture or as synthesis from hydrogen and nitrogen. Note that ammonia in combination with other materials play an important role in the manufacture of nitrogen fertilizers.**

### *Nitrogen Solutions*

Nitrogen solutions are used in many Western countries. They are composed of ammonium nitrate and urea which are dissolved in water and contain about 30 per cent of nitrogen. Nitrogen solutions are safe to handle and easy to apply directly to ecosystems because they are not under pressure. They are also used in the preparation of liquid fertilizers.

### *Urea (Carbamide)*

It is the most extensively used nitrogen fertilizer and contains about 46 per cent of nitrogen. Ammonia reacts with carbon dioxide under a pressure of about 400 atmospheric pressure and at temperature of 160-200°C to form urea.

$$2NH_3 + CO_2 \longrightarrow \underset{\text{Ammonium carbamate}}{NH_2\text{–}COONH_4} \longrightarrow \underset{\text{Urea}}{NH_2\text{–}CO\text{–}NH_2} + H_2O$$

Since the ammonium carbamate is unstable at pH 7.0, ammonia is released to the atmosphere in gaseous form. Moreover, urea is hydrolyzed to ammonia at a temperature of 26.7-32.2°C and relative humidity of 50-75 per cent. This ammonia is highly toxic to eggs, fry and fingerlings of different species of fish. Therefore, it should be properly used to prevent fish mortality.

1. *Reaction in Soil :* When urea is added to the soil, it is decomposed by the enzyme urease produced by micro-organisms to ammonium carbonate. This ammonium carbonate is oxidized to nitric acid. This nitric acid again reacts with free calcium carbonate (if it is present in the soil), humic acid, and clay micelle.

$$\underset{\text{Urea}}{NH_2\text{–}CO\text{–}NH_2} + 2H.OH \xrightarrow{\text{Urease}} \underset{\text{Ammonium carbonate}}{(NH_4)_2CO_3}$$

$$(NH_4)_2CO_3 + 3O_2 \longrightarrow 2HNO_2 + 3H_2O + CO_2$$

$$2HNO_2 + O_2 \longrightarrow 2HNO_3$$

$$2HNO_3 + CacO_3 \rightleftharpoons Ca(NO_3)_2 + CO_2 + H_2O$$

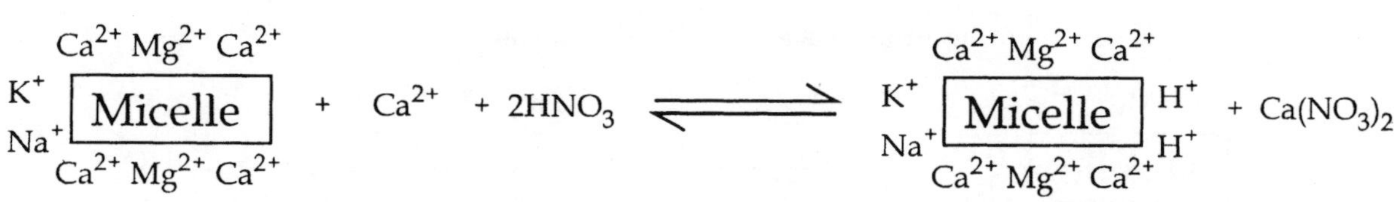

This calcium nitrate is leached in the soil when the above reaction proceeds in the forward direction. Therefore if urea is applied for prolonged period, the clay and humic micelle is saturated with hydrogen ions and the soil becomes acidic if calcium carbonate is not present in adequate amount. However, soil acidity can be neutralized by adding 80 kg of limestone for each 100 kg of urea.

### *Ammonium Sulfate*

Like urea, ammonium sulfate is also considered as the most important and widely used nitrogen fertilizer that contains about 21 per cent of nitrogen. When hydrogen and nitrogen in the ratio of 3:1 are passed over heated (500°C) platinized asbestos or powdered iron under 250 atmospheric pressure, ammonia is formed which is passed through a suspension of gypsum in water through which carbon dioxide is also passed.

$$N_2 + H_2O \xrightarrow{\text{250 atmosphere}} 2NH_3$$

$$2NH_3 + CO_2 + H_2O \longrightarrow (NH_4)_2CO_3$$

$$(NH_4)_2CO_3 + CaSO_4 \longrightarrow (NH_4)_2SO_4 + CaCO_3$$

1. *Reaction in Soil* : Ammonium sulfate generally reacts with mono-calcium phosphate of single superphosphate, potassium chloride (muriate of potash) and calcium carbonate as shown in the following equations :

$$(NH_4)_2SO_4 + Ca(H_2PO_4)_2 \longrightarrow 2(NH_4)H_2PO_4 + CaSO_4$$

$$(NH_4)_2SO_4 + 2KCl \longrightarrow 2NH_4Cl + K_2SO_4$$

$$(NH_4)_2SO_4 + CaCO_3 \longrightarrow (NH_4)_2CO_3 + CaSO_4$$

$$(NH_4)_2CO_3 \xrightarrow{\text{Heat}} 2NH_3 + CO_2 + H_2O$$

When ammonium sulfate is added to soil, ammonium ions replace basic cations from clay and humic micelle.

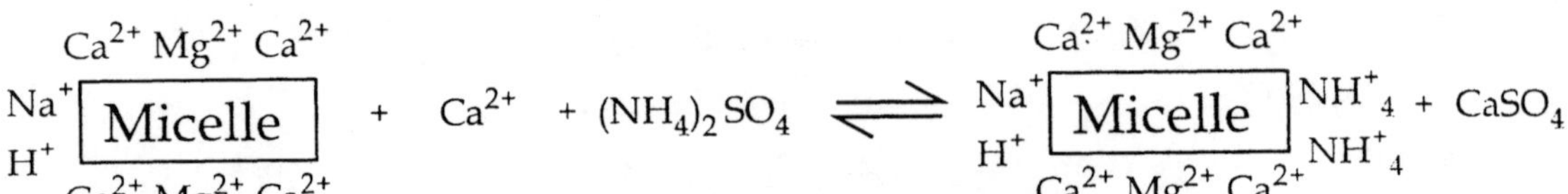

These absorbed ammonium ions are oxidized to nitrate.

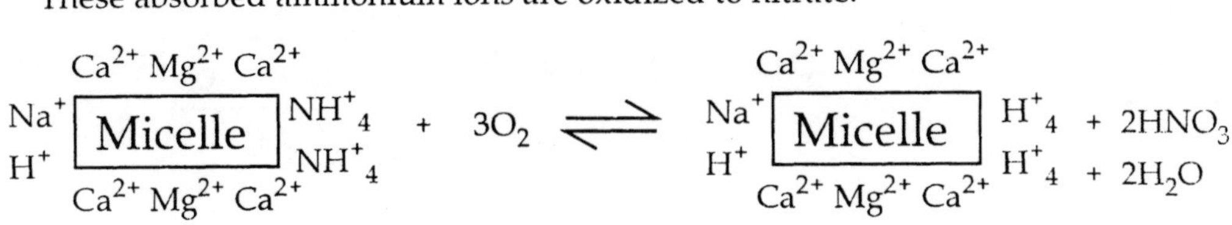

$$2HNO_2 + O_2 \longrightarrow HNO_3$$

This nitric acid reacts with the calcium carbonate and the clay and humic micelle.

$$CaCO_3 + 2HNO_3 \longrightarrow Ca(NO_3)_2 + CO_2 + H_2O$$

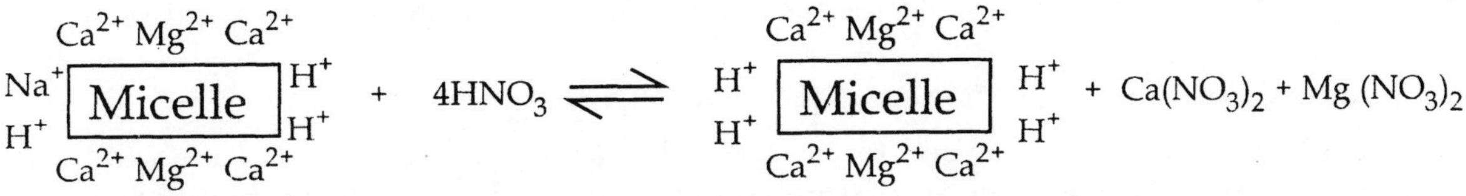

When the clay and humic micelle is saturated with hydrogen ions, the soil becomes acidic if the soil does not contain calcium carbonate. Soil acidity can, however, be rectified by adding 110 kg of limestone for each 100 kg of ammonium sulfate.

### *Calcium Ammonium Nitrate (or Nitrochalk)*

When hydrogen and nitrogen in the ratio of 3:1 are passed over heated (500°C) platinized asbestos or powdered iron under 250 atmospheric pressure, ammonia is formed. A part of ammonia is then mixed with air and passed over a red hot platinum guage and nitric acid is formed.

$$4NH_3 + 5O_2 \longrightarrow 4NO + 6H_2O$$

$$4NO + 2O_2 \longrightarrow 4NO_2$$

$$4NO_2 + O_2 + 2H_2O \longrightarrow 4HNO_3$$

Summing up : $4NH_3 + 8O_2 \longrightarrow 4HNO_3 + 4H_2O$

This nitric acid then reacts with the remaining ammonia to form ammonium nitrate.

$$NH_3 + HNO_3 \longrightarrow NH_4NO_3$$

This ammonium nitrate solution then reacts with limestone to remove nitric acid.

$$2HNO_3 + CaCO_3 \longrightarrow Ca(NO_3)_2 + H_2O + CO_2$$

$$Ca(NO_3)_2 + NH_4NO_3 \longrightarrow Ca(NO_3)_2 NH_4NO_3$$

Nitrochalk

Nitrochalk is a brown or light grey in color and contain 25 per cent nitrogen, one forth of which is ammoniacal and three forth is nitrate.

1. *Reaction in Soil* : Nitrochalk reacts with soil as in the case of ammonium sulfate, but the soil does not become acidic even if there is no reserve calcium carbonate. The nitrate which is produced after oxidation of absorbed ammonium ions by clay and humic micelle, reacts with calcium carbonate to generate calcium nitrate. If the soil does not contain free calcium carbonate, this nitric acid reacts with the clay and humic micelle to form calcium carbonate. Since the soil contain an appreciable amount of calcium nitrate, the reaction does not become acidic.

### *Ammonium Phosphates*

These are also important nitrogen fertilizer because they contain both nitrogen and phosphorus. Ammonium phosphates are prepared from ammonia and phosphoric acid. Since these fertilizers are highly soluble in water, they have good demand for fish culture. Ammonium phosphates are used in liquid form because liquid fertilizers are most effective than solid forms of fertilizers.

### *Ammonium, Potassium, Sodium, and Calcium Nitrates*

Generally ammonia is oxidised to produce nitric acid. This nitric acid is used to make ammonium salts, potash, sodium, and calcium nitrates. Ammonium nitrate contains both nitrate and ammonia and the percentage of nitrogen in ammonium nitrate ranges from 20 to 33 per cent. Potassium nitrate is made by combining nitric acid and potassium chloride. It supplies both potash and nitrate. Sodium nitrate is obtained from salt beds in Chile. Because of its high cost per unit of nitrogen, sodium nitrate is a minor supplier of nitrogen. Calcium nitrate is prepared by combining nitric acid and calcium carbonate. It is also a by-product of the manufacture of nitrophosphate fertilizers.

## 2.3 Phosphorus Fertilizers

Several phosphorus carriers are also widely used to supply phosphorus in fish ponds as shown in Table 2.2. Table indicates that like nitrogen carriers, there is a wide range of phosphorus contents of phosphorus carriers – from 15 to 76% in different types of carriers. The chief source of phosphorus carriers is rock phosphate (Figure 2.2), the important component of which is the mineral apatite. The phosphorus in apatite is slowly available. This material should be treated with sulphuric, phosphoric or nitric acid to change the phosphorus into more available forms such as $CaHPO_4$ and $Ca(H_2PO_4)_2$.

**Table 2.2 : Phosphorus Carriers**

| Fertilizer | Chemical form | Available $P_2O_5$(%) | Phosphorus (%) |
|---|---|---|---|
| Superphosphates | $Ca(H_2PO_4)_2$; $CaHPO_4$ | 16-50 | 7-22 |
| Ammoniated superphosphate | $NH_4H_2PO_4$; $CaHPO_4$ | 16-18 | 7-8 |
| Monoammonium phosphate | $NH_4H_2PO_4$ | 48-55 | 21-24 |
| Diammonium phosphate | $(NH_4)_2HPO_4$ | 46-53 | 20-23 |
| Ammonium polyphosphate | $(NH_4)_3HP_2O_7$; $(NH_4)_3H_2P_3O_{10}$; $NH_4H_2PO_4$ | 58-60 | 25-26 |
| Basic slag | $(CaO)_3.P_2O_5.SiO_2$ | 15-25 | 7-11 |
| Rock phosphate | Fluoro-, chloro-, and hydroxyapatites | 25-40 | 11-16 |
| Calcium metaphosphate | $Ca(PO_3)_2$ | 62-63 | 27-28 |
| Superphosphoric acid | $H_3PO_4$; $H_4P_2O_7$; $H_5P_3O_{10}$ | 68-76 | 29-33 |
| Nitrophosphate | $Ca(H_2PO_4)_2.H_2O$ | 20-35 | 10-15 |

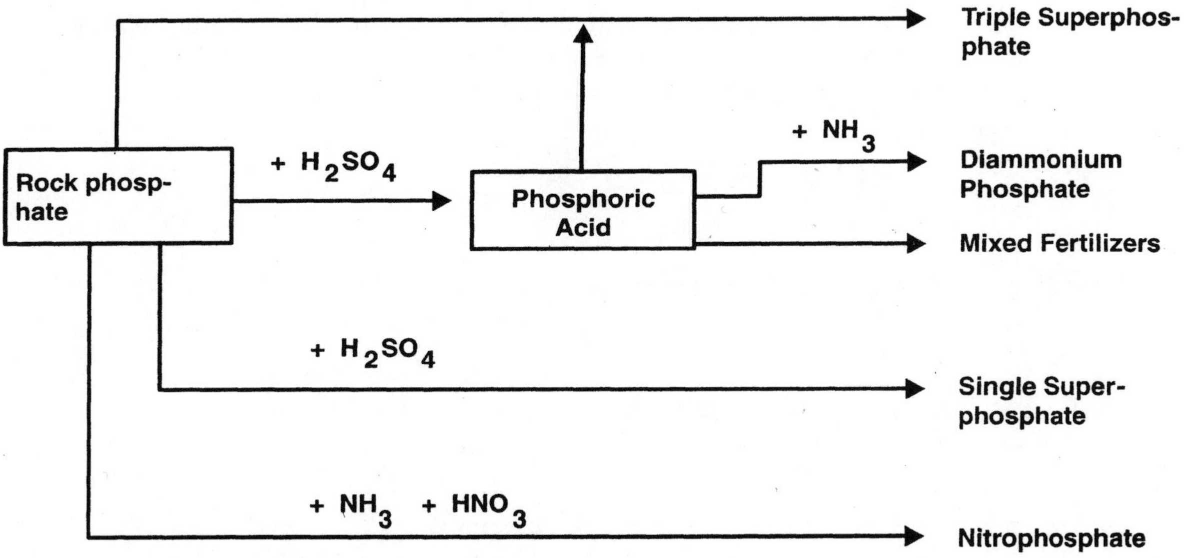


**Fig. 2.2 : Synthesis of various phosphatic fertilizer materials from rock phosphate. Rock phosphate is obtained from naturally occurring mineral apatite. Rock phosphate and other synthetic materials have supplied several important phosphate fertilizers.**

Various types of phosphorus compounds are present in phosphatic fertilizers and are classified as *water-soluble* and *water-insoluble* compounds. Water-soluble compounds are $Ca(H_2PO_4)_2$. $H_2O$, $NH_4H_2PO_4$, and $(NH_4)_2HPO_4$ whereas water-insoluble compounds are called *phosphate* ($[3(Ca_3(PO_4)_2]$. CaX) (X = OH, Cl or F). Generally the phosphorus content of phosphatic fertilizers is expressed as *phosphorus pentaoxide* ($P_2O_5$) rather than as *phosphorus* (P). It is to be noted that $P_2O_5$ is not present in any phosphatic fertilizers and the term oxide is only used as a means of phosphorus content.

## *Single Superphosphate*

When finely ground insoluble rock phosphate is treated with calculated amount of sulphuric acid, water-soluble monocalcium phosphate is formed. This reaction is carried out in a pit den where the temperature rises to about 100°C. The mixture of monocalcium phosphate (27%) and gypsum (50%) is pulverized to powder and marketed as single superphosphate.

$$Ca_3(PO_4)_2\ 3CaF_2 + 7H_2SO_4 \longrightarrow \underbrace{3Ca(PO_4)_2 + 7CaSO_4}_{\text{Single superphosphate}} + 2HF$$

Single superphosphate is an ash color powder containing about 16 per cent water-soluble $P_2O_5$, 20 per cent calcium, and 12 per cent sulphur. However, single superphosphate should not be mixed up with lime because monocalcium phosphate is converted to water-insoluble tricalcium phosphate.

$$Ca(H_2PO_4)_2 + CaCO_3 \longrightarrow Ca_3(PO_4)_2 + 2CO_2 + 2H_2O$$

1. *Reaction in Soil* : Acidic pond soils contain excessive amounts of aluminium and ferric ions and also aluminium and ferric hydrous oxides which react with monocalcium phosphate and convert it into insoluble aluminium and ferric hydroxy phosphate when single superphosphate is applied to acidic ponds.

$$2Al \left\langle \begin{matrix} OH \\ OH \\ OH \end{matrix} \right. + 2Ca\,(H_2PO_4) \rightleftharpoons 2Al \left\langle \begin{matrix} OH \\ OH \\ OH \end{matrix} \right. + 2Ca^{2+} + 2(OH)^-$$

$$2Fe \left\langle \begin{matrix} OH \\ OH \\ OH \end{matrix} \right. + 2Ca\,(H_2PO_4) \rightleftharpoons 2Fe \left\langle \begin{matrix} OH \\ OH \\ OH \end{matrix} \right. + 2Ca^{2+} + 2(OH)^-$$

$$2Al^{+3} + Ca\,(H_2PO_4)_2 + 2H_2O \rightleftharpoons Al \left\langle \begin{matrix} OH \\ OH \\ OH \end{matrix} \right. + Ca^{2+} + 2H^+$$

$$2Fe^{+3} + Ca\,(H_2PO_4)_2 + 2H_2O \rightleftharpoons Fe \left\langle \begin{matrix} OH \\ OH \\ OH \end{matrix} \right. + Ca^{2+} + 2H^+$$

Alkaline pond soils contain excessive amounts of calcium carbonate and calcium ions which react with the monocalcium phosphate and converted it into water-insoluble tricalcium phosphate.

$$Ca(H_2PO_4)_2 + 2Ca^{2+} + 4(OH)^- \rightleftharpoons Ca_3(PO_4)_2 + 4H_2O$$

$$Ca(H_2PO_4)_2 + CaCO_3 \longrightarrow Ca_3(PO_4)_2 + 2CO_2 + 2H_2O$$

## *Triple Superphosphate*

Triple superphosphate is manufactured by treating high-grade rock phosphate with phosphoric acid.

$$Ca_3(PO_4)_2\ 3CaF_2 + 14\,H_3PO_4 \rightleftharpoons 10Ca(H_2PO_4)_2 + 2HF$$

Triple superphosphate (also called concentrated superphosphate) contains 43 per cent water-soluble $P_2O_2$. The reaction of triple superphosphate (TSP) in soil is similar to that of single superphosphate. TSP is widely used in many countries as a good phosphatic fertilizer in fish ponds.

## *Nitrophosphates*

As the name indicates, nitrophosphates (NP) contain both nitrogen and phosphorus. They also contain both ammoniacal and nitrate form of nitrogen and therefore, can be used in wide range of soil conditions. However, the most important features of NP are : (1) it has high water-solubility of the phosphorus components, (2) Nitrophosphates are acidic in their initial reaction with soil. Acidification has two beneficial effects : first, it improves trace element availability and second, it pushes the ammonia : ammonium equilibrium more towards ammonium, thus reducing any tendency to ammonia loss and (3) the synergism between ammonium and phosphate helps to increase phosphate uptake by plankton – essential natural food organisms for fish.

These fertilizers are manufactured by using nitric acid instead of phosphoric or sulphuric acid to increase the solubility of the phosphorus in rock phosphate.

$$2Ca_3(PO_4)_2 + 6HNO_3 + H_2O \longrightarrow 2CaHPO_4 + Ca(H_2PO_4)_2 \cdot H_2O + 3Ca(NO_3)_2$$

Nitrophosphate is a granulated fertilizer containing a stabilizer which prevents the reversion of citrate soluble phosphate to insoluble tricalcium phosphate. Generally three type of nitrophosphates are manufactured. The first type of nitrophosphate contains 20% nitrogen, 20% phosphorus and 2% potash; the second type contains 18% N, 18% $P_2O_5$ and 9% $K_2O$ and the third type contains 18% N, 15% $P_2O_5$ and 15% $K_2O$. The trade name of nitrophosphate is called '*Suphala*'.

### *Rock Phosphate*

Rock phosphate is the trade name of mineral phosphates which denotes the products obtained from metallurgical processing of phosphorus-containing ores. Rock phosphate contains complex apatites which occur in different geological periods. Such complex apatites are : (1) Sedimentary deposit in Uttar Pradesh, (2) Metamorphic-cum-sedimentary deposit in Madhya Pradesh, Bihar and West Bengal, and (3) Igneous deposit in Andhra Pradesh and Rajasthan. However, among different types of rock phosphates, sedimentary rock phosphate is the most effective due to its marine in origin and recent geological age. When finely ground, it is beneficial on acid and mineral soils having considerable organic matter. Rock phosphate is a natural phosphatic fertilizer and contain 25-40% available $P_2O_5$ (12-17%P). Because it is natural, they contain other elements such as calcium (37-45%), magnesium (4-7%), sulphur (2-6%) and trace amounts of zinc, copper and molybdenum. Although India has a vast reserve of about 128 million tonnes of phosphate rock, the consumption of this fertilizer has reached a level of 3.8 lakh tonnes per year during 1995-1996. However, phosphate rock holds a great promise in fish culture ponds mainly due to easily available and much cheaper than any other phosphorus carriers.

### *Diammonium Phosphate*

This compound is widely used as phosphorus-containing fertilizer in agriculture and aquaculture programs. Diammonium phosphate (DAP) contains 16-18% nitrogen and 46-48% $P_2O_5$ (20-21% P). DAP is manufactured by producing phosphoric acid from rock phosphate and reacting the acid with ammonia.

Other ammonium phosphate-containing material is ammonium polyphosphate which is widely used in the manufacture of liquid fertilizers. This type of fertilizer is superior because it prevents the precipitation of impurities present in it and also keep micronutrients in solution. Polyphosphates contain orthophosphates (salts of phosphoric acid), purophosphates (salts of $H_4P_2O_7$), and tripolyphosphates (salts of $H_5P_3O_{10}$).

## 2.4 Potassium Fertilizers

Potassium is obtained by mining underground salt beds. Sometimes, they are leached from the mines by pumping the hot water or hot unsaturated solutions of salts and taking out the saturated solutions. The solutions are then either cooled to crystallize the dissolved potassium salts or subjected to thermal evaporation to get various fractions of solid raw concentrates of potassium salts. Brine from salt lakes are also an important source of potassium salts. Sylvite, Carnalite, Kainite,

Langbeinite; Leonite, and Picronite are also the most common of the crude potash sources. The most important potash carriers are shown in Table 2.3. All potash salts are highly soluble in water and hence are easily assimilated in ecosystem. Although potassium fertilizers are not used in fish culture, they are important so far as the growth of aquatic plants (particularly macrophytes and phytoplankton) are considered.

## 2.5 Micronutrients in Fertilizers

Generally the amount of micronutrients in fertilizers should be carefully controlled than the macronutrients. This is due to the fact the micronutrients are very toxic and required amounts of these elements are extremely low. Therefore, micronutrients should be supplied only when their need is certain. Salts of micronutrients commonly used in fertilizers are shown in Table 2.4. Among different types of micronutrients, zinc, sulfate, cobalt chloride, copper sulfate, and magnesium sulfate are used in fish feed at very low application rates. Copper sulfate is also used to eradicate snails and algal bloom from fish culture ponds.

**Table 2.3 : Potassium Carriers**

| Fertilizer | Chemical form | K (%) | $K_2O$ (%) |
|---|---|---|---|
| Potassium sulfate | $K_2SO_4$ | 40-42 | 48-50 |
| Potassium nitrate | $KNO_3$ | 37 | 44 |
| Potassium magnesium sulfate | Double salt of K and Mg | 19-25 | 25-30 |
| Potassium chloride | KCl | 40-50 | 48-60 |
| Manure salts | KCl | 15-25 | 20-30 |
| Kainit | KCl | 10-15 | 12-15 |

**Table 2.4 : Salts of Micronutrients Commonly Used in Fertilizers**

| Compound | Chemical formula | Nutrient content (per cent) |
|---|---|---|
| Copper sulfate | $CuSO_4.5H_2O$ | 35 |
| Ferrous sulfate | $FeSO_4.7H_2O$ | 19 |
| Ferric sulfate | $Fe_2(SO_4)_3.4H_2O$ | 23 |
| Cupric oxide | CuO | 75 |
| Manganous oxide | MnO | 41-68 |
| Sodium borate (borax) | $Na_2B_4O_7.10H_2O$ | 11 |
| Zinc oxide | ZnO | 80 |
| Zinc sulfate | $ZnSO_4.7H_2O$ | 23 |
| Sodium molybdate | $Na_2MoO_4.2H_2O$ | 39 |
| Ammonium molybdate | $(NH_4)_6Mo_7O_{24}.2H_2O$ | 54 |
| Manganous sulfate | $MnSO_4.4H_2O$ | 26-28 |

*Source :* Tisdale *et al.* (1985).

## 2.6 Mixed Fertilizers

Mixed fertilizers mean a mixture of two or three single fertilizers. For example, single superphosphate and ammonium sulfate may be mixed to get a mixed fertilizer. Application of mixed fertilizers has several advantages such as fertilizer elements can be uniformly applied to soil and less labour requirement for the application.

In order to prepare mixed fertilizers, several materials are required. Although straight fertilizers are the basic materials used to make the mixed fertilizers, some low grade organic materials (such as paddy husk and peat) are added at the rate of 45.5 kg per tonne to the mixed fertilizer to maintain them in a good physical condition. Moreover, dolomite is also added to neutralize acidity of the mixed fertilizer. The following points should be kept in mind before preparing mixed fertilizers :

1. Single superphosphate and ammonium phosphate should not be mixed with fertilizers that contain free lime because water-soluble phosphorus would be converted to water-insoluble phosphorus.

   $$Ca(H_2PO_4)_2 + 2CaCO_3 \longrightarrow Ca(PO_4)_2 + 2CO_2 + 2H_2O$$

2. Urea and sulfate of potash should be mixed only at the time of application to prevent lump formation.
3. Urea should be mixed with mono or dicalcium phosphate.
4. Ammonium sulfate and ammonium nitrate should not be mixed with rock phosphate because they may decompose to liberate ammonia.

   $$(NH_4)_2\ SO_4^{-} + CaCO_3 \longrightarrow Ca(PO_4)_2 + 2CO_2 + 2H_2O$$

   $$(NH_4)_2CO_3 \longrightarrow 2NH_3CO_2 + H_2O$$

## 2.7 Speciality Fertilizers

Various multi-national companies such as Roten-Anfert-Negev Ltd. and Haifa Chemical, Isreal are being produced different types of fertilizers for modern agriculture under the brand names Multi-K, Multi-K + Magnesium and Multi-NPK potassium nitrate enriched with soluble phosphate. These fertilizers are being offered as a economical, fully water-soluble PK fertilizers for fertigation. These fertilizers can be used in agricultural fields with technical monoammonium phosphate or potassium nitrate. All these fertilizers are said to be solid salts, fully water-soluble, chloride free and making the most concentrates commercial fertilizers.

Although these fertilizers are designed for use in intensive agriculture, their application in aquaculture should be considered because of the fact that they contain high percentage of $P_2O_5$ and due to their high solubility power, phosphorus is easily available to overlying water. It may also be used either as direct application or as an intermediate for compound NPK mixture to give the desired solid mixture or liquid formulations. However, the above-mentioned companies are claimed to be environmentally friendly and are said to give plants a balanced supply of chloride-free nutrients without residue or waste.

## 2.8 Recommendation of Fertilizers

Nutrients are generally exhausted from pond soil and water by the development of plankton populations which ultimately consumed by fish. Therefore, nutrients of aquatic ecosystems would be at least partially withdrawn by the time of fish harvesting. The fish farmers should know the accurate level of depletion of nutrients before the next fish culture program is accomplished in order to compensate for their deficiency after the previous harvesting and leaching.

## 2.9 Rapid Test for Soil and Water Nutrients

The group of test widely used for nutrient availability are the rapid tests. As the name implies, the individual determination is rapidly made. Sometimes, highly experienced fish farmers may fail to predict the deficiencies of a particular aquatic ecosystem. However, scientific methods have been developed to test water and soil samples and to provide reliable information. Detailed analysis of soil and water and their interpretations are difficult and only well-organized laboratory can be able to analyse different samples. But it is important to note that it is possible to obtain valuable informations about water and soil samples by carrying out simple and rapid tests which are generally given in several hand-books. Required chemicals are kept inside water and soil testing kits and the kits are so well designed that anybody can do the tests without any difficulty. However, the interpretation of rapid test data is accomplished by experienced and trained personnel, who fully understand the scientific principles underlying the common field procedures.

## 2.10 Aspects of Fertilizer Practice

Fertilizer practice usually involves several factors such as water, soil, fish, and fertilizers. Since variations of these factors depend on places, it is difficult to arrive at definite conclusions for fertilizer use. For phytoplankton growth, nitrogen fertilization is the most important issue. Phosphorus fertilization is made to balance the nitrogen supply. The main aspect of fertilizer practice relates to the effects of fertilizer use other than fish production. Application of correct dose of fertilizer can result in increased soil and water fertility over a period of time, if fertilization rates exceed the removal rates of nutrient by the fish. On the other hand, excessive use of fertilizer is not only adversely affect the quality of the environment but wasteful of the nutrients. Excessive levels of nitrate, nitrite, ammonia, and phosphate are harmful to aquatic life. Therefore, to determine fertilizer practice, consideration of environmental quality should not be ruled out.

## 2.11 Economy of Fertilizer

The relative costs of nitrogen and phosphorus indicate that the latter is more costly per unit of phsophorus than a similar unit of the former. Generally, there is considerable variation in the cost of different nutrient carriers for a given element. As a result, the cost of manufacture of nitrogen in solid fertilizers (such as urea and ammonium nitrate) is double that in anhydrous ammonia.

## 2.12 Guarantee of Nutrient Carriers

Stringent law should be adopted so that every nutrient carrier has a guarantee so far as the content of nitrogen (N), phosphorus ($P_2O_5$), potash ($K_2O$), and other elements (if necessary) is considered. In any nitrogen fertilizer, however, the total nitrogen is expressed in its elemental form (N). Similarly in a phosphatic fertilizer, the phosphorus is marked in terms of available phosphoric acid ($P_2O_5$) or available phosphorus (P). Thus, 46-0-0 or 0-16-0 fertilizer contains 46 per cent of total nitrogen or 16 per cent of available phosphoric acid (7 per cent available phosphorus). In fish culture, fertilizers that contain either N or $P_2O_5$ are generally used. Mixed fertilizers, however, are not usually recommended under Indian conditions.

According to nutrient ratio, commercial fertilizers are classified into groups. For example, 0-16-0, 0-48-0, 22-0-0 and 44-0-0 respectively have 0-1-0, 0-3-0, 1-0-0 and 2-0-0 ratio. These nutrient carriers should give essentially the same result when used in equivalent amounts. Thus, 250 kg of

0-48-0 furnishes the same amount of phosphorus as does 750 kg of 0-16-0. Similarly, 100 kg of 44-0-0 provides the same amount of nitrogen as does 200 kg of 22-0-0.

It should be recalled that this grouping is very important when the comparative costs become the deciding factor as to which fertilizer should be used.

## 2.13 Lime*

The term lime is,assigned to the calcareous matter present in nature which finds wide use as agricultural fertilizers, in various manufacturing processes, laboratories, in medicines, and in fish culture ecosystems. It mainly represents the carbonate, oxide, phosphate, sulfate, nitrate, and hydroxide with which it is sold in bulk quantity. The chief element of lime is calcium and its total amount in these compounds is expressed as percentage.

Calcium is abundant element of Group IIA of the periodic table and occurs in a wide range of materials in combination with other elements. It is widely distributed in soil, water, plants and animals as calcium and plays an important role in different physiological activities of plants and animals. Moreover, biotic communities can easily assimilate the calcium from water and soil to great extent.

### *Types and Sources of Calcium*

In India, the main deposits of calcium occur as stratified bodies and contain a very few calcium materials such as calcium oxide (CaO) or quicklime, calcium carbonate ($CaCO_3$) or marble, and calcium sulfate ($CaSO_4.2H_2O$) or gypsum which are commercially important for recovery of calcium. Calcium is mixed with hydrogen, oxygen, iron, chlorine, bromine, iodine, sulphur, nitrogen, carbon, and phosphorus. Calcium is found in chalk, limestone, marble, calspar, coral, etc. and other important minerals such as dolomite ($CaCO_3.SiO_2O_8$), fluospar ($CaF_2$), gypsum, apatite [$Ca(PO_4)_2$. $CaF_2$], and calcium phosphate $Ca_3(PO_4)_2$. Calcium carbonate occurs in Madras, Madhya Pradesh, Rajasthan and Punjab; limestone in Madhya Pradesh and Punjab; fluospar is found in Rajasthan and Jabalpur; and gypsum occurs in Bikaner and Jodhpur.

### *Chemistry of Liming Materials*

Of the various liming materials used in fish culture, the followings are very important and hence their physical and chemical properties will be briefly discussed :

1. *Calcium Oxide :* It is a white amorphous powder and its melting point is 2570°C. The reaction with water is vigorous, attended by the evolution of much heat and formation of slaked lime. The lime swells up, cracks and form a fine powder.

$$CaO + H_2O \longrightarrow Ca(OH)_2$$

   It is employed as a drying agent. It is widely used in furnace linings, in fish ponds, and in the production of calcium carbide.

2. *Calcium Hydroxide :* It is also a white amorphous powder and sparingly soluble in water. The solubility in water decreases with increase in temperature of water. On heating, it is decomposed into oxide. The suspension of slaked lime in water is known as the *milk of lime,* while the clear solution *lime water,* is strong alkaline in nature. It absorbs carbon

* For further study, see Tisdale *et. al.* (1985).

dioxide from air or produce calcium carbonate and consequently, lime water turns milky. With excess of carbon dioxide, this clears out due to the formation of calcium bicarbonate.

$$Ca(OH)_2 + CO_2 \longrightarrow CaCO_3 + H_2O$$

$$CaCO_3 + H_2O + CO_2 \longrightarrow Ca(HCO_3)_2$$

It is used in different industries and as a general disinfectant in pond water.

3. *Calcium Carbonate* : It is a white powder and sparingly soluble in water but soluble in acids and in carbon dioxide water. In acid, it forms soluble bicarbonate. On heating, calcium carbonate generates carbon dioxide and lime and the reaction is reversible. It is widely used in different industries, laboratories, medicines, and in fertilizer.
4. *Calcium Sulfate* : It is also sparingly soluble in water but dissolves in a hot concentrated solution of ammonium sulfate and calcium-ammonium sulfate is formed. Precipitate finds application in the preparation of ammonium sulfate as a fertilizer.
5. *Calcium Nitrate* : It is highly soluble in water and alcohol. On heating, it decomposes into oxide. It is also used as a fertilizer.
6. *Calcium Phosphate* : It is sparingly soluble in water but dissolves in dilute mineral acids and in aqueous solution of phosphoric acid. The latter solution yields crystals of calcium dihydrogen phosphate, generally termed as *monocalcium phosphate.* It is a very important fertilizer and is converted into *superphosphate of lime* $Ca(H_2PO_4)_2$. It is the mother substance for the preparation of phosphorus, phosphoric acid, and superphosphate of lime.
7. *Calcium Carbide and Calcium Cynamide* : Calcium carbide is an important industrial compound and is primarily used as a source of acetylene and in the manufacture of calcium cynamide. The latter is used as a fertilizer.
8. *Dolomite* : It is a mixture of carbonates of calcium and magnesium. It is a greyish powder, soluble in pure water, and gives effervescence with dilute hydrochloric acid.

## 2.14 Quality Evaluation of Liming Substances

Fish production largely depends on the development of bottom organisms in pond soils. Therefore, it is a very important task to keep the bottom mud in good condition throughout the culture period. Preparation of pond before and during fish culture plays a very key role for existence of bottom-dwelling organisms. To serve the purpose, liming fish ponds is one of the most important criteria that should be followed with regularity.

Prior to application of liming substances into the pond for higher productivity, neutralization value (or calcium carbonate equivalent), fineness factor, and percentage of effective calcium carbonate should be determined. Otherwise, it would not be possible to recommend the accurate quantities of liming substances required for neutralization of acid ponds.

### *Neutralization Value (NV)*

The term neutralization value (NV) or calcium carbonate equivalent (CCE) refers to the relative power of lime substances to neurtalize acidity. Generally pure calcium carbonate possess 100% NV and it is the standard against which different types of liming substances are compared. Hence, the neutralizing power of any liming substance is considered as a strength with reference to calcium carbonate or its CCE.

### *Fineness Factor (FF)*

For calculation of fineness factor (FF), liming material is allowed to sieve through 60 (0.25 mm) or 8 (0.03mm) mesh sieves. Lime particles are passed through 60 mesh sieve are rated 100% efficient, those passing through 8 mesh sieve are rated 50% efficient and those retained in 8 mesh sieve are rated 20% efficient. In order to determine the FF of lime substances (Table 2.5), six samples are collected from different sources and is allowed to mix all the samples thoroughly. Two hundred and fifty gram of the mixture is sieved through 60 mesh through constant shaking and the remaining material over the sieve is then passed through 8 mesh sieve. The finer the liming substance, quicker is the reaction with the soil. Various types of lime substances are available in the market and vary in their particle size. Therefore, fineness of liming substances should be considered very carefully before or during fish culture operations.

**Table 2.5 : Estimation of Fineness Factor of Liming Substances**

| Sieve number (in mesh) | Quantity of lime passed (g) | Residue (g) | Percentage of particles | Efficiency factor | Fineness factor (%) |
|---|---|---|---|---|---|
| 60 | 235 | - | 94 | 1.0 | 94.0 |
| 8 | 10 | - | 4 | 0.5 | 2.0 |
| 0 | | 5 | 2 | 0.2 | 0.4 |

*Source* : Gupta *et al.* (1997).

### *Percentage of Effective Calcium Carbonate (PECC)*

The value of PECC is obtained by multiplying the estimated CCE with fineness factor. With the adoption of standard methods, the values of CCE, FF (%), and PECC of commercially available lime substances have been calculated (Table 2.6).

**Table 2.6 : Efficiency of Liming Substances Used by Farmers**

| Lime material | NV or CCE | Fineness factor (%) | PECC |
|---|---|---|---|
| Agricultural lime | 93.55 | 96.4 | 90.2 |
| Calcite | 99.63 | 90.0 | 89.7 |
| Dolomite | 89.19 | 99.0 | 88.3 |
| Hydrated lime | 86.39 | 68.2 | 58.9 |
| Hydrated granules | 123.37 | 52.4 | 64.6 |
| Quick lime | 127.45 | 84.1 | 108.5 |
| Shell powder | 110.14 | 71.0 | 78.2 |

*Source* : Gupta *et al.* (1997).

## 2.15 Calculation of Lime Requirement for Ponds

Problems with acid-base relationship in fish ponds can often be solved by liming. Though it is not a form of fertilization, a remedial procedure to improve conditions for fish survival, growth and reproduction. For synergistic effects of fertilizers, bottom mud must not be highly acidic. Acidic muds adsorb phosphate, benthic organisms do not grow well, and phytoplankton do not have adequate carbon and calcium for growth. Although lime requirement techniques have been developed for agricultural purposes, it is also applicable for fish ponds.

The requirement of lime substances of a pond is defined as the quantity of lime substance that should be applied to enhance the soil pH from 4.0 to 7.0. However, the calculation of lime requirement is made through the following two steps :

1. *Lime Requirement :* It is calculated on the basis of the actual pH of pond soil and the extent of the area to be applied. This can be derived from the following formula :

$$\text{Lime required} = \frac{\dfrac{\text{Desired pH} - \text{Actual pH}}{0.1} \times 0.5}{\text{Efficiency of lime}} \times \text{Area}$$

2. *Recommended Rate (RR) :* The recommended rate (RR) is calculated by dividing the value of lime required as pure calcium carbonate (100%) with the PECC value of that particular lime substance using the following formula.

$$\underset{\text{(Tonne/ha)}}{\text{Recommended Rate (RR)}} = \frac{\text{Liming rate as pure calcium carbonate (Tonnes/ha)}}{\dfrac{\text{PECC}}{100}}$$

Table 2.7 shows the quantities of different types of lime substance required to raise the soil pH upto 7.0.

**Table 2.7 : Amount of Different Types of Lime Needed to Raise the Soil pH Upto 7.0**

| Soil pH | Pure $CaCO_3$ | Amount of lime substances required | | | | | | |
|---|---|---|---|---|---|---|---|---|
| | | Agricultural lime | Calcite | Dolomite | Hydrated lime | Hydrated granules | Quick lime | Shell powder |
| 6.5 | 2.5 | 2.8 | 2.8 | 2.8 | 4.2 | 3.9 | 2.3 | 3.2 |
| 6.0 | 5.0 | 5.5 | 5.6 | 5.7 | 8.5 | 7.7 | 4.6 | 6.4 |
| 5.5 | 7.5 | 8.3 | 8.4 | 8.5 | 12.7 | 11.6 | 6.9 | 9.6 |
| 5.0 | 10.0 | 11.1 | 11.1 | 11.3 | 17.0 | 15.5 | 9.2 | 12.8 |
| 4.5 | 12.5 | 13.9 | 13.9 | 14.2 | 21.2 | 19.3 | 11.5 | 16.0 |
| 4.0 | 15.0 | 16.6 | 16.7 | 17.0 | 25.5 | 23.2 | 13.8 | 19.2 |
| Efficiency percent | 100 | 90.2 | 89.7 | 88.3 | 58.9 | 64.6 | 108.5 | 78.2 |

*Source :* Gupta *et al.* (1997).

Another technique for the determination of lime requirement of soil is that about 20 g of air dry soil is mixed with 20 ml water and after one hour, the pH of solution is determined. The wet soil sample is mixed with a buffer solution (pH 7.0). The solution is then shaked for one hour and the pH is determined. Each 0.1 reduction in pH denotes need for approximately 500 kg calcium carbonate.

## 2.16 Chemical Reactions of Liming Substances in Water and Soil

Liming materials usually react with acidity as follows for dolomite :

$$CaMg(CO_3)_2 + 4H^+ \longrightarrow Ca^{2+} + Mg^{2+} + 2H_2O + 2CO_2$$

This reaction neutralizes acidity and pH is increased. The $Ca^{2+}$ and $Mg^{2+}$ concentration in the water increases accordingly, total hardness (the sum of the calcium and magnesium ions expressed as equivalent calcium carbonate) also increased. Dolomite and other liming materials are sparingly soluble in water. Carbonate or hydroxide ions resulting from their dissolution reacts with carbon dioxide to form bicarbonate :

$$CO_3^{2-} + CO_2 + H_2O \longrightarrow 2HCO_3$$

$$OH^- + CO_2 \longrightarrow HCO_3$$

If all the carbon dioxide is removed, pH will increase above 8.5 and as a result of which hydroxide of carbonate will accumulate. Higher pH (above 10.0) results from use of calcium hydroxide or calcium oxide to form calcium in water.

It has been be noted that liming also increased total alkalinity of pond water because of higher concentrations of bicarbonate, hydroxide and carbonate. The effects of liming on total hardness, pH and total alkalinity are shown in Table 2.8.

**Table 2.8 : Water and Soil pH, Concentration of Hardness and Alkalinity in Ponds Treated With Limestone and in Control Ponds. Values are Means of Four Ponds**

| Variables | Limed ponds | Control ponds |
|---|---|---|
| pH of water | 8.0 | 7.0 |
| pH of soil | 6.5 | 5.9 |
| Total hardness (mg/1) | 45.7 | 12.9 |
| Total alkalinity (mg/1) | 185.0 | 67.5 |

When liming materials are added to a fish pond, it is deposited in pond mud where the calcium and magnesium compounds react with carbon dioxide and with the acid colloidal complex. Generally in acid soil, liming materials react with carbon dioxide and water to form bicarbonate due to high partial pressure of carbon dioxide in the soil. The following chemical reactions take place :

$$CaO + H_2O + 2CO_2 \longrightarrow Ca(HCO_3)_2$$

$$Ca(OH)_2 + 2CO_2 \longrightarrow Ca(HCO_3)_2$$

$$CaCO_3 + H_2O + CO_2 \longrightarrow Ca(HCO_3)_2$$

Liming materials react with acid soil, the magnesium and calcium replacing hydrogen and aluminium on the coiloidal complex. Assuming that hydrogen ions are replaced, the adsorption of calcium compounds to soil colloidal may be indicated as follows :

$$\begin{matrix}H^+\\H^+\end{matrix}\boxed{\text{Micelle}} + CaCO_3 \rightleftharpoons Ca^{2+}\boxed{\text{Micelle}} + H_2O + CO_2$$

$$\begin{matrix}H^+\\H^+\end{matrix}\boxed{\text{Micelle}} + Ca(OH)_2 \rightleftharpoons Ca^{2+}\boxed{\text{Micelle}} + 2H_2O$$

These reactions are completed with the formation of carbon dioxide. Moreover, adsorption of magnesium and calcium increases the percentage base saturation of the colloidal complex and as a result soil pH significantly increased.

In limed soil, three forms of calcium and magnesium are found such as exchangeable bases adsorbed by the colloid matter, dissociated cations in the soil solution, and undissociated solid calcium and Ca-Mg carbonates. As the reaction goes on, calcium and Ca-Mg carbonates completely dissolved and involves the exchangeable cations and those in soil solution.

In fish culture ecosystems where the leaching takes place, soluble calcium and magnesium compounds are eliminated from the soil resulting in gradual reduction of the percentage base saturation (soil acidity or alkalinity) and pH. Therefore, further application of lime is necessary.

## 2.17 Role of Chemical Composition of Liming Materials

Different forms of liming materials contain essential calcium and magnesium which affects the reaction rate with soil. For examples, limestone (calcite or dolomite) reacts more slowly for longer period to react fully with soil. On the other hand, hydrated and burned limes react very rapidly and soil pH changes within a very short time. Calcium and magnesium contents in four liming materials are shown in Table 2.9. Table shows that dolomitic limestone has a slightly higher neutralizing power than calcitic limestone with the same degree of purity. On the contrary, burned and hydrated limes have greater acid neutralizing power. This is important for determining the type of lime to be used and the actual amount to be applied for maximum benefit.

**Table 2.9 : A Comparison of Different Means of Expressing the Calcium and Magnesium Contents in Four Liming Materials**

| Types of material | Actual chemicals (%) | Means of expressing composition | | | | | | |
|---|---|---|---|---|---|---|---|---|
| | | Element (%) | | Conventional oxide (%) | | CaO equivalent | Total Carbnates | $CaCO_3$ equivalent |
| | | Ca | Mg | CaO | MgO | (%) | (%) | (%) |
| Hydrated lime | $75Ca(OH)_2$ | 40.5 | - | 58.6 | - | 78.9 | - | 140.8 |
| | $23Mg(OH)_2$ | - | 9.6 | - | 15.9 | | | |
| Burned lime | 77CaO | 55.0 | - | 77.0 | - | 102.2 | - | 182.5 |
| | 18MgO | - | 10.9 | - | 18.0 | | | |
| Calcitic limestone | $95CaCO_3$ | 38.0 | - | 53.2 | - | 53.2 | 95 | 95 |
| Dolomitic limestone | $35CaCO_3$ | (14) | - | (19.6) | - | (19.6) | 95 | (35) |
| | $60CaMg(CO_3)_2$ | (13) | (7.9) | (18.2) | (13.1) | (36.4) | | (65) |
| | | 27 | 7.9 | 37.8 | 13.1 | 56 | | 100 |

*Source* : Brady (1991).

### *Particle Size of Limestone*

If liming material is finer, it will rapidly react with soil and water. Generally calcium oxide and calcium hydroxide are available as powdered form. But in some limestones, particle size vary considerably. Agricultural limestone is prepared by pulverizing calcium carbonate or dolomite. It was discovered long ago that soil acidity was neutralized faster by fine than by coarse particles of limestone. However, fine limestone is better than coarse limestone for use in fish ponds. The fineness of limestone is measured by passing soil sample through a series of standard screens with openings of designated size. Limestone particles of different size classes for use in efficiency tests are shown in Table 2.10.

**Table 2.10 : Particle Size Distribution of Two Agricultural Limestones**

| Classification of materials by screen | Particle size (mm) | Particle distribution (%) | |
|---|---|---|---|
| | | Dolcito | Southern stone |
| Retained on 10 mesh | > 1.7 | – | 7.82 |
| Retained on 20 mesh | 1.69-0.85 | 0.04 | 14.03 |
| Retained on 60 mesh | 8.84-0.25 | 5.29 | 24.57 |
| Retained on 100 mesh | 0.25-0.15 | 44.36 | 16.25 |
| Retained on 270 mesh | 0.14-0.053 | 9.31 | 16.32 |
| Passed through 270 mesh | < 0.052 | 41.00 | 21.01 |

*Source :* Boyd and Hollerman (1982).

## 2.18 Liming in Acid-sulfate Soils

Ponds with low pH (below 4.5) generally receive drainage from acid-sulfate soils. These soils contain pyrites that oxidize to form sulphuric acid. Although liming reduces the acidity of ponds with acid-sulfate soils, the acidic conditions will rapidly appear. For successful fish culture in such ponds, the acid run-off and seepage should be diverted, or the pyrite-containing soil should also be treated to reduce the potential for release of sulphuric acid.

## 2.19 Organic Manures*

In addition to the use of inorganic fertilizers, organic manures have been employed in freshwater fish culture ponds. Generally manures are obtained from either plant or animal sources and they contain small quantity of nutrients such as nitrogen, phosphorus, and potash. Organic manures are added to fish ponds to augment fish production by increasing productivity through increase in the concentration of nutrients. It should be remembered that inorganic fertilizers pollute the ecosystem in such a way that the existence of aquatic life becomes impossible. In contrast, organic manures are less hazardous than chemical fertilizers. Organics are important for improving and maintaining physical conditions of soil and also effective in soils where water and nutrients are leached considerably. Therefore, organic manures are called as *soil conditioners*. Although nutrient contents of different organics are very low in contrast to chemical fertilizers, the former types are widely practised in rural areas because it is easily available and more cheaper. Nutrient contents of various types of organic materials are shown in Table 2.11.

### *Oil Cakes*

Oil cakes are the residual left after extraction of oil from oil seeds. Different types of oil cakes are used in fish culture to increase the fertility of soil and water. Oil cakes have great demand as fertilizer in agricultural soil as well as fish culture ecosystems. Among different types of oil cakes, mahua, mustard, groundnut, and neem cakes are very important so far as fish production potential is concerned. After addition of oil cakes to soil and water, they are decomposed by soil micro-organisms.

1. *Groundnut Cake :* Groundnut plant (*Arachis hypogea,* Family : Leguminecae) is known as an annual branched herb with pinnate compound leaves and widely distributed in tropical, subtropical, and temperate areas of the world. The thick, oblong and wrinkled

* For further study, see Yawalkar *et. al.* (1986), Hall and Smith (2002).

fibrous pods contain 2 to 4 red or brown ovoid nutritious seeds with white fresh and are harvested from below soil level. The average yield of unshelled pods is about 1,600 kg/ha. The protein-rich seed residue (groundnut cake) is widely used in animal feed, agriculture and in aquaculture as fertilizer and feed. Prior to application in fish culture, the cake is powdered and then uniformly broadcast over pond water. Groundnut cake quickly decomposes in pond mud and make available the nutrients present in it for the growth of plankton and bottom fauna.

**Table 2.11 : Different Types of Organic Manures Used in Fish Ponds and Their Percentage of Nutrient Composition**

| Manure | Nutrient composition (%) | | |
|---|---|---|---|
| | N | $P_2O_5$ | $K_2O$ |
| I. *Animal Origin* | | | |
| 1. Cattle dung | 0.5 | 0.4 | 0.2 |
| 2. Pig dung | 0.6 | 0.6 | 0.4 |
| 3. Poultry dropping | 1.7 | 0.9 | 0.7 |
| 4. Duck dropping | 0.9 | 0.5 | 0.6 |
| 5. Farm Yard Manure | 0.6 | 0.5 | 0.1 |
| II. *Plant Origin* | | | |
| 1. Groundnut cake | 6.5 | 1.0 | 1.0 |
| 2. Mustard cake | 4.5 | 1.5 | 1.6 |
| 3. Cotton-seed cake (decorticated) | 6.4 | 2.9 | 2.2 |
| 4. Linseed cake | 5.6 | 1.4 | 1.3 |
| 5. Mahua cake | 2.5 | 0.8 | 1.9 |
| 6. Neem cake | 5.2 | 1.1 | 1.5 |
| 7. Coconut cake | 3.0 | 1.9 | 1.8 |
| 8. Rape-seed cake | 5.2 | 1.8 | 1.2 |

2. *Neem Cake* : Neem tree (*Azadiracta indica*) is a member of the Mahogany family (Meliaceae). The tree grows is Asia, Africa, Australia, and other tropical and subtropical areas. A fully grown tree (about 10 years old) can yield about 50 kg of seeds. The oval fruits (1.4-2.4 cm long) are produced usually once and sometimes twice a year. However, neem has attracted interest because of its manurial value, pesticidal products for controlling rice pests, its fuelwood, shade value, and as a component of reforestation. Neem cake is a residue of the plant *A. indica*. This residue is produced after neem oil is extracted. Neem cake is powdered before use in fish culture and it is broadcast over pond water.

Nimbidin and azadiractin are considered as the most active ingredients of the neem tree. All parts of the tree contain these ingredients, but is more concentrated in the seed. These are tarpenoid compounds which might responsible for toxic property of the neem cake. For this purpose neem cake should be detoxified using cold water. The process involves the soaking of 3 kg neem cake in 90 litres of water for 24 hours. Supernatant is decanted and the sowlen cake is broken into finer pieces by rubbing in between the palms. The pot is filled again with same quantity of water, allowed to settle for half an hour and the supernatant is decanted. The process is repeated ten times. The cake mass is then filtered through a calico to remove water and dried in the sun. In this process all toxic compounds are removed.

3. *Mahua Cake :* The mahua (*Madhuca indica* and *Bassia latifolia*) seed cake is very bitter and poisonous and is mainly used as manure. Mahua cake is also widely used for removing unwanted aquatic life during fish pond preparation. The cake contains 4-6% saponin (mowrin) which is soluble in water. The toxicity of mahua cake is due to presence of saponin which enters the blood stream through gills and buccal epithelium of fish. Besides saponin, other toxic compounds such as resins, tannins, and sapoglucoside are also present in considerable amounts. Similar to neem cake, expeller-processe mahua cake obtained from market, is detoxified using cold water treatment and as a result, all toxic compounds are completely removed from the cake. Nutrients of mahua cake becomes available to soil and water after two months of application. Therefore, a period of 45-50 days should be allowed after application of the cake before stocking the ponds with fish. This will ensure the removal of toxic effects of the cake to get released into soil and water for building up of suitable ecosystem for fish culture.
4. *Mustard Cake :* Mustard cake is a residue of seeds of the plant *Brassica nigra* (Family : Cruciferiae). This residue is obtained after the extraction of oil from seeds. The cake is extensively used in fish culture as feed and fertilizer. It is easily available in the market in rural areas.

## 2.20 Green Manure

It is the fresh green plant material which is mixed up with the soil when the pond is dried up during summer. Green plants are cut into small pieces and spread over pond soil and then ploughed just before the onset of monsoon. The plant material is slowly decomposed by soil micro-organisms and as a result nutrients are released to overlying water.

Generally green leaf manuring plants must grow rapidly and easily decompose in the soil. Green foliage of some important green leaf manuring plants is added to the soil. Hence, organic matter and nutrients are added to water from outside sources. Among different plants, *Sesbania bispinosa, Pongamia glabra* and *Gliricida maculata* are important. The first type of plant produces about 2,500 kgs of fresh green leaves per hectare, the second type yields 130 kgs of fresh green material while the third type of plant produces 20-25 kgs of green material per hectare. After complete decomposition of green manures, it improves the structural and textural properties of soil due to formation of humus compounds. Humus are more or less stable fraction of the soil organic matter remaining after the major portion of added plant residues have decomposed. It is dark in color.

Green manuring in fish culture ponds means a mechanical preparation of the bottom mud of the pond followed by growing of rye, oats or barley, which is mowed prior to filling the pond with water during monsoon season. Grain crops are suitable as they decompose slowly. On solid parts of plant materials, a large number of fish food organisms are developed and consumed by the fish. Green manuring plants are grown at the edges of the ponds and are utilized during the fish-rearing periods. Cattle being allowed to graze there, affording additional manure.

### *Role of Green Manuring in Fish Culture*

Green manuring is practiced in ecosystems where there is excessive loss of water and nutrients through leaching. Because the cost of chemical fertilizers are relatively high and have got pollution potential, it is better to use green manure to avoid water pollution and nutrient loss. Different species of fish have response differently to green manuring under Indian conditions. In rural areas where poor farmers fail to procure costly chemical fertilizers for fish cultivation, use of green manure is beneficial to farmers.

The role of green manuring is manifold, such as (1) it supplies organic matter – it not only acts as food for soil organisms but also includes microbial activity in the soil and as a result, productivity is increased, (2) it restores soil nitrogen if leguminous plants are used, (3) green manure increases the solubility and availability of lime, (4) after decomposition of green manure, it provides inorganic nutrients and increase the concentration of magnesium, potassium, and iron, (5) on decomposition of green manure, humus compounds are formed which increases the adsorption capacity of the soil and as a result soil aeration, drainage, and granulation is triggered, and (6) green manure forms substratum in the pond bed for development of algae, worms and insect larvae.

## 2.21 Compost

It is an organic residue, or a mixture of organic residue that have been piled, moistened, and allowed to undergo biological decomposition. Compost is also called *'artificial manure'* or *'synthetic manure'* if produced from plant residues. The compost is also considered as a natural product of the decomposition of dead plant materials. The most familiar compost occurring naturally is present in nature and deciduous woodland as leaf-mould on the forest floor. However, compost is used in fish ponds to provide a supply of humus and nutrients.

Plant materials are composed of lignin, cellulose, hemicellulose, and proteins. These compounds are gradually decomposed in presence of available nitrogen, phosphorus, and potash. If the plant materials contain 1.2% of nitrogen, there is no need to add extra nitrogen. Generally plant materials contain about 0.5% of nitrogen and therefore, 0.8 parts of soluble nitrogen should be added to every 100 parts of the plant material for proper decomposition by micro-organisms.

### *Preparation of Compost*

Compost is prepared by dumping alternately with layers of waste plant materials and cattle dung along with urine-soaked litter. There are three processes for preparation of compost : (1) Indore method, (2) Bangalore method, and (3) Coimbatore method.

In Indore method, about 25 kg of dry waste plant material is spread in a pit under each pair of animal (1m deep x 2m wide x 5m length) as bedding. In next day, the mixture of cattle dung and urine-soaked litter is spread at the bottom. A *slurry* (3.5 kg of urine + 4.5 kg of fungal inoculation + 0.5 kg of wood ash + 4.5 of cattle dung in 4 gallons of water) is added over the bed. The fungal inoculation is the material which is collected from an old compost pit. The old compost material contains micro-organisms. Freshwater is spread over the slurry. In this way, the compost pit is filled up gradually till it rises about 35 cm above the ground. When the pit is filled up, water is spread over it, it is mixed and replaced in the pit. After one month it is taken out and heap is made below the shade. It is ready in 3-4 months. In this way well-decomposed manure is obtained within very short time although some nitrogen and organic matter is lost due to rapid decomposition.

In Bangalore method, about 20 kg of waste plant material is spread under each pair of animal in a pit (1m x 2m x 5m). In next day, the mixture of cattle dung and urine-soaked litter is filled in sections of 60 to 120 cms at a time. A bamboo partition is placed across the pit at a distance of 60 to 120 cms from one end. Water is sprayed over it. Mixture of cattle dung and urine-soaked litter is sprayed daily over the previous layer in the pit. After filling this section, bamboo partition is taken out and a similar section is made next to it. In this way, manure pit is filled up section-wise. When the pit is completely filled up, it is covered with mud. The manure is ready in about 6-months. Since the compost material is slowly decomposed, less amount of nitrogen and organic matter is lost.

In Coimbatore method, waster plant material is spread over the bottom of a pit. A mixture of 10 kg of cowdung, 50 litres of water = 1 kg of bone and meal is sprinkled over it. A layer of plant material is again placed over the previous layer and the mixture of cowdung, water and bone meal is spread. In this way, the pit is filled up until the material rises about 60 cms above the ground. After two months, compost material is taken out, water is applied to it and it is dumped below the shed. It takes about 6-months to make the compost more effective.

## 2.22 Vermi-compost

Solid organic waste material can be processes in several ways to make it a useful product. Waste materials are collected from agricultural washes and urban areas and utilized by converting them into organic manures. These waste materials are processes through simple technique with the aid of some species of earthworms such as *Elisenia fetida, Edurilus engenaia* and *Perionix excavator.* This process is generally termed as *vermi-compost*.

### *Composting Method*

Waste materials are collected and kept in a tank. Cattle dung is thoroughly mixed with the waste. The surface layer is covered with a 3 cm thick soil layer and 2 cm diameter holes at 25 cm apart for proper aeration in the chamber. The earthworms are released on the surface of the tank at the rate of 1,500-2,000 $m^2$. Worms will slowly enter into the organic matter through the holes on the soil covering. In this way, the entire tank is kept without disturbing for six weeks.

The soil covering from the tank is then removed and the manure compost is kept on the floor, a pyramid is made for ten hours. While making this, earthworms are found to move down at the base of the pyramid of the manure compost. The pyramid is then pressed from the top to spread the compost. The entire material is then dried in the shed for 3 days and passing through a 3 mm sieve. Small worms and cocoons are separated from the vermi-compost.

### *Application of Vermi-Compost*

Because the nutritional status of vermi-compost is very high (Table 2.12), it can be utilized as an alternative animal protein substitute for aqua feeds. Fish feed can be substituted by complementing the vermi-compost with fish feed. Besides, it is also used as a good organic manure in aquaculture.

## 2.23 Farm-Yard Manure

Farm-yard manure (FYM) or farm manure is the material which is produced from decomposition of the mixture of animal dung. Fresh FYM should not be used directly because micro-organisms assimilate all the available nutrients in the soil and therefore, fish culture ecosystems adversely affected by the lack of nutrients. FYM should be thoroughly mixed with the soil by ploughing when the water bodies become dried up during summer months. It decomposes in about a month and improves the soil structure and texture to significant extent.

**Table 2.12 : Chemical Composition of Vermi-Compost**

| Constituent | Range |
|---|---|
| Organic carbon (%) | 9.15 - 17.98 |
| Total nitrogen (%) | 0.5 - 1.5 |
| Available phosphorus (%) | 0.1 - 0.3 |
| Available potassium (%) | 0.15 - 0.56 |
| Calcium and Magnesium (mg/100g) | 22.67 - 45.00 |
| Copper (mg/l) | 2.0 - 9.5 |
| Iron (mg/l) | 2.8 - 9.3 |
| Zinc (mg/l) | 5.7 - 11.7 |
| Sulphur (mg/l) | 128 - 548 |

*Source :* Kumaraiah *et al.* (1997).

Farm-Yard Manure is composed of two parts : liquid and solid parts. The liquid manure is a very important fertilizer and triggers the concentration of plankton and macrophytes which ultimately accelerates fish growth. The solid part of FYM is widely used in Asia and South-East Asian countries in fish culture management strategies. Among different types of FYM, cow, pig, goat and horse dung as well as chicken litter are equally important because these are cheaper and easily available for the purpose. When FYM is applied to water, it undergoes decomposition and improves the soil and water conditions. Because nutrients from FYM are easily lost and contain very low amounts of nitrogen and phosphorus, it is better to apply nitrogen and phosphorus fertilizers along with FYM to obtain satisfactory results in fish culture.

## 2.24 Sewage and Sludge

These are the settled sewage solids combined with varying amounts of water and dissolved materials, which are removed from sewage by screening, sedimentation, chemical precipitation or bacterial digestion. They are considered to be a rich fertilizer with nitrogen and phosphorus as the principal ingredients. Besides these ingredients, they also contain lots of objectionable colloidal materials and pathogens and therefore, it is diluted with freshwater before use in fish culture.

The sewage is treated with either septic tank or filter bed or activated sludge methods. All these processes depend on the biological oxidation of urine by micro-organisms either anaerobically or aerobically. Part of it is converted to ammonia, hydrogen and methane. Some portion of it remains is solution and some other portion is settled down to the bottom which is called as *sludge* while liquid part is called as *effluent*. In the septic tank, the faecal matter is anaerobically decomposed. In the filter bed method, the sewage is gradually passed through filters and the faecal matter is aerobically decomposed.

In the activated sludge methods, air is passed through the sewage and all the faecal matter is digested and only a very little amount of it settles down to the bottom as sludge, leaving the effluent which is as clear as water. The effluent part of the sludge is generally used in fish culture. In many places, sewage is periodically used in ponds in the proportion of 4:1 (water : sewage). Sludge is a very important source of nutrients and was found to increase yield of fish and fish food organisms. But it has been found that periodic application of sewage to fish culture ecosystems increases fish mortality and the life span of plankton becomes very short. Moreover, the colloidal material in sewage may sometimes choke the soil pores and therefore, continuous or frequent application of sewage should be avoided.

## 2.25 Conclusion

Chemical fertilizers and organic manures are very important in supplementing the ability of fish culture ponds and lakes to provide nutrients for fish and fish food organisms. At present, a large variety of nutrient inputs are available to the market to meet the demands of fish farmers.

Application of solid fertilizers and organic manures are very common in large-scale fish culture programs. Moreover, liquid fertilizers are also becoming popular because the cost of labour is very cheap.

Among different organic manures, oil cakes, green manure, compost, vermi-compost, and farm yard manure are of special interest not only because these are less hazardous to water and soil than chemical fertilizers but also improve the physical properties of soil to significant degree. Moreover, they are capable to retain the loss of nutrients from soil through leaching.

Sewage and sludge are rich in nutrients and after proper treatments, they are being used in fish culture. Of course, periodic application of sewage to fish culture ecosystems degrades the environmental quality to lesser or greater degree.

Before application of fertilizers in a particular situation, it is necessary to know the factors which determine the correct application rates of fertilizers and manures. Several factors involve the relationship between the amount and type of fertilizers and productivity of ecosystem. The type of fertilizers is determined by the ability of aquatic ecosystem to provide its nutrients for fish and fish food organisms. The cost of the nutrient inputs used in fish culture and the value of increased production obviously help to determine the amount and kind of inputs to be applied. Moreover, the quality and quantity of fish biomass in relation to the effects of nutrient inputs on water and soil properties should be considered.

It is important to note that the use of chemical fertilizers in excessive amounts be avoided, not only because of the potential to damage the quality of aquatic ecosystem but also because of the potential for the loss of nutrients. The challenge to the fish culture experts is to determine the amount and kind of fertilizers required to produce fish biomass while helping to keep the fish culture ecosystem free from contaminants.

## *References*


Boyd, C.E. and W.D. Hollerman. 1982. Influence of particle size fo agricultural limestone on pond liming. *Proc. Annu. Conf. South-east. Assoc. Fish and Wild1.* 36 : 196-201.

Brady, N.C. 1991. *The Nature and Properties of Soils*. MacMillan Company, New York.

Collings, G.H. 1962. *Commercial Fertilizers.* Tata Mc-Graw-Hill Publishing Co. Ltd., New Delhi, India.

Gupta, B.P., K.O. Joseph, and M. Muralidhar. 1997. Optimum use of lime materials in brackish water aquaculture pond preparation. *Fish. Chim.* 17 : 34-36.

Hall, A.F. and A.M. Smith. 2002. *Fertilizers and Manures*. Daya Publishing House, Delhi, India.

Kumaraiah, P., N.M. Chakrabarty, and S.L. Radhavan. 1997. Integrated aquaculture with agricultural (animal and plant crop) components including processing of organic waste and their utilization. *Fish. Chim.* 17(8) : 19-20.

Tisdale, S.L., W.L. Nelson, and J.D. Beaton. 1985. *Soil Fertility and Fertilizers.* Mac-Millan Company, New York.

Yawalkar, K.S., J.P. Agarwal and S. Bokde. 1986. *Manures and Fertilizers.* Agri-Horticultural Publishing House, Nagpur, India.


## Questions

1. What are the nutrient carriers? Why they are very important in connection with fish culture?
2. In fish culture ecosystems, nitrogen and phosphorus fertilizer are widely used than potassium fertilizers? What are the reasons for the wide use of these materials?
3. When nitrogen fertilizers are used in fish culture ponds, pond soil becomes acidic. How do these fertilizers bring about soil acidity?
4. Why phosphorus fertilizers along with the nitrogen fertilizers are used in fish culture?
5. Why nitrogen fertilizers should be used in fish ponds most efficiently?
6. Assume you have 150 kg of single superphosphate (18% of $P_2O_5$) and you want to make 500 kg of a 10:20 fertilizer. How much of each of the following materials would you need to formulate this fertilizer : diammonium phosphate (18 : 46 : 0), muriate of potash (0 : 0 : 60), and a nitrogen solution (32 : 0 : 0)?
7. A fish culture expert advise a fish farmer to apply 300 kg of a 10 : 10 : 0 fertilizer on his pond. After two months you have come to know from soil tests that the nitrogen and phosphorus levels are low and hence nitrogen and phosphorus fertilizers should be applied. How much urea (46 : 0 : 0) and single superphosphate (0 : 16 : 0) would you need to provide the same amount of nitrogen and phosphorus you applied before two months?
8. Why environmental quality should be considered at the time of fertilizer application?
9. What is the importance of soil tests as guide to determine the rates and kinds of fertilizers to be used?
10. What are the types and sources of calcium essential for fish growth? Why lime is essential for fish culture?

11. How the quality of liming materials is evaluated?
12. Describe how lime requirement for a fish pond is determined?
13. What chemical reactions take place when liming materials are applied to ponds?
14. Distinguish between organic manure and green manure. Why they are called soil conditioners?
15. Why mahua oil cake is widely used in fish culture.
16. State the role of green manure in fish ponds.
17. What is artificial manure? Describe different methods for the preparation of compost and vermi-compost.
18. What are the advantages of organic manures compared to chemical fertilizers when they are added to ponds?
19. Suppose you have been applying 100 kg of 10 : 10 : 0 fertilizer to a fish culture pond and you want to apply oil cake or compost instead of the chemical fertilizer. How much of oil cake/compost would you apply to provide the same amount of nitrogen and how much single superphosphate (16% of $P_2O_5$) would you need to supplement the oil cake/compost to provide the same nutrient balance as present in the 10 : 10 : 0 ?

# 3
# Pond and Lake Ecosystems

The term ecosystem was first advocated by A.G. Tansley in 1953. The *eco* means environment and the *system* denotes a complex coordinated units. However, the system is a collection of interdependent parts of events that make up a complete structure. For example, the main components of a computer are inputs, outputs, and central processing unit. Though each unit has its own specific function, the entire system fails to function unless all units work smoothly. The ecosystem also functions in a similar way. Energy from the sun, the input, is fixed by green plants and transferred to animals. Nutrients are taken up by plants and animals where they are deposited, cycled from one feeding group to another, liberated by decomposition to the water, soil, and air (output). Thus, the nutrients and energy that comprising an integral part of the entire system, keeps the biosphere functioning.

It is needless to mention that water is a most essential and important abiotic factor of pond and lake ecosystems. It also forms the habitat for a large variety of plants and animals. These ecosystems are called as *freshwater lentic ecosystems.* Freshwater pond and lake ecosystems involve a gradient from lakes to ponds to bogs, marshes and swamps. Freshwater lentic ecosystems have low percentage of dissolved salts and subjected to the physical and chemical factors. Wide or slight fluctuations of factors significantly affect the fauna, altering their distribution, concentration, growth and existence.

## 3.1 Physico-Chemical Nature of Freshwater

Fish has its own specific and unique set of environmental requirement for survival, growth and reproduction potentials. Unless environmental requirements are uniformly distributed, it is beyond possible to live fish in any ecosystem. Some important factors of pond and lake ecosystems are density, pressure, buoyancy, temperature, light, oxygen, carbon dioxide, salinity, and pH of water which interacts together with fish and the situation becomes more complicated.

### *Density*

The density of water varies inversely with temperature and directly with the concentration of dissolved substances. Generally dissolved inorganic and organic salts increase the density of water. Thus, the density of a freshwater bodies is much less than that of the sea. When evaporation of water from ponds and lakes takes place during summer season, the concentration of salts becomes higher – a condition termed as *hypersalinity.* In contrast, salt concentration during rainy season becomes lower which is termed as *hyposalinity.*

### *Pressure*

In different freshwater environments, pressure is very less than in the sea, and aquatic animals appear to adjust them readily. The pressure of water on pond and lake-dwelling animals is combination of the weight of the water column and weight of the air.

### *Buoyancy*

The buoyancy of aquatic animals is equal to the weight of the water it displaces. It varies with the density of water and is influenced by viscosity and temperature of water. Fish keep stations by swimming movements by fins or by decreasing the specific gravity of the body with the help of air bladder.

### *Temperature*

Temperature of ponds and lakes exhibits an unique fluctuations and diurnal as well as seasonal fluctuations of temperatures are evident than in marine ecosystems. For example, a diurnal variation range of 4.8-5.0 °C is evident in tropical fish ponds with an average depth of 3 metres. On the other hand, in a polluted shallow moat with an average depth of 1.5 metres, the lowest and highest temperature at night and day times were 26.6 and 32.0°C, respectively.

On the basis of thermal stratification, ponds and lakes are classified into three types :

1. *Temperate lakes/ponds* : Surface temperatures differ below and above 4 °C.
2. *Tropical lakes/ponds* : Surface temperatures are always maintained above 4 °C.
3. *Polar lakes/ponds* : Surface temperatures never go above 4 °C.

Thermal stratification phenomenon is controlled by the seasons and has a significant impact on the survival and existence of aquatic life. Increasing and decreasing temperatures cause a fall and rise in metabolism, respectively, resulting in decrease or increase rate of food consumption. The extremes of higher and lower temperature have lethal effects on aquatic animals. For example, temperature above 35 °C or below 14 °C are reported to be lethal for *Macrobrachium malcolmsonii* and they prefer at temperature range below 29 and 31°C.

### *Light*

The influence of light on ponds and lakes is of considerable importance because light influences the productivity of these lentic environments. Since waters often have a lot of suspended materials, it is obvious that these materials obstruct the light and hence cannot able to reach the bottom. Thus, shallow lentic water bodies receive light to the bottom resulting in an abundant growth of phytoplankton and macrophytes. These ultimately form an excellent food chain. Moreover, light also controls the changes in position, nature of growth, and diurnal migration of different species of plankton.

### *Oxygen*

The oxygen is a most essential chemical composition of life processes and remain dissolved in freshwater environments. The freshwater ecosystems which remain in close proximity with air, contain an abundance of oxygen. This oxygen is mixed with the water either by wave action or water circulation or by diffusion. In this process, freshwater lentic water bodies dissolve about 5 per cent oxygen.

The concentration of oxygen in freshwater ecosystems exhibits diurnal variation. Oxygen concentration remain very low at dawn and at peak between 12 noon and 4 PM. Oxygen is consumed by aquatic animals and decomposition of dead organisms in accelerated. Generally, the oxygen content of surface water remains high than in the deeper zone due to presence of phytoplankton population in significant amounts. Ponds and lakes have a lot of decaying vegetation and oxygen content remains in a stage of complete depletion. Reduction in the level of oxygen can be observed by the release of several gases due to decomposition of dead and decaying matters.

### *Carbon Dioxide*

Carbon dioxide gas diffuses directly from air and dissolved in water thus forming carbonic acid ($H_2CO_3$) which ultimately affects the pH of water. It is also present in freshwater ecosystems as carbonates and bicarbonates of magnesium and calcium. Phytoplankton and aquatic plants require carbon dioxide for photosynthetic activity. Aquatic animals and decomposition process release and appreciable amounts of carbon dioxide. The high saturation levels of oxygen and carbon dioxide exhibit toxicity to aquatic life.

### *Other Gases*

In cases where freshwater lentic ecosystems have decaying vegetation, show an abundance of hydrogen sulphide. Nitrogen, ammonia, and sulphur dioxide are also found in freshwaters. All these gases are highly toxic to aquatic biota. Gradual increase in the concentration of these gases by any means results in complete or partial elimination of bottom fauna.

### *Salinity*

Salinity of freshwater bodies is mainly due to presence of common salt (sodium chloride) in significant amounts and the degree of salinity depends on the amount of salt present in dissolved condition. Since the concentration of salt in freshwater lentic water bodies is very low (about 2-5 ppt.), aquatic animals face the problem of osmo-regulation. The salt concentration within the cell remains higher than the freshwater environment and therefore, the water tends to enter the body which should be removed. Fishes are above to remove extra amount of water by osmo-regulation. Many aquatic invertebrates possess excretory organs for elimination of excretory products along with water.

### *Dissolved Salts*

Freshwater ecosystems contain various types of dissolved salts such as compounds of nitrogen, phosphorus, calcium, magnesium, manganese, iron, sodium, zinc, sulphur, and potassium. Dissolved salts accumulate within the ecosystem by surface run-off and decay of aquatic organisms. Although ammonium salts, nitrates and nitrites are important for the growth of algae and aquatic macrophytes, the compounds are highly toxic to fish and fish food organisms. Phosphorus compounds are also important because plankton utilize this phosphorus as source of food. Thus, nitrogen and phosphorus concentrations gradually decreased. Compounds of salts greatly influence the composition of fauna. For example, deposition of calcium carbonate takes place by the activity of plants. This calcium carbonate is essential for external covering of arthropods and the shell of molluscs.

### *Hydrogen Ion Concentration or pH*

The pH of freshwater bodies is a determining factor for aquatic life and has a limiting factor. The pH values in a particular ecosystem fluctuate seasonally and annually. Although the range of pH is species specific, higher aquatic species (such as fishes) responded quickly to slight pH change, while lower aquatic species exhibited little reaction to changes in pH of water.

## 3.2 Characteristic Features of Ponds and Lakes

Ponds are considered as shallow standing water bodies of small or medium dimensions. They are so shallow that rooted plants are able to grow most of the bottom. On the other hand, lakes are inland depressions containing standing water. Although lakes may vary in size from less than acre to large seas covering thousands of square kilometres, the size of freshwater lakes generally vary from one to several hectares. Ponds and lakes possess outlet streams and are temporary features on the landscape.

The freshwater pond and lake habitats are vertically stratified into five distinct zones depending upon the intensity of light, temperature, hydrostatic pressure, and wave length absorption (Figure 3.1). These are as follows :

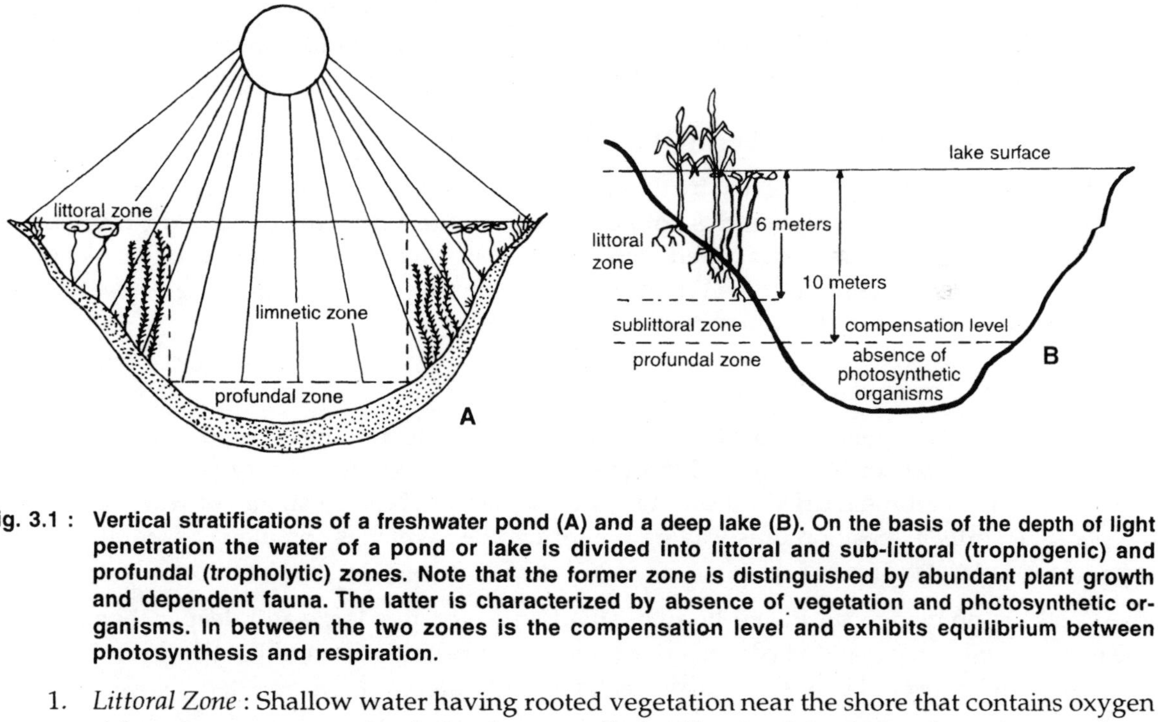


**Fig. 3.1 : Vertical stratifications of a freshwater pond (A) and a deep lake (B). On the basis of the depth of light penetration the water of a pond or lake is divided into littoral and sub-littoral (trophogenic) and profundal (tropholytic) zones. Note that the former zone is distinguished by abundant plant growth and dependent fauna. The latter is characterized by absence of vegetation and photosynthetic organisms. In between the two zones is the compensation level and exhibits equilibrium between photosynthesis and respiration.**

1. *Littoral Zone* : Shallow water having rooted vegetation near the shore that contains oxygen rich and upper warm circulating layer – called *epilimnion*. It begins at about 6 metres from the surface.
2. *Sublittoral Zone* : This zone extends from littoral zone to non-circulating poor oxygen cold water zone – called *hypolimnion*. This zone begins at about 10 metres from the surface.

3. *Limnetic Zone* : It is an open water zone away from the littoral zone and extends upto the depth where the rates of respiration and photosynthesis are equal. This zone begins at about 50 metres from the surface.
4. *Profundal Zone* : This zone is situated beneath the limnetic zone that extends beyond the depth of effective light penetration.
5. *Abyssal Zone* : This zone is found only in the deep water lakes. It begins at about 2,000 metres from the surface of water.

In freshwater pond ecosystem, littoral zone is larger than the limnetic and profundal zones. Small ponds do not have the limnetic and profundal zones. Therefore, ponds have little vertical stratification. In general, freshwater ponds have three zones such as littoral zone, limnetic zone, and profundal zone.

On the basis of the depth of light penetration, lentic freshwater lakes or ponds are also classified into two zones such as trophogenic zone and tropholytic zone.

1. *Trophogenic Zone* : This zone includes littoral and sublittoral zones and is distinguished by growth of plants.
2. *Tropholytic Zone* : It includes the upper part of profundal zone and is characterized by the absence of aquatic plants. However, in between two zones, there is a compensation level which forms a boundary between the two zones. In this compensation level, the rates of respiration and photosynthesis remain in a state of perfect equilibrium.

## 3.3 Water Circulation in a Thermally Stratified Lakes

The epilimnion is always heated by the sunshine and mixed by the wind and water currents. On the other hand, the hypolimnion is not heated and circulated by the wind and currents. The transition area between the two is called *thermocline* or *metalimnion*. The circulation layer of lake water is maintained by temperature. Whenever a metalimnion is formed, exchange of water does not takes place between hypolimnion and epilimnion.

## 3.4 Origin of Ponds and Lakes

In most regions, ponds occur due to rainfall in adequate amount. Permanent ponds have adequate depths and contain water all the year round, while the temporary ponds are shallow, seasonal and most of them dried up during summer seasons. When a stream of water is flooded the plains, it shifts its position thus leaving the formed bed which is completely separated in the form of a lentic water area.

A lake is defined as an inland body of standing water, it is large and deeper than a pond. The term includes an expanded part of a river, a reservoir behind a dam, and a lake basin formerly or intermittently covered by water. Lakes are formed in different ways as noted below :

1. By the deposition of silt, driftwood etc. in the beds of slow-flowing streams.
2. Gradual accumulation of rain and ground water in a volcanic crater or caldera.
3. Glacial abrasions of slopes in high mountain valleys impress deeply resulting in the formation of depressed areas with no surface outlets, which is filled with water from rain and melting snow.

4. Retreating valley glaciers are left behind as crescent-shaped ridges of rock debris and filled up with water.

The above-mentioned lakes are natural in origin. On the other hand, man-made lakes are called *artificial lakes* which is constructed by damming rivers for power, irrigation and water storage or by constructing small marshes and ponds for wild life and fisheries. Artificial lakes are also called *impoundments.* Natural lakes differ from artificial lakes in thermal and oxygen variations and are characterized by high percentage of bottom organisms.

### 3.5 Coldwater Lakes and Ponds

So far as the fisheries point of view is concerned, the term coldwater fish refers to those species which prefer thermal regime between the snow-melt waters of the greater Himalaya as the North and watersheds draining the Southern slopes of Himalaya. These waters are generally characterized by the dissolved oxygen contents, high transparency, low carbon dioxide, inorganic soil and sparse biotic communities.

Assessment of coldwater fishery resources in the Indian uplands of the Himalaya and Deccan plateau and developmental strategies for their conservation and management is a recent issue. The Indian uplands are divided into two main regions such as (1) Himalayan and (2) Peninsular. The former is about 2,500 km in length and 300-500 km in breadth which covers the watersheds of the three river systems : the Ganga, the Indus and the Brahamaputra, while the latter covering the watersheds of the Cauvery and the Krishna river systems. Besides these, a large number of lakes, ponds and reservoirs are found that has indigenous and exotic coldwater fish fauna.

### 3.6 Benthic Communities of Lakes and Ponds

On the basis of the dependence of water and soil, different organisms of the ponds and lakes can be classified into two forms such as *Pedonic forms* and *Limnetic forms.* The former types depend on the substratum and the latter forms are free from any substratum. Moreover, according to the size of habitats, aquatic organisms are also divided into following types :

1. *Plankton* : These are small animals and plants that possess limited power of self-locomotion. Plankton constitutes phytoplankton and zooplankton and can move in such a way that they are able to control their vertical distribution. Some species of planktons are very active, move great distances but are very small and their distribution is controlled by water currents. Such planktons are termed as *nektoplankton.*
2. *Nekton* : These are large size animals, powerful and active swimmers. The size vary from about 2 mm long (swimming insects) to the largest animal (the blue whale).
3. *Neuston* : These are found in all freshwater ponds and lakes at the air-water interface. They include free-floating plants (such as *Lemna, Azolla, Salvinia, Pistia, Spirodella, Eichornia,* and *Wolffia*) and animals. Animals that spend most of their times below the water-air interface (such as back swimmers, diving beetles), are called *hyponeuston,* while others (water striders) live on the top of the air-water interface are called as *epineuston.*
4. *Benthos* : The term benthos means the organisms living at the bottom of the freshwater ecosystem. They either live above the soil-water interface-termed as *benthic epifauna* or within the soil-termed as *infauna.*

## *Organisms of Littoral Zone*

Aquatic organisms are most prolific in this zone. Emergent vegetation is the most characteristic feature of the shore proper. These plants are deeply rooted in the shore but their tops remain exposed such as *Sagitaria* (arrow heads), *Scripus* (bulrushes), *Typha* (cattails) etc. (Figure 3.2). Some rooted plants with floating leaves such as *Nelumbo, Nymphaea, Trapa,* and *Marsilea* are found just slightly deeper part of the zone (Figure 3.3). Still deeper part of the zone, there are occurrence of rooted, submerged and thin stemmed aquatic weeds. These are *Vallisneria, Mycrophyllum, Hydrila, Chara, Elodea* etc. (Figure 3.4)

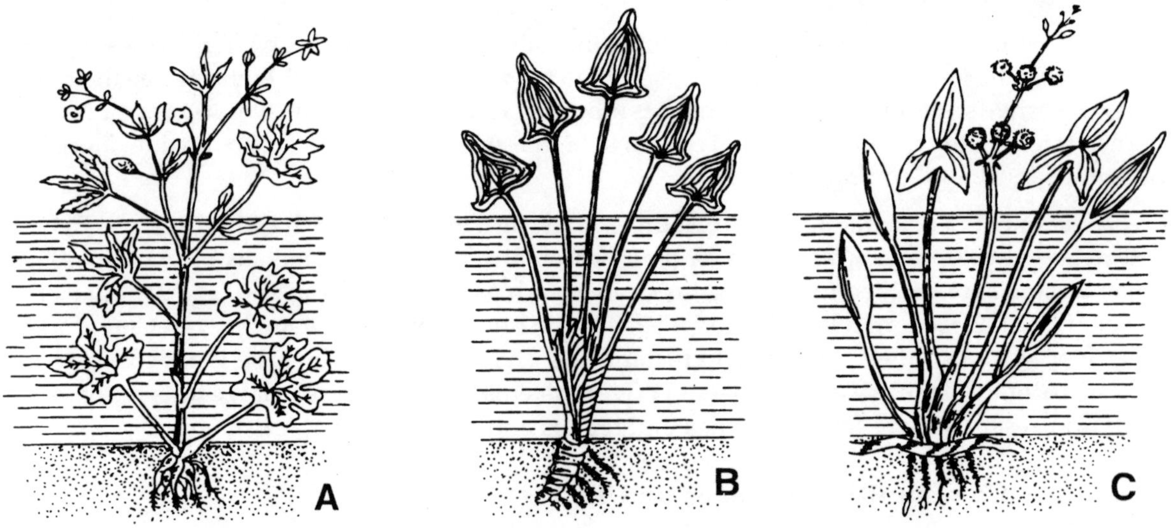


**Fig. 3.2 : Some rooted emergent hydrophytes grown in a freshwater ecosystem : (A) *Rananculus*, (B) *Monochoria*, (C) *Sagittaria***

The zooplankton of littoral zone consists of Rotifers, Ostracods and Daphnids. The nekton includes *Euglena, Paramoecium, Corixa, Ranatra, Dytiscus,* larvae of *Gyrinus, Culex* and *Chaoborus* (Figure 3.5). Neuston includes water spiders, protozoa, whirling beetles etc. Various types of pond fish such as sunfish, bass, pike, and top minnows spend much of their time in this zone for various purposes.

Different species of animals of littoral zone live in association with plants or other objects. These include larvae of *Dytiscus, Glossophonia* (leech), Dragon fly and Damsel fly nymphs, *Brachionus,* Bryozoa, *Lymnaea, Vorticella,* and *Chironomus.*

## *Organisms of Limnetic Zone*

This zone has autotrophs in abundance and because this zone is characterized by rapid variation of temperature, oxygen and water level, some species (such as *Rotatoria, Philodina, Daphnia, Cyclops,* Snails, *Volvox, Euglena,* and fishes) which are capable of encystment during unfavourable situations, live in this zone.

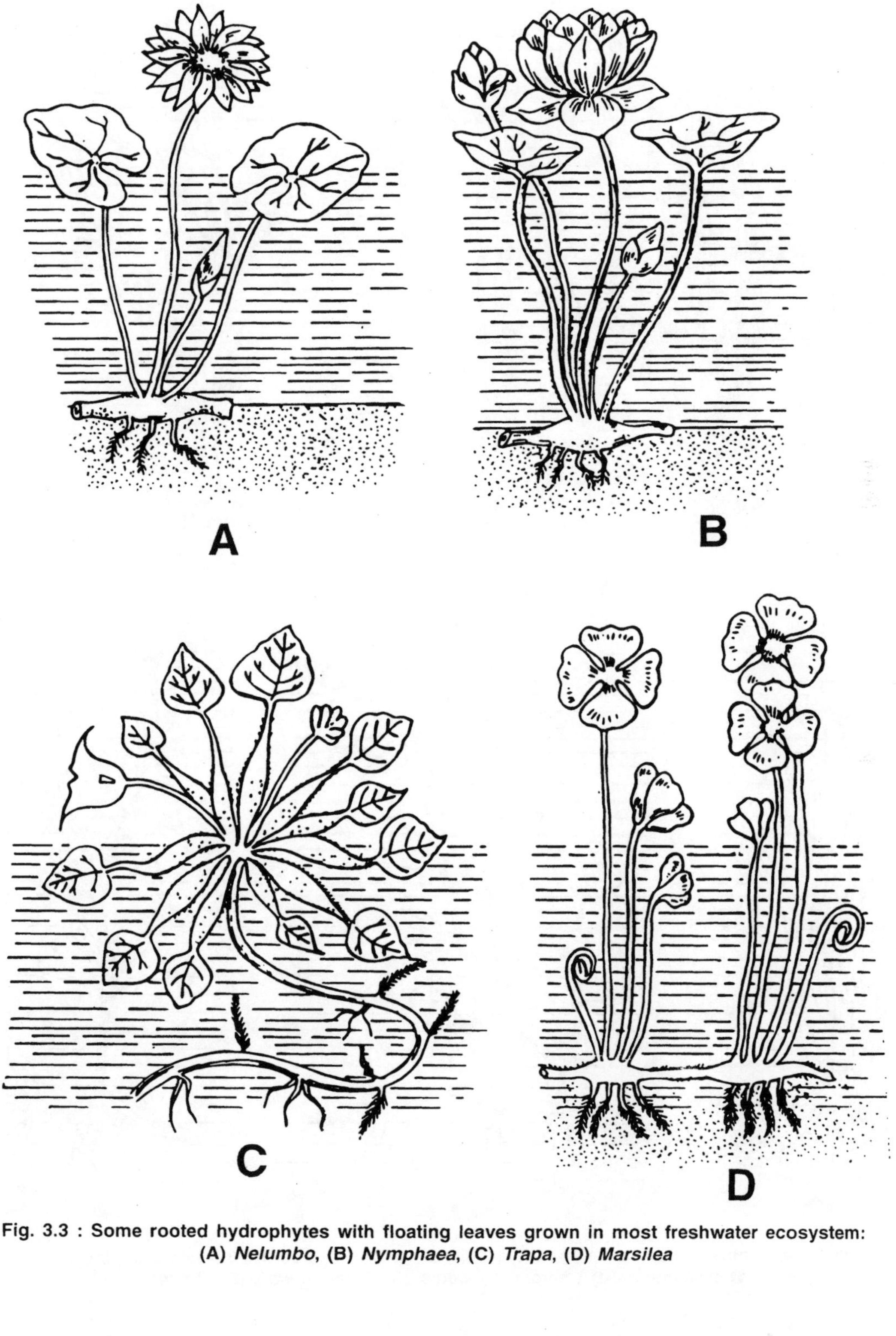


Fig. 3.3 : Some rooted hydrophytes with floating leaves grown in most freshwater ecosystem: (A) *Nelumbo*, (B) *Nymphaea*, (C) *Trapa*, (D) *Marsilea*

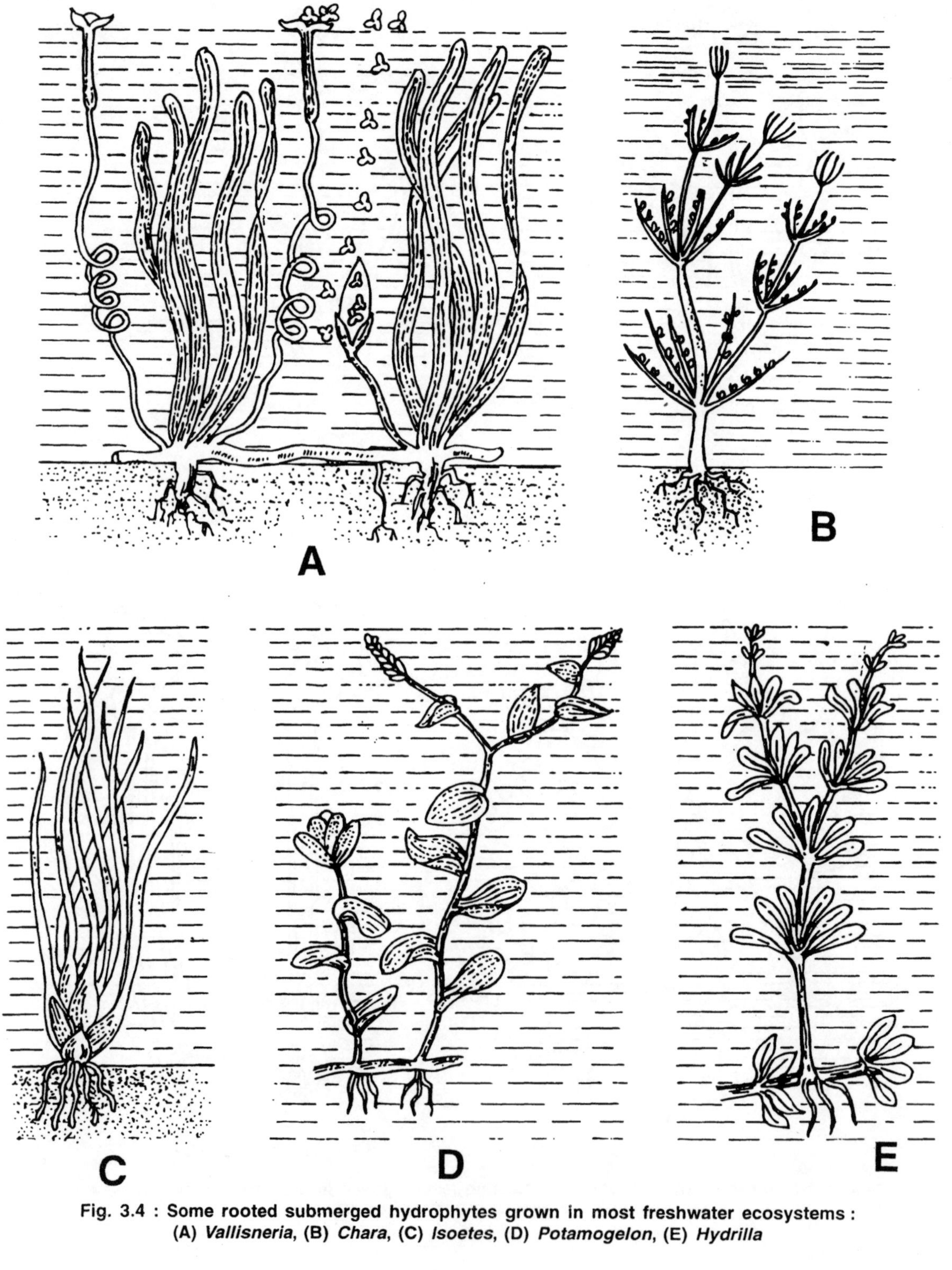


Fig. 3.4 : Some rooted submerged hydrophytes grown in most freshwater ecosystems : (A) *Vallisneria*, (B) *Chara*, (C) *Isoetes*, (D) *Potamogelon*, (E) *Hydrilla*

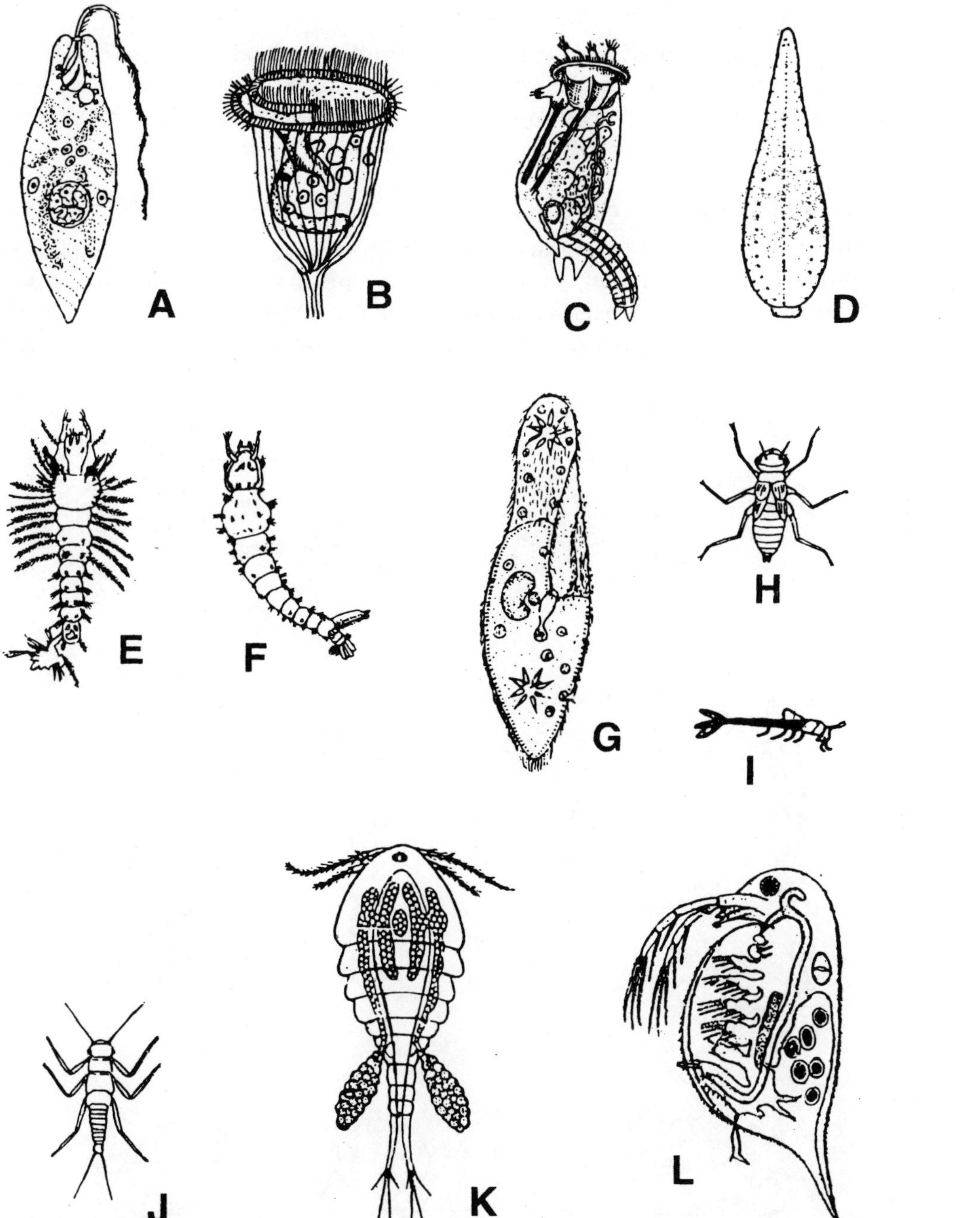


**Fig. 3.5 : Fauna of littoral zone comprises an extremely diverse animal population ranging from microscopic unicellular organisms to multicellular ones as shown in this figure. Most of them are used as natural food for fish. (A) *Euglena*, (B) *Vorticella*, (C) *Brachionus*, (D) *Glossophonia*, (E) *Anophelis* larva, (F) *Culex* larva, (G) *Paramoecium*, (H) Dragon fly nymph, (I) Damsel fly nymph, (J) Stone fly nymph, (K) *Cyclops*, (L) *Daphnia*.**

## *Organisms of Profundal Zone*

Generally autotrophs in this zone are not able to synthesize food and therefore detritus is the main source of energy for existence of several types of organisms such as clams, bacteria, fungi, annelids etc. that are capable of living in water having low oxygen and little light. Organisms are heterotrophs in nature and they are either carnivores or detritus feeder.

## *Organisms of Pond and Lake Bottom*

Most of the pond and lake bottoms are covered with sediment to form a uniform substratum of sand or mud. The most common benthic organisms include several species of insect larvae (*Chaoborus, Chironomus* etc.) oligochaetes (*Aulodrilus, Branchiura* etc.), bivalves (*Pisidium, Parreysia, Indonaia,' Lamellidens*), gastropods (*Viviparus, Melanoides, Gyraulus, Planorbis*) and crustaceans (*Macrobrachium, Scylla* etc.) (Figure 3.6)

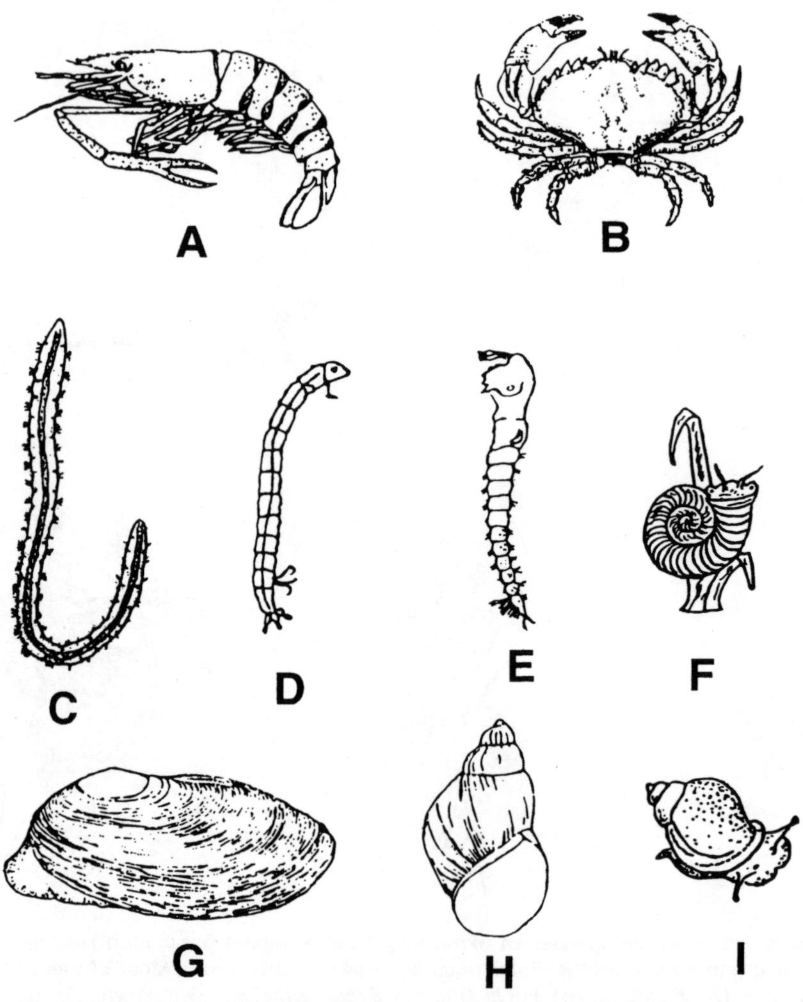


**Fig. 3.6 : Fauna of benthic region of ponds and lakes : (A) *Prawn*, (B) *Scylla*, (C) *Branchiura*, (D) *Chaoborus*, (E) *Chironomus*, (F) *Planorbis*, (G) *Lamellidens*, (H) *Viviparus*, (I) *Lymnaea*.**

## 3.7 Sustainable Management of Freshwater Ecosystem

Water is the light of life. While water is an important necessity, politics continue to shape the water as market. It is difficult to develop a regulatory system that can save the freshwater resources. The challenging of managing freshwater resources and facilitating the adoption and use of environmentally sound technologies, governments, private sector, and farmers are all significant stake-holders.

Active involvement of private sector and farmers' community is essential for successful prevention and management of freshwater resources, the development of ecosystem-related projects, and the maintenance of associated infrastructure. It is better to involve and educate the fish farmers/fishermen, rather than to wait for a disaster due to water pollution.

Private sector and public utilities both play important roles in ensuring that freshwater resources should keep free from pollution for their present and future interests. As indicated in Figure 3.7, a number of policy considerations may exist when addressing freshwater ecosystems and their management strategies.

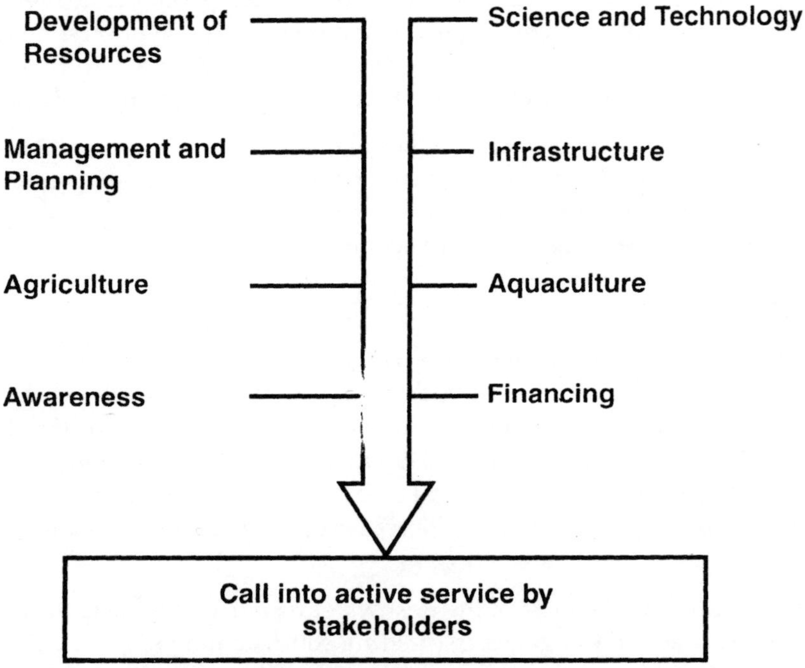


**Fig. 3.7 : Freshwater and its technology : Important policy considerations [Modified after UNEP (1998)]**

### *Development of Resources*

Development of freshwater bodies require sophisticated technology, human and financial resources. Where such resources are not available, training and education to aquaculturists should be adopted which is intimately associated with the development of fisheries as a whole. In cases where approaches to development of resources differ, the following features should not be overlooked.

1. Active participation by stakeholders at various levels of planning, evaluation, and development.
2. Effects of interactions among various groups of fish farmers.
3. Inter-relationships between planning and development.
4. Proper evaluation of different management system.
5. Implementation of development and management assistance programs.
6. Realistic expectations about the scope of developmental program relative to local capabilities.

### *Management of Planning*

Planning on regional basis are important where physical constraints in the management of freshwater fish culture ecosystems (FFCE) exists. Management and planning strategies of FFCE depends on the geographical condition and standard of living of fish farmers. Therefore, a coordinated framework for ecosystem management and pollution prevention should be considered. Although fishery management strategies vary in terms of resources and quality of water, the main object should be to achieve and sustain improvement of FFCE for maximum yield of fish biomass.

Financial assistance must be received by fishermen from different sources for implementation of management strategies pertaining to waste management, pesticide management, water quality protection, and other associated programs. However, by involving different groups of fishermen, fish culture objective should be integrated with economic goal.

### *Infrastructure*

In many countries, serious problems arise when utilization of freshwater ecosystem, where fish culture operation is contemplated, is considered for sustainable development of aquaculture. In under-developed countries, development of infrastructure takes place or if not, must be taken into consideration when rapid population build-up makes water highly polluted and waste-water treatment systems not so adequate. Withdrawal of large volume of water from fish culture ponds and lakes for agricultural purposes drastically decrease the water level, thus, the entire fish culture management strategies are jeopardized.

In many urban areas of India, most of the freshwater resources are being destroyed/abolished by those persons who are actively engaged in building construction programs. Undoubtedly, such infrastructure development damages the aquaculture strategies.

In many rural areas, poor people do not use septic tanks for disposal of domestic sewage. Consequently, freshwater resources become highly polluted and not only unfit for fish culture but also causes water-borne diseases as well as unhygienic living conditions.

Although prevention is the best solution for pollution control of FFCE, problems have persisted because of limited funding, absence of effective policies, planing and management strategies. The potential for degraded infrastructure to jeopardize fish culture operations, often goes unrecognized.

For example, where unpolluted waters are available for viable fish culture, and water is supplied by a sub-standard distribution network, the water can be degraded to the point where it is not safe for fish. This can be alleviated by introducing another alternative which is characterized by treating water and waste water in close connection to the fish farm. This approach eliminate the possibility of degradation of unpolluted water.

Since the operational costs of conventional polluted and/or waste water treatment systems are high and adequate fund cannot afford by fish farmers, a large quantities of waste/polluted waters are discharged untreated. Therefore, there is growing interest to develop low cost systems which will combine waste water purification and nutrient recycling.

### *Financing*

In order to keep the freshwater resources free from polluted materials, adequate financing for the development of freshwater infrastructure is necessary. Where private and co-operative sectors fail to afford fund, local fish farmers are becoming the likely source of future funding. Often, a lack of integrated planning and management give rise to developmental inconsistency relevant to management aptitude. Infrastructure expenditures should be considered with respect to developmental and financial issues. Financial mechanisms (such as adequate developmental fund, risk management insurance etc.) need to be improved to encourage technological cooperation, use of environmentally sound technologies for aquaculture and the development of local capacity for supporting the implementation and management of sustainable preferences.

Actions should be directed towards appropriate financial support to encourage fish farmers for debit avoidance and self-reliance and also for efficient use of inputs for sustaining yield of fish. Prevention of pollution and environmentally-driven process alterations can offer economic advantages while reducing pollution.

### *Science and Technology*

Science and technology play a very important role in understanding and resolving freshwater issues. The growing scale of fish culture development poses severe threats to water quality criteria and hence aquatic science must be integrated with adequate planning and development. Scientific research and new technologies, if effectively applied during fish culture, can contribute to protect freshwater resources and also to explore the following facts :

1. Sources of water pollution and causes of pollution.
2. Rate of degradation of water.
3. Safe limits of different kind of pollutants.
4. Mode of toxicity, toxic actions and the effects of pollutants on fish.
5. Effects of pollutants on soil, water, and fish food organisms.
6. Disease and disease-producing organisms.
7. Physiological alterations in fish.

Aquaculture experts are actively engaged in developing freshwater and waste water management and treatment technologies, and sound production of aquatic resources so that nation can use aquatic food items more effectively. Although better management practices are required to treat contaminated aquatic ecosystems, such development is in response to more stringent regulatory requirements and the need to address and expanding range of complex pollutants. However, the adoption and use of technologies endogenous to developing countries should be carefully considered for long-term success of ecologically sustainable options.

### *Awareness*

Latest and scientific informations regarding development of environmentally sound technologies should be shared among stakeholders. For adoption and use of appropriate technologies, actions to educate stakeholders and establish integrated database information network should be targeted. However, to protect ecosystem from pollution, emphasize should be given for planning and development, the establishment of ecosystem objectives and the implementation of management options.

## 3.8 Benthic Food Chains and Food Web in Pond and Lake Ecosystem

On the basis of the feeding habits, zoobenthic invertebrates may be divided into four types as noted below :

1. *Carnivores* : These include some species of protozoa (such as *Polomyxa* sp., *Acanthocystis* sp. etc.), rotifera, hirudinea and Odonata nymphs. Protozoa and rotifera consume ciliates, zooflagellates, and organic matter; hirudinea consume snails and tubificids; while Odonate nymphs consume rotifers, protozoa, chironomids and crustacea.
2. *Omnivores* : Different species of protozoa such as *Paramoecium* spp., *Amoeba* sp., and *Stentor* sp. consume sediment algae and bacteria.
3. *Herbivores* : A large number of protozoan species and chironomids consume sediment algae, microbes, and organic detritus. Molluscan species depends on sediment algae and decomposed tissues of macrophytes.
4. *Detrivores* : Several species of tubificid oligochaetes and zooflagellates consume organic debris and bacteria-organic detritus complex, respectively.

From the above discussion, it is evident that the food web of benthic community is very complex. The basic source of food constitutes the rich microbe-organic detritus complex system which is transferred to different trophic levels. The trophic factor decides about the occurrence and biomass of benthos in any freshwater ecosystem. However, the basic source of food in the form of detritus is supplied by organic matter and animal wastes that are added to the ecosystem (Figure 3.8). Second, bacteria and protozoa act as primary decomposers and form an organic detritus-microbe complex. Secondary decomposers – the zoobenthos – are also equally important. Both types of decomposers together constitute the source of food for benthic predators. Thus, the detritus food chain is completed.

### *Trophic Structure of Benthic Community*

The trophic structure of the bentic community principally depends upon the degree of eutrophication of any freshwater ecosystem. In an eutrophicated ecosystem, where primary production level exceeds consumers, the residues of primary producers gradually accumulate at the bottom mud. Soon after their accumulation, decomposition process is initiated that results in

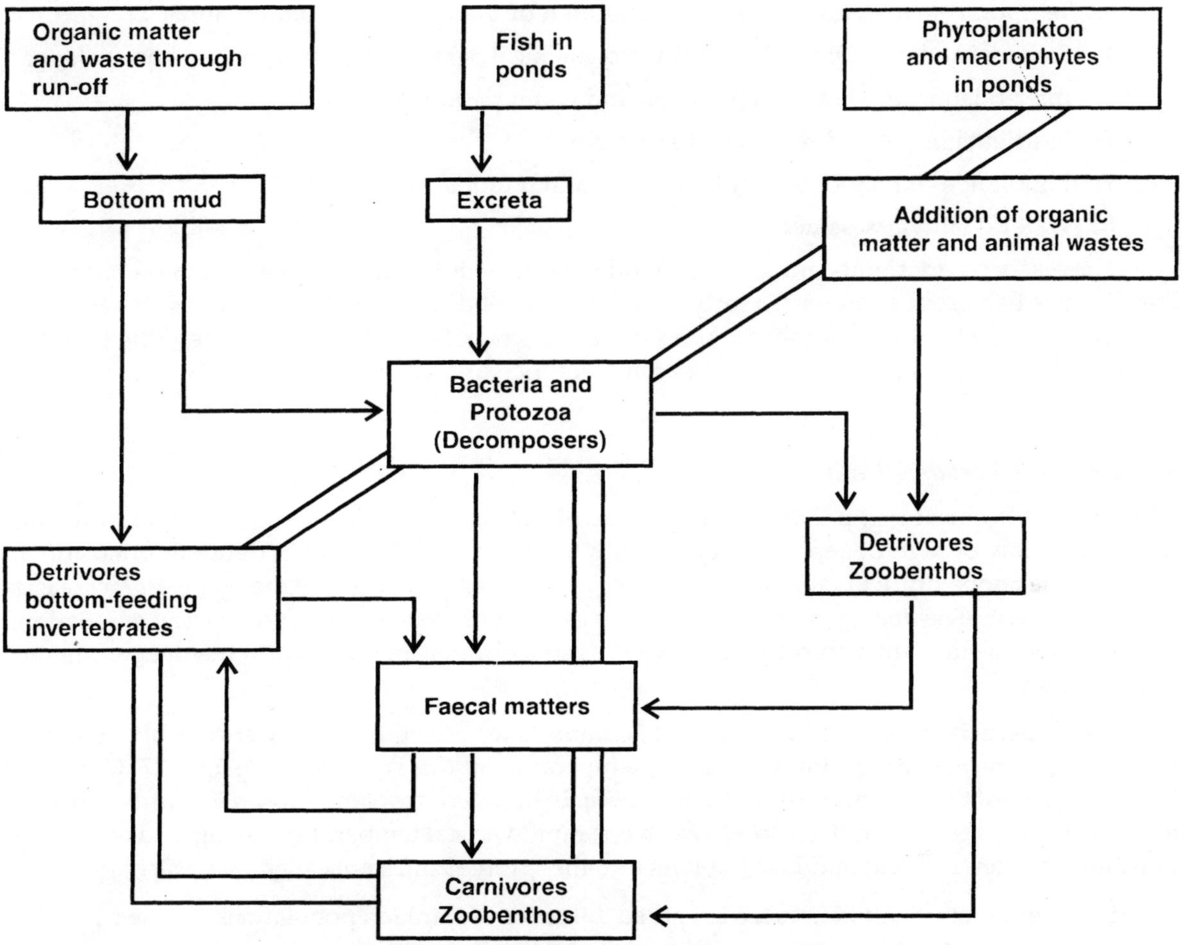


**Fig. 3.8 : Schematization of a general model showing flow of materials in a benthic food chain. [(Used with permission of A.K. Pandit (1996)]**

the shortening of food chain. Shortening of the food chain is mainly due to the switching over of primary consumers (such as snails, chironomids, and protozoa) from algae to bacteria and detritus. The increase in concentration of soil invertebrates is due to the breakdown of organic matter and animal wastes which is influenced by the existence of primary decomposers. However, the pyramid of biomass (on the basis of mean biomass, gram/square metre) of benthos can be determined by herbivores/primary producers and carnivores/omnivores + herbivores ratios.

## 3.9 Interactions Between Abotic and Biotic Factors in Ecosystem

Elaborate investigations on the limnobiology of ponds and lakes reveal the interactions of various types of factors of soil and water which influence the productivity of aquatic ecosystem.

Investigations on various aspects of pond and lake ecosystem include the following :

1. Density and diversity of plankton population and bottom fauna.
2. Nutrient status.

3. Temporal, spatial, and seasonal distribution of plankton and bottom fauna.
4. Seasonal and diurnal rhythm of physico-chemical parameters, plankton and bottom fauna.
5. Inter-relationship between plankton and water parameters.
6. Nutrient budget and primary productivity.
7. Population dynamics of plankton and bottom fauna.
8. Water quality assessment.

These aspects are significant to understand the interactions between abiotic and biotic factors that initiates fish production particularly in lentic water bodies. Though it is not possible to make a scientifically justifiable generalization on these aspects, a few examples will describe how the relationship between abiotic and biotic parameters increases the productivity of pond and lake ecosystems.

### *Temperature Versus Plankton*

In many tropical fish ponds, the amplitude of seasonal changes of water temperature was found to be low or high depending on the depth of water. The increase in depth of ponds often increases the temperature fluctuations. In shallow ponds, temperature of water generally increases to a great extent. The deeper layers of water also get a lot of heat as a result of vertical circulation, wherein, due to circulatory movement of water, surface water are brought to the deeper regions and vice versa.

The superficial layers of freshwater ponds and lakes are continuously stirred by wind and called as *epilimnion.* During summer months, temperature of this layer may rise upto 27 °C and the stagnant water of the bottom constitute *hypolimnion.* In between hypolimnion and epilimnion, occurs an intermediate zone called *thermocline* which has rapid vertical temperature changes. The process of differentiations of pond and lake habitats into these three zones are called *thermal stratification.*

Such temperature differences have marked influence of plankton populations. While a positive correlations of temperature with Cyanophycease and Bacillariophyceae exist in deep water ponds, the reverse situations have been observed in shallow waters. Furthermore, higher concentration of zooplankton populations in many tropical fish ponds is primarily due to increase in temperature of water. However, several species of algae can thrive in pond and lake waters as warm and cold as 42° and 7 °C, respectively.

### *Dissolved Oxygen Versus Plankton*

In general, a direct relationship among dissolved oxygen and different groups of algae always exist in pond and lake ecosystems. Figure 3.9 indicates that dissolved oxygen concentration is dependent upon phytoplankton concentration and the regression line is linear.

### *Range of Temperature Tolerance*

In plankton organisms, all metabolic activities start at a certain *minimum temperature* and increase with rise in temperature until they attain the maximum level at a temperature called *optimum temperature*. Further rise in temperature beyond optimum brings about decrease in metabolic activities, until it ceases at a temperature called *maximum temperature.* Therefore, the favourable temperature range for any particular species is determined by the prevailing temperature – suitable for normal activities of the animal. Many species of Rotifera, Copepoda, and Cladocera, for example, are known to disappear from ponds and lakes in the tropics when the temperature is below 23 °C and appear when the temperature rises above 32 °C. However, their abundance reach the peak during summer months (March to June).

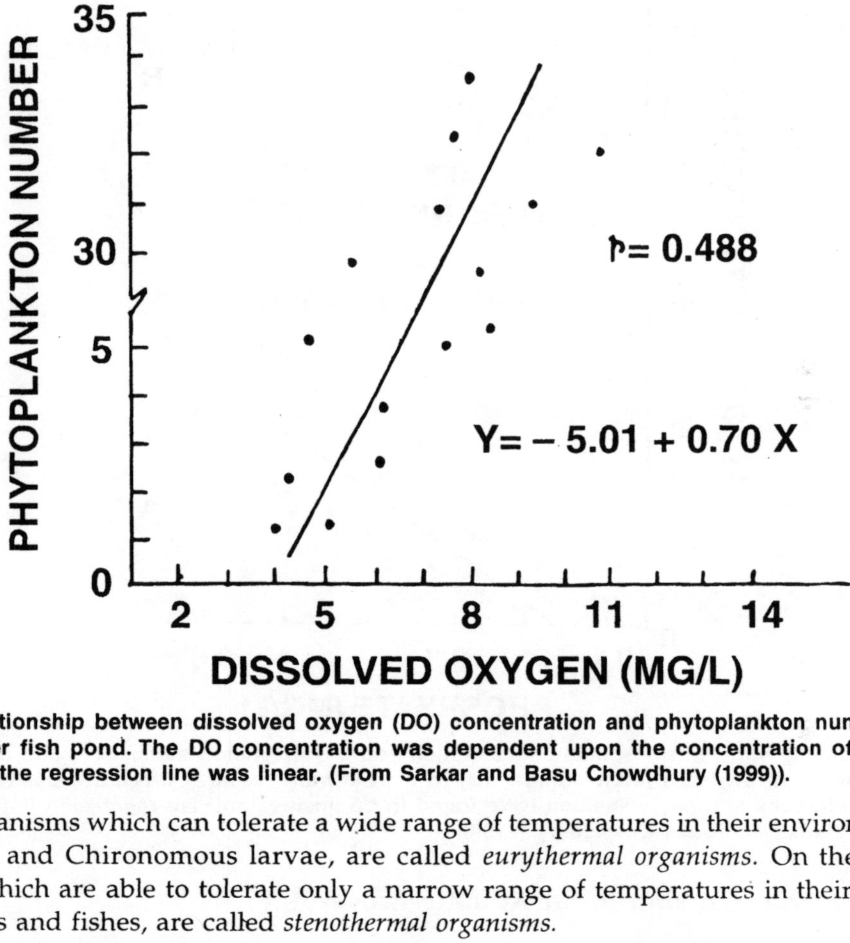


**Fig. 3.9 : Relationship between dissolved oxygen (DO) concentration and phytoplankton number in a freshwater fish pond. The DO concentration was dependent upon the concentration of phytoplankton and the regression line was linear. (From Sarkar and Basu Chowdhury (1999)).**

The organisms which can tolerate a wide range of temperatures in their environment such as *Cyclops* spp. and Chironomous larvae, are called *eurythermal organisms.* On the other hand, organisms which are able to tolerate only a narrow range of temperatures in their environment such as snails and fishes, are called *stenothermal organisms.*

## *Nutrient Versus Plankton*

The presence of nutrients in any freshwater ecosystem are the key for development and existence of plankton populations to great extent. This statement points to an obvious conclusion that a distinct positive correlation between total ammonia nitrogen, phosphate, and phytoplankton concentrations exists and their general relationships are shown in Figure 3.10. The nutrient level of water, however, usually exhibits a great variation in different ponds and lakes depending upon the soil type, floral and faunal compositions, depth of water, human activities, and nature of productivity. Whenever the relationship between nutrient levels and biotic factors is established, these aspects should be kept in mind.

* For further study, see Kramer *et. al.* (1979)

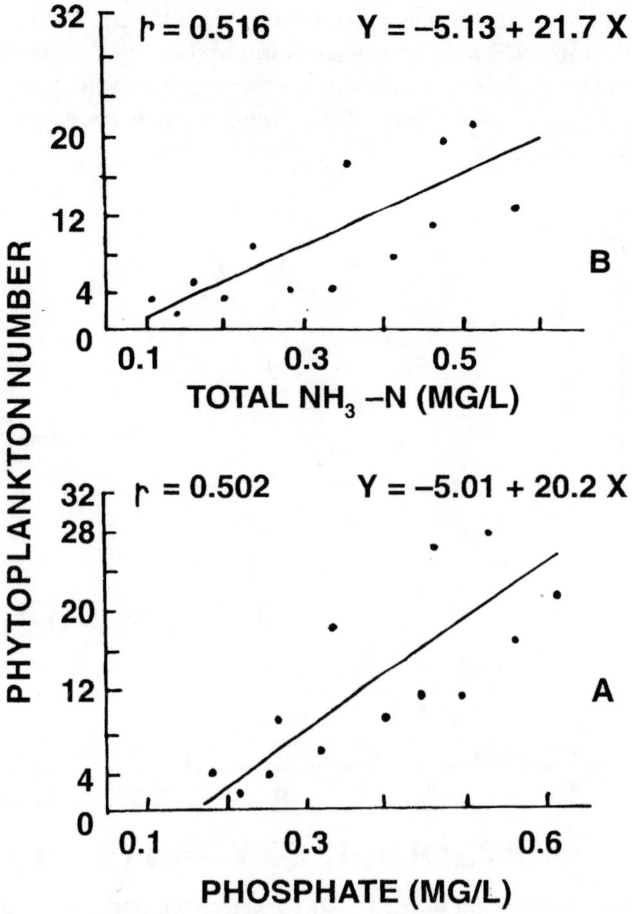


**Fig. 3.10 : Relationship between phosphate concentration and phytoplankton number (A) and total ammonia-nitrogen and phytoplankton number (B) in a freshwater pond. Correlations between nutrient concentrations and phytoplankton were found to be positive and the regression lines were linear. (From Sarkar and Basu Chowdhury (1999)).**

## 3.10 Structure and Function of an Aquatic Ecosystem

A typical aquatic ecosystem has two major parts such as structure and function. The former part indicates the followings : (1) the composition of biological community including number, species, biomass, life history and distribution of plants and animals and (2) the amount and distribution of water and nutrients, and (3) the fluctuations of temperature and light. The function denotes (1) the rate of flow of biological energy such as production and respiration of the community, (2) the rate of nutrient cycles, and (3) biological/ecological direction including direction of organisms by environment and vice-versa. Since fish production is directly related to the ecosystem productivity, it is necessary to understand the structure and function of any aquatic ecosystem where fish culture is undertaken on commercial basis. The following brief discussion will provide an overall conception about the structural and functional aspects of an aquatic ecosystem. A generalized model of structural and functional aspects of an ecosystem is shown in Fig. 3.11.

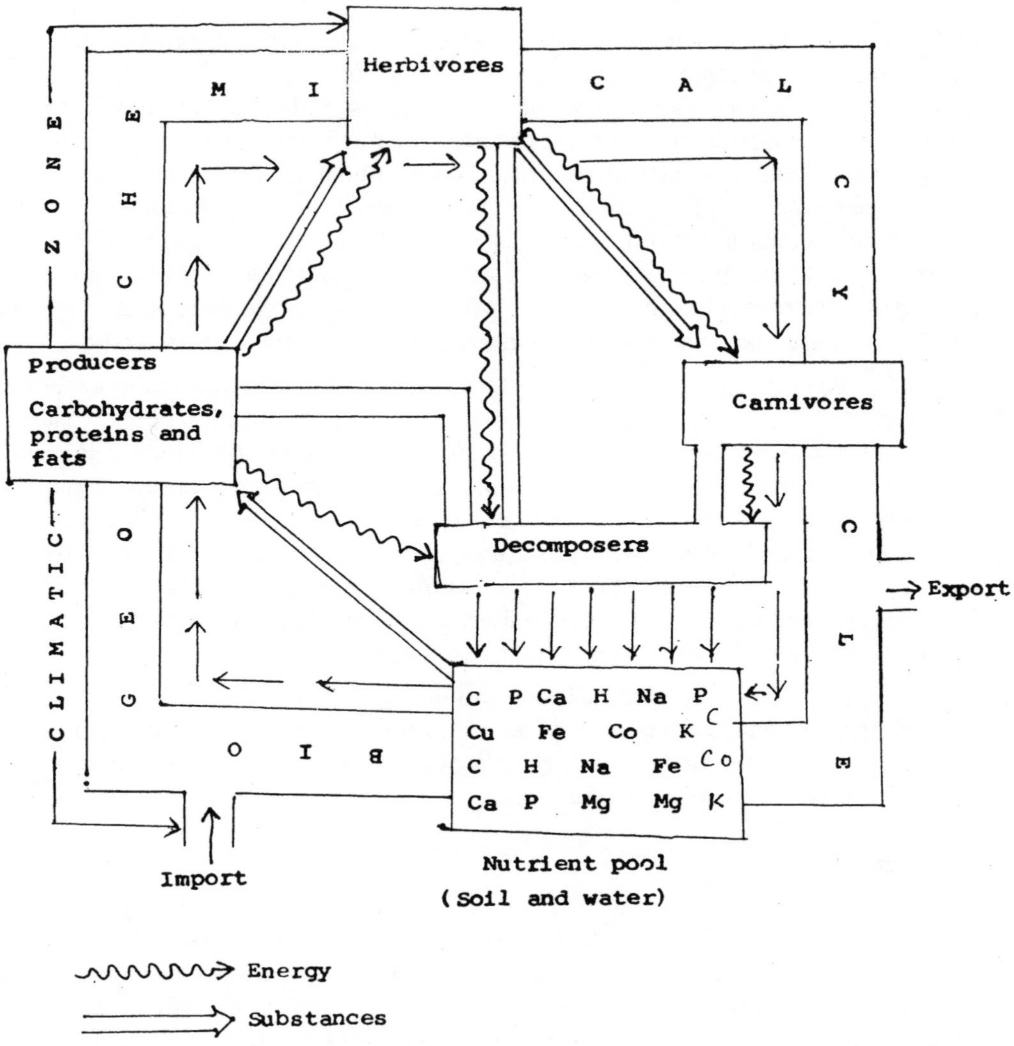


**Fig. 3.11 : A universal diagram of an aquatic ecosystem exhibiting its structure and function. Observe the biogeochemical cycles overlapping on different components of the ecosystem. Movement of energy is uni-directional (non-cyclic) and that of substance is cyclic.**

## 3.11 Structure of an Ecosystem

An aquatic ecosystem has two basic components : (1) abiotic component and (2) biotic component. The former type embraces (i) the amount of inorganic elements such as nitrogen, hydrogen carbon, oxygen, phosphorus, and sulfur – responsible for mineral cycles. The amount of these elements present in ecosystem at any given time is termed as *standing attribute*, (ii) the amount and allotment of chlorophylls, carbohydrates, proteins, and lipids which are present either in the environment, or in the biomass and (iii) climatic factors. This biochemical structure is closely related to the abiotic and biotic components of the ecosystem.

### *Abiotic Components*

1. *Inorganic Components* : Though various life processes of living organisms necessitate the involvement of, at least, forty elements in any aquatic ecosystem. Carbon, oxygen, hydrogen, nitrogen, potassium, calcium, phosphorus, and cobalt are, however, necessary for productivity. Fish food organisms obtain some nutrients such as calcium, phosphorus, nitrogen, and cobalt (and also some other micronutrients) for their growth and metabolism in the form of $Ca^{2+}$, $K^+$, $PO^3{}_4{}^-$ and $NO^-{}_3$. The availability of each of these elements depend on its chemical form in the soil, temeprature, and pH of water. Deficiency of these elements entirely overbalance the structure of ecosystem, resulting in poor production of fish food organisms and consequently, limits fish production. Some elements such as phosphorus and calcium are taken up in amounts much higher than their concentrations.
2. *Organic Components* : The most important organic substances include carbohydrates, proteins, lipids, and their derivatives resulted from the denaturation of the waste products of animals and plants. Organic fragments are moiety of various compounds produced as a result of decomposition of organic residues – collectively termed as *organic detritus*. Decomposing organic matter liberates essential nutrients along with the formation of a dark, colloidal, and amorphous substances – termed as *humus* which play a critical role in ecosystem fertility. Through the process of mineralization, this humus gets converted into mineral elements. Therefore, both these processes (humification and mineralization) proceed simultaneously and ceaselessly in an aquatic ecosystem.
3. *Climatic Factors* : These factors include rainfall, temperature, intensity of solar radiation, diurnal and seasonal fluctuations of temperature and water currents. These factors are not insignificant in the survival and growth of fish and under adverse circumstances exert their limiting effects on fish production.

### *Biotic Components*

Biotic components constitute the nutritional structure or trophic structure of any ecosystem where living organisms are celebrated based on their nutritional relationships. Biotic components of a fish culture ecosystem are, however, diverse that range from micro-organisms to large-sized fish. From nutritional stand-point and mode of procurement of energy and nutrients, they may be categorized into three groups such as producers, consumers, and decomposers. Such an assortment is termed as *trophic assortment*.

1. *Producers* : Producers, primary producers, or autotrophs as the name implies, largely include the chlorophyll-bearing green plants that can produce their own food from simple inorganic substances. Majority of the autotrophs of any fish culture ecosystem are microphytes or several species of algae including photosynthetic bacteria. To lesser extent, a few chemosynthetic bacteria contribute to the build up of organic matter.
2. *Consumers* : Consumers or heterotrophs include the organisms that cannot manufacture their own food but derived their energy and nutrients from autotrophs. In heterotrophic component, utilization, rearrangement, and decomposition of complex materials predominate. Some consumers depend directly on the autotrophs for food and are termed as *herbivores* or *primary consumers*. In a fish culture pond, primary consumers include

zooplankton, a few insects, molluscs, and some herbivorous fishes. Some of these organisms such as copepods and mussels are filter feeders, extracting microscopic primary producers from water.

Aquatic animals which depend on herbivores are known as *secondary carnivores*. These types of consumers include insect larvae and carnivorous fishes.

In some aquatic ecosystems, there is a third category of consumers that feed on secondary consumers. They are also carnivorous in nature and are termed as *tertiary consumers* such as game fishes.

3. *Decomposers* : These are largely microscopic organisms and include protozoa, bacteria, fungi, and actinomycetes. Complex compounds of dead plants and animals and waste products of fishes are broken down by these micro-organisms and subsequently absorb the breakdown or decomposition products and liberate inorganic nutrients in ecosystem, making them useful again to producers.

   The most important role of the decomposers in the ecosystem is to permit recycling of nutrients through mineralization (conversion of dead organic materials into simple forms).

## 3.12 Trophic Structure of Ecosystem

The biotic components of any ecosystem may be recognized as the functional kingdom of nature, since biotic components are based on the type of nutrition and the energy source used for the purpose. The foregoing discussion on the structure of aquatic ecosystem based on the nutritional relationships, has acknowledged us to assort various organisms into different levels, termed as *trophic levels*. The trophic structure is one type of autotroph-heterotroph agreement where each food level is called as *trophic level*. Such arrangement of different trophic levels constitutes what is termed as *food chain*. The amount of living component in different trophic levels is known as *standing crop*. The total quantity of fish present in a particular pond may be considered as standing crop. It may be expressed in terms of (1) number of organisms including fish per unit area and (2) biomass per unit area and can be measured as living weight, dry weight, or any other suitable units.

In most aquatic ecosystems, autotrophs occupy the position of the first trophic structure. The second trophic level position is the secondary consumers belong to the third trophic level and finally, the tertiary consumers to the fourth trophic level. The number of trophic levels in most of the fish culture ecosystems varies from 3 to 4 and very infrequently exceeds five. A sound idea of the trophic structure is an essential prerequisite for study of the functioning and dynamics of an ecosystem. In other words, the investigation of trophic levels provides an opportunity to get acquainted with the process of energy transformation in an ecosystem. Trophic structure affords an useful basis to comprehend all organisms that allot the same mode of feeding into one rank and thus belongs to the same trophic level.

### *Ecological Pyramids*

As mentioned earlier, trophic structure is characteristic of each type of ecosystem. The presentation of trophic structure and function at successive trophic levels is made graphically in the form of a pyramid and thereby referred to as *ecological pyramids* where the base of each pyramid exhibits the autotrophs while the apex represents the tertiary consumers. The other consumer trophic levels subjugate their ranks in a hierarchic manner. Since this well-known model is formed in association with biotic community, it is also termed as *biotic pyramid*. However, the pyramid shows how an aquatic ecosystem works. It looks like a series of blocks stacked one on top of another, each depicting an organism or group of organisms depending upon one below for sustenance. Ecological pyramids are of three types as noted below.

1. *Pyramid of Numbers* : These types of pyramids exhibit the relationship between autotrophs, herbivores, and carnivores at different trophic levels in terms of their numbers. Thus, the graphical representation of the total number of individuals of different species belonging to each of the trophic level in an ecosystem is known as *pyramid of numbers*. The pyramid of numbers in a pond ecosystem is shown in Fig. 3.12. Figure indicates that the number of autotrophs such as algae and photosynthetic bacteria is maximum and forms the platform of the pond ecosystem; the herbivores – small fish, are lesser in number than the former. The secondary consumers such as larger fish are still lesser in number than the herbivores. Lastly, the number of tertiary consumers – the larger fish are least. Thus in most pyramid of numbers, the number of individuals in each trophic level gets smaller as you go up the pyramid and consequently, furnishes the diagrammatic representation of a typical pyramid shape.

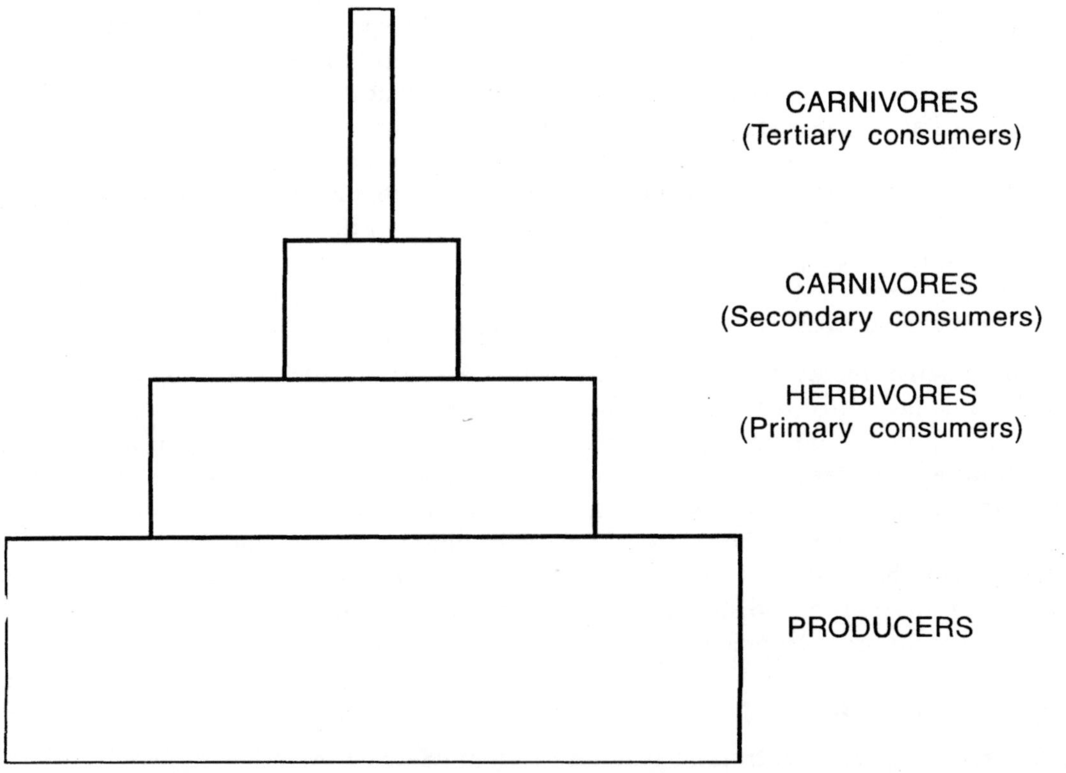


**Fig. 3.12 : Pyramid of numbers (individual/unit area) in a pond ecosystem.**

Though the pyramid of numbers contribute many useful informations, it suffers from certain limitations such as the fluctuation in numbers of different trophic levels in often so extensive that it becomes arduous to depict the pyramids to scale. It also fails to differentiate size alterations of the autotrophs (algae and photosynthetic bacteria in case of pond ecosystem). For this reason, a true pyramid is not always obtained.

2. *Pyramid of Biomass :* In contrast to pyramid of numbers where the pyramid is upright, the amount of biomass gradually exhibits and increase towards the apex and consequently, the pyramid in such ecosystems forms an inverted shape (Fig. 3.13).

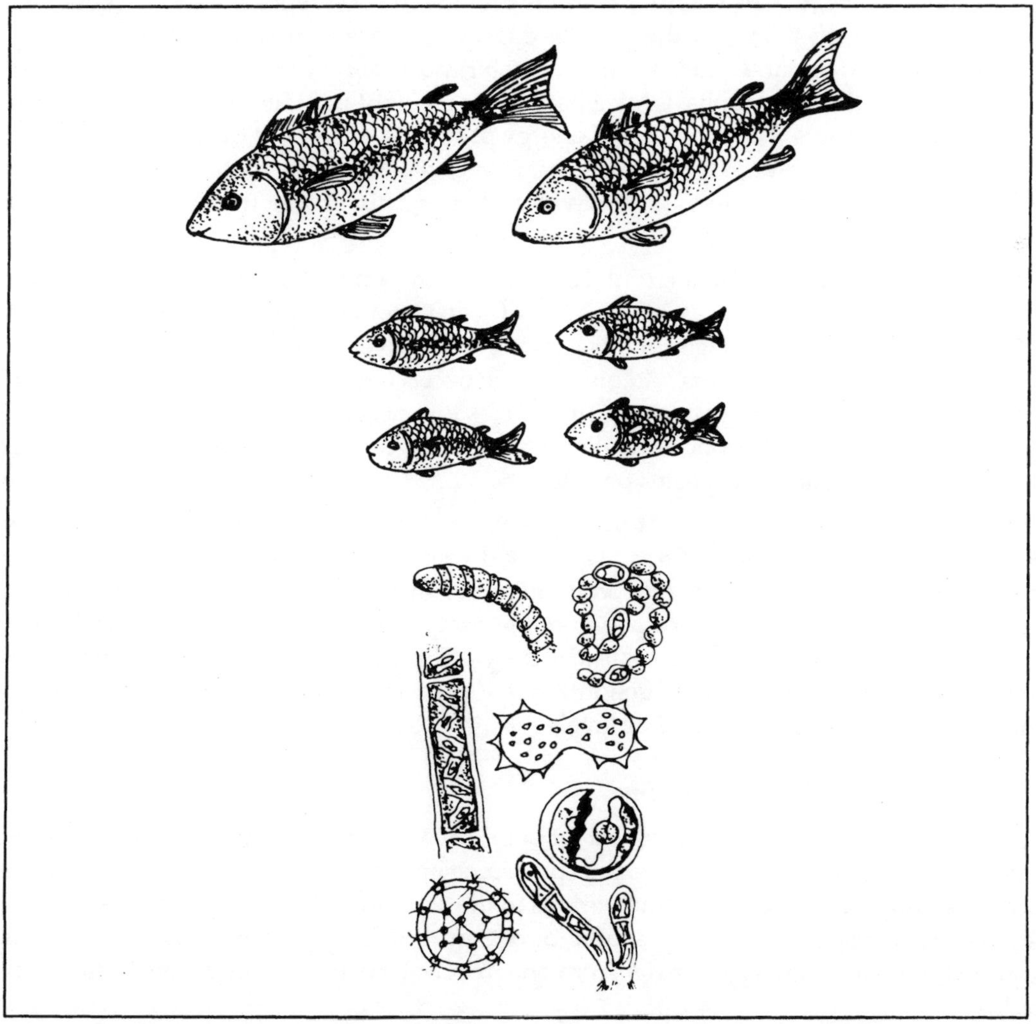

**Fig. 3.13 : Pyramid of biomass in a pond ecosystem. In large lakes and ponds, phytoplanktons constitute the autotrophic layer. Due to their short life cycle, they do not concentrate much biomass. Zooplankton and fish rapidly consume phytoplankton populations. As a result, the standing crop of phytoplankton considerably decreased than that of the herbivores or other consumers. Two reasons such as density of different populations and their growth rates are accountable to form such a pyramid.**

In a fish culture pond ecosystem, a pyramid of biomass appears to be upside down, with more primary consumers than primary producers. At first glimpse, this appears to be impracticable. But the upside down pyramid is observed in conditions in which the autotrophs reproduce very rapidly, in spite of being eaten by the consumers. The trophic levels represent the rate at which biomass is being produced. Therefore even though there may be few autotrophs at a given time, the rate of their production is much greater than the rate of biomass production for the primary consumers. Biomass is an estimate of the total amount of living material.

3. *Pyramid of Energy* : The energy pyramids provide the best representation of general nature of a pond ecosystem and the flow of energy in ecosystem can be represented by such pyramids. Algae and photosynthetic bacteria use only a fraction of incident light and trapped by the autotrophs. Only a part of the incident light, in turn, becomes assimilated by the carnivores. Thus the flux of energy primarily decreased during its course through different trophic levels. It obviously overcomes the limitation of the pyramids of numbers and biomass and ponders the behaviour of energy. In shape, therefore, such pyramid is always upright.

   In energy pyramid, each horizontal bar denotes the amount of energy at each trophic level in a given time and expressed in terms of Kilo Calories/square metre/year (K Cal/$m^2$/yr). Some of the advantages of this pyramid are (i) weight and number of total organisms in the pyramid of numbers and biomass are deceiving because two species do not always possess the same energy content, (ii) it takes into account the productivity of ecosystem, and (iii) this pyramid not only allows a collation between two ecosystems but also the relative importance of species populations.

The above discussion follows that there is a greater amount of energy at the producer level than at the other trophic levels, and this gradually decreases at each successive trophic level. The efficacy of conversion of the energy embodied in food substances to animal protoplasm increases with the trophic level. The rate of production of the components of a pond ecosystem may thus be compared to a pyramid – the so-called *pyramid of energy*. This pyramid is represented by the synthesis of green plants at the base with production rates of primary consumers and those of secondary consumers laying at successively higher levels.

### *Limitations of Ecological Pyramids*

Though ecological pyramids provide a lot of informations regarding the structure and energy relationships of an ecosystem, they fail to interpret some important expressions : (1) They are unable to indicate the diurnal and seasonal fluctuations in the activities of organisms, (2) Detritus greatly affects the ecosystems by acting as a source of energy, but the pyramids cannot illuminate this aspect, and (3) the rate of energy transfer from one trophic level to another is not mediated in these pyramids.

Different types of pyramids are, of course, elementary conceptions. The original pyramid of any vast aquatic ecosystem is a complex tangle of food chains dependent on assistance and contest among the links. If there is an alteration in any one component, many other components must accommodate harmoniously.

## 3.13 Function of an Ecosystem

A second important measurement of energy flow is an aquatic ecosystem is its functional aspects. It is expressed as how an ecosystem operates under natural conditions.

The network of living and non-living components of an ecosystem is so intricate that their separation from each other often becomes practically arduous. The way of movement of nutrients and energy in an ecosystem is shown in Fig. 3.11. The autotrophs along with the elements synthesize carbohydrates but not energy and because they transform radiant energy into chemical form, it should be better to call the producers as *converters*. The term converter is, however, not common and generally not used in many text books.

The flow of energy and cycling of minerals within an aquatic ecosystem involve complex interactions between the biotic components and the physico-chemical environment should be regarded as the nucleus of the dynamics of ecosystem. While energy flow is an uni-directional (non-cyclic) and proceeds from the sun to the decomposers via producers and consumers, the minerals flow in a cyclic manner. The mineral cycling is highlighted by different cycles known as *biogeochemical cycles* which is an important aspect so far as the flow of energy is considered. Moreover, the energy not only flows in a non-cyclic way but also substantial amount of energy is lost from the system in various ways. At the same time, the minerals also exhibit a significant loss. Functional aspects of an ecosystem are characterized by the ecosystem productivity, food chain, and food webs. These processes will be given brief consideration.

### *Productivity of Ecosystem*

The amount of organic substances concentrated in the living components of a pond or lake ecosystem in unit time is referred to as the *productivity of ecosystem*. The productivity of a fish culture pond is very important whereby living substances are fabricated through the interactions of ingredients of the natural environment. Ponds and lakes are essentially self-supporting. The basic methods in the operation of such an ecosystem are : (1) the acquittance of energy, (2) synthesis of organic substances by producers, (3) use of organic substances by consumers, (4) decomposition of organic and inorganic substances, and (5) their transformation into suitable forms for the nutrition of autotrophs. Productivity can be divided under the following heads.

1. *Primary Productivity* : Primary productivity is connected with the autotrophs, most of which are algae, photosynthetic bacteria, and green plants. These are the common autotrophs in a fish culture pond and constitute an important aspect in the strategy of ecosystem productivity. Primary productivity can be defined as the rate at which radiat energy is stored by the activity of autotrophs.

   Primary productivity is evaluated by light and dark bottle method. The bottles are immersed in pond water and are incubated for 4 hours at mid-day light and temperature conditions. The dissolved oxygen concentration of pond water sample before the start of experiment and that of samples of light and dark bottles after 4 hours incubation is measured. The primary productivity has three basic components such as gross productivity (GP), net productivity (NP), and respiration.

$$\text{Gross productivity (GP) (mg of oxygen liberated per litre/hour)} = \frac{\text{Dissolved oxygen of light bottle} - \text{Dissolved oxygen of dark bottle}}{\text{Hours of incubation}}$$

$$\text{Net productivity (NP) (mg of oxygen liberated per litre/hour)} = \frac{\text{Dissolved oxygen of light bottle} - \text{Dissolved oxygen of initial bottle}}{\text{Hours of incubation}}$$

$$\text{Respiration (R) (mg of oxygen consumed per litre/hour)} = \frac{\text{Dissolved oxygen of light bottle} - \text{Dissolved oxygen of dark bottle}}{\text{Hours of incubation}}$$

(i) *Gross primary productivity* : It is the total rate of photosynthesis including the organic matter used up in respiration. Generally, it depends on the concentration of chlorophyll in autotrophs. Greater the amount of the chlorophyll content the higher is the gross primary productivity.

(ii) *Net primary productivity* : It is the rate at which organic matter is stored in tissues in excess of the respiratort utilization by plants. The net primary productivity refers to the balance between the gross primary productivity and respiration. The net primary productivity is essentially a statement of growth.

The rates of gross and net production are measured on the basis of organic carbon present in the unit volume or underneath unit area of the pond surface and are expressed as mg $C/m^2/day$. The word day usually refers to day-light hours during which gross or net photosynthesis takes place in the surface water.

2. *Secondary Productivity* : Collation of rates of gross and net primary productivity in an ecosystem furnish important results on the utilization of plant food by the plants themselves. In most cases, only a small portion of net primary production is consumed by organisms at second trophic level or by decomposers during a given period of time and the unutilized portion is taken up by the standing crop. Overgrazing of the autotrophs decreases the total standing crop. The secondary productivity, however, ascribes to the consumers and it is the rate at which energy is stored at consumer levels. Because consumers utilize food substances during respiration, secondary productivity is not divided into gross and net amounts. Therefore, the term assimilation is used instead of production at the consumers level.

3. *Net Productivity* : It is defined as the rate of storage of organic matter unutilized by consumers. In other words, it is the equivalent to net primary production less consumption by the consumers during the unit period. It is, however, expressed as production of g $C/m^2/day$ which is then converted into month or year.

## 3.14 Classification of Ponds on the Basis of Productivity

Water quality variables generally determine the productivity of any fish culture ponds and accordingly, they are classified as productive, poor productive, unproductive, and high productive, several *abiotic* and *biotic factors* such as pH, total alkalinity, dissolved oxygen, inorganic nitrogen, filtrable orthophosphate, plankton, and bottom fauna play a key role for productivity of fish ponds. Investigations on the water quality and soil conditions of fish ponds in relation to fish production have provided significant data (Table 3.1) to fish farmers so that they can manipulate their own ponds if they feel it is necessary.

**Table 3.1 : Classification of Ponds Based on the Abiotic and Biotic Factors of Water and Soil**

| | Parameter | Type of pond | | |
|---|---|---|---|---|
| | | Unproductive | Low productive | High productive |
| I | **Water** | | | |
| | pH | 6.0-7.0 | 7.0-7.5 | 7.5-8.5 |
| | Total alkalinity (mg/l) | 10-30 | 30-70 | 70-170 |
| | Inorganic nitrogen (mg/l) | Below 0.1 | 0.1-0.3 | 0.3-0.7 |
| | Orthophosphate (mg/l) | Below 0.05 | 0.05-0.2 | 0.2-0.6 |
| | Dissolved oxygen (mg/l) | 2-4 | 4-6 | 6-9 |
| | Plankton concentration[1] (Number/l) | 5-35 | 35-70 | 70-250 |
| II | **Soil** | | | |
| | pH | 6.0-7.0 | 7.0-7.5 | 7.5-8.5 |
| | Available nitrogen (mg/100g) | 5-15 | 15-35 | 35-90 |
| | Available phosphorus (mg/100g) | 3-6 | 7-20 | 21-50 |
| | Exchangeable calcium (mg/100g) | 20-50 | 51-150 | 151-300 |
| | Bottom fauna[2] ($g/m^2$) | 1-15 | 15-20 | 20-50 |

1. Plankton includes different groups of zooplankton (Cladocera and Rotifera) and phytoplankton (Cyanophyceae, Chlorophyceae, and Bacillariophyceae).
2. Bottom fauna commonly encountered with the following genera : *Chironomus*, Dragon-fly nymph, *Chaoborous*, *Branchiura*, Earthworms, and *Viviparus*.

Prior to fish culture on commercial basis, water bodies are arranged according to their productivity in relation to different physico-chemical and biological features and hence, intensive hydrobiological studies on such water bodies deserve special attention. To bring unproductive water bodies under fish cultivation, however, treatment of ponds with nutrient carriers at recommended rates is necessary to make the ponds more effective.

## 3.15 Food Chain in Ecosystem

The food chain is a series of organisms in an ecosystem through which food energy is transferred from the autotrophs to the top carnivores with repeated swallowing and being swallowed. Autotrophs utilize solar energy which is transferred to chemical form (in the form of adenosine triphosphate) during photosynthesis. Therefore, green plants form first nutritional level – called the *primary producers*. The energy stored within the producers is then utilized by the second trophic level – the *primary consumers* which, in turn, are swallowed by the third trophic level – called the *secondary consumers*. These, in turn, are further devoured by tertiary consumers. Omnivorous fishes, however, depend on the producers and consumers at their lower level in the food chain. Taxonomically different species may inhabit the same trophic level because they have homogeneous function in the food chain. Several genera such as *Volvox, Euglena, Nostoc*, Diatoms, *Vaucheria, Chlamydomonas, Nymphaea, Chara*, and photosynthetic bacteria, for example, belong to the same trophic level but they are taxonomically different and depends on direct solar energy as well as the addition of organic matter. The degradation of fallen leaf remnants which are teemed with benthic algae and phytoplankton, are engulfed and re-engulfed by a group of small but large number of animals such as copepods, insect larvae, crabs, mysis, amphipods, and bivalve molluscs (Fig. 3.14). These animals are called as *detritus consumers* and consume plant detritus. These animals are then consumed by small carnivores such as game fish and minnow which finally acts as the principal food for those carnivores which are situated in the vertex of the pyramid.

In opposition to the grazing food chain (see Section 3.16), the detritus consumers in terms of trophic levels are a mixed assamblage. These include herbivores, omnivores, and primary carnivores. Detrivorous animals obtain energy, to some extent either from plant materials (primary food material) or from micro-organisms (secondary food material).

## 3.16 Food Web – Interlocking Arrangement of Organisms

In an aquatic ecosystem, none of the food chains is detached from each other or as simple as we have already described earlier. In fact, most aquatic ecosystems embrace a general function. A food chain in a pond starts with aquatic plants and algae, proceeds through copepods (*Daphnia, Cyclops* and others) followed by smaller fish, larger fishfish, and birds/man.

Based on the type of organisms that constitute the first trophic level, food chain may be classified as (1) Grazing food chain and (2) Detritus food chain.

### *Grazing Food Chain*

Progressive fish culturists are quite familiar with the grazing food chains. Such chains are characterized by autotrophs (phytoplankton and algae), grazing animals (zooplankton and fish), and their predators (Birds). Most aquatic ecosystems have this type of food chain which includes phytoplankton, zooplankton, small fish, large fish, and game fish (Fig. 3.15). Aquatic food chains are usually larger than those of terrestrial ones because of the small size of the phyto- and zooplanktons that constitute the first two trophic levels. This type of ecosystem depends on autotrophic energy capture and the captured energy is proceeded to top carnivores via the herbivores. Grazing food chains are predominant in most fish culture ponds.

### *Detritus Food Chain*

This type of food chain begins with dead organic matter into micro-organisms and then to organisms feeding on detritus and their predators. Enormous quantities of organic matter are contributed by the remains of animals and plants as well as their excretory products. Such food chains are found in several ecosystems but predominant mostly in grassland, forest, and in shallow water communities. Such ecosystems are less multitudinous inter-connected nutrient interactions. These nutrient interactions between the members belonging to different food chains are referred to as a *food web* (Fig. 3.16). Because it is composed of several food chains, a food web explains all conceivable transfer of energy and nutrients among the organisms in an ecosystem, whereas a food chain delineates only a single pathway in the food web. In most ecosystems, the food web exhibits very confused relationships involving various types of living organisms.

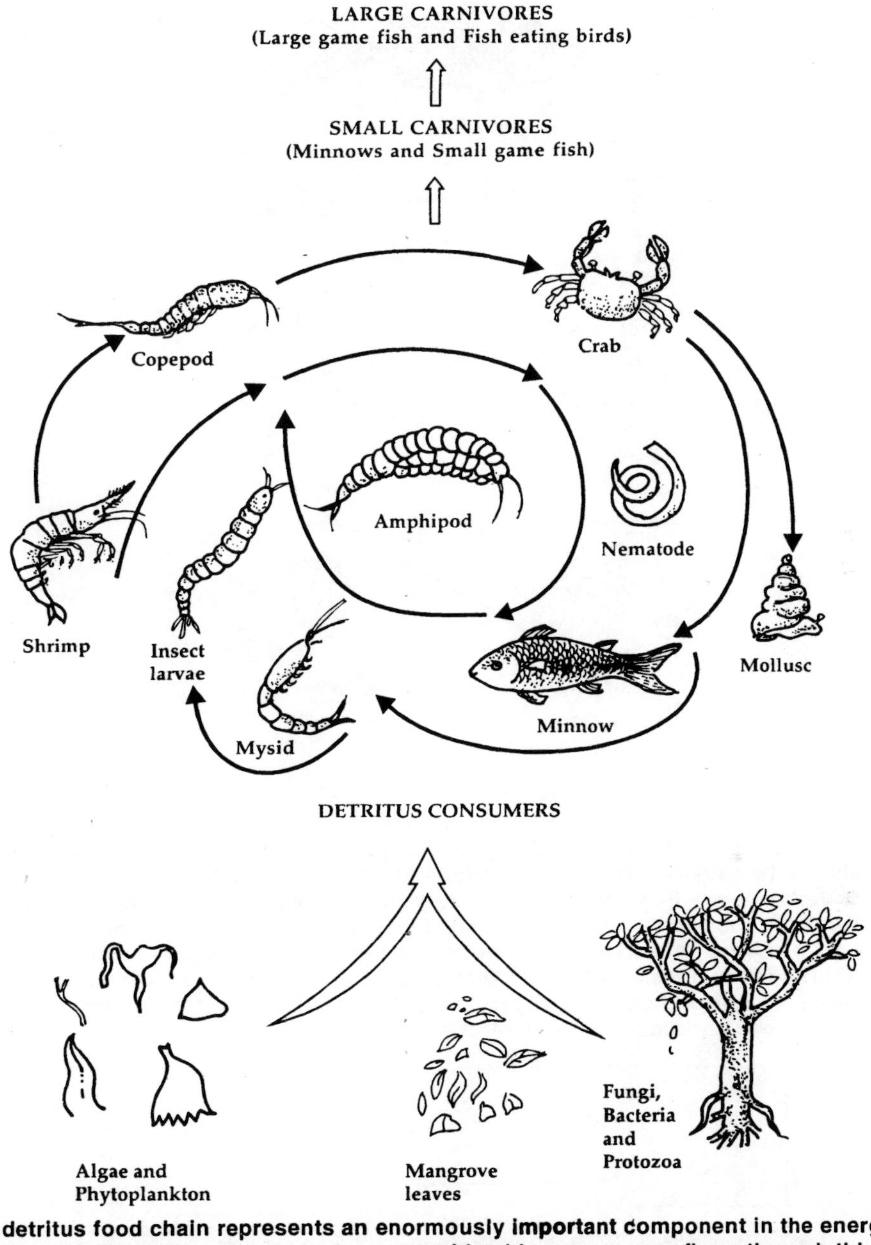


**Fig. 3.14 : The detritus food chain represents an enormously important component in the energy flow of an aquatic ecosystem. In some ecosystems, considerably more energy flows through this type of food chain. The organisms in this chain are many and include algae, bacteria, fungi, insects, protozoa, crustacea, molluscs, rotifers, annelid worms, and fish. Some species consume almost anything and some are specific in their food requirements. Many protozoa, for example, require certain vitamins, specific organic acids, and nutrients. Detritus organisms consume pieces of decomposed organic matter and digest them partially. After extracting some of the chemical energy from the food, undigested materials are excreted in the form of simpler organic molecules. The wastes from one organism can be utilized by a second and thus repeats the process. The complex organic molecules present in the dead organisms are broken down to much simpler compounds and consequently, some organic substances – humic acids or humus – are formed which forms an essential part of the soil.**

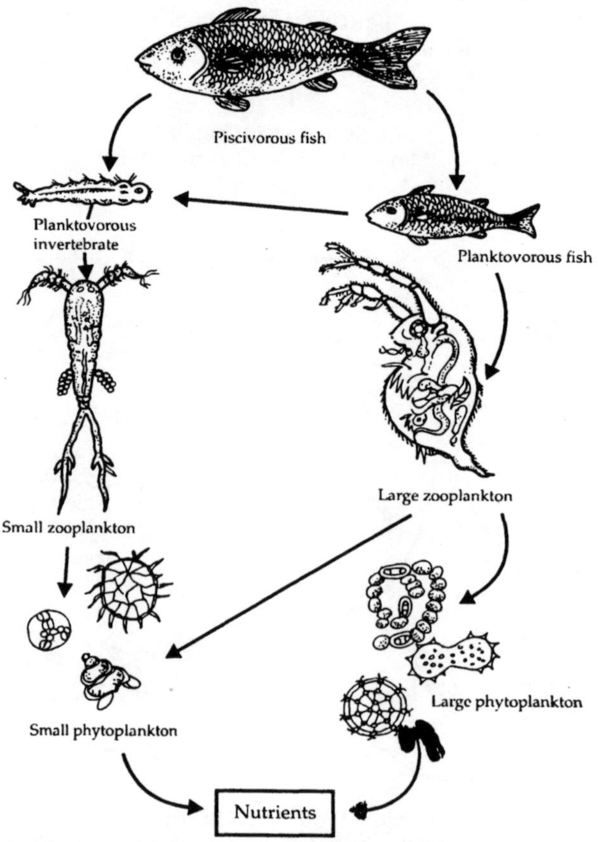


**Fig. 3.15 : The grazing food chain in a typical aquatic ecosystem begins from phytoplankton and terminates to carnivores by passing through herbivores. Thus the gross production of phytoplankton may meet three destinations – it may be oxidized in respiration, it may die and decay, and it may be consumed by herbivores. In herbivores, the assimilated food can be stored as carbohydrates, proteins, or fats either transferred into simple substances or reconstruct by the animals into complex organic molecules. These transformations require energy that is supplied by respiration. The ultimate flow of energy in herbivores occurs by (i) respiration, (ii) decay of organic matter, and (iii) other decomposer organisms and consumption by carnivores.**

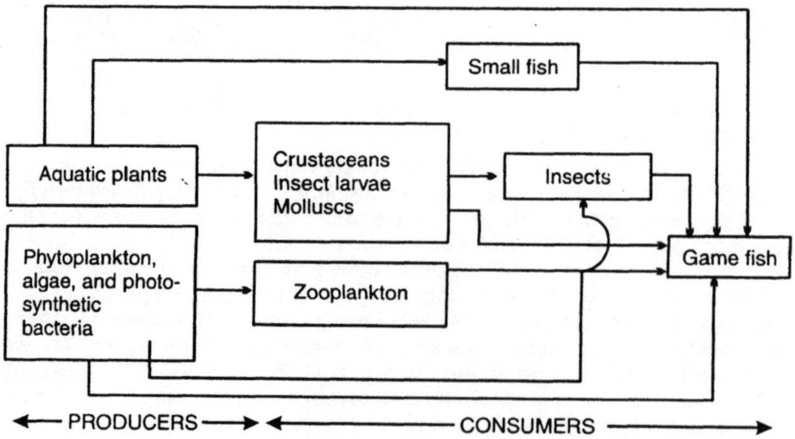


**Fig. 3.16 : Diagramatic representation of a food web in a freshwater pond.**

The food webs are very consequential in sustaining the stability of an ecosystem to great extent. Decrease in the population of herbivores, however, would obviously cause destruction of the producers due to competition. Besides, the survival of primary consumers is also connected with the secondary consumers and this sequence goes on. Therefore, an equilibrium condition of the ecosystem is indispensable for the survival of living organisms.

## 3.17 Dynamics of a Freshwater Ecosystem

Organisms within a freshwater ecosystem which interact with one another either directly or indirectly through feeding relationships is termed as *trophic interactions*. This involves transformations of food energy from individual to the next through a process called *consumption*. At the same time, substances move within the ecosystem and the pathways of such movements are closely associated with the flow of energy.

Aquatic organisms are not able to utilize full energy for various metabolic functions. They, however, obtain energy either through photosynthesis or through consumption. According to the law of thermodynamics (to be discussed later), organisms dissipate energy in two ways such as (1) Movement of aquatic animals (including fish) results in the transfer of kinetic energy to the surrounding water and (2) Loss of energy at each biochemical step in the transformations required for their activity.

Inspite of these losses, deficiency of energy is not generally observed because the autotrophs extract energy from the sun. Thus, an uninterrupted input of energy into the ecosystem takes place. Aquatic macrophytes and algae take an important part in the transfer of energy. These aquatic plants convert some of the energy from sunlight into forms usable by the fish. Fish use some energy for their metabolic activities.

### *Behavior of Energy in an Aquatic Ecosystem*

Energy can be defined as the ability to perform different activities of fishes for their survival and growth. The movement of energy, however, into and out of the organisms and also the movement of materials permit aquatic life to survive in any ecosystem. Ecosystems are, therefore, characterized by the movement of energy and the materials between the system and the environment generally.

It is important to emphasize that energy does not available like non-energy producing materials. It is, rather used once by a given organism and converted into heat, soon lost from the ecosystem, and again regenerated through autotrophs. In fact, the behaviour of energy can be explained by the fundamental concepts of physics termed as the *Laws of Thermodynamics* in the following way.

1. *First Law or the Law of Conservation of Energy* : This law states that energy may be transformed from one type into another but is neither created nor destroyed. Light energy of the ecosystem, for example, is converted into potential or kinetic energy of the food by producers.
2. *Second Law or the Law of Entropy* : According to the second law, no process involving an energy transformation will occur unless there is a degradation of energy from concentrated form into a dispersed form. Only a fraction of the energy conserved under the first law is, however, available to perform different activities within the system. In other words, the

energy accumulated by the producers is passed to the consumers through food materials (carbohydrates) with a gradual loss of energy from the base to the apex of the food pyramid of an ecosystem.

It appears, therefore, that aquatic organisms particularly fish, plankton, and bottom fauna exhibit the essential thermodynamic features. Thus the behavior of energy in aquatic ecosystem can be conveniently perceived from its flow since energy flow is uni-directional. The interaction of material and energy in the ecosystem is, however, a primary concern to ecologists.

## 3.18 Four Approaches to the Energy Flow of an Ecosystem

For convenience of the study and understanding, the energy flow of an aquatic ecosystem can be grouped into four categories as mentioned below.

### *Energy Fixation by Autotrophs*

We have learnt that ecosystems will cease to function unless there is a constant input of energy from sunlight. Solar energy makes its passage into the ecosystem through autotrophs. The process by which energy is fixed in autotrophs is called *photosynthesis*. It is the utilization of light energy from the sun to produce food in the form of carbohydrates. The process of photosynthesis is carried out by green plants – called *chlorophylls,* which can absorb only a fraction of radiant energy of the sun and convert it to chemical energy. This chemical energy is used to make organic matter (in the form of glucose) from the gas carbon dioxide and water. The overall reaction may be simplified as :

$$6CO_2 + 6H_2O + \text{Solar energy} \xrightarrow[\text{Enzyme}]{\text{Chlorophyll}} \underset{\text{Glucose}}{C_6H_{12}O_6} + 6CO_2$$

From this basic reaction the amount and rate of energy fixation by autotrophs can be determined. If the amount of carbon dioxide used in unit time is calculated, the amount of glucose produced can easily be determined.

Among the autotrophs able to synthesize organic matter are certain photosynthetic bacteria such as *Chlorobium* sp., *Chromatium* sp., and *Rhodopseudomonas* sp. In the presence of light, these organisms are able to yield various organic substances and simultaneously fix energy. The photosynthetic bacteria possess bacteriochlorophyll which can absorb light energy. The chemical reaction of the various photosynthetic bacteria may be presented by a generalized formula.

$$CO_2 + \underset{\substack{\text{Hydrogen}\\\text{sulphide}}}{2H_2S} + \xrightarrow[\text{Bacteriochlorophyll}]{\text{Light energy}} \underset{\text{Carbohydrate}}{(CH_2O)} + 2S + H_2O$$

Several different types of bacteria such as *Nitrobactor, Nitrosomonas,* and *Beggiatoa* are able to yield organic matter. The synthesis of organic matter is brought about by the energy derived from oxidation reactions – independent of the light energy and this process is known as *chemosynthesis.*

The amount of organic matter synthesized by these bacteria varies greatly with the pH, nutrient level, intensity of light, and the photosynthetic pigments. In general, the rates of organic matter production by these bacteria are thought to be much lower than those synthesized by blue-green algae. The contribution of these bacteria is, however, insignificant but that of blue-green algae is thought to be of several importance for energy transfer especially in well-managed fish culture ponds.

### *Estimation of Energy Accumulation by the Autotrophs*

To know the amount of energy accumulated by the autotrophs, measurement of biomass is necessary. A number of methods such as carbon dioxide assimilation, oxygen production method, harvesting, and radio-isotope process are involved in the measurement of pond biomass. Although these processes are very complex and beyond the scope of discussion in this book, following brief notes on the overall process will provide an idea of how the amount of energy accumulated by autotrophs is estimated.

According to an estimate, one kilogram of aquatic plant contains about 0.446 kilogram of carbon. This amount of carbon enters into the plant body through photosynthesis which is equivalent to 1.115 kilogram of glucose. It is termed as *net production* (NP) and the amount of carbon released during respiration (R) is 0.136 kilogram. Therefore, the gross production (GP) is NP + R = 1.115 + 0.136 = 1.252 kilogram. It is also known that 3,760 Kilo Calories are required for the production of 1 kilogram of glucose. Therefore, for the production of 1.251 kilogram of glucose, 4,704 Kilo Calories will be required and this requires 2.912 Kilo Calories of solar energy.

### *Energy Flow from Producers to Primary Consumers*

A number of data regarding the flow of energy from producers to primary consumers has been obtained from different lake and pond ecosystems. Raymond Lindeman in 1942, for example, analyzed the energy flow in Cedar Bog Lake ecosystem of Minesota and calculated the production in terms of gram calorie per square centimetre per year (g Cal/cm$^2$/yr). The fundamental truth of food chains and functioning of the laws of thermodynamics can better be explained by means of the following diagram.

The Fig. 3.17 represents a very simplified energy flow model as described by R. Lindeman in 1942 having only three trophic levels. The description of the energy flow model was based on the following data : Incident solar radiation – 118,872 g cal/cm$^2$/yr. Net production (NP) – 86.9, Utilization in respiration (R) – 24.1, and gross production (GP) – 111. The data indicates that out of the total input of solar radiation (118,872 g cal/cm$^2$/yr), substantial amounts of radiation (118,761 g cal/cm$^2$/yr) remain unutilized and the gross production (NP + R) by autotrophs is 111 g/cal/cm$^2$/yr i.e. 0.10 per cent energy is trapped. Out of 111 g cal/cm$^2$/yr energy, 23 g cal/cm$^2$/yr or 20.7 per cent of energy is used by the autotrophs for growth, respiration, and reproduction. It may further be observed that 15 g cal/cm$^2$/yr energy are utilized by primary consumers which feed on autotrophs and this amounts to 17 per cent of net autotroph production. Decomposition (3 g cal/cm$^2$/yr) considers for about 3.4 per cent of net production. The remaining plant material – 70 g cal/cm$^2$/yr or 63.1 per cent of net production is unutilized and deposited in the soil.

### *Energy Flow from Primary to Secondary Consumers*

Fig. 3.17 also indicates that the primary consumer accumulates 15 g cal/cm$^2$/yr of energy. Of this energy, 30 per cetn or 4.5 g cal/cm$^2$/yr is used for metabolic activities and the remaining

10.5 g cal/cm$^2$/yr is available to the secondary consumers. It is, therefore, evident that 30 and 20.7 per cent energy is lost through respiration by herbivores and autotrophs, respectively. At the same time, substantial amounts of energy are available for the carnivores (10.5 g cal/cm$^2$/yr or 70 per cent) which is not fully utilized; only 3 g cal/cm$^2$/yr or 20 per cent of the net production proceeds to the carnivores. In the carnivore level, about 60 per cent energy is utilized in metabolic activities, the remaining part of energy is deposited in the soil, and trace amount of energy undergoes decomposition.

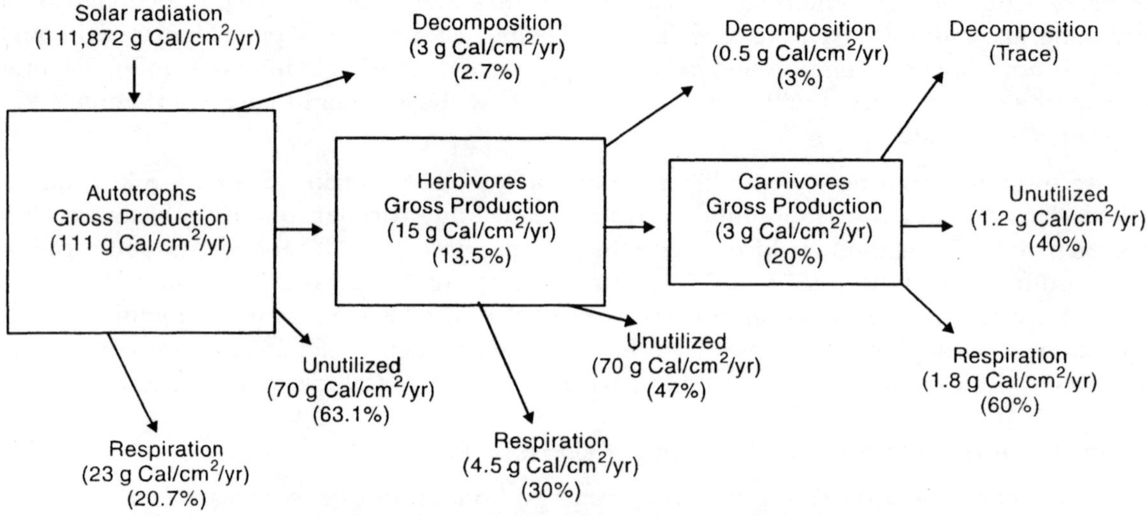


**Fig. 3.17 : Energy flow diagram in a freshwater lake.**

Discussion regarding the flow of energy in a pond ecosystem indicates that the amount of energy that is arrested by the producers does not return to solar input. Because it moves gradually through different nutrient levels, it is not procurable in the preceding level. Therefore, the energy flow is a one-way path and the entire system will cease to function if the principal source of energy – the sun – is not allowed to enter into the system. On the other hand, there occurs a gradual decrease in energy at each nutrient level. This is considered by the energy disseminated as heat in metabolic activities and estimated as respiration.

It should be borne in mind that the above-mentioned model indicates the principal pattern of energy flow in an aquatic ecosystem. But under natural situations, organisms are so inter-related that several food chains become interlocked and consequently, a complex food web is developed. Interlocking arrangement of a number of food chains in a food web obviously exhibit a multi-channel flow of energy.

A well-managed fish culture ecosystem generally considered as one of the most important and economic value that makes a substantial contribution to the food chain which is related to fish production – an important economy in regions where fish culture is the main source of income for the farmers and hence must take these conditions into account.

## 3.19 Nutrient Cycle in Aquatic Ecosystems

The conversion of carbon in photosynthesis and respiration are adequate enough to make a vivid picture that the flux of energy depends on the availability and movement of nutrients in ecosystems. Energy flows involve nitrogen, phosphorus, carbon and many other elements of the ecosystem. Where energy transfer requires an uninterrupted supply of solar energy to the ecosystem for high productivity, nutrients in contrast, are held within the ecosystem where they are ceaselessly recycled among different types of organisms as well as between the organism and the physical environment. Chemical fertilizers and organic manures serve as the sources of different important nutrients – essential for ecosystem productivity.

Brief discussions on the structure and function of an ecosystem have shown that energy and nutrients move through abiotic and biotic components of the system. Though nutrient pathways are closely related to the flow of energy, it is significant to note that there is an important difference between the energy flow and the nutrient recycling. While nutrient cycle is moderate with elements being extracted from restricted compartments or pools and reserved within the ecosystem, the flow of energy is shameless in the perception of being prosecuted by immeasurable supply of energy from the sun. The elements are, however, withdrawn from the soil and water by the autotrophs and again released to the ecosystem. A number of organisms play an important role in returning and extracting phenomena of minerals and at the same time some physico-chemical manifestations are also involved that make together a methodical cycle in a well-furnished aquatic ecosystem. Since the movement of nutrients is executed by various chemical cycles that assist to flow the nutrient elements forward and backward between the environment and organisms. Therefore, movement of nutrients undertake unique cyclical pathways that involves biological, chemical, and physical exchanges and for this reason nutrient cycles are also referred to as *biogeochemical cycles*. For smooth running of these cycles, constant supply of several micronutrients (such as iron, copper, and cobalt) and macronutrients (such as carbon, oxygen, potassium, nitrogen, and phosphorus) is necessary. A variety of biogeochemical cycles operate the connections between unavailable and available forms of nutrients in water and soil. Among different cycles, however, nitrogen and phosphorus cycles are extremely important for constant productivity of fish culture ecosystems.

## 3.20 Two Approaches in an Aquatic Ecosystem – Energy Flow and Biogeochemical Cycle

Two non-living functions such as energy flow and biogeochemical cycles are not unimportant that make an aquatic ecosystem operational and productive. energy flows from the sun into the aquatic ecosystem where organic matter is synthesized through photosynthesis. Energy is stored and used later, but once used, energy cannot become sunlight again to synthesize food. If ecosystem productivity is to continue, sunlight must allow to enter into the system.

Since some energy is continually squandered as disordered heat energy called as *entropy*, no transformation such as from light to food, can be 100 per cent efficient. To survive fish culture ecosystems for sustained production, need high-quality energy, storage capacity for periods when input is not sufficient, and the methods to convey heat from entropy. If energy flow is reduced, the ecosystem itself becomes irregular.

In contrast to energy, material can be reused. Biogeochemical cycles are the courses of chemical substances proceeding between organisms and the environment. More than twenty types of nutrients important to organisms are distributed throughout ecosystem. Nutrients present in large, non-easily available pools and in small, active and available pools and usually cycle from unavailable to available forms.

## 3.21 Conclusion

Fish culture ponds and lakes usually emphasize the importance of the structural and functional ecology termed as *ecosystem*. An ecosystem is the organizational unit which includes a community of non-living substances and living organisms. Ponds and lakes are called *natural* and *limnetic ecosystems* – differ from each other not only in their species composition but also in productivity, all of them have a similar plan of their gross structure and function. The structure of an ecosystem is principally a description of different species of plants and animals. While the structure includes abiotic substances (inorganic and organic compounds), producers, consumers, and decomposers, the function describes the energy flow and nutrient cycle, the latter two constitute the *food chain.*

The availability of freshwater resources under suitable conditions has become critical, with an increasing proportion of population unable to access adequate use for fish culture. And as demand for freshwater resources increases, aquatic ecosystems suffer as water is either diverted to agriculture or gets polluted. Therefore, sound management of water is a key step.

## *References*

Kramer, J.R., S.E. Herbes, and H.E. Allen. 1979. Phosphorus analysis of biomass and sediment. *In* : *Nutrients in Natural Waters* (H.E. Allen and J.R. Kramer *eds*), pp. 52-100. Wiley Interscience Publications, New York.

UNEP (United Nations Environmental Programs). International Environmental Technology Centre's Insight News Letter, September, 1998.

Sarkar, S.K. and P. Basu Chowdhury. 1999. Role of some environmental factors on the fluctuation of plankton in a lentic pond at Calcutta. *In* : *Limnological Research in India* (*Ed*: K.D. Mishra), pp. 108-132, Daya Publishing House, New Delhi, India.

Verma, S. and R.C. Mohanty. 1995. Phytoplankton of Malyanta pond of Laxmisagar and its correlation with certain physico-chemical parameters. *Poll. Res.* 14 : 243-252.

Verma, P.S. and V.K. Agarwal. 1985. *Principles of Ecology.* S. Chand and Company, New Delhi.

## Questions

1. State the physico-chemical nature of a freshwater pond/lake.
2. How ponds/lakes are classified on the basis of thermal stratification?
3. What are the characteristic features of ponds?
4. Distinguish the following terms : (a) hyposalinity and hypersalinity, (b) trophogenic zone and tropholytic zone, (c) epilimnion and hypolimnion.
5. Describe the benthic communities of a pond/lake.
6. How sustainable management of freshwater resources is executed for fish production? State the role of science and technology in understanding freshwater issues in fisheries sector.

7. How a benthic food chain or food web is formed in a pond/lake ecosystem?
8. How abiotic and biotic factors interact with each other in an aquatic ecosystem?
9. Give an account of the structure and function of an aquatic ecosystem.
10. Discuss the abiotic and biotic components of a freshwater ecosystem.
11. Discuss how does energy flows through different trophic levels.
12. What is food chain? Give a comparative account of grazing and detritus food chains.
13. Explain the following :

    (a) Productivity; (b) Trophic structure, (c) Standing crop, (d) Ecological pyramids, (e) Food webs, (f) Secondary productivity, (g) Nutrient cycling.

# 4

# Aquatic Plants in Pond and Lake Environments

Different types of plants and animals which live in water, form the complex biotic community and exhibit inter-relationship among themselves. Fish ponds treated with organic manures and chemical fertilizers are occasionally infested with various types of aquatic plants which disturb physiological activities of fish and other aquatic animals and, as a result, fish culture operations become impossible. Aquatic plants limit the living space for fish and their reduction in survival, growth and reproduction, upset the equilibrium of physico-chemical factors of pond soil and water, cause imbalance of biochemical oxygen demand and disbalance of balanced relationship between plants and animals, promotes siltation of ponds and lakes and ultimately these ecosystems become unfit for human use. But one cannot ignore beneficial effects of aquatic plants in fish ponds such as (1) production of biogas, paper pulp, animal feed and organic manures, (2) pollution control, (3) absorption of hevy metals such as chromium, mercury, nickle, zinc, iron, and lead and (4) absorption of essential nutrients such as nitrogen and phosphorus from water that has already been treated with chemical fertilizers.

## 4.1 Units of Aquatic Vegetation

Aquatic plants are not uniform in any pond or lake. Depending on climatic and edaphic factors aquatic plants of a pond or lake may be divided into a number of natural units whose composition is different and distinct. A large unit of aquatic vegetation in a pond under identical climatic conditions is called a *plant formation.* A formation is said to be a *climax* when it is dominated by one or more species which are abundant in it. For example, in floating vegetation there may be certain dominant marginal plants.

A major sub-division of a plant formation is termed as *association.* An association is similar in general outward appearance floristic composition, and ecological structure such as marsh association or hydrophytic. A plant association may have one or more dominant species. Within the association there may be *communities* of plants closely associated.

## 4.2 Types of Aquatic Plants

Aquatic plants are defined as those vegtation which grow and reproduce in small and large bodies of water. From the edge of pond or lake to the bottom , different types of aquatic plants or hydrophytes grow in different depths of water. According to their distribution in a freshwater pond/lake ecosystem, hydrophytes have been classified as follow :

1. *Floating Plants* : These plants float above the surface of water with roots hanging from underneath. They remain in contact with air and water but not soild such as *Eichornia* sp. (water hyacinth), *Wolffia* sp., *Lemna major* and *Lemna minor* (duck-weeds), *Azolla* spp. (water fern), *Pistia* sp. etc. (Figure 4.1).

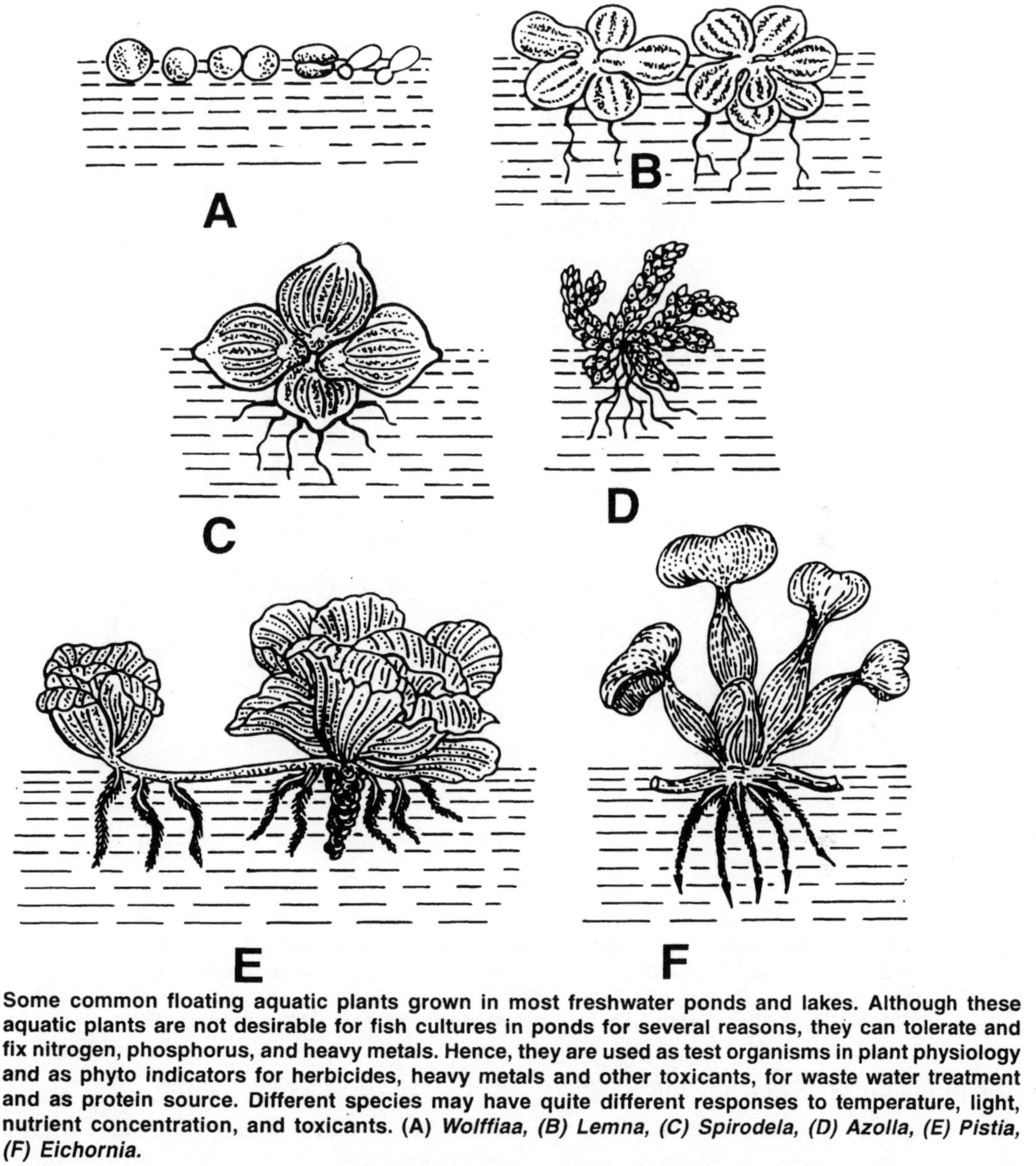


**Fig. 4.1 : Some common floating aquatic plants grown in most freshwater ponds and lakes. Although these aquatic plants are not desirable for fish cultures in ponds for several reasons, they can tolerate and fix nitrogen, phosphorus, and heavy metals. Hence, they are used as test organisms in plant physiology and as phyto indicators for herbicides, heavy metals and other toxicants, for waste water treatment and as protein source. Different species may have quite different responses to temperature, light, nutrient concentration, and toxicants. (A) *Wolffiaa, (B) Lemna, (C) Spirodela, (D) Azolla, (E) Pistia, (F) Eichornia.***

2. *Rooted Emergent Plants* : These plants are partly or completely exposed to air. The root system is completely under water and fixed with soil such as *Nelumbo* sp. (lotus), *Nymphaea* ap., *Sparganium* sp., *Sagittaria* sp. etc.
3. *Rooted Submerged Plants* : These plants remain submerged in water but rooted in soil such as *Hydrilla* sp., *Vallisneria* sp., *Ottelia* sp., *Lagarosiphon* sp., *Myriophyllum* sp., etc.
4. *Marginal Plants* : They are mostly rooted and infest the shallow water body such as *Typha* sp., *Cyperus* sp., *Panicum* sp., *Ipomoea* sp. etc.

It is important to note that except a few species of rooted emergent plants (such as *Chara* sp.), varieties of rooted emergent species are developed in different depths of water (Figure 4.2). However, some authors claim that in fish ponds and lakes, aquatic vegetation (both algae and phanarogams) is very important because of the following reasons :

1. Different species of algae form the food of minute creatures which in turn, are the food of small animals, and again these small animals become food for still larger animals followed by large-sized fish until the cycle is completed.
2. Phanarogams are the higher plants generate oxygen they bind the bottom soil by their roots and hence keep the water transparency, some fish species obtain food organisms from roots, leaves and stems; also shade the fish from bright sun and provide sanctuary to young fish from their enemies.

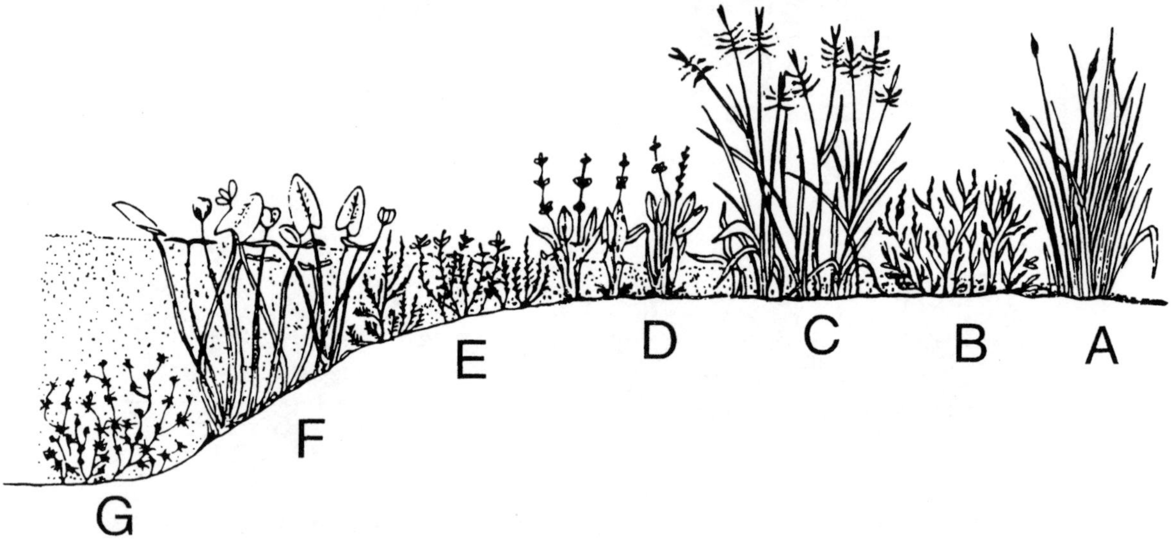


**Fig. 4.2 : Flora of the littoral zone of a lentic pond or lake ecosystem. Lentic aquatic life is most prolific in this zone upto the depth to which effective light penetration is possible. Note that from the shore proper to the deeper part in the littoral zone, the following types of vegetation occur. (A) *Typha* (Cattail), (B) *Sporangium* (Burred), (C) *Zazania* (Wild rice), (D) *Sagittaria* (Arrowhead), (E) *Myriophyllum* (Water milfoil), (F) *Nuphar* (Spatter dock), (G) *Chara* (Muskgrass). [After Smith (1977)]**

## *Combination of Aquatic Plants*

For different groups of fish, it is important to the fish farm to have the exact combination of plants for the ponds. While garden ponds need emergent plants, fish culture ponds require submerged plants. This is due to the fact that the submerged plants are the most useful because they not only provide an assembly place for fish food organisms but good oxygenators. Marginal and floating plants can be used in the rearing ponds for shade purposes. Though aquatic plants are important for real use in fish ponds and lakes to greater or lesser degree, some water weeds suddenly develop in a pond from seeds or fragments dropped by any means or carried out by the wind. These weeds, as soon as recognized, should be destroyed. A list of some common aquatic plants in fish ponds have been compiled in Table 4.1

**Table 4.1 : Some Common Auqatic Plants in Ponds and Lakes**

| Common name | Genus | Type |
|---|---|---|
| Sweet sedge | *Acorus* | Marginal |
| Floating plantain | *Alisma* | Submerged |
| Starwort | *Callitriche* | Submerged |
| Tasselled Sedge | *Carex* | Marginal |
| Hornwort | *Ceratophyllum* | Submerged |
| Rosebay | *Epilobium* | Marginal |
| Water grass | *Glyceria* | Marginal |
| Mare's tail | *Hipparis* | Submerged |
| Frogbit | *Hydrocharis* | Floating |
| Water poppy | *Limnocharis* | Floating |
| Lobelia | *Lobelia* | Submerged |
| Money wort | *Lysimachis* | Marginal |
| Purple loosestrif | *Lythrum* | Marginal |
| Water mint | *Mentha* | Emergent |
| Monkey musk | *Minulus* | Marginal |
| Watercress | *Nasturtium* | Marginal |
| Drop wort | *Oezantha* | Emergent |
| Pickerel weed | *Potederia* | Marginal |
| Curled pondweed | *Potanogetons* | Submerged |
| Water crowfoot | *Rananculus hederaceus* | Marginal |
| Water buttercup | *R. fluitans* | Floating |
| Lizard's tail | *Satrus* | Marginal |
| Water soldier | *Stratiotes* | Submerged |
| Water chestnut | *Trapa* | Floating |
| Bladderwort | *Utricularia* | Submerged |
| Speed well | *Veronica* | Marginal |
| Horned pondweed | *Zannicbellia* | Submerged |
| Arrowhead | *Saqittaria* | Emergent |
| Lotus | *Nelumbo* | Emergent |
| Lily | *Nymphaea* | Emergent |
| Water fern | *Azolla* | Floating |
| Duckweed | *Lemna* | Floating |
| Water hyacinth | *Eichornia* | Floating |
| | *Pistia* | Floating |
| Tape grass | *Vallisneria* | Submerged |
| Stonewort | *Chara* | Rooted emergent |
| Water milfoil | *Myriophyllum* | Submerged |
| Spatter-dock | *Nuphar* | Emergent |
| Wild rice | *Zazania* | Emergent |
| Reed mace | *Typha* | Emergent |
| Arun | *Colocasia* | Emergent |
| | *Cyperus* | Emergent |
| | *Limnophila* | Emergent |
| Morning glory | *Ipomoea* | Emergent |
| Oriental pepper | *Polygonum* | Emergent |
| | *Cardenthera* | Rooted emergent |
| | *Hydrilla* | Submerged |
| Water ferm | *Marsilea* | Rooted emergent |
| | *Najas* | Submerged |
| | *Hydrolea* | Submerged |
| | *Ottelia* | Submerged |
| | *Caldesia* | Submerged |
| | *Aponogeton* | Submerged |
| | *Salvinia* | Free-floating |
| | *Spirodela* | Free-floating |
| | *Wolffia* | Free-floating |

It has been observed that about three hundred species of aquatic macrophytes are distributed all over the world. Some species grow anywhere and under most conditions. Some thrive only in hard water, others in soft water. Some species are exclusively indigenous to a particular country or region. Generally, aquatic plants are flourished in different lentic and lotic environments in tropical and sub-tropical fish ponds. Therefore, their abundance, variations, and distribution in different soil types and freshwater ecosystems are extremely variable.

### 4.3 Efficiency Evaluation of Aquatic Plants

In order to examine the efficiency of floating aquatic plants in the treatment of toxic compounds in fish ponds, long-term experiments (365 days) were carried out in ponds under Indian conditions. Ponds were treated with mustard oil cake, single superphosphate, and ammonium sulfate at levels which are toxic or non-toxic to aquatic life (Table 4.2). These chemicals were added to ponds in several instalments. Long-term experiments were also carried out in other countries using triple superphosphate and ammonium nitrate at the rates of 15.5 and 11.2 kg/hectare area, respectively. These fertilizers were broadcast over pond surfaces at two-week intervals. In these ponds, initial crops of duckweed and water hyacinth were 30.0 and 21.5 kg/hectare area, respectively. Adequate space for their constant growth was provided by sampling half to two-fifth of the plant at 15 day intervals. Water samples were collected at 15 days intervals and analyzed for determination of chemical properties of water.

**Table 4.2 : Specification of Experimental Ponds and Application Rates of Single Superphosphate (SSP), Ammonium Sulfate (AS) and Mustard Cake (MOC). Rates are in Terms of kg per hectare**

| Treatment | Water area (hectare) | Depth (metre) | MOC | AS | SSP | Aquatic plants (kg) |
|---|---|---|---|---|---|---|
| P 1 | 0.03 | 1.00 | 1,000 | - | - | - |
| P 2 | 0.05 | 1.00 | 1,000 | 437 | - | - |
| P 3 | 0.04 | 1.30 | 1,000 | - | 375 | - |
| P 4 | 0.03 | 0.95 | 1,000 | 437 | 375 | - |
| P 5 | 0.04 | 1.40 | 1,000 | - | - | 2.0 |
| P 6 | 0.03 | 0.90 | 1,000 | 437 | - | 2.0 |
| P 7 | 0.04 | 1.20 | 1,000 | - | 375 | 2.0 |
| P 8 | 0.03 | 1.30 | 1,000 | 437 | 375 | 2.0 |

*Source :* Sarkar (1997).

Several research workers also collected water hyacinth and duckweed fornightly from different types of polluted water bodies (Table 4.3). These water bodies exhibited distinct variations in their habitats and levels of pollution. It was possible to understand the quality and quantity of elements removed by these floating aquatic plants from different habitats.

In another types of experiments, water hyacinth and some species of algae (referred to as simple, microscopic and single-celled plants) such as *Microsystis* sp., *Scenedesmus* sp., *Chlorella* sp., and *Chlamydomonus* sp. were also used for evaluation of their efficiency of domestic waste-water purification. These experiments were conducted for 14 day (Table 4.4).

### *Results*

It has been reported that the wet weight of plants increased by 377 kg/hectare per 3 months and 197 kg/hectare per 5 months for duckeweed and water hyacinth plants, respectively, and

**Table 4.3 : Composition of Some Elements in *Eichonia crassipes* (E) and *Lemna minor* (L) Collected from Moderately Polluted (MP), polluted (P) and Unpolluted (UP) Areas**

| Element | Plant | UP | P | MP |
|---|---|---|---|---|
| Nitrate-nitrogen (mg/l) | E | 18.6 | 29.6 | 24.4 |
| | L | 8.5 | 16.8 | 12.3 |
| Phosphate (mg/l) | E | 3.8 | 6.5 | 4.7 |
| | L | 0.6 | 1.5 | 0.8 |
| Calcium (mg/l) | E | 8.1 | 15.3 | 10.3 |
| | L | 10.0 | 18.7 | 15.2 |
| Magnesium (mg/l) | E | 4.4 | 8.8 | 6.2 |
| | L | 2.3 | 4.3 | 3.2 |
| Potassium (mg/l) | E | 20.9 | 35.5 | 27.3 |
| | L | 14.6 | 27.3 | 22.5 |
| Sodium (mg/l) | E | 1.7 | 3.3 | 2.7 |
| | L | 1.6 | 2.9 | 2.0 |

*Source* : Tripathi et al. (1990).

**Table 4.4 : Treatment of Domestic Wastewater by Water Hyacinth (A) and Algae (B)**

| Parameter | Domestic waste water | A | | B | |
|---|---|---|---|---|---|
| | | 14 day retention time | % reduction | 8 day retention time | % reduction |
| Dissolved oxygen (mg/l) | – | 2.0 | – | 16.7 | – |
| Nitrate-nitrogen (mg/l) | 3.2 | 0.8 | 76.3 | 0.4 | 87.4 |
| Phosphate (mg/l) | 12.7 | 3.5 | 72.8 | 1.1 | 91.7 |
| Total alkalinity (mg/l) | 656.0 | 176.0 | 42.7 | 94.5 | 85.6 |
| Biochemical oxygen demand (mg/l) | 290.0 | 94.7 | 67.3 | 30.7 | 89.4 |
| Chemical oxygen demand (mg/l) | 570.0 | 370.5 | 35.0 | 290.0 | 49.1 |
| Phytoplankton (Number/l) | – | 624 X $10^2$ | – | 1,465 X $10^1$ | – |
| Zooplankton (Number/l) | – | 275 | – | 7,935 | – |

*Source* : Shukla and Tripathi (1989).

concentration of plantkton also increased by 20.0-40.5 per cent. The level of nutrients of water significantly decreased from 2.15 to 0.15 mg/l for ammonia nitrogen, 1.80 to 0.20 mg/l for nitrate nitrogen, and 3.0 to 0.15 g/l for phosphate. These represent absorption and accumulation of nitrogen and phosphorus from water. Generally *Azolla pinnata* (a water fern) and some blue-green algae (such as *Gleocapsa atrata, G. punctata, G. calcarea, Nostoc commune, N. hatei, Scytonema fritschii, Anabaena spirodes, A. ambigera, and A. gelatinicolà*) are considered as the most important agent for fixation of nitrogen from water and their excessive growth is due to absorption and accumulation of nitrogen and phosphorus at a faster rate from nutrient-enriched waters. Regularly fertilized fish ponds have high content of nutrients and hence favourable to plant growth particularly macrophytes and algae than species of zooplankton and other animals. Other experiments have shown that (1) the water hyacinth and duckweed can grow over a wide spectrum of water quaity and may be used as tools for removing nutrients from water, (2) the sewage may be treated with the aid of water hyacinth and algae, and (3) a good quality of water may be obtained within 20 day retention time.

### *Nutrient Assimilation*

Several species of floating plants (such as *Azolla pinnata*) assimilate oxygen in association with algae. Moreover, they contribute abut 20-30 kg N/hectare per season through nitrogen fixation and addition of phosphorus fertilizers at the 250 kg/hectare rate increased nitrogen level upto 69 kg N per hectare per season in many ponds of temperate, tropical and sub-tropical zones. It has been reported that combined effects of nitrogen and phosphorus fertilizers stimulate plant growth and nutrient assimilation rate also increased. In presence of some chemical fertilizers like diammonium phosphate and nitro-phosphate, blue-green algae secrete considerable amount of growth promoting substances into the surrounding water. Therefore, blue-green algae may exhibit synergistic influence on the potentiality of nutrient assimilation by aquatic plants from water. Of equal interest is the phosphorus fixation capacity by phytoplankton and algae. They can able to store ten times as much phosphorus as found in pond water. Therefore, occasional growth of aquatic plants of particular species for several weeks may be considered which will undoubtedly lower the levels of metabolites of fertilizers and manures below the toxic concentration in fish culture ponds where manuring and fertilization porgrams are undertaken for improving physical properties of soil and constant fertility of water.

### *Nutrient Removal*

Removal of toxic compounds by different species of floating plants such as *Eichornia crassipes, Azolla pinnata, Lemna minor, and Pistia stratiotes* depends upon the days of retention of plants and nutrient levels in ponds. For example, water hyacinth removes upto 200 kg nitrogen per hectare per year and 322 kg phosphorus per hectare per year from water. This plant also removes about 10-95 per cent nitrogen and 25-90 per cent phosphorus within 2 to 50 days. Marsh (a swamp-loving plant) successfully removes phosphorus at the 0.949 g per squre metre from effluent-receiving water. It has also been established that *Hydrilla verticillata* is able to eliminate about 1.329 g phosphorus per square metre from eutrophicated water. Nutrient content of aquatic plants was found to be significantly positive correlation with nutrient concentrations of water. Generally, the intensity of nutrient removal by aquatic plants depends upon the plant density, bacterial activity, amount of fertilizers and manures used in ponds and their interactions with various abiotic-biotic factors of ecosystem such as temperature, season, wind, and rainfall.

Experiments have shown that like macrophytes, mixed algal forms in ponds also have many advantages in removing ammonia-nitrogen, nitrate-nitrogen, and nitrite-nitrogen within 30 days. Algal forms also exhibited dramatic reduction in the concentration of hardness, potassium, mercury, iron, zinc, and copper from water. It has been suggested that different species of algae such as *Anabaena variabilis, A. cylindrica, Aulosira* sp., *Calothrix elenkenii, Tylopothrix tenuis,* and *Nostoc muscorum* imparted a notable reduction in pollution load from eutrophicated ponds rather than a single species of algae. Scientists have advocated about the usefulness of macrophytes and algae and their use in fish culture ponds to control pollution.

### *Quantity of Pollutants Absorbed by Macrophytes*

A number of field and laboratory experiments have been made to treat wastewaters using different types of macrophytes. Macrophytes are so efficient that water hyacinth, for example, covering one hectare water area is sufficient to recyclr the waste produced by 2,500 persons. However, regulation of harvesting schedules are necessary for maximum absorption of pollutants. When intensive and semi-intensive fish culture systems become eutrophicated, immediate removal of pollutants (particularly ammonia and nitrate) should be undertaken on emergency basis. The efficiency of water hyacinth, for example, in reducing excess quantities of ammonia, nitrate, and phosphate from pond water is quite dramatic. In sewage-fed fisheries where diluted sewage (sewage : freshwater = 1 : 1 or 1 : 2) is used, may increase the level of nutrients beyond threshold concentrations. Therefore, three-forth of the water area is covered with single species of young macrophyte for short duration. Experimental studies have shown that water hyacinth absorbed highest quantity of nitrate-nitrogen followed by total phosphorus and ammonia-nitrogen (Fig .4.3).

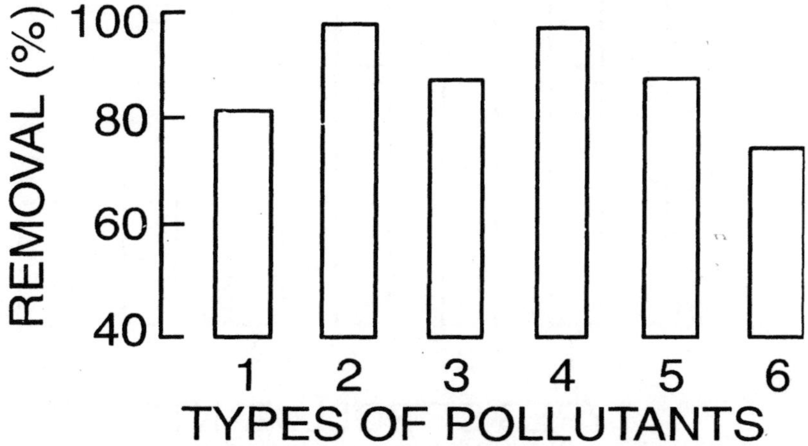


**Fig. 4.3 : Removal of pollutants from sewage effluents by water hyacinth. Note that in all the cases, more than 70 per cent of pollutants was removed within 14 days. (1) Ammonia-nitrogen, (2) Nitrate-nitrogen, (3) Total phosphorus, (4) Biochemical oxygen demand, (5) Chemical oxygen demand, (6) Suspended solids. (Drawn from data as proposed by Trivedy and Gudekar (1985)).**

Since a number of inorganic compounds are found in sewage sludge, they were studied for absorption by water hyacinth and found that one hectare of the plant can efficiently remove significant quantities of heavy metals from sludge (Figure 4.4).

## 4.4 Mechanism of Pollutant Removal

When macrophytes grow over ponds profusely, concentration of dissolved oxygen in water is depleted that create anaerobic conditions. This condition favours the process of denitrification and nitrate removal is initiated. Some species of macrophytes are able to transport oxygen from the foliage to the root that results in the formation of oxidized environment in water. Such environment is principally due to the diffusion of oxygen which is consumed by bacteria. A large number of anaerobic bacteria are associated with the roots of plants. Organic compounds are degraded by aerobic bacteria. The degraded products remain in suspended condition in water which strike the root surface. Consequently, their electrical charges are lost.The presence of enzyme dehydrogenase in the root system of plants facilitate the removal of suspended particles from waters.

### *Factors Affecting Removal of Pollutants*

The removal of pollutants by aquatic plants depends on a number of factors such as temperature, pH, depth of water, nutrient composition, nature and quantity of plants and their stages of development. For example, young plants are able to absorb pollutants more rapidly than the older ones. While increase in water temperature favours the rate of absorption of pollutants, alkaline ponds resist it.

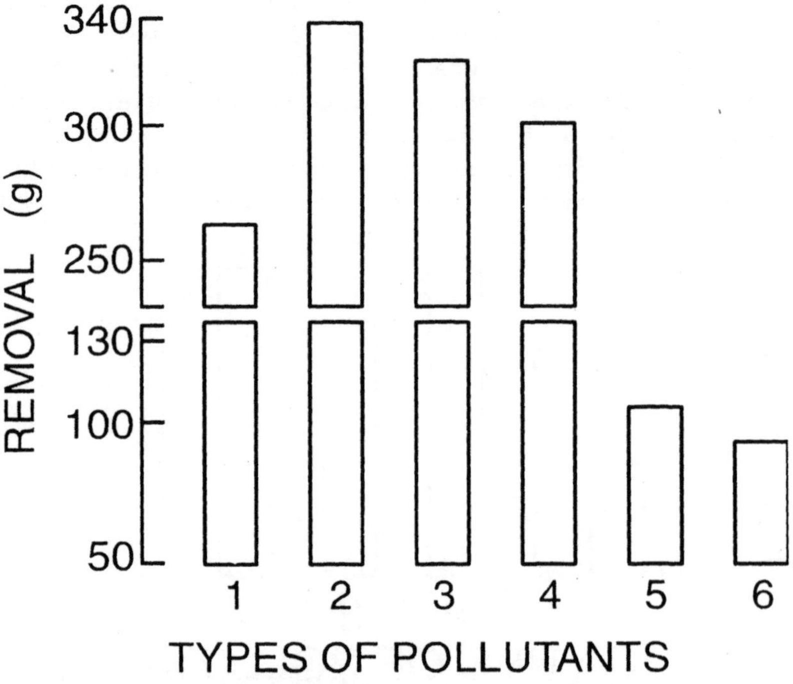


**Fig. 4.4 : Removal of metals from sewage sludge by water hyacinth. Note that an appreciable quantities of metals were removed within one day. (1) Silver, (2) Copper, (3) Strontium, (4) Cadmium, (5) Lead, (6) Mercury. (Drawn from data as proposed by Wolverton and McDonald (1975)).**

## 4.5 Accumulation of Nutrient Elements

Floating aquatic plants such as water hyacinth and duckweed are also able to accumulate calcium, magnesium, potassium, and sodium from different types of water bodies (Table 4.3). Such accumulation is significant for plant growth because these elements are utilized by plants for various physiological processes.

## 4.6 Utilization of Aquatic Plants

Absorption of nitrogen and phosphorus by aquatic plants is significant to fish farmers. Because these plants are of floating nature, their collection from water bodies is not difficult. However, there is an ample scope for controlling fertilizer pollution by triggering the dynamics of nutrient pool through soil-water-plant relationship and different feeding status of plant-animal-man. Such relationship makes the innovative way to recycle the nutrients to greater extent. Nutrient-enriched plants are used for biogas production and preparation of animal feed as well as compost. The compost is a natural product of breakdown of dead plant material. It is used in fish culture to provide nutrients to soil and water. The compost is particularly used in fish ponds where there is severe loss of water through soil. It has been reported that the annual biogas production was highest in water hyacinth (44.381 litres) and somewhat lower in water fern (17.186 litres). On the other hand, single layer of water fern covering a hectare of water area weighing about 150 quintal of green matter will ensure about 30 kg of nitrogen after the breakdown of dead plants and animals by bacteria and fungi. The nitrogen is slowly released and utilized by aquatic ecosystem. Dehydrated and chopped water hyacinth contains 10-20% crude protein, 20-30% carbohydrate, 20-30% ash free crude, and 12% ash. Similarly, dehydrated azolla also contains 5-12% crude protein, 7% ash and 10-12% carbohydrate. Therefore, water hyacinth and azolla are very palatable food for pigs, ducks, sheep, cattle and some species of fish. Such utilization of aquatic floating plants can be realized only when integrated approach would be undertaken as a village industry thereby increasing the resource management, employment generation and improving the quality of polluted waters for adoption of fish culture management techniques (Figure 4.5).

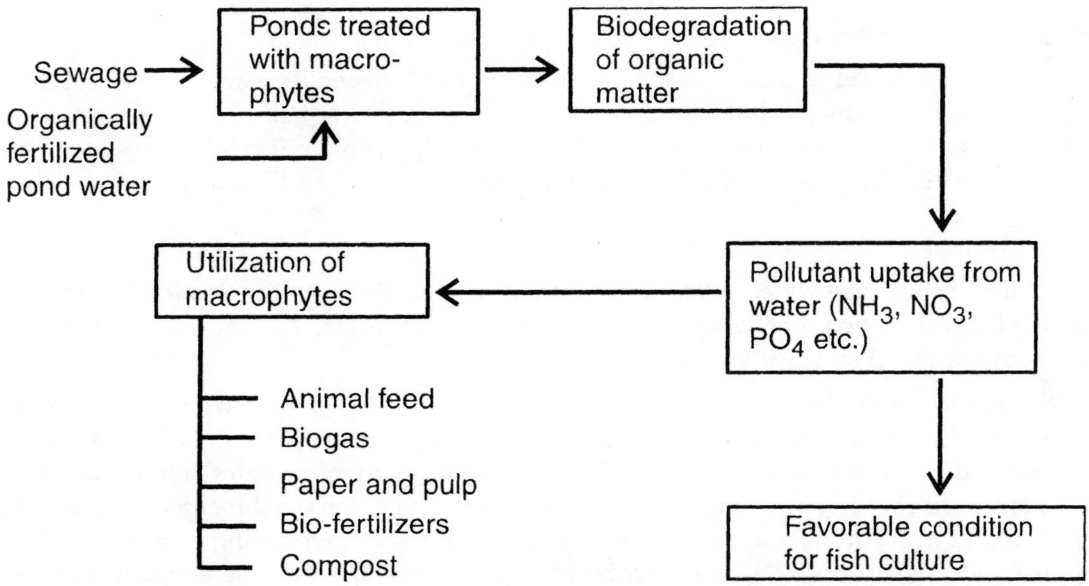


**Fig. 4.5 : Layout for removal of pollutants from eutrophicated pond ecosystem. Macrophytes absorb nutrients which are further utilized for various purposes and the pollutant-free waters are utilized for fish culture.**

## 4.7 Removal of Decaying Vegetation

It is important to note that some aquatic plants die away in a particular season and again appear after several months. The dead and decaying plants remain settle into the bottom of the pond and decomposition process starts. Therefore, it is necessary to remove the dead plant and avoid toxicity. Aquatic vegetation that has been treated with chemicals should not be allowed to remain in the pond, otherwise it will be rotted, thus de-oxygenating the water. Dead weeds should be properly shifted from the edges of ponds and lakes and if it is not done or just dragged out on the margins of ponds and lakes, small broken pieces of seeds of weeds might be back into the water resulting in further occurrence of weeds all over the water surface.

## 4.8 Algal Blooms

Optimum level of algal blooms is beneficial for fish production. They are the food for fish food organisms and therefore, considered as an important link in the food chain. With high rainfall and temperature the algal blooms rapidly cover the ponds. The diatoms appear at first and they feed on dissolved substances present in the water. After death and decay of diatoms, blue-green algae (such as *Anabaena* sp. and *Oscillatoria* sp.) appear. However, in freshwater ecosystems about nineteen hundred species have been reported so far. They are either attached to other plants and substratum in the pond bed or free-floating. They require nutrients for normal growth and development. The nutrients are derived from the decomposition of dead and decaying plants and animals present in the mud. Heavy growth of these algal species indicates that the ponds and lakes are productive. Algal mats which are formed over pond surface by the growth of some species of blue-green algae such as *Spirogyra* sp., *Oedogonium* sp., *Cladophera* sp., etc. may cause the fish to seek the pond bottom because fish gills are clogged by these filamentous algae. After several days fish obtain food from the tiny animal life to be found on the algal mats.

## 4.9 Control of Aquatic Plants

Generally older and fertilized ponds and lakes succumb to aquatic vegetation which is very important for aquatic ecology. But it becomes necessary to remove aquatic vegetation if they grow rapidly. Therefore, in order to keep fish in good health and to maintain fish farms an economically viable, adequate control of several types of aquatic plants should be adopted as noted below.

### *Manual Control*

This is done in small water areas simply either by hand picking or uprooting emergent and marginal plants or cutting them with sickles. Of course, these practices are executed in countries where labour is cheap and easily available.

The shallow submerge plants are removed mechanically by (1) raking with weeders made of wood which is fitted with iron spikes and wire, (2) rotating bamboo poles fitted with basal cross attachment and (3) cutting the weeds with long-handled sickles. In many states of India, mechanical lifts or power winches have been used extensively to eradicate dense submerged weeds. Motor boats equipped with cutting knives are also used. Sometimes a weed cutting launch with its bow connected with V-shaped sickles that have reciprocal action, is used for the purpose. In order to check the migration of floating plants into ponds and also to control marginal plants, creation of barriers and grazing of harbivorous animals, respectively, are necessary. For controling algal blooms, ponds and lakes are shadded by some means. Such work should be done in monsoon or winter seasons. If this is done in summer, deoxigenation of water may result and thus fish will suffer.

### *Chemical Control*

Now-a-days, a variety of herbicides are used to control aquatic plants. The basic consideration of the use of herbicides in fish culture is that (1) herbicides should be used as sublethal levels for weed control strategies and will not adversely affect the aquatic ecosystem as a whole, (2) the rate of application should not be exceeding 10 mg/l and 10 kg/ha for those applied in water and on surface area basis, respectively, (3) herbicides should be non-toxic to warm-blooded animals, and (4) easily available and should not require costly equipment to spray herbicides. Some commonly used chemical and their application rates are shown in Table 4.5.

**Table 4.5 : Chemical Control of Some Aquatic Plants**

| Types of Plant | Chemical | Application rate |
|---|---|---|
| A. ***Floating*** | | |
| 1. *Eichornia* | 2, 4-D | 4.5-6.7 kg/ha |
| | 2, 4-D + Detergent | 6 kg/ha + 0.25 per cent |
| | Simazine | 5 kg/ha |
| 2. *Pistia* and *Salvinia* | Paraquat | 0.02 kg/ha |
| | Aqueous ammonia | 1-2 per cent |
| | 2, 4-D Ester (MCPA) | 9-13 kg/ha |
| | Taficide-80 + Simazine | 5 + 5 kg/ha |
| B. ***Marginal*** | | |
| 1. *Cyperus* | 2, 4-D Sodium | 10-12 kg/ha |
| 2. *Eleocharis* | 2, 2-Dichloropropionic Sodium | 10-15 kg/ha |
| 3. *Ipomoea* | 2,4-D Sodium | 5 kg/ha |
| 4. *Panicum* | 2,2-Dichloropropionic Sodium | 10-15 kg/ha |
| C. ***Emergent*** | | |
| 1. *Nymphaea, Nelumbo* and *Nymphoides* | 2,4-D | 5-10 kg/ha |
| D. ***Submerged*** | | |
| 1. *Chara* | Copper sulfate + Ammonium sulfate | 18 + 36 kg/ha |
| 2. *Ottelia* | Copper sulfate + Sulphuric acid | 250 mg/l + 10 per cent |
| | Urea | 50-100 mg/l |
| 3. *Hydrilla* | Simazine | 6-10 mg/l |
| | Aquathol | 10 mg/l |
| | Sulphur dioxide in acid medium | 40 mg/l |
| | Urea | 250-300 mg/l |
| E. ***Algal Blooms*** | | |
| 1. *Microcystis* | Simazine | 1 mg/l |
| | *Diuron* | 0.3 mg/l |
| | *Copper sulfate* | 2 per cent |
| | *Bleaching powder* | 2 kg/1,000 gallons of water |

*Source :* Compiled from Thomas and Srinivasan (1949), Banerjee and Mitra (1954), Philipose (1957), Ramachandran and Ramaprabhu (1968), Ramachandran and Reddy (1975), Jhingran (1988).

For controlling aquatic plants in large ponds and lakes it is necessary to use chemicals, but use of chemicals in small water areas should be avoided. The chemicals are lethal to fish and other aquatic life and if they are not used at safety limits, the fish will be killed.

Action of chemicals differs in hard and soft water, structure and texture of bottom soil. However, the pond may be emptied half-way before use of chemicals to reduce the cost of treatment. The pond is filled again with water to its desired level. Of course, it is better to use the chemical before the onset of monsoon season.

1. *Use of Arsenic* : Sodium arsenite is widely recognized in some western countries more that fifty years ago and it is so successful that the use of this chemical became an established

factor in weed control management strategies. This compound is also practiced in many Asian countries and the safest effective concentration is 1.0 to 2 mg/l.

2. *Use of Copper Sulfate* : Copper sulfate is highly effective in controlling algal blooms, *Chara* sp. and *Ottelia* sp. at the rates of 2.0%, 18 kg/ha, and 250 mg/l, respectively.
3. *Use of 2,4-D and Simazine* : Simazine and 2, 4-D are extensively used in India to control water hyacinth from large ponds and lakes. In such cases, the spreading of the chemical should be continued for several days. The surface water should be roughly divided into several equal sections to ensure proper distribution and safeguard water from being treated twice or even left untreated.

### *Biological Control*

Some species of aquatic plants can be controlled by culturing some herbivorous fishes such as *Oreochromis mossambicus, Ctenopharyngodon idella, Puntius javanicus, Cyprinus carpio, Osphronemus gorany* etc. These fishes are hardy, does no interfere with other fishes and easily consume weeds.

In different parts of the world, grass carp, *C. idella,* has been found to be effective in controlling some species of aquatic weeds such as *Azolla* sp., *Salvinia* sp., *Hydrilla* sp., *Wolffia* sp., *Ottelia* sp., *Lemna* sp., *Spirodella* sp., *Myriophyllum* sp., *Trapa* sp., *Limnophila* sp., Guinea grass, leaves of cabbage and *Ipomoea* sp. These plants are highly palatable and hence accepted by the fish. Some other species such as *Eichornia* sp., *Pistia* sp., *Nymphoides* sp., etc. are also consumed by the fish but the fish does not appear to feed voraciously.

The common carp, *C. carpio,* is also equally important in the uprooting of aquatic plants at the time of digging and burrowing in search of food. This species eliminates filamentous algae and *Myriophyllum* sp. from fish ponds successfully and did not interfere with other species.

The culture of different species of tilapia has also been found suitable for weed control management strategies. This species consumes filamentous algae and submerged vegetation such as *Najas* sp., *Chara* sp. etc.

It is important to note that the feeding rates of different species of herbivorous fish depends on the environmental temperature, fish species to be reared, availability of natural food organisms, various life stages of fish (such as juveniles, fingerlings and adults), concentration of aquatic vegetation, stocking density of fish and geographical as well as climatic conditions of a particular area. For example, stocking of grass carp in ponds under Indian conditions at the rate of 79-173 kg/ha and 175-225 kg/ha consume hydrilla and lemna, respectively within a month. In China, this fish when stocked in ponds at the rate of 100 per hectare, consumes a wide variety of aquatic plants. In Israel and USA, stocking of *Cyprinus carpio* in ponds at the rate of 8,000 and 400 per hectare, respectively, successfully controlled aquatic plants of the species *Ceratophyllum* sp., *Myriophyllum* sp., and filamentous algae. Stocking of tilapia at the rate of 2,470-4,940 per hectare in large Indian impoundments controlled a variety of aquatic plants. Therefore, no definite conclusion can be drawn regarding the effective control of aquatic plants so far as the above-mentioned factors are considered.

In many Asian and European countries, geese and ducks are used in fish culture ponds for dual purpose. First, they effectively destroy some species of floating plants such as azolla and lemna and second, manuring the fish ponds. About 2,500 ducks are stocked per hectare water area having 6,000 fish fingerlings of indigenous and exotic species. In this case, manuring and aeration of pond is dispensed with, because of duck faecal washing and paddling in the pond. However, once aquatic vegetations have been completely removed from ponds and lakes, periodical maintenance should be undertaken to prevent regeneration of aquatic plants. Control of some aquatic plants through biological means have been briefly summarized in Table 4.6

**Table 4.6 Biological Control of Some Aquatic Plants**

| Types of plant consumed by fish | Name of fish | Stocking rate of fish (kg/hectare) | Consumption rate of plant (kg/month) |
|---|---|---|---|
| *Lemna* | Grass carp | 175-225 | 54 |
| *Hydrilla* | Grass carp | 79-173 | 50 |
| *Chara* | | | |
| *Potamogeton* | Grass carp | 40-80 | 43 |
| *Eleocharis* | | | |
| *Myriophyllum* | | | |
| *Ceratophylla* | Common carp | 400-8,000 | 40 |
| Filamentous algae | | | |
| Filamentous algae | Tilapia | 2,470-4,940 | 55 |

*Source* : Chattopadhyay (1951), Jhingran (1988).

## 4.10 Conclusion

Different types of aquatic plants grow in ponds and lakes and in association with aquatic organisms, they form a complex biotic community. Fertilizers and manures trigger the development of aquatic plants (particularly macrophytes and algae) and as a result, physiological activities and fish culture strategies are jeopardized.

Though aquatic plants upset the equilibrium conditions of physico-chemical qualities of water, their beneficial effects in fish culture (such as pollution control, production of animal feed and compost) should not be ignored. Efficiency of floating plants in nutrient assimilation, removal of toxic elements, accumulation of excess nutrients, and their utilization for myriad purposes must be emphasized. In spite of various beneficial impact of aquatic plants, there are some harmful effects and therefore, these plants should be controlled when fish culture management strategies are contemplated.

Treatment of eutrophicated ponds with macrophytes is a cheapest and common method through which pollutants can easily be removed from water within a very short period. In cases where organic manures are widely used for pond productivity, occasional spreading of macrophytes can be regarded as a more common solution for drastic reduction of pollutants.

### *References*

Banerjee, R.K. and E. Mitra. 1954. Observations on the use of copper sulphate to control submerged aquatic weeds in alkaline waters. *Indian J. Fish.* 1 : 204-216.

Chattopadhyay, S.B. 1951. Control of *Chara,* an algal weed in paddy fields of West Bengal. *Proc. Ind. Sci. Cong.* 38 : 141-142.

Jhingran, V.G. 1988. *Fish and Fisheries of India.* Hindustan Publishing Corporation, New Delhi.

Philipose, T.M. 1957. A report on the present status of aquatic control in the Indo-Pacific countries. *IPFEC,* 25 p.

Ramachandran, V. and T. Ramaprabhu. 1968. Investigations on the aquatic weed control with special reference to the use of chemicals. *FAO Fish Rep.* 44 : 92-108.

Ramachandran, V. and P.V.K.G. Reddy. 1975. Observations on the use of ammonia for the control of *Pistia straitotes. J. Inland Fish. Soc. Ind.* 7 : 124-130.

Sarkar, S.K. 1997. Seasonal influence of the plant *Lemna major* in the treatment of eutrophicated ponds. *Poll. Res.* 16 : 247-249.

Smith, R.L. 1977. *Elements of Ecology and Field Biology.* Harper and Row Publishers, New York.

Shukla, S.C. and B.D. Tripathi. 1989. Biological treatment of domestic wastewater by water hyacinth and algal culture. *Sci. Cult.* 55 : 209-210.

Thomas, K.M. and A.R. Srinivasan. 1949. Weed killers. *Indian Fmg.* 10 : 101-106.

Tripathi, B.D., J. Srivastava and K. Misra. 1990. Impact of pollution on the elemental composition of water hyacinth and duckweed in various ponds of Varanasi. *Sci. Cult.* 56 : 327-329.

Trivedy, R.K. and V.R. Gudekar. 1985. Water hyacinth for wastewater treatment – A review of the progress. *In : Current Pollution Researches in India,* Environmental Publications, India. pp. 109-145.

Wolverton, B.C. and R.C. McDonald. 1975. Water hyacinth and alligator weeds for removal of lead and mercury from polluted waters. NASA Tech. Men. TM-X-72723.

## Questions

1. Name some common aquatic plants found in freshwater lakes/ponds. Add a brief note on different types of aquatic plants.
2. How the efficiency of aquatic plants are evaluated in the treatement of toxic compounds in freshwater ecosystems?
3. State the role of aquatic plants in the removal of pollutants from an ecosystem.
4. How aquatic plants are utilized for fish culture?
5. How algal blooms affect fish culture?
6. Describe different methods adopted for controlling aquatic plants.

# 5

# Actions of Pond Soil Organisms

Since the inception of plant and animal life on earth, their residues are continuously being accumulated on soil surface. This accumulation is observed particularly in tropical and temperate regions where herbs, shrubs and trees drop their leaves on soil. Animal residues such as excreta, dead and decaying animals are also deposited on the soil surface. On careful examination of the underlying soil layers, it will be observed that a portion of it remains as soil *humus* where most of the plants and animals have been incorporated. However, plant and animal residues are either decomposed by soil organisms, yielding water and carbon dioxide, or degraded and synthesized into compounds that comprise soil humus or nutrient elements are released that are utilized by fishes for their growth.

Most pond soil organisms belong to plant life (flora) rather than animals (fauna). Soil organisms are classified as micro-organisms (can be seen only with the microscope) and macro-organisms (seen with the naked eyes). Both types of organisms have significant role on physical properties of soil. They are also important in the soil biological processes associated with the plant and animal life. A general classification of important groups of pond soil organisms are shown in the Figure 5.1.

## 5.1 Actions of Soil Organisms

Various activities of soil plants and animals are closely related. Different types of live aquatic plants are subject to infect by soil organisms called as *herbivores.* For example, insect larvae and parasitic plant nematodes attack plant roots; termites, ants and beetles devour plant materials. Figure 5.2 depicts how aquatic plant tissues are gradually broken-down. Since plant tissues capture carbon dioxide and solar energy, they are called as *primary producers.* After the death and decay of producers, they are attacked by soil fauna and microflora of primary, secondary and tertiary consumers. These organisms liberate carbon dioxide and energy and produce humus.

### *Primary Consumers*

Consumers are referred to organisms which feeds upon another organism or consumes food, the nature of which indicates the position of organisms in food chain. Immediately after dropping the dead plants and animals to the pond mud, it is subjected to attack by microflora and detrivores such as mites, woodlice, earthworms etc. These animals gradually chew to facilitate quick attack by the microflora. The plant-eating animals such as snails and mussels together with microflora, are called *primary consumers.*

While the actions of the bottom fauna of a fish culture pond are both chemical and physical, those of the microflora are mostly chemical. Herbivores (such as floating zooplankton) chew the plant parts and move them from different places either on the soil/water surface or into the soil or both. Earthworms convert the plant residues into the mineral soil by passing them through their bodies along with mineral soil particles.

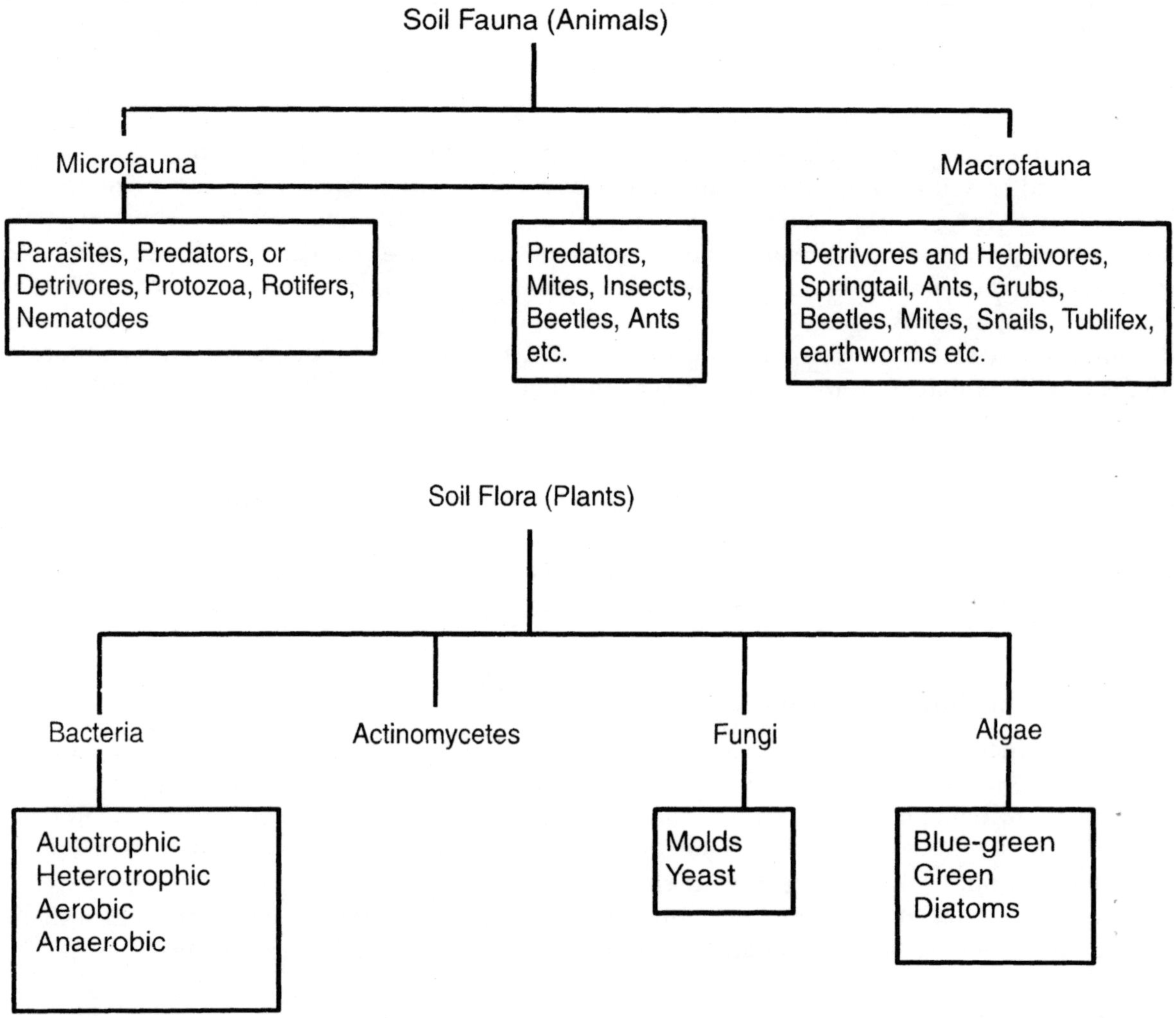


**Fig. 5.1 : A general division of some important groups of a fish culture pond and lake soil organisms.**

### *Secondary and Tertiary Consumers*

The primary consumers are used as food by bacteria, fungi, algae, and lichens. These are termed as *secondary consumers.* For examples, carnivorous fishes consume small insects and snails; some species of small fish depends upon tubifex and earthworms; termites and protozoa use the microflora as source of food.

The secondary consumers prey for still larger carnivorous fishes, they are called *tertiary consumers.* The soil microflora are closely involved in the decomposition of organic material associated with the animals.

### *Decomposers*

Decomposers are organisms which relies upon the dead tissues of other organisms as an energy source. By using this energy, it liberates nutrients from those tissues into the ecosystem.

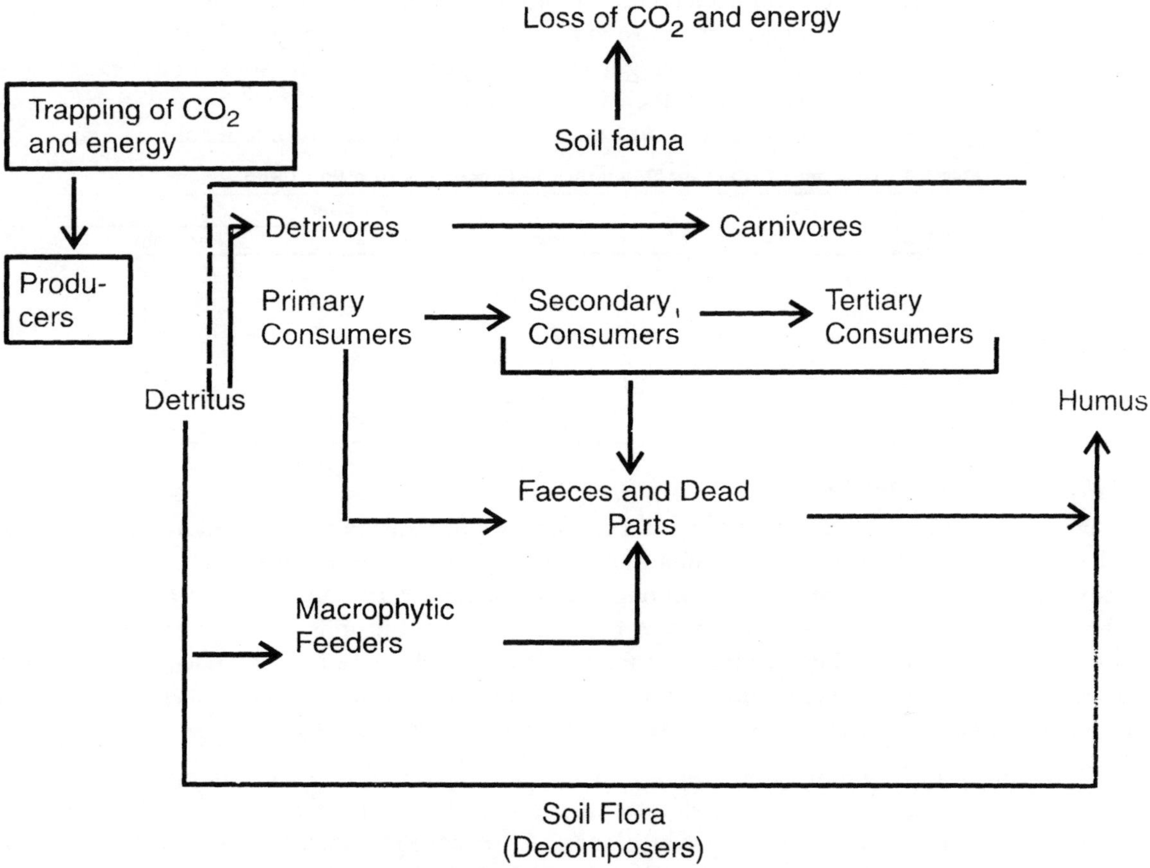


**Fig. 5.2 : A general pathway showing the breakdown of aquatic plant tissues and their subsequent transformations.**

Soil microflora are also found to be active in the digestive systems of some aquatic animals where they attack the finely organic material in animal faeces and later decompose the dead animals to simpler forms as nutrients which remain in the pond for some time and then taken up by the producers for further use.

## 5.2 Numbers and Biomass of Soil Organisms

In general, the numbers and biomass are greatly influenced by the physical and chemical factors of soil and water. It is needless to mention that the species composition in a pond ecosystem will obviously be different from that of a lake or reservoir ecosystem. Moreover, alkaline pond soils are populated by quite different species from those in acid soils. Similarly, the number of species and their diversification are also different in tropical pond from those in temperate or subtropical ponds.

While generalization are difficult to establish, the numbers and biomass of soil fauna in undisturbed ponds are usually low than in cultivated ponds where the species diversity always exist following the application of nutrients to the systems.

The number of soil fauna and flora commonly found in surface soils is shown in Table 5.1. It is expected that micro-oraganisms are the most numerous and exhibited highest biomas. Microflora along with earthworms and protozoa trigger the biological activity in soils. Some detrivores particularly digest organic residues and their excreta undergoes microfloral degradation. Some soil fauna simply live in the bottom mud and improves physical conditions of soil.

**Table 5.1 : Number of Different Types of Soil Organisms in a Pond Ecosystem**

| Organisms | Number | |
|---|---|---|
| | Per m$^2$ | Per gram |
| Bacteria | $10^{10} - 10^{11}$ | $10^5 - 10^6$ |
| Actinomycetes | $10^8 - 10^9$ | $10^5 - 10^6$ |
| Algae | $10^6 - 10^7$ | $10^3 - 10^4$ |
| Fungi | $10^8 - 10^{10}$ | $10^3 - 10^4$ |
| Protozoa | $10^7 - 10^8$ | $10^2 - 10^3$ |
| Earthworms | **10 – 35** | – |

### *Energy and Carbon Source*

On the basis of the source of energy and carbon, soil organisms are divided into autotrophic and heterotrophic. Autotrophic organisms are those which manufacture carbohydrates and proteins from water, carbon dioxide and mineral salts. The most common and important autotrophs are the green plants which use energy derived from sun (photoautotrophs) to fuel the manufacturing process, photosynthesis. Other autotrophs, mostly bacteria, make use of energy released by various chemical reactions (from the oxidation of nitrogen, sulfur, and iron) to convert inorganic molecules into food (chemoautotrophs). This process is known as *chemosynthesis*.

The heterotrophic organisms are those which obtain carbohydrates and proteins ready-made from other organisms. All animals are heterotrophs, as many bacteria and fungi. Many heterotrophs are herbivores, feeding directly on green plants. Carnivorous heterotrophs may feed on the herbivores. Parasites are heterotrophs which obtain their food from other living organisms. On the other hand, saprophytes are heterotrophs which derive their food from the dead remains and waste products of animals and plants. The heterotrophic soil organisms obtain their energy and carbon from the breakdown of soil organic materials.

## 5.3 Earthworms

Farmers consider earthworm as their friends and hold them in high esteem as nature's ploughmen. Some species, however, play a key role in decomposing organic matter and mineral cycling, which is an important criterion for selecting these organisms in waste utilization and vermi-composting.

It is well known that burrowing habit of earthworms leads to soil aeration and significant involvement in its physico-chemical conditions. There are about 1,800 species of earthworms known world-wide today, of which over 40 species are found in India. Pheretima is the largest genus of the Class Oligiochaeta and has 500 species of which 13 species are Indian. It is regularly found in

South-East Asia, Malaya, Australia, Japan, and United States. Some forms are deep-boring (such as *Lumbricus* sp.) while other forms are shallow-boring (such as *Allolobophora* sp.). The length of earthworm vary from 25 cm to 3 m.

### *Role of Earthworms on Soil Fertility*

Earthworms ingest soil and organic matter. As these materials pass through their alimentary canal, they are subjected to digestive enzymes. The weight of the material (cast) passing through their alimentary canal each day may equal to the weight of the body. In the tropics as much as 275 tonnes/hectares of casts may be produced annually mainly due to extensive earthworm activities.

In cases where the concentration of earthworms is higher in fish culture ponds, soil organic matter is changed dramatically. For example, *Metaphire posthuma* increased the organic matter by 1.13% in soil having initial pH 7.8, while *Eutyphoseus waltoni* by 0.18% in soil having initial pH 5.9 after one month (Table 5.2). Nitrogen content of soil increased by *M. posthuma* and *E. waltoni* was 1.64 and 1.13 fold, respectively.

### *Role of Earthworms of Pond Productivity*

Owing to their burrowing behavior, soil nutrients are released into water thereby increasing the levels of nitrogen and phosphorus, an important consideration in natural food organisms and ecosystem development. They are also important for various reasons : some carnivorous fishes are used as food as earthworms's body contains nutrient materials in-significant amounts. Moreover, the worms increased soil porosity from 1.69 to 4.28% (Table 5.2).

**Table 5.2 : Changes Caused in Physico-Chemical Parameters of Three Soil Samples by *Metaphire posthuma* and *Eutyphoseus waltoni***

| Parameter | Sample 1 | | | Sample 2 | | | Sample 3 | | |
|---|---|---|---|---|---|---|---|---|---|
| | pH | OM (%) | P (%) | pH | OM (%) | P (%) | pH | OM (%) | P (%) |
| Initial value of soil samples | 5.9 | 1.34 | 74.4 | 7.8 | 1.75 | 71.5 | 8.4 | 0.27 | 74.3 |
| Altered values due to *M. posthuma* | 5.9 | 1.9 | 78.7 | 7.2 | 2.9 | 74.2 | 8.0 | 0.42 | 76.09 |
| Altered values due to *E. waltoni* | 5.9 | 1.43 | 75.1 | 7.4 | 1.9 | 73.5 | 8.1 | 0.35 | 75.3 |

*Source :* Singh (1998), used with permission of S.M. Singh. Data were based on the laboratory earthen pot experiment. OM – organic matter; P – porosity.

## Development of Earthworms in Soil

In order to improve soil conditions, investigations have been carried out to produce earthworms using organic wastes such as kitchen waste (soil 1 : waste 3), cowdung (soil 1 : cowdung 3) etc. These are mixed with soil in the group with cowdung, hatching percentages, cocoon production and number of young were found to be less in both the species than biogas slurry (Table 5.3). Although cowdung exhibited less response than biogas slurry for earthworm activities, the former is widely used in fish culture and therefore should be recommended for development of earthworm in bottom mud because it is easily available and more cheaper animal waste.

**Table 5.3 : Effects of Various Wastes on the Egg-laying and Hatching Capacities of Earthworm Species**

| Waste | Ratio of Soil : water | *Lampito mauritii* | | | *Perionyx excavatus* | | |
|---|---|---|---|---|---|---|---|
| | | Cocoon production | Number of young | Hatching (%) | Cocoon Production | Number of young | Hatching (%) |
| Control | 1 : 0 | 10 ± 2 | 5 ± 2 | 48 ± 1 | 7 ± 2 | 3 ± 1 | 40 ± 2 |
| Kitchen | 1 : 2 | 15 ± 2 | 11 ± 2 | 73 ± 3 | 12 ± 1 | 8 ± 2 | 66 ± 3 |
| Cowdung | 1 : 2 | 20 ± 3 | 17 ± 2 | 15 ± 2 | 15 ± 3 | 12 ± 2 | 75 ± 2 |
| Bio-gas | 1 : 2 | 32 ± 4 | 29 ± 4 | 23 ± 3 | 23 ± 2 | 20 ± 2 | 85 ± 2 |

*Source* : Ponnuraj, *et al.* (1998).

### *Factors Affecting Earthworm Activity*

Earthworms must have organic matter as a source of food. They also prefer moist habitat. Consequently, they live where organic manures have been added to the soil. Most earthworm species thrives best where the soil is not too acid, but a few species are tolerant to low pH.

Temperature of pond ecosystem affects earthworm numbers and their distribution. For example, a temperature of about 10 and 30°C appears optimum for *Lumbricus* sp. and *Pheretima* sp., respectively. The sensitivity of temperature probably account for the maximum activity of earthworm observed in winter and monsoon seasons in tropical and temperate regions.

## 5.4 Soil Protozoa

Protozoa are possibly the simplest form of animal life in pond soil. Although unicellular organisms, they are larger than bacteria, ranging from 4 to 120 μm. They include amoeba, flagellates and ciliates. They generally thrive in moist soils and are concentrated in surface layers and therefore, protozoa can be regarded as a major factor in organic matter decay and release of nutrients.

## 5.5 Soil Fungi

Soil fungi are known to play an important role in the transformation of soil constituents. Since fungi contain no chlorophyll, they must depend for their carbon and energy on the soil organic matter. Fungi may be divided into three groups : moulds, mushroom fungi and yeasts. Only the first is important in pond soil.

Molds are filamentous, microscopic and develop in acid, neutral or alkaline soils. Therefore, molds are abundant in acid soils, where actinomycetes and bacteria offer mild competition. In acid soils, molds are very important in decomposing the organic residues.

In affecting the processes of humus formation, molds are very important than bacteria. More than 50 per cent of the dead residues of pond ecosystem are decomposed by molds, compared to about 20 per cent for bacteria. On the other hand, soil fertility depends on the concentration of molds, since decomposition of organic materials continues after actinomycetes and bacterial activities have ceased.

## 5.6 Soil Algae

Several hundred species of algae have been reported from pond soils. Most algae bear chlorophyll and therefore, exhibit photosynthetic activity. Because light is necessary for photosynthesis, they are abundant near the surface of the soil. Their activity are more in shallow ponds where light is able to penetrate at the deepest zone. In many ponds and lakes where turbidity of water is very less, light can be reached even at the bottom of water. Such water bodies are highly productive because photosynthetic and other activities of bottom soil organisms are more pronounced.

Soil algae are divided into four groups such as green, blue-green, yellow green, and diatoms. They are best developed in soils having low pH. Blue-green algae are able to fix nitrogen – significant for pond productivity.

## 5.7 Soil Actinomycetes

Actinomycetes are similar to bacteria and molds that they are unicellular and filamentous. They are sensitive to acid soils.

The number of actinomycetes in soil sometimes reached hundreds of millions, about one tenth of the bacterial population. Their numbers are numerous in soils rich in humus. Moreover, actinomycetes are very important for the decomposition of soil organic matter and the release of its nutrients.

## 5.8 Bacteria

Bacteria are unicellular organisms, one of the smallest (0.002-0.003 mm in length) and simplest forms of life known. Their capacity for rapid reproduction allows bacteria to increase their numbers and activities in aquatic habitats resulting in the change of fish pond ecosystem. The shape of bacteria varies : They are round, rod-like and spiral.

Although the numbers of bacteria are highly variable, the numbers are very high, ranging from a few billion to 3 trillion per kilogram of soil. In the soil, they exist as filamentous, clumps and mats called *'colonies'*. Many soil bacteria are able to produce resting and growing phases. The latter forms are very important because they are able to survive even in unfavourable conditions.

Bacteria are of two types : autotrophic and heterotrophic. the former types obtain their energy from the oxidation of inorganic substances such as iron, sulfur, and ammonium. Their carbon come from carbon dioxide. Their abundant is very low, but play very important roles in controlling the availability of nutrients. Heterotrophic bacterial populations are numerous and very important, their energy and carbon generally comes from soil organic matter.

## 5.9 Role of Bacteria in Ponds

Different autotrophic bacteria are able to oxidise (or reduce) some elements in sediments. Thus, through nitrification (nitrate oxidation), selected bacteria oxidize ammonium compounds to the nitrite and nitrate forms. Since nitrate is absorbed by most plants, this transformation is very important for aquatic plants. Some other bacteria oxidize sulfur, yields sulfate ions that are absorbed by plants. Oxidation and reduction of manganese and iron by bacteria also help to determine the color of the soils.

Heterotrophic bacteria play an important role not only to determine the trophic level of ponds and lakes, but also the nutrient condition of fish culture ecosystems. Generally bacterial populations have no direct relationship to fish growth, rather they are used as food source by some species of fish.

Some species of bacteria can able to fix atmospheric nitrogen to form ammonia which is then incorporated into organic nitrogen compounds utilized by plants. This type of bacteria is associated with plant roots. For example, water fern (*Azolla* sp.) can able to fix nitrogen is significant amount. This plant, when die, undergoes decomposition at the pond bottom and nutrients are thus available to overlying water.

## 5.10 Factors Governing the Availability of Bacteria

A large number of factors of freshwater ecosystems are responsible for the growth and development of bacteria. Among the most important are the seasons, organic loading, geochemical properties of the pond, oxygen, temperature, pH, different forms of nitrogen and phosphorus as well as stocking density of fish in ponds. These factors are briefly noted below :

1. Appearance of large number of bacteria in ponds treated with fertilizers and manures are possibly due to prevalence of specific nutrient conditions.
2. Some bacteria use oxygen gas (aerobic), combined oxygen (anaerobic) and free or combined oxygen (facultative).
3. Bacterial activity is greatest at 20-40°C; below 10°C, there is occurrence of temporary suppression of bacterial activity.
4. In contrast to autotrophic bacteria, heterotrophic bacteria use organic matter as an energy source.
5. pH values from 6 to 8 and high calcium concentration are best for most bacteria. However, some bacteria function at pH $< 3.0$ and at pH $< 8.5$.
6. Different forms of nitrogen such as ammonia nitrogen, nitrite nitrogen, and nitrate nitrogen show strong influence on the population growth of heterotrophic bacteria.
7. The highest peak of bacteria is found in post-monsoon months (July to September).
8. Development of heterotrophic bacteria is higher in mono- and polyculture ponds than in traditional system of farming.

## 5.11 Detrimental Effects of Pond Organisms on Fish

It is needless to mention that some pond organisms are injurious to fishes. For example, different species of snails and slugs act as vector for transmission of diseases in fish. Some species of monogenic and digenic trematodes, cestodes, nematodes, hirudinea, and crustaceans are very important fish ectoparasites. Since soil and water are easily infested with disease-borne organisms through fish farm implements and manuring from animals that were fed infected plant materials.

Large-scale mortality of fish occurs in ponds due to bacterial, fungal, viral, and protozoan infections. In ponds where fishes are stocked beyond normal stocking densities, fishes are easily succumbed to various diseases because heavy stocking provides favourable conditions for rapid spread of infections among the fish. However, symptoms of fish diseases are characterized by lethargic, ulceration of the skin followed by haemorrhage, blindness, congestion at the base of anal and pectoral fins, red flecking on the gill filaments (fungal diseases) ; ulcers on the skin, accumulation of a body fluid, exophthalmic condition, opaque eyes (bacterial diseases) and presence of small white cycsts on the gills and fins, emeciation, falling on scales and loss of chromatophores (protozoan diseases).

### *Competition for Nutrients*

Pond soil organisms may be detrimental temporarily due to competition for available nutrients. Soil organisms rapidly absorb nutrient and as a result phytoplankton use only what is left. In some situations, bottom fauna also use very low amount of nutrients for growth. Consequently, relationship between natural food and fish yield is jeopardized. However, competition for nitrogen is greatest, although similar competition occurs for potassium and calcium.

## 5.12 Disease Control by Pond Management

Prevention is the best defence against microbial diseases. Removal of the vectors from the infested pond should be done at regular intervals. Moreover, strict quarantine systems will check the transfer of micro-organisms from one pond to another. Disease-free and healthy fish should be stocked in ponds.

Agricultural lime, due to its toxic and caustic actions, kills pathogenic bacteria as well as fish parasites in different stages of their life history. Thus, the fishes may become tolerant to diseases. It is well known that fishes are succumbed to diseases in acidic ponds than in alkaline one and hence, application of lime to ponds in suitable rate is necessary. However, for thorough disinfection of ponds, an application rate of about 10,000 kg/hectare is recommended.

Stabilized chlorine dioxide has been known to be a very powerful disinfectant and effective against wide variety of bacteria, fungi and molds. To treat gill rot, fungal and protozoan infections in fishes, 625 ml per hectare feet once a day and alternative day for treatment is recommended. Besides chlorine dioxide, calcium hypochlorite and benzalkonium chloride are also recommended to fish farmers for disinfecting the pond water during pond preparation at the rates of 16 and 64 mg/l, respectively.

## 5.13 Benefits of Soil Organisms

Different species of soil organisms play an important rôle in nutrient cycle both in water and soil. They are broken down the natural organic compounds into simpler components. While soil organisms have many beneficial effects on fish production, only the most important should be discussed as under.

### *Inorganic Transformation*

Presence of nitrate, phosphate and sulfate ions in soil is due to the actions of soil organisms. Organic forms of nitrogen, phosphorus and sulfur are converted by the microbes into plant-available forms. Besides these, iron and manganese elements are oxidized by autotrophic organisms to their higher valency states, in which forms their solubilties are very low. As a result, these elements remain in insoluble and non-toxic forms in both acidic and alkaline conditions.

### *Decomposition of Organic Matter*

Decomposition of organic matter by soil flora and fauna results in release of nutrients (particularly different forms of nitrogen) for use by plankton and bottom fauna. Decomposition also enhances the stability of soil aggregates and humus.

Soil organisms must require nutrients and energy for efficient function. For this purpose, organic matter is broken down by organisms which ultimately release compounds – utilized by algae and macrophytes in ponds.

### *Nitrogen Fixation*

Some blue-green algae, actinomycetes, and bacteria can fix nitrogen. Worldwide, enormous amounts of nitrogen are fixed into forms usable by plants. After fixation of nitrogen by these organisms, it is transferred from organisms to organisms. For example, some bottom fauna, due to their brushing behavior in the sediment, engulf the nitrogen-fixing bacteria which stimulates the growth and development of bottom fauna. Bottom-dwelling fishes consume such bottom fauna and thus fish production is accelerated.

An enzyme system of particular importance is that it promotes the fixation of atmospheric nitrogen. This is of considerable interest for a variety of reasons. It is a very important step in the nitrogen cycle, providing available nitrogen for plant nutrition. It is an intriguing process since it occurs readily in various bacteria, blue-green algae, and in symbiotic bacteria-legume associations under mild conditions.

1. *In-Vivo Nitrogen Fixation* : In this case, both free-living and symbiotic species are involved. They are the strictly anaerobic *Clostridium pasteurianum,* facultative aerobes like *Klebsiella pneumoniae,* and strictly aerobes like *Azotobactor vinelandii.* Even in the aerobic forms it appears that the nitrogen fixation takes place under essentially anaerobic conditions. The most important nitrogen-fixing species are the mutualistic species of *Rhibobium* living in root nodules of various species of legumes.

The active enzyme in nitrozen-fixation is nitrogenase. It is not a unique enzyme but appear to differ somewhat from species to species. Nevertheless, the various enzymes are similar. Generally two proteins having high molecular weights (57,000-73,000 and 220,000-240,000 for small and high proteins, respectively) are involved. Recombination of a soluble protein-free co-factor containing molybdenum and iron with inactive nitrogenase restore the activity.

2. *In-Vitro Nitrogen Fixation* : In case of *in-vitro* nitrogen fixation, high temperature and pressure is necessary. It is efficient and entrenched, and it can produce large volumes of product (nitrogen fertilizers) in short time period.

## 5.14 Role of Aquacultural Practice on Soil Organisms

Changes in environment dramatically affect both the kind and number of soil organisms. Removal of aquatic plants and algal mats by herbicides and algaecides temporarily reduce the food for the organisms. Moreover, the species diversity and the total population of organisms also reduced due to application of pesticides and industrial sewage. Other practices such as manuring, fertilization, and liming generally increase the activities of the microflora.

## 5.15 Competition Among Soil Micro-organisms

Soil micro-organisms exhibit an intense intermicrobial rivalry for food. When organic manures are applied to pond water heterotrophic soil organisms complete with each other for food. Bacteria, due to high reproductive capacity, dominate initially and always consume simple compounds. When simple compounds are broken down, the fungi and actinomycetes become more competitive. Obviously, such competition among micro-organisms in fish pond soil for food is the general rule.

## 5.16 Conclusion

In any fish culture ecosystem bottom organisms are very important to the cycle of life in soil. They incorporate animal and plant residues into the soil and digest them liberating carbon dioxide which is utilized by higher plants. Soil organisms also produce humus which is vital to improve physical and chemical conditions of pond soil. At the time of decomposition of organic matter, soil organisms release essential inorganic nutrients in water that can be absorbed by aquatic plants and plankton or leached from the mud. Soil organisms control the color of pond mud by stimulating oxidation-reduction reactions.

Soil bacteria, fungi, and actinomycetes are considered as decay organisms. Algae and bacteria play a role in providing essential elements for improving the productivity of fish culture ecosystem. The success of pond and lake management systems depends upon the microbial activity in soil and very important because it not only improves the overall conditions of ecosystem but also triggers the production of fish biomass.

Earthworm is generally considered as an important bottom animal because it increases soil organic matter, nitrogen, phosphorus, and soil porosity. Some bottom-dwelling carnivorous fishes consume earthworms as food because they are highly palatable.

### *References*

Ponnuraj, P., A.G. Murugesan, and N. Sukumaran. 1998. Effect of organic wastes on fecundity of two earthworms, *Lampito mauritii* and *Perionyx excavatus*. *J. Environ. Biol.* 19 : 57-61.

Singh, S.M. 1998. Variable activities of two earthworm species in altering physico-chemical parameters of soils. *Geobios.* 25 : 54-57.

## Questions

1. At which stage in the breakdown of organic matters do the microflora active? Describe.
2. Describe why earthworms are considered as the most important soil animal.
3. Explain the role of protozoa, bacteria, and algae in pond productivity.
4. State the factors responsible for the development of bacteria in a fish culture pond.
5. State the detrimental effects of organisms on fish.
6. How pond soil organisms are related to inorganic transformation, decomposition of organic matter, and nitrogen fixation?

# 6
# Organic Matter in Pond Soil

It is needless to mention that organic matter [For review on the subject, see Stevenson (1982) and Tate (1987)] influences the chemical and physical properties of soils. Since organic matter increases the cation exchange capacity of soil, it is very important for the stability of soil aggregates and physical as well as chemical conditions of soil. Organic matter also supplies energy and nutrients which are responsible for the growth and development of micro-organisms.

## 6.1 Sources of Organic Matter

Plants and animals are recognized as the chief sources of organic matter. Under natural conditions, leaves, herbs, and aquatic plants annually supply large quantities of organic matter. These plant tissues are decomposed and digested by soil organisms and become part of the soil either by physical incorporation or infiltration. In this way, the residue of plant material supply nutrients for soil organisms and this soil organic matter is maintained throughout the year.

Animals consume the plant material, digest and contribute waste products and thus improves the soil organic matter. Earthworms, ants and some forms of insects play an important role in the translocation of soil residues. In integrated fish farming systems, a large amounts of organic matter are added from various sources such as duck, poultry, cattle etc. A variety of organic manures (plant and animal origin) are also applied during fish culture management programs which adds organic matter for high productivity.

## 6.2 Constituents of Plant and Animal Residues

The water content of plant and animal residues is very high, varying from 70-90 per cent. Besides this, hydrogen, oxygen, and carbon also represent the bulk of organic tissue in the soil. The other essential elements such as nitrogen, sulfur, phosphorus, potash, calcium, and magnesium are present in small quantities and they play an important role in plant and animal nutrition.

The carbohydrates are present either in the form of simple sugars or starch, lignin and cellulose. Sugars are found in plants and animals while starch and cellulose are the most prominent organic compounds in plants. In contrast to animal carbohydrates, plant carbohydrates are very resistant to decomposition. Proteins, in addition to carbon, oxygen and hydrogen, contain nitrogen, sulfur, manganese, copper, and iron.

### *Composition of Compounds*

The proteins contain carbon, oxygen, hydrogen, nitrogen and small amount of sulfur, manganese, copper, and iron. The complex proteins are resistant to breakdown but the simple proteins are easily decomposed.

The carbohydrates such as sugars, starch, and cellulose contain carbon, oxygen and hydrogen and considered as the most prominent of the organic compounds found in plants and animals. Like proteins, simple forms of carbohydrates are readily decomposed while the more complex forms are very resistant to decomposition. Lipids are somewhat more complex than carbohydrates, are found in some parts of plant and animal tissues.

### *Decomposition of Organic Substances*

When plant and animal body is deposited in the pond bottom, different types of organic compounds begin to decompose simultaneously. The rate of decomposition of organic substances generally vary depending upon their structural composition. Simple proteins, sugars and starches undergo rapid decomposition while fats, cellulose, hemicellulose, and crude proteins exhibit very slow decomposition. During breakdown processes, three reactions take place as noted below :

1. Organic compounds undergo oxidation in presence of enzymes and release water, carbon dioxide, energy, and heat.

$$-(C, 4H) + 2O_2 \xrightarrow[\text{Oxidation}]{\text{Enzyme}} CO_2 + 2H_2O + \text{Energy} + \text{Heat}$$

2. Nitrogen, phosphorus and sulfur are released by a series of reactions which is specific for each element.
3. Organic compounds which are very resistant to microbial action, constitute soil humus.

Proteins are succumbed to microbial action and yields carbon dioxide, water, and amino acids. Amino acids are further decomposed and produced ammonium nitrite and nitrate.

### *Organic Decay in Pond Soil*

When a new pond is constructed, no readily decomposable substances are present in the soil and the number as well as activity of micro-organisms (such as bacteria, fungi and actinomycetes) are low. When organic substances in the form of manures are added to ponds, they sink in to the pond mud, micro-organisms act on sugars, starches etc., releasing water and carbon dioxide. As a result, micro-organisms are developed dramatically and new organic compounds are formed.

When the level of decomposed food substances is exhausted, the concentration of micro-organisms begin to decline. After death and decay of micro-organisms, their bodies are utilized by living micro-organisms and simultaneously water and carbon dioxide are produced. On further reduction in the concentration of decomposed food substance, general activity of micro-organisms continues to decline which is associated with release of sulfates, nitrates and nitrites. The remaining portion of organic matter is more or less stable left in the soil after the major portions of added plant and animal residues have decomposed. It is dark, heterogenous colloidal mass, and newly formed resistant compounds, generally termed as *humus*.

## 6.3 Carbon Cycle

It is known to all that carbon is a common constituent of all organic matter and involved in all life processes. Decomposition of organic matter yields carbon dioxide and the transformation of this element called *carbon cycle*. It is interesting to mention that carbon dioxide and humus are stable components of this cycle. Changes of carbon in a pond environment is shown in Figure 6.1.

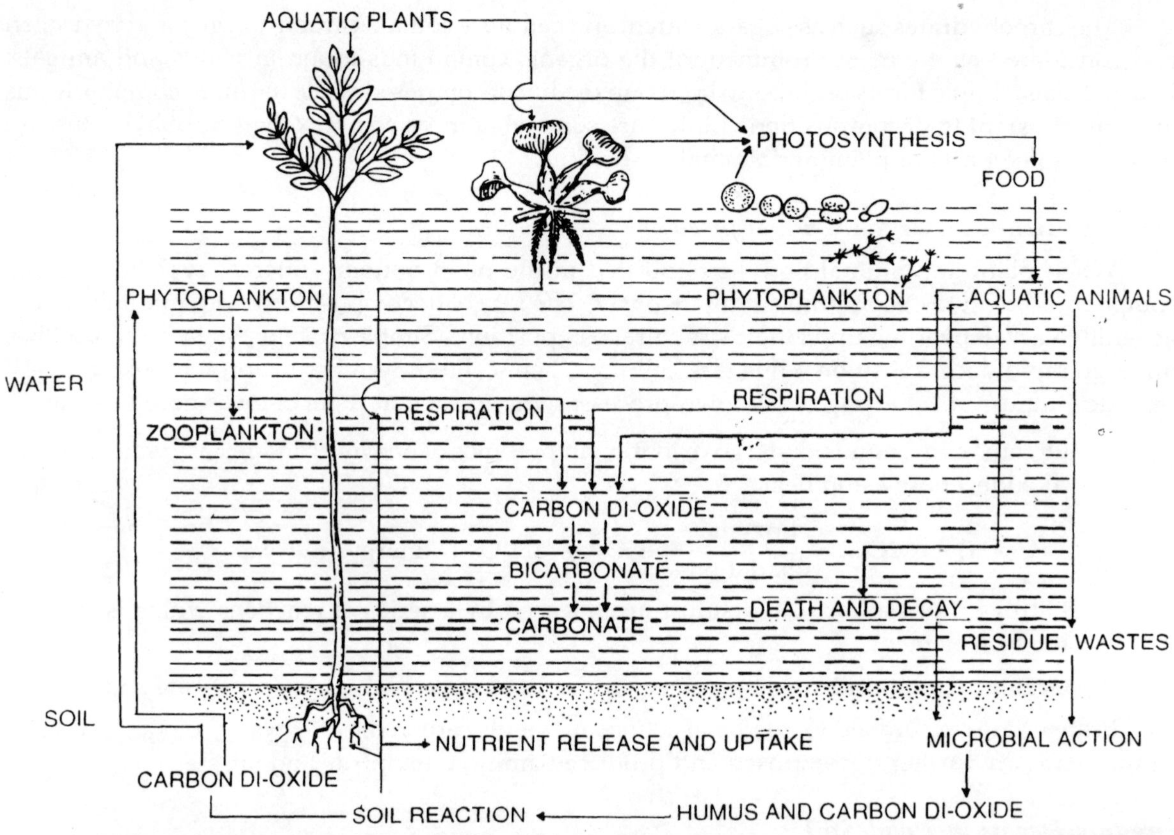


**Fig. 6.1 : Carbon cycle in an aquatic environment. It is the name given to one of the most important biogeochemical cycles, in which the element carbon circulates between living organisms and the nonliving environment. Carbon dioxide is used by plants for photosynthesis, produce food (Carbohydrates), and stored by the plant. Aquatic animals use carbon compounds from plants for growth and respiration. The carbon compounds in dead animals and plants are broken down by microorganisms. Micro-organisms degrade the complex carbon compounds into simpler ones and used in respiration, ultimately releasing carbon dioxide. In an environment, there is a balance between the amount of carbon dioxide absorbed by plants in photosynthesis and the amount returned to the environment by respiration of living organisms. Environmental pollution due to human activities can upset this balance.**

In an aquatic ecosystem, phytoplankton fix the carbon dioxide. The primary consumers, zooplankton, feed upon phytoplankton. Zooplanktons are consumed by small animals which, in turn, provide food for fish.

During photosynthesis, carbon dioxide is assimilated by phytoplankton and aquatic plants and converted into different types of organic compounds. In the pond mud, these organic compounds undergo enzymatic oxidation and carbon dioxide is evolved. Although soil microbial activity is the main source of carbon dioxide, some amounts of carbon dioxide come from respiration of plant roots and rain water. However, carbon dioxide is released which is further utilized by plants.

Appreciable amount of carbon dioxide reacts in the soil and water, producing carbonates and bicarbonates of calcium, magnesium, potassium etc. and carbonic acid. Carbonates and bicarbonates are highly soluble in pond water and can be removed either by leaching or aquatic plants.

## 6.4 Decomposition Products in Pond Soil

As result of decomposition of organic matter in pond mud, sulfur, nitrogen and phosphorus are produced. For example, proteins are broken down by soil microbes to produce ammonium compounds, sulphides and inorganic phosphorus and finally nitrates, sulfates and phosphates. Such conversion of element from an organic form to an inorganic state as a result of microbial decomposition is called *mineralization*.

Most of the inorganic ions such as nitrogen, phosphorus, sulfur, calcium, potassium, sodium, and magnesium released by decomposition are used by algae, macrophytes, and micro-organisms. Nitrates are subject to leaching, but phosphates are either tend to be retained in insoluble aluminium, iron and calcium compounds. Nitrogen is lost as gaseous form. Other cations are dissolved in soil solution and adsorbed by negatively charged soil colloids. Finally they are either removed by leaching or utilized by plants.

### *Energy of Organic Matter*

High quantity of energy is used by soil micro-organisms for decomposition of humus and plant and animal residues. It has been estimated that the addition of 10 metric tonnes of farm yard manure containing 2,500 kg of dry matter contain about 13 million kilocalories of latent energy. This is equivalent to the energy in more than 1.5 metric tonnes of coal. However, most of the energy is lost as heat and a small amount of energy is used by micro-organisms.

## 6.5 Humus

Humus in pond soil is formed by the decomposition of dead animals and plants [For detailed discussion of humus, see Haynes (1986) and Frimmel and Christman (1988)]. It is a heterogenous mixture of complex organic matter in different states of decay and the organisms that have caused the decomposition (mainly fungi and bacteria) and synthesis.

Plant and animal residues are decomposed by soil organisms and produce simpler organic compounds which are ultimately metabolized into new compounds in the soil by micro-organisms. The new compounds undergo further modification and synthesis as the micro-organisms are attacked by other micro-organisms.

On the other hand, synthetic reactions also occurs in soil which is characterized by the breakdown of lignin of aquatic plants and produced phenols. This product is called *monomer* which is transformed into *polymer* in presence of colloidal clays and as polyphenols. These compounds react with nitrogen-containing compounds and forms stable humus.

### *Humic and Non-humic Groups in Humus*

The humic substances are characterized by aromatic and ring-type structures that includes polyphenols. These are formed by decomposition, synthesis and polymerization. They comprise about 65-80% of the soil organic matter.

On the basis of resistance to degradation, humic substances are of three types : (1) *Humin* : insoluble in both acid and alkali and most resistant to microbial attack, (2) *Humic acid* : insoluble in acid but soluble in alkali and intermediate in resistance to microbial attack, and (3) *Pulvic acid* :

soluble in acid and alkali and most susceptible to microbial attack. However, depending on the ecosystem, it takes 10-100 years to decompose different types of humic substances in the soil.

Non-humic groups make up 20-30 per cent of the organic matter in soils. These are less complex and less resistant to microbial attack. Among different types of non-humic substances, polysaccharides, organic acids and protein-like materials are worth mentioning. These substances are present in pond soil in very small quantities and affect the availability of nutrients such as nitrogen and phosphorus.

### *Interactions Among Different Humic Substances*

Different types of humic substances interact with each other. For example, nitrogen substances and proteins react with humic acids and polysaccharides. This interaction is very important for storage of nitrogen in soils because protein nitrogen is protected from microbial attack. If this reaction does not occur, the protein substances would succumb to decomposition and hydrolysis and as a result nitrogen is released as soluble ammonium and nitrate forms. These are quickly lost from the soil through leaching.

Besides this, humic substances also react with silicate clays. These clays bind with amino acids, proteins and peptides and form complexes and thus nitrogen-containing compounds are protected in the soil from microbial degradation. Moreover, different types of organomineral complexes (such as organic acids) are linked with silicate clays or aluminium oxides. Huge amount of organic acids in soil is linked with clay and inorganic constituents and as a result, high organic matter content of clay soils is found.

### *Colloidal Features of Humus*

1. The water-holding capacity of humus is 5-6 times that of the silicate clays.
2. The black color of humus can be distinguished from the other colloidal constituents in soils.
3. Cation exchange reactions with humus are similar to that of silicate clays.
4. Colloidal humus particles (micelles) are composed of hydrogen, carbon and oxygen.
5. The colloidal surfaces of humus are negatively charged and the extent of negative charge is pH dependent.
6. The surface area of humus colloids is very high than silicate clays.

### *Effect of Humus on Nutrient Availability*

Due to continual microbial attack, humus undergo mineral breakdown and as a result different types of nutrients become available. For example, polysaccharides form stable organomineral complexes with metallic ions such as copper, zinc and iron. Minerals are attracted by the cations and forms complex by the organic molecule. They are then used up by plants or take active part in the synthesis of inorganic constituents. Another way of nutrient availability in soil from humus can be explained in the following way :

Humic acids react with minerals, undergo decomposition and as a result, base-forming cations are formed :

$$H^+ \boxed{\text{Micelle}} + CaAlSi_3O_8 \longrightarrow HAlSi_3O_8 + Ca \boxed{\text{Micelle}}$$

| Humic acid | Microcline | Acid silicate | Adsorbed Ca |
|---|---|---|---|

$$H^+ \boxed{\text{Micelle}} + KAlSi_3O_8 \longrightarrow HAlSi_3O_8 + K \boxed{\text{Micelle}}$$

The calcium and potassium are absorbed by algae and macrophytes.

### *Stability of Humus*

Because humus is always subject to microbial attack, it is obvious that the level of soil organic matter gradually reduced unless plant and animal tissues are added. But inspite of the continual microbial attack of humus, some humic materials which are formed from organic carbon, are very resistant to degradation by micro-organisms. This resistance is very important as nitrogen, phosphorus, calcium, and organic matter levels in soils are maintained for constant productivity of fish culture ecosystems.

## 6.6 Role of Soil Organic Matter on Soil Properties

1. Nitrogen, sulfur, phosphorus, and micronutrients are held in organic forms.
2. Cations present on humus colloids are easily replaced.
3. Organic matter has high cation exchange capacity (10-30 times as great as mineral colloids).
4. Water-holding capacity is greatly increased.
5. Nutrients are released from minerals by acid humus.

## 6.7 Carbon/Nitrogen Ratio

It is known that there exists a close relationship between nitrogen and organic matter contents of soils. Because carbon is present in organic matter in large and definite proportion, the C/N ratio of soils is fairly constant. This is important for controlling the total organic matter, available nitrogen and also for the development of sound fish culture pond management schemes.

### *Ratio in Microbes and Plants*

The C/N ratio in micro-organisms is much lower, ranging from 4:1 to 9:1. But in organic manure, the C/N ratio is higher than microbes, ranging from 20:1 to 30:1. However, organic residues contain small amounts of total nitrogen and large amounts of carbon. Consequently, the C/N ratio is very high.

### *C/N Ratio in Fish Ponds**

Although there is no definite correlation between fish production and carbon content of fish ponds, organic carbon was found to be less than 0.5 that can be considered as very low for a fish pond. The range 0.5 to 1 per cent organic carbon exhibited average fish production and 1.5 to 2.5 per cent seemed to be optimal. In general, C/N ratio less than 5 indicates poor fish production.

* For further details, see Banerjee (1967).

Although high production is found in the ratio varied between 5 and 10, the range 10 to 15 indicates ideal condition. However, the C/N ratio more than 15 considered as the less favourable for fish production.

### 6.8 Significance of C/N Ratio

The C/N ratio in fish ponds is significant for two reasons : (1) competition among soil micro-organisms for available nitrogen in soil when organic matter having a high C/N ratio (50 : 1) is added to ponds, and (2) since the C/N ratio is constant in soils, the maintenance of soil organic matter is constrained by the soil nitrogen level.

If a composite fish culture pond is fertilized with inorganic materials in suitable proportions, it will be seen that nitrates are present in adequate amounts, the C/N ratio is low and the decay organisms are at a low level of activity as evidenced by the evolution of carbon dioxide at low level. Furthermore, if such pond is further treated with large amounts of organic materials with moderate or high C/N ratio (20:1 or 40:1), various types of micro-organisms such as fungi and bacteria become active, increase their number and produce carbon dioxide (Figure 6.2). At this stage, micro-organisms will consume nitrate-nitrogen and ultimately disappear from the soil. As a result, little or no ammonia and nitrate-nitrogen is available; but this availability is temporary. As the process of decay takes place ceaselessly on, the C/N ratio of the remaining organic material decreases because of nitrogen is being stored and carbon dioxide is lost.

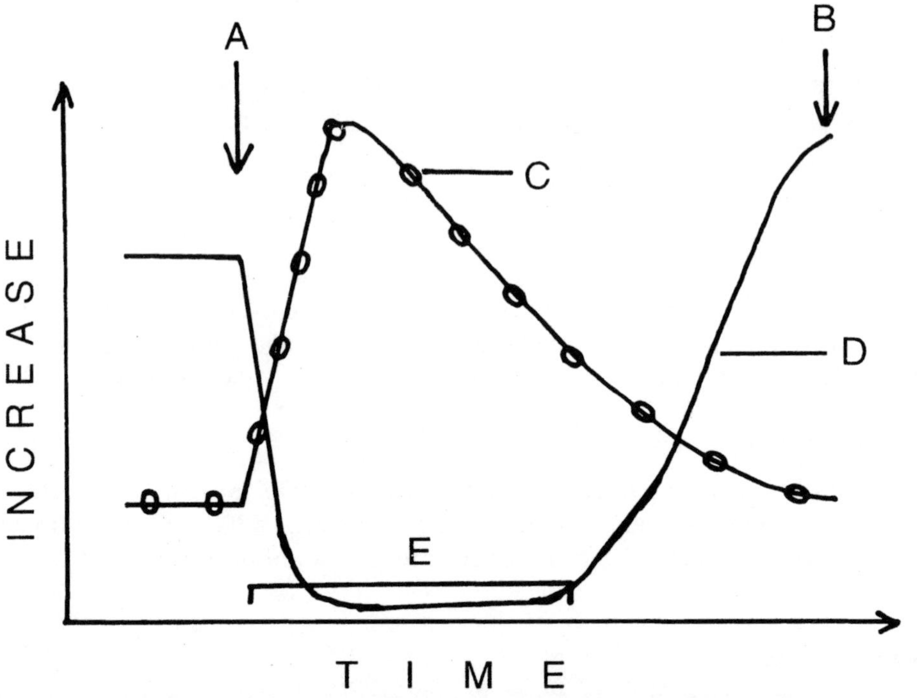


**Fig. 6.2 : Relationship between the level of nitrate-nitrogen and decay of organic matter. When organic matter with high C/N ratio is added to soil (A), micro-organisms are dominant and the nitrifying bacteria are not in active stage. Consequently, nitrate depression occurs (E). During depression period, phytoplanktons receive little nitrogen. The period of nitrate depression depends on the C/N ratio, (C) Activity of decay organism and production of carbon dioxide, (D) Nitrate-nitrogen level of soil.**

This situation is generally observed so long as activities of micro-organisms subside. The number of micro-organics and formation of carbon dioxide gradually decrease, and nitrification (release of nitrates) can proceed.

### *Nitrate Depression Period*

The time of interval of nitrate depression may be only a week or more. The rate of decomposition of organic residues either shorten or lengthen the period depending on the quantity of decomposable residues applied (Figure 6.2). For example, the greater the quantity of decomposable residues added, the longer nitrate formation will be blocked and the lower the C/N ratio of the residues added, the more rapidly the cycle will run. Mature organic residues of plant origin (legume and non-legume) have a higher C/N ratio than that of younger residues (Figure 6.3). This is very important and should be considered at the time of addition of organic residues to fish ponds.

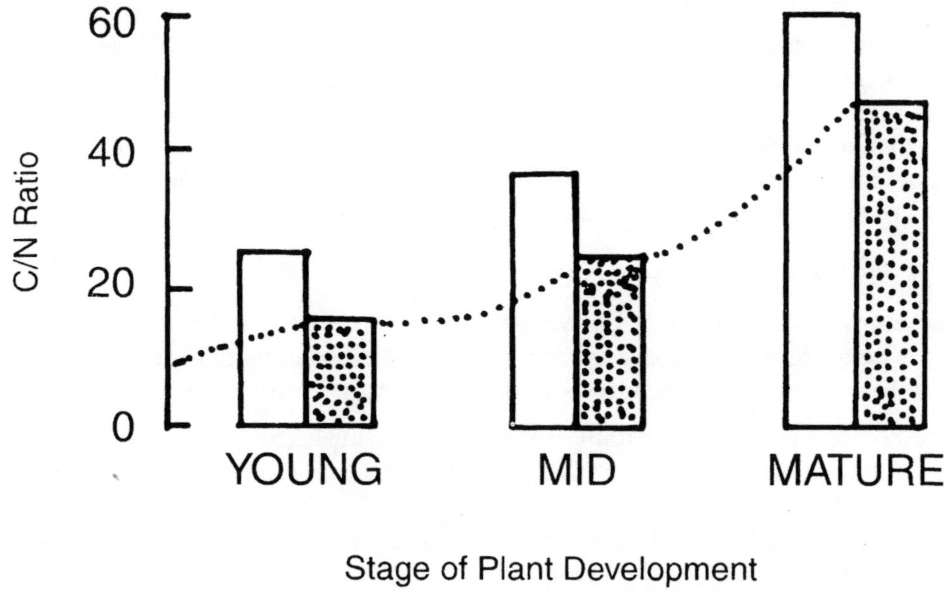


**Fig. 6.3 : Relationship between the stage of plant development and C/N ratio. Note that mature plants have high C/N ratio than the younger ones and hence the nitrate suppression period will be longer. Leguminous plant tissues have an advantage over non-leguminous plant tissue because it triggers a rapid organic turnover in soils. Hollow columns indicate non-leguminous plant and dotted columns indicate leguminous plant.**

## 6.9 Factors Affecting Soil Organic Matter and Nitrogen

1. *Soil Texture* : The texture of soil influences the amount of humus and nitrogen present. For example, soils high in silt and clay have higher organic matter than the coarser-textured soils. Furthermore, in cases where organomineral complexes are formed in soils having high clay content, soils are able to protect nitrogen and thus soil organic matter is protected from microbial decay.

2. *Nutrients* : The application of lime, manure, and phosphorus help maintain higher organic matter levels. Therefore, fish culture ponds that are kept productive throughout the growing season by application of lime, fertilizers, and manures are likely to have more organic matter than in unfertilized ponds.
3. *Influence of Temperature* : In general, the decomposition of organic matter is accelerated in warmer areas and a lower rate of decomposition occurs in cooler areas.

## 6.10 Pond Soil Organic Compounds as Chelators

As early as the late seventies, the importance of chelation in soil has been recognized. Chelators are the collection of chemicals which are cheaper and easier for soil fertility management. Chelators are of interest because they are able to bind with calcium, magnesium, potassium, iron, copper etc. and can make these elements as well as phosphorus more available to water.

Microbial activity in the soil produces a great amounts and variety of metal-binding substances (chelators). Decomposition of organic matter in soil leads to the formation of humic substances as well as different organic acids which have important chelation properties due to the presence of functional groups in their molecules. Humic substances act as strong complex forming and chelating agents better than or comparable to the synthetic chelating agents. This property of humic substances is very important for increasing the availability of phosphatic fertilizers which are rendered insoluble in the soil by fixation mechanisms.

### *Examples of Chelating Complexes*

Chelate compounds are more stable when they contain a system of alternate double and single bonds. This is better represented as a system in which electron density is delocalized and spread over the ring. Examples, of this includes acetylacetone and porphyrin complexes with metals.

Chetale complexes are more stable than similar complexes with unidentate legands, as dissociation of the complex involves breaking two bonds rather than one. The more rings that are formed, the more stable the complex is. Chelating agents with 3, 4, and 6 donar atoms are known and are termed tridentate, tetradentate and hexadentate legands. An important example of the latter is *ethylenediaminetetracetic acid (EDTA)*. This bonds though two nitrogen and four oxygen atoms to the metal, and so form five rings. Due to this bonding EDTA can form complexes with most metal ions. Even complexes with large ions such as calcium ($Ca^{2+}$) are relatively more stable (the $Ca^{2+}$-EDTA complex is only formed completely at pH 8, not at lower pH).

The most important example of a simple well-known copper-glycine chelate is shown in Figure 6.4. Copper acts as an important component of many enzymes (such as uricase, tyrosinase etc.). It is an essential trace element for many micro-organisms and animals. Free glycine although microbiologically unstable, does occur in small amount in soil where it is excreted by micro-organisms. Glycine is also a part of organic matter in different soil types. Its carboxyl and amino groups are the most important legands in chelation. The chemistry involved in the formation of copper-glycine chelate illustrates how organic constituents in organic manures or in soil may chelate metals.

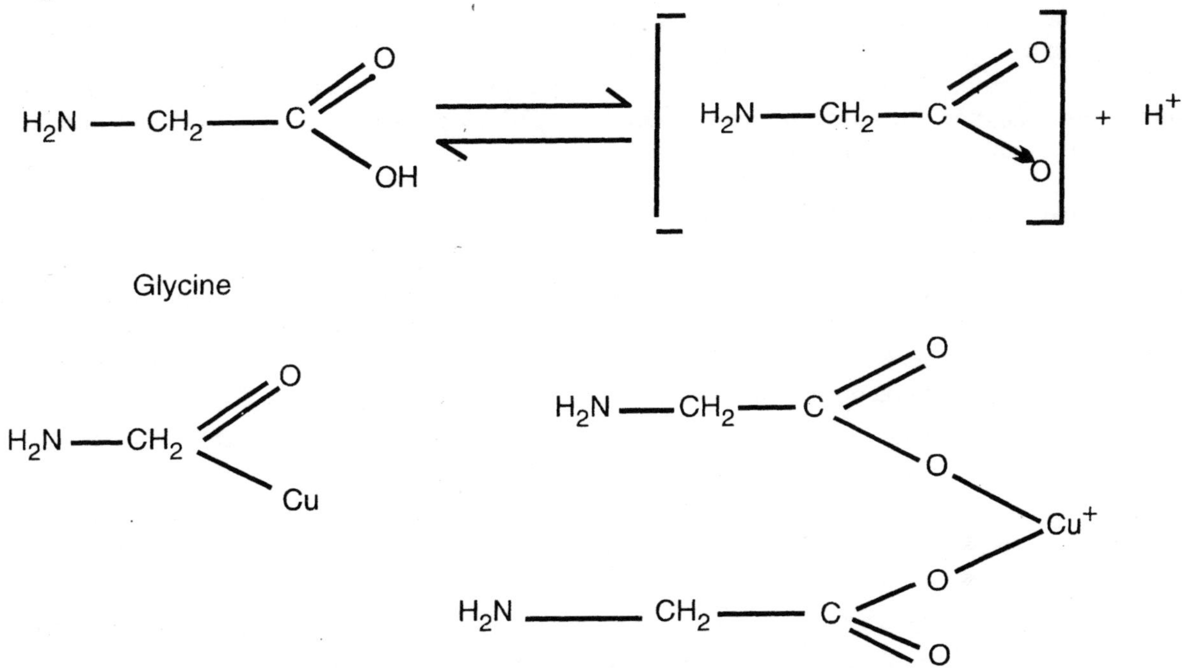


**Fig. 6.4 : Copper-glycine chelation.**

## *Beneficial Effects of Chelators*

Organic matter plays an important role in soil fertility by virtue of its chelating or complexing property. This beneficial effect was attributed to the mineral nutrients contained in it which were returned to the soil. However, many organic substances formed by decomposing organic matter are excellent chelators whose most important function is to make trace elements available to plants.

Organic matter counteracts or overcomes certain toxic substances in the soil. The accumulation of insecticides, herbicides, fungicides and other types of pesticides in aquatic environments has resulted in toxic effects on fish and other aquatic animals. However, the adverse effects of toxic substances can be overcome by amending the soil with good quality humus having high cation exchange capacity and sources of energy and nutrients to the soil micro-organisms taking part in the detoxification process of these substances. Humus also has the ability to chelate the lead, mercury, copper, arsenic, and zinc of certain pesticides.

Much attention has been given to a few synthetic chelates while the vast reservoir of natural products and easily available arouses less interests. This cannot be attributed to the complexity of humus. Many organic compounds in the soil are simple, fairly stable chemically and microbiologically, and highly active metal-binders. In these respects, they compare with the EDTA. The total amount of synthetic chelates is only a very small fraction of the quantity of naturally-occurring organic matter such as compost material, humus, farmyard manure etc. that are produced in or applied to the soil.

The use of synthetic chelates in several countries is actually existing in spite of the fact that they are quite expensive. They are used to meet micronutrient deficiencies of several species of plankton. Although chelates may not replace the conventional methods of supplying most micronutrients, they offer possibilities in special cases. Research will, however, likely continue to increase the chances for their use in fish culture.

## 6.11 Control of Organic Matter in Soil

1. If soil organic level in fish ponds is to be maintained, the ponds must receive a continuous supply of organic substances. Compost, organic waste and animal manures are the main sources of organic materials. However, experiments have shown that additions of about 6 metric tonnes organic materials per hectare per year are necessary to maintain soil organic matter at optimum level.
2. To minimize the constraints to fish yield from chemical toxicites, moderate application of fertilizers and lime should be used.
3. Since there is a strong correlation between nitrogen and organic matter, management practices include the application of organic and inorganic materials. This results a higher organic matter levels.
4. Biological and mechanical stirring of the ponds mud is necessary to control organic matter. Stirring of pond bottom helps to release nutrients which is dissolved in water.

## 6.12 Conclusion

Organic matter is very important in enhancing the fertility of pond ecosystem for high fish production. It supplies nutrients and has water-holding capacity. Organic matter can be regarded as a source of food for micro-organisms and consequently, their activities significantly increased. The optimum level of organic matter should always be maintained for constant fertility of fish ponds. Through intensive fish cultivation, abundant residues can be returned to the pond ecosystem. The importance of organic matter as a natural chelating material has been fully recognized. The uptake of phosphorus fertilizers, increase in solubility of nutrients, and counteracting certain toxic substances in the soil is of special interest in soil science.

### *References*

Banerjee, S.M. 1967. Water quality and soil condition of fish ponds in some states of India in relation to fish production. *Ind. J. Fish* 14 : 115-144.

Frimmel, F.H. and R.F. Christman. 1988. *Humic Substances and Their Role in The Environment.* Wiley-Eastern Ltd., New York.

Haynes, R. J. 1986. The decomposition process : mineralization, immobilization, humus formation, and degradation (*Ed.* R.J. Haynes). *In : Mineral Nitrogen in Plant-Soil System.* Academic Press, New York.

Stevenson, F.J. 1982. *Humus Chemistry – Genesis, Composition, and Reaction.* Wiley-Eastern Ltd., New York.

Tate, R.L. 1987. *Soil Organic Matter : Biological and Ecological Effects.* Wiley-Eastern Ltd., New York.

## Questions

1. What are the chemical constituents of plant and animal residues? What are the sources of organic matter for pond productivity?
2. A fish culturist explains that humus is a mixture of compounds that originally was found in plants and that resisted microbial decay. Do you agree or disagree with the fish culturist?
3. Distinguish between humic and non-humic organic compounds in pond soils.
4. State how humus increases the processes of nutrient availability.
5. How carbon-nitrogen ratio (C/N ratio) is related to pond productivity? State its significance.
6. How soil organic compounds act as chelator? State the beneficial effects of chelators.
7. What are the sources of chelates and of what importance are they in increasing micronutrient availability? Under what conditions would they most likely be useful?

# 7

# Micro-organisms in Relation to Fish Culture Ecosystem

A micro-organism is any organism of microscopic dimensions and it is considered as a part of the animal and plant populations which consists of individuals so small that it is not able to distinguish clearly without the aid of microscope. Micro-organisms are, however, closely associated with the health of fish and other aquatic animals. Some micro-organisms are detrimental and others are beneficial. Most microbes are unicellular. In unicellular organisms, all the life processes are performed by a single cell. In higher forms of life, organisms are composed of many cells that are arranged in tissues and organs to perform specific functions.

Microbiology of aquatic ecosystems is the study of micro-organisms and their activities in springs, lakes, ponds, rivers, bays, and seas. Various types of micro-organisms such as bacteria, protozoa, algae, viruses, and microscopic fungi inhabit these natural waters. Some of these micro organisms are indigenous; others are transient, entering the water from soil or air or from industrial or domestic waters. Micro-organisms make a number of changes in soil and water. For examples, in pond soils, organic matter is transformed into simple inorganic substances that provide the nutrient material for the fish food organisms. Sewage and waste water treatment procedures and decay of organic wastes, both natural or artificial, are largely dependent upon microbial activity to eliminate or greatly reduce the development of objectionable or hazardous situations.

### 7.1 Historical Background of the Study of Micro-organisms

The discovery of micro-organisms by scientists came after development of microscopes and used them to examine droplets of natural fluids. Millions of microbes were found from a variety of samples. During the period from 1600 to 1800, considerable information accumulated about the occurrence of these microscopic forms of life. At the time, there was great controversy regarding the origin of micro-organisms. The controversy centered on the question of whether micro-organisms arose from spontaneous generation (i.e. from non-living substances). Simultaneously with the development of evidence over hundred years to discard spontaneous generation was the growing acceptance of the concept that micro-organisms were the cause of many adverse or favourable conditions that occurs in daily life. In the latter part of the nineteenth century, a problem confronted investigators searching for evidence to prove that a specific type of micro-organisms was responsible for spoilage or a disease. Questions arose how could they isolate the micro-organisms suspected of causing the change and prove that it was the causal agent. A solid nutrient substance was necessary upon which a specimen could be spread so that individual microbial cells would keep at a distance from each other. Upon incubation, each cell would reproduce, resulting in a mass of identical cells (a colony). A small portion of the colony could then be transferred to a fresh medium and be maintained as a pure culture.

Robert Koch (1843-1885) was concerned with the need to develop a technique for the isolation of micro-organisms in pure culture to establish the causative agent of a disease. During his experiment, he faced a problem for several reasons. However, the solution to the problem was provided in 1883 by a German housewife, Fannie E. Hesse who worked with her husband, Walther Hesse, a German physician. She worked on agar and found that *agar,* a polysaccharide of algal origin, may be used as a substitute for gelation in microbial media. His experiments on this aspect with agar were so successful that Walther Hesse reported the experiments with agar to Robert Koch. Robert Koch recognized the importance of agar as a solidifying agent for microbial media. Agar turns into solution (1.5 per cent) at 100 °C and solidifies at 45 °C. Upon jelling at 45 °C, it remains solid at elevated temperatures (just below 100 °C). This feature makes it possible to incubate the innoculated media at any desired temperature and still have the medium remain solid.

## 7.2 Groups of Micro-organisms

Until the eighteenth century, the classification of living organisms placed all organisms into plant and animal kingdoms. Some organisms are predominantly plant like, others are animal like, and some others share characteristics common to both animals and plants. Because there are organisms that do not fall into either the plant or the animal kingdom, it was proposed that new kingdom be established to include those organisms which are neither animals nor plants.

An earliest proposal was made in 1866 by a German Zoologist, E.H. Haeckel. He suggested a third kingdom, *Protista,* be formed to incorporate those micro-organisms that are typically neither plants nor animals. The Protists, however, include algae, bacteria, fungi, and protozoa. Some features of these microbes are shown in Table 7.1.

**Table 7.1 : Some Characteristic Features of Major Groups of Micro-organisms and Their Significance**

| Group | Size | Features | Significance |
|---|---|---|---|
| Bacteria | 0.2 to 100 μm | Unicellular, procaryotic, simple internal structure, reproduction asexual, grow on culture media | Some cause disease; some perform important role in cycling of elements and contribute to soil fertility; some make foods and spoil foods |
| Fungi (Moulds) | 2-10 μm to several mm | Eucaryotic; multicellular, with many structural features; cultured in laboratory culture media; reproduction : sexual and asexual | Causes diseases of humans and other animals; useful for production of many chemicals; responsible for decomposition of many materials |
| Fungi (Yeast) | 5-10 μm | Eucaryotic; unicellular; grow on laboratory culture media; reproduction : budding, sexual, and asexual | Some cause disease; Some used as food supplement and production of beverages |
| Protozoa | 2-200 μm | Eucaryotic; unicellular; some cultivated in laboratory; reproduction : asexual and sexual | Some cause disease; food for aquatic animals; some cause degradation of cellulose, sewage and sludge |
| Algae | 1 μm to several feet | Eucaryotic; unicellular and multicellular; contain chlorophyll; reproduction : sexual and asexual; mostly found in aquatic ecosystems | Important to food production in aquatic ecosystems; some algae produce toxic substances; used as food supplement; source of agar for microbiological media |

## 7.3 Distribution and Interactions Among Micro-organisms in Aquatic Ecosystem

Micro-organisms occur nearly everywhere – starting from the mountain top to the greatest depth of the ocean. Fertile soil and water are teemed with them. They are transported by streams and rivers into lakes, ponds, etc. Human and other organic wastes are discharged into streams, lakes and ponds, diseases may be spread from one place to another. Micro-organisms grow abundantly where they find food and a temperature suitable for their growth and reproduction. It is fortunate that most micro-organisms are harmless to fish and other aquatic animals. A variety of micro-organisms are beneficial to fish culture ecosystem and the productivity of ecosystem dramatically increased. It is also important to note that micro-organisms are found to exist under harsh environmental conditions.

Micro-organisms in different types of freshwater environment may occur at all depths. The surface flim and the bottom sediments, however, harbor the higher concentrations of micro-organisms.

### *Plankton*

The aggregation of drifting and floating microbial organisms in the surface region of the aquatic ecosystem is termed as *plankton*. Plankton is composed of algae (phytoplankton) and minute animal life (zooplankton). Phototrophic micro-organisms are regarded as the important plankton since they are the primary producers of organic matter through photosynthesis. Several physical conditions which influence the plankton population qualitatively and quantitatively are shown in Figure 7.1.

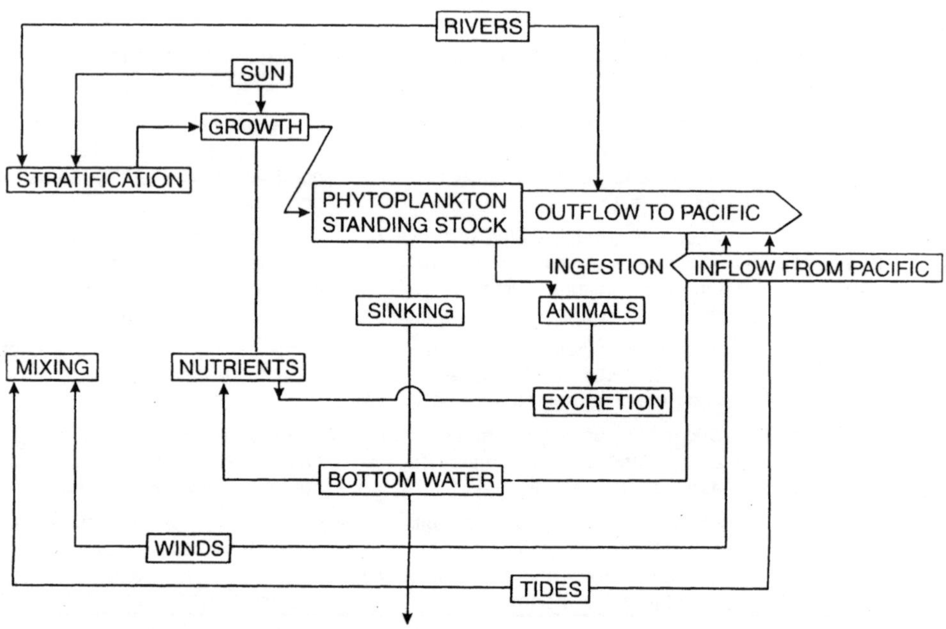


**Fig. 7.1 : Physical conditions generate both negative and positive effects on the growth of phytoplankton and balanced physical conditions produce blooms. Consequently, phytoplankton standing stock is increased by nutrient uptake and photosynthesis. These are controlled by mixing, stratification, and sunlight. Standing stock is decreased by sinking, animal consumption, and flushing by runoff and winds. [From Strickland (1983)]**

### *Benthic Micro-organisms*

Micro-organisms which inhabit the bottom region of any aquatic ecosystem are termed as *benthic organisms.* A large number of benthic micro-organisms perform various function – essential for the overall productivity of ponds and lakes.

### *Mixing of Waters*

The movement of water by currents and winds accomplished redistribution of micro-organisms. Mixing of water (also called upwelling) takes place when water rises from a deeper to a shallow depth due to divergence of winds or currents. During upwelling, the bottom water carries with it a rich supply of nutrients that are delivered to the surface water. Consequently, productivity of water significantly increased.

### *Interaction among Micro-organisms*

Highly productive aquatic ecosystems are occupied by numerous numbers of micro-organisms. However, the conditions as influencing the growth of micro-organisms to aquatic ecosystems can be briefly summarized as follows : (i) hydrogen ion concentration (pH), (ii) temperature, (iii) dissolved oxygen, (iv) amount of nutrients, and (v) application of chemical fertilizers and manures.

Interactions between and among microbial species undoubtedly has an important effect on the numbers of the population and consequently, and extremely complex situation results. Since the bacterial population generally exceeds the population of all other groups of micro-organisms in both variety and number, they are responsible for the transformation of many organic and inorganic compounds and as a result, bacterial population play an important role in building fertility of fish culture ecosystems.

The micro-organisms exhibit many different types of interactions. Some of the associations are *neutral* (neutralism), some are *beneficial* (mutualism and commensalism) and others are *detrimental* (antagonism, competition, and parasitism). These interactions are, however, play a significant role in both productivity of water and fish production generally. Although studies on the complexities of different interactions among or between micro-organisms are beyond the scope of this section, their interactions with fish in relation to fish production and fish diseases should be considered.

## 7.4 Importance of Micro-organisms in Aquatic Ecosystem

The fungi are heterotrophic organisms and they require organic compounds for nutrition. When they feed on dead and decaying organic matter, they are termed as *saprophytes.* Saprophytes decompose complex animal and plant remains, breaking them down into simpler chemical substances that are returned to the soil, and consequently, soil fertility is increased.

When fungi live in or on fish, they are known as *parasites* and they cause diseases in fish; of course, fungal diseases are less commonly encountered than bacterial diseases.

### *Algae*

Different algal species occur in great abundance in the seas, oceans, salt lakes, freshwater ponds and lakes, rivers, and streams. Many are found in damp soil, on stones, rocks, and trees bark. Some species of algae are subject to bacterial decomposition and the decomposition products are made available for enrichment of aquatic ecosystems.

When dispersed in aquatic environments algae increase the carbon dioxide concentration in water. Heavy growth of some algae reduces hardness of water and removes salts which are the cause of brackishness.

Some planktonic algae produce toxins which are lethal to fish. Several toxic substances are liberated from the algae (see Section 7.10) by bacterial decomposition of water blooms and in serious cases respiratory failures in fishes can result.

There is increasing interest in the use of the smaller forms of blue-green algae (Group : Cyanophyceae) especially *Chlorella* and *Spirulina,* as food for humans, domestic animal (cattle and poultry), and fish. When these organisms are grown under suitable conditions, they provide a rich source of proteins, fats, carbohydrates, and other essential ingredients.

### *Protozoa*

Protozoa serve as an important link in the *food chain* of communities in aquatic ecosystems. Also, the particular importance in the balance of wetlands and aquatic ecosystems are the bacteria-feeding and saprophytic protozoa. They make up of the substances produced and organisms involved in the final decomposition stage of organic matter. This can be represented by the following stages :

| Excretory products Dead bodies of plants and animals | → | Decomposition of dead bodies by bacteria and fungi | → | Bacterial ingestion by bacteria |
|---|---|---|---|---|

Although bacteria serve an important role in the degradation of sewage, the role of protozoa is becoming more completely understood and appreciated. Biological sewage treatment involves both an aerobic digestion and/or aeration. Aeration and flocculation include the aerobic protozoa such as *Paramoecium, Bodo, Vorticella, Aspidisca* while those treatment steps requiring an aerobic digestion include an aerobic protozoa such as *Saprodinium, Metopus,* and *Epalsis.*

In the treatment of industrial wastes, where nitrates and phosphates are accumulated in appreciable quantities, the settling tanks are exposed to sunlight to hasten the growth of protozoa and algae. These micro-organisms remove the nitrates and phosphates for their own synthesis. The quality of water is significantly improved, used for fish culture and crop production. At the same time, the autotrophs are skimmed from the water surface, dried, and used as fertilizer.

## 7.5 Natural Waters

The moisture of the earth surface undergoes a continuous circulation. It is a process generally termed as *hydrologic cycle* or *water cycle* (Figure 7.2). It has been estimated that about 24,000 cubic kilometres of water from lakes and land surfaces and 1,28,000 cubic kilometres from oceans evaporate annually. Various types of micro-organisms are present at different stages of this cycle process in surface water, atmospheric water, and ground water. Because the types of aquatic environment are extremely variable it is obvious that different species of micro-organisms are considered to be indigenous to specific habitats.

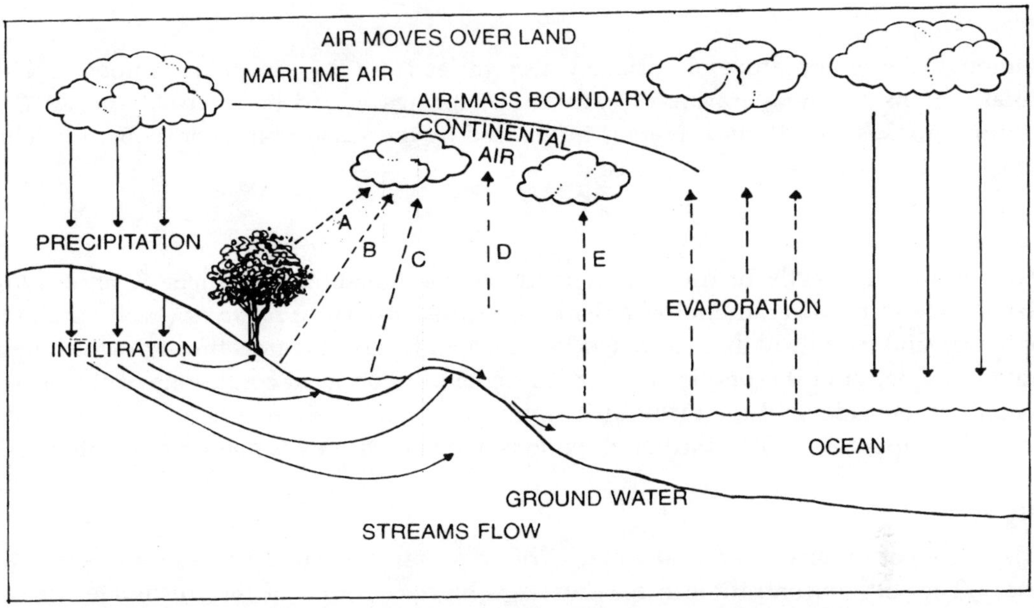


**Fig. 7.2 : Diagramatic representation of water cycle. Water returns to dry continental air through transpiration (A) evaporation from soil, (B) lakes and ponds, (C) streams, (D) Continental air moves over ocean to become moist, (E) with conversion to maritime air with precipitation. [From Holzman, (1982)]**

### *Surface Water*

Rivers, lakes, ponds, oceans etc. represent surface water. These water bodies are, however, susceptible to contamination with micro-organisms from precipitation (atmospheric water), the surface runoff from soil, and from dumped waste materials. In general, microbial populations vary in both type and number with several conditions such as composition of the water in terms of microbial nutrients, biological and climatic conditions, and the source of water.

### *Groundwater*

Groundwater occurs where soil pores or rock-containing materials are saturated. Suspended particles and bacteria are removed by filtration, in varying degrees, depending on the permeability characteristics of the soil. Springs consist of groundwater that reaches the surface through exposed porous soil or rock fissures. Wells are made into the ground to penetrate the groundwater level. If cautions are taken, springs and wells produce water of good quality.

## 7.6 Conditions for Existence of Micro-organisms

The population of micro-organisms in different types of natural water is determined by the chemical and physical conditions in that particular area. Although this generalization applies to all aquatic environments, it is apparent that these conditions vary greatly when lakes, ponds, rivers, oceans etc. are compared. The most common and important conditions which are responsible for development and existence of micro-organisms are briefly described below.

### *Temperature*

Although the temperature of surface water varies from 0°C in polar regions to 42°C in equatorial regions, the temperature of natural hot springs varied between 40 and 80 °C. The temperature in lakes, ponds, and rivers is influenced by the seasons and consequently, there are corresponding shifts in the micro-organisms.

### *Light*

Aquatic life, indirectly or directly depends on the metabolic products of photosynthetic organisms. These photosynthetic organisms are called *primary producers* (such as algae and macrophytes), and their growth is restricted to the upper layers of waters through which light can penetrate. The depth of the photic zone varies depending on local conditions of the ecosystem such as season, latitude, and the turbidity of the water. However, the photosynthetic activity is confined to the upper 40-120 m. Carbon dioxide is available from gas and bicarbonate ions.

### *Salinity*

The degree of salinity in natural water varies between zero in freshwater and saturation in salt lakes. The salt concentration in sea water is high than that of the freshwater ecosystem (Table 7.2) which is remarkably constant. The concentration of salts varies in different regions of water. For examples, it is usually less in shallow offshore regions and near river mouths but in estuary, it is moderate. Most micro-organisms grow best at salt concentration of 2 to 4 per cent whereas those from lakes, rivers, and ponds are salt sensitive and do not grow at a salt concentration of more than 1 per cent.

**Table 7.2 : Composition of some Natural Waters. Values are in Terms of g/litre**

| Water | $Na^+$ | $K^+$ | $Ca^{2+}$ | $Mg^{2+}$ | $Cl^-$ | $SO_4^{2-}$ | $CO_3^{2-}$ |
|---|---|---|---|---|---|---|---|
| Seawater | 10.7 | 0.39 | 0.42 | 1.34 | 19.3 | 2.69 | 0.073 |
| Freshwater | | | | | | | |
| Soft | 0.016 | - | 0.010 | 0.00053 | 0.019 | 0.007 | 0.012 |
| Hard | 0.021 | 0.015 | 0.065 | 0.014 | 0.04 | 0.025 | 0.119 |

*Source :* Baldwin (1948).

### *Turbidity*

There is a marked variation in the turbidity of surface waters. The suspended material responsible for the turbidity includes : (1) detritus, predominantly particulate organic material, (2) particles of mineral material which originate from land, and (3) suspended micro-organisms. Turbidity of the water influences the penetration of light, which in turn affects the photosynthetic zone. Particulate matter also serves as a substrate to which micro-organisms adhere or as substrates that are metabolized.

### *Inorganic and Organic Materials*

A variety of inorganic and organic substances present in the aquatic ecosystem are important in determining the microbial flora. Phosphates and nitrates are important organic substances for the growth of algae. Organic compounds are required for the growth of saprophytic bacteria and fungi. Waters that receive domestic wastewater, are subject to intermittent variations in their nutrient load. Industrial wastes may contribute antimicrobial substances to estuaries and coastal waters. Certain heavy metals in small concentrations may inhibit growth of some micro-organisms while simultaneously permitting the growth of resistant forms.

## 7.7 Freshwater Micro-organisms

The micro-organisms of freshwaters constitute a part of the science of limnology which deals with the study of the flora, fauna and conditions for life in lakes, ponds, and rivers.

### *Ponds and Lakes*

Ponds and lakes have a characteristic zonation and stratification (Figure 7.3). Along the shore, there is a large *littoral zone* – enriched with rooted vegetation and includes regions where light penetrates to the bottom. In open areas, the *limnetic zone* is determined by the light-compensation level (depth of effective light penetration). Photosynthetic activity decreases progressively in the deeper regions of the open water – called the *profundal zone*. The *benthic zone* is composed of mud

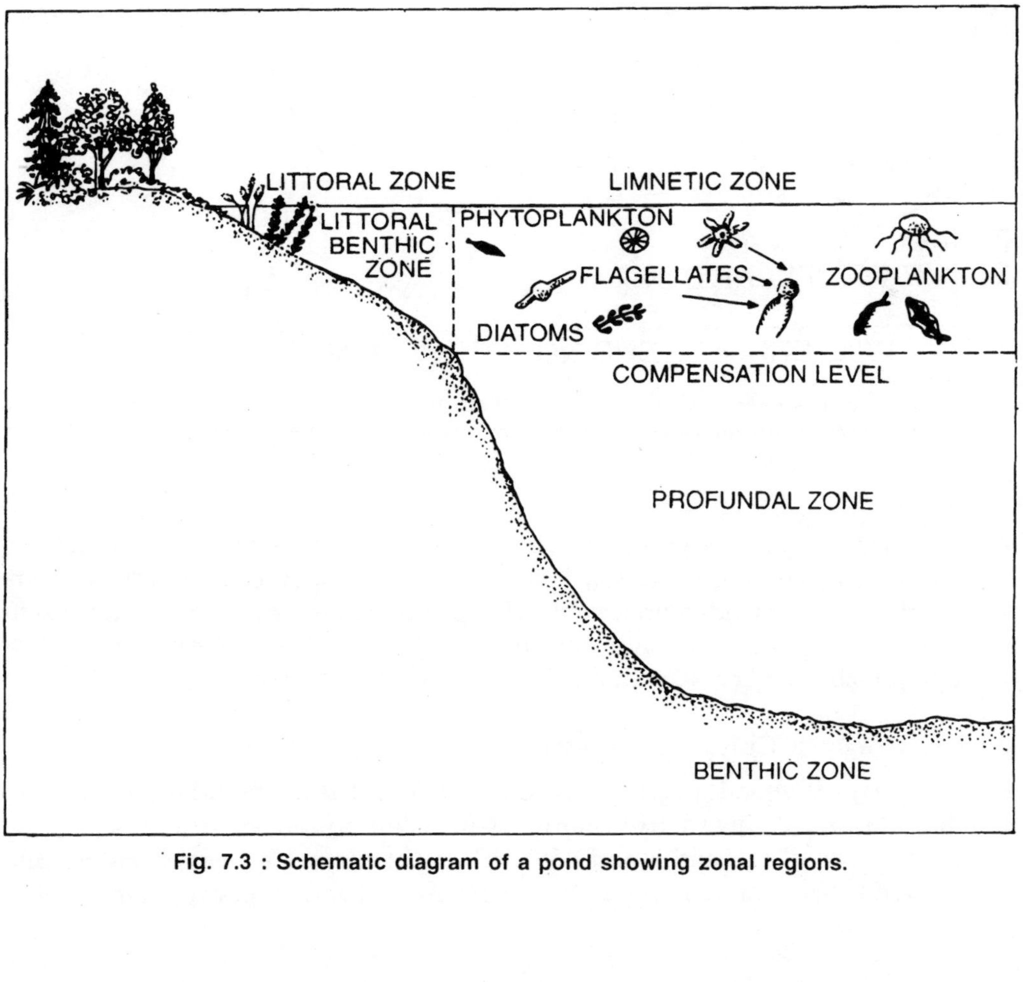


**Fig. 7.3 : Schematic diagram of a pond showing zonal regions.**

at the bottom. This zone is highly populated by heterotrophic organisms. When the benthic zone is enriched with organic materials, the majority of micro-organisms will be an aerobic decomposers. In the littoral and limnetic zones, greatest variety of physiological types are found and consequently, these zones are the most productive one. Of course, productivity is affected by the nature of imported substances from rivers and streams and the chemical nature of the basin. Ponds and lakes of the temperate region, due to stratification of the water as a result of temperature differences, exhibit seasonal fluctuations in their microbial populations. Such stratification acts as a barrier to oxygen and nutrient exchange. In the summer season, the top layers of water become warmer than the lower regions; but in the winter season, the top layers become too cold and as a result, a reversal of temperature and mixing occurs, often resulting in massive growth of algal bloom (Figure 7.4). Ponds and lakes enriched with nitrogen and phosphorus, are likely to support excessive algal bloom.

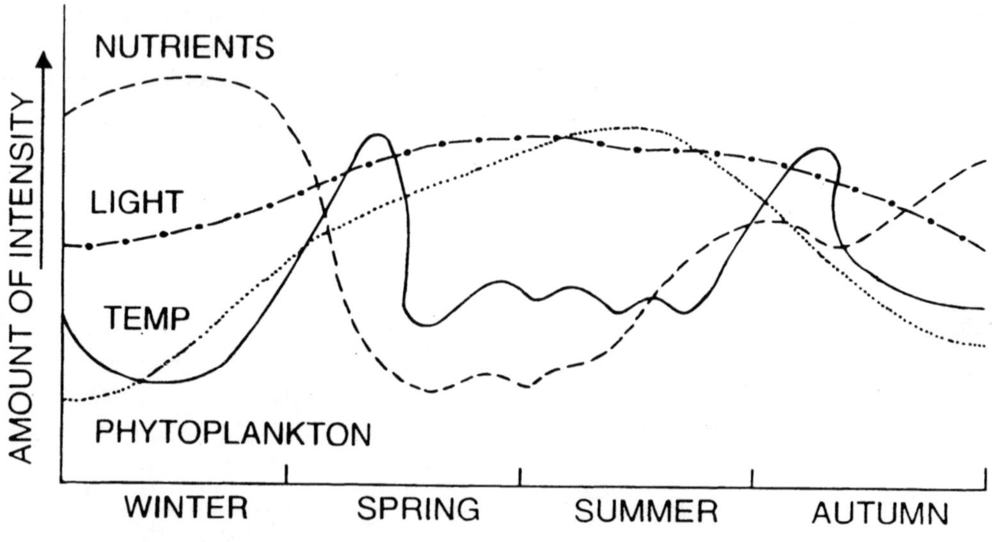


**Fig. 7.4 : Development of phytoplankton "pulses" in northern temperate lakes in springs and autumn. Mixture of nutrients level, temperature and light encourages this phenomenon. [From Odum (1971)]**

*Streams and Rivers*

Streams and rivers receive large amounts of nutrients from the flow of organic and inorganic materials from the surrounding terrestrial system. The micro-organisms reflect the immediate terrestrial conditions. The drastic environmental changes in the streams and rivers due to effects of agricultural and industrial practices on the one hand and the changes in farming practices on the other make it impossible to generalize upon characteristic micro-organisms.

### 7.8 Productivity of Fish Culture Ecosystem

Since most algae are aquatic organisms, it is probable that as much carbon is fixed (captured as carbon dioxide and changed to organic carbon compound such as sugars) through photosynthesis by algae. Tiny benthic and floating algae of freshwater ecosystems constitute the phytoplankton and serve as an important food source for fish and other organisms. These algae form the beginning of most aquatic food chains because of their photosynthetic activities and are therefore called *primary producers* of organic matter. The biological activity of a fish culture ecosystem is undoubtedly dependent upon the rate of primary production. In shallow freshwater pond and lake ecosystems, the role of photosynthetic organisms is considerably reduced.

### *Food Web and Food Chain*

Food chains are a way of expressing the feeding relationships between organisms in an ecosystem. They describe how energy is locked into food by plants and other autotrophic organisms (organisms are those which manufacture complex organic molecules, such as carbohydrates and proteins, from simple inorganic substances including water, carbon dioxide and mineral salts) and passes from organism to organism. The most simple food chain, if such a thing existed, would consist of just three organisms : a plant, a herbivore, and a carnivore. The position of each occupies in the food chain is called a *trophic level* and classifies them into producers (such as algae) and consumers (such as fish). The herbivore, which eats the plant, is called the *primary consumer;* the carnivore, which eats the herbivore, is called the *secondary consumer.* The larvae of many kinds of animals and zooplankton (such as water flea and other crustacea) feed on microscopic plants (such as algae and diatoms). These small animals are consumed by surface-feeding fish, the fish, in turn, forms part of the diet of man (Figure 7.5).

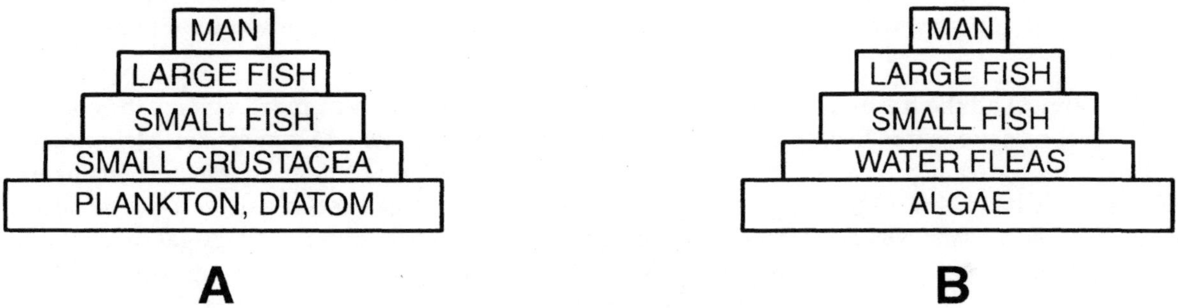


**Fig. 7.5 : Examples of food chains**

It is important to realize that links in a food chain do not all involve similar numbers of individuals. The organisms at the first stage of a chain are usually small and very numerous; for example, in the food chain illustrated in Figure 7.5A, the diatoms in the phytoplankton are microscopic and very abundant. The crustaceans which feed on the phytoplankton, somewhat larger than diatoms, and each of these small animals consumes great number of diatoms during its lifetime. The fish which consume are larger still, and far less numerous. Finally large numbers of fish are needed to feed one man.

When the population of one of the animals or plants is altered, all the other species in the chain are affected. For example, if all the fish could be removed from a pond, the number of fish in the pond would rise as the food requirements of the fish population increased, the number of water fleas would fall.

## 7.9 Role of Micro-organisms in the Productivity of Fish Culture Ecosystem

Fish and other aquatic animals and plants exhibit a vast complex interactions among micro-organisms. Micro-organisms, especially protozoa and algae, occupy a key role in the food chain of the aquatic ecosystem. Bacterial populations perform a variety of biochemical changes in various substrates that allow the recycling of elements and nutrients – the most important fish production potential.

The metabolic activities of micro-organisms solubilize phosphate from insoluble calcium, iron, and aluminium phosphates. Phosphates are also released from organic compounds by microbial degradation. Bacteria are able to change insoluble oxides of iron and manganese to soluble maganous and ferrous salts. These examples of biogeochemical changes that take place in an aquatic ecosystem, clearly indicate that micro-organisms do perform numerous and essential functions that contribute to the productivity of ecosystem.

## 7.10 Detrimental Impact of Micro-organisms in Fish Culture Ecosystem

Apart from the non-pathogenic micro-organisms which are beneficial to aquatic ecosystem and are responsible to establish food chains, there are pathogenic micro-organisms present in the aquatic environment and create problems to fishery industries. Some of them many cause diseases in fish and hence production losses. Presence of some human pathogens in fish ponds such as *Vibriocholerae* spp., *Salmonella* spp., *Listeria* sp. etc. lead to rejection and consequent post-harvest losses.

In almost all the freshwater ponds and lakes in tropical and sub-tropical regions, the color of water becomes green due to excessive development of certain species of Cyanobacteria (commonly called as *blue-green algae*) caused by surface run-off waters laden with detergents and fertilizers. However, it has been established that species of blue-green algae (such as *Anabaena flosaquae, Microcystis aeroginosa, Nodularia spumigena* etc.) generate some toxic substances. Algal cells release toxins into the water only when they die or when the algal bloom is damaged using copper sulfate as chemical. The most important toxins produced by toxic species of blue-green algae are neuro toxins (Anatoxin-A), hepatotoxins, and cytotoxins. The first type of toxin interferes with the functioning of the nervous system of fishes followed by fatigue and paralysis. Hepatotoxins cause destruction of liver cells in fishes and are termed as *hepatocytes.* However, the primary consumers of an aquatic ecosystem do not consume toxic algae; but in some situations, they are compelled to consume these algae as food. Consequently, the concentration of primary consumers drastically reduced and may leads to their complete disappearance from the water.

## 7.11 Micro-organisms in Relation to Soil Aggregation

Soil particles held in a single mass by physical and biological processes are called *aggregation* that has great practical significance. Any action that will emphasis contracts between soil particles, will encourage aggregation. Mechanical binding action of soil micro-organisms, particularly the algae, bacteria, the thread-like filaments of fungi, and other soil organisms also stimulate aggregate formation. These effects are, however, more pronounced within 1-2 weeks after the application of organic matter to pond soils. Therefore, micro-organisms are important in improving the physical properties of soil by aggregating particles adding organic matter.

## 7.12 Transformation of Elements

A major biochemical activity of micro-organisms is the dissimilation and mineralization of organic matter. In the process of dissimilation, organic compounds are converted to water, carbon dioxide, and various inorganic salts. Organic substrates are originated from metabolic processes of the plant and animal life. Under aerobic conditions the main products resulting from dissimilation of organic compounds are ammonia, carbon dioxide, sulfate, and phosphorate. These are considered as nutrients for plant growth, including the phytoplankton. Anaerobic degradation yields methane, hydrogen, and hydrogen sulphide in addition to ammonia, carbon dioxide, and phosphate.

### *Biogeochemical Cycles*

The cycling of chemical constituents such a as organic carbon, nitrogen, sulfur, phosphorus, and other compounds in an aquatic environment through a biological system is referred to as *biogeochemical cycle.* Micro-organisms play an important role for the conversion of complex organic compounds into simple inorganic compounds or elements and reutilization occurs in the aquatic ecosystem. The most important cyclic processes occur in a fish pond ecosystem are : nitrogen cycle, phosphorus cycle, and carbon cycle.

A number of microbial transformations of nutrient elements involved in these cyclic processes have been described in Sections 6.3, 16.2 and 17.3. In each case, micro-organisms play an important role in releasing the element (such as nitrogen) from an organic compound through a series of reactions mediated by various species of mico-organisms, especially bacteria. The element in the form of an inorganic compound, becomes available again as a plant nutrient.

## 7.13 Micro-organisms as Food for Fish

In contrast to the diversity of species consumed by humans, blue-green algal food for fish has centered around six genera such as *Spirulina, Chlorella, Scenedesmus, Cladophora, Oscillatoria,* and Diatoms (Figures 7.6 and 7.7). These species are marketed in the form of biscuits, capsules, and lozenges. However, these should be used cautiously as their excessive use may cause stone formation in kidney and gout in man.

Microbiologists have created a biological engineered micro-organism called *Cyclop-eeze* which they claim is more superior nutritionally than any other live feed organisms for use in larval and grow-out culture of fish, prawn, and shrimp. Biologically altered micro-organisms have the capacity to produce high levels of highly unsaturated fatty acids which are known to be essential to the growth and development of high-priced fish. This is cultured in hypersaline lakes near the Arctic circle. It is available in deep frozed and freeze dried form.

### *Single Cell Protein*

The term *single cell protein* (SCP) refers to dried cells of micro-organisms grown in large quantities for use as animal feed and protein supplements. These micro-organisms include some species of algae, yeast, and fungi. The SCP is characterized by high percentage of nutrient ingredients. For example, the nutrient composition of *Spirulina* powder as suggested by Michaelson Laboratory, California, U.S.A., are shown in Table 7.3. Therefore, it points to an obvious conclusion

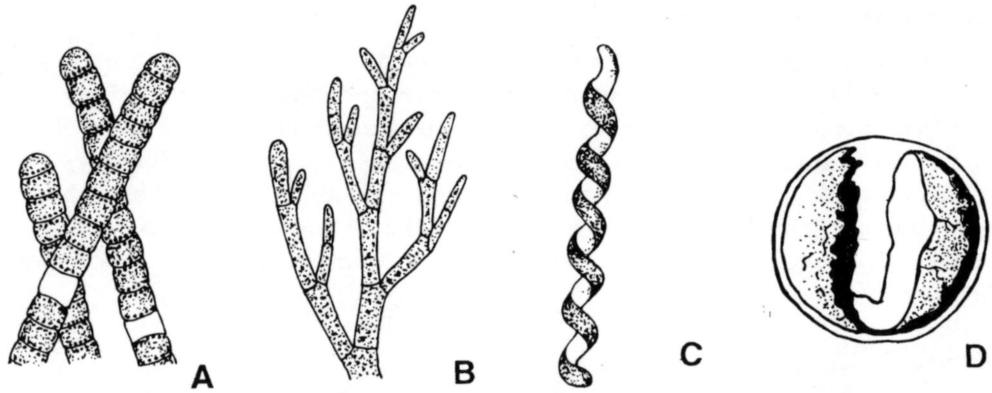


Fig. 7.6 : Though phytoplankton is comprised of algae which are uniquely adapted in the freshwater and marine environments, many species of algae are used as food. They occur in myriads of shapes, sizes, and arrangements. The drastic increase in their numbers due to rapid growth result in the appearance of dense algal blooms or carpets. There is increasing interest in the use of some smaller forms of algae as food for fish, domestic animals, and humans. *(A) Oscillatoria, (B) Cladophora, (C) Spirulina, (D) Chlorella.*

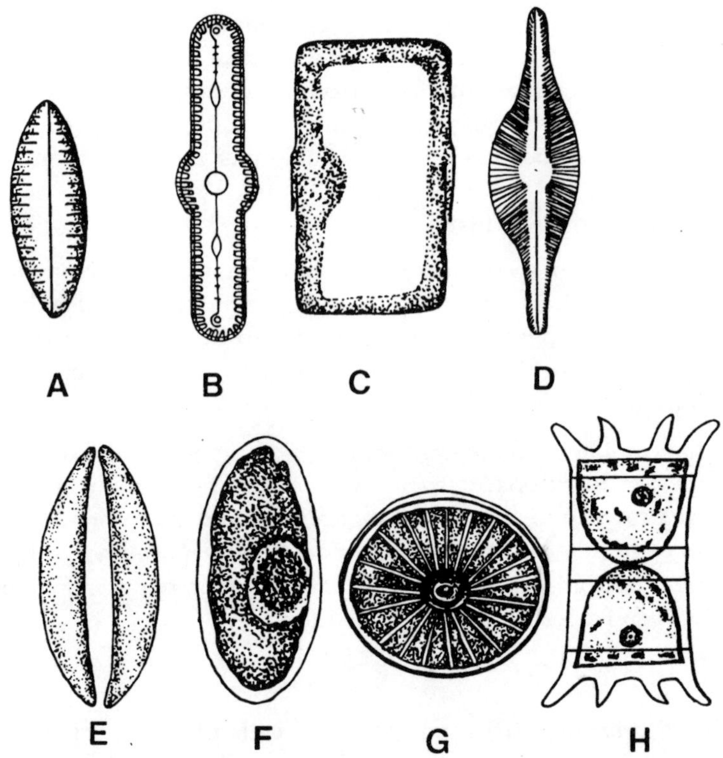


Fig. 7.7 : Diatoms are unicellular, colonial or filamentous algae occurs in a variety of shapes found abundantly in fresh and sea waters. They are abundant in cold waters. Their hard silica-containing walls consist of two halves which fit together, many with beautiful surface designs. The thousands of species of diatoms provide an ever present and abundant food supply for aquatic animals. A few species of diatoms are shown in this figure. (A) *Gomphonema*, (B) *Pinnularia*, (C) *Melosira*, (D) *Navicula*, (E) *Cymbella*, (F) *Cocconies*, (G) *Coscinodiscus*, (H) *Biddulphia*.

**Table 7.3 : Nutrient Composition of *Spirulina fusiformis* Powder as Analyzed by Michaelson Laboratory, California, USA**

| | Constituent | Composition (per 100g of powder) |
|---|---|---|
| I. | **Major constituents (percentage)** | |
| | Total protein | 64.6 |
| | Carbohydrate | 16.1 |
| | Fat | 6.7 |
| | Fiber content | 9.3 |
| | Bile pigments and chlorophyll | 6.0 |
| | Caloric value | 3,46,000 |
| | Ash | 3.0 |
| II | **Vitamins** | |
| | Beta-carotene (IU) | 3,20,000 |
| | Botin (mg) | 0.22 |
| | Cyanocobalamin (mg) | 65.76 |
| | Folic acid (mg) | 17.6 |
| | Riboflavin (mg) | 1.78 |
| | Thiamin (mg) | 0.118 |
| | Tocopherol (IU) | 0.773 |
| III | **Minerals (mg)** | |
| | Calcium | 6.58 |
| | Phosphorus | 977.00 |
| | Iron | 44.7 |
| | Sodium | 796.00 |
| | Potassium | 1.28 |
| IV | **Essential amino acids (percentage)** | |
| | Lysin | 2.99 |
| | Cystine | 0.474 |
| | Methionine | 1.38 |
| | Phenylalanine | 2.87 |
| | Therionine | 3.04 |

*Source* : Chaturvedi and Habib (1999)

that the extracts of micro-organisms could be valuable alternative to some traditional sources such as soyabean fish meal. Micro-organisms which are considered as sources of essential nutrients for fish, must possess certain characteristic features such as non-pathogenicity to fish and other animals, good nutritional value that are acceptable as food, low production cost, and absence of toxic ingredients.

## *Cultivation of Algae*

Mass cultivation of algae has been undertaken by several countries of the world such as Germany, Mexico, USA, Canada, Japan, and India. Methods for cultivation of these plants, using waste products and sewage for their nutrition have ben developed. However, for the production of algae, rectangular cemented tanks are generally used. Each tank contains about 100 litres of water and the ratio of the inoculum and the medium should be around 1 : 4. The temperature of the culture medium should be maintained between 25 and 30°C and that of pH between 8 and 10. For better production of algae, the culture medium should be stirred regularly. Harvesting can be started from the 10th day when large floating filaments of the algae are seen as mats over the surface water. The algae are collected and washed in freshwater to remove debris, then chopped and spread on frames to dry into thin sheets. The yield of algae is highly significant. The yield of *Spirulina* powder, for example, is about 20 tonnes/hectare/year.

After algae have been grown on waste products, the residues can be disposed of in streams, lakes, and without causing pollution. Although it can be stated in favour of using algae in place of higher plants for human food, acceptance of the use of algae in some countries should not be expected until food from higher plants is in short supply. At the same time algae will obviously find wide application as feed supplements or animal feeds.

A variety of micro-algae are now being used as for finfish and shellfish because of the presence of growth-promoting factors. They are cultured either in indoor hatcheries or laboratory conditions. The growth of micro-algae largely influenced by salinity, rain, light intensity, temperature, and certain nutrients such as silica, sulfur, nitrogen, and phosphorus. A relatively few species are able to grow profitably in mass culture. Micro-algae provide essential nutrient for fish growth and development. Different types of micro-algae have variable nutritional compositions and the types of nutrients required for different species of farmed fish also differ. Hence the most suitable genera of micro-algae should be selected for fish nutrition and health management. This is a subject of another chapter and has been described in Volume 2.

### 7.14 Micro-organisms as Indices of Water Quality

Interactions between micro-organisms and different factors in an aquatic ecosystem is of considerable importance in assessing the ecological impact of environmental alterations. The species composition in a particular ecosystem often reflects the environmental conditions. Species that are sensitive, may be eliminated when environmental conditions become adverse. Such fluctuations in the species composition persist for sometime that helps in monitoring the adverse environmental conditions which makes several groups of aquatic organisms more reliable than the chemical indicators.

It is known that some species of bacteria and algae that gain entrance into bodies of water arrive there via intestinal discharges of humans and other animals. Bacterial species, designated as *coliforms,* faecal streptococci, and *clostridium perfrigens,* are normal inhabitants of the large intestine of humans and other animals and are consequently present in faeces. Therefore, the presence of any of these bacterial species in water is evidence of faecal pollution.

### Micro-organisms Other Than Coliform Bacteria

1. *Faecal Streptococci* : Besides coliform bacteria, some micro-organisms could also be used as indicators of faecal contamination of water such as faecal streptococci (*Streptococcus faecalis, S. bovis, S. equinus, S. faecium* etc.) and *Bacillus coli.* Some other micro-organisms are regarded as *nuisance organisms,* because they create problems of color, odor, and taste of water. Since these bacteria are primarily responsible for causing diseases, their presence in any water body clearly indicate that the water is not fit for fish culture.

2. *Algae* : When nutrient-enriched pond water is exposed to sunlight, a variety of algal species (such as *Spirogyra, Oscillatoria, Scenedesmus, Euglena, Navicula, Chlorella, Chlamydomonus, Nitzschia* etc.) are developed. Algae are developed in all natural aquatic environments. Their nuisance characteristics involve odor, discoloration, and taste in water as well as production of turbidity. Therefore, these micro-organisms are, however, very important in determining the degree of water quality.

Considerable research on this aspect is under way for determination of the indices of water quality. At the same time more attention is being given to the assessment of water quality for fish culture.

## 7.15 Regeneration Capacity of Aquatic Environment

Waters from lakes, ponds, rivers, oceans, etc. evaporate to form cloud which after condensation, pure water is formed and it is dropped in the form of rain. Rain waters before reaching the earth's surface, acquire a variety of impurities. This water then travels through rocks, vegetation, soils etc. and consequently, more impurities are added to the running waters that ultimately form natural sources of surface waters. These natural waters are generally fit for myriad purposes. When these waters pass through various regions, more impurities are added to this due to human activities and as a result, the water becomes unsuitable for fishery activities.

Micro-organisms have the capacity to regenerate the natural water. This process involves the activities of protozoa, algae, and bacteria and transform the deleterious substances into innocuous products.

Algae and cyanobacteria are photoautotrophs. They use light as their source of energy and carbon dioxide as their source of carbon. Thus, these micro-organisms remove carbon dioxide through photosynthesis and generate oxygen. The bacteria absorb and digest organic substances from solution and can thrive well in most of the polluted waters. Protozoa consume algae and bacteria as food. Microscopic invertebrates (such as Crustacea) also ingest algae as food and ultimately, a food chain principle is followed. Several bacteria such as manganese bacteria, sulfur bacteria, iron bacteria, nitrifying bacteria, and denitrifying bacteria have been found to remove manganese, sulfur, iron, and nitrogen respectively from the polluted waters most efficiently.

## 7.16 Conclusion

Fish culture ecosystems comprise the conceal habitat of micro-organisms and they are the building blocks of pond productivity, occupying and being essential to fish production. In some situations, micro-organisms cause fish diseases making them disagreeable for human consumption.

Micro-organisms and their activities are of great importance in many ways such as (a) they occupy a key position in the food chain by providing rich nourishment for the next higher level of aquatic life, (b) they are instrumental in the chain of biochemical reactions which accomplish recycling of elements, and (c) they may affect the health of fish.

Micro-organisms affect fish and other aquatic animals in a great many ways. They occur in large numbers in almost all aquatic environments and bring about many changes, some undesirable and others desirable. The diversity of their activities ranges from causing diseases in fish to the enhancement of productivity of fish culture ecosystem.

Some species of algae are effectively utilized as food for humans and animal feed. At the same time micro-organisms are too important to improve the physical properties of soil by aggregating soil particles and adding organic matter.

Urbanization and consequently the growing demand for water by human populations, the importance of aquatic ecosystem as a major food source and other developmental activities have resulted in the establishment of several Government and Non-Government agencies which exercise jurisdiction over many aspects of aquatic ecosystem.

### *References*


Baldwin, E. 1948. *An Introduction to Comparative Biochemistry.* Cambridge University Press, London.

Chaturvedi, U.K. and I. Habib. 1999. Cyanobacteria as a source of food. *Proc. Acad. Environ. Biol.,* 8 : 241-245.

Holzman, B. 1982. *Encyclopedia of Science and Technology.* McGraw-Hill Co., New York.

Odum, E.P. 1971. *Fundamentals of Ecology.* Saunders Publishers, Philadelphia.

Strickland, R.M. 1983. *The Fertile Fjord,. Plankton in Puget Sound,* Puget Sound Books, University of Washington Press, Seattle.


## Questions

1. Why micro-organisms are important for study in relation to fish culture ecosystem?
2. State the characteristic feature of some groups of micro-organisms.
3. What is plankton, zooplankton, and phytoplankton? Name some organisms that belong to each category.
4. Discuss some of the properties of the freshwater fish culture ecosystem in terms of their role on microbial growth.
5. State the importance of micro-organisms in an aquatic ecosystem.
6. What is meant by the term upwelling? What effect does upwelling have on productivity of a deep aquatic ecosystem?
7. What is referred to by the term primary producers?
8. What contributes to the fertility of a freshwater ecosystem?
9. How productivity of a fish culture ecosystem is established?
10. What role does micro-organisms play in creating problems to fish culture?
11. How micro-organisms help to improve the physical properties of soil?
12. State the activity of micro-organisms on the process of transformation of elements.
13. Name some important micro-organisms used as food by fishes. Why these micro-organisms are considered as sources of essential nutrients for fish? What is single-cell protein?
14. How algae are cultivated for fish food production?
15. How micro-organisms are related to regeneration of aquatic ecosystem?
16. How micro-organisms are considered as indices of water quality?

# 8

# Natural Food in Fish Ponds

Different types of natural food forms a definite relationship among themselves in a pond ecosystem. This relationship describes how energy and nutrients pass from organism to organism. Energy is locked into food by plants (mainly phytoplankton) and then transferred to primary consumers (mainly zooplankton and bottom fauna) and ultimately consumed by fish (secondary consumers). Thus, a food chain or web is formed within the aquatic ecosystem. Food web describes situations closer to reality, where each organism may feed at several different trophic levels and produce a web of feeding interactions. Food chain begins with a single producer organism, but a food web may involve several producers.

Progressive fish farmers use a number of small ponds for cultivation of natural food for fry and fingerlings of fishes. They use natural food organisms rather than artificial ones particularly for young fish. Therefore, question may arise how different types of natural food are developed in fish culture environment. The answer is that natural food should be cultivated in ponds using different scientific techniques.

## 8.1 Classification of Natural Food

Natural food organisms which are used by carps, are comprised of phytoplankton, zooplankton, diatoms, some Protozoa, and Annelids which are well adapted to freshwater environments (Figure 8.1). In general, however, natural fish food organisms are classified into two types such as (1) plankton and (2) bottom fauna. The former types are present in water as suspended condition and the latter types remain on the bottom mud. Plankton is defined as free-floating microscopic or sub-microscopic plant and animal and possess feeble locomotory power. Neustons are organisms which are related to surface water either by hanging upon the surface flim (Gyronidae, Gerridae etc.) or by hanging against the lower side (insect larvae, protozoa, algae etc.). True plankters belong to the euplankton and are classified as : (1) macro-plankton (> 3mm in size such as mysids, euphausids etc.), (2) micro-plankton (< 3mm in size) and (3) nano-plankton (> 60μm in diameter).

On the basis of the quality, plankters can be divided into two types. (1) *phytoplankton* and (2) *zooplankton*. Phytoplanktons are chlorophyl-containing organisms while zooplanktons are plankters of animal origin. Furthermore, on the basis of site of occurrence, planktons are of various types such as : (a) *lake plankton* (or limnoplankton), (b) *running water plankton* (or heoplankton), (c) *pond plankton* (or heleoplankton), (d) *salt-water plankton* (or helioplankton), and (e) brackish-water plankton (or hypalmyroplankton).

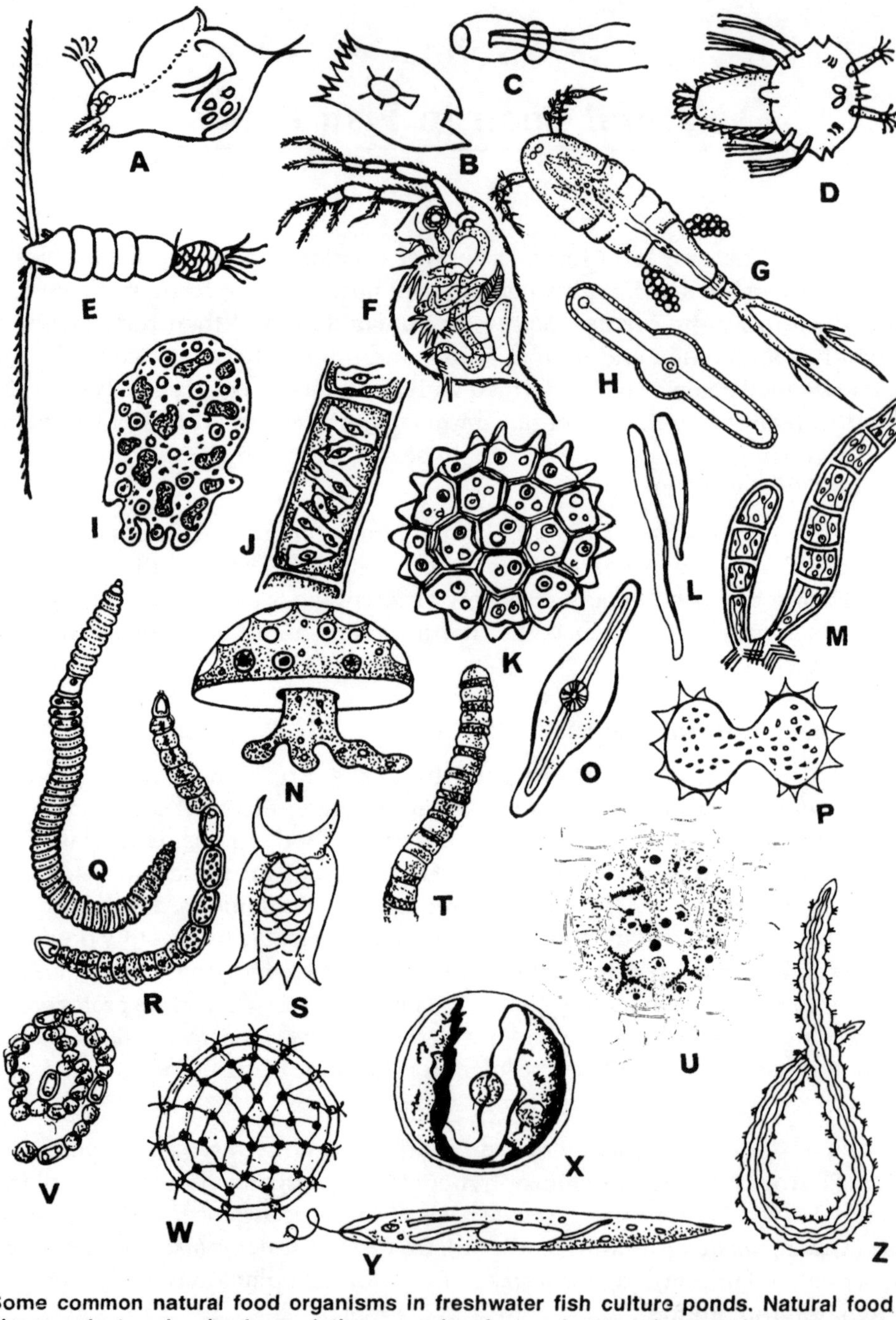


Fig. 8.1 : Some common natural food organisms in freshwater fish culture ponds. Natural food is extremely diverse plant and animal populations ranging from microscopic organisms to multicellular metazoans. (A) *Moina*, (B) *Keratella*, (C) *Filinia*, (D) *Nauplii*, (E) *Diaptomus*, (F) *Daphnia*, (G) *Cyclops*, (H) *Pinnularia*, (I) *Difflugia*, (J) *Spirogyra*, (K) *Pediastrum*, (L) *Synedra*, (M) *Ulothrix*, (N) *Arcella*, (O) *Navicula*, (P) *Staurastrum*, (Q) *Pheretima*, (R) *Anabaena*, (S) *Brachionus*, (T) *Oscillatoria*, (U) *Pandorina*, (V) *Nostoc*, (W) *Volvox*, (X) *Chlorella*, (Y) *Euglena*, (Z) *Tubifex*.

## 8.2 Plankton in a Fish Pond

A large number of plankton populations occur in any pond and lake ecosystem which determines their productivity – an obvious link with fish growth and production. This biological diversity of pond ecosystem is very important so far as fisheries management strategy is considered. Planktons are unpredictable in a pond and the data available at present are not adequate. Some important planktons which are used as food by fishes are shown in Table 8.1.

**Table 8.1 : Some Common Freshwater Planktonic Organisms Consumed as Food by Fishes**

| Group | Species | Range per litre of water | | |
|---|---|---|---|---|
| | | Abundant (200 and above) | Frequent (101-199) | Rare (1-100) |
| **Phytoplankton** | | | | |
| 1. Chlorophyceae : | *Chlorella vulgaris* | 235-260 | 120-160 | 30-50 |
| | *Chlosteridium lunula* | 245-270 | 137-170 | 20-45 |
| | *Clasterium setaceum* | - | 120-150 | 10-30 |
| | *C. acerasum* | - | - | 25-40 |
| | *Coelestrum spharicum* | - | 135-160 | 30-55 |
| | *C. microporum* | - | 115-160 | 10-30 |
| | *Desmidium swartzii* | 230-280 | 140-170 | 20-35 |
| | *Pandorina morum* | 220-245 | 120-145 | 10-40 |
| | *Pediastrum boryanum* | 230-250 | 135-160 | 15-40 |
| | *P. simplex* | - | 140-170 | 25-40 |
| | *P. tetrus* | - | 120-150 | 20-45 |
| | *Scenedesmus obliquas* | 220-260 | 130-160 | 15-40 |
| | *S. abundanus* | - | 110-140 | 10-30 |
| | *S. falcatus* | - | 130-160 | 30-50 |
| | *Selenestrum westi* | - | 135-145 | 25-50 |
| | *S. gracile* | - | – | 10-30 |
| | *Tetradesmus cumbricus* | - | – | 10-45 |
| | *Cosmarium granatum* | - | – | 20-35 |
| | *Volvox globator* | - | 120-150 | 20-40 |
| | *V. areus* | - | 135-160 | 5-25 |
| | *Actinastrum sp.* | - | 125-145 | 5-15 |
| 2. Cyanophyceae : | *Anabaena affinis* | 210-260 | 115-160 | 15-50 |
| | *A. spirodes* | 205-245 | 130-165 | 10-30 |
| | *Aphanocapsa pluchrum* | - | 120-140 | 5-30 |
| | *Calothrix fusca* | - | - | 10-30 |
| | *Gleocapsa rupestris* | - | - | 5-30 |
| | *Nostoc spharium* | - | 120-140 | 15-50 |
| | *Oscillatoria limosa* | 240-350 | 130-190 | 20-50 |
| | *O. tenuis* | 220-300 | 120-150 | - |
| | *O. lacustris* | 200-260 | 101-140 | - |
| | *Tetrapedia* sp. | - | 120-140 | 10-30 |
| | *Spirulina gigantaea* | 230-330 | 130-170 | 10-25 |
| | *S. major* | 220-300 | 140-180 | 10-20 |
| | *Microsystis aeroginosa* | 230-400 | 130-190 | 30-50 |

*contd...*

**Table 8.1** ***contd...***

| Group | Species | Range per litre of water | | |
|---|---|---|---|---|
| | | Abundant (200 and above) | Frequent (101-199) | Rare (1-100) |
| 3. Bacillariophyceae | *Synedra* sp. | - | - | 20-40 |
| | *Stephanodiscus* sp. | - | 120-170 | 10-30 |
| | *Nitzshia* sp. | - | 130-185 | 10-20 |
| | *Fragillaria* sp. | - | 130-190 | 20-40 |
| | *Cyclotella* sp. | 220-290 | 120-175 | 10-20 |
| | *Navicula* spp. | 220-400 | 120-180 | 20-40 |
| 4. Euglenophyceae | *Euglena* spp. | 220-350 | 115-180 | 5-20 |
| | *Phacus* spp. | 200-370 | 120-170 | 5-30 |
| **Zooplankton** | | | | |
| 1. Rotifera | *Brachionus calyciflorus* | 220-400 | 120-180 | 10-30 |
| | *B. rubens* | 230-340 | 130-170 | 15-30 |
| | *B. florficula* | 210-300 | 110-160 | 5-30 |
| | *B. bidentata* | 200-300 | 110-140 | 10-25 |
| | *Keratella tropica* | 210-390 | 120-170 | 10-40 |
| | *K. valga* | 220-350 | 110-140 | 10-25 |
| | *Filinia longiseta* | 210-400 | 120-170 | 5-25 |
| | *Rotatoria* sp. | 210-450 | 110-150 | 10-30 |
| | *Asplanchna periodonta* | 210-300 | 120-170 | 10-30 |
| | *Trichocera longiseta* | 230-390 | 120-160 | 5-20 |
| 2. Cladocera | *Daphnia magna* | 220-500 | 110-175 | 40-90 |
| | *D. carinata* | 200-450 | 115-170 | 30-60 |
| | *Moina micrura* | 200-470 | 120-170 | 25-70 |
| | *Cariodaphnia* spp. | 210-350 | 110-150 | 10-30 |
| | *Bosmina longirostris* | 200-290 | 115-150 | - |
| 3. Copepoda | *Tropocyclops* sp. | - | 110-160 | 5-25 |
| | *Mesocyclops sp.* | - | 120-150 | 5-15 |
| | *Diaptomus* sp. | - | - | 30-50 |
| | *Cyclops viridis* | 200-500 | 120-180 | 15-30 |

### *Fluctuations of Plankton in Ponds**

In any fish culture ponds, there occur qualitative and quantitative variations in plankton populations. Some species of plankton reappear at specific periods and disappear during others. The concentration of phytoplankton is always greater than that of zooplankton. Moreover, the concentration of plankton in different months and seasons also varied greatly. Fluctuations of

* For further study, see Sarkar and Basu Chowdhury (1999).

plankton depend upon the geographical and climatic conditions of a particular ecosystem and different abiotic factors of water such as pH, dissolved oxygen, total alkalinity, hardness, temperature, ammonia and nitrate nitrogen and therefore, it is not possible to make a scientifically justifiable generalization on various aspects of pond ecology.

## 8.3 Culture of Natural Food

Addition of inorganic and organic fertilizers such as urea, ammonium sulfate, lime, oil cakes and cattle dung in freshwater ponds increase the nutrient levels which supports excessive growth and development of algal blooms and macrophytes. Nitrogen and phosphorus are the key for the occurrence of algal blooms in ponds while calcium increases the absorption of phosphorus by phytoplankton. Most species of phytoplankton such as *Spirulina nordestedtii, Trachelomonas vaolvocina, T. charkoweinsis, Phromidium fragile,* and *Microsystis aeruginosa* are found in abundance in freshwater bodies that have tendency of either light scale depostion of calcium carbonate or have a property of corrosion.

It is needless to mention that fertilization of ponds accelerated the growth and development of phytoplankton upon which zooplankton is dependent. Zooplankton concentration is developed when pond is fertilized with organic materials particularly cowdung and poultry litter. In fish culture ponds, *Daphnia* sp., *Moina* sp., *Cyclops* sp., *Diaptomus* sp., *Brachionus* sp., *Filinia* sp., *Asplanchna* sp., *Mesocyclops* sp. etc. of the groups Rotifera, Cladocera and Copepoda are developed. Therefore, if plankton populations are collected from water bodies and introduced into the culture ponds, the ponds will exhibit plankton swarms. Zooplankton feeds on the small green algae and diatoms and also on dead animal and decaying vegetable matter. Generally light and temperature hasten the life processes of various species of planktons. Summer, monsoon and winter makes little difference in their concentration, except that the greatest abundance is in summer. Highest growth of phytoplankton was recorded during October-December and March-May.

Pond fertilization with inorganic and organic materials is initiated the development of first trophic level (such as phytoplankton) and as a result greenish color of the water first appear. As soon as zooplankton populations are inoculated or developed, the greenish color gradually disappears due to consumption of plant nutrients by zooplankon. This cycle should always be maintained during fish culture operation for sustaining yield of fish biomass.

### *Mass Culture of Moina and Chaetocaros*

The natural live food organisms supply minerals, micronutrients, proteins, fats, and carbohydrates resulting in healthy growth of fish fry and therefore, the culture of live food species is necessary. However, mass culture of *Moina*, sp. and *Chaetocaros* sp. can be done using organic and inorganic fertilizers. At Central Institute of Fisheries Education, Mumbai, these species are cultured using slurry prepared with groundnut cake, single superphosphate and poultry litter or cowdung in the ratio of 2:1:4, respectively. Generally 10 kg of groundnut cake, 5 kg of single superphosphate and 20 kg of chicken litter is mixed with 500 litres of freshwater. This mixture is called *slurry* and is kept under continuous stirring to remove harmful gases developed during the process of degradation of organic manures. The slurry is added at the rate of 4 ml per litre for the first three days. Then the application rate of slurry is reduced to 2 ml per litre per day. For continuous culture, it is necessary to replace one third of the water from the culture system with filtered freshwater.

### *Mass Culture of Brachionus*

For culture of this species in the laboratory, 100 ml capacity test tubes are filled with 16 ppt of filtered brackishwater and inoculated with one individual per ml. *Brachionus* sp. is fed with yeast suspension at the rate 200 ml per litre once in a day and about 250 individuals per litre is obtained within four days. These test tube cultures are used as stock cultures. A series of 30 litre plastic jars are filled with 12 ppt filtered brackishwater and inoculated with 300 individuals per litre. *Brachionus* sp. is fed with yeast suspension at the rate of 400 ml. The species will multiply until a production of about 250 individuals per litre is obtained. The stock culture is used as inoculum of mass culture using slurry at the rate of 4 ml per litre for the first three days followed by 2 ml per litre per day. After one week of inoculation, the total concentration ranges from 2.0 to 2.5 lakhs per litre. If the species is harvested at peak densities and replaced at regular intervals, the culture can be continued upto 60 days.

### *Mass Culture of Daphnia and Cyclops*

For mass culture of *Daphnia* spp. and *Cyclops* sp. freshwater is used as medium. Inoculation of this two species is done separately at the rate of 50 individuals per litre. After a week of inoculation, they reach peak density varying from 15,000 to 20,000 per litre. Continuous mass culture can be maintained by harvesting and replacement at regular intervals.

## 8.4 Continuous Mass Culture of Zooplankton

Fish nursery management practices for mass culture of the larvae of commercially important fishes require the provision of excessive quantities of zooplankters comprising principally rotifers, cladocerans and copepods at field levels which enhanced seed production to the extent of about 80 per cent by phased fertilization techniques. Phased fertilization is a method of adding adequate amounts of organic manures along with phosphatic fertilizer at fixed time intervals so that zooplankton populations can be generated. The biomass of planktonic bacteria produced due to the breakdown of organic manures, is consumed by zooplankton. The zooplankton populations will grow with equal rate leading to mass culture.

### *Method of Culture*

In the laboratory, stock culture of zooplankton is maintained in one litre beakers. Manuring with groundnut cake solution is made every 48 hrs by introducing 1 ml of manuring solution adjusted to 4 ml/litre rate on the basis of studies on its solubility in freshwater. It has been observed that 0.26 g is dissolved in 1 litre of water out of 1 g of the gross oil cake. On the basis of this information, the strength of stock solution is adjusted to 4 mg/litre so that 1 ml of this solution is enough every 48 hrs to maintain cultures in 1 litre beakers.

For mass culture, a plastic pool with a capacity of 2,404 litres is set up in the field. The pool is filled up with filtered freshwater. Crude fertilizer solution is prepared by grinding 37 g of groundnut cake and producing the soluble part at 4 mg/l rate. This soluble part is used in plastic pool. On the second day, 110 g of calcium oxide is added, thereby increasing the pH to nearly 8.2. On the third day, the cultured zooplankton from one litre beaker is introduced into the pool. During this first cycle the population of zooplankton is allowed to grow for a total number of 15 days from the first day. A second application rate of fertilizer solution is applied during the first cycle at the end of 15 days to boost zooplankton biomass production. Thereafter following observation is made every day :

1. *Standing Biomass* : By plankton net operation (made of 120 mesh cloth with a diameter of the opening ring around 27.5 cm), the total quantity of 340 litres of pool water is filtered. Before collection of zooplankton, the pool water is agitated for uniform distribution of zooplankton biomass. The standing biomass per litre of plastic pool and for the entire pool is determined.
2. *Total Debit per Day* : The standing biomass is collected twice or thrice per day. Net weight is determined so as to work out the total debit weight per day. By using this data, the closing balance of zooplankton biomass is determined by weight.

By comparing the closing balance of pervious day and the opening balance of the following day, the zooplankton biomass added or deleted due to continuous production/declining process is determined per day. Quantitative details of zooplankton culture in a plastic pool are shown in Table 8.2.

**Table 8.2 : Quantitative Details of Zooplankton* Culture in a Plastic Pool**

| Number of cycle | Number of days | Standing biomass (g) | | Debit weight of biomass (g) | | Amount of biomass produces per day (g) | | Minus amount of biomass on the last day (g) |
|---|---|---|---|---|---|---|---|---|
| | | Range | Mean | Range | Mean | Range | Mean | |
| 1. | 15 days for initial growth + 6 days of observation and exploitation total | 17.74-27.18 | 21.3 | 2.52-3.86 | 3.11 | 3.77-6.90 | 3.5 | 0.73 |
| 2. | 15 | 15.84-45.77 | 29.29 | 2.25-10.25 | 4.72 | 0.99-20.24 | 6.73 | 2.32 |
| 3. | 27 | 29.73-85.90 | 61.66 | 4.22-19.14 | 11.26 | 1.6-30.98 | 12.78 | 7.15 |
| 4. | 22 | 30.91-72.00 | 52.50 | 4.40-14.82 | 9.67 | 0.72-32.10 | 10.66 | 6.92 |
| 5. | 28 | 13.07-53.09 | 33.94 | 1.86-10.14 | 6.57 | 0.31-16.86 | 8.5 | 7.27 |

* This species of Zooplankton is *Moina micrura*

*Source :* Shirgur and Indulkar (1987)

## Observations

1. As indicated in the Tabe 8.2, it is clear that the first cycle lasted for total of 21 days. For other cycle, the duration varied from 15 to 28 days. However, the average number of days per cycle is 22.6 days. The standing crop of zooplankton and the total debit per day increased from first cycle to the third cycle and thereafter it declined.
2. For a total of 113 days, nearly 798.458 g of zooplankton is able to exploit and at the end of the last day of exploitation, 32.794 g of biomass still remains in the pool. Therefore after 113 days, the total quantity of zooplankton biomass produced in a pool of 2,404 litres capacity is about 834.252 g.
3. The day on which the less production is observed indicates that the phase fertilizer at the standard rate should be introduced so as to start the new cycle.
4. For production of 834.252 g biomass, only 222 g of groundnut cake is used.
5. The decline of biomass from the fourth cycle onwards is possibly due to loss of water quality resulting from cumulative effect of metabolites. Of course, this problem can be alleviated by changing 50 per cent of water from the pool at the end of fifth cycle which will help to maintain the steady mass culture.

## 8.5 Micro-nutrients and Abiotic Factors in Mass Culture

In addition to organic and inorganic fertilizers, use of some trace elements are necessary for enhancing the growth and reproduction of different species of zooplankton. Population of *Moina micrura* in cement cistern of 185 ml capacity has been found to increase in freshwater treated with 0.2 and 0.4 mg of zinc sulfate per litre in combination with Farm Yard Manure (FYM) at the rate of 500 mg/litre (Table 8.3). Use of copper sulfate and zinc sulfate at the rates of 0.08 and 0.6 mg/litre respectively in combination with FYM or cowdung was also found to increase the average zooplankton population. Among different trace elements, cobalt chloride, ferrous sulfate, zinc sulfate, copper sulfate and magnesium sulfate are very effective in combination with manures and fertilizers for mass culture of zooplankton.

It has been estimated that for successful culture of zooplankton, water temperature ranging between 21 and 28$^0$C was found to be suitable because at this range of water temperature, growth and reproductive potential of different species of zooplankton proceeds in an unhampered manner. Increase of pH (8.2-14.0) and total alkalinity (170-250 mg/l) of water were also found to be superior for production of zooplankton.

**Table 8.3 : Population of *Moina* sp. in Water Treated with Zinc in Combination with Farm Yard Manure at the Rate of 500 mg/l**

| Concentration of Zn (mg/l) | Average population of *Moina* (Number/one week) Initial | First | Second | Third | Fourth | Average population for 28 days (Nos./1) |
|---|---|---|---|---|---|---|
| Control | 450 | 2,093 | 1,300 | 1,920 | 926 | 1,560.00 |
| 0.2 | 450 | 25,100 | 8,460 | 16,133 | 13,800 | 15,873.25 |
| 0.4 | 450 | 22,760 | 7,753 | 14,786 | 10,800 | 14,024.75 |
| 1.6 | 450 | 986 | 706* | 820* | 693* | 801.25 |
| 3.2 | 450 | 726 | 613* | 700* | 593* | 658.00 |

* After 15 days these were refertilized with farm yard manure

*Source* : Gupta (1997)

## 8.6 Bottom Fauna

Although benthic macrofauna form either predatory animals or essential food items for various commercially important species of fish, very little importance has been given on the population structure and biomass of benthic animals in fish culture ponds. Bottom fauna is significant to understand the relationship between benthic animals and fish production and also to exploit biological productivity in fish ponds. Generally the dynamics of benthic fauna in ponds largely depends upon the topographical conditions of a particular area, nutrient levels, abiotic factors of soil, and pollution status of ecosystem where they live temporarily of permanently.

### *Bottom Organisms*

The term bottom organisms or benthos means the community of several species of plants and animals which live in the bottom of a water body. Many bottom organisms possess self movement from one place to another and hence they are responsible for increasing soil fertility and ultimately triggers the yield capacity of fish and other aquatic animals. They form essential food items of many commercially important species of some aquatic fauna including fish.

### *Existence*

Although all ponds are heavily infested with various species of bottom organisms in varying quantity, their absence or presence principally depends upon the physico-chemical properties of water, nutrient and trace element loads in water bodies, soil structure and texture, seasons and geographical areas of ponds. In ponds where fertilization and manuring programs are extensively undertaken for constant fertility, their existence is perpetual and therefore, there is excessive growth of bottom organisms. However, the above-mentioned factors accelerate the potential behavior of bottom organisms for their successful existence in soil of any ecosystem.

### *Development*

Manuring and fertilization of ponds play a key role for the growth and development of bottom organisms. Recently it has been reported that several agricultural chemicals at sublethal levels exert their synergistic influence on the potential level of nutrients in soil which are essential for the growth of micro-organisms, act as metabolic activator and triggers to increase the number and growth of soil organisms. Generally partially or completely degraded molecules of chemicals stimulate growth of different species of micro-organisms (protozoa, bacteria and fungi) in pond soil leading to their higher numbers in significant amounts. Several bottom organisms due to their browsing behavior in the sediments ingest contaminated micro-organisms as food which stimulates to increase their growth and other physiological activities. As a result fish food species consume such types of organisms and exhibit significant increase in fish yield. Thus, a food web is formed in the benthic region itself of ponds among micro-organisms, bottom fauna and fish. The relative importance of growth of bottom organisms is likely to vary according to the levels and types of nutrients in aquatic ecosystem.

### *Classification*

On the basis of their habitat, the benthos has been classified as (a) endobenthos (boring in the solid substrate), (b) herpobenthos (growing through mud), (c) psammon (growing through sand),

(d) haptobenthos (attached to immersed solid surface), and (e) rhizobenthos (rooted and extending into water). In general, the most commonly encountered bottom organisms in fish ponds have been divided into Zoobenthos and Phytobenthos.

1. *Zoobenthos* : Larval and adult forms of different groups of fauna such as Oligochaeta, Insecta (Diptera, Collembola, Odonata, Coleoptera, Ephimeroptera), and Mollusca (Bivalvia, Gastropods) are considered as an important constituent in fish ponds. The dominant forms of zoobenthos include *Tubifex, Lumbriculus, Pheretima, Branchiura, Dero, Chaetogaster, Aeolosoma, Mesenchytraeus, Culicoides, Chironomus,* May fly nymph, Dragon fly nymph, *Dytiscus, Cybister, Gyrinus, Pila, Lamellidens, Pisidium, Viviparus, Lymnaea, Melanoides, Digoniostoma, Cypris,* and *Eucypris.*
2. *Phytobenthos* : In many ponds, various species of algae form a green scum over pond bottom. Different groups of flora such as Chlorophyceae, Myxophyceae, and Bacillariophyceae comprise various species of *Oedogonium, Oscillatoria, Botrydium, Ulothrix, Coleochaete, Closterium, Microsterias, Nostoc, Navicula, Amphora, Spirogyra* and others. As the eutrophic zone in many ponds is limited, an appreciable amount of phytoplankton sinks below and forms the food of many benthic communities. Therefore, a magnitude of phytoplankton population could sustain a very rich in zoobenthos of ponds.

### *Management*

The main aspect of bottom organisms involves their adequate management and maintenance of a high quality ecosystem for their success. Because of differences in ecological conditions among localities, the management procedure may vary greatly and therefore, these differences should be considered precisely. It has been demonstrated many times that there is a positive correlation between bottom organisms and fish production. Generally nitrogen, phosphorus, organic matter and lime determines the productivity of fish ponds through development of bottom organisms. Newly constructed ponds have insignificant amount of benthic organisms than older ones due to presence of nutrients in very low amounts. Of course, the progressive fish farmers do not consider the age of ponds during pond management program. The bottom mud should not be highly acidic because bottom organisms do not grow well at low pH and organic matter. Therefore, liming and fertilization have more than doubled the production of benthic organisms in ponds.

### *Role of Abiotic Factors and Toxicants*

A number of bottom organisms appear at specific period while other forms disappear. The periodic appearance and disappearance of bottom organisms are controlled by nitrate, ammonia, phosphate, pH, temperature, salinity, total alkalinity, hardness, dissolved oxygen, and trace elements present in water. Correlations between different groups of bottom fauna and abiotic factors of soil exhibited distinct significant relationship (Table 8.4).

During rainy season, aquatic ecosystems are contaminated with a variety of toxic chemicals generated from diverse sources such as domestic, agricultural and industrial. Among different types of pollutant, pesticides, heavy metals and detergents are the major cause of concern for aquatic environment because of their toxicity, persistency and tendency to accumulate in bottom mud and organisms. The impact of these pollutants on bottom organisms is due to the movement of toxicants from various diffuse or point sources which gives rise to coincidental mixtures in the ecosystem, thus, posing a great threat to fish and fish food organisms. Acute and sublethal concentrations of different types of toxicants exhibit direct and/or indirect effects on bottom organisms. In some cases, no significant alteration in the concentration of bottom organisms was observed at sublethal concentrations of combined pollutant exposure and there may be some indirect effect which can bring about disbalance in the food chain. Continuous or repeated exposure to low concentration of toxicants can lead to accumulation of high residual concentrations without any mortality of bottom fauna. Comparatively less effect on bottom fauna might be due to negative interaction between constituent toxicants on survival, growth and reproduction. Therefore, toxicant mixture at lethal concentration if allowed to interact in natural water bodies, the consequent interaction might produce higher toxicity (synergism) to fish food organisms.

**Table 8.4 : Correlation Coefficients Between Different Groups of Bottom Fauna and Abiotic Factors of Soil and Water. Significance is Shown at P< 0.01**

| Relationship | | "r" | "t" |
|---|---|---|---|
| 1. *Soil* | | | |
| Mollusc Vs | pH | 0.346 | 0.738 |
| | Free calcium carbonate | 0.989 | 13.338* |
| | Available nitrogen | 0.955 | 6.440* |
| | Available phosphorus | 0.942 | 5.614* |
| Arthropod Vs | pH | 0.315 | 0.668 |
| | Free calcium carbonate | 0.996 | 22.282* |
| | Available nitrogen | 0.967 | 7.590* |
| | Available phosphorus | 0.963 | 7.149* |
| Annelid Vs | pH | 0.332 | 0.704 |
| | Free calcium carbonate | 0.995 | 19.900* |
| | Available nitrogen | 0.962 | 7.045* |
| | Available phosphorus | 0.948 | 5.975* |
| 2. *Water* | | | |
| Mollusc Vs | Alkalinity | 0.985 | 18.308* |
| | Dissolved oxygen | 0.986 | 18.956* |
| | Temperature | 0.970 | 12.609* |
| Annelid Vs | Alkalinity | 0.983 | 16.919* |
| | Dissoved oxygen | 0.995 | 31.485* |
| | Temperature | 0.979 | 15.175* |
| Arthropod Vs | Alkalinity | 0.996 | 35.245* |
| | Dissolved oxygen | 0.983 | 16.919* |
| | Temperature | 0.991 | 23.408* |

*Source :* Sarkar (1989, 1992).

## 8.7 Abundance and Variations of Bottom Fauna

The productivity of benthos is closely associated with the fish production and exchange processes in the profundal zone. Therefore, abundance and fluctuation of different species of bottom fauna in fish culture ponds are very important so far as the indicator of water quality and eutrophication are considered. The abundance of bottom fauna is seasonal and irregular (Table 8.5). For example, molluscan species strongly dominated all the year round (49.3-80.5 per cent) than that of other groups of bottom fauna. Oligochaetes ranked second (9.8-31.9 per cent) followed by Insects (8.5-26.9 per cent). However, in many lentic pond ecosystems, a different trend of the mean concentration of bottom fauna in different months is a common occurrence (Figure 8.2) and their numbers also varied. Seasonal abundance of various forms of bottom fauna revealed that insect populations were abundant during summer -monsoon seasons followed by oligochaetes (summer- winter) and molluscs (monsoon-winter) (Table 8.6). Monsoon season seems to be the most important factor influencing the abundance of insect fauna. Therefore, seasonal variations must be considered so that fertilization and manuring programs can be manipulated in such a way that production strategy of bottom fauna may proceed in an unhampered manner.

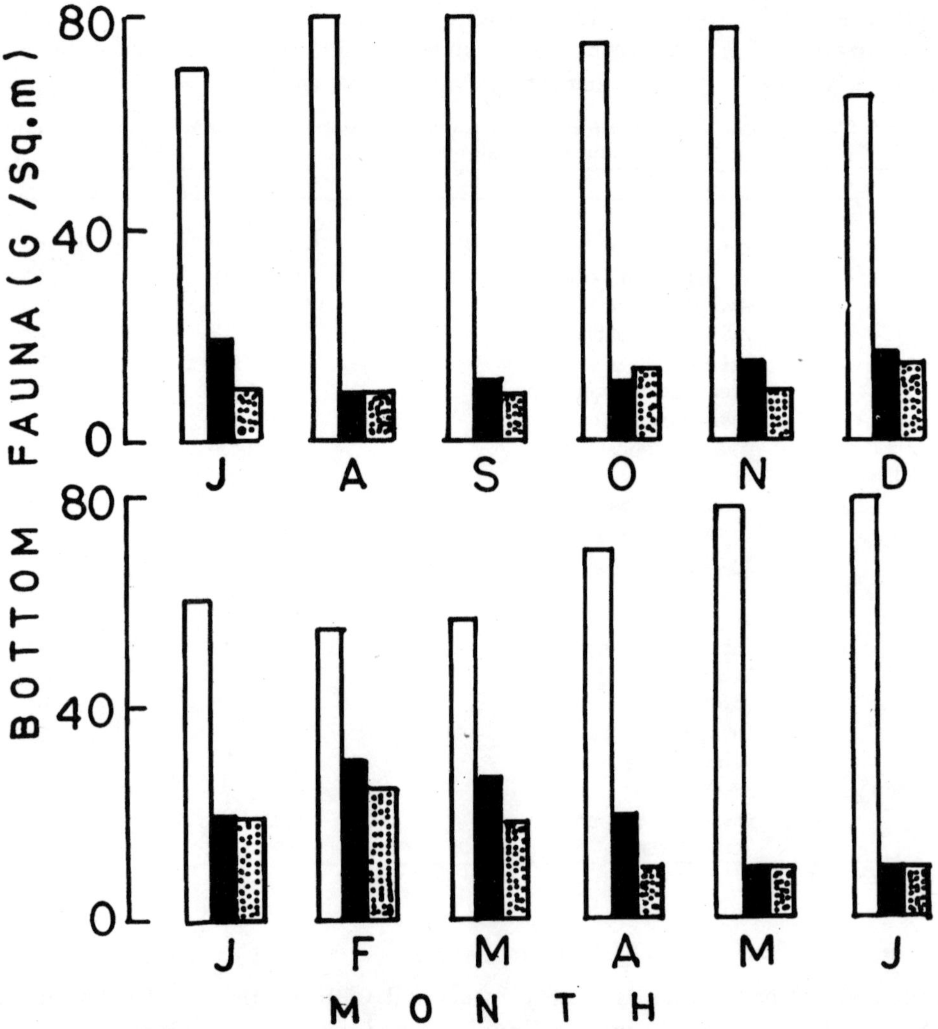


**Fig. 8.2 : Mean concentrations of bottom fauna in a pond ecosystem in different months of the year 1990. Bottom fauna constitutes molluscs (hollow columns), arthropods (solid columns), and annelids (dotted columns). The pond was perennial with an average depth of 2.8 metres. Allochthonous organic matter was found to be higher due to human activities on the edge of the pond. [From Sarkar (1992)]**

Quantitative and qualitative compositions of various taxa of macro-invertebrates recorded from fish ponds and sewage-irrigated tanks (Table 8.7) revealed the presence of meagre number of *Tubifex* sp. and larvae of *Dicrotendipes* sp., the former being dominant constituting 86 per cent and the latter 12 per cent of the total invertebrates in fish ponds compared to those of the sewage-fed irrigation tanks inhabited by *Tubifex* sp., larvae of *Dicrotendipes* sp., *Bellamya dissimilis*, and *Indoplanorbis exustus*.

**Table 8.5 : Occurrence of Some Benthic Macrofauna in a Freshwater Pond During the Year 1984. D, Dominant (>150-200); A, Abundant (>100-150); P, Present (>50-100); R, Rare (>1-50); N, Absent. J-D, January to December**

| | Month | | | | | | | | | | | |
|---|---|---|---|---|---|---|---|---|---|---|---|---|
| Species | J | F | M | A | M | J | J | A | S | O | N | D |
| Insects (per cent) | 40 | 60 | 68 | 20 | 16 | 70 | 68 | 15 | 13 | 18 | 23 | 18 |
| *Chironomus* | R | R | R | N | R | R | P | N | N | R | R | N |
| Culicoides | N | N | R | R | N | N | N | N | N | N | N | N |
| Dragon fly nymph | N | R | P | R | R | P | N | N | R | P | N | R |
| *Notonecta* | N | N | N | N | N | R | R | R | N | N | N | N |
| *Dytiscus* | P | R | N | P | R | P | N | R | N | R | P | P |
| *Belostoma* | N | R | P | R | R | N | N | P | R | N | R | R |
| *Cybister* | N | N | N | N | N | R | N | N | R | N | N | R |
| Oligochaetes (per cent) | 25 | 17 | 10 | 20 | 10 | 17 | 5 | 8 | 4 | 17 | 17 | 19 |
| *Tubifex* | R | N | R | N | P | P | R | N | R | D | D | D |
| *Chaetogaster* | R | N | P | P | A | R | P | R | N | N | P | R |
| Molluscs (per cent) | 35 | 40 | 22 | 60 | 67 | 25 | 24 | 81 | 79 | 65 | 60 | 63 |
| *Lemellidens* | N | R | R | N | N | R | R | R | N | P | P | R |
| *Viviparus* | R | R | R | P | P | P | A | A | D | A | A | P |
| *Lymnaea* | N | R | R | N | P | R | P | P | A | P | N | R |
| *Pila* | R | N | P | A | P | P | D | D | A | A | P | A |
| *Planorbis* | R | N | R | N | R | N | R | P | N | N | R | N |

*Source* : Sarkar (1989)

**Table 8.6 : Percentage of Abundance of Various Groups of Macrofauna in Different Seasons of the Year 1984**

| | Groups (per cent) | | |
|---|---|---|---|
| Season | Insect | Oligochaet | Mollusc |
| Summer (Feb-May) | 33.6 | 36.4 | 30.0 |
| Monsoon (June-Sep) | 33.0 | 30.0 | 37.0 |
| Winter (Oct-Jan) | 28.4 | 36.6 | 35.0 |

*Source* : Sarkar (1989)

## 8.8 Interactions Between Bottom Fauna and Nutrients

Nutrient carriers such as nitrogen, phosphorus and oil cakes are the most important compounds regulating biological productivity of water bodies, and their cycles are the basis for management of fish culture systems. But their accumulation in excessive amounts cause deleterious effects on fish food organisms particularly bottom organism. However, optimum levels of nutrient trigger survival, growth and reproduction of algae and bacteria, which in turn, sustain populations of zooplankton and bottom organisms resulting in utilization of these organisms by different species of fish.

Although there is no clear explanation for the direct influence of chemical fertilizers and organic manures on fish growth in ponds, a partial explanation involves the relationship between nutrient load and fertilizer treatment. Fertilizers exert their synergistic influence on the potential level of nutrients, which are essential for the growth of micro-organisms, by acting as metabolic activators that trigger the growth and reproduction of soil organisms. Because the concentration of micro-organisms depends upon the amounts of nutrients in the soil, it is not unrealistic to expect fertilizers to affect the growth dynamics of soil organisms.

**Table 8.7 : Population Abundance of Different Taxa of Benthic Macro-invertebrates in Sewage-Irrigated Tanks and Fish Ponds. Values are mean ± SE and expressed in Number/m²**

| Zoobenthic group | Sewage-fed tank | Fish pond |
|---|---|---|
| 1. *Oligochaeta* | | |
| *Tubifex* sp. | 2,036.4±521.3 | 610.9±3.2 |
| 2. *Diptera* | | |
| Larva of *Dicrotendipes* sp. | 880.7±317.2 | 712.0±237.6 |
| 3. *Mollusca* | | |
| *Bellamya dissimilis* | 5,192.8±742.6 | 560.0±152.7 |
| *Indoplanorbis exustus* | 8,619.9±748.6 | 4,632.1±384.2 |

*Source* : Selected data form Reddy and Rao (1989)

Chemical fertilizers and organic manures are degraded by bacterial, and as a result, a variety of metabolites are released into the sediments and ultimately act on the biomass of bottom fauna. The residual as well as the direct toxicity of some toxic compound (such as 2, 4-D, atrazine, simazine etc.) causes mass mortality of plankton populations, particularly in the early stages of devèlopment. On degradation, the result is a release of nutrients for the growth of bottom organisms. However, partially or completely degraded fertilizers in the sediments stimulate bacterial growth rates; these molecules are used in microbial cell synthesis and lead to higher numbers of bacteria. It is possible that nutrients of fertilizers at optimum rates are absorbed or absorbed by bacteria, and hence removed from the soil by different species of bottom organisms. Some benthic animals ingest contaminated bacteria/algae as food from the sediment. Fertilizers and manures also enhance the metabolic capacities of bottom organisms after entering their bodies. The increased accumulation of nutrients following induction of various abiotic factors of the ecosystem is more likely to result from fertilizer affinity for cellular structures. In ponds where fertilizers and manures increase the number of bacterial and algal species, the bottom fauna engulf these contaminated bacteria and as a result the concentration of bottom fauna is increased.

## 8.9 Ingredients of Natural Food

Different species of fish consume different types of natural food for their nutrition. Generally protein, fat, carbohydrate, and vitamins are present in appreciable quantities and therefore, it is most important for fish health. The most common and principal types of natural food consumed by fish and their ingredients are shown in Table 8.8.

### *Energy Contents of Natural Food*

Determination of calorie is very essential so far as the bioenergetics are considered. Measurement of energy in fish food organisms of an ecosystem provides the reliable information of the functional nature of the communities. A comparative estimates of energy content of some common fish food organisms have been made using wet oxidation method and bomb calorimetry and the results obtained in the latter method were only within 3 per cent higher than the former. The highest calorific value (7.12 K cal./g) was recorded in tubificid worm while the *Cypris* has the minimal value (3.62 K cal/g), energy content of the other organisms (such as *Asplanchna, Heliodiaptomus, Nauplius* larva, *Moina, Daphnia,* Chironomid larva and *Lymnaea*) lying in between (Figure 8.3).

**Table 8.8 : Ingredients of Some Natural Food Organisms. (1) Water, (2) Crude Protein, (3) Fat, (4) Carbohydrate, (5) Calorific Value (K Cal/g), (6) Ash, (7) Total Digestible Nutrient.**

| Type | Ingredient (per cent) | | | | | | |
|---|---|---|---|---|---|---|---|
| | 1 | 2 | 3 | 4 | 5 | 6 | 7 |
| *Mysis* | 77.8 | 16.3 | 3.3 | 2.8 | 3.62 | - | 19.6 |
| Small shrimp | 17.0 | 55.5 | 5.5 | 17.6 | - | - | 58.9 |
| *Viviparus* | 83.4 | 14.1 | 0.5 | 2.1 | - | 1.5 | 12.9 |
| *Tubifex* | 87.2 | 8.1 | 2.0 | - | 7.12 | 1.0 | - |
| *Chironomus* | 87.7 | 8.2 | 1.9 | - | 5.05 | 0.9 | - |
| *Planorbis* | 73.0 | 10.6 | 0.7 | - | 4.86 | 7.0 | - |
| *Daphnia* | 91.6 | 3.5 | 0.6 | - | 5.53 | 1.6 | - |
| Locust | 65.9 | 25.5 | 2.0 | 2.2 | 25.0 | - | 25.9 |
| *Cyclops* | 90.3 | 2.6 | 0.9 | - | 5.44 | 1.4 | - |

*Source :* Selected data from Tamura (1961), Mann (1961), Jana and Pal (1981)

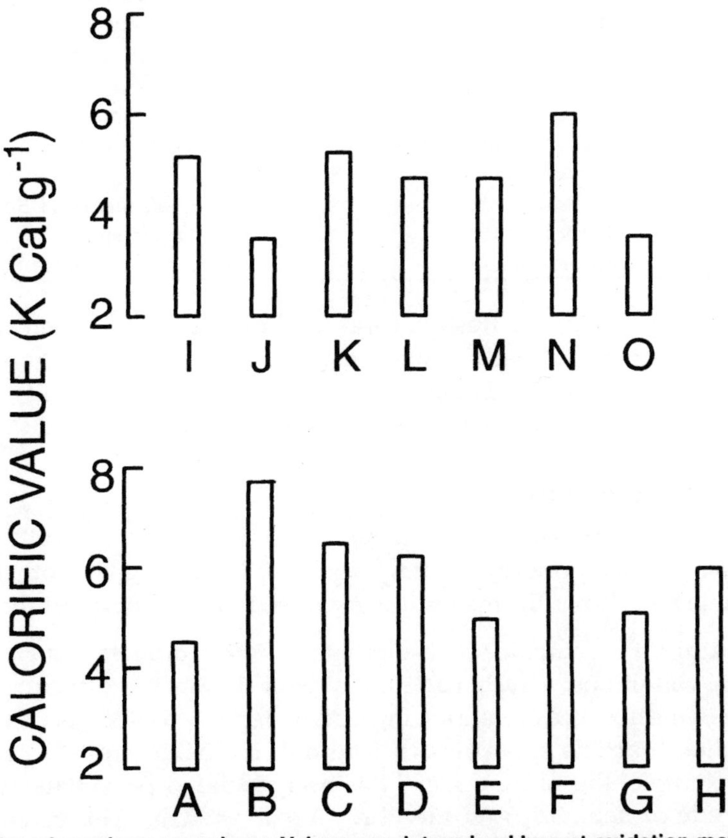


**Fig. 8.3 : Calorific values in various organisms. Values are determined by wet oxidation method. (A) *Asplanchna*, (B) *Tubifex* worm (large), (C) *Tubifex* worm (small); (D) *Heliodiaptomus* (large), (E) *Heliodiaptomus* (small), (F) *Nauplius* larva, (G) *Moina*, (H) *Daphnia* (large), (I) *Daphnia* (small), (J) *Cypris*, (K) Chironomid larva (large), (L) Chironomid larva (small), (M) Chironomid pupa, (N) *Lymnaea* (large), (O) *Lymnaea* (small) [Redrawn from the data as proposed by Jana and Pal (1981)]**

## 8.10 Natural Feed Ingredients

The main point of a discussion on feed ingredients for fish should be included an overview of the food preferences of the fish in question, the latter serving as a guide for the identification of potential feed ingredient sources for use by fish farmers.

Generally the bottom-dwelling fish have benthiophagic omnivorous feeding habits. The main food items consumed include crustaceans, molluscs, annelids, insects, and associated detritus-based micro/meiofauna and flora. An important fact has come out from the survey on natural feed ingredients is that cultured freshwater fish generally feed lower on the aquatic food chain and a large variety of food organisms than their marine counterparts. However, due to natural feeding habits of different species of freshwater fish, a wide variety of natural food items have been used by fish farmers as exogenous feed sources for their culture operations (Table 8.9).

**Table 8.9 : Major Natural Feed Ingredients Employed as Supplementary Feed for Freshwater Fish**

| Group | Common name | Species |
|---|---|---|
| Crustaceans | Water flea | *Daphnia* sp. |
| Molluscs | Snail | *Viviparus bengalensis* |
| | | *Lymnaea leuteola* |
| Annelids | Bloodworm | *Americonuphis* sp. |
| | | *Pheretima* spp. |
| Insects | Aquatic insect larvae | |
| Fish | Tilapia | *Oreochromis mossambicus* |
| Vertebrates | Slaughterhouse waste | *Azolla* sp. |
| Plants | Water velvet | |
| | Green algae | *Cladophera* and *Spirogyra* |
| | Duckweed | *Lemna* spp. |
| | Sweet potato | *Ipomoea batatas* |

The nutritional value of live or fresh natural food organisms has been found to be higher than that of processed food or artificially compound diets. However, to a large extent this has been due to their superior water stability and higher biological value and digestibility; the latter resulting from a well-balanced essential amino acid (EAA) and essential fatty acid (EFA) profiles.

## 8.11 Processed Feed Ingredients

These include all plant and animal food items which have been physically processed prior to feeding either by composting, grinding, cutting, drying or by mixing with other food items into a compound diet. Table 8.10 shows the major processed feed ingredients used for pond fish culture.

Although it is difficult to compare the results obtained from different experimental trials under varied agro-climatic conditions, a systematic analysis of the data is difficult to draw a definite conclusion because of the absence of definite ingredient definitions such as : (1) international feed number, (2) ingredient particle size prior to feeding or mixing, and (3) proximate chemical composition. For example, the 'shrimp meal' has been declared proximate composition – 3.2% crude fat, 71.8% crude protein, 3.14% crude fibre, and 16.9% ash. This cannot be nutritionally compared with 'shrimp meal' which has been declared proximate composition – 4.5% crude fat, 43.3% crude protein, 10.7% crude fibre, and 34.5% ash.

**Table 8.10 : Major Processed Feed Ingredients Used as Feed for Freshwater Fish**

| | Ingredient | Species |
|---|---|---|
| 1. | *Animal by-product* | |
| | Earthworm meal, Fish meal, Fish silage, Trash fish, Poultry by-product meal, Blood meal, Meat meal, and Feather meal | |
| 2. | *Plant by-product* | |
| | Soyabean meal | *Glycine max* |
| | Cotton seed meal | *Gossypium* sp. |
| | Mustard seed meal | *Brassica nigra* |
| | Rice bran and Rice polished | *Oryza sativa* |
| | Wheat meal | *Triticum* sp. |
| | Corn/Maize meal | *Zea mays* |
| | Turnip leaves | *Brassica nepus* |
| | Lucerne/Alfalfa | *Medicago sativa* |
| | Single-cell proteins (SCP) : Bacterial SCP, Algal SCP and Fungal SCP | |

On a nutritional basis the best feed ingredients for fish and shellfish are those whose biochemical composition approximates closest to that of their own body composition. Since protein constitute the largest chemical group within the fish body, it follows therefore that the nutritive value of a feed ingredient will be governed to a large extent by its amino acid composition, and upon the availability and profile of the EAA present.

The nutritional value of a fee ingredient is also dependent upon its physical properties and characteristics, including particle size, density, water solubility, stability, and pelletability. For example, feeding ingredients in compound diets should be grind to a particle size of 200μm for maximum water stability and feed utilization efficiency in *Paenius indicus*. By contrast, the reverse was true : fine grinding did not improve feed digestibility within cold pelleted diets, although it improved pellete stability in *Paenius vannamei*.

## 8.12 Future Prospects

Whilst there is no doubt of the high nutritional value of fish meal for cultured fish and shellfish, they represent an expensive item which are also utilized within livestock feeds or used for human consumption. Moreover, increasing raw ingredient and feed manufacturing costs necessitates that the fish farmer reduce production costs so as to maintain profitability. As a result, since food and feeding represents the largest single operating cost item of intensive fish farming, clearly further research emphasize must be placed on the development of improved feed management strategies or through the development of feeding strategies aimed at reducing feed costs as noted below :

1. Development of low-cost supplementary diets, as compared to high-cost complete diets, for use within intensive fish pond farming systems at moderate stocking densities, should be adopted. For example, studies have shown that there was no difference in the growth of pond reared *Paenius monodon* (10 animal per square metre) fed a high-cost nutrient dense complete pelleted diet or a locally produced ball.

2. Maximizing the role played by naturally available pond food organisms in the overall nutritional budget of the farmed fish through improved pond fertilization, manuring and water management techniques. For example, chicken litter (applied at a rate of 200 kg/ha/week) could profitably substituted for high cost pelleted shrimp feeds during the first eight to nine weeks of the pond grow-out phase at a stocking density of 5 shrimp per square metre.
3. Replacement of feed ingredients with less expensive protein sources and overcoming nutrient imbalance through dietary supplementation with feeding attractants or/and stabilized free amino acids so as to obtain the desired dietary EAA and nutrient profile.

## 8.13 Conclusion

There are two methods of feeding fish ponds : (a) through artificial feeding and (b) with the natural food production of the pond through fertilization and manuring. In the latter case the pond provides a natural grazing ground. The fish finds its nourishment along the edges, on the bottom or in the water. Natural fish foods are classified into plankton and bottom fauna. A large number of plankton and bottom fauna occur in pond and lake ecosystems which determines their productivity. The concentration of plankton in different months and seasons varied greatly. Plankton fluctuations usually depend upon the climatic conditions and different abiotic factors of ecosystem.

Since natural food supplies minerals, micronutrients, proteins, fats and carbohydrates for fish growth, mass culture of different species of animals used as food is necessary for constant supply for fish particularly in nursery ponds. For culture of zooplankton and bottom fauna, use of nitrogen and phosphorus fertilizers, organic manures and micronutrients are very effective. The most commonly encountered bottom organisms can be divided into zoobenthos and phytobenthos.

Natural food contains relatively high content of protein than artificial food. It is known that the relationship between the nutrients and fish food organisms plays an important role in the growth of fish. For this reason the ratio between protein and other nutrients has already been indicated in food charts. In the natural feeding of the fish, it is mostly around 1:1 up to 1:1.8, while in artificial food lower ratios are encountered. Since protein constitute the largest chemical group within the fish body, the nutritive value of a feed ingredient will be governed by it amino acid composition.

The periodic appearance and disappearance of natural food organisms in pond and lake ecosystems are triggered by various abiotic factors and toxicants. The impact of toxic compounds on natural food species is due to the movement of metabolites of toxicants which gives rise to coincidental mixtures, thus posing a great threat to various forms of aquatic life. Therefore, caution must be taken so that fish culture ecosystem may not be contaminated with toxicants.

Energy contents of fish food animals must be evaluated to understand the functional nature of the ecosystem. Among different natural food, the highest and lowest calorific values were recorded in blood worm and *Cypris* sp., respectively and the energy content of the common natural food species varied between 7.12 and 3.62 K cal per gram.

### *References*

Gupta, A.K. 1997. Observation on fertilization of water with zinc for increased production of *Moina micrura*. *Geobios*. 24 : 231-235.

Jana, B.B and G.P. Pal. 1981. Energy contents of some common fish food organisms. *Biol. Bull. India*. 3 : 93-100.

Mann, H. 1961. Fish culture in Europe. *In* : *Fish As Food* (*Ed.* G. Borgstrom). Vol. I, pp. 77-102. Academic Press, New York.

Reddy, M.V. and B.M. Rao. 1989. Community structure of benthic macro-invertebrates of fish ponds and sewage-irrigated tanks in an urban ecosystem. *Environ. Ecol*. 7 : 713-716.

Sarkar, S.K. 1989. Seasonal abundance of benthic macro-fauna in a freshwater pond. *Environ. Ecol*. 7 : 113-116.

Sarkar, S.K. 1992. Composition and changes of benthic macro-invertebrates of a lentic pond in Calcutta. *Geobios*. 19 : 10-14.

Sarkar, S.K. and P. Basu Chowdhury. 1999. Role of some environmental factors on the fluctuation of plankton in a lentic pond at Calcutta. *In* : *Limnological Research in India* (*Ed.* S.R. Mishra), pp. 108-132, Daya Publishing House, New Delhi.

Shirgur, G.A. and S.T. Indulkar. 1987. Continuous mass culture of Cladoceran, *Moina micrura* in plastic pool by phased fertilization techniques. *Fish. Chim.* May, pp. 19-23.

Tamura, T. 1961. Carp cultivation in Japan. *In* : *Fish As Food* (*Ed.*G. Borgstrom). Vol. I, pp. 103-120, Academic Press, New York.

## Questions

1. Name some common natural food organisms used as food by carp fish. Why they are so important in the production of fish?
2. Classify the natural food organisms. What are the conditions for fluctuations of plankton in a fish pond?
3. Name some common zooplankton cultured in commercial scale. How they are cultured?
4. Describe the process of continuous mass culture of zooplankton.
5. Trace the role of some abiotic factors in mass culture of zooplankton.
6. State the role of some factors and toxicants in the management of natural food organisms.
7. Define the terms plankton and bottom fauna.
8. Discuss how nutrients are related to the development of natural food organisms in fish culture ecosystems.
9. Distinguish between natural and processed feed ingredients.

# 9

# Artificial Feed in Fish Culture

Fishes are widely distributed in various lakes, rivers, ponds, and reservoirs. Most of the freshwater fishery resources are very productive and fish growth depends on the availability of natural food organisms. Since reproductive potential of fish is very high, they grow fast in polluted-free water bodies. For this purpose only commercial species of fish are scientifically cultured using quality and balanced artificial feed fortified with vitamins, minerals and other organic compounds.

Sustainable and successful freshwater fish culture on scientific basis principally depend upon the use of nutritionally adequate, economically viable and environmentally friendly artificial feeds. Since the feed costs vary between 40 and 65 per cent of the total managerial expenditure in freshwater fish culture systems, artificial feeds must be scientifically formulated and adequately supplied so that the entire amounts of nutritional inputs in the feeds may be utilized by the fish.

## 9.1 Importance of Artificial Feed

Though natural food contributes to the nutrition (For fish nutrition, see De-Silva and Anderson 1992) of the cultured fish in semi-extensive and extensive fish culture systems, the exogenous supply of artificial food needs to supply nutrients which may be deficient in natural food. Fish production from these systems depends on the natural biogenic potential on the quality and quantity of natural food production in the pond and artificial feed supplied and on the physico-chemical factors of water and soil.

In intensive fish culture systems, where very high stocking densities are maintained (such as cage culture, recirculatory and pond culture systems), the cultured fish has to rely on the artificial feeds. At high densities, insufficient feeding leads to poor growth and nutritional diseases and succumbed to infection. Therefore, it is necessary to design and prepare artificial feeds in balanced proportions with all essential nutrients so that the metabolic activities of fish may accelerate immediately after feeding.

## 9.2 Energy in Artificial Feed

The basic function of fish food is to supply adequate energy for growth and reproduction. There are two advantages in energy utilization of fish such as (1) the excretion of waste nitrogen requires less energy in fish than in land animals and (2) fish do not expend energy to maintain body temperature different from that of their environment. In case of warm-blooded animals, energy converts ammonia to less toxic metabolites. But in fish, ammonia diffuses into the water through gills. However, it has been found that deficient or excess dietary energy levels results in reduced growth of fish.

Generally energy is obtained from carbohydrates, proteins and fats and the gross energy levels of these food items are 4,100, 5,650 and 9,450 K cal/kg, respectively. The digestible energy is defined as the gross energy of the food consumed by the fish and is expressed as K cal/kg. However, the digestible energy values of non-leguminous carbohydrate, leguminous carbohydrate, plant protein, animal protein, and fat for fish are 3.0, 2.0, 4.25, 3.8 and 8.0 K cal/g, respectively. If the proximate compositions of feed ingredients are known, the digestible energy (DE) of feed can be calculated for balancing energy levels in artificial feed.

*Example*

Suppose a fish feed has to be prepared by mixing with three types of feed ingredients (such as fish meal, soyabean and rice bran) in the ratio of 3:3:4, respectively) having the percentages of their proximate composition as shown in Table 9.1 From these data, it is possible to calculate the digestible energy level in 1.0 kg of fish feed in the following way :

**Table 9.1 : Nutrient Composition In Three Feed Ingredients Used for the Preparation of Fish Feed**

| Nutrient composition | Per cent in Feed ingredients | | |
|---|---|---|---|
| | Rice bran | Soyabean meal | Fish meal |
| Crude proteins | 10 | 40 | 65 |
| Lipids | 4 | 4 | 6 |
| Soluble carbohydrates | 50 | 38 | 5 |

Compiled by the author from different published literature

DE of protein : (195 x 4.25) + (120 x 3.8) + (40 x 3.8)

= 828.8 + 456 + 152 = 1,436.8 K cal

DE of lipids : (18 x 8) + (12 x 8) + (16 x 8)

= 144 + 96 + 128

= 368 K cal

DE of carbohydrate : (15 x 3) + (114 x 2) + (200 x 3)

= 45 + 228 + 600

= 873 K cal

Therefore, the total DE in 1.0 kg of fish feed will be 1,436.8 + 368 + 873 = 2,677.8 K cal.

## 9.3 Artificial Feed Versus Water Quality

Application of artificial feed affects water quality criteria more than any other management factors. As a result, freshwater bodies produce excessive plankton blooms which use dissolved oxygen. Decomposition of uneaten food also results in oxygen demand. Therefore, the amount of fish that can be produced in lentic water bodies is dependent upon the amount artificial feed to be added while maintaining proper water quality enough for fish growth. However, the application of artificial feed in fish culture is the most important and expensive item and therefore, the supply of optimum quantity of feed under various environmental factors (such as temperature and dissolved oxygen) is necessary.

Artificial feeds are used for culture of different species of fish to increase production above that obtained with chemical fertilizers and organic manures. Metabolic wastes from feed reach the water, exert an oxygen demand, and provide nutrient for phytoplankton. The phytoplankton also exerts an oxygen demand. Therefore, as feed rates increase, the concentration of phytoplankton increased and consequently water quality deteriorates.

Maximum daily feeding rates of 0, 28, 56, 84, 112, 168 and 224 kg/hectare have been established in some fish ponds of U.S.A. When concentrations of dissolved oxygen fall below 2 mg/l, aeration was supplied at 6.1 kilowatt per hour. Strong correlations (r = 0.85 to 0.98) were observed between concentrations of dissolved oxygen and feeding rates particularly at dawn. It has been observed that little aeration is necessary at feeding rates of 56 kg/ha or less. Constant aeration at night in ponds with feeding rates of 112 kg/ha/day and above was found to be effective for fish production in significant levels. However at higher feeding rates, feed conversion ratios (feed applied/net fish production) increased and fish production steadily declined.

Deterioration of water quality obviously limits the amount of feed that can be applied with a given aeration regime. Since the cost of aeration is very high for poor farmers, the amount of aeration must be limited. If excessive aeration is applied without any economic considerations to check low dissolved oxygen at high feeding rates, fish production will obviously be limited by large concentrations of ammonia.

## 9.4 Artificial Feed Versus Fish Health

In different types of farming systems fish obtain dietary nutrients from available pond food organisms and nutritionally complete diet throughout their life cycle. Dietary imbalances may arise from the presence of disproportionate levels of amino acids, fatty acids, minerals, and vitamins in foodstuffs and have resulted in manifestations of several nutritional pathological conditions such as scoliosis, eye cataract, fatty liver, exophthlmous and skin/fin haemorrhage. However, good morphological conditions of fish depend on the quality and quantity of feed available, nature of farming systems, variations of water and soil parameters and other management schedules. In fish culture ponds where stocking densities exceed the availability of food, addition of supplemental feed is very significant because both natural and artificial food provides different types of nutrients to fish in adequate quantities. Deficiency of feed in ponds leads to stunted/diseased fish while excess diet pollutes the fish culture environments and thus enhance the cost of fish production. Furthermore, excess diet has led to various abnormalities in fin-fish such as vertebral deformities, scale loss, haemorrhages on skin and fin bases, exopthalmia, anorexia, depigmentation or changed in body color, muscle damage, anaemia, lethargy and reduced white blood cell count. Of course, these abnormalities also depend on the absence of several amino acids in the diet.

## 9.5 Types of Diet

Artificial feeds are generally considered as the dawn of a new era in the strategy of fish culture industry. In order to maintain optimum fish health and growth, balanced diet in adequate amounts should be supplied to fish culture ponds. Generally poor fish farmers use locally available diet for fish and these diets are often poor in nutrient composition and not palatable to fish although they contain appreciable amounts of proteins, carbohydrates, fats and trace amounts of calories.

Generally two types of diets are used for fish culture : (1) pelleted diet and (2) non-pelleted diet. Experiments have shown that the non-pelleted diet was conducive to survival and production of tilapia fish. In contrast, it has also been observed that there was significant benefit to feed conversion for catfish and carps if the food stuffs were pelleted. These indicate that selection of fish species is very important so far as the types of diets and their conversion efficiency are concerned.

Recently various types of artificial feeds such as Shirimin, Agrimin, and Fishmin have been developed by Shirimin Minerals and Allied Products, Hyderabad, Glaxo India Limited, Mumbai and Agro-Vet Industries Limited, Mumbai, respectively for aquacultural management practices. These are multi-micronutrient mineral fertilizers with essential balanced mixture of amino acids, macro and micro elements. They are mixed with artificial feed in suitable proportions. Generally, 10 kg of agrimin or shirimin or fishmin is mixed with one tonne of feed. However, use of these nutrients is very important because (1) they are active stimulating agents for cell development, (2) stimulate to increase the rate of photosynthesis, (3) enable for active participation of both nitrogen assimilation and fixation, (4) enhance production of phytoplankton, zooplankton and bottom organisms, (5) reduce annual production cost, and (6) help bone and muscle formation in fish.

A wide variety of feed ingredients are used to prepare artificial feeds. The simplest fish feeds are prepared by using locally available raw materials such as corn or rice bran and rice mill sweepings as sources of carbohydrates. These are mixed with animal proteins such as fish meal, trash fish and snail meat. However, commercial feed preparations and their formulations depend on culture system (such as semi-intensive, intensive, and traditional), species of fish and protein requirement. Thus, supplemental feeds for tilapia are prepared using 80 per cent rice bran and 20 per cent fish meal; for prawn/shrimp, supplemental feeds include fresh raw materials such as snail/ mussel meat and other slaughter-house left overs. Fish feeds are marketed in various forms such as starter, grower, and finisher. Starter feeds are added on the first month of culture, finisher feeds on the last month, while the grower feeds in between finisher and starter feeds. Starter and finisher feeds have the crude protein contents of about 40 and 30 per cent, respectively.

## 9.6 Overfeeding

Although supply of high quality diet to fish ponds in necessary, overfeeding may sometimes be harmful to fish and ecosystem as a whole. An experiment was conducted at Auburn University (USA) to determine precisely how much artificial food in earthen ponds would consume daily from small channel catfish fingerlings to harvest-size fish. Result which represents the maximum amount of supplemental feed that the catfish would accept, indicate that food decreased markedly as the growing season progresses.

## 9.7 Prepared Food in Container

There are several types of prepared feeds in container such as biscuits or combined meals, bio-enriched feeds and Artificial Plankton Rotifers (APR) which are used as fish food. Using advanced processing technology and excellent quality control standards, it has paved the way for making industry leading feeds for various developmental stages of some finfish and shellfish which are designated as biofeeds. The process of bio-enrichment involves in organisms by feeding or adding different types of materials such as micro-diets, micro-encapsulated diets, baker's yeast an emulsified lipids together with fat-soluble vitamins. Generally the following organisms are considered for bio-enrichment : Rotifers (50-150μ), Artemia (200-350μ), Moina (400-1,500μ), Daphnia (200-350μ), and Microalgae (2-20μ). Ocean Star International Inc (USA) developed fish feeds that include Brine Shrimp Flakes, a micro-encapsulated, high fat diet called APR and Spirulina powder. These diets greatly enhance fish and shrimp growth and survival rates at the same time reduces the feed costs.

## 9.8 Feeding of Fish

Regular feeding of fish with high quality diets is necessary; of course, feeding intensity of fish largely depends upon the climatic conditions, geographical and environmental conditions, availability of feed stuffs, local prices, local technical infrastructure, seasons, physiological conditions of fish, age and sex of fish, fish species and their combinations in a particular water body where they are cultured. These parameters determine "optimal" working pattern of farmers. Therefore, there is no universally applicable solution for optimal feeding of fish. However, several reputed companies have specialized in balancing these parameters and finding the solution that optimizes yield of fish in ponds.

Artificial feeding is not advisable if natural food is available or can be cultivated on commercial basis. Moreover, there is no universal feeding diet for all species of fish. The feeding rate is calculated as a percentage of the estimated fish biomass in the pond. Higher rations are given when fishes are small and starts at 5 per cent and 10-15 per cent of estimated biomass of fish and shrimp, respectively, and decreases to a low of 2 and 5 per cent for fish and shrimp, respectively at the end of culture period.

### *Feeding Devices*

In many countries, compressed-air feeding machine has become very popular for fish feeding; of course, this method is very useful for feeding where ponds are stocked with large numbers of fish. In India, fish feeding by hand is the most common method, although it takes more time to accomplish the practice. This method is generally employed in case of small ponds. But in larger ponds and lakes, small wooden boats are float on water and feed is distributed by throwing handful into the water. However, feeding of fish should be done in such a way that all the fish may consume feed uniformly for achieving more or less uniform growth.

In semi-intensive and intensive fish culture systems, feeding trays (Figure 9.1) are submerged into the water at certain points along the periphery of the pond. Trays are provided with known quantities of feed and the feeding tray is lifted two or three hours after the feed was supplied to check how much of it has been consumed by fish and to see if the fishes are healthy and feeding. Empty feeding trays may indicate that the quanitity given is not adequate and may have to be increased. Conversely, full or slightly touched trays indicate excessive feed quantities. The feeding ration is subsequently adjusted accordingly to optimize feed utilization. Adequate number of feeding trays (30 to 40 per hectare water area) should be used to create maximum opportunities for the fish to feed.

A peculiar feeding device is adopted in many carp farming areas in India where feed mixture (rice bran and groundnut or mustard oil cake in the ratio of 1:1) is kept in perforated bags (20 to 30 bags per hectare) tied to bamboo poles. Feed mixture is transported in a small raft and equally distributed in the feeding bags. Fish browse on the feed through the perforation in the bags and within a few hours, most of the feed kept in the bags is consumed by fish.

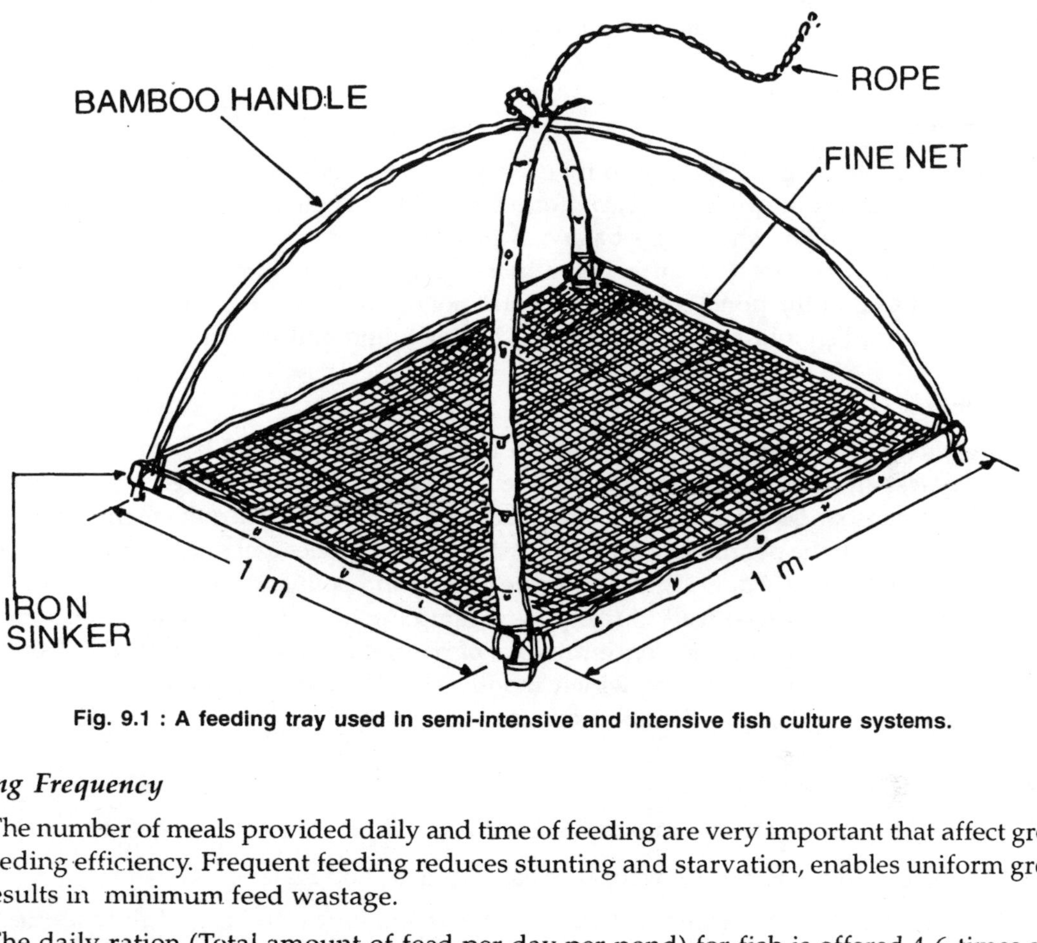


**Fig. 9.1 : A feeding tray used in semi-intensive and intensive fish culture systems.**

### *Feeding Frequency*

The number of meals provided daily and time of feeding are very important that affect growth and feeding efficiency. Frequent feeding reduces stunting and starvation, enables uniform growth that results in minimum feed wastage.

The daily ration (Total amount of feed per day per pond) for fish is offered 4-6 times a day and as the fish grow the number of meals are reduced to either one or two per day. Feeding of fish fingerlings is done between 6 AM and 4 PM.

### *Size of Feed*

The size of the feed is also important for the growing stages. Under-or over-sized particles in the feed, if present, should be removed through screening. Fine particles often clog the gills of the fish and provides a medium for propagation of a variety of micro-organisms.

Generally the number of feed particles in the daily ration should correspond to the estimated number of fish present in the pond. For example, suppose a pond has been stocked with the fingerlings of *Labeo rohita* at the rate of 1,50,000 per hectare. Obviously, each fish should get a feed particle. Assuming that the total biomass of fish is 1000 kg and the feed offered at the rate of 6 per cent of the biomass. Therefore, 60 kg of feed to be offered daily. If this total amount of feed is equally distributed at 4 feeding times, each meal size will be 15 kg and each of this 15 kg feed contains about 1,50,000 particles for all the fish to get a pellet or granules. Therefore, artificial feeds should contain adequate number of particles to provide almost the total surviving population of fish.

### *Quality of Feed*

Fish feed should be good quality so that fish can easily consume and the following points should be considered for the purpose. (1) Uniform size, (2) Feed pellets should be good smell, taste and fresh, (3) Feed pellets should be balanced in nutrient constituents and stable in water, (4) Attractants must be present in feed for increasing the consumption rate to avoid the accumulation of feed at the pond bottom and water pollution, (5) Food conversion ratio must be low, and (6) Fish feed should not contain fungus, bacteria, and anti-nutritional factors.

### *Time of Feeding*

It is not advisable to feed fish all the year round. It has been found by experience that fish consume very less amount of artificial feed in winter season but consume a good amount of feed in summer and monsoon seasons when the temperature of water becomes quite high and fluctuates between 24 and 36$^0$C. If the ponds are loaded with natural food species, very little amount of artificial feed is necessary. If feeding is done, there is a waste of food and pollute the water bodies. Therefore, prior to application of artificial feed to fish ponds, sampling of water and soil is done once in a month to check the abundance and concentrations of natural food species in ponds. Experience has shown that feeding of fish during winter months should be done at noon while in summer and monsoon months, feeding in the afternoon is the best time.

## 9.9 Divisions of Artificial Feed

Feeding stuffs can be divided into three heads :

1. Vegetable products, flour, soyabean-meal, oat-meal, and maize-meal.
2. Fresh meats of cattle, sheep and pig liver, heart and lungs, and fresh fish (both fresh and sea water fish).
3. Ground and dried products such as fish meal, meat meal and others.

Generally fish farmers use the cheapest possible ingredients provided they give better survival and yield of fish biomass. Different combinations of the foods have been made with varying degrees of success. Carnivorus fish species and prawn consume meat meal, fish meal, fresh fish and fresh meat; prawns consume fresh meat of mussels and clams; grass carps prefer vegetable products while major and exotic carps consume soyabean meal and fish meal. When dried vegetable products are mixed together and fed to trout fish or when used separately, fairly successful results have been observed. On the other hand, if any one of the vegetable products is mixed with fresh meat/ fresh fish, the fish will consume only the meat/fish leaving the vegetable parts of feed.

## 9.10 Nutrient Requirement of Fish Feed

Natural food constitutes an important source of nutrient for extensive fish culture but for intensive and semi-intensive fish cultivation, artificial feed is necessary. In intensive fish culture operations, feeds are the more expensive and accounting for about 60 per cent of the total input costs. Therefore, the development of feed quality is important for fish farming. Moreover, consumer awareness and choice strongly dictate the susceptibility of fish as food. In general, nutrient requirements of fish feed include proteins, carbohydrates, lipids, vitamins, and minerals.

### *Proteins*

Proteins are molecules composed of long chains of amino acids linked by peptide bonds; they make up about half of the dry weight of the bodies of fish. The 'primary' structure of a protein is the order, or sequence, of amino acids in its polypeptide molecules. Proteins differ in the sequence of amino acids in their polypeptide chain; the precise sequence determines the final three-dimensional shape of the protein molecule. The stages in the formation of a three-dimensional shape from a straight polypeptide chain are called *secondary, tertiary,* and *quaternary* levels. However, proteins in the diet are essential to repair or replace existing molecules, or for growth, but may be exploited as an energy source.

Proteins are considered as the best nutritional and most expensive components in fish diet. Fish consume proteins for proper maintenance, growth, and reproduction. Fish growth is determined by the level of protein and it constituent amino acids. The diversity in feeding habits exhibited by the fish is reflected in the variation in their protein and essential amino acid requirements. If excess protein is supplied, only part of it will be used as growth while the remainder part will be converted into energy. Inadequate supply of protein to fish feed leads to loss of weight. Studies on the growth response of Indian Major Carp have shown that consumption of protein is high in their early life stages (Table 9.2).

**Table 9.2 : Protein Requirement of an Indian Major Carp, *Catla catla***

| Stage | Protein in diet (Per cent) |
|---|---|
| Fry | 47.17 |
| Fingerlings | 45.78 |
| Advanced fingerlings | 41.28 |

*Source :* Devaraj and Seenappa (1988)

Fish requires a balanced mixture of essential and non-essential amino acids. The minimum amount of dietary protein essential for optimum fish growth varies between 24 and 55 per cent. These variations are due to various reasons such as life stages and age of fish, quality of protein, species variations, palatability of diet, culture technology, and availability of natural food organisms.

To obtain maximum growth and production of cultured fish, optimum dietary protein level should always be maintained (Table 9.3). In general, protein requirement of fish is higher than that of terrestrial animals. When protein is mixed with the diet at high rate, a substantial amount of it is deaminated* and the carbon is burned. It has been shown that protein requirements are highest in fry and fingerlings and that decreases along with the increase in fish size (Table 9.4).

*Refers to removal of amide ($NH_2$) group from an amino acid and leading to the formation of ammonia.

**Table 9.3 : Protein Requirement of Some Common Important Commercial Fish**

| Common name | Species | Protein content (g/kg feed) |
|---|---|---|
| Rohu | *Labeo rohita* | 400-450 |
| Catla | *Catla catla* | 470 |
| Mrigal | *Cirrhinus mrigala* | 400-450 |
| Silver Carp | *Hypophthalmicthys molitrix* | 370-420 |
| Catfish | *Clarias* spp. | 357 |
| Murrel | *Channa* spp. | 520 |
| Tilapia | *Oreochromis mossambicus* | 520 |
| Grass Carp | *Ctenopharyngodon idella* | 360-420 |
| Common Carp | *Cyprinus carpio* | 330-480 |

*Source :* Selected data from Paulraj (1995).

**Table 9.4 : Dietary Protein Requirement of Tilapia**

| Fish weight (g) | Protein requirement (Per cent) |
|---|---|
| 1.0 | 35-50 |
| 1.0-5.0 | 30-40 |
| 5.0-25.0 | 25-30 |
| More than 25.0 | 20-25 |

1. *Essential Amino Acids* : Requirement of protein is highly affected by the quality of dietary protein. The quality of protein depends upon the balanced levels and bio-availability of essential amino acids (EAA). The most important amino acids necessary for fish diet are isoleucine, leucine, lysine, methionine, histidine, arginine, phenylalanin, therionine, tryptophan, and valine. The amino acid profile of a protein, however, indicates the relative proportions of the EAA within that protein.
2. *Diet Supplemented with Amino Acids* : Fish feed having inadequate amounts of amino acids significantly limit the growth of farmed fish. Therefore, fish feed should be supplemented with amino acids; of course, shellfish and finfish effectively utilize free amino acids at varying degrees. Trout and salmon are able to utilize free amino acids for growth. Soyabean meal supplemented with several amino acids was superior source of protein to the meal. Diet containing fish meal, yeast, bone meal, and soyabean meal could further be improved by adding cystine and tryptophan at the rate of 10 and 5 g/kg of feed, respectively. But a diet for common carp having free amino acids (such as cystine, arginine, tryptophan, or methionine) had little effect on growth.
3. *Utilization of Non-Protein Nitrogen* : Carp diet containing 2.4 or 9.5 per cent of urea or ammonium citrate respectively has been found to utilize the feed most efficiently. In some species of catfish, addition of urea (3 per cent of the feed) substituted one fifth of the dietary protein, did not affect the performance of the fish.

Protein having an EAA profile have high nutritive value. Therefore during preparation of fish feed, this factor should be considered precisely. However, when a balanced diet fortified with essential amino acids are used in fish culture, a substantial amounts of amino acids are lost. Various substances and techniques have been developed to bind amino acid with proteins so that such loss can be reduced. Through improved technology, it has also been possible to increase the retention time of amino acids within the alimentary canal of fish for utilization more efficiently.

### *Carbohydrates*

Carbohydrates are one of the major classes of natural organic compounds with the general formula $C_X(H_2O)_Y$. They include sugars, starch, and cellulose. The simplest carbohydrates are sugars (such as ribose and glucose). These are called *monosaccharides* and are the basic units from which all other carbohydrates are built. When two of these simple sugars are bonded together, they form compounds called *disaccharides*; these include sucrose and moltose. Compounds known as *polysaccharides* are formed by joining together ten or more monosaccharides. These are very important biological compounds and include storage materials such as starch and glycogen.

Carbohydrates are an essential component of the diet of fish and may be used as a source of energy (respiration) or modified by being combined with fats or proteins. Although fish utilize carbohydrates as energy sources, but their efficiency of utilization is variable depending on the level and type of carbohydrate used in the diet. Experiments have shown that different forms of carbohydrates are better source of energy due to rapid absorption in the digestive tract. The most important source of carbohydrate in fish feed is wheat/rice bran, oil cakes, grasses, wheat and maize flours.

> *Carbohydrate Utilization in Fish* : Since carnivorous fishes have poor ability to digest carbohydrates, the formulated feed for such fishes must contain carbohydrate level less than 20 per cent. Carnivorous fishes produce very low amount amylase and for this reason, they are not able to utilize food containing carbohydrates. In contrast, omnivorous and herbivorous species (such as Indian Major Carp, Tilapia Channel catfish and others) are able to utilize more than 45 per cent carbohydrate in the form of cooked starch or mixture of cereal brans and oil cakes. Glucose, maltose, and sucrose are, however, utilized by fish at varying degrees.

### *Lipids*

Lipids are a major group of natural organic compounds comprising fats, oils, phospholipids, and sterols. The molecules of all these substance contain a high proportion of $CH_2$ groups and therefore not very soluble in water, but are readily soluble in organic solvents such as ethanol and chloroform.

A high lipid content in fish meal has been considered to lower the value of the meal for feeding purposes. The lipid content of a fish meal may contribute to the overall nutritive value of the product as they are source of energy, fatty acids, sterols, fat soluble vitamins, and phospholipids. Detail information regarding the nutritive value of fat derived from fish feed is not available and research conducted so far did not provide satisfactory answer. Fish meal generally contains about 40 per cent of free fatty acid. It has ben observed that free fatty acid is utilized by many freshwater fish. It seems probable that the fat content of the fish meal may exerts an indirect effect on the nutritive value of the meal. Generally freshwater fish do no tolerate as high levels of dietary lipid as do marine fishes. Lipid level of 4-80 per cent appears to be optimum for fish. Presence of high level of lipid (>40 per cent) in fish meal has been found to reduce fish growth. Fatty acids of the linoleic (18 : 2 n-6), linolenic (18 : 3 n-6, 20 : 5 n-3, 22 : 6 n-3 etc.), stearic (18 : 0), palmitic (16 : 0) and oleic (18 : 1 n-9) series are probably important for better growth of fish. It has been

**Table 9.5 : Requirement of Lipid of Some Freshwater Fish**

| Common name | Lipid level (g/kg feed) |
|---|---|
| Indian carps | 50-80 |
| Chinese carps | 50-80 |
| Common carp | 80-100 |
| Tilapia | |
| (a) Upto 0.5g | 100 |
| (b) Upto 35g | 80 |
| (c) More than 35g | 60 |
| Rainbow trout | 120 |
| Catfish | 80-120 |
| Eel | 100 |

*Source* : Selected data from Paulraj (1995)

experimentally proved that the nutritional values of different types of oil seeds which is high n-9 fatty acids, promoted better growth of prawns and shrimps. The total lipid level recommended in fish feeds is shown in Table 9.5.

In addition to fatty acids, phospholipids (called lecithin, choline etc.) are also equally important for dietary requirement. Fish meal fortified with phospholipids was found to be highly effective so far as survival and growth of fish is considered. The chief source of phospholipids is soyabean.

*Lipid Utilization in Fish* : Since carnivorous fishes utilize lipids more effectively, more than 25 per cent lipid level may be maintained in their diets. Less amounts of lipids are utilized by herbivorous fishes. Dietary lipids are digestible source of energy and when lipids are incorporated in feed at optimal levels, about 90 per cent of lipid is digested by fish.

### *Minerals**

The constituents of food can be divided into organic nutrients, inorganic nutrients, and the minerals, serve a variety of functions in animals including fish. The essential minerals are probably common to all animals but their identity and the amounts needed have not been studied in detail. The minerals may be required in such small amounts that they are difficult to detect but, with improved chemical analytical techniques, trace elements are still being added to the list of essential minerals. It is customary to distinguish between minerals and trace elements, but there is no sharp delineation between the two groups. The amounts of essential minerals decrease gradually over many orders of magnitude. Trace elements are present in the body in concentrations smaller than 1 in 2,00,000. Iron, however, occupies a transitional position.

Minerals are very important for several regions such as :

1. Minerals maintain acid-base balance and osmotic relationship with the aquatic environment.
2. They provide rigidity and strength to bones of fish.
3. Minerals are the main components of blood pigments, enzymes, and organic compounds in tissues and organs.
4. Minerals are responsible for normal interactions between endocrine and nervous systems.

About 21 recognized mineral elements are known which perform essential functions in fish. Minerals such as calcium, phosphorus, magnesium, sodium, potassium, sulfur, and chlorine are

* For further study, see Hilton (1989) and Lall (1989).

required relatively in higher amounts and are termed as *macronutrients* whereas copper, iron, zinc, cobalt, manganese, iodine, selenium, nickel, florine, vanadium, tin, silicon, chromium, and molybdenum are required in very small amounts and therefore, these minerals are termed as *micronutrients.* Generally good quality of diet is the main source of minerals, although some minerals are absorbed from the environment. The mineral content of fish feed varies depending on the raw materials used during preparation of feed. Non-availability of minerals in fish feed in adequate amounts affect survival, growth, and reproduction of fish and may also cause deficiency diseases.

Fish can absorb minerals directly from the environment through body surfaces and gills. Therefore, the dietary requirement of minerals principally depends upon the mineral concentration of the fish culture environment. Some mineral requirements of fish are satisfied through absorption from the water. If fish culture ecosystems are not fertilized with phosphorus-containing materials and agricultural lime, phosphorus and calcium levels of water will obviously be lowered and hence, these minerals should always be incorporated in the diet. However, different species of fish exhibit considerable variations in the utilization of minerals.

1. *Phosphorus and Calcium* : In extensive culture systems, fish obtain phosphorus and calcium from the water and natural food. But in intensive and semi-intensive culture systems, these elements should be supplied through diet. Since calcium and phosphorus levels in natural waters are very low, balanced diet is the main source of these elements. Generally animal ingredients are considered as the chief source of calcium and phosphorus. Plant sources of these elements are poorly available because they are bound with plant phytin. However, the availability of phosphorus varies with the source of raw materials and fish species (Table 9.6).

**Table 9.6 : Availability Factors for Phosphorus**

| Ingredient | Per cent availability factor (to apply to the total phosphorus in the feed) | |
|---|---|---|
| | Fish containing no stomach | Fish containing stomach |
| Plants | 30 | 30 |
| Plant products | 57 | 58 |
| Animal products | 97 | 90 |
| Yeast and Bacteria | 90 | 90 |
| Monobasic sodium, potassium or calcium phosphate | 95 | 95 |
| Dibasic calcium phosphate | 45 | 70 |
| Tribasic calcium phosphate | 15 | 65 |
| Rice bran | 25 | 19 |
| Fish meal | 10-33 | 60-72 |

*Source :* Paulraj (1995).

Most phosphorus is concentrated within the nucleus. It combines with proteins and form phosphoproteins which initiate muscle action. Phosphorus also binds with lipids and form phospholipids essential for lipid metabolism. It should be remind that phosphorus occurs in fish in the form of high-energy phosphates, nucleic acids, and phosphoproteins which limits the value of total phosphorus determination. Organic phosphorus compounds are, however, very important for various metabolic processes.

2. *Zinc* : It is involved in the synthesis of nucleic acid and also a component of metallo-enzymes. Phytic acid, a plant protein, binds with zinc rendering it unavailable. Since fish can able to tolerate relatively high concentration of zinc (upto several hundreds mg) without any harmful effects, it may supply at high levels in the diet to prevent calcium antagonism.
3. *Iron and Copper* : Iron is a component of haem found in haemoglobin, peroxidases, and in cytochromes. It is also involved in respiratory processes including electron transport and oxidation-reduction reactions. Ferrous ions are better absorbed by fish than that of ferric ones.

   Copper is a co-factor in tyrosinase and ascorbic acid oxidase. Addition of copper in the diet improves the growth performance of fish.

   Copper and iron together enters into the constituent of a complex of iron-copper nucleoproteins from which they are liberated by trypsic digestion. The complex contains about 40 per cent ionizable iron which is rapidly absorbed by the intestinal membrane. Due to the iron content in feed, fish can able to regenerate haemoglobin in case of anaemia.
4. *Iodine* : Iodine is an essential constituent of thyroxine that is intimately associated with the mortality of farmed fish. Since freshwater contains very small amounts of iodine, this should be provided with feed to prevent fish goiter.
5. *Cobalt* : It is an essential component of vitamin $B_{12}$. Addition of cobalt salt in the diet increases the growth of fish.
6. *Manganese* : It is a co-factor in the enzyme asrginase and certain metabolic enzymes which are responsible for bone formation and erythrocyte regeneration in fish. However, manganese is very important for the growth of different species of fish such as *Cyprinus carpio, Cirrhinus mrigala, Ictalurus punctatus* and other freshwater fishes. This mineral is mixed with fish feed at rates varying between 0.5 and 12 mg/kg of feed in combination wit oil cake and rice bran (1:1 ratio).
7. *Magnesium* : It plays an important role in enzyme co-factors, structural components of cell membrane, and in extracellular fluids. Fish has the capacity to absorb this element from the environment. Since freshwater ecosystems contain very low amounts of magnesium, fish entirely depends upon the dietary sources.
8. *Selenium* : It is the main component of the metallo-enzyme glutathione peroxidase and plays an important role in the defence mechanism of the fish. It has been shown that the dietary selenium and vitamin E function synergistically to prevent oxidative damage. High dietary levels (10 mg/kg of feed and above) result in uncoordinated swimming movement of fish and death occurs within 12-24 hours.
9. *Sulfur* : Sulfur-containing amino acids are the chief carriers of this mineral in fish feed. The sulfur content in fish feed is low but other sulphurous compounds are absent. The sulfur or protein is centred in the following amino acids : Cysteine, Methionine, and Cystine. Other organic forms of sulfur such as glutathione and taurocholic acid are found in fish body. However, the mineral ingredients used for mineral premix in fish feed and their requirement in diet (per kg of feed) are shown in Tables 9.7 and 9.8.

**Table 9.7 : Different Types of Ingredients Used for Mineral Premix**

| Mineral | Types of ingredients |
|---|---|
| Phosphorus | Di-calcium phosphate, Mono-calcium phosphate, Mono-sodium phosphate, Mono-potassium phosphate |
| Calcium | Calcium carbonate, Calcium phosphate, Di-calcium phosphate |
| Magnesium | Magnesium sulfate, Magnesium carbonate |
| Sodium | Sodium chloride |
| Potassium | Potassium sulfate, Potassium chloride |
| Manganese | Manganese sulfate |
| Copper | Copper sulfate |
| Zinc | Zinc sulfate |
| Iron | Ferrous sulfate, Ferrous carbonate |
| Iodine | Potassium iodide |
| Cobalt | Cobalt sulfate, Cobalt chloride |

**Table 9.8 : Requirement of Minerals in Fish Diet**

| Mineral | Requirement (per kg feed) | | | |
|---|---|---|---|---|
| | Trout and salmon | Common carp | Indian and chinese carp | Other fish |
| Phosphorus (g) | 7-8 | 6-7 | 5-7 | 1-2 |
| Calcium (g) | 0.2-0.3 | 0.28 | 5-18 | 5 |
| Magnesium (g) | 0.5-0.7 | 0.4-0.6 | - | 0.5 |
| Sodium (g) | 1-3 | 1-3 | 1-3 | 1-3 |
| Potassium (g) | 1-3 | 1-3 | 1-3 | 1-3 |
| Sulfur (g) | 3-5 | 3-5 | 3-5 | 3-5 |
| Copper (mg) | 3 | – | 4 | 1-4 |
| Manganese (mg) | 12-13 | 4 | 4 | 20-25 |
| Zinc (mg) | 15-30 | 15-25 | 10-20 | 20-25 |
| Cobalt (mg) | 2-4 | – | 3-6 | 5-10 |

*Source* : ADCP (1983), New (1987).

## 9.11 Vitamins

The enzymes are proteins, but many metabolic enzymes require non-protein components for their function. These components may be small organic molecules, vitamins, or inorganic ions. Thus, vitamins are organic nutrients which are required in small amounts in the food for proper function and growth of the fish. Vitamins are very diverse group of substances, not related structurally or chemically. All other needed as coenzymes in chemical reactions within the fish. There is little obvious connection between the known biochemical roles of the vitamins and the enzymes of which they are constituents on the one hand and the diverse structural and functional symptoms of deficiencies on the other hand. These symptoms are mainly known from humans, cattles, chickens, and fishes.

Although green plants and most micro-organisms synthesize their own vitamins, fish must obtain most types of vitamins in their diets. Absence of short supply of any of these substances causes a specific vitamin deficiency disease.

Vitamins are essential for myriad purposes such as disease resistance immune response, (for comments, see Hardie *et.al.* 1990, 1991, Swain and Das, 1996), maintenance, growth and reproduction of fish. In extensive fish culture system where fish are stocked at very low densities, natural food species may provide a substantial amount of vitamins to fish. But in case of high stocking densities (in intensive and semi-intenisve systems), the concentrations of natural food items are drastically reduced and therefore, addition of vitamins to the diet is necessary to meet their requirement.

The vitamin content of fish meals may vary greatly according the processing procedure and the raw materials. Vitamin requirements for commercially important species of fish have been well defined. Although three fat-soluble vitamins (such as vitamin A, D and K) and the water-soluble vitamins (such as vitamin C, choline, inositol, folic acid, biotin, nicotinic acid, pyridoxine, thiamin, vitamin $B_{12}$, and riboflavin) have been shown to be dietary important for growth and metabolism and their exact requirement depends upon fish weight and deficiency symptoms.

Very few reports have been obtained to throw scientific outlook on the specific amounts of vitamin requirements for important species of cultured fish. Still more obscure is the methodological constraints and biological significance of vitamins in the metabolism of fish. The data obtained so far for vitamin requirement in fish diet is still inadequate as it is desirable to undertake experiments on fish. However, major obstacles discouraging such experiments are frequent failures in meeting the requirements of vitamins for experimental raising of fish and technical difficulties involved in working with fish as they live in water.

**Table 9.9 : Recommended Levels of Vitamins in Fish Feed**

| Vitamin | Recommended levels (per kg of feed) | | | |
|---|---|---|---|---|
| | Trout and salmon[1] | Carps[1] | Catfish[2] | Tilapia[2] |
| Vitamin A | 1,000-2,000 | 2,000 | 1,500 | 5,000 |
| Vitamin $D_3$ | 1,600-2,000 | – | 750 | 600 |
| Vitamin E | 30-50 | 100 | 125 | 25 |
| Vitamin K | 10 | 4 | 6 | 4 |
| Ascorbic acid (mg) | 300 | 30-50 | 10 | 16 |
| Cyanocobalamin (mg) | 0.01-0.02 | ** | 0.01-0.9 | 125 |
| Choline (mg) | 800 | 500-600 | 250-470 | 200 |
| Biotin (mg) | 0.4 | 0.6-1.0 | 0.9 | 0.6 |
| Folic acid (mg) | 5-10 | 15 | 6 | 8 |
| Niacin (mg) | 150 | 28 | 25-40 | 13 |
| Inositol (mg) | 200-400 | 440 | 170-250 | 18 |
| Pyridoxine (mg) | 10-20 | 5-10 | 3-8 | 5 |
| Pantothenic acid (mg) | 60 | 30-40 | 15-25 | 30 |
| Riboflavin (mg) | 20 | 4-7 | 6-15 | 22 |
| Thiamine (mg) | 10 | 2-3 | 4-7 | 4 |

*Source* : 1, Paulraj (1995); 2, Compiled by the author from different published literature
** Gut microbes synthesize it in adequate levels

The optimum dietary requirements of vitamins are affected by the age, growth rate, health status, nutrient compositon of the diet, size of the fish, physiological conditions, availability of vitamins from natural food, and environmental conditions. During processing and storage of feed ingredients, a substantial amounts of vitamins are lost which also affect the availability of vitamins to the cultured fish. Dietary requirement of vitamins (per kg of feed), their sources and deficiency symptoms are shown in Tables 9.9 and 9.10.

**Table 9.10 : Source of Vitamins and Their Deficiency Symptoms in Fish**

| Vitamin | Source | Deficiency Symptoms |
|---|---|---|
| Vitamin A | Fish liver oil (shark, cod etc.) palmitate acetate, beta carotene | Poor growth, reduced survival, cataract xerophthalmia, haemorrhage in eye and fin |
| Vitamin $D_3$ | Cod and shark liver oil | Poor appetite, reduced growth, lethargy, lower body phosphate |
| Vitamin E | Synthetic tocopherol in estarified acetate or phosphate form | Poor growth, clubbed gills, xerophthalmia and muscular dystrophy |
| Vitamin K | Soyabean, green vegetables, menadione sodium bisulphate complex (synthetic compound) | Haemorrhage in gills and eyes, anaemia |
| Ascorbic acid | Kidney tissues, fresh insects, coated ethyl cellulose, gelatin-coated ascorbic acid | Metabolic disorders, fin erosion, bacterial infection, reduced calcium absorption, haemorrhage, gill deformities |
| Cyanocobalamin | Fish meal, fish liver and kidney, slaughter-house waste and poultry by-product | Poor appetite, growth and food conversion, fragmented erythrocytes. |
| Choline | Wheat meal, beans, choline chloride solution (70 per cent) | Poor growth and feed conversion, loss of appetite, enlarged livers, haemorrhagic kidneys |
| Biotine | Yeast, liver, milk products, egg yolk | Loss of appetite, skin disorder, muscle dystrophy, fragmentation of erythrocytes, depigmentation |
| Folic acid | Yeast, liver, kidney, green vegetables, fish viscera | Poor growth and loss of appetite, lethargy, fragile fins, anaemia, dark skin pigmentation |
| Niacin | Yeast, liver, kidney, gut microbial synthesis, nicotinic acid | Loss of appetite, skin and fin erosions, poor feed conversion and haemorrhages |
| Inocitol | Yeast, liver, green vegetables, mesoinocitol | Poor growth and feed conversion, edema and dark color in skin |
| Pyridoxine | Yeast, egg yolk, liver, pyridoxine hydrochloride | Affect protein metabolism, gasping breathing with flexing opercules, poor appetite, anaemia and nervous disorders |
| Pantothenic acid | Cereal bran, yeast liver, fish flesh calcium pantothenate | Impairment of reproduction, poor growth, clubbed gill disease (clubbing together of gill filaments and lamillae), distended opercules, necrosis, sluggishness |
| Ribolfavin | Liver, Kidney, yeast, groundnut, soyabean, riboflavin tetrabutyrate | Poor appetite and growth, increased mortality, cataract, short-body, fin necrosis, photophobia and haemorrhages |
| Thiamine | Yeast, cereal bran etc. thiamine mononitrate and thiamine hydrochloride | Impaired carbohydrate metabolism, poor growth and appetite, nervous disorders and increased sensitivity to shock |

## *Role of Vitamins in Fish Flesh*

The amount of vitamins in fish flesh is influenced by the quality and quantity of diet and also the relative amount of flesh. Also, the amount of flesh in fish varies with size, species, sex, and season of capture. It has been estimated that the amount of flesh constituted 50 to 60 per cent of the total weight of the fish. Therefore, it may be concluded that the relative amount of muscular tissues is generally higher in fish than in man or domestic animals.

Fresh fish flesh is, in general, contains a very poor amounts of vitamin, A, D, and E. The muscle tissues of "lean" fish contain little or none, but in case of "fatty" fish contain rather more. However, it is improbable that provitamin D in the muscle tissues converted to vitamin D by the action of ultraviolet light on the body surface, as in mammals. The activating rays of the sun penetrate only to a depth of one metre, and fish normally inhabit much deeper waters.

Fish probably ingest provitamins D, since these occur in many natural food species. At, present, there is no convincing evidence that fish can able to convert provitamins to vitamin D, since the energy required for this conversion is derived from ultraviolet light. If conversion does occur, there is, furthermore, no evidence to decide whether the fish employs provitamins from an external source, or whether it forms its own provitamin by dehydrogenation of cholesterol. Though the origin of vitamin D is a matter of speculation, it has already been suggested that fish can able to synthesize vitamin D.

The origin of vitamin A in fish is beta-carotene, which is present in the diatoms on which small crustacea subsist. The latter contains the preformed vitamin and consumed by small fishes, which in turn forms the food of the larger fishes. It has been demonstrated that fish can convert beta-carotine to vitamin A.

Vitamin E exerts an influence through controlling peroxidation in lipids. Fishes are not able to oxidize lipid mainly due to the presence of polyunsaturated lipid. This causes the flesh susceptible to spoilage. Taste of fish flesh can be improved through increased feeding of vitamin E to a large number of fishes such as trout, channel catfish, salmon, African catfish and carps. Experiments have shown that muscle tissue accretion was dependent on vitamin E-containing diet (Figure 9.2). With the

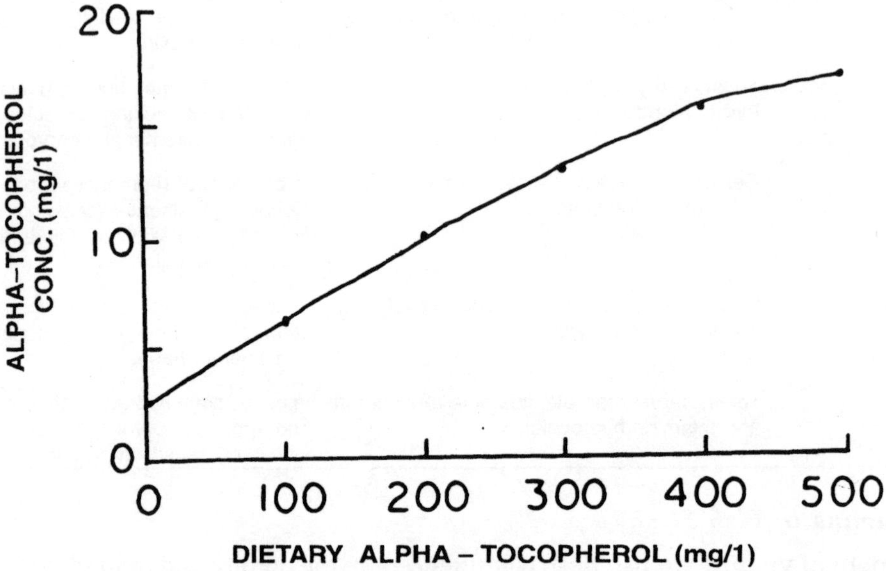


Fig. 9.2 : Fillet vitamin E concentration from catfish diets varying in vitamin E dose. Muscle tissue accretion of vitamin E was dependent on dietary supply. Vitamin E improves fish flesh quality. [Redrawn from Baker (1997)]

gradual increase in the dietary alpha-tocopherol level, accumulaiton of alpha-tocopherol increased within the muscle and thus fish flesh quality is enhanced to a greater degree. Moreover, increase in the level of vitamin E in the diet suppresses lipid oxidation in fish flesh. Vitamin contents of some freshwater fish flesh are shown in Table 9.11.

**Table 9.11 : Vitamin Contents of Some Freshwater Fish Flesh**

| Common name | Vitamin | | | | | | | |
|---|---|---|---|---|---|---|---|---|
| | A[1] | D[1] | Thiamine[2] | Riboflavin[2] | Niacin[2] | Ascorbic acid[2] | Pantothenic acid[2] | Folic acid[2] |
| Common carp | 170 | 10,000 | 0.9 | 0.08 | 0.8 | 166 | 7.5 | 0.004 |
| Rohu | 145 | 7,500 | 0.05 | 0.07 | 0.7 | 171 | 1.4 | 0.009 |
| Mrigal | 100 | 3,800 | 0.07 | 0.05 | 0.5 | 166 | 2.1 | 0.03 |
| Catla | 165 | 6,800 | 0.02 | 0.10 | 0.9 | 176 | 4.2 | 0.07 |
| Catfish | 30 | 38 | 0.08 | 0.17 | 0.9 | 161 | 5.2 | 0.12 |
| Eel | 4,500 | 4,700 | 0.004 | 0.03 | 0.2 | 100 | 2.5 | 0.10 |

Compiled from different published literature

1. i.u./100g, 2. mg/g

Table 9.12 indicates that vitamin A and D contents of carp and eel are very high. Though vitamin A content in sturgeon is much higher, vitamin D content is very negligible. The Indian major carps also contain negligible amounts of vitamin A and D. However, in cases where fish contain high amounts of oils, they are relatively rich in tocopherol. They also contain large amounts of highly unsaturated fatty acids, which acts as antagonists to vitamin E. Therefore, their actual biological potency may be much less than would be expected from the alpha-tocopherol content.

**Table 9.12 : Vitamin A and D Contents in Fish Oils**

| Common name | Vitamin (i.u./g oil) | |
|---|---|---|
| | A | B |
| Common carp | 5,000-7,000 | 10,000-11,000 |
| Rohu | 39 | 50 |
| Mrigal | 12 | 9 |
| Eel | 5,700-16,500 | 6,450-8,870 |
| Sturgeon | 26,000 | Less than 1 |

Compiled from different published literature

## 9.12 Factors Affecting the Nutritive Value of Fish Feed

It is difficult to generalize with respect to the relative nutritive value of fish feeds made from various types of raw materials. Types of raw materials affect the composition of the final product, not only because of differences in the composition of the entire materials, but because of differences in the proportions of different materials which are utilized by the fish. However, the information regarding fish feed and fish meal available so far is confined to the laboratory under well-defined controlled environmental conditions. The information though well-defined, cannot be directly applied to commercial fish production mainly due to differences in species and size, feeding practices and culture conditions.

Much of the reports available on the nutritional quality of fish feed is difficult to assess because it is based on systematic studies. Furthermore, in drawing useful conclusions from the results regarding the nutritive value of fish meal on the relative merits of fish meals produced from different types of ingredients, it is important that careful consideration should be given to the nutrients supplied by the diet.

In present day, with a knowledge of nutrition available, an appreciation of essential nutrients and the interactions among nutrients, there can hardly be a general statement of the value of any fish meal when composition of the diet and the amount as well as availability of fats, proteins, carbohydrates, and minerals present in the fish meal have to be taken into consideration in any evaluation of nutritive worth. A clearcut distinction between vegetable and animal proteins should also be made : the latter assumed to be of higher nutritive value. The emphasis on the importance of animal protein grew out of the fact that prior to development of processing techniques for vegetable protein concentrates, it was difficult to compensate for the amino acid shortage in the cereal protein of the diet without recourse to some protein derived from animal sources. Animal protein concentrates possess vitamins which were short in the vegetable protein concentrates. This is true with regard to vitamin $B_{12}$. Due to the rapid development of synthetic processes for manufacturing amino acids and vitamins, the formulation of feed is based on the analysis of feed-stuffs for all the nutrients.

## 9.13 Ingredients of Fish Feed and Their Nutritive Values

A variety of fish feed ingredients have been known in different parts of India. But it is very important to identify suitable ingredient for use in the preparation of fish feed. Though adequate information on the nutritive value of various feed ingredient have received from various experiments, their biological significance for cultured fish are less clearly understood. The most important common and widely available feed ingredients are briefly summerized below.

### *Animal Ingredients*

1. *Fish Meal* : It is the most abundant animal protein source commercially produced as pulverised dried miscellaneous fish and crustanceans, often termed as *trash fish*. The quality of fish meal depends upon the size, species, freshness of the material, and the process employed in the manufacturing of the meal. Fresh single fish species is the best ingredient for preparation of meal. Fish meal is prepared by steam cooking fish, pressing to remove oil and water contents of the body, drying and then pulverising the whole body of fish. However, some problems which are associated with the quality of dry fish, are noted below :

    (a) variation in species and size compositions of the fresh fish.

    (b) high levels of acid-insoluble ash (about 16 per cent).

    (c) trash fish is spoiled before processing due to microbial attack.

    (d) the salt content is very high (more than 12 per cent) which is more than the required level.

    (e) when crustacean ingredients are used, the level of chitin becomes very high which is, at all, not desirable.

    (f) variable moisture content.

Good quality fish meal contains the following nutrients : Protein (50-65 per cent), ash (17-30 per cent), lipid (10 per cent), crude fibre (1-4 per cent), calcium (2-7 per cent), phosphorus (2-4 per cent), vitamins and amino acids, sodium chloride (3 per cent), moisture (10 per cent), and antitoxidant (200mg/l).

2. *Fish Solubles* : It is the water-material obtianed after the removal of oil from the liquid pressed out during the preparation of fish meal. It is used in fish feed as a good attractant. It contains vitamin B-complex groups.
3. *Fish Silage* : It is prepared from several species such as trash fish, waste fish, viscera and crabs. These are crushed together, mixed with a mixture of hydrochloric acid or sulphuric acid and formic or propionic acid to prevent bacterial decay. Lactic acid bacteria are also added to prevent the fish from mortality.
4. *Poultry by-product Meal* : It is prepared from poultry viscera, heads, and feet which has high crude protein (50-60 per cent), all essential amino acids, minerals, vitamins, lipid (16-35 per cent), and ash (9-20 per cent).
5. *Slaughterhouse Offals* : Bone meal is prepared from cattle wastes that contains fat (10 per cent), ash (30 per cent), phoshorus (5 per cent), calcium (10 per cent), and protein (50 per cent). Dried and pulverised blood meal is prepared by drying the meal. It also contains high crude protein (85 per cent), iron (2784 mg/kg dry matter), and trace amounts of minerals, niacine and cyanocobalamin.
6. *Milk By-products and Chicken Eggs* : Skimmed and whole milk powder, albumen and yolk parts of eggs contain all essential amino acids, minearls, vitamins, crude protein (46-50 percent), and fat (40-45 pecent). These by-products, though costly, can be used as ingredients in nursery fish feed.

### *Plant Ingredients*

1. *Soyabean Oil Cake* : It is obtained after the extraction of soyabean oil from the seeds. The cake is very rich in protein (48 per cent), fat (2-7 per cent), ash (5-7 per cent), crude fibre (6-8 per cent), amino acids, vitamins, and minerals. The nutritive value of full fat soya will be discussed latter on.
2. *Cotton Seed Oil Cake* : It is also obtained for cotton seeds. The cake contains lipid (4-8 per cent), protein (30-40 per cent), ash (6-9 per cent), crude fibre (6-22 per cent), therionine, methionine, lysine, and tryptophan.
3. *Groundnut Oil Cake* : It is obtained from groundnut kernals after decortication of pods. The cake is widely used in the preparation of fish feed. It contains crude protein (35-40 per cent), lipid (3-8 per cent), crude fibre (6-8 per cent), ash (4-8 per cent), lysine, tryptophan, methionine, therionine, and some vitamins.
4. *Sunflower Oil Cake* : After extraction of oil from the seeds, the cake is obtained. It contains lipid (4-6 per cent), crude protein (40-45 percent), crude fibre (14-16 percent), ash (6-7 per cent), cystine, methionine, vitamin B-complexes, and carotine.
5. *Linseed Oil Cake* : Linseed oil cake, though available in India, has very poor amino acids profile and low levels of lysine and methionine. Vitamin contents are also very low.

6. *Coconut Oil Cake* : This cake is also available in India and used, though not extensively, in fish feed as it contains protein (22-28 per cent), ash (20 per cent), lipid (6-9 per cent), and fibre (14 per cent). It is rich in iron, potassium, choline, niacine and vitamin A.
7. *Mustard Oil Cake* : The oil cake is easily available in the market because it is used as a good organic manure both in agriculture and aquaculture. It contains crude protein, fat, trace amounts of some minerals and amino acids.
8. *Cereal Product* : Generally rice, wheat and maize are used as fish feed ingredients considering their cost, availability and carbohydrate levels. Cereal products are the main sources of energy. Rice bran contains crude protein (12 per cent), crude fibre (12-18 per cent), lipid (7-12 per cent), ash (8-12 per cent), and vitamin B-complexes.

   Wheat bran contains crude protein (10-14 per cent), crude fibre (12-18 per cent), ash (6-18 percent), lipid (6-8 per cent), trace amounts of potassium, phosphorus, maganese, zinc, biotin, pantothenic acid and niacine.

   Maize bran contains crude protein (10-12 per cent), fat (4-6 percent), crude fibre (4-8 per cent), ash (6-9 per cent), arginine, lysine, niacine, and vitamin E.
9. *Yeast and Alfalfa* : Brewers yeast is the unextracted yeast (*Saccharomyces* sp.) produced as a by-product from the brewing of beer. It contains crude protein (40-45 per cent), crude fat (1 per cent), crude fibre (2-7 per cent), ash (6-8 per cent), vitamin B-complexes, phosphorus, iron, and potassium.

   Dried powdered leaf of alfalfa is also used in fish feed as it contains crude protein (15-25 per cent), fat (2-4 per cent), crude fibre (15-30 per cent), ash (10-15 per cent), B-group vitamins, vitamin E, calcium, iron, zinc, potassium, and manganese.
10. *Other Ingredients* : Dried and powdered sea weeds, fruit wastes, algae, silage of plant origin, ipil-ipil leaf, and leaves of aquatic weeds are good sources of proteins, vitamins and minerals. The levels of these nutrients vary according to their habitats, species, size and mode of extraction procedures. Though these nutrients are costly and not easily available in the market, they exhibit synergistic effect on fish if used in small amounts; of course, extensive biological potency should be evaluated before use in fish feeds.

Composition of feed ingredients for Indian Major Carp fry, common carp and tilapia fingerlings are shown in Tables 9.13 and 9.14 for better understanding of the relationship between fish species and composition of feed ingredients.

## 9.14 Soyabean – The Chief Source of Nutrients

Fish meal is generally used as a protein source in fish feeds. Due to increase in price of fish meal with concomittent decrease in quality, full fat soya flour which is derived from soyabean, is being considered as an alternative source of fish meal as it contains essential amino acids, fatty acids, lecithin, and antitoxidants.

**Table 9.13 : Composition of Feed Ingredients Used for Indian Major Carp Fry**

| Ingredient | Quantity (g/kg dry feed) |
|---|---|
| Groundnut cake | 200 |
| Sesame cake | 350 |
| Fish meal | 100 |
| Rice bran | 340 |
| Dicalcium phosphate | 5 |
| Sodium chloride | 3 |
| Mineral premix | 1 |
| Vitamin premix | 1 |
| **Mineral premix** | **mg/kg in the final diet** |
| Copper sulfate | 10 |
| Ferrous sulfate | 100 |
| Zinc oxide | 50 |
| Cobalt chloride | 0.05 |
| Manganous sulfate | 50 |
| Potassium iodine | 1 |
| **Vitamin premix** | **mg/kg of diet** |
| Vitamin A (i.u.) | 5,000 |
| Vitamin D (i.u.) | 600 |
| Thiamine | 10 |
| Riboflavin | 20 |
| Pantothenic acid | 30 |
| Ascorbic acid | 200 |
| Niacin | 50 |

*Source* : Chow (1982a).

**Table 9.14 : Composition of Feed Ingredients Used for Tilapia (A) and Common Carp Fingerlings (B)**

| Ingredient | Per cent composition | |
|---|---|---|
| | A | B |
| Extracted cottonseed meal | 53.0 | - |
| Extracted soyabean meal | - | 50.0 |
| Extracted rice bran[1] | 43.0 | 46.2 |
| Bone meal[1] | 3.0 | 3.5 |
| L-lysine | 0.4 | - |
| Sodium chloride | 0.2 | 0.2 |
| DL-methionine | 0.3 | - |
| Vitamin premix | 0.1 | 0.1 |
| **Vitamin premix** | **mg/kg of diet** | |
| Vitamin A (i.u.) | 5,000 | |
| Vitamin D (i.u.) | 600 | |
| Niacin | 50 | |
| Pyridoxine | 2 | |
| Ascorbic acid | 200 | |
| Pantothenic acid | 30 | |
| Thiamine | 10 | |
| Riboflavin | 20 | |

1. These two ingredients are boiled with water equivalent to 100 per cent of the whole diet to accelerate the binding capability. Other ingredients are then added and dried in the sun.

*Source* : Compiled from Chow (1984b), New (1987)

### *Probelms*

In recent years, it has been known that the quality of fish meal is altered when it is used in ponds due to interaction with the soil, thus resulting in gradual degradation of protein and fat present in it. Furthermore, fish meal is contaminated with the fungus of the genus *Salmonella* sp. due to improper processing. The production of quality fish meal, however, depends upon the avilability of raw materials.

### *Nutritive Value of Full Soya Powder*

Using advanced processing techniques, it has been possible to produce soya flour from soyabeans and consequently, the potency of different types of minerals, nutrients, vitamins, amino- and fatty acids have been retained. On comparison of chemical composition between fish meal and full fat soya flour, it is evident that the latter contains high amounts of ingredients (Table 9.15).

**Table 9.15 : Chemical Composition of Full Fat Soya Flour and Fish Meal (Values for 100 g)**

| Composition | Full fat soya flour | Fish meal |
|---|---|---|
| Moisture (g) | 6-8 | 7 |
| Protein (g) | 38-40 | 60 |
| Oil (g) | 18-20 | 7 |
| Carbohydrate (g) | 20-21 | - |
| Fibre (g) | 3-4 | 0.6 |
| Calcium (g) | 0.24 | 4.14 |
| Iron (mg) | 10-14 | 43 |
| Magnesium (mg) | 175 | 320 |
| Copper (mg) | 1.12 | 11.2 |
| Manganese (mg) | 2.11 | 27 |
| Carotene (Vitamin A) (i.u.) | 426 | - |
| Thiamine (mg) | 0.73 | 0.6 |
| Niacin (mg) | 3.2 | 49.8 |
| Pantothenic acid (mg) | 1.68 | 7.8 |
| Riboflavin (mg) | 0.39 | 6.8 |
| Biotin (mg) | 0.055 | - |
| Folic acid (mg) | 8.65 | - |
| Inositol (mg) | 200 | - |
| Arginine (%) | 7.2 | 4.5 |
| Isoleucine (%) | 5.4 | 3.6 |
| Leucine (%) | 7.7 | 5.4 |
| Methionine + Cystine (%) | 3.1 | 2.6 |
| Phenylalanine + Tyrosine (%) | 8.1 | 3.1 |
| Therionine (%) | 3.9 | 2.9 |
| Tryptophan (%) | 1.4 | 0.9 |
| Valine (%) | 5.3 | 3.8 |
| Histidine (%) | 2.4 | 1.6 |
| Lysine (%) | 6.3 | 5.7 |
| Lenoleic acid (%) | 58.8 | - |
| Lenolenic acid (%) | 8.1 | - |
| Lecithin (%) | 3.6 | - |
| Calories (K Cal) | 410 | 303 |
| Nitrogen solubility index (%) | 50-60 | 10 |
| Digestive efficiency (%) | 80 | 60-70 |

*Source :* Ganesh and Joseph (1997), used with permission of N. Ganesh

### *Use of Soyalecithin in Fish Feed*

In the recent trend of freshwater fish culture system, qualitative aspects of lipid have attracted the attention to fish nutritionists. The important role of two fatty acids – cholesterol and phospholipid – essential for a healthy diet of fishes – has been observed by many aquaculturists.

Soyalecithin, which is a by-product of soyabean processing industry, can be utilized as an ideal source of phospholipid, fatty acids, and cholesterol.

### *Effects of Soyalecithin on Survival and Growth of Fish*

Soyalecithin is the major source of metabolic energy throughout the embryonic development of fishes. In nursery and rearing systems, fish larvae in a number of species require phospholipids in their diets since they cannot synthesize it *de novo*. The Eicosapentaenoic acid (EPA) and Docosahexaenoic acid (DHA) ratio and their individual content in absolute terms is very important for high survival and growth of fish.

## 9.15 Agrimin and Fishmin

A variety of artificial feeds such as Agrimin, Fishmin, and Shirimin are being manufactured by various companies for aquacultural management practices. These are multi-micronutrient mineral fertilizers with several essential balanced mixture of amino acids, macro-and micronutrients (Table 9.16). These are mixed with the artificial feeds in suitable rates. Generally, 10 kilogram of Agrimin or Fishmin or Shirimin is mixed with one tonne of feed. Use of these nutrients is important because of the following reasons :

(1) These are active stimulating agents for cell development.

(2) Increase rate of photosynthesis.

(3) Active participation in both nitrogen and phosphorus fixation and assimilation.

(4) Enhance production of phytoplankton, zooplankton, and benthic organisms.

(5) Reduce annual production cost.

(6) Stimulate bone and muscle formation in fish.

**Table 9.16 : Composition of Agrimin[1]**

| Composition | Quantity used (per kg of feed) |
|---|---|
| Copper (mg) | 312 |
| Cobalt (mg) | 45 |
| Magnesium (g) | 2.114 |
| Iron (mg) | 979 |
| Zinc (g) | 2.130 |
| Iodine (mg) | 156 |
| DL-Methionine | 1.920 |
| L-Lysine Mono HCl (g) | 4.4 |
| Calcium (%) | 30 |
| Phosphorus (%) | 8.25 |

1. Recommended by Agrivet Farm Care, Glaxo India Limited, Mumbai

### *Efficiency of Agrimin*

Experiments on the efficiency of mineral mixture agrimin at different rates (1.25, 2.50, 3.75, and 5.0g/kg of feed) in relation to the growth and food consumption of fingerlings of *Labeo rohita* have shown that although the fish daily consumed minimum food containing 3.75 g agrimin (Table 9.17), the food consumption ratio was highest. Also, the application of agrimin at the rate of 3.75 g/kg of feed (rice bran : oil cake = 1 : 1) exhibited better growth of fish (Table 9.18). This result is not convincing because the experiment lacked replicates and therefore needs elaborate reassessment of the efficiency of agrimin on the growth dynamics of fish in ponds.

**Table 9.17 : Daily Consumption of Food in Percentage Body Weight of *Labeo rohita* Fingerlings**

| Diet Number | Daily food consumption (g) | | | |
|---|---|---|---|---|
| | Upto 15 day | Upto 30 day | Upto 45 day | Upto 60 day |
| A | 1.93 | 1.88 | 1.81 | 1.75 |
| B | 1.88 | 1.78 | 1.70 | 1.71 |
| C | 1.84 | 1.81 | 1.81 | 1.72 |
| D | 1.91 | 1.80 | 1.79 | 1.86 |
| Control | 1.87 | 1.88 | 1.88 | 1.92 |

*Source :* Sharma and Rai (1997), used with permission of O.P. Sharma

**Table 9.18 : Growth (per cent gain in body weight) of *Labeo rohita* Fingerlings**

| Diet number | Per cent gain in body weight | | | |
|---|---|---|---|---|
| | First day | Second day | Third day | Fourth day |
| A | 7.54 | 21.87 | 41.59 | 70.51 |
| B | 13.78 | 41.52 | 77.33 | 101.51 |
| C | 17.10 | 39.43 | 61.80 | 101.11 |
| D | 9.62 | 32.57 | 54.88 | 69.79 |
| Control | 14.67 | 29.00 | 44.34 | 55.02 |

*Source :* Sharma and Rai (1997), used wth permission of O.P. Sharma

## 9.16 Anti-Nutritional Factors in Fish Feed

The constituents of feed ingredients that are responsible for the deleterious effects to fish health are called as *anti-nutritional factors*. These factors are toxic and depress digestion of protein (such as lactones, tannins, saponins, and protease inhibitors), reduction in the solubility or interfering the utilization of mineral elements (such as oxalic acid, phytic acid, gossypols and glucosinolates), and inactivating the requirements of vitamins and hormones (such as anti-vitamins; vitamin A,D,E and K). Some compounds (such as moulds, cyanogens, and mycotoxins) are also known to exhibit toxic effects on fish.

1. *Protease Inhibitors* : As the name implies, these substances inhibit the proteolytic activity of some enzymes and found in leguminous plant and soyabean. For example, these react with trypsin and chymotrypsin and thus inactivate the enzymes. Because protease inhibitors are heat-labile, these can be destroyed when cooked.
2. *Lactins* : Lactins are heat-labile proteins and has an affinity for sugar molecules. They cause agglutination of red blood corpuscles and hence it is also called *haemagglutinins.* Lactins reduce fish growth by reducing nutrient absorption in the alimentary canal.

3. *Saponins* : Saponins have been identified in about 550 plant species belonging to 90 families. Saponins are plants such as lentil, alfalfa, soyabean, jackbean, etc. Saponins can easily be removed from food stuffs by soaking saponin-containing materials in water for one day.
4. *Tannins* : These are polyphenolic compounds and found in rape-seed, mustard seed, bean, sorghum, and sunflower. They affect digestion of crude fibres, loss of mucus, irritation and damage to digestive tract which increases tannin absorption. The toxicity of tannins can be remved by heating and soaking of tannin-containing materials in water.
5. *Oxalic Acid* : It is widely distributed in animal and plant kingdom, in salt and free forms. It causes increased respiration, irritation, depression, and weakness. It can be removed from feedstuff by soaking it in water.
6. *Phytic Acid* : It is found in all plants particularly in rapeseed, sesame and soyabean in connection with proteins. Phytic acid is a salt which forms complexes with zinc, iron, and manganese, thus resulting in the formation of insoluble compounds in the digestive tract. Addition of phytic enzymes in the feed can reduce the level of phytic acid.
7. *Glucosinolates* : These are heat-labile and found in mustard and rapeseeds. They suppress thyroid synthesis and reduce fish growth.
8. *Anti-vitamins* : These are some factors which causes the efficiency of vitamins A,D,E,K and anti-pyridoxine present in the diet. For example, presence of raw soyabeans at 30 per cent in fish feed produces a sharp decline of vitamin A and carotene. This is mainly due to the presence of lypoxygenase enzyme in raw soyabean which catalyzes carotene oxidation. This enzyme is considered as the precursor of vitamin A. Linseed meal contains I-amino-D-proline that suppresses pyridoxine activity. All anti-vitamins can be destroyed using heat processing of the food ingredients.
9. *Mycotoxins* : These are fungal metabolites causing physiological or pathological changes in fish. Vomitoxin, aflatoxin and ochratoxin A are the three important types of mycotoxins. These are produced by the fungi of the genus *Aspergillus parasiticus, A. flavus,* and *Penicillium* sp. They produce tumours and liver cell carcinomas. Afflatoxins have been recorded from groundnut meal, copra meal, maize products, sorghum, oats, soyabean, and sunflower. They produce health hazards such as anaemia, slow growth, liver damage, impaired blood clotting and susceptibility to various infectious diseases.

Although anti-nutritional factors are present fish feeds, ingredients are added to some commercial feeds. However, with adequate pre-treatment of unconventional ingredients in feeds, it can be used to replace about 25 per cent of fish meal protein in feeds without any health hazards.

## 9.17 Steroids in Fish Feed

Steroids are the non-saponifiable fraction of lipid extracted by the fat solvent. It has cyclopentano-perhydrophenanthrene nucleus in its molecular structure. Manipulation of several types of hormones in the diet is very significant so far as the acceleration of fish growth is taken into consideration. Although the use of hormones in fish culture management strategies is of recent issue, yet it is a matter of controversy whether they can be used without any detrimental effect on fish biomass.

On the basis of physiological functions of fishes, steroids can be classified into four categories such as (1) Progesterone, (2) Estrogen, (3) Androgens, and (4) Corticosteroids.

1. *Progesterone* : These are female sex hormones secreted by the corpus luteum of the ovary. These hormones trigger implantation of the fertilized ovum.
2. *Estrogen* : Estrogens are females sex hormones secreted by the ovary and responsible for the development of eggs or oocytes within the ovary. Among different types of estrogens, estrone and estradiol are important. They are anabolic in function.
3. *Androgens* : Androgens are male sex hormones secreted by the matured testes. They have growth promoting effect on sex organs.
4. *Corticosteroids* : These are produced by adrenal cortex or internal gland in fishes of similar nature. They perform various physiological functions such as protein, carbohydrate and lipid metabolism and maintain electrolytic as well as body fluids.

Although these hormones are found in all higher vertebrates, they are available in a numbr of fishes. Except corticosteroids, all other hormones are anabolic in function. The term anabolic effect is defined as an increase in nitrogen retention by increased nitrogen intake through feed.

### 9.18 Accumulation of Steroidal Hormones into Fish Body

It is in general agreement that excessive use of steroids is detrimental to fish and hence not recommended for the purpose. In spite of this, some amounts of hormones are accumulated gradually within the body of fish through feed.

(1) Sturgeons consume steroids through plankton.

(2) Steroids are also entered into the body of fish through fish meal that contains significant amount of steroids.

(3) Fish oils are the source of steroids. Oils are used to prepare fish feed through which steroids are accumulated into the body of fish.

(4) Fish meal is pepared from liver and viscera of mammals through which steroids are entered into the body of fish.

Several reproductive steroids have been reported in fish meal and fish feed. These are androgens, estradiol and estrone (in fish meal) and 17, 20-dihydroxy progesterone, androstendione, estradiol, testosterone, 11-ketotestosterone etc. (in fish feed). The estimated amount of sex steroids in fish meals and fish diets are shown in Table 9.19.

### 9.19 Use of Steroids in Fish Feed*

In many developed countries like USA and Canada, use of a large number of synthetic steroids in fish feed is a recent issue in aquaculture programs. Application of different types of synthetic steroids in fish feed either alone or in combinations have been recommended. These are diethylstilbestrol and estradiol benzoate, diethylstilbestol plus progesterone, estradiol benzoate plus testosterone propionate, melengestrol and others. Use of some synthetic steroids at different rates in fish feed have shown that the growth rate of fish significantly enhanced (Table 9.20). It is to

* For further report, see Santandrev and Diaz (1994).

note that androgen and estrogen at the rates of 1-10 mg/l and 0.6-150 mg/l, respectively, exhibited significant impact on fish survival and growth. Table also indicates that estradiol is more tolerant to fish than diethylstilbestrol. Therefore, the level of seroids to be used in fish feed depends on the chemical structure of setroids, their mechanism of action and fish species to be administered.

**Table 9.19 : Amounts of Some Sex Steroids in Fish Meals and Fish Diets**

| Source | Steroid | Concentration (mg/100g) | |
|---|---|---|---|
| | | Lowest | Highest |
| Fish diet | Testosterone | 40 | 700 |
| | 11-Keto testosterone | 240 | 850 |
| | Androstendione | 480 | 900 |
| | Androgens | 280 | 1,100 |
| | Estradiol | 240 | 615 |
| | Progesterone | 1,280 | 2,690 |
| | 17, 20 Dihydroxy-progesterone | 900 | 8,200 |
| Fish meal | Androgens | 240 | 1,340 |
| | Estradiol | 40 | 1,510 |
| | Estrone | 350 | 940 |

**Table 9.20 : Effects and Application Rates of Some Steroids Used as Feed Additives**

| Fish | Steroid | Dose | Result |
|---|---|---|---|
| Coho salmon | 17-L-Methyl-tes-tosterone | 10 mg/kg | Higher growth and reduced mortality |
| | | 50-100 mg/kg | Reduced growth and increased mortality |
| Rainbow trout | 17-L-Methyl-tes-tosterone | 1 mg/kg | Higher growth |
| | | 2 mg/kg | Reduced growth |
| Plaice | Diethylstil-bestrol | 0.6 mg/kg | Higher growth |
| | | 1.2 mg/kg | Lower growth |
| | | 1-150 mg/kg | Higher growth |

## 9.20 Role of Biotechnology in Fish Feed

Fish culture industries in South and South-east Asian countries are the fastest growing in the world. Their success is a story of growing awareness among people : driven by a need to supply adequate protein and employment for a large population, different fish farmers' development organizations and Government made strong decision to improve their fish culture industry more viable through use of recent techniques. To keep the rate of production constant, the Indian fish industry is going to have to rely on biotechnology products. Feed efficiency can be enhanced by adding enzymes and help minimize the feed costs.

Biotechnology has played an important role in the fish culture industry for many years. It amalgamates the principles of biochemistry, microbiology, molecular biology and immunology, and a biological process by which biological material is produced, or by which the nutritive value of feed is improved. Different feed additives are produced by methods based on biotechnology.

### *Immunostimulation through Diet*

Adequate nutrition of fish is necessary to maintain several physiological functions such as cell division, cell activity and repair and production of substances that are protective in function such as lectins, precipitins, and antibodies. These factors are important to sustain fish survival and growth. It has been reported that some vitamins such as A, C, D, and K are important for immunostimulating effects when used at high rates (about ten times than normal rates).

Immunostimulants hold great promise in increasing the resistance of fish under culture to microbial diseases. Some immunostimulants such as Bio-MOS, Bio-MOL, PK-565, levamisole, glucans, lectoferrin, EF-203, crude bacterial extracts, lipopolysaccharides, and zymozan have shown to increase the innate protective functions of fish. Following the administration of several immunostimulants along with artificial diets, augmentation of defence mechanisms of fish take place and it has been fully explained to evaluate the mechanisms of enhanced protection.

### *Amino Acids*

Excess amounts of amino acids in fish feed may lead to high losses of nitrogen through faecal matter and hence responsible for degradation of pond water. This situation is generally predominant in intensive and semi-intensive fish culture systems. For this reason, more attention is paid to the amino acids pattern in fish diet. Amino acid patterns composed from plant materials can be improved using synthetic amino acids which are produced by fermentation with bacteria. This will results in a better utilization of dietary protein and the content of nitrogen in the faecal matter will be less.

### *Enzymes*

Enzymes are types of protein found in all living organisms. They allow all the chemical reactions or metabolism to take place, regulate the speed at which they progress, and provide a means of controlling individual biochemical pathways. Enzymes owe their activity to the precise three-dimensional shape of their molecules. The substances upon which an enzyme acts (substrates) fit into a special space in the enzyme molecules, called the active site. A chemical reaction takes place at the active site and the appropriate products are released, leaving the enzyme unchanged and ready for re-use. Enzymes are very specific in relation to the substrates with which they work, and normally effective only for one reaction or a group of closely related reactions.

Certain enzymes have been developed for use as feed additives. They are produced by fermentation based upon different microbes such as *Bacillus subtilis, Trichoderma reesei,* and *Aspergillus niger.* The products contain one enzyme or a mixture of enzymes. However, glycanases are enzymes available in the market and designed to cleave the non-starch polysaccharides which exhibit anti-nutritive activity in the diet. Anti-nutritive activity is characterized by poor digestibility of nutrients. This problem can be alleviated by using enzyme or enzymes which triggers partial cleavage of the non-starch polysaccharides and thus increase the digestion of fat, starch, protein, and metabolizable energy in the diet.

Most cereal grains and their by-products consist of large amounts of arabinoxylan and cellulose as the main non-starch polysaccharides. The structure of these polymers is well established and the enzyme technology is available for complete breakdown of these polymers.

### *Use of SP-604, C-Care 500, and Nutripro in Fish Feed*

Recently several research organizations have manufactured and introduced a range of products to improve aquaculture productivity, farm management, and disease control in fish. Among different products, SP-604, C-Care 500 and nutripro are very important. These are balanced nutrients having high efficiency power for survival, growth and reproduction of fish. The recommended rates of SP-604, C-Care and nutripro are 1.0, 1.0 and 10 kg per tonne of feed, respectively.

### *Use of Antibiotics in Fish Feed**

Antibiotics are substances excreted from micro-organisms such as fungi, and used to destroy or inhibit the growth of others, such as infective bacteria and fungi. However, the most commonly used antibiotics include streptomycin, cephalosporins, tetracycline, and penicillins. These natural extracts are often chemically modified to increase their stability and specificity. Inadequate or inappropriate use often leads to the development of resistant strains of the damaging micro-organisms.

Feed is a good source of bacterial and fungal development which in turn, produces the toxins and hazardous to fish. Several antibiotics are used for effective treatment of various diseases such as vibriosis, epizootic ulcerative syndrome, hepatic speticemia, acute dermatitis, and oriental fish sores. The recommended rates of some antibiotics in fish feed are shown in Table 9.21.

**Table 9.21 : Application Rate of Some Antibiotics Used in Fish Feed**

| Antibiotic | Application rate (per tonne of feed) |
|---|---|
| Neodox forte | 5 kg for 5 days |
| Neochlor forte | 1 kg for 5-7 days |
| Colidox | 2-3 kg for 5 days |
| 3-Care | 1 kg for 10 days |
| G-Probiotic | 1-2 kg for 10-15 days |
| UTP-5 | 5-8 kg for 7-10 days |
| Virginianycin | 80 g for 10 days |
| Terramycin | 100 g for 7days |

*Source :* Sarkar (1997).

### *Importance of Astaxanthin in Fish Feed*

Astaxanthin is generally considered as a fish feed additive belonging to the carotenoid group. It occurs either free or in the form of a carotenoid protein couples. Since *de novo* synthesis of carotenoids is confined to fungi, algae and plant, fish entirely depends on dietary intake. As the contribution of the food chain to the nutrient intake of fish in ponds is negligible, astaxanthin should be added to feed.

1. *Physiological Role of Astaxanthin* : The physiological role of astaxanthin has recently centered around finfish and shellfish. It has been proposed that astaxanthin acts as antioxidants, and reactive oxygen forms (such as free redicals and singlet oxygen) are deactivated. These findings stimulated the interest on the effects of astaxanthin on fish production and other biological functions.

---

* For further detail, see Ahmad and Matty (1989).

During oxidation reaction, highly reactive oxygen species and oxygenated free radicals are formed in the body of fish. These compounds encourage massive destruction of cell components, especially through degradation of phosphos lipids of cell membranes. Addition of astaxanthin to the diet has a strong preventive effect against these harmful free radicals. The preventive activity of astaxanthin against the action of free radicals increased with increasing concentrations of astaxanthin (Figure 9.3). However, it is also found to be an effective free radical scavanger (antioxidants) and consequently, the cell defence mechanisms are initiated.

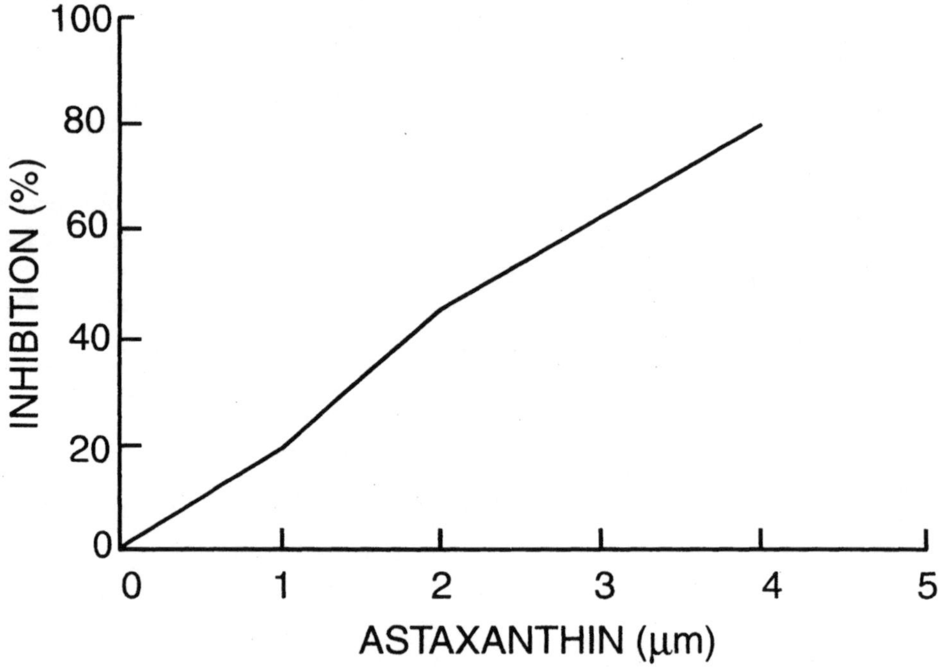


**Fig. 9.3 : Inhibitory activity of astaxanthin against the action of free radicals on membranes of red blood cells. Redrawn from Miki (1991).**

Experiments have shown that fish eggs developing in an environment with poor oxygen in water contain more astaxanthin than eggs developing under high oxygen content. This suggests that the utility of astaxanthin may be considered as an adaptive process to a hypoxic situation in which the carotenoid serves as an oxygen reserve for embryological development. Evidence also indicates that astaxanthin enhances macrophase activity and lymphocyte proliferation, natural killer cells, protect white blood cells against autooxidation, increases *in vitro* antibody production to T-cell dependent antigens and also prevent oxidative damage of cell membranes.

2. *Role of Astaxanthin on Fish Growth* : Laboratory studies have shown that survival and growth of red tilapia and Atlantic salmon significantly increased when astaxanthin-supplemented diet was added. However, about 63.6 and 75 per cent survival and growth of *Penaeus japonicus* respectively were observed after a 3- month period when fed 50 mg astaxanthin/

kg feed (Figure 9.4). These observations indicate a synergistic influence of the potentially of astaxanthin on fish and shellfish production. Further systematic research on this aspect using other types of freshwater fish is needed for evaluation of interactions between astaxanthin and fish production.

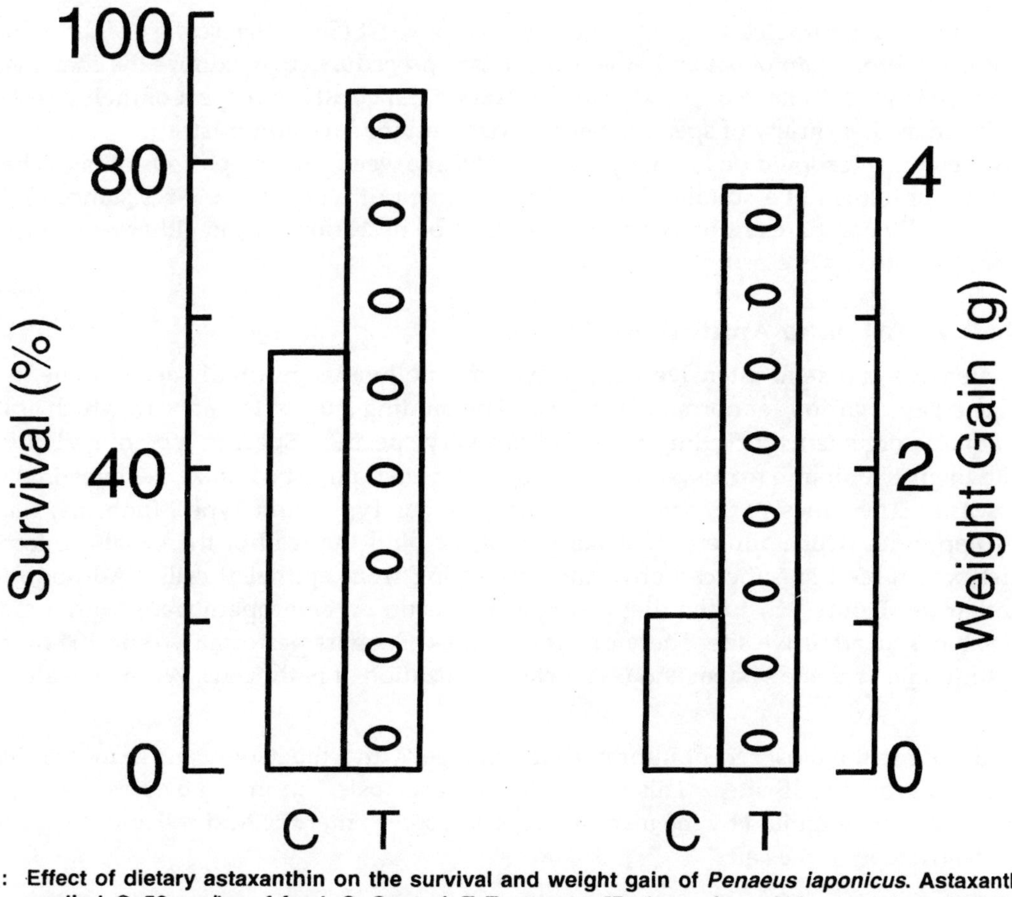


**Fig. 9.4 : Effect of dietary astaxanthin on the survival and weight gain of *Penaeus iaponicus*. Astaxanthin was applied @ 50 mg/kg of feed. C, Control; T, Treatment. [Redrawn from Chien and Jeng (1992)]**

### *Use of Color and Flavor in Fish Feed*

Prior to selection of artificial feed for fish culture, it is necessary to look into the components which provide nutrition to fish. Better survival and yield can be obtained if fish diet is fortified with such compounds. For this purpose, fish feed should be of good shape, color, taste and texture. Moreover, color, flavor, antibiotics, and hormones are also important. Color can alter food intake in carps. Dark colored feeds are prefered by the major carps such as *Labeo rohita* and *Catla catla*.

Flavors are chemical qualities of a substance and their addition affect the taste or gratifies the palate thereby influencing food consumption although it is also possible to render nutritionally correct food of low palatability acceptable to the fish. It has been found that Atlantic salmon and *Tilapia zilli* when fed with different flavors such as squid extract, amino acids, fried fenugreek seeds – *Trigonell* sp., seed and root extracts of murrya plant etc. exhibited significant consumption of food. Addition of these Havors, Lowever, generally makes the feed more palatable to the cultured fish.

### *Use of Vetregard in Fish Feed*

Vetregard is a natural feed ingredient comprised of yeast (*Saccharomyces cerevisiae*) which has been processed through autolysis and other proprietary procedures to maximize the content of Beta 1.3 glucan. Beta glucans have been extensively studied, especially in respect of their effect on the immune systems of a variety of species. Vetregard is designed for administration at a rate of 1 kg per tonne of feed and should be fed daily for at least three weeks during periods of high challenge or risk. Although it is not a suitable for good managment, it will assist the resistance to disease and, in fact, will potentiate antibacterials should they be necessary. It can either be incorporated during pelleting to top dressed on to finished feed.

## 9.21 Mode of Action to Antitoxindants

In intensive and semi-intensive culture systems, pollutants gradually accumulate in water and soil and cause various abnormalities in fish. The binding sites of toxins with which epithelial wall of aquatic organisms including fish bind are very specific. Specific type of carbohydrates bind with toxins. Similar to toxins, several pathogenic bacteria must colonize the intestinal wall to cause diseases. The adhesive fimbriae of bacteria (that has Type 1 and Type 2 fimbriae) *E. coli and Salmonella* sp. with which alimentary canal pathogens bind the wall of the canal are lectin-like structures which bind specific carbohydrates projecting from epithelial cells. Adding mannan oligosaccharide (Figure 9.5) to the diet provides the lectin on enteropathogens with a mannose residue to block an adhesive site. Therefore, these types of sugars perform two specific functions : (1) they stimulate immune system and (2) blocks colonization of pathogens. Within the alimentary canal.

According to the most recent information, pathogens attach beta glucans to macrophages at specific sites, termed iC3B sites. This stimulates phagocytosis, but in some cases the pathogens attach to but are not engulfed by the macrophages and some enter the host cell and divide rapidly to cause destruction of the cell.

An immune stimulant will generally activate the macrophages, giving them ability to react quickly to destroy the invading organism. Research has shown that immune stimulants such as beta glucans of a particular molecular size have the ability to attach themselves to specific sites on the macrophages outside the iC3B sites. When attched the beta glucan, stimulate the macrophages to activate a process whereby destructive enzymes are released capable of destroying the invading pathogen and finally it is phagocytized.

Recent research work in fish confirms that action of beta 1.3 glucan in stimulating macrophage activity and speeding up the general immune response. However, after stimulation with beta 1.3 glucan at the $CR^3$ sites, there is a release of cytokinase, particularly interferon Y and interleukin-1, which turn can activate T-cells in the immune system. The T-cells then produce lymphokines which activate B-cells and then release a high level of antibodies against the invading pathogens.

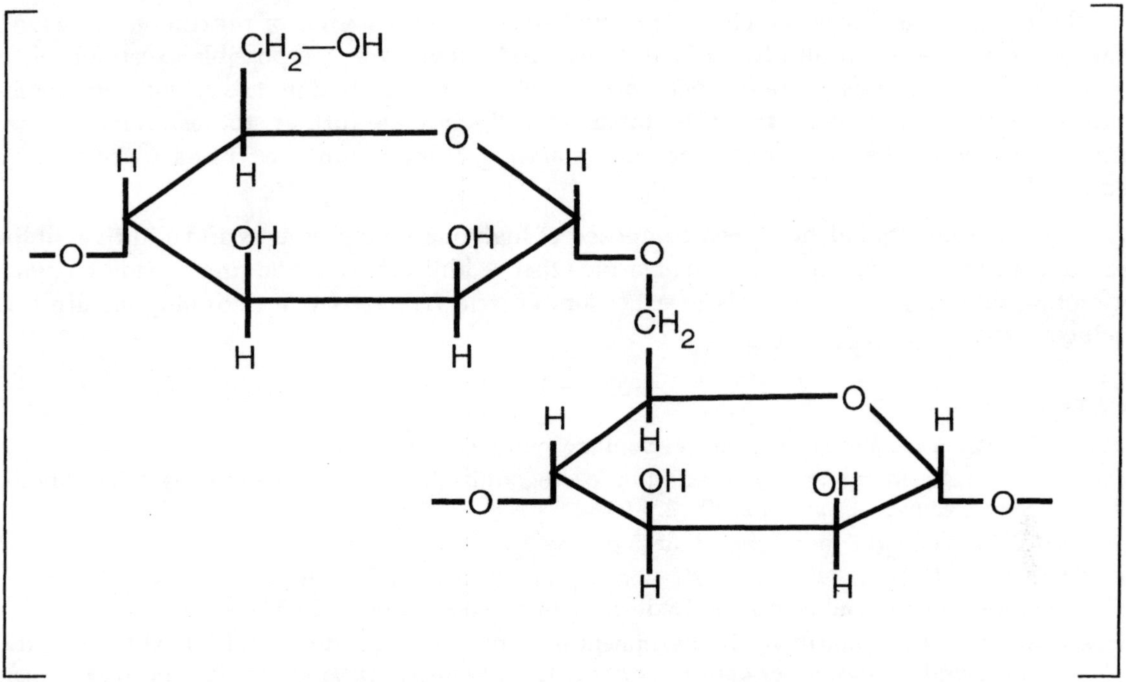


**Fig. 9.5 : Chemical environment of a mannan sugar (Mannan oligosaccharide).**

In nature, the immune system remains crucial in fish. The question is whether immune system can be "tweaked" improve performance. However, research with catfish have shown improved feed conversion and less scouring using mannan oligosaccharide with feed.

## 9.22 Conclusion

The main development in production systems has been intensificaion of feeding with high nutritional quality feed ingredients and increasing thanks to a steady rise in yield capacity of fish culture ecosystems. Further development of low-cost feeds depends on the informaiton on nutrient requirements and bioavailability of different feed ingredients. This information for different species of freshwater fish is very limited. During the last fifteen years, however, progress has been made in the field of fish nutritional research.

Since nutrient requirements for fish depend on various factors (such as age or size, composition of basal diet, dietary nutrient levels, physical quality of diet, culture conditions, and feeding practices), it is very difficult to detect exact quantity of nutrients as it varies considerably even for the same species. The requirement values of finfish for some nutrients are low compared to shell-fish. Though lower levels of minerals and vitamins have been recommended for finfish, it is not exactly known how much quantity of nutrients is lost into the water during consumption. However, the results obtained in the laboratory under controlled environmental conditions are well-defined, but can never be directly applied to large-scale commercial production. Therefore, recommendation of correct amounts of feed with balanced ingredients pertaining to different life stages of fish reared under large-scale production systems are very important for the formulation of cost-effective feeds.

The dietary imbalance which has resulted in the manifestation of nutritional deficiency symptoms cause severe health hazards and low production. With the possible exception of the studies in some countries, there has been no systematic attempt to date to various pathological conditions in fish. It is hoped that this statement will stimulate further studies under practical farming conditions so as to increase the understanding of the importance of balanced artificial feed in fish culture.

The availability of balanced feed composed of high quality ingredients and additives at low cost stimulate growth and reproduction of farmed fish. Addition of steroids, amino-acids, enzymes, antibiotics, vitamins, and minerals in exact amounts in fish feed will definitely ensure high production of fish.

### *References*

ADCP. 1983. Fish Feeds and Feeding in Developing Countries. FAO, Rome.

Ahmad, T.S. and A.J. Matty. 1989. The effect of feeding antibiotics on growth and body composition of *Cyprinus carpio. Aquaculture.* 77 : 221-230.

Baker, T.M. 1997. Vitamin E improves fish flesh quality. *Fish. Chim.* 16 : 19-20.

Chien, Y.H. and S.C. Jeng. 1992. Pigmentation of Kuruma prawn, *Penaeus japonicus* (Bate) by various pigment sources and levels and feeding regimes. *Aquaculture.* 102 : 333-346.

Chow, K.W. 1982a. Carp nutrition : Establishment of a nutrition laboratory and intimation of a diet. *Development programme for Carp Polyculture,* FAO, Rome, FI-DP/IND/75/031, Field Document, Number 4.

Chow, K.W. 1982b. *Diet Formulation for Selected Species for Aquaculture in Mexico.* Project Report Number ME/77/002, FAO, Rome.

De-Silva, S.S. and T.A. Anderson. 1992. *Fish Nutrition in Aquaculture.* Chapman and Hall, New York.

Devaraj, K.V. and D. Seenappa.1988. Studies on the nutritional requirements and feed formulations of cultivable fishes. *Ann. Res. Rep.* of the Pl-480. Fish Nutrition Project (1987-1988). 29pp.

Ganesh, N. and M.A. Joseph. 1997. Use of full fat soya flour as a fish meal substitute. *Fish. Chim.* 16 : 21 : 22.

Hardie, L.J. T.C. Fletcher, and C.J. Secomber. 1990. The effect of dietary vitamin E in the immune response of the Atlantic Salmon, *Salmo salar. Aquaculture.* 87 : 1-13.

Hardie, L.J., T.C. Fletcher, and C.J. Secomber. 1991. The effect of dietary vitamin C on the immune response of the Atlantic Salmon, *Salmo salar. Aquaculture.* 95 : 201-214.

Hilton, J.W. 1989. The interactions of vitamins, minerals and diet composition in the diet of fish. *Aquaculture.* 99 : 223-244.

Lall, S.P. 1989. The minerals. *In : Fish Nutrition* (*Ed.* J.E. Halver), pp. 216-255, Academic Press, New York.

Miki, W. 1991. Biological functions and activities of animal carotenoids. *Pure and Appl. Chem.* 63 : 141-146.

New, M.B. 1987. Feed and feeding of fish and shrimp : A manual on the preparation and presentation of compound feeds for shrimp and fish in aquaculture, FAO, Rome, ADCP-REP/89/26.

Paulraj, R. 1995. *Aquaculture Feed.* The Marine Products Export Development Authority, Kochi, India.

Santandrev, I.V. and N.F. Diaz. 1994. Effect of 17-Methyl Testosterone on growth and nitrogen excretion in mass salmon, *Oncorthynchus* spp. *Aquaculture.* 214 : 321-333.

Sarkar, S.K. 1997. Comments on artificial feed used in fish culture. *Fish. Chim.* 16 : 15-17.

Sharma, O.P. and H.O. Rai. 1997. Efficacy of Agrimin in the growth of Indian major carp, *Labeo rohita* fingerlings. *Fish. Chim.* 17 : 15-16.

Swain, S.K. and B.K. Das. 1996. Effects of micronutrients in immune response and disease resistance of fish. *Fish. Chim.* 16 ; 21-23.

## Questions

1. Why artificial feeds are so important in fish culture?
2. Suppose a farmer wants to prepare fish feed by mixing two feed ingredients (such as fish meal and rice bran) in the ratio of 2 : 2 respectively having the percentages of their composition as shown in the Table 9.1. Calculate the digestible energy in 2 kg of fish feed.
3. State how artificial feeds affect the quality of pond water.
4. State how artificial feeds are related to fish health.
5. What are the qualities of feed required for good health of fishes?
6. State the method adopted for feeding of fish in intensive fish culture system. State the frequency and time of feeding of fish.
7. What are the types of artificial feed. Why multi-micronutrients are so important in the formulation of fish feed?
8. Discuss the role of some nutrients required for the preparation of fish feed.
9. Why minerals are so important in fish feed?
10. What is vitamin? Why vitamins are essential for fish growth?
11. What are the factors that affect the nutritive value of fish feed?
12. Name some ingredients used in the preparation of fish feed. Briefly summarize the widely available feed ingredients.
13. Why soyabean is considered as the main source of nutrients?
14. What are anti-nutritional factors? How these factors affect the general health of fish?
15. Define the terms steroids, mycotoxins, antibiotics, astaxanthin, vetregard, and antitoxidants.
16. Discuss the role of steroids in fish feed.
17. How immunostimulation affects fish growth?
18. State the importance of astaxanthin in fish feed.
19. State the mechanisms of action of antitoxidants in fish.

# 10

# Management of Water in Fish Culture

The success of Indian fish culture schemes depends on the use of a variety of manures and fertilizers. Although these inputs play an important role for boosting the yield of fish in ponds, their detrimental effects in relation to toxicity should not be overlooked. Application of nutrient carriers also cause depletion of dissolved oxygen and eutrophication of toxic metabolites which cause histopathological changes of fish, high mortality, suppress fish growth and deteriorate water quality variables. Therefore, to improve productivity through good farm management including disease control (For further study, see Warren (1983), Sindermann (1986), and Das (1997) in fishes, a clean pond environment is essential. In this context, there is an urgent need to ameliorate the deteriorating aquatic ecosystem of fish farms through additional management inputs. The importance of water quality in relation to fish production should be strongly emphasized. High production of commercially important species of fish in suitable aquatic ecosystem necessitate the use of some water-added agents to increase the water quality, particularly ammonia control.

## 10.1 Products Used for Cleaning Pond Ecosystem

Several pharmaceutical organizations such as Teragon Chemie Ltd., India and Altech Biotechnology Centre, USA, have manufactured and introduced a variety of research products for aquaculture to improve productivity of fish ponds. To keep the pond environment clean and free from toxic metabolites as well as micro-organisms that are harmful to fish, use of their products on scientific basis – that are related to aquaculture, is significant. Among different products, De-odorase, Aquazyn, Chlortech, Geolite, Sokrena, Calcium hypochlorite and Benzalkonium chloride are very important. A short description of these products and their impact on fish and shellfish production is given below.

## 10.2 De-Odorase

Ammonia is a product of nitrogen metabolism which is hazardous for fish. Its accumulation leads to low survival, poor growth and feed conversion efficiency due to the need for energy-requiring mechanisms to eliminate the ammonia. The concentration of ammonia increases when pH of pond water increases and decreases the dissolved oxygen concentration.

De-odorase is a product which is extracted from the plants *Yucca shidigera* and *Y. aloifolia.* The extracted product contains certain enzymes and active components (glycocomponents) such as sapogenin, smilogenin, tigogenin, gitogenin, chlorogenin, 'CI' and 'CII'. These glycocomponents selectively bind the ammonia and other noxious gases. The active component actively binds with ammonia molecules to produce non-toxic nitrogenous compounds. These compounds are used by pond bacteria as a nitrogen source. While 'CI' fraction of glycocomponent binds with ammonia, indol and sketol groups, 'CII' binds with hydrogen sulphide. Moreover, De-odorase also improves dissolved oxygen level in water and decreases chemical oxygen demand as well as biochemical oxygen demand. Therefore, the ammonia-binding capacity of De-odorase is of special interest in aquaculture. This product is brown in color and highly soluble in water.

## *Effects of De-Odorase on Shellfish*

1. *Survival and Yield* : Experiments on freshwater prawn (*Macrobrachium rosenbergii*) culture ponds have shown that application of De-odorase for day 1 and day 60 significantly increased survival rate (Table 10.1). Total yield of prawn also increased in another series of test where application De-odorase did not start until day 60 and repeated every 15 day till harvest. At this rate of treatment, prawns exhibited best feed conversion, a difference of 7 per cent over the control experiment.

**Table 10.1 : Layout of the Experiment for Prawn Production in Ponds Under Three Management Regimes**

| Particular | Series | | |
|---|---|---|---|
| | A | B | C |
| Pond size ($m^2$) | 200 | 400 | 500 |
| Number of juveniles | 7,500 | 15,000 | 18,750 |
| Density of prawn/$m^2$ | 5 | 5 | 5 |
| Initial length (mm) | 10±0.06 | 9.7±0.07 | 8.9±0.06 |
| Initial weight (g) | 6.5±0.07 | 5.8±0.06 | 6.0±0.05 |
| Days to harvest | 105 | 110 | 117 |
| Survival rate (%) | 92±3 | 102±4 | 75±3 |
| Total yield (kg) | 150±3.7 | 150±4.8 | 280±3.9 |
| Feed conversion ratio | 1.4 | 1.3 | 1.4 |
| Total feed (kg) | 208 | 465 | 394 |
| Treatment of De-odorase | Day 1 and every 15 day | Day 60 and every 15 day | Control |
| Application rate of de-odorase (mg/l) | 6 | 6 | 6 |

*Source* : Sarkar (1998b)

The pond which has been treated with De-odorase from day 1, exhibited lower yield of prawn biomass (less than half from the control pond) but the yield was highest in pond where De-odorase was used at day 60 (more than half from the control pond).

Studies on the efficacy of De-odorase on survival and growth rates of freshwater prawns in ponds have shown significant results. In this case, total amount of De-odorase was added to pond water at the rate of 1.5 kg/hectare from day 1 and repeated once every 15 days until harvest.

## *Effects of De-Odorase on Pond Water*

Addition of De-odorase in water every 15 day interval resulted in 66 per cent lower mean ammonia and nitrate levels (Figure 10.1). Moreover, ammonia and nitrate levels gradually decreased as application of De-odorase progressed. Following the application of De-odorase to water, ammonia and nitrate levels exhibited decreasing trend after 45 day (Figure 10.2). Mean dissolved oxygen, total alkalinity, and pH of water significantly increased (Figure 10.3).

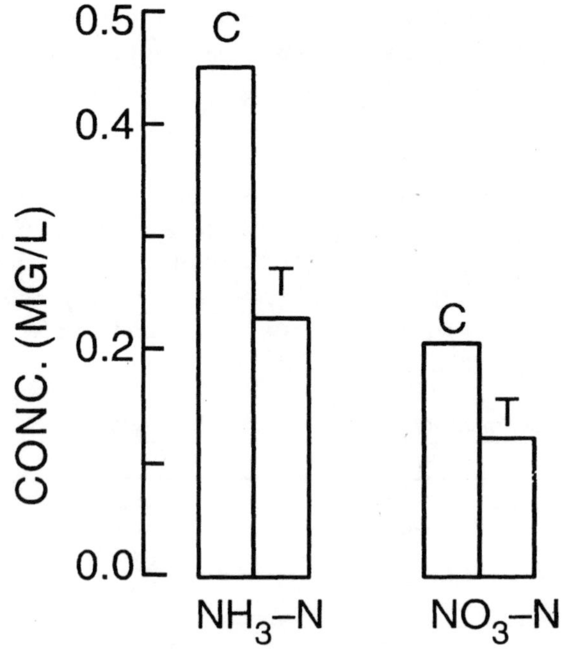


Fig. 10.1 : Effect of De-Odorase on the mean concentration of ammonia-nitrogen and nitrate-nitrogen in pond water. C – Control; T – Treatment. [From Sarkar (1998b)]

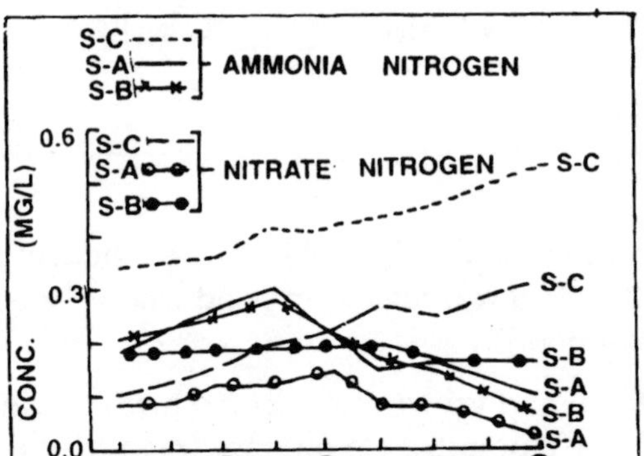


Fig. 10.2

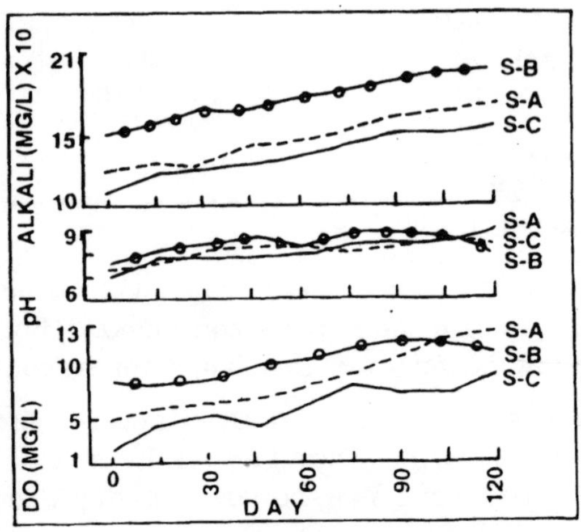


Fig. 10.3

Fig.10.2 : Effects of De-Odorase (De-O) on the fluctuations of ammonia-nitrogen and nitrate-nitrogen in pond water. S-C, Control; S-A, Series-A; S-B, Series-B. Series-A received an application of De-O @ 6 mg/l at day 1 and every 15 day till harvest. Application of De-O in Series-B did not start until day 60 and was repeated evey 15 day till harvest. Note that ammonia and nitrate levels decreased after 45 days. [From Sarkar (1998b)]

Fig. 10.3 : Effects of De-Odorase (De-O) on the fluctuations of dissolved oxygen (DO), pH, and total alkalinity (TA) of pond water. S-C, Control; S-A, Series-A; S-B, Series-B. The mode and rate of application of De-O was similar to that as described in Figure 10.2. The DO and TA significantly increased. [From Sarkar (1998b)]

### *Effect of De-Odorase on Plankton*

Mean phytoplankton and zooplankton concentrations (Number per litre) in ponds treated with De-odorase (at the rate of 6 mg/l and additional 6 mg/l every 15 day till harvest) significantly increased (Table 10.2).

**Table 10.2 : Concentration of Phytoplankton and Zooplankton in Ponds. Values are Means and Ranges of 9 Observations**

| Item | Plankton concentration (number per litre of water) | |
|---|---|---|
| | Phytoplankton | Zooplankton |
| Control | 417 | 366 |
| | 360 – 474 | 302 – 430 |
| Treatment | 623 | 461 |
| | 527 – 719 | 390 – 532 |

*Source :* Sarkar (1999)

### *Effects of De-Odorase on Bottom Fauna*

Concentrations of bottom fauna before treatment with De-odorase were comparable among different series of tests. At the end of test (120 days), bottom fauna concentrations increased in ponds which have been treated with De-odorase. Monthly observations in the concentrations of bottom fauna in ponds exhibited decreasing trend from the 60th day. It is possible that prawns consumed different species of bottom fauna resulting in increase in the growth and yield of prawn biomass thereby reducing the concentrations of bottom fauna.

## 10.3 Interactions of De-odorase and Fertilizers of Fish

For evaluation of major carp spawn (72h after hatching, 6.95 mm in total length, 5 mg weight), one month experiment was carried out in earthen vessels (0.30 sq.m. area each) that had been fertilized with lime, urea and single superphosphate at different application rates (Table 10.3). Vessels were received an initial treatment of 3 mg/l De-odorase and an additional 3 mg/l every 10 day.

**Table 10.3 : Application Rates of Fertilizers and De-odorase (De-O) During Rearing of *Labeo rohita* Eggs in Vessels and physico-Chemical Parameters of Water During the Period of Experiment. N, Nitrogen From Urea; P, Phosphorus from single Superphosphate; L, Lime. N and P are in terms of Nutrients only.**

| Treatment | Denoted as | Application rate (kg/ha) | Parameter | | | |
|---|---|---|---|---|---|---|
| | | | pH | Dissolved oxygen | Total alkalinity | Total ammonia nitrogen |
| L (C) | T 1 | 550 | 8.0 | 6.4 | 160 | 0.24 |
| L + De-O | T 2 | 550 + De-O | 8.3 | 6.8 | 197 | 0.16 |
| L + N (C) | T 3 | 550 + 80 | 8.0 | 7.4 | 210 | 0.41 |
| L + N + De-O | T 4 | 550 + 80 + De-O | 8.3 | 8.6 | 245 | 0.29 |
| L + P (C) | T 5 | 550 + 40 | 8.5 | 7.2 | 200 | 0.24 |
| L+P+De-O | T 6 | 550 + 40 + De-O | 7.8 | 8.5 | 253 | 0.18 |
| L + N + P (C) | T 7 | 550 + 80 + 40 | 7.8 | 7.5 | 214 | 0.28 |
| L + N + P + De-O | T 8 | 550 + 80 + 40 + De-O | 8.0 | 8.9 | 249 | 0.22 |

C, Control. De-O was added at 3 mg/l at day 1 and every 10 days. Except pH, all parameter values are in terms of mg/l. Values are means of 5 observations. Experiments were conducted for 30 days in earthen vessels. In each vessel, 2 kg of unfertilized soil was added to make 1 cm thick layer at the bottom. The capacity of water to each vessel was 60 litres. Each vessel was stocked with one hundred eggs.

*Source :* Sarkar (1998a)

### *Survival and Yield*

Though application of lime, urea and single superphosphate to vessels increased the mean survival and yield of *Labeo rohita* fry, addition of De-odorase to fertilizer combinations further enhanced the survival rate and yield (Figure 10.4). Obviously, De-odorase exhibited synergistic influence on the potentiality of nutrient combinations in survival and production of carp fry.

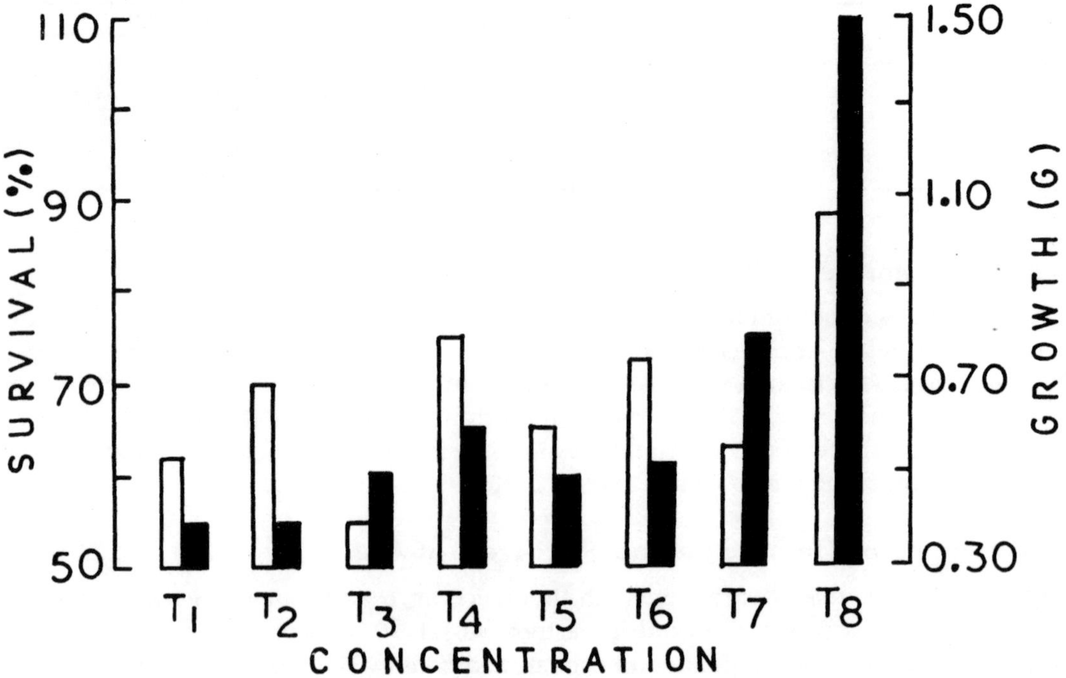


**Fig. 10.4 : Influence of De-Odorase (De-O) and fertilizers on survival and growth of *Labeo rohita* egg reared in earthen vessels. Experiments were conducted for 30 days in earthen vessels (0.3 sq.m. area each). In each vessel, 2 kg unfertilized loamy soil was added to make 1 cm thick layer at the bottom and 60 litres of borehole water was added to each vessel. Each vessel was stocked with *L. rohita* eggs @ 100 per vessel (25 lakhs/ha). De-O was added at the rate of 3 mg/l at day 1 and every 10 days. T1, Lime (control) 500 kg/ha; T2, De-O + Lime; T 3, Lime + nitrogen from urea @ 80 kg/ha (Control); T4, Lime + nitrogen + De-O; T5, Lime + phosphorus from single superphosphate @ 40 kg/ha (Control); T6, Lime + phosphorus + De-O; T7, Lime + phosphorus + nitrogen (Control); T8, Lime + phosphorus + nitrogen + De-O.**

Yield of major carp fingerlings (*Labeo rohita, Cirrhinus mrigala* and *Catla catla*) in ponds treated with De-odorase have shown that the mean growth rate of fish gradually increased (Figure 10.5). However, mean yield of fish also increased by 15 per cent of control yield. Total yield of *Catla catla* was highest (200 kg/hectare) followed by *Cirrhinus mrigala* (168 kg/hectare) and *Labeo rohita* (160 kg/hectare) (Figure 10.6).

At different application rates of fertilizers and De-odorase (Table 10.4) survival and yield of tilapia (*Oreochromis mossambicus*) increased. But when application rates of fertilizers were doubled, fish production did not further improve inspite of the addition of De-odorase during the culture period (Figure 10.7). The addition of De-odorase to fish culture ponds clearly indicates that the application rates of fertilizers may be greatly reduced which will not only reduce the cost of fish production but also will check the pollution of pond ecosystems.

**Table 10.4 : Application Rates of Calcium Ammonium Nitrate (CAN), Lime (L), Single Superphosphate (SSP), and De-Odorase (De-O). Except De-O and L, CAN and SSP are in Terms of Nutrients**

| Treatment | Denoted as | Application rate (kg/ha) |
|---|---|---|
| L + CAN | T 1 (C) | 300 + 80 |
| L + CAN + De-O | T 2 | 300 + 80 + De-O |
| L + CAN | T 3 (C) | 300 + 160 |
| L + CAN + De-O | T 4 | 300 + 160 + De-O |
| L + SSP | T 5 (C) | 300 + 40 |
| L + SSP + De-O | T 6 | 300 + 40 + De-O |
| L + SSP | T 7 (C) | 300 + 80 |
| L + SSP + De-O | T 8 | 300 + 80 + De-O |
| L + CAN + SSP | T 9 (C) | 300 + 80 + 40 |
| L + CAN + SSP + De-O | T 10 | 300 + 80 + 40 + De-O |
| L + CAN + SSP | T 11 (C) | 300 + 160 + 80 |
| L + CAN + SSP + De-O | T 12 | 300 + 160 + 80 + De-O |

C, Control. De-O was applied to water at 1 mg/l at day 5 after initial treatment of fertilizer(s) and an additional 1 mg/l every 15 day till the experiment was over. Each pond was stocked with fry of tilapia (*Oreochromis mossambicus*) at the rate of 6,000 numbers/hectare. Experiments were conducted for three months.

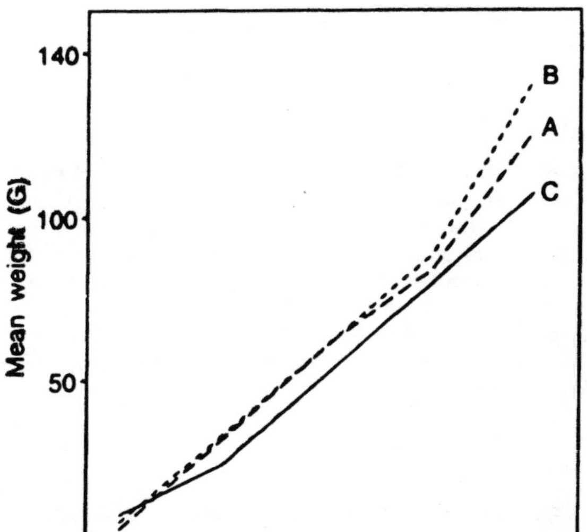


**Fig. 10.5 : Mean weight gain of *Labeo rohita* (A), *Catla catla* (B), and *Cirrhinus mrigala* (C) in ponds treated with De-Odorase (De-O). The mode and rate of application of De-O was similar to that as described in Figure 10.2. Note that the mean growth of fish gradually increased in ponds receiving De-O. [From Sarkar (1999)]**

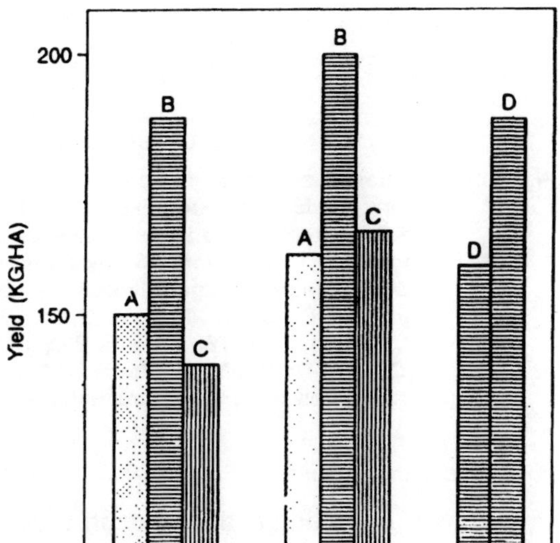


**Fig. 10.6 : Net yield of *L. rohita* (A), *C. catla* (B), and *C. mrigala* (C) in ponds treated with De-O. C, Control; T, Treatment; D, Mean total yield. Mean yield of fishes in treated ponds increased by 15 per cent of control. [From Sarkar (1999)]**

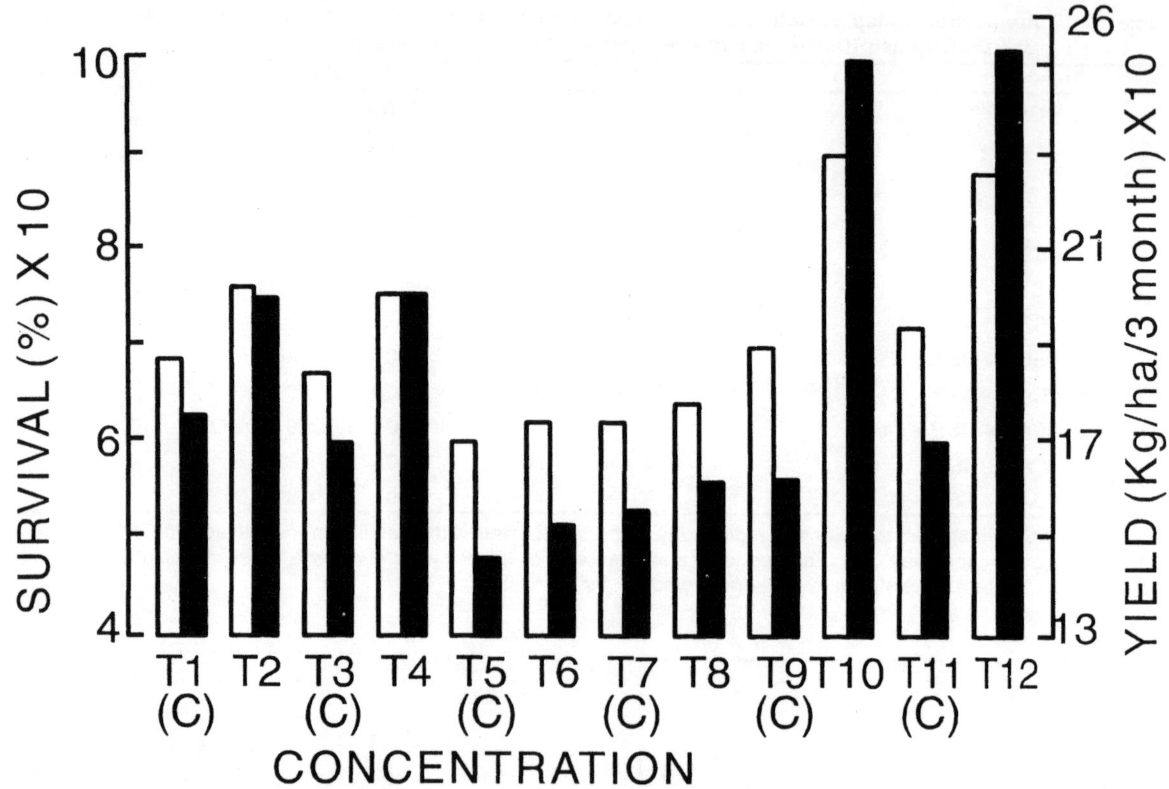


**Fig. 10.7 :** **Influence of nitrogen (from calcium ammonium nitrate), lime (L, as calcium oxide), phosphorus (from single superphosphate), and De-O on the survival and yield of *Oreochromis mossambicus*. De-O was applied to water at the rate of 1 mg/l at 5 day after initial treatment of fertilizer(s) and an additional 1 mg/l every 15 day till harvest. Each pond was stocked with fry of *O. mossambicus* @ 6,000 number/ha. Experiments were conducted for three months. (C), Control; T1, 80 kg/ha N; T2, 80 kg/ha N + De-O; T3, 160 kg/ha N; T4, 160 kg/ha N + De-O; T5, 40 kg/ha $P_2O_5$; T6, 40 kg/ha $P_2O_5$ + De-O; T7, 80 kg/ha $P_2O_5$; T8, 80 kg/ha $P_2O_5$ + De-O; T9, 80 kg/ha N + 40 kg/ha $P_2O_5$; T10, 80 kg/ha N + 40 kg/ha $P_2O_5$ + De-O; T11, 160 kg/ha N + 80 kg/ha $P_2O_5$; T12, 160 kg/ha N + 80 kg/ha $P_2O_5$ + De-O. Lime was added in all treatment combinations at the 300 kg/ha rate.**

## *Aquatic Ecosystem*

Drastic reduction in the concentration of total ammonia-nitrogen from pond water (reduction varied between 50 and 62 per cent) indicates that the ammonia-nitrogen level was reduced by the use of De-odorase. Although uptake of total ammonia-nitrogen by algal mats and blooms is the most important mechanism for complete/partial elimination from water, their appearance also upset oxygen budget of the water causing fish mortality. Fish mortality was found to be higher in ponds with the highest nitrogen input (1.2 g $N/m^2$) and low dissolved oxygen levels and therefore, not safe for fish culture. Since the safe concentrations of ammonia-nitrogen for fish culture are difficult to establish, the application of De-odorase will obviously lower the concentration of total ammonia-nitrogen and thus exhibited antagonistic influence on ammonia toxicity. Use of De-odorase in composite fish culture ponds where fertilization and manuring programs are undertaken, thus appears to be essential for active management of water quality.

## 10.4 Stabilized Chlorine Dioxide (or Chlortech)

Chlorine dioxide ($ClO_2$) has been known for many years to be a very powerful disinfectant in the gaseous form. Until recently, it has not been possible to stabilize this gas in liquid form and hence, its use has been limited. But now it has been paved the way for to produce the complex chlorine dioxide in liquid form and with activation, this product is one of the most powerful and safe versatile disinfectant available to day. This product has Environmental Protection Agency (EPA) and United States Development Agency (USDA) clearance for many applications.

### *Production of Chlorine Dioxide*

Production of $ClO_2$ involves two major processes : (1) Solvey process and (2) Altech process. In the first process, $ClO_2$ is generated by reacting chlorate, methanol and acid. The resultant gas, $ClO_2$, is unstable and in solution, quickly produces chlorides and chlorites.

In the Altech process, $ClO_2$ gas is stabilized in a precursor form in solution. Chlorine dioxide is stable for 18 months waiting to be activated. By activation with some selected acids, this precursor releases $ClO_2$ at levels of 2 to 5 per cent (2,000-5,000 mg/l) (Figure 10.8).

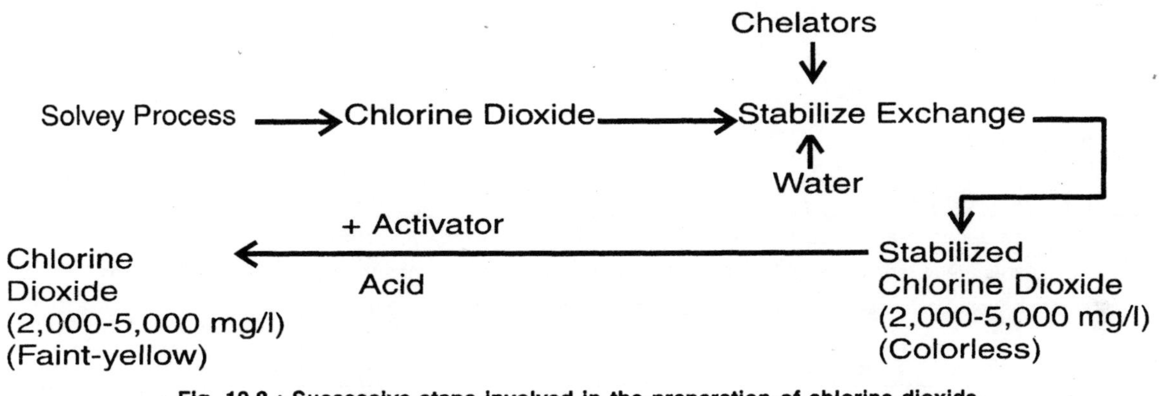


**Fig. 10.8 : Successive steps involved in the preparation of chlorine dioxide.**

### *Features of Chlorine Dioxide*

Active chlorine dioxide pushes aside the most commonly used disinfectants such as lodophores, quaternary ammonium compounds, phenolic compounds, formaldehyde, chlorine, hypochlorides and other traditionally used disinfectants because of its various advantages and superiorities as noted below :

1. Low toxicity to finfish and shellfish.
2. Effective bactericide, viricide, slimicide, and protocide. It is 2.5 times more potent than chlorine and 50 times more effective than hypochlorites.
3. It is less corrosive than chlorine, hypochlorites or ammonium compounds.
4. It is effective over broad pH ranges (5-10) whereas the activity of other disinfectants is limited to a very narrow pH range (5.0-7.5).

### *Action of Chlorine Dioxide Against Bacteria*

Chlorine dioxide is a very powerful and effective disinfectant against wide variety of bacteria such as *E. coli*, *Salmonella* sp., *Staphylococcus* sp., *Streptococcus* sp., *Pseudomonous* sp., *Listeria* sp. and other pathogenic bacteria and some species of fungi such as *Candida* sp., *Aspergilus* sp., *Trichophyton* sp. and others. It is also effective against various molds and viruses. Studies have shown that chlorine dioxide is most superior than other disinfectants (Table 10.5) and antimicrobial tests also indicate that a large number of micro-organisms are killed effectively (100 per cent) at 50-100 mg/l in 60 seconds (Table 10.6).

**Table 10.5 : Comparison of Chlortech with Other Disinfectants and Their Effects on Various Micro-organisms.**

| Disinfectant[1] | Micro-organisms | | |
|---|---|---|---|
| | A | B | C |
| Chlorine compound (chlorox) | 1,000 | 1,000 | 1,000 |
| Chlorine dioxide | 310 | 1,300 | 640 |
| Stabilized chlorine dioxide (chlortech) | 48 | 93 | 95 |
| Iodophor | 440 | 440 | 450 |
| Glutaraldehyde-phenol peroxide | 2,300 | 1,200 | 620 |
| Ammonium compound | 580 | 140 | 740 |
| Peroxide | 36,000 | 68,000 | 270,000 |
| Acidified quarternary | 150 | 1,200 | 300 |
| Phenol | 1,500 | 380 | 190 |
| Acid glutaraldehyde | 6,600 | 2,200 | 18,000 |

A, *Pseudomonas aeruginosa;* B, *Staphylococcus aureus;* C, *Saccharomyces cerevisiae*

1. All disnifectants are in terms of mg/l to achieve hundred per cent kill of micro-organisms within 60 seconds

*Source :* From a leaflet published by Vetcare, Division of Tetragon Chemie Ltd., Bangalore

**Table 10.6 : Effects of Chlortech on Various Micro-organisms**

| Micro-organisms | Contact time | Concentration (mg/l) | Per cent kill |
|---|---|---|---|
| *Eschericha coli* | 30 sec. | 100 | 99.99 |
| *Legionella pneumophila* | 60 sec. | 25 | 99.99 |
| *Listeria monocytogens* | 30 sec. | 100 | 99.99 |
| *Salmonella choleraesuis* | 10 min. | 500 | 100.00 |
| | 1 hr. | 50 | 100.00 |
| *Trichophyton mentagrophytes* | 5 min. | 500 | 99.99 |
| *Pseudomonas aeruginosa* | 60 sec. | 100 | 99.99 |
| *Staphylococcus aureus* | 60 sec. | 100 | 99.99 |
| *Saccharomyces cerevisiae* | 60 sec. | 100 | 99.99 |
| *Streptococcus faecium* | 60 sec. | 100 | 99.99 |
| *Proteus mirabilis* | 60 sec. | 100 | 99.99 |
| *Candida albicans* | 60 sec. | 100 | 99.99 |
| *Streptococcus faecalis* | 60 sec. | 100 | 99.99 |
| *Aspergillus fumigatus* | 60 sec. | 200 | 99.99 |
| *Bacillus cereus* | 5 min. | 200 | 99.99 |

*Source :* From a leaflet published by Vetcare, Division of Tetragon Chemie Ltd., Bangalore

### *Activation Before Application*

This product should be activated before use in fish culture ponds. However, activation involves the following steps :

1. Activation is initiated by adding 5g activator crystals for every 50 mg of chlortech in a clean container.
2. The container should be allowed to stand for 5 minutes to dissolve and color change from colorless to faint yellow.
3. After changing the color, it is diluted with water to required concentration depending on the purpose of application (Table 10.7).

**Table 10.7 : Concentration of Active Chlorine Dioxide Required for Various Purpose**

| Purpose of application | mg/l | ml of chlortech per litre of water |
|---|---|---|
| Drinking water treatment | 10 | 0.5 |
| Fish and shellfish processing plants | 50 | 2.5 |
| | 80 | 4.0 |
| | 100 | 5.0 |
| | 200 | 10.0 |
| For sensitization and odor control | 400 | 20.0 |
| | 500 | 25.00 |

*Source :* From a leaflet published by Vetcare, Division of Tetragon Chemie Ltd., Bangalore

It is very active because each litre of chlortech contains 2 per cent stabilized chlorine dioxide (20,000 mg/l).

### *Indication for Use*

In fish and shrimp culture ponds, this sanitizer is used at the rates varied between 150 and 250 ml per acre feet every 15 day once for the following cases :

1. To remove or destroy floating microbes.
2. To sanitize the pond water for better survival and growth of fish.
3. To treat gill rot, fungal, and protozoan infections.
4. To destroy the external parasites such as *Vorticella* sp. and *Zoothamnium* sp.
5. To remove the algae developed in larval rearing ponds.

## 10.5 Aquazyn

The Quinn India Limited, Hyderabad, has presented some scientifically blended concentrations of selected, adapted and cultured bacterial formultations grown on a cereal and mineral substrate along with enzymes and special buffers that are designed to degrade organic waste materials emanating from dead phyto and zooplankton, excreta and unconsumed feed. The product is considered as the right blend for every pond water treatment requirement.

### *Features of Aquazyn*

1. It increases average body weight and survival of fish.
2. It reduces biochemical oxygen demand and pond bottom sludge.
3. It also improves feed conversion ratio.
4. Organic and faecal waste materials are broken down rapidly.
5. It increases nutrient concentrations for plankton growth.
6. It prevents various infectious diseases.
7. It is non-pathogenic, non-toxic, non-hazardous, salmonella-free, biodegradable and environment friendly.

## 10.6 Sokrena

It is a liquid substance, generally used as pond water sanitizer and helps prevent fish diseases to a greater extent. This material is used at the rate of 1 mg/l or 10 litres of Sokrena is added in a pond of 1 hectare with an average depth of 1 metre. However, it has the following advantages over other disinfectants :

1. Broad spectral activity against bacteria, fungi and viruses.
2. It is effective in the prevention and treatment of all cases of *Vibriosis, E. turda, Aeromonas* sp., *Pseudomonas* sp., and white spot diseases.
3. It is most effective in alkaline pond water.
4. It is effective in presence of organic matter and pollutants.
5. It does not settle on the pond mud.

## 10.7 Geolite

It is a natural aggregate of inorganic salts of silicon, aluminium and calcium. It is also regarded as a health stone which is manufactured by Guybro Chemical, Mumbai and distributed in various maritime states for use in fish and shellfish culture. At present, geolite is mixed with neem oil to make the material more effective. This compound is used at the 10 kg/hectare rate with an average depth of 1 metre. It is a very important material for the following reasons :

1. It improves growth and survival of fish.
2. Pond water quality is improved.
3. The pH of water and soil is normalized throughout the culture period.
4. It brings down the clay turbidity of water.
5. It is also useful in prawn and fish culture to absorb harmful substances such as ammonia, nitrite and hydrogen sulphide.
6. It removes bad odor from pond water.
7. It improves soil conditions and re-activates soil.

## 10.8 Application of Gypsum

The quality of water for fish culture in relation to gypsum ($CaSO_4.2H_2O$) application is extremely important in the management of the water. Pond water high in suspended particles can bring about harmful effects unless these particles are reduced by soluble salts. Knowledge of the quality of water is a requisite for good management of fish culture ponds.

Two types of general management practices have been recommended to reduce the turbidity of water. The first is a *conversion* of suspended clay particles and the second is designated *control.*

### *Conversion*

The use of gypsum is a general practice that has been recommended for reducing the turbidity of water. Gypsum helps to exchange $Ca^{2+}$ for silica on the micelle and removing silica particles from the water.

$$SiO_2 + CaSO_4 + H_2O \rightleftharpoons CaSiO_2 + H_2SO_4$$

Sulphuric acid can be used to advantage on salty waters, particularly where sodium carbonate abounds. Since sulphuric acid decreases the alkalinity, the reactions of sulphuric acid with sodium carbonate may be shown as follows :

$$Na_2CO_3 + H_2SO_4 \rightleftharpoons \underset{\text{(leachable)}}{Na_2SO_4} + H_2O + CO_2$$

The application rate of gypsum depends on the quantities of suspended clay particles. The higher the quantities of clay particles, the higher is the amount of gypsum to be used. During rainy season, the turbidity in most of the ponds and lakes normally rises upto 1,000 mg/l and values even as high as 2,500 mg/l are on record. In case of medium and high turbidity of water, several hundred kilograms of gypsum per hectare are necessary. To hasten the reaction, gypsum should be thoroughly mixed into the water so that the entire amounts of clay particles may be precipitated.

### *Control*

The retardation of surface run-off into the pond ecosystem, human and cattle interferences are important features in the control of turbidity. Surface run-off is extremely important on the turbidity, particularly during the rainy season. Run-off waters carry huge amounts of clay particles to fish culture ecosystems thereby increasing the turbidity of water.

Human and cattle activities in fish culture ponds may sometimes create serious hazards by increasing the turbidity of water. These activities should be avoided.

### *Effects of Turbidity on Fish*

Most of the culturable fishes tolerate high ranges of turbidity. For example, fishes did not show any behavioral reaction under laboratory conditions until the turbidity approaches 20,000 mg/l (as silicon dioxide). Some experimental fishes tolerate turbidity higher than 1,00,000 mg/l for a week or month, but finally succumbed to gill clogging at turbidities of 1,75,000-2,25,000 mg/l. Generally, mechanical effect of suspended mineral particles on fish occurs when the water contains upto 4 per cent (by volume) of solid particles.

Continuous secretion of mucus by the skin of some species of fish (Order : Anguilliformes, Family : Ophichthidae) (snake eels) helps protect the gills from being chocked by suspended particles. For example, secretion of a few drops of mucus by the snake eels (*Pisoodonophis chilkensis*, *P. boro*, and *P. cancrivorus*) to turbid water has been found to precipitate within a few minutes. This indicates how mud-dwelling fishes exhibit a number of adaptations. These fishes have narrow gill openings and the mucus secreted by their skin exhibits the property of precipitating suspended particles.

## 10.9 Historical Background of Water Disinfection

Sustainable fish culture requires water that must be free from organisms which are detrimental to fish and fish food organisms. Presence of several disease-producing micro-organisms cause serious fish health hazards. Therefore, supervision and operation of fish culture ponds should be made to eliminate micro-organisms.

Disinfection of water was first developed in 1879 in France and England where chlorinated lime was used for treatment of sewage effluents (For further review, see White (1978). In 1893, USA and Germany used chlorine for treatment of pathogenic bacteria in sewage effluents. Later on, in 1909 and 1915, liquid chlorine was developed on commercial basis. In 1961, a chlorine residual controlled system for wastewater effluent was established in California. In 1976, use of chlorine for removal of nitrogen from wastewater became very popular. In 1986, it has been established that besides chlorine, a number of compounds have been identified and implemented as disinfectants all over the world.

## 10.10 Types of Disinfectants

Recently, the problem of pathogens in fish culture ecosystems has received great attention. Each disinfectant has its own *efficiency* for removal of pathogens. Therefore, on the basis of efficiency, it is considered what type of disinfectant should be used. The following disinfectants are generally used.

### *Chlorine and Chlorine Compounds*

Many compounds of chlorine are available which can be handled more conveniently than free chlorine and under proper conditions of use, they are equally effective as disinfectants.

1. *Chlorine* : The chlorine is a good disinfectant and its efficiency in water vary depending on the presence of ammonia (inorganic nitrogenous compound) and proteins (organic nitrogenous compound). Chlorine reacts with ammonia and forms monochloramine which is then converted into chloro-compounds. Monochloramine is much less reactive than free chlorine and persists in water over a long period of time. On the other hand, organic nitrogenous compounds react with chlorine and form organic-chloramines. These compounds, thus, reduce the germicidal efficiency of chlorine.
2. *Chlorine Dioxide* : Chlorine dioxide is more superior than the chlorine because the former does not produce chloro-organic compounds. Moreover, chlorine dioxide molecules remain unchanged in strong acidic (pH 4.0) and alkaline (pH 8.0-9.0) conditions. This compound does not hydrolyze in water and hence do not reduce its efficiency. It has been confirmed

that chlorine dioxide reacts with peptone of bacteria and virus coats through absorption that causes the efficiency of disinfectant.

3. *Chloramines* : These are the most important chlorine compounds (such as Chloramine-T, $NH_2Cl$, Monochloramine and Azochloramide). However, the most advantage of the chloramines is their stability. They are more stable than the hypochlorites in terms of prolonged release of chlorine.

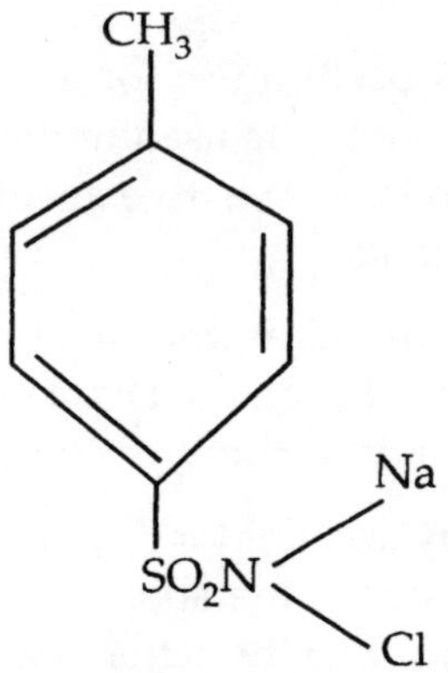


Chloramine–T

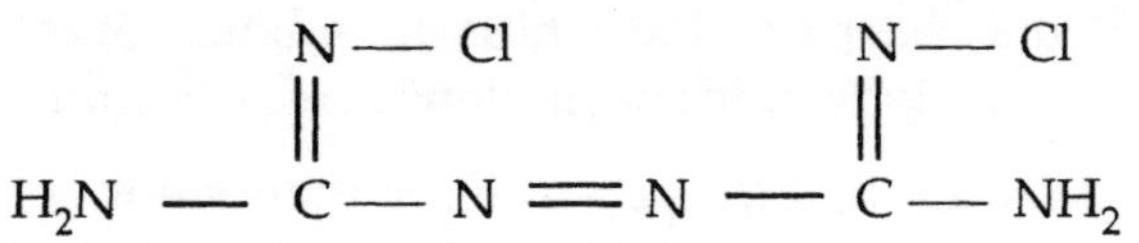


Azochloramide

4. *Hypochlorites* : The most widely used chlorine compounds are hypochlorites. Generally, three types of hypochlorites are extensively used as disinfectants such as (a) *sodium hypochlorite flakes*, (b) *liquid sodium hypochlorite*, and (c) *granular calcium hypochlorite*. These compounds react with water and generate hypochlorous acid which acts as an disinfectant.

$$\underset{\text{Sodium hypochlorite}}{NaOCl} + H_2O \longrightarrow \underset{\text{Hypochlorous acid}}{HOCl} + NaOH$$

## *Mode of Action of Chlorine*

The action of chlorine and chlorine compounds is due to the formation of hypochlorous acid when free chlorine is added to water.

$$Cl_2 + H_2O \longrightarrow HCl + HClO$$

Chloramines and hypochlorites also undergo hydrolysis with the formation of hypochlorous acid which is further decomposed as follows :

$$HClO \longrightarrow HCl + O$$

The oxygen thus released is called *nascent oxygen* which is a strong oxidizing agent. This agent reacts with the cellular constitutents of micro-organisms and consequently they are completely destroyed. The action of chlorine compounds is also due in part to the direct combination of chlorine with proteins and enzymes of micro-organisms.

### *Bromine*

In alkaline medium, bromine ($Br_2$) is converted into dibromamine that has germicidal efficiency. Bromine has been considered as a most potent germicide but its germicidal efficiency is least compared to other halogens. However, it has been demonstrated that 100 mg/l of bromine is as effective as 13.3 mg/l of chlorine.

### *Iodine*

Iodine is the oldest and most effective germicidal agent. It has been in use for more than a century. It is only slightly soluble in water but readily soluble in alcohol. It is used in the form of substances called as *iodophors.* Iodophors are mixture of iodine with surface-active agents which act as carriers for the iodine. Iodine is slowly released from these agents.

Elemental iodine ($I_2$), though inferior than chlorine, is considered as the most effective bactericidal agent and is unique in that, it is effective against all kinds of bacteria. However, about 0.2 mg/l of iodine has been found to destroy various species of bacteria within three minutes.

1. *Mode of Action :* Iodine is an oxidizing agent, and this fact may account for its antimicrobial action. Oxidizing agent can oxidizes and this inactivate essential metabolic compounds such as proteins with sulphydryl groups. It has been suggested that the action may involve the halogenation of tyrosine units of enzymes.

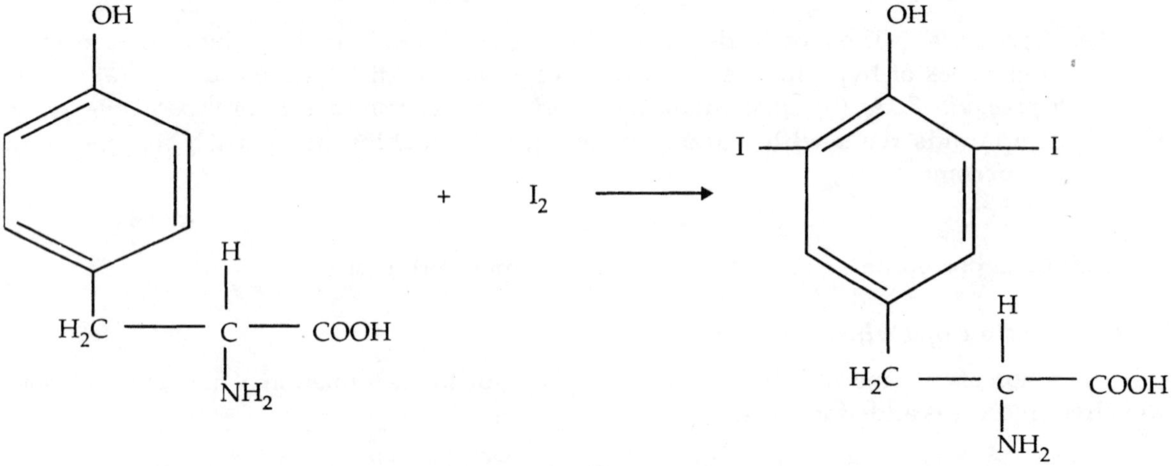


Tyrosine Diiodotyrosine

### *Heavy Metals*

Although some heavy metals (such as mercury, silver etc.) have antimicrobial activity, the most effective is copper sulfate. It is more effective against algae and molds than bacteria and 2 mg/l in water is sufficient to prevent algal growth in water.

1. *Mode of Action :* Copper sulfate combines with cellular proteins of micro-organisms and inactivating them. Note below that the metal reacts with the sulphydryl group of enzyme.

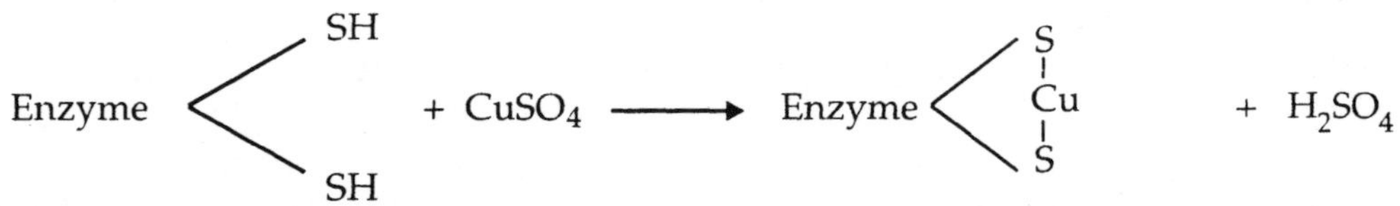


Active enzyme Copper sulfate Inactive enzyme

## *Quaternary Ammonium Compounds*

Compounds of the cationic-detergent class are called *quaternary ammonium compounds.* However, different quaternary ammonium compounds have been synthesized and evaluated for their antimicrobial activity. Some compounds are available commercially as effective antimicrobial agents used for myriad purposes (Figure 10.9).

The bactericidal power of these compounds is very high and have the ability to manifest bacteriostatic action far beyond their bactericidal concentration. The limit of bacteriostatic action for a given compound may be a dilution of 1:30,000; but it may be bacteriostatic in dilutions as high as 1:200,000. The action of these compounds demonstrates the need to distinguish between lethal and static activity for the evaluation of disinfectants.

The combined properties such as germicidal activity, low toxicity, high solubility in water and stability in solution have resulted in many applications of quaternary ammonium compounds as sanitizing agents and disinfectants in the dairy and fishing industries to control microbial growth on equipments and the aquatic environment in general.

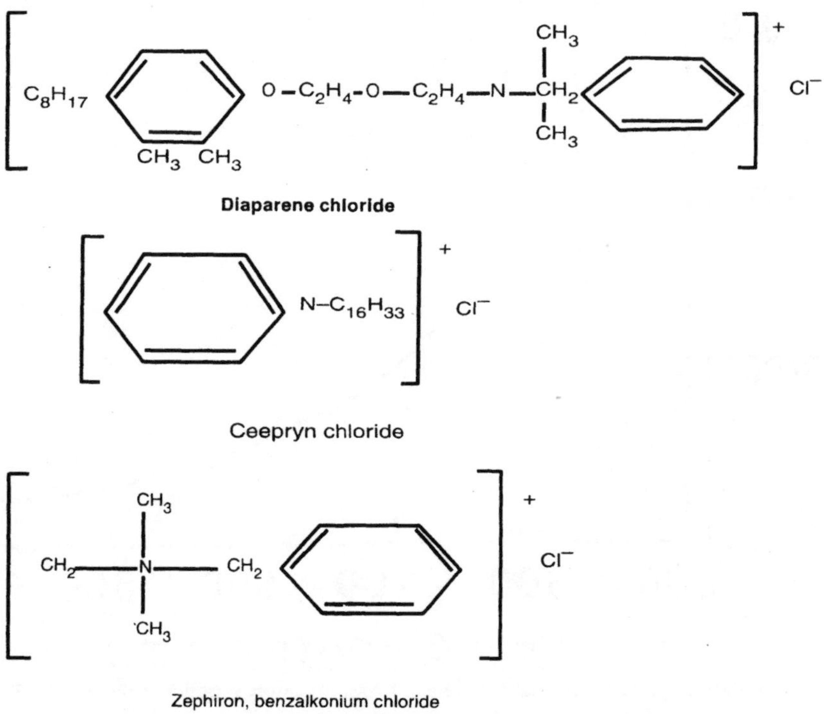


**Fig. 10.9 : Some examples of quaternary disinfectants.**

### *Ultraviolet Light*

Energy transferred through space in a variety of forms is termed as *radiation*. The most significant type of radiation is possibly *electromagnetic radiation* of which light is the most common example. However, in contrast to Gamma rays and X-rays, ultraviolet light is considered as less energetic radiation; it is absorbed specifically by various compounds because it excites electrons and raises them to higher energy levels, thus encouraging a variety of chemical reactions in the environment.

The ultraviolet portion of the spectrum includes all radiations from 100 to 3900 Å. It has been established that wavelengths between 2000 and 7000 Å have the highest bactericidal efficiency (Figure 10.10). Although the radient energy of sunlight is composed of ultraviolet light, most of the shorter wavelengths are filtered out by the earth's atmosphere. As a result, the ultraviolet radiation is greatly reduced from about 2670 to 3900 Å. Therefore sunlight, under certain conditions, has

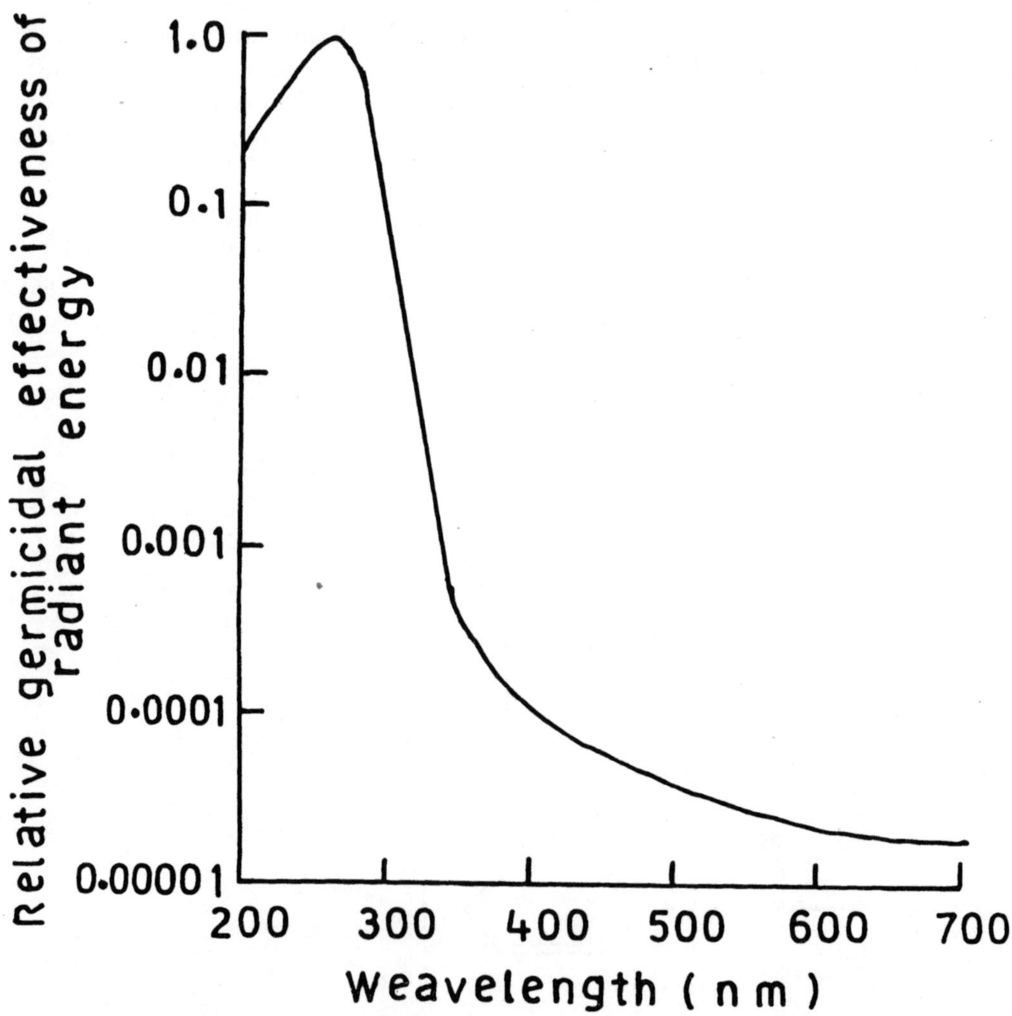


**Fig. 10.10 : Relative germicidal effectiveness of radiant energy between 2000 and 7000 Å.**

germicidal activity. It has been detected that ultraviolet light is so effective that different species of bacteria (such as *Escherichia coli*, *Salmonella* sp., *Bacillus* spp. etc.) are destroyed within 60 seconds.

Because of the microbicidal effect to a limited degree, the ultraviolet light is attractive for practical consideration in fish culture. To serve the purpose, fish culture ponds should always be constructed/selected in such places that maximum amount of sunlight can penetrate into the water.

## 10.11 Other Management Schedules

Pond water management techniques for finfish and shellfish culture, while varying slightly depending on the specific biological requirements of the culture species and the type of the culture environment, are similar in that they involve several activities such as pond preparation, stocking, feeding and fertilization, manuring, pond management, water management, and harvesting.

Intensively managed ponds generally require full artificial feeding and substantial water management to ensure optimum culture conditions for the species being reared. On the other hand, extensively managed systems require the least management, with no supplemental feeding and minimum water exchange on account of the low stocking density used.

### *Oxyguard*

Oxyguard International, Denmark, manufacture water quality measurement and control equipment that not only guards against dangerously low oxygen levels, high temperature or wrong pH levels, but also helps the fish farmers to optimize production. The Oxyguard-8 is very popular with smaller farms. This unit can be fitted with the number of measurement channels needed up to a maximum of 8. Nursery and rearing pond culture requires this type of equipment and found to be useful since the condition of the smaller quantities of water usually involved can fluctuate very rapidly.

This system is highly accurate, reliable and easy to use. It is so designed that fish farmers can operate it without difficulty. Oxyguard systems are now widely used all over the world.

### *Oxyflow*

For sustained flow of oxygen during nursery and rearing pond management, oxyflow is recommended for farmers. It is a product of Guybro Chemical, Mumbai, India, and though it is not widely used by farmers, it has several advantages over paddle-whell aerator such as (i) it increases dissolved oxygen levels rapidly (from 3.1 to 7.6 mg/l) and the dissolved oxygen levels are maintained for more than 48 hours, (ii) it eliminates toxic gases from the ponds, (iii) it is effective even in the presence of organic matter, and (iv) it increases survival, growth, and maturity of fishes. Further investigation on this aspect is badly needed.

### *Depth of Pond Water*

Water in the pond is kept at certain levels for optimum fish growth. In general, a pond water depth of 1 metre is considered best for culture of carps, tilapia, prawns, and shrimps. Such ponds are highly productive because enough sunlight can penetrate into the water to be effective in photosynthesis.

### *Exchange of Water*

Pond water is not just maintained at a certain depth; its quality must also be kept high to ensure optimal growth of fish. This is very important in semi-intensive and intensive culture systems where large amounts of metabolites are continuously accumulated into the pond and where excess, unconsumed feeds add to the bottom load and serve to pollute the water.

To prevent the deterioration of the pond environment, pond water should be continuously freshened by the entry of new water and the old water is drained through the outlet.

A flow-through systems of water managemet that allows the simultaneous entry and exit of water into and out of the pond is very important in any high-density culture systems. This is effected by the provision of separate inlets and outlets for all the ponds, each inlet regulating the flow of water from the source of water to the pond and each outlet controlling the discharge of water out of the pond. Outlet and inlet systems are so designed that as to bring water into and drain water out of the lower levels of the pond, where water quality tends to get poorer very rapidly as a result of accumulation of wastes and their subsequent decomposition.

Frequent replenishment of pond water can be made by the use of pumps. Although there is no definite rule as to the rate of water change necessary for medium to high density fish culture, semi-intensive culture systems change water at the rate of 15 per cent per day for an equivalent total replacement of water every 10 days or three times per month. Intensively managed ponds require greater water exchange in view of the much higher organic load on the pond bottom, particularly toward the latter phase of the culture period when fishes excrete more wastes.

### *Aeration*

Intensive fish culture ponds require to provide aeration facilities/equipment to prevent anoxia that may lead to poor growth and mass mortalities. Oxygen depletion in high-density fish ponds results the faster rate of ultilization of dissolved oxygen. Proper aeration of pond water accelerates the decomposition at faster rate at the pond bottom by oxygen-consuming micro-organisms.

Generally, paddlewheels are used in the ponds to effect the introduction of greater quantities of oxygen into the water. The aerators are operated at periodic/regular intervals for certain fixed durations in the early morning hours when the concentration of dissolved oxygen is known to be lowest (due to the absence of photosynthetic activity in the pond). At the end of the culture period when oxygen demand is highest, continuous aeration should be provided. At that time, water pumps need to be run for longer periods to effect greater water exchange. However, the number and type of aerators depends on the pond depth, shape and size, stocking density, and culture systems.

### *Measurement of Water Parameters*

In general, poor growth and mass mortalities of fish have been shown to be associated with deteriorating water qualities. Most of the problems encountered by pond fish culturists come through several means such as discharge of waste materials, run-off water, and different farming activities feeding, fertilization and others. The main problems include the following : (1) Development of toxic algal species; (2) Agricultural and industrial chemicals; (3) Urban wastes; (4) Development of pathogenic bacteria; (5) High oxygen loading; (6) Extreme pH ranges.

Whether the way of contaminating the fish culture ecosystems, steps though strenuous, should be taken to root out the causes of deterioration of water. Strict measures are taken to improve the quality of water for the production or quality and healthy fish. For this purpose, pond water is regularly sampled and basic/essential parameters such as pH, dissolved oxygen, ammonia, salinity and phosphate should be measured. This is very important for the purpose of determining the need for remedial action to bring water quality to optimum levels good fish production.

Problems of acidity are corrected by liming. Dissolved oxygen levels are kept above 6 mg/l by aeration. Salinity is an important parameter for prawn and shrimp culture and should be maintained within a range of 3-8 and 15-25 ppt., respectively. During summer months, high-salinity water can be diluted by mixing with freshwater. Since nitrogen and phosphorus are very important which determines the productivity of pond water, it is necessary to determine their optimum levels in water. However, deficiency of these nutrients can be rectified by fertilization at recommended rates. Other techniques of management of pond water in relation to the productivity of fish culture ecosystems have been discussed in chapter 11.

### *Role of Alum in Pond Cleaning*

These are double salts, generally of alkali metal sulfates and aluminium sulfate and have the empirical formula, M $Al(SO_4)_2.12H_2O$, where M is the alkali metal. On doubling, the empirical formula gives the molecular formula of alum. Common alum is generally manufactured from powdered bauxite. This compound is colorless, octahedral or cubic crystals and highly soluble in water. Because of high solvent property, it is widely used in fish ponds to remove colloidal impurities from water and thus the turbidity of pond water significantly decreased. The compound is also effective in removing toxic gases from pond mud. However, the effective recommended rate of alum varies between 400 and 500 kg/hectare. Prior to application, this compound should be powdered for uniform distribution over water.

## 10.12 Management of Water in Tropical and Sub-Tropical Regions

Wise use and management of water in tropical and sub-tropical countries are the most critical factors to increase fish production. At the same time, agricultural and industrial requirements for water have expanded dramatically and have created strong competition for fishery activities. This competition forces increased emphasis on efficiency of water use in fish culture. Considerations of water management will undoubtedly follow those concerning fish management.

Management of water is much influenced by several factors. In fact, the most important factors imposed by the environment are highly significant. A brief description of some factors follows.

### *Evaporation of Water from Aquatic Environment*

In tropical and sub-tropical countries where the temperature of water is extremely high for most of the year, vapor losses of water occur by evaporation. The loss resulting from this process is termed as *evaporation*. As a result of this phenomenon, water level gradually decreased to such an extent that an unfavourable environment for the fish is created.

### *Temperature*

A rise in temperature increases the vapor pressure at the water surface but has less effect on the vapor pressure of the air. This temperature difference obviously enhances the rate of evaporation.

### *Wind*

If a dry wind flows continuously over the water surface, moisture vapor will sweep away. Therefore, high winds intensify evaporation from aquatic environments.

### *Sunlight*

It has been estimated that 540 calories energy are necessary to evaporate each gram of water. On cloudy days the sunlight striking the water surface is reduced. Consequently, evaporative potential is not as great as found on cloud-free days. However, shading of water surface by floating aquatic plants reduces evaporation although a substantial amounts of water are lost through transpiration from the leaf surface during fish culture period.

## 10.13 Leaching Losses of Nutrients

Leaching losses of water are high in ponds having sandy soils. Even in sandy-loam soils, leaching losses are also of significance. Therefore, losses of water through leaching depend on the character of the soil. Since most of the nutrients remain in dissolved conditions, such losses of water rob the pond soil much of its fish production potential.

The loss of nutrients through leaching along with water is determined by soil-nutrient interactions. Soil properties have a definite effect on nutrient-leaching losses. Sandy soils permit greater nutrient loss than do clays because of the higher rate of percolation and lower nutrient-absorbing power of the sandy soils. For example, soluble phosphorus is rapidly bound chemically in fine-textured soils with appreciable amounts of iron and aluminium oxides. As a result, little amount of phosphorus is lost by leaching from these soils. Nitrates react by anion exchange with iron and aluminium hydrous oxides. For these reasons, nitrates are less prone to leach from soils high in iron oxides.

The leaching of cations is affected by the cation-exchange capacity of the soil. Soils with a high cation-exchange capacity tend to hold the nutrients and prevent their leaching. Because such soils are often high in exchangeable cations, they provide a reservoir of nutrients; of course, a small portion of these cations is subject to leaching.

### *Control of Leaching Losses*

The most effective practices aimed at controlling leaching of water to a lesser extent, however, are those that provide some cover to the bottom mud. The most important water-saving materials are organic manure, straw and litter. These are highly effective in reducing water from fish culture ecosystems. Even when following these suggestions, some losses will occur. The aim should be to minimize these losses both for the sake of the farmer and for ecosystem in general.

## 10.14 Conclusion

Routine applications of De-odorase, chlortech, geolite and aquazyn in composite fish culture ponds where manuring and fertilization programs are undertaken, appears to be essential for successful management of water quality. Due to high solubility of De-odorase in pond water, it rapidly binds with the ammonia, thus the former exhibit antagonistic effect on the potentiality of the toxicity of ammonia to fishes.

Chlortech (stabilized chlorine dioxide) is a very powerful disinfectant which has Environmental Protection Agency and United States Development Agency clearance for many applications including aquaculture farms and finfish as well as shellfish processing units. It is also extremely effective against a wide range of micro-organisms.

Aquazyn is a blended concentrations of selected bacteria, enzmes and buffers that are designed to remove ammonia, nitrate, nitrite, reduce Biological Oxygen Demand, degrade organic wastes and enhance survival as well as growth of finfish and shellfish. It is possible to adopt these techniques by fish farmers for fish culture management because these products are easily available and can be used at very low rates.

To prevent deterioration of the pond environment, pond water is continuously freshened by the entry of new water. The amounts of water to be replenished entirely depend on the type of culture systems. Bacterial and organic loadings as well as phytoplankton density widely fluctuate in low, medium and high volume water exchange systems. Other methods of water management include aeration of water by aerators at regular/periodic intervals and management of water and soil parameters.

### *References*


Das, D.N. 1997. Aqua-hygiene and its management. *Fish. Chim.* 17 : 35-37.

Sarkar, S.K. 1998a. Effects of interactions of De-Odorase and fertilizers on the survival and growth of carp spawn. *Geobios.* 25 : 137-140.

Sarkar, S.K. 1998b. Effects of plant-extracted glycocomponent De-Odorase on *Macrobrachium rosenbergii* in ponds. *Fish. Chim.* 18 : 13-15.

Sarkar, S.K. 1999. Role of De-Odorase on water parameters in fish ponds. *J. Environ. Biol.* 20 : 131-134.

Sindernann, C.J. 1986. Role of pathology in aquaculture. *In* : Realism in Aquaculture (*Ed.* M. Bilio, H. Rosenthal and C.J. Sindernann). pp. 395-419. European Aquaculture Society, Bredene.

Warren, J.W. 1983. Synthesis of a fish health management programme. *In* : Guide to integrated fish health management in the Great Lakes Basin. (*Eds.* F.P. Meyer, J.W. Warren and T.G. Carey), pp. 151-158, Special Publication, 83(2), Great Lakes Fishery Commission.

White, G.C. 1978. Disinfection of wastewater and water for reuse, Van Nostrand Reinhold Co., New York.


## Questions

1. What is de-odorase? Why it is important in fish culture?
2. State the effects of de-odorase on fish and prawn as well as natural food organisms.
3. List some products that are used for the management of water in fish culture. What are the essential features of these products?

4. Why application of gypsum is important in fish culture management?
5. List several compounds of halogens that are used to control micro-organisms. What are their mode of action upon micro-organisms?
6. Describe some management schedules (other than disinfectants) that are related to fish culture.
7. How losses of nutrients through leaching affect fish production? How it is controlled?

# 11

# Management Techniques in Fish Culture

Fish culture management [For further detail on the subject, see Agarwal, (1990)] can be defined as the science and technology of producing annual crop of fish mainly for commercial use from a well-maintained water areas. However, depending on the geographical and climatic conditions, the management procedures vary from region to region and even country to country. Therefore, it is difficult to generalize the management strategies. Some of the techniques that may be effective are described in a general way.

Aquaculture is one of the most important technique for healthy environment because aquatic ecosystems are able to utilize a lot of undesirable substances through various processes. Man is always looking for development which involves various management procedures of aquatic ecosystems, and the impact of human activity on environment includes both detrimental as well as beneficial effects on aquatic life. For improvement of aquaculture, the use of fertilizers, manures, algicides, and herbicides are important but there is every possibility of adverse impact from their indiscriminate use and hence should not be ignored.

## 11.1 Importance of Management

It is needless to say that use of fertilizers, oil cakes, organic manures, and lime at optimum rates are important for constant productivity of fish culture ecosystems. Of course, their use in excessive amounts causes fish mortality and poor production. The levels of these nutrient ingredients and their interactions with various biotic and abiotic factors of water and soil significantly limit fish production. However, prior to filling the ponds with water before the start of the monsoon, nitrogen, phosphorus, and calcium-based fertilizers should be thoroughly mixed with pond water for (1) balanced interactions among them, (2) the cultivation of natural food species, and (3) maintaining good quality ecosystem in which fish can grow.

One of the production techniques identified by the fish culture development and coordination experts was the need for new long-term approaches to increase production of low-cost fish through fish culture techniques. This need is greatest in those countries and sub-regions where population growth is rapidly accelerating the gap between market demand for fish and all potential sources of supply. Special priority shoud be given to the promotion of adequate production systems, which are feasible from a socio-economic and environmental point of view, to enhance production in water bodies which are not considered unsuitable. Generally culture-based fisheries are seen as potentially useful tools for resource enhancement, in those countries where vast open-water areas are available to use for fisheries development.

A fish farmer has to manage his fish culture pond efficiently to ensure production and profit. The importance of management is very critical because inadequate management can undoubtedly upset the balance inspite of the application of the latest and cheapest technologies. Generally, intensive and semi-intensive fish culture systems faced a lot of biological and organizational problems by fish farmers and hence extreme care is essential for the purpose.

## 11.2 Pond Management

Production capacity of Fish Culture Ecosystem (FCE) can be enhanced through fertilization, manuring, liming, and adequate feeding. Furthermore, suitable ecosystem, stocking and other management are the most important means of intensifying fish yield.

### *Fertilization*

Fertilization aims at greater production of aquatic food species through increased growth of plankton and bottom fauna. Fish culture ponds are regularly fertilized with inorganic fertilizers such as urea, ammonium sulfate, single superphosphate, rock phosphate, ammonium sulfate nitrate, calcium ammomium nitrate, diammonium phosphate, and dicalcium phosphate are important for maintaining plankton populations. Intensive and semi-intensive culture systems do not require fertilization since they are not natural food-based, except for those which grow as plankton-feeders.

Nitrogen compounds are of little importance as they are consumed by bacteria. Moreover, nitrogen carriers do not exhibit any effect unless they are mixed with phosphorus fertilizers. It is important to note that fertilizers should be used at very low rates, fish culture ponds should be fertilized during one growing season (12 months), fertilizers should be divided into several lots and in suitable combinations so that maximum effect of fertilizers can be obtained under different climatic conditions.

Fertilization of fish ponds in the United States usually consists of 10-12 periodic applications of inorganic fertilizers, with the first application being made in late winter or early spring. In India, however, pond fertilization is made during one growing season. It has been established that there is a positive correlation between temperature and fertilizer toxicity to fish and fish food organisms. Therefore, environmental temperature should be considered before the onset of fertilization. In other words, fertilization should be discontinued during summer and monsoon seasons when water temperature becomes high and fluctuates between 30 and 38$^0$C.

In general, liquid and solid fertilizers are used in fish culture but both are not equally effective in increasing fish production. In South-eastern United States, for example, liquid fertilizers were found to be more effective in increasing nitrogen and phosphorus concentrations as well as phytoplankton production than broadcast applications of solid fertilizers.Therefore, it may be concluded that liquid fertilizers appear potentially useful in fish culture. But it should be remembered that liquid fertilizers are expensive, not easily available everywhere and Indian fish farmer's are not accoustomed to use these fertilizers very carefully. Obviously, the application of liquid fertilizers at different rates for fish production in ponds remain outside the realm of Indian fish farmers.

In contrast to the efficiency of NPK- fertilizer combinations the use of only N and P has been found to be more superior to fish production. The use of potassium fertilizers in fish culture has been abandoned as it did not bring a higher return although a trace amount of potassium is accumulated in water and soil through artificial feeding and organic manuring which is essential for tilapia culture but toxic to Indian major carps. The combinations of nitrogen and phosphorus should be such that the total fertilizer should have good aquacultural values. Several nitrogen fertilizers such as ammonium sulfate and ammonium sulfate nitrate are 10 to 15 times more toxic than other fertilizers and therefore, use of such fertilizers should not be encouraged. Moreover, use of these nitrogen fertilizers far excess than requirement tend to increase in acidity of soil which is not desirable for fish farming.

### *Organic Manuring*

1. *Green Manures* : In addition to the use of chemical fertilizers, organics have also been used in fish culture. One technique for this is green manuring, which means a preparation of the bottom soil of FCE, which is mowed prior to monsoon season or to filling with water and the stocking with fish. Generally grain and leguminous plants are suitable for it because the former decomposes slowly and the latter one is able to fix atmospheric nitrogen into the soil.
2. *Animal Manures* : Animal manures have the advantage to release nutrients as they slowly disintegrate and become available to the water. Release of nutrients generally depends upon the type of manures. For example, release of soluble salts from cowdung is not as high as observed in case of poultry manure. On the other hand, cowdung particles sink at the rate of 2.3 cm per minute as against 4.4 cm per minute in the case of pig manure. This provide sufficient time to release nutrients from the cowdung. However, daily addition of poultry manure and occasional racking of bottom maintain pH, alkalinity, specific conductivity, nitrogen, and phosphorus at desired levels to stimulate fish production. Animal manures increase the concentrations of plankton and bottom fauna, physical properties of soil and are more suitable in sandy-loam soil where leaching of water is a common phenomenon.

   Although high production was observed from ponds treated with poultry manure and feedlot cattle manure, management of fish ponds is accomplished with little organic fertilization principally because of concern over high biochemical oxygen demand loadings and the conventional knowledge gained from experience that a high quality diet is necessary for good production. However, use of organic manures from animals could result in different growth of fish because manures collected from different animals vary in quality.
3. *Oil Cakes* : Although a variety of oil cakes such as neem cake, mahua oil cakes, mustard oil cake, and groundnut cake have long been practiced in Indian fish farming systems along with the chemical fertilizers, little or no systematic investigations to optimize their use have been accomplished. Because of recent increase in feed ingredient costs, fish farmers have been able to consider that the fish production cost could be decreased by using animal manures and/or oil cakes along with the chemical fertilizers. Fish is well-suited in conversion of organic matter into body protein and lipid by the use of oil cakes in combinations with the inorganic fertilizers. Therefore, the interaction between oil cakes and chemical fertilizers is significant to pond productivity.

   Prior to application of oil cakes, they are powdered for uniform distribution in water. Application of mahua oil cake at the rates varying between 250 and 350 kg/hectare in fish culture is important because it not only removes predatory fishes from ponds but also increases the fertility of water and soil when its toxic action is over. Studies have, however, shown that the application rate of mahua oil cakes can be reduced to 220 kg/hectare and thus the cost of fish production is also reduced.

Fish production from ponds has been found to be more economical when oil cakes at different rates are used. Application of mustard and groundnut cakes, for example, at the rates of 1,775 and 1,550 kg/hectare/year exhibited superior results than cow manures. Of course, use of oil cakes along with the chemical fertilizers has been found to reduce the rates of oil cakes and the cost of fish production.

### *Liming*

Liming of fish ponds is important for obtaining alkaline condition of water and soil. Acidic mud has several disadvantages such as (1) abosrb phosphate, (2) bottom organisms do not grow well, (3) fish will not survive and do not have adequate growth and reproduction, and (4) insufficient carbon and calcium for growth of plankton.

The rate of liming depends upon the soil condition of the ecosystem and type of liming materials. It the total alkalinity and pH of water remain more than 20 mg/l and 7.5-8.0, respectively, liming is not essential but in many situations (such as dystrophic condition of ecosystem), liming is necessary when alkalinity and pH drops below these levels. Therefore, mud samples should be collected from fish ponds for determination of lime requirement which has already been described in Chapter 2.

*Collection of Soil Samples for Liming* : Pond mud samples are collected from several sampling areas of ecosystem, thoroughly mixed to provide a single sample. This sample must be air dried, powered, passed through a 20 to 25 mm mesh size and pH is measured in a 1 : 1 mixture of dry, pulverized soil, and distilled water. This sieve analysis is necessary to assign an efficiency rating of a liming substance on the basis of particle size. Generally, liming rates for diffferent forms of lime are expressed as a calcium carbonate equivalent with a neutralizing value and efficiency rating. For example, if a fish pond is limed at the rate of 1,000 kg/hectare with a calcium oxide having a neutralizing value of 92 per cent and an efficiency rating of 85 per cent, the amount of calcium oxide required can be calculated as follows :

$$\text{Application Rate} = \frac{1{,}000 \text{ kg/ha}}{\frac{92}{100} \times \frac{85}{100}} = 1{,}278 \text{ kg/hectare}$$

Different forms of lime such as calcium carbonate, calcium oxide, calcium hydroxide and basic slag possess 90-105, 180, 135, and 55 per cent neutralizing vlaues, respectively. However, soil acidity is neutralized at faster rate by fine particels than by coarse particles of limestone. Generally, liming substances are composed of particles of differnt sizes. The different size classes for use in efficient test prepared by sieve analysis for agricultural purposes are also used in fish culture.

## 11.3 Application Methods of Fertilizers and Lime

### *Fertilizers*

Application of chemical fertilizers to fish culture pond is difficult because they should be sprayed over water. For small ecosystems, difficulties may not arise because relatively a small amount of materials is needed. In case of larger ecosystems, special devices are necessary for uniform application of fertilizers. However, prior to application of fertilizers, they are dissolved in water and then sprayed over water. The following may be considered as effective methods for the purpose :

1. *Pump* : A centrifugal irrigation pump with a capacity of 400 litres per mintue is used to discharge fertilizer into the ecosystem at a single site.
2. *Power sprayer* : It is operated by a 5-HP engine and the ferilizer is sprayed from one site from the edge of ecosystem.
3. *Hand sprayer* : A compression type hand sprayer is used to supply fertilizer. The operator is moved around the edges of ecosystem.
4. *Broadcasting* : In many cases, fertilizers are broadcast in solid form from different sites of ecosystem.
5. *Boat method* : The boat is moved over water surfaces and the solid forms of fertilizers are broadcast. Organic manures are also applied by this method.
6. In some situations, cattle manures are dumped at shallow corners of ecosystems.
7. Prior to application of oil cakes, they are detoxified using cold water. The process consists of soaking of oil cake in water for several days, after which it is again diluted with water and then broadcast over water.

### *Liming Methods*

As the solution of calcium compounds is removed from the soil by leaching, the per cent base saturation and pH are gradually reduced and therefore application of lime is necessary. While losses vary in different location of ponds, it is well to recognize that liming of pond is not a one-way venture, rather it has to be repeated with somewhat regularly. This type of cyclic activity should be adopted by fish farmers. However, the techniques for applying bulk agricultural limestone is to construct a platform on the edge of a large boat (18 feet in length) which can carry 700 kg limestone and apply 30-40 tonnes of material per day (8 hours) with the help of 3-4 workers. Limestone is sprayed with a shovel as the boat moves over water. Pumps and sprayers are also used to spray powdered limestone and hydrated lime. In some cases, hydrated lime is kept in a tank situated near the ecosystem and pump is used to discharge the lime. Liming materials of small piles are sometimes dumped along shallow edges to make the method more effective.

## 11.4 Plankton Density

Another important aquacultural management technique is the development of phyto-and zooplankton through fertilization and manuring. Plankton concentrations determine the productivity of fish ponds. Phytoplankton groups such as Chlorophyceae, Cyanophyceae, Bacillariophyceae, and Englenophyceae and that of zooplankton groups such as Copepoda,

Cladocera, and Rotifera are important. Some species of phytoplankton such as *Volvox* sp., *Anabaena* sp., *Spirogyra* sp. and others form a green scum over water surface which is not desirable for fish culture. Plankton abundance is the most important natural fish food organisms and simple methods (direct count and wet weight methods) for determination of their concentration and abundance in ponds should be followed by fish farmers.

## 11.5 Stock Monitoring

The cultured fishes are monitored closely and regularly to determine their rate of growth and the general condition of the stock. In the first few months of culture, the feeding tray is a good tool for stock monitoring. As fishes grow in size, cast-netting is used as a sampling tool, with those caught in the throw of the cast providing an indication as to size and weight of stocks. On the basis of sample weights and daily feed consumption, it is possible to predict the available biomass and make projections on volume of harvest.

## 11.6 Bottom Fauna

Bottom fauna forms essential natural food items of many commercially important species of fish. Fish culture ponds are infested with various species of bottom fauna in varying quantities, their presence or absence principally depends upon the quality of water and soil. In ponds where manuring and fertilization programs are undertaken for constant fertility, their existence is perpetual and there is excessive growth and development of bottom fauna.

In semi-intesive and intensive culture systems where moderate or high stocking densities are maintained, fishes are not dependent on bottom fauna. In spite of the abundance of bottom fauna, fish entirely depends upon the supplemental feeds. In contrast, extensive systems use low stocking densities and no supplemental feeding, but fertilization and manuring may be done to stimulate the growth and production of bottom fauna. Generally, the growth of bottom-dwelling fishes depends entirely on the availability of bottom fauna.

## 11.7 Biofertilizers*

These are biological origin which serve as manure. Biofertilizers have attracted greater attention in developing countries as a substitute for costly chemical fertilizers. The importance of cyanobacterial biofertilizers was recognized as early as 1939. Since then a good deal of literature on these aspects has appeared. However, the advances made in biofertilizer research reveal that different strains of bacteria present in the nodules of leguminous plants use sunlight as the source of energy for fixation of carbon and nitrogen. They are able to release nitrogenous products spontaneously. Field trials have shown that about 30 kg Nitrogen/hectare can be saved by the use of cyanobacterial biofertilizers. Cyanobacteria have advantage over grain legumes, forage, and water ferns in that they are easily attacked by various diseases and pests. Although biofertilizers are now being used in agricultural fields, specially under water-logged conditions, it can unequivocally be stated that biofertilizers may also be used in fish culture management practices because it is a potential substitute for chemical fertilizer and serve as an efficient, pollution free agent as well as cheaper than fertilizers.

---

* For excellent discussion, see Desmukh (1998)

### *Azolla as Biofertilizers**

*Azolla* spp. is a heterosporous water fern and it has as siginificant role as a biofertilizer. It is a free-floating aquatic plant which fix atmospheric nitrogen through *Anabaena azollae* present in dorsal leaves. Due to its high nitrogen fixing capacity and decomposition ratio, release of nitrogen is very rapid and hence *Azolla* is considered as an ideal nutrient input in fish culture ponds.

### *Culture System of Azolla*

Azolla can easily be cultivated separately for periodic application in fish ponds. The system of cultivation involves a network of earthern raceway (20 x 3.0 x 0.6 metre) with water supply and drainage facilities. In each raceway, 6 kg of *Azolla* is incorporated. Sixty five gram of single superphosphate is added having water depth of 10-15 cm. *Azolla* is harvested weekly at the rate of 20-25 kg/raceway. It has been estimated that about 1 tonne of *Azolla* can be harvested every week from a water area of 650 $m^2$ with nitrogen and phosphorus fertilizer ratio of 5 : 1. It has also been reported that application of *Azolla* at the rate of 20 tonne/ha/year supply about 50 kg of N, 13 kg of $P_2O_5$ and 45 kg of $K_2O$. Besides these nutrients, organic matter at the rate of 750 kg/ha/ year is also added. Obviously, *Azolla* is a recent aquacultural input with high potential in both trophic enrichment and fertilization.

### *Benefits to Fish Culture Ecosystems*

The cultivation of green manure crop where fixed nitrogen is supplied to the ecosystem is important. Green manures improve the physical properties of pond soil. Most legumes are able to fix about 100 kg N/ha/year. Grain legumes tend to fix lower level of nitrogen than forages. Therefore, cultivation of forage plants seems to be advisable in fish culture ecosystem. Generally lucerne can fix about 600 kg N/ha/year than other types of legumes (Table 11.1). Among different forms of legumes, the members belonging to the genera *Aeschynomene aspera, Sesbania rostrata, S. aculeata,* and *Leucaena* sp. are important for nitrogen fixation.

**Table 11.1 : Rate of Nitrogen Fixation (kg/hectare/year) in some Common Green Manure Crops**

| Species | Amount of nitrogen fixed by plants |
|---|---|
| *Sesbania aculeata* | 60 |
| *Medicago sativa* | 150-250 |
| *Leucaena* sp. | 350 |
| *Glycine max* | 300 |
| *Vigna unquiculata* | 50-100 |
| *Lucerne* sp. | 600 |
| *Phaseolus vulgaris* | 30-50 |
| *Pisum* sp., *Lens* sp. | 40 |
| *Azolla* spp. | 150-300 |

*Source :* Selected data from Brady (1991) and Chakraborty (1990).

## 11.8 Pond Conditioning

A pond is scientifically prepared before stocking of fish seeds. The pond is totally drained and the pond bottom is dried. Tobacco dust, Nagdona, Tuba, Balanu, mahua oil cake, teaseed cake, and rotenone along with the lime are used at very low rates to remove predators and obnoxious gases from the pond. If complete drainage is not possible, the pesticidal plant products should be

* For further detail, see Talley and Raims (1982).

avoided because it may suppress the growth and production in the later phase of culture. Because of the toxic action and manurial value, mahua oil cake is, however, more effective and therefore widely used by Indian fish farmers for eradication of undesirable fishes from their ponds.

Ponds with acid-sulfate soils are repeatedly dried and flushed. Agricultural lime is then applied to correct the soil pH and bring it upto at least 6.5 or 7.0.

To maintain and stimulate the growth of natural plankton, organic or inorganic fertilizers are applied to the pond ecosystem. After fertilizer application, water is let in to a depth of about 20-40 cm and gradually increased to 1 metre a week after fertilization. Intensively managed ponds where artificial feeding shall be given, do not need to be fertilized. Extensive ponds need regular fertilization during the culture period to maintain the growth of natural food. Semi-intensive ponds may use a mix of fertilization and supplementary feeding.

## 11.9 Feeding and Feed

Fish grown in semi-intensive and intensive culture ponds are given supplementary and full artificial feeds, respectively; the former augments the natural food in the pond, and the latter replaces the natural organisms in the water as a source of nutrition.

Since artificial feed is most expensive and may not available all the times, it is necessary to calculate the food consumption rate of fish to avoid the loss of feed. Food consumption is expressed in terms of milligram (mg) dry food consumed per gram live fish per day and growth as mg dry substance gained per gram live fish per day. The gross conversion efficiency (GCE) can easily be expressed in terms of percentage of food converted into fish flesh.

$$\text{GCE} = \frac{\text{Growth (mg/day)}}{\text{Food consumed (mg/day)}} \times 100$$

Various types of artificial feeds and feeding methods have been developed by different organizations for fish culture management practices. A wide variety of feed ingredients are used to prepare supplemental/artificial feeds. The simplest fish feeds are prepared using rice or corn bran, wheat bran and oil cake. These are mixed with trash fish/fish meal in suitable ratios.

Commercial feed preparations are also available now in a wide range of brand names. These commercial diets consist of a number of ingredients for high fish production. However, detailed description regarding the importance of feeds, their composition, ingredients used in feed, and nutrition to fish have been given in Chapter 9.

### *Relationship Between Feeding and Abiotic Factors*

Relationship between the amount of feed added to ponds and dissolved oxygen as well as temperature is significant. At high feeding rate (> 112 kg/ha/day), the net fish production decreased. Excess feeding to fish drastically reduced the concentration of dissolved oxygen in pond water.

Fish growth under different fixed daily feeding rate (Table 11.2) at 30$^0$C was found to be higher (3.7 per cent) than those at 35$^0$C (3.1 per cent). The differences in growth rate between fish at 35$^0$C and other temperatures increased with 10 per cent per day feeding, and growth at 35$^0$C was less than that of fish at other two temperatures. Conversion efficiencies were highest at 10 per cent and lowest at 4 per cent per day feeding. The lowest and highest conversion efficiencies of fish were observed in 35$^0$C on 4 per cent per day feeding and 30$^0$C on 10 per cent per day feeding, respectively. Therefore, the behavioral responses of fish to temperature should always be taken into consideration. The consequences of environmental temperature variations to growth and feeding of fish in relation to other abiotic factors of pond ecosystem should not be rule out.

**Table 11.2 : Growth Rate and Conversion Efficiency of *Ctenopharyngodon idella* Fed With Aquatic Plant *Vallisneria* sp. Under Different Environmental Temperatures**

| Temperature ($^0$C) | Fixed daily feeding (per cent) | | | | | |
|---|---|---|---|---|---|---|
| | 4 | | | 6 | | |
| | A | B | C | A | B | C |
| 25.5 | 2.5 | 112.4 | 52.7 | 2.5 | 184.7 | 58.3 |
| 30.0 | 2.8 | 102.0 | 60.0 | 2.8 | 181.6 | 66.5 |
| 35.0 | 2.4 | 107.6 | 45.3 | 2.4 | 171.8 | 53.3 |
| | 8 | | | 10 | | |
| | A | B | C | A | B | C |
| | 2.5 | 22.6 | 64.8 | 2.5 | 271.5 | 103.4 |
| | 2.8 | 207.5 | 73.1 | 2.8 | 254.4 | 87.4 |
| | 2.4 | 218.6 | 59.3 | 2.4 | 226.7 | 78.0 |

*Source* : Sarkar (1989)

A, Initial weight of fish (g); B, Final weight of fish (g); C, Conversion efficiency.

Experiments were set in different seasons of the year 1983. Growth rate of fish was estimated in earthen vessels (area 0.4m$^2$; mean depth 35 cm) containing 60 litres of water (dissolved oxygen, 7.6 mg/l; pH 7.9; total alkalinity, 200 mg/l as calcium carbonate). Experiments were run for 15 days.

## 11.10 Significance of Liming in Pond Management

Liming materials provide dramatic and obvious examples of pond management as a factor of various functions to the production potential of fish culture ecosystems. Liming causes overall improvement of fish culture environment by increasing their effectiveness to survive fish in better condition and by altering the pond dynamics in ways that render fish more resistant to disease.

Examination of fish cultured under intensive farming system with adequate management lead inevitably to the conclusion that liming must undoubtedly and seriously increase favourable chances for production in ponds. Liming has, however, a great significance in fish culture for several reasons as follows.

(1) Lime materials are responsible for demineralization of the pond soil; (2) Liming triggers mineralization of organic matter; (3) Liming helps to precipitate excessive organic matter and consequently nitrification of ammonium compounds is controlled; (4) Lime substances reduce the concentration of colloidal organic matter. Thus, the turbidity of water greatly reduced; (5) Lime substances supply calcium to water that stimulate the growth of phyto-and zooplankton; (6) Concentrations of calcium, mangnesium and phosphorus significantly increased; (7) Lime substances act as disinfectant and as a result pathogens are destroyed from fish ponds; (8) Lime substances release ions in the water column and therefore, the concentration of total alkalinity increased; (9) Liming substances increase the availability of carbon for photosynthesis.

## 11.11 Turbidity

Turbidity of pond and lake waters is a very important factor for existence of the fish. Turbidity of water is principally due to presence of numerous inorganic substances such as silt, caly or planktonic organisms which limit the productivity of water. These particles remain in suspended conditions and obstruct the penetration of sunlight through water. Turbidity of water, due to planktonic organisms, indicates high productivity but that caused by silt or mud beyond permissible limit, is detrimental to fish and fish food organisms. The suspended particles adsorb considerable amounts of nutrient elements such as phosphorus, nitrogen, and potash in their inorganic forms making them unavailable for plankton production.

High turbidity in fish culture pond ecosytems drastically affects the plant and animal life to a greater extent. Since visible light is indispensable for photosynthetic activity, phytoplankton populations gradually disappear. Zooplankton populations also reduced and thus fish ponds become unproductive. Occasional racking of bottom mud, human and cattle interferences, rainfall, and algal blooms are responsible for high turbidity. Turbidity of water is measured by Secchi Disc and Jackson's Turbidimeter which are considered as the best and simple method (Figure 11.1).

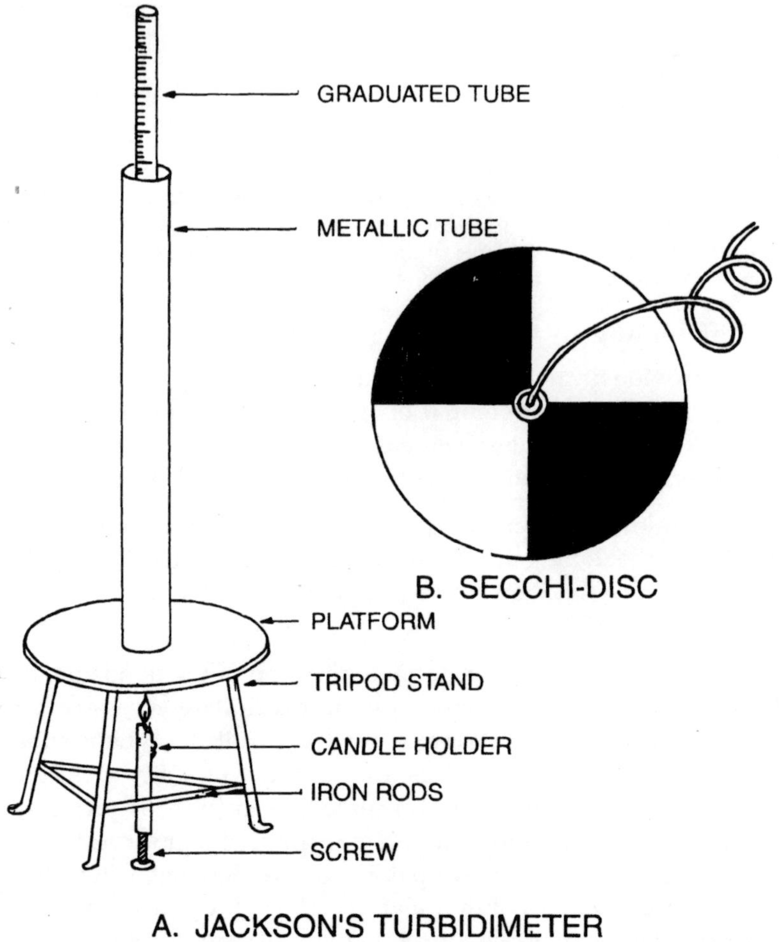


**Fig. 11.1 : Jackson's turbidimeter (A) and Secchi disc (B) are widely used to determine the transparency of pond water. These methods are very simple and easy to operate within a very short time.**

### *Secchi Disc*

It is composed of a large and a small disc both of which are fitted together by a central iron bar. The smaller disc is more heavier than the larger one. A hook is fitted at the central point on the larger disc. The upper surface of the larger disc is marked by four equal and alternate black and white quadrants. A nylon rope is tied with the hook.

Operation of the disc is usually carried out by a boat that floats on water surface. The disc gradually immersed into the water from the boat. As soon as the quadrants disappear from view, the rope is immediately stopped from further lowering.

The disc is pulled up from the water and measured with a centimetre scale from upper surface of the larger disc to the mark of the rope. The reading in centimetre is the amount of light penetration in that particular zone of the pond or lake. For best result, the same operation is carried out at different zones of the pond and the average value is considered for determination of the turbidity of water.

### *Jackson's Turbidimeter*

It is made up of a tripod stand – fitted with three horizontal iron rods. In the middle of the stand, a candle holder is fitted that can move up and down with a screw system. At the upper surface of the stand, a round platform is attached over which an elongated metallic tube is vertically fixed. A graduated glass tube is inserted within the metallic tube. One end of the glass tube is closed and the other end is open, the closed end is directed towards the platform of the tripod stand. The base of the platform is provided with an aperture. The arrangement of these parts is such that the aperture, candle flame, metallic stand, and graduated tube lie in a straight line.

At the time of operation, the glass tube is carefully placed within the metallic tube. A lighted candle is kept in the candle holder. The water sample, of which the turbidity will be measured, is gradually poured into the glass tube until the flame of the candle just disappears. Thus, the turbidity of the sample is measured in centimetre from the reading of the graduated glass tube.

Measurement of turbidity of the water with Secchi Disc has some advantages over that of Jackson's Turbidimeter because the disc is cheap, unbreakable, easily operated without any difficulty and hence widely used.

## 11.12 Ecosystem Treatment with Chemicals

Treatment of ecosystems with chemicals before or after stocking with fish has also an important role in the strategy of management techniques. Chemicals check various diseases of fishes and disinfect the ecosystem, because prevention is always the best possible control of diseases and parasites. Generally, disease transmission can occur through four routes : (1) water containing infected intermediate host such as Copepods or Oligochaetes, (2) water containing infected intermediate or final host, (3) wet pond bottom contaminated with pathogens, and (4) infected amphibia that have appeared to water containing fish. If any of the four routes of transmission are blocked by management techniques, medicaments for the fish may not be necessary.

Infection can also be avoided by obtaining fish fry and fingerlings from sources certified free from any diseases. Fish from non-certified sources can be disinfected upon arrival and subsequent release into the ponds. Fishes are disinfected by their submersion for several minutes in suitable chemicals at recommended rates. A variety of chemicals are known to control or check various diseases in fish and suitable dose of chemicals for the purpose have been developed (Table 11.3).

**Table 11.3 : Some Common Important Diseases and Parasites in Carp Fish and Their Control**

| Name of disease/ parasite | Chemical | Application rate | Treatment Schedule |
|---|---|---|---|
| A. Fungus | Malachite green | 5 mg/l for eggs and 4 mg/l for adults | As a daily 10 minutes bath for one hour |
| B. Protozoa | | | |
| 1. *Icthyopthiriasis* | Sodium chloride | 2-5 percent | For 7 days |
| | Copper sulfate | 8 kg/hectare | Three day interval in 3 instalments |
| | Malachite green + Formalin | 0.1 mg/l + 25 mg/l | At 3 day interval |
| | Copper sulfate | 0.5 mg/l for hardness 50 mg/l 2 mg/l for hardness 200 mg/l | Weekly in ponds |
| | Methylene blue | 2 mg/l | Daily for one hour |
| | Acriflavin | 10 mg/l | 2-20 days |
| 2. *Myxosporidia* | Quicklime | 1 tonne/acre | Drained but moist pond bottom should be limed prior to filling with water |
| 3. *Coccidiosis* | Bleaching powder | 500 kg/hectare | Applied at the time of draining of pond |
| | Stovarsol | 1 mg/g of feed | – |
| C. Bacteria | | | |
| 1. *Furunculosis* | Acroflavin | 1,000 mg/l | Applied for 30 minutes |
| 2. *Septicemia* | Terramycin | 50-75 mg/kg of feed | For 10 days |
| 3. Fin and gill rot | Terramycin | 50 mg/l | Sprayed over ponds |
| | Copper sulfate | 1:2,000 | Bath for 2 minutes |
| 4. Peduncle | Copper sulfate | 50 mg/kg of feed/day | For 10-15 days |
| 5. Ulcer | Potassium per-manganate | 5 mg/l | Dip treatment for 5 minutes |
| 6. Dropsy | Chloromycetin | 60 mg is mixed with 4.5 litres of water | Dip treatment for 5 minutes |
| D. Helminth | | | |
| 1. *Gyrodactylus* | Formalin | 25 mg/l | For 10 days |
| 2. *Dactylogyrus* | Potassium per-manganate | 10 mg/l | For 10 days |
| 3. Tapeworm | Dibutyl tin dioxide | 5 per cent mixed with feed | For 5 days |
| 4. Fluke | Di-n-buryl tin oxide | 5 per cent mixed with feed | For 5 days |
| E. Predator | | | |
| 1. Insect | Floating oil | 5 gallons/acre | |
| | Dylox (Dipterex, Neguvon, Chlorophos, Trichlorofon) | 0.5 mg/l | |
| F. Snail | Copper sulphate | 2 kg/hectare | To be applied for 1,000 sq. feet of pond bottom Removal of detritus by shovel |

*Source :* Selected data from Hoffman (1984), Cipriano and Bullock (1983, 1984), and Jhingran (1988)

Prior to stocking of ponds with fish, treatment of pond bottom by drying during summer or chlorination, or by applying hydrated lime is necessary. Pond bottom can be disinfected by drying during summer or winter. If the pond cannot be drained, the water should be chlorinated. The most commonly used disinfectants with extreme caution include slaked lime, hydrated lime and calcium hydroxide at rates depending upon the quality of soil and water. Generally, 5.5 tonnes/hectare is recommended for the purpose. Dosage of 1.0 tonne of 20 per cent ammonia water is also recommended for disinfection program in some situations. Calcium hypochlorite can also be applied at the rate of 1.0 kg/100 cubic metre.

## 11.13 Disease Protection Through Management

Fish diseases are the result of interactions of the etioloic agents such as the fish and the environment. Although facultative fish pathogens are present in water supplies, epizootic seldom occur unless environmental quality and the host defence systems of the fish also deteriorate.

In intensive fish culture, fish are continuously affected by environmental fluctuations and by management techniques such as handling, crowding, hauling, chemical treatement, drug treatment and water chemistry. Fish diseases commonly considered to be stress-mediated and certain environmental factors predisposing to epizootic are characterized by low oxygen concentration (less than 4 mg/l), overcrowding of fish, high temperature (more than 30$^0$C), wet diet, elevated ammonia nitrogen (1mg/l), inadequate pond cleaning, low hardness (less than 100 mg/l as calcium carbonate), non-supply of water to pond by sand gravel filter, chronic sublethal exposure to heavy metals and pesticides, excessive size variation among fish in ponds, nutritional imbalance, particulate matter in water, and rough handling of fish. However, environmental stress and resulting disease problems can be minimized by maintaining high water quality standards. Table 11.4 presents guidelines for water chemistry which are conducive to optimum fish health.

**Table 11.4 : Suggested Criteria for Water Quality Required for Optimum Health of Warmwater Fish Ponds**

| Water Quality | Upper limits for continuous exposure |
|---|---|
| pH | 6.0-9.0 |
| Alkalinity | 20 mg/l (as calcium carbonate) |
| Ammonia nitrogen | 0.02 mg/l |
| Phosphate | 0.2 mg/l |
| Hardness | >15 mg/l |
| Calcium | 0.004 mg/l in soft water (< 100 mg/l alkalinity) |
| | 0.003 mg/l in hard water (> 100 mg/l alkalinity) |
| Chromium | 0.03 mg/l |
| Copper | 0.006 mg/l in soft water, 0.03 mg/l in hard water |
| Lead | 0.03 mg/l |
| Mercury (organic/inorganic) | 0.05-0.2 mg/l |
| Nitrogen | Maximum total gas pressure 110 per cent |
| Carbon dioxide | 10-15 mg/l |
| Dissolved oxygen | 9.2 mg/l at 20°C, 7.5 mg/l at 30°C |
| Manganese | 0.073 mg/l |
| Polychlorinated biphenyl | 0.002 mg/l |
| Total suspended solids | 80 mg/l or less |

*Source* : Selected data from Environmental Protection Agency (1973), Jhingran (1988)

## 11.14 Floor Space of Ecosystem

In extensive and semi-intensive systems where fertilization and manuring programs are undertaken, pond bottom is covered with various natural food organisms. Experience has shown that if floor space is increased by keeping tree branches over bottom mud, fish production significantly increased. Therefore, the floor space is the limiting factor for pond productivity. Phytoplankton populations generally grow over tree branches which is directly related to fish growth.

While considering the floor space of ecosystem, several problems are encountered. For examples, complete or partial harvest of the stocked material may be difficult due to infestation of pond bottom with tree branches. High cost of repeated removal of the branches at the time of netting may prove uneconomical. In such cases, harvesting of fish may be done at the end of culture period.

## 11.15 Control of Nutrient Loss

Soil and water fertility of ponds and their management practices are greatly influenced by various factors such as weather, season, temperature, metals, and salts. Because nutrient dynamics depends upon the fluctuation for these factors, it is logical to assume that prior to application of fertilizers these factors should be considered. For examples, the loss of nitrate-nitrogen is highest in early summer and lowest in monsoon seasons. In some situations, however, the level of nitrate-nitrogen is elevated in late summer and depressed in winter. Nitrogen uptake by plankton is higher in summer and lower in winter.

Although it is not possible to check nutrient loss completely, the aim is to minimize nutrient loss for the sake of farmers. Several guidelines may be mentioned for the purpose : (1) only commercially important species of fish should be kept in ponds, (2) in case where loss of nutrients is severe, the use of nutrient carriers is totally dispensed with, (3) fertilizer application rates should not be higher than what can be justified by water and soil tests, and (4) different types of nutrient carriers should be applied to pond water at the time of nutrient utilization by fishes.

## 11.16 Artificial Fish Habitats

An artificial fish habitat (AFH) is an external object or stable structure placed in fish culture ecosystem to attract fishes. Shelter and food provided by the artificial fish habitats are mainly the attractors of fish to such habitats. There are three types of artificial fish habitats; (1) AFH is set on the bottom, (2) Mid-water AFH anchored in the water column, and (3) AFH is floated on the surface by bamboo rafts.

Artificial fish habitat is a very imporant device for increasing fish biomass because there is good growth of algae and bottom organisms on these reef structures. This lead to aggregation of different species of bottom-dwelling fish.

## 11.17 Computer Vision in Fish Culture Management

The industry of fish culture and fisheries in general, includes extensive diversity of fish size, species, shape, methods of catching, and their processing. Fishery industries, by their characteristic features, require labour force for various activities such as freezing and storage plants, fishing crafts and gears and in fish farms. The employment of labour requires skilled personnel and involves heavy cost so far as their remuneration is concerned. Moreover, labour troubles obviously jeopardise the functional activities of the industry.

It is needless to mention that the present day will be known as the computer application for myriad purpose because it is witnessing a remarkable growth and development of computer technology and application. The computer is a sensitive and information processing machine, a tool for storing, manipulating, and correlating data. Recently, it has been well documented that computer can ensure consistent accuracy, less fatigue, and to expedite the activities of the industry. Therefore, the development of fisheries and aquaculture has closely been associated with the evolution of computer and undoubtedly has paved the way for undertaking studies that involve compilation and analysis of large masses of data with the aid of computer particularly when standard or packaged software programs are available.

### *Detection of Fish Products*

At present, quality of fish products are generally detected simply through naked eyes and chemical testing. Besides this, optical techniques and combinations of optical method with X-rays and acoustics are also used. Other methods involve optic fibre*, use of UV monitor for viewing and a picture of the fillet. To develop a machine for sorting fillets into defect-free and spoiled, it is necessary to obtain better pictures and process them in the same method.

### *Concept of Computer Vision*

The concept is based on the transmission of light through a fish fillet. In this case, a substantial proportion of visible light is transmitted. Light transmission not only depends on the thickness of samples but also the wavelength of visible light. However, the phase and direction of a light wave is changed randomly on its way through the fillet. Information is then coded into the light beam. Thus an image of a defect is smeared out. The scattering of light is very less at longer wavelengths. During transmission, some light is absorbed and converted into heat. The combined effect of absorption and scattering of light results in gradual attenuation. The direct transmission of light is characterized by the fraction of the original light beam that is neither scattered nor absorbed on its way.

During transmission, the amount of light which is absorbed within the sample, is extremely variable and this variation depends on the type of sample to be used for detection of quality. For example, fish flesh having a moderately blood clot would provide a more pronounced absorption peak than the slightly blood-stained fish flesh.

> *Image Analysis System* : This system consists of an illuminating source emitting any part of the electromagnetic sprectum (such as Ultra-Violet, visible light, X-ray and infra-red), image sensor, and digitizer. While digitizer helps to digitize analog from the image sensor using an A/D convertor, computer machine processes the digitize image. The output from the image processor can then be used to operate with precision.

---

*There are three parts in a typical optical fibre communication system : (a) the potical fibre as transmission medium, (b) the optical transmitter, and (c) the optical receiver. Some passive optical components and interconnection elements such as connectors and couplers are also used. Electronic systems are used as repeater, multiplexer, coder, decoder, and supervisory circuits.

### *Potential Uses*

Recently, computer vision is being used for various fisheries strategies such as spotting for fish catches, quality inspection of fish and fish products, fish sorting, fish farm management, hatchery management, and detection of fish parasites.

1. *Management of Fish Farm* : Using vision technique, it is possible to sort fish in ponds and sea enclosures, count fish, estimate their length and weight, and also to estimate the growth curve of fish. Measurement of length and weight is very important to calculate length-weight regressions, and the condition factor which is significant to predict production of fish and their value. Moreover, monitoring of water quality variables by particle diffuse scattering, control of hatching and farming units and thermal control of water in fish ponds to regulate water renewal rate, and fish dynamics through sonar systems are also possisble for successful management of fisheries.
2. *Fish Hatchery Management* : Generally the counting of juvenile fish is undertaken when they are being transferred from one point to another or before they leave the farm. The problem of counting of fish is not severe in small-sized hatcheries but in cases where large quantities of juvenile larvae or fish are counted regularly in large-sized hatchery farms, application of computer vision is necessary. In many hatchery farms, counting is done either by manually or photoelectric chambers. The former method, though common in rural areas, is a very stupendous, monotonous, and time-consuming job. Manual counting often results in fish diseases and death due to severe stress.

   For counting of juvenile fish, the juveniles are allowed to pass through a chamber in which juveniles are counted. The transfer of juveniles is executed either manually or under the control of a computer. The chamber is closed but with valves which are responsible for allowing the fish to enter. A camera is fitted on the hood which covers the basin and takes the image and trasmit it to the computer. As soon as the counting is completed, the computer opens the output valves and as a result, the fish along with the water is migrated towards the receiving basin thus continued the cycle.

   The Microsoft's Assembler and C, and Fox-base plus are used as programming languages. Furthermore, User Interface, Data Analysis, and Image Processing are the software used for this purpose.
3. *Spotting for Fish Catches* : To transmit satellite image and detect potential fishing grounds, the micro-computer-cellular telephone link is used. By this technique, it has been possible to move fishing vessels to that particular area where fishing can be done successfully. National Marine Fisheries Service, Mississipi, has developed display software that allowed satellite transmission of compressed image from receiving centres to the fishing vessels. However, this technique is principally adopted in marine fisheries management.

## 11.18 Use of Ozone in Fish Culture Management

Ozone is an allotrophic form of oxygen produced by passing dry oxygen or air into it through an electric discharge (5,000-20,000 volts). It is unstable, highly toxic blue gas with a pungent odor. It is a powerful oxidizing agent. Generally a large number of chemicals are used for controlling fish diseases. But most of the chemicals are absorbed by soil organic matter in the pond mud, thereby reducing the production capacity of ponds and at the same time exhibits direct residual effects that are harmful to fish, their food organisms and water quality variable generally. For this reason ozone has been gaining popularity in respect to water sanitation mainly due to easy application, high reactivity and low cost. Because ozone is slightly soluble in water and unstable, it has no residual effects. It is toxic to pathogenic bacteria and viruses and oxidizes ammonia, nitrite, nitrate and dissolved organic matter.

The Central Institute of Fisheries Education, Mumbai, has developed an ozonizer (B10-KLEN-108-R) to discharge ozone gas into the water. This equipment produces a total ozone of 3.389 mg/minute. In order to evaluate the effect of ozone in reducing bacterial population, dissolved oxygen and ammonia level of ponds water, an experiment was conducted in the Central Institute of Fisheries Education, in three tanks using poultry litter, groundnut cake and single superphosphate as fertilizer. The initial bacterial counts in the three tanks were $13.5 \times 10^3$, $29 \times 10^4$ and $50 \times 10^4$ per ml before use as control, aerated and ozonated tanks simultaneously. After four hour treatment of tanks with ozone, the bacterial load was $15.4 \times 10^3$, $38 \times 10^4$ and $45 \times 10^4$ per ml respectively in control, aerated and ozonated tanks. When tanks are ozonated for 4 days, the bacterial density was very low (140 per ml) compared to control and aerated tanks. Furthermore, ammonia level in ozonated tanks was zero but quite high in control tank (0.1 mg/l).

## 11.19 Aeration in Fish Culture Ponds

In semi-intensive and intensive type of culture systems, where stocking densities are high, dissolved oxygen and ammonia form the most important limiting factor affecting the growth of the cultured organisms. The metabolites and other water pollutants adversely affect fish growth as these are oxidized in the presence of oxygen depletion in dissovled oxygen concentration which upset environmental conditions for fish growth.

The artificial aeration in fish culture ponds is done for absorption of atmospheric oxygen into water by increasing air-water interfacial area. Aeration improves water quality which supports higher density culture for increasing fish production. Aeration not only increases dissolved oxygen content in water but also provides mixing of water by continuously moving highly oxygenated surface water to the bottom by breaking stratification.

### *Aeration Device*

Aeration in fish ponds is done by means of electric and diesel power aerators. The mechanical aerators, such as paddle wheels, invariably need electricity and in most cases they need the three phase electric supply. Therefore, the paddle wheel aerators have limited scope for large scale utilization in rural areas.

At present, bamboo basket cascade type aerator (Figure 11.2) is being operated by Indian fish farmers with the help of a 3-HP portable diesel pump set. Diesel pump sets are usually used for agricultural purposes having capacity of about 10 litre/second used for pumping water.

An aeration unit is made of 4 numbers of 1.5 metre diametre bamboo baskets vertically at 0.3 metre spacing with 4 numbers of bamboo poles giving the effect of vertical cascade. The design of this low-cost aeration device has been constructed by the Central Institute of Freshwater Aquaculture, Orissa, for aeration of 0.4 hectare semi-intensive fish culture pond. The cost of fabrication of this aerator is only Rs. 400/- and is could be used for more than one year. The pump set is used for pumping water from the bottom layer of the pond to the top-most basket of the cascade. The water is then allowed to fall gradually from top-most basket to the pond surface travelling through all the baskets one after another.

Field trials have shown that this aeration device is able to diffuse oxygen in water from atmospheric air, a quantity of about 15.6 kg of oxygen in three hours with an oxygen transfer rate of about 5.2 kg/hour at an efficiency of 2.3 kg/Kw-hour. Moreover, this device is as effective as the paddlewheels.

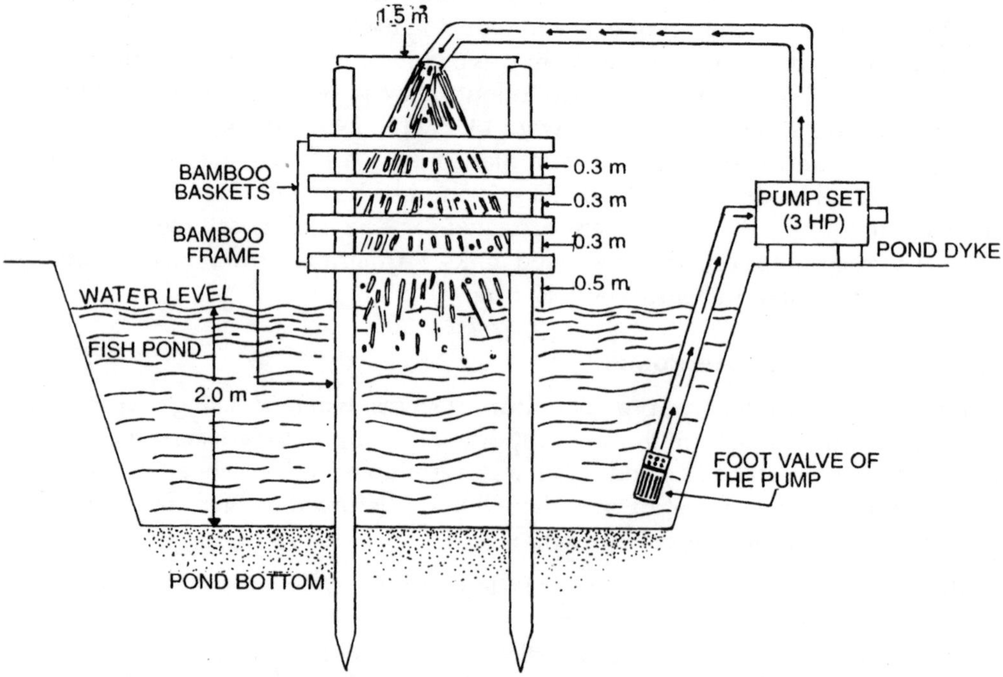


**Fig. 11.2 : Installation of a bamboo basket aerator in a fish culture pond. It is a low-cost aerator device for use in fish culture. This aerator may be used for more than one year without any maintenance cost. Note that the water falls from top-most basket to the pond surface through all baskets that causes increase in air-water interfacial area for maximum diffusion of oxygen into water. [Re-drawn from Saha (1998)]**

### *Period for Aeration*

The critical period for aeration in a fish pond is just before sunrise when over night demands have depleted the oxygen shortage in water and resupply by photosynthesis has not been activated during the night. At this critical period, the pump set is used for aeration during the night and during early morning hours. During the day time, fish farmers can use it for agricultural purposes when fish ponds receive oxygen from the process of photosynthesis by phytoplankton.

## 11.20 Integrated Fish Farming*

In a number of countries in Asia and in some parts of Africa, freshwater fish culture is integrated with the farming of crops mainly rice, vegetables, fruits, and animals (usually pigs, ducks, chickens

---

* For further discussion, see NACA (1989) and Jhingran et. al. (1998).

and cattles). This leads to overall efficiency of the farming system as by-products/wastes of one component are used as inputs in another. For example, poultry or pig manure can be used to fertilize the fish pond and the vegetable garden and the waste vegetables can be fed to the fish and the pigs.

The integrated fish farming system is defined as the linking together of two or more normally separate farming systems which become subsystem of a whole farming system. It is the horizontal diversification of agriculture with the development of a fish pond and as a subsystem on a farm with existing crop and /or livestock subsystems. This type of farming is a complementary farming in which one practice is indirectly or directly connected with the other and both are benefitted simultaneously. The objective of integrated fish farming is to obtain maximum food production from various corners of cultivation such as fish-poultry, fish-duck, fish-cattle, and fish-paddy. The pond embankment is utilized efficiently for rearing duck, poultry, and pig and also for cultivation of vegetables, fruits, and fodder. Because these animals are succumbed to various diseases, adequate sanitation and health care are important for their survival. Therefore, integrated fish farming system is broad in perspective with several models and can produce more material for the people (Figure 11.3)

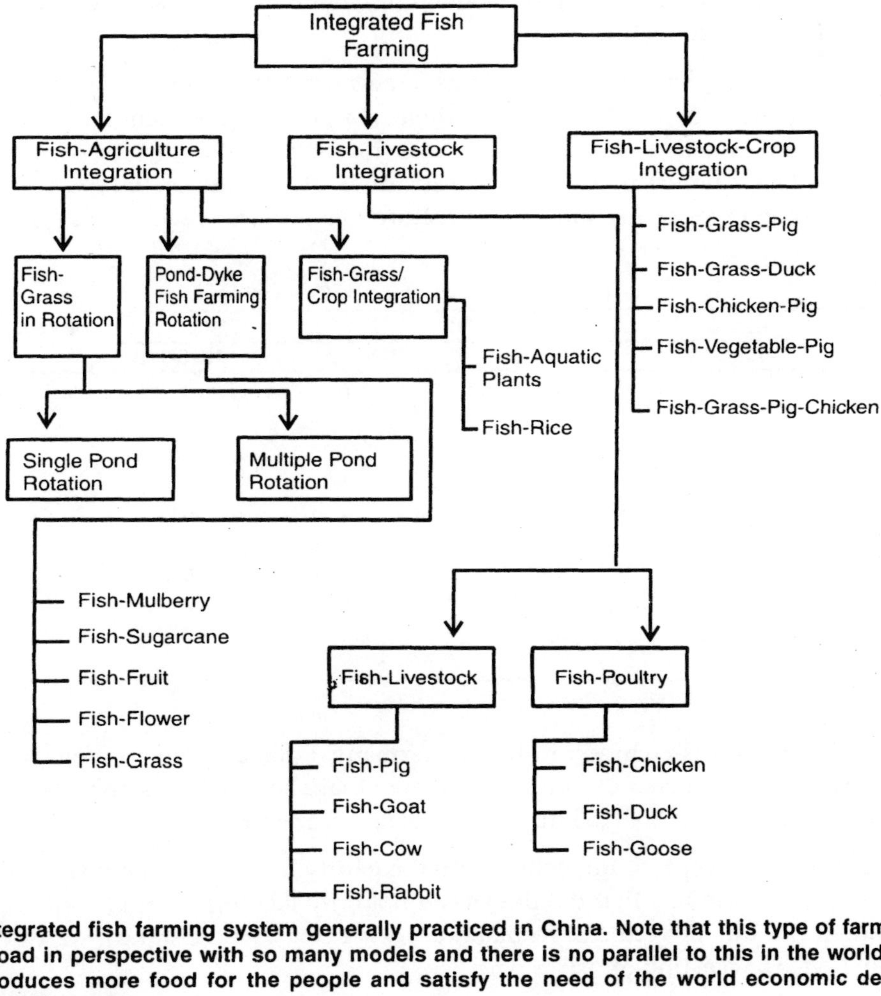


**Fig. 11.3 : Integrated fish farming system generally practiced in China. Note that this type of farming system is broad in perspective with so many models and there is no parallel to this in the world. This system produces more food for the people and satisfy the need of the world economic development on several aspects.**

### *Fish-Duck Integration*

Despite being a favourite in Asia for their meat and eggs, ducks have remained the poor relations within the poultry industry. In the Philippines, for example 1998's estimate duck population of 9 million was over-shadowed by nearly 79 million chickens. Yet ducks offer untapped benefits. In some countries, increased duck farming could not only boost per caput meat consumption, but also provide extra income for many farmers. Apart from meat and eggs, demand for fish is believed to offer a secure market for the farm.

In the duck-fish farming system, raising ducks for ducklings (broiler rearing) or raising ducks for producing eggs (layer ducks) are considered. As ducks choose water, the most suitable place could be near to a water source such as fish ponds.

For fish-duck farming, only improved breeds such as Indian Runner, Sylhet, Khakhi Campbell, and others are kept in fish ponds. An average sized duck excretes about 150 g droppings per day and accordingly, 250-230 ducks per hectare is required to yield carp fish of about 100 kg/hectare. Ducks are brought to the pond at a late stage and are ready for slaughtering after 60 days at a weight of 2 kg. Duck house is made up of wooden or bamboo materials. The size of the shed is generally 25 feet long and 20 feet wide.

Duck droppings on dry basis contain 80 per cent moisture, 0.9 per cent nitrogen and 0.4 per cent phosphorus. Ducks obtain about 70 per cent of their total feed requirement from the pond in the form of insects, aquatic weeds, and molluscs. Besides this, balanced poultry feed and rice bran are mixed in the ratio of 1:2 which is used as food for duck at the rate of 100 g/duck/day in two instalments. The economics of fish-cum duck culture per hectare water area is shown in Table 11.5.

**Table 11.5 : Economics of Duck Production**

| Particular | Year | | | | Average production |
|---|---|---|---|---|---|
| | 1992-1993 | 1993-1994 | 1994-1995 | 1995-1996 | |
| Duck cultivated (Nos.) | 100 | 200 | 200 | 150 | 187 |
| Egg production (Nos.) | 6,500 | 12,000 | 22,000 | 1,000 | 12,625 |
| Income (Rs. in lakh) | 0.1 | 0.2 | 0.4 | 0.24 | 0.23 |
| Expenditure (Rs. in lakh) | 0.13 | 0.27 | 0.53 | 0.16 | 0.27 |
| Net income (Rs. in lakh) | 0.03 | 0.07 | 0.13 | 0.08 | 0.07 |

*Source :* Shukla et al (1996)

Duck fish integration is common in India, Germany, Poland, China, Hungery, and Russia. However, experiments conducted in India have shown that about 4,325 kg fish/hectare/year can be obtained from a pond with 100-150 ducks per hectare water area.

Layer ducks are also kept in connection with the paddy field and it is considered as an ideal farming. This is due to the fact that the ducks can not only feed off insects, earthworms, snails, weed seeds, and pests, but also reduces feed and pesticide costs. At the same time, ducks also find food in the field once harvesting of fish is over.

1. *Types of Duck-Fish Farming* : There are three main types of duck-fish farming usually followed by the farmers such as (i) grazing type of duck rearing, (ii) rearing of ducks near the fish ponds, and (iii) rearing of ducks on the surface of fish ponds. In the first type of farming system, the ducks are allowed to graze on the ponds during the day time and are maintained in enclosures ('pens') at night. This method is not suitable for fish culture because duck droppings, may not be effectively utilized. Through this type of farming, only egg and meat are obtained.

   In the second type of farming, a large duck shed is constructed in the vicinity of fish culture ponds having a cemented portion of wet and dry runs outside. The average stocking density of 3 ducks per square meter is generally recommended. The duck runs are clean daily and thus duck manure is flushed into the pond.

   In cases where ducks are reared on pond surface, the embankments of fish ponds are partly fenced to form a dry run and a part of the water area is fenced to form a wet run.

   The enclosures should be installed by about 50 cm above and below the water surface. Consequently, fishes are able to enter the wet run for food but the ducks cannot escape under the net. The average stocking density of duck is 4.5 individuals per square meter. Under this system, both egg-laying and meat ducks are cultured in fish ponds.

2. *Benefits to Fish and Ponds* : Due to the browsing behavior of ducks at the pond bottom, nutrients are released from the soil which increase the pond productivity and consequently, fish production significantly increased. The feed spilled by ducks and their droppings, act as a substitute to fish feed. Moreover, droppings are spread by ducks themselves over the pond surface. Thus, the cost of labour for pond manuring can be saved.

   The duck droppings provide a continuous supply of organic matter, nitrogen and phosphorus to pond soil and water. This triggers biological productivity of ponds and increases the natural fish food organisms.

3. *Demerits of Fish-Duck Farming*

   (i) Ducks may damage the embankment of the pond.

   (ii) There is every possibility of eutrophication of pond water due to excessive addition of duck manure.

   (iii) Eutrophication of water causes algal blooms.

   (iv) There may be depletion of dissolved oxygen and accumulation of ammonia.

   (v) Since there is every possibility to consume small-sized fish by ducks, they should be kept away from the nursery ponds. Therefore, ducks should be reared in rearing and stocking ponds.

4. *Economic Benefits of Duck-Fish Integration* : The introduction of ducks to a fish pond clearly indicates that the space of pond surface is fully utilized and no additional space is required for duck rearing. This culture results in high production of eggs, meat, and fish from a common place. Thus, more animal protein may be obtained with less capital investment.

### *Fish-Poultry Integration**

A very simple and economically viable system of fish-poultry farming is increasing day by day. In this system of farming, poultry droppings are utilized and fully built-up poultry litter is recycled into fish ponds with fish production level to the tune of 4,500-5,000 kg/hectare. In fish-poultry integration, high egg-laying varieties of birds such as Rhode Island Red, Leghorn etc. are reared. Generally broilar birds are kept near the pond embankment through deep-litter system. However, it is better to rear broiler birds than egg-laying ones because the former type becomes ready to sell the market after eight weeks of rearing. In contrast, egg-laying birds take 6-7 months to lay eggs if adequate feeding is maintained.

Since broilar production provides good and immediate returns to the farmers, it is necessary to study the market demand where the products will be sold. The success of poultry farming depends on the efficiency of the farmer, aptitude, procurement of high quality brood stock, housing, feeders, water tarys, prevention and control of diseases.

The residual poultry feeds and droppings are used in fish pond to stimulate the biological productivity of water. Generally poultry litter is used in pond water at the rate of about 50 kg/hectare/day; of course, the application of litter should be discontinued as soon as the algal blooms appear in water. It has been estimated that one adult bird produces about 25 kg of poultry manure in one year and 500-600 birds are necessary to fertilize one hectare of water area.

### *Pig-Fish Integration*

Pig-fish integration is practiced in Taiwan, China, Thailand, Hong Kong, Malaysia, Hungary and other European countries. A number of exotic breeds of pigs such as white york shine, Berkshine and others have been introduced in India to cultivate them in close connection with the fish pond so as to avoid the transportation of manure for the purpose. Local or hybrid varieties of pigs are either cultured in captivity or they are allowed to graze in open area. A run for the pigs adjacent to the pig house is essential. The size of the pig house depends on the number of pigs to be reared. The pig house should be constructed at the pond site. The washing of the run containing dung and urine are directly drained into the pond or composted before they are used. About 4 square meter area should be provided for each pig weighing 80-90 kg. Generally an adult pig liberates about 2 kg of faeces per day.

Conversion of pig manure into fish flesh is highly significant. It has been estimated that about 50 kg of pig manure results in the production of 2.5-3.0 kg of fish flesh. The amount of dung which is produced by about 40 pigs was found to be adequate to fertilize one hectare of pond water. Fish production ranges from 6,000 to 7,000 kg/hectare/year along with 4,200-4,500 kg pig meat.

Since pigs are fed on kitchen wastes, aquatic plants and crop wastes, the quality and quantity of pig dung depends on the feed provided and also the age of pig. It has been estimated that 30 pigs excrete about 15,000 kg dung annually which is equivalent to 1 tonne of ammonium sulfate.

### *Fish-Cattle Integration*

Fish-cattle farming system is considered as an excellent practice for judicious recycling of organic materials and production of fish at very low cost. In this farming system, organic wastes may be disposed and utilized more efficiently.

---

* For comments, see Banerjee *et. al.* (1979).

Application of cattle manure in fish culture is an age-old practice all over the world. Among different cattle manures, cow manure is the most abundant and widely used in rural areas. A cow weighing about 450 kg liberates 12,000 kg of dung and 1,095 litres of urine annually. Cow dung and urine are beneficial to *Catla catla* (Catla) and *Hypophthalmichthys molitrix* (Silver carp). About 4 cows can able to provide adequate amount of manure for one hectare water area. Cow sheds should be constructed very close to fish ponds so as to discharge the entire amounts of manure and urine into the water. The edible part of the dung is consumed by many species of fish. In this integration, fish farmers simultaneously obtain fish and milk.

An adult cow consumes about 10,000 kg of grass per year and the remaining quanitites of grass can be used as fish feed. Each cow can provide manure for a pond area of 0.13 hectare if fish culture subsists on wasted cow feed and manure having a net production target of about 250 kg.

### *Fish-Mushroom Integration*

Mushrooms are fleshy fungi and are considered as one of the most important nutritious food items. Their consumption as food and medicine has been written in the religious books such as Bible and Vedas. Although more than 1,000 species of edible mushrooms are reported all over the world, about 200 species have been recorded in India. In India, only three types of mushrooms are being cultivated in commercial basis. The species includes *Agaricus bisporus* (European white button), *Volvoriella* spp. (paddy straw mushroom), and *Peurotes* spp. (oyster mushroom). The cultivation of these species require high degree of humidity (80-85 per cent) and wide temperature fluctuations (24-37°C) and hence their cultivation should be done along with fish culture. Mushrooms are rich in protein (25-30 per cent), Vitamin B, C, D, and K, minerals, and some amino acids such as niacine, pantothenic acid, riboflavin, and nicotinic acid. For mushroom cultivation, paddy straw is used as substrate that can further be utilized as cattle feed and for the production of natural food in fish ponds.

Mushroom cultivation in different countries such as China, India, Japan, Taiwan, and South Korea has an advantage because plant residues and straw are easily available and the environment is conducive to the cultivation of mushroom. Mushroom cultivation on pond embankment is being carried out with success because of low capital investment with proportionately higher production.

In fish culture practices, about 60 per cent of the total expenditure of this integration is utilized as fish feed. Thus, fish-mushroom integration drastically reduces fish production cost because the spent straw, after harvesting the crop, is teemed with nutrients. This nutrient-loaded straw could be used for pond productivity.

### *Fish-Horticulture Integration*

Horticulture and related enterprises are also integrated along with pond fish culture. In some fish farms, the dykes are devoted to economic crops such as sugarcane, fruit trees, vegetables and fodder crop. Adequate management of pond dykes involves utilization of top areas of dykes for cultivation of fruit trees while the outer and middle areas of dykes are used for intensive vegetable cultivation. Residues of the vegetables could be recycled within the fish ponds. The pond water, if necessary, can be used for irrigation of crops. It has been demonstrated that culture of four species of carps such as Rohu (*Labeo rohita*), Catla (*Catla catla*), Mrigal (*Cirrhinus mrigala*), and Grass carp

(*Ctenopharyngodon idella*) in the ratio of 15:20:15:50 at a stocking density of 5,000 fish per hectare could yield about 3,000 kg/hectare/year; of course, this production may vary depending on the intensity of crop cultivation and crop rotation. Fish-horticulture integration fetches about 20 per cent higher returns compared to fish culture alone.

### *Fish-Rice Integration**

Fish culture in rice field is an age-old practice because rice is considered as a major crop and staple food for billions of peoples of the world. Fish-rice integration is extensively practiced in India, China, Bangladesh, Korea, Africa, Phillippines, America, Hungary, Indonesia, Thailand, Vietnam, Malaysia, Japan, and Italy. However, fish production in rice field in India has been obtained as 200-800 kg/hectare. Production of deep water rice is very low due to a number of abiotic and biotic factors such as drought, floods, pest and disease incidence, lack of sufficient risk management technology, difficulties in good water management, and poor economic condition of the farmers. For this purpose, cultivation of some species of fish in the deep water rice field is carried out to increase the productivity of rice field. Generally, fish-rice culture can increase rice production due to the following reasons :

(i) Fish culture in rice fields can reduce the loss of nutrients in the field.

(ii) Fish in rice fields consume larvae of stem borrer, mosquito, rice beetle, and brown plant hopper.

(iii) Some bottom-dwelling fish due to their food searching activities, loosen and aerating the soil of the paddy fields which helps to release nutrients from the soil.

(iv) Excretory products of fishes act as fertilizer of rice fields.

1. *Environment in Rice Field for Fish Culture* : The characteristic features of water in rice fields are quite different from those of tanks or ponds. The depth of water in low-land paddy field is very low and for this reason a high rate of exchange of oxygen from the air may takes place and the temperature of water is greatly influenced by the air temperature. During photosynthesis, a large quantities of oxygen are released that results in high concentration of oxygen. This indicates that the environment of rice fields are not suitable for fish culture. But research on deep and semi-deep water rice field in some states of India such as West Bengal has shown that the physico-chemical parameters of rice-fish system are favorable for fish farming (Table 11.6).

**Table 11.6 : Physico-Chemical Parameters of Soil and Water in a Rice-Fish Field Ecosystem**

| Treatment | Water | | | | Soil | Plankton and benthos | | |
|---|---|---|---|---|---|---|---|---|
| | pH | Temp. (°C) | Dissolved oxygen (mg/l) | Specific conductivity (μ mhos/cm) | pH | P[1] | Z[2] | B[3] |
| Rice+ Fish | 6.8-7.3 | 24.8-32.6 | 3-9 | 119.1-400 | 7.0-7.5 | 500-2,400 | 60-400 | 200-500 |

*Source :* Deepwater Rice-Fish Farming System Bulletin, 1991, 1, Phytoplankton; 2, Zooplankton; 3, Benthos

2. *Rice Varieties* : It is recommended to cultivate specific varieties of rice which possess special characteristic features such as good elongation and submergence tolerance, strong and

* For further discussion, see CIFRI (1985), Sollows and Dela (1992) and Sinhababu and Venkateswarlu (1998).

deep rooting system as well as photoperiod sensitivity, tolerance to insects, pest and diseases, strong seedling vigour and drought tolerenace, and good quality of grain. Under deep and semi-deep water conditions, several varieties of rice such as NC 492 (Sabita), CN 704-7-3, NC 491, CN 705-18, CN 450 (Suresh), Patnai 23, Jaladhi - 1 and 2, Chakia, Jalamagna, Sudha, Kasai, Maghi and others are suitable for rice-fish culture system under Indian conditions. Rabi paddy is harvested after 3-4 months of sowing and a production of about 2.5 tonnes/hectares of fish and 4.5 tonnes/hectares of rice can be achieved.

3. *Species of Fish* : In India, Rohu, Catla, Mrigal, Tilapia, and Common carp have been recommended for culture in rice field. These species thrive in shallow waters, withstand high trubidity, tolerate high temperature, and grow to marketable size in a very short period. Some species of air-breathing carnivorous fishes such as *Clarias* sp., *Heteropneustes* sp., and *Channa* spp. are also cultured in rice fields.

4. *Rice Cultivation* : Cultivation of Kharif rice principally depends on the monsoon season. The geographical location, advent of monsoon, and harvesting period of rice vary greatly from country to country and even region to region. Deep and semi-deep water paddy is sown in April-May and June-July. Before sowing, urea, single superphosphate and muriate of potash each at the rate of 60 kg/hectare along with compost or Farm Yard Manure at the rate of 10 tonnes/hectare should be applied to the field. For semi-deep water rice fields, application rates of these inputs should be doubled. Harvesting of deep and semi-deep water rice is done in the month of December-January and October-November, respectively. Two application rates of nitrogen (100 and 50 kg/hectare) are recommended one after 15 days of transplantation of paddy and again at the time of flowering.

5. *Use of Pesticides* : As soon as infestation of paddy pests are detected, pesticides which are less toxic to fish, low residual effect and high efficacy, should be used. It has been experimentally verified that carbamates (carbaryl and methonyl) are less toxic than organochlorine (endrine and endosulfan) and organophosphates (dichlorvos and fenitrothion). However, use of pesticides is one of the major limitations to rice-fish farming as chemicals may damage the fish, causes environmental health hazards, secondary pest problem, biological magnification through food chain, and development of resistance species of pests to pesitcides. Therefore, it is advisable that before or during application of chemicals, the field water must be drained to drive fishes to the trenches, volume of field water should be increased and the water should be changed after the use of pesticides.

6. *Fish Culture* : Stocking of deep and semi-deep water rice fields with different species of fish (3-4 inches in length) at the rate of 6,000 number/hectare should be done in the month of July when the depth of water reaches 35 cm and rice seedlings are firmly attached to the soil. The selection of fish species depends entirely on the availability of fish seed in areas where rice-fish integration is adopted. A mixture of rice bran and mustard oil cake (1:1 ratio) is applied as supplementary feed at the rate of 3 per cent of the body weight of fish. Complete and partial harvesting of fish is done by dewatering the rice fields and with the help of dragnet, respectively. Fish harvesting is done after the harvesting of rice. Fish production and economics of rice-fish farming systems are most significant (Table 11.7).

**Table 11.7 : Productivity and Economics of Rice-Fish Seed Farming in Rainfed Lowlands[1]**

| Treatment | Rice | | Fish Yield (Kg/hectare) | Net return (Rs./hectare) | Profit to turnover (per cent) |
|---|---|---|---|---|---|
| | Grain yield (tonne/hectare) | Straw yield (tonne/hectare) | | | |
| Rice | 3.3* | 7.5 | - | 5105* | 42.3* |
| | 3.8 | 7.6* | - | 5857 | 43.5 |
| Rice + Fish | 3.6* | 7.9 | 81.0* ± 9.6 | 7055* | 46.7 |
| | 4.0 | 8.1* | 88.0 ± 10.9 | 7442 | 47.8* |
| Rice + Fish + FYM @ 2 tonnes/ hectare | 4.1 | 7.6 | 104.3 ± 9 | 7802 | 47.6 |
| Rice + Fish + Fish feed @ 0.5 tonne/ hectare | 3.9 | 7.4 | 143 ± 24.4 | 6572 | 39.7 |

*Source* : Sinhababu (1993)

1 Under trench refuge system covering 5 per cent of the field area

* Direct seeded rice; Without asterisk, transplanted. FYM, Farm Yard Manure

## 11.21 Advantages and Disadvantages of Integrated Fish Farming

### *Advantages*

1. The scope of intergrated fish farming is wide because this type of farming not only provide fish but also eggs, meat, milk, vegetables, and fruits.
2. It is also possible to utilize animal manures and vegetable wastes which supply nutrients to pond water.
3. Socio-economic condition is upgraded due to self employment generation in rural areas.
4. Integrated fish farming systems exhibit excellent synergistic effects where plant and animal products significantly increased.
5. Nutrient status in soil and water as well as fish nutrition is highly improved.
6. Integrated fish farming system is considered as an artificial ecosystem with no wastes (Figure 11.4).

### *Disadvantages*

1. In many situations, construction of animal sites on the pond embankment may not be possible because this type of farming is constructed away from farmers' residence where there is every possibility of theft.
2. Piglets, day-old chick and ducklings may not be available all the time due to lack of breeding farms.
3. Pig farming based on organized feeding system with concentrated feed, is not profitable and therefore, feeding cost may be reduced by reducing the composition of concentrated feed with the addition of agricultural wastes.

4. It is nct possible to continue integrated fish farming in dry areas where there is no adequate rainfall. If the ponds are seasonal, farming system should not be adopted.
5. Villagers are not interested for pig farming system because of their religious superstition.
6. Construction of an integrated fish farming involves heavy costs which is not possible for small and marginal farmers.

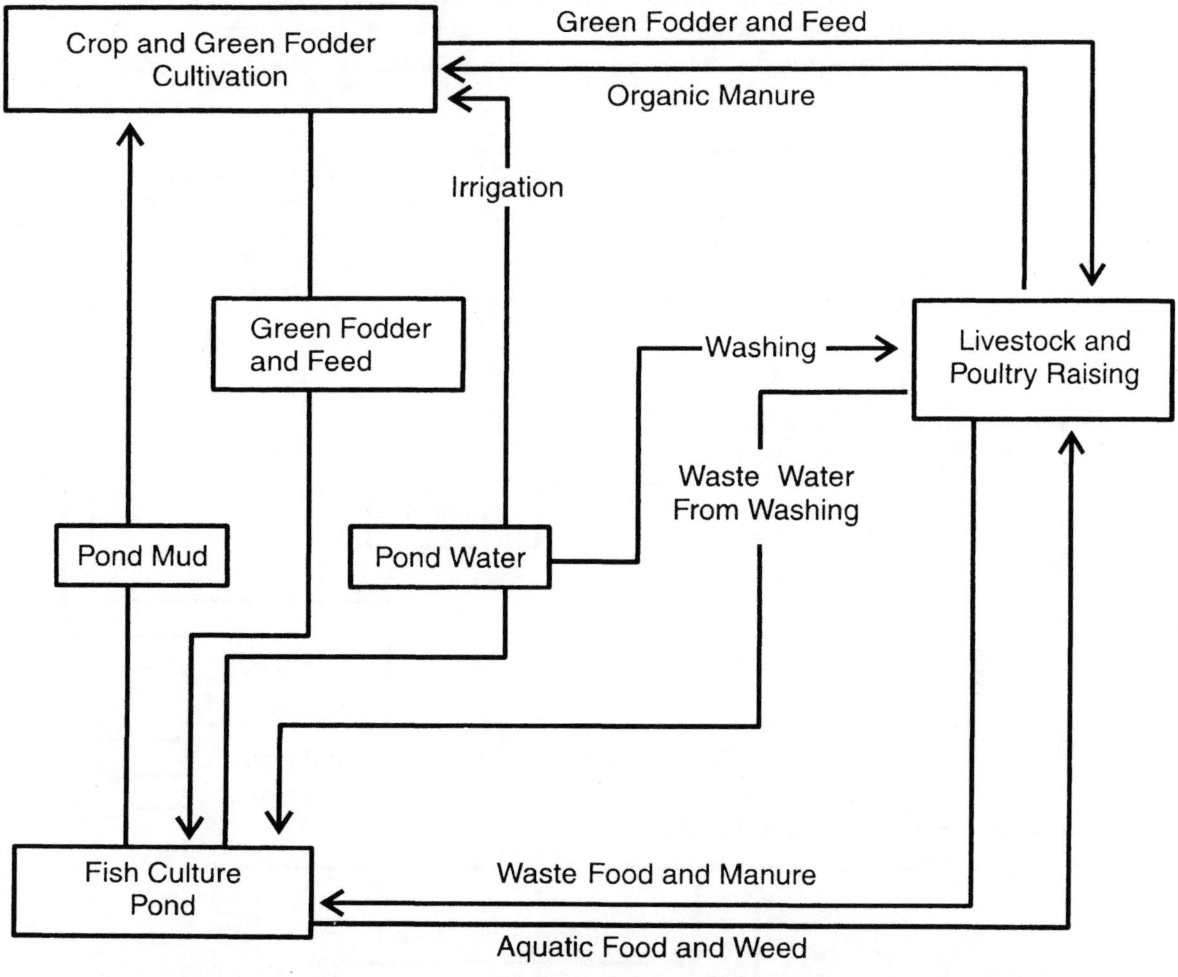


**Fig. 11.4 : Cycling and reutilization of different agricultural materials in integrated fish farming system**

## 11.22 Multilevel Waste Utilization in Integrated Fish Farming

In many countries where famers cultivate fish through the application of latest technologies, have established complex integrated farming and management systems (Figure 11.5). This combination can increase the level of comprehensive utilization of natural resources, the rate of energy utilization, and the production of livestock, fish as well as poultry resulting in job opportunities and income. With such complex integrated farming system, fish production to the tune of 10-12 tonnes/hectare/year is achieved.

## 11.23 Cage Culture*

Cage culture involves the rearing of fish within floating or fixed net enclosures supported by frameworks made of wood, bamboo or metal and set in shallow portions of ponds, lakes, rivers, and bays.

While fish pond culture has its 4,000 year tradition, cage fish culture is of recent origin. It seems that as early as 1920, cage culture has developed independently in at least two countries such as Kampuchea and Indonesia where bamboo cages and baskets have been used. Since then, this type of fish culture has spread throughout the world to about 40 countries in Asia, Europe, and the Americas.

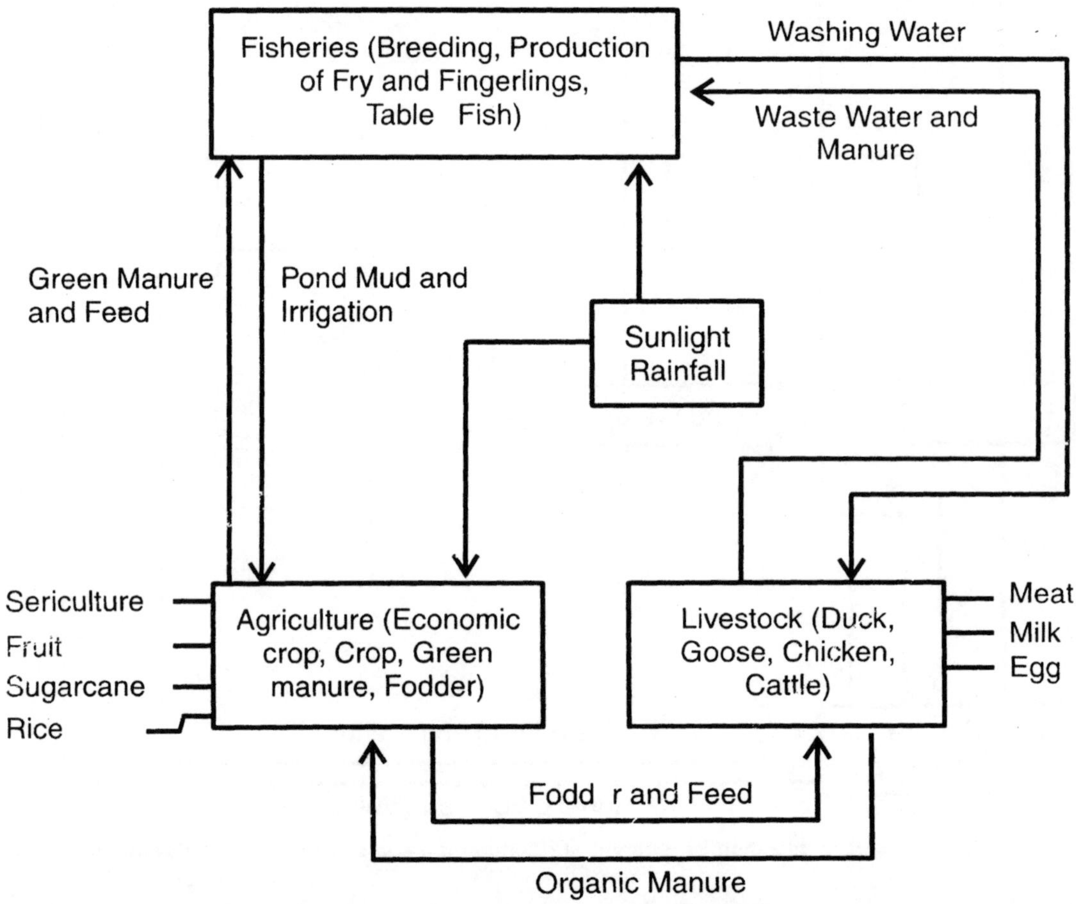


**Fig. 11.5 : Network of integrated fish farming practiced in China. Through such network, trades develop on both the input and the output ends. While feed processing industries can be set up on the input end, the products are processed and marketed on the output end. This network obviously increases the utilization of natural resources, energy, and production of fish, meat, egg, and milk.**

The wide popularity of cage culture is due to its greater flexibility in terms of siting the cage. For example, the cages may be installed in ponds, lakes, and rivers as long as they are protected

* For excellent discussion on the subject, see Beveridge (1996).

from adverse environmental conditions. However, cage culture has expanded rapidly over the past three decades vis-a-vis the decreasing availability of land-based resources for fish culture and an increasing awareness of their merits over traditional fish culture, such as (a) their applicability in different types of water bodies, (b) their high productivity (as much as 10 times that of ponds of comparative sizes) with a minimal inputs and the lower costs to develop and operate, and (c) the greater socio-economic opportunities they provide to low-income families in the rural areas.

### *Floating Cage System for Pond Fish Culture*

Team Aqua Corporation, Taiwan, has developed a pond floating cage system which greatly enlarges the available space of fish ponds. The system increases the pond utilization for fish culture and enhance fish production apart from shortening the growing period. This system involves the following important aspects :

1. Considering a depth of 2 meters, it is necessary to deepen the pond 4 meters to allow for better circulation of pond water (Figure 11.6). The size of each cage should be 6m (length) X 5m (width) X 2m (height). However, different mesh sizes of cages may be available as required by the farmer.

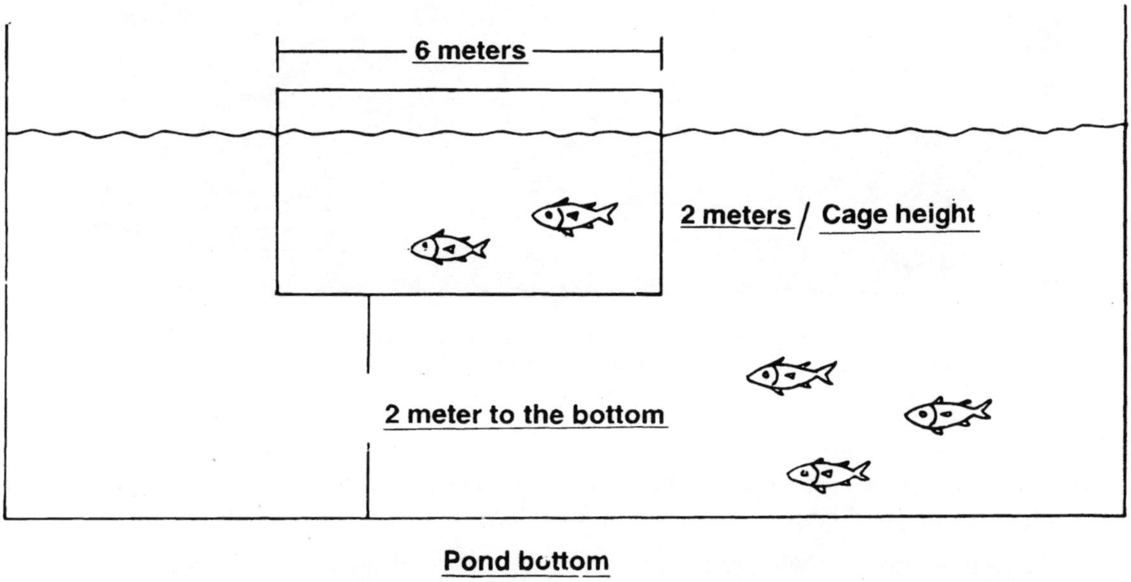


**Fig. 11.6 : The basic aspect for pond floating cage system. Considering a depth of 2 meters, there is a need to deepen the pond to 4 meters to allow for better pond water circulation. The size of the cages is stated to be 6 meters (length) x 6 meters (width) X 2 meters (height).**

2. The system allows the flexibility of raising at least two different species of fish; one inside the floating cage and another outside or in the pond for different sizes of the same species.
3. The system gives the advantage of stocking new fishes without any time-lag. In other words, a farmer would not have to wait for the harvest of the fish in the ponds for use of the floating cages. The cages will be helpful particularly when all the ponds are simultaneously under the culture. After stocking the fry, pond cages can be used to get them together for taming.

4. Several cages can be connected together inside the pond and the fish farmer can easily walk along the top perimeter of the cage (Figure 11.7).
5. If the cage is fixed in the centre of the pond during the period, it is possible to move the cage towards the side of the pond embankment so that harvesting can be done more efficiently.

**Fig. 11.7 : Diagramatic representation of a view of the fish farm with floating cage system. Note the walk ways of cages on which fish farmers can move around with ease for effective management.**

### *Culture Species*

The choice of species for stocking and rearing in cages is governed by the same criteria as in species selection for pond culture. The criteria for selection of fish species is noted below :

1. High tolerance to a wide range of environmental conditions.
2. Fast growth in confinement.
3. Resistance to diseases and parasites.
4. Ready supply of fish seed for stocking.
5. Ease of culture and management.

There are about ten species of freshwater fish which are commercially cultured in cages in both tropical and temperate waters, including tilapias, carps (Indian, Chinese and Common varieties), air-breathing fishes, milkfish, catfishes, snakeheads, and salmonoids (Salmon and Rainbow trout). Commercially important species of freshwater cage farming in different countries are shown in Table 11.8.

**Table 11.8 : Commercially Important Species in Inland Water Cage Farming**

| Species | Countries | Climate | Farming type | Lentic/Lotic |
|---|---|---|---|---|
| **Carps** Chinese, silver, grass and bighead | Asia, Europe, North America | Tropical and temperate | Mainly semi-intensive; also extensive (Asia) and intensive (Europe and North America) | Lentic and lotic |
| Indian Major carps | Asia | Tropical and Sub-tropical | Semi-intensive | Lentic |
| Common carp | Asia, Europe, North America, South America | Temperate and tropical | Semi-intensive and intensive | Lentic |
| **Tilapias** | Asia, Africa North America South America | Tropical and subtropical | Semi-intensive and intensive | Lentic |
| **Snakeheads** | | | | |
| *Ophiocephalus* spp. | South-east Asia | Tropical | Semi-intensive and intensive | Lentic and lotic |
| *Pangasius* spp. | South-east Asia | Tropical | Semi-intensive | Lentic |
| **Catfishes** | | | | |
| Channel catfish | North America | Sub-tropical | Intensive | Lentic |
| *Clarias* spp. | Africa and South-east Asia | Tropical | Semi-intensive | Lentic |
| **Salmonoids** | | | | |
| Rainbow trout | Europe, North America, Japan, high altitude tropics (such as Columbia, Bolivia, Papua New Guinea) | Temperate | Intensive | Lentic |
| *Salmon* spp. | Europe, North America, South America | Temperate | Intensive | Lentic |

*Source* : Beveridge (1984)

## *Site Selection*

The selection of site for fish cage culture should be guided by the following criteria :

1. Good water quality such as adequate dissolved oxygen, stable pH, low turbidity and absence of pollutants.
2. Firm bottom mud to allow cage framework to be driven deep into substrate for better support.
3. Freedom from natural hazards and predators.
4. Water exchange that will enable the flow of water containing nutrient through the cages.
5. Protection from high winds.
6. Accessibility to source of inputs.

It is important to note that the selection of a suitable site is a very important criteria to the success of the culture system because a good site solves much of the management problems of cage culture. The following factors, however, must be considered in selecting sites for cages in different freshwater environments.

1. *Protection of Natural Factors* : Of different natural factors flood and water current are the most common phenomena that, to large extent, cause harm to cage culture systems. Floods cause great damage to fish cages. Consequently, the entire fish population is destroyed by partially or completely damaging the cages and fishes are swept away. Though a flood is a natural calamity, steps should be taken to prevent fish loss.

   Water currents are also equally important that interfere the normal growth and production of caged fish. Occasionally high currents of water break the structural framework of the cages and at the same time removes almost all the natural food organisms.

2. *Water Circulation* : Fish and their food organisms are distributed unevenly throughout the ecosystem. Their distribution is largely determined by light penetration, temperature, and oxygen profiles. Similar to other fish culture systems, cage culture environments also have the following characteristic features : (i) a layer of free circulating surface water (epilimnion), (ii) a deep and cold layer of dense water (hypolimnion), (iii) a steep and rapid decline water temperature (metalimnion), and (iv) a layer of bottom mud. These stratifications have marked influence on the production potential of fish stocked in cages. Temperature fluctuations in different seasons of the year develop currents which bring about a mixing of surface water and with water richer in nutrients. These natural forces push away surface water and cause upwelling of colder and denser subsurfaces water generally richer in nutrients. Though these forces (and also certain other environmental factors) within the reach of technological possibility, only in some regions can the basic production be increased by means of using quality fish seed, artificial feeding, and avoiding pollution, disease as well as parasites and thus increase fish yields. Only if we have sound knowledge about those production-oriented factors, which in each individual region are limiting the basic yield, may effective feeding and stocking of suitable species on a large scale be feasible.

3. *Water Quality* : A number of factors are primarily responsible for survival, growth, and production potential of fish in cages. Though certain environmental factors are directly related to cage fish production, other factors are also essential for drastic reduction in fish production. For sustained production, however, the following factors must be taken into consideration before implementing the cahe culture operations : Ammonia level, pH, biological oxygen demand, hardness, salt water intrusion, temperature, siltation, turbidity, floating objects, depth variations, algal blooms, predators, aquatic vegetation, natural productivity, disease and parasites, pollution, and pollutants.

4. *Spacing* : Adequate space between the cages should be provided for well circulation of water to eliminate the waste products of fishes and to increase the dissolved oxygen concentration in water.

5. *Access and Security* : These include supply of materials, feeds and fingerlings, labour availability, easy assess for regular monitoring, efficient precautions, and security from interference of all sorts.

The guiding principles on which the cage culture development programs would be based are judicious utilization of water and other resources related to this culture system and the promotion of a source in a total panorama accepting the fulfilment of engaging special attention of both the cage and pond culture systems. Specific plans, projects, and schemes would have to be drawn up keeping in view the regional needs such as evolving viable solution to the problem of cage cultivation, ecological and environmental protection, reduction in infrastructural bottle-necks, and generation of manpower as well as productive employment.

### *Design and Construction*

A fish cage is like an inverted mosquito net with the cage bottom of the netting material used for its four sides. Moreover, fish cages cannot exceed 1,000 $m^2$ in area for reasons of the quantity of material necessary for cage construction and manageability of operation. Fish cage may be either fixed or floating and they are individual units for either seed production or grow-out and they are installed in clusters with a common framework (Figure 11.8).

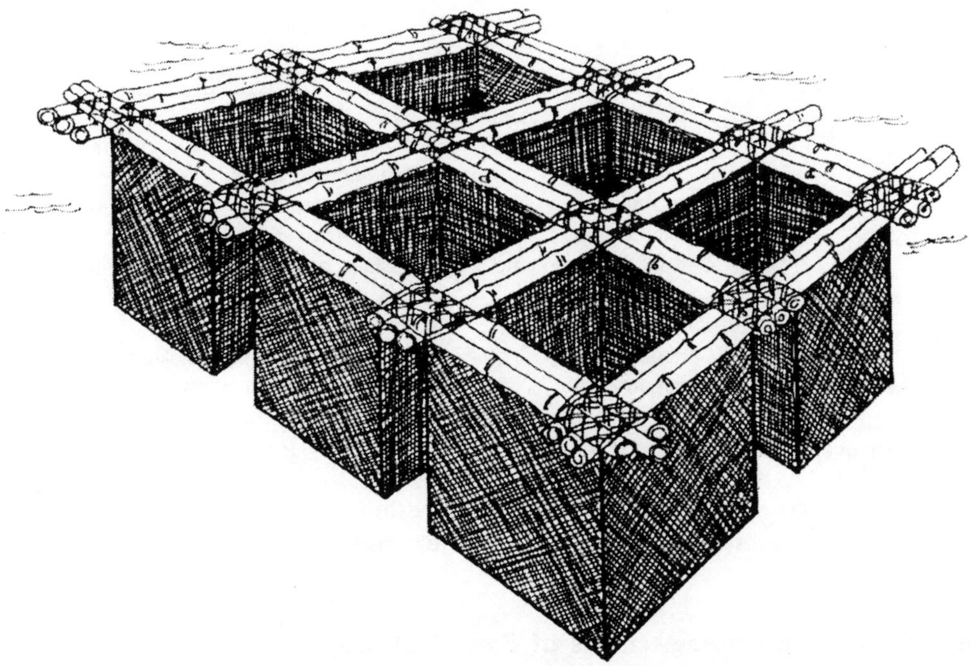

**Fig. 11.8 : Design of the structures of cages. Cages are individual units for either grow-out or seed production. They are installed in clusters with a common framework. Fish farmers have higher control in the operation for most effective management as that of inland fish culture. Inland pond cage is a historical breakthrough to effectively enlarge the aquacultural space.**

Fish cages have various shapes and sizes and are composed of different types of materials. Most cages are rectangular or square although some may be circular or cylindrical. Rectangular cages are preferred for easy operation and management.

Wire mesh, polythylene and nylon monofilament twine are widely used for fabricating cages. The structural frame work is made up of bamboo or wood. Cage floating materials include polyvenyl chloride pipes (PVC), bamboo, plastic or steel drums, and aluminium floats. The type of anchor for floating cages varies depending on the nature of bottom, depth of water and currents.

### *Cage Operation*

The operation procedures involved in the management of cage culture are very much similar to those in pond culture, starting with complete construction and preparation of the culture facilities for stocking, rearing, and harvesting. Variations in specific activities exist as the result of very nature of the system. For example, there is no need to apply lime, fertilizers, and organic manures as in the case of pond fish culture since the cage culture system has open water exchange between the outside environment and the inner compartment.

After construction of the cage, fry/fingerlings are stocked at the rate of 20,000-40,000/hectare. Cage-reared fish may or may not be fed artificial diet depending on the stocking density used and the level of technology in the country. Current feed practices in freshwater cage culture involve the provision of artificial feeds using rice bran/wheat bran and poultry feeds. However, different countries use artificial feeds based on simple diets (Table 11.9) prepared in pelleted form for best results.

**Table 11.9 : Types of Feed Applied to Cage-Reared Fish**

| Country | Fish Species | Types of feeds |
|---|---|---|
| India | Indian major carps | Soyabean powder, groundnut, rice bran and oil cakes |
| Indonesia | *Tilapia niloticus* | Aquatic plants (*Lemna, Chara, Hydrilla*) |
| Nepal | Common carp | Wheat flour, oil cakes, rice bran |
| Thailand | Catfish, Common carp and Tilapia | Pellets containing ground fish meal, Soyabean, peanut, and rice bran |
| Hungary | Carp | Pelleted feed |
| Russia | Common carp | Mixture of trash fish, mollusc, crayfish |
| German Democratic of Republic | Common carp | Formulated feed/pellets, 33.7 per cent crude protein |

*Source :* Proceedings of the SEAFDEC/IDRC International Workshop on Pen and Cage culture of fish, 1979.

At the end of the culture period, the fish are harvested using nets (such as gill nets and cast nets) or by lifting the cage and to collect the fish at one corner of the cage for scooping out using a pail.

## 11.24 Applications and Characteristics of Pond Fish Cage

Recently, inland pond fish cages have been modified to some degree for effective management. However, inalnd pond fish cages are made up of nylon thread with mesh size of 1 cm. A single cage unit has the net dimension of 5.25 X 5.25 X 2 metre (Figure 11.9) having 32 number of floats (four floats in 4 corners X 4 = 16 and four floats in 4 sides X 4 = 16; 16+16 = 32). The total number of units consist of eight sets. The cage net may require thorough cleaning after each crop by flushing with freshwater. The cages does not require any maintenance during the crop but care should be taken not to clean or scarp the floats with sharp-edged tools. The pond fish cage has the following applications and characteristics :

### *Polyculture*

1. In different cages of the same pond, one can raise fishes with different feeding habits (carnivores, omnivores or herbivores) without the fear of mortalities caused by cannibalistic behavior among the fishes.
2. The farmer's utilization of the working capital can be increased by raising fish species with different grow-out period in the same pond.
3. It is possible to establish a balanced pond water environment by properly growing different species of fishes in the same pond.
4. It is highly effective in culturing the pond and oyster in the cage, or raise the shrimp in the pond and fish in the cage.

**Fig. 11.9 : Clusters of inland pond fish cages. The pond fish cage has several applications for high fish production.**

### *Fry Taming and Conditioning*

1. Pond fish cage is more efficient in pond utilization.
2. A variety of fishes can be stocked in different cages to be raised in the same pond so as to minimize the need to occupy different ponds.

### *Aquaculture Operation*

1. Pond fish cage helps to grade the size of the fishes more efficiently. By using the correct mesh net and properly manipulating the net during feeding, grading the fishes can be easily undertaken.
2. It is so easy to operate the cage that only two persons can harvest the fish inside the cage without difficulty thus the cost of labour is reduced.
3. When fishes inside the pond are not yet to be harvestable in size, one can easily start a new crop by raising the same or different species inside a cage.
4. If required, the entire fish stock can be transferred from one pond to another without any mortality.
5. Observations for the development of diseases, feeding and growth rates can be easily recorded.
6. Diseased or sick fishes can be segregated inside the floating cage for treatment.

### *Culture Quality*

1. Feed conversion ratios are more efficient that reduces feed cost. Feed can be efficiently converted to meat due to the limited space in the cage.
2. Marketable sized fish can be harvested from cage at a time to increase the survival rates of the live fishes during transport.
3. Prior to harvest, fishes are not fed with supplemental feed. Under normal condition of the pond, starved fishes have a tendency to go to the pond bottom to feed on mud or even detritus. Consequently, a muddy flavor to the body is introduced. This condition reduces the quality of fishes. This situation does not arise under cage conditions.

## 11.25 Closed Fish Culture System

Recently, closed water circulation systems for fish culture are being considered for high production. This system cuts down the quantity of water needed in terms of weight of fish produced per unit space, production of over 100 tonnes/hectare/year can be obtained compared to about 6 tonnes/hectare/year that can be achieved through intensive culture in open systems. Heavy feeding has to be resorted to for achieving the high rate of production in a closed system, which may push up costs, but the returns will equally high. The remnants of heavy feeding and the resulting biological wastes cause no damage because water circulation removes all wastes.

Closed circulation system is no doubt a supplemental means to augment fish production. Vulnerability of the fishes towards diseases and pollution of water can be controlled. In India, the availability of open lentic waters is about 10 lakhs hectare and should be brought under fish production through this technology. However, this technology cannot easily be adopted by all fish farmers, because it calls for greater investments compared to open water system; of course, the fish grown in closed systems will be cheaper because of the high rate of production which may counteract the higher costs.

## 11.26 High-Tech Super-intensive Fish Culture

The recent development in high-tech fish culture is gaining rapid popularity for high remuneration and multiple benefits both for the farmers and the nation. Tremendous advancements have been made throughout the world in the recent past in fish culture tehnologies to achieve quite high rates of fish production. Therefore, there is a great scope for increasing fish production through the use of advanced techniques of biotechnology. The biotechnological applications permit direct utilization of new products by the use of genetic engineering. Biotechnology also helps in obtaining recalcitrant species for culture which will offer opportunities for gene selection and manipulation.

The success of fish culture principally depends on the manipulation technologies of both the stocked material and the environment. High fish production depends on stocking density, supplementary feeding, fertilization, and aeration of the environment. In super-intensive system of culture, water circulation and aeration are applied along with the balanced pelleted feed for the fish. Genetical manipulation techniques have produced gynogenesis, polyploidy and more sophisticated transgenic fish with rapid somatic growth factor help to culture unisex, sterile or profitable species in a water area to yield maximum production without loss of growth during maturation of both the sexes of fish species. Through this culture system, several hundred times higher yield of fish (25-40 tonnes/hectare/year) is possible than that in still water because of the elimination of growth inhibiting metabolites and continuous supply of oxygen.

## 11.27 Multiple Stocking and Harvesting

Most of the fish farms generally adopt rotational stocking and harvesting methods for fish particularly where about 90 per cent fingerlings of several species are stocked. In this case, it has been found that fish production is increased in an unhampered manner because multi-production activities provide more waste materials to be recycled into the ponds for sustaining productivity of fish biomass. Multi-production activities in addition to fish production per hectare water area include rearing of cows, pigs, ducks, and chickens with the annual production of 5,60,000 litres milk, 1,000 pigs, 30,000 ducks, and 50,000 chickens. Therefore, multi-production activities increased both ecological and economic efficiency of the farm and thus there is every possibility to increase employment generation per unit farm area.

Though Indian and exotic carps have dominated the fish farming scenario in the twenty years, the other species such as tilapia and/or air-breathing fishes are also being cultured. Modern freshwater fish culture management techniques involve multiple stocking and harvesting of fish for continuous returns. But all the farmers do not adopt this type of management strategies because of the variability in temperature and rainfall patterns which produces a great diversity in the production capacity of most tropical fish ponds. It is, therefore, necessary to develop fish culture according to the suitability of each cultivable species on the basis of the profitability and needs to the individual farmer.

## 11.28 Induced Breeding

The recent growth in fish culture industry for significant production in Asia and South-east Asian countries is striking and the introduction of induced breeding techniques is making fish production still more effective. To meet the needs of the increasing human population, fish seed prodcution will rapidly expand further, mainly through brood-stock management and induced breeding.

Though fish culture industry is rapidly expanding, shortage of fish seed is one of the most important constraints in the development of this industry. To obtain pure seed of cultivable fishes from lentic freshwater bodies or from controlled conditions, the techniques of using pituitary hormone or *hypophysation* has been developed.

### *Hypophysation**

The technique consists of removing pituitary glands from gravid fishes of both sexes belonging to the same or closely related species. Pituitary glands are kept in absolute alcohol for dehydration and parmanent storage. At the time of injection, glands are crushed with a tissue homogenizer containing distilled water or 0.3 per cent saline solution. The gland suspension is then centrifuged and the clear supernatant water is collected with the aid of a hypodermic syringe. The extract is then injected in required amounts usually in two doses. This technique is very simple and has become very popular in all parts of India. Though the technology of pituitary extract injection is fairly well standardized and is being widely employed, the quality, potency, and availability of pituitary glands have become undependable. Consequently, spawning failure in several fish farms is very common.

### *Gonadotopins*

As a whole the anterior lobe of pituitary gland is the most active endocrinal portion, secreting at least six distinct hormones. Of them, two hormones (such as follicle stimulating hormone and luteinizing hormone) participate in the control of reproductive system in fishes and are referred to as *gonadotropins* because of their influence on the growth and function of the gonads. One gonadotropin, called *follicle stimulating hormone,* regulates the production of eggs and sperms whereas luteinizing hormone, also called *interstitial cell stimulating hormone,* affect the production of progesterone and testosterone.

Extensive studies on induced breeding of fishes have shown the relative effectiveness of pituitary extracts over mammlian pituitary hormones and steroids. For this reason, fish pituitary extract has been widely used all over the world. However, among different mammalian hormones, *corionic gonadotropin* (CG) and *human chorionic gonadotropin* (HCG) have been found to stimulate spawing in many species of fish such as *Ictalurus punctatus* (channel catfish), *Ictalusus furcatus* (blue catfish), *Salmo gairdnerii* (salmon fish), *Roccus saxatilis* (stiped bass), *Polydon spathula* (paddle fish), *Pylodictis olivaris* (flathead catfish), *Schaphirhynchus platorynchus* and *Acipenser fulvenscens* (sturgeons). Moreover, a mixture of CG and HCG and mammalian pituitary extract in combination with fish pituitary has also been found to be effective in inducing the fish to breed. A number of other possible substitute such as 'SZhK' (prepared from pregnant mare serum), 'Estrovest' and 'Hypophysine Forte' have successfully tried on trouts and carps.

### *Use of Ovaprim*

It is a drug manufactured by Syndal Laboratories, Canada, USA, and each millilitre of ovaprim contains an analogue of salmon gondotrophin releasing hormone (sGnRH) (20 mcg) and dopamine

* For further detail, see Jhingran (1988).

antagonist, domperidone (10 mg). This material is, at present, considered as the world's most advanced spawning technology and a wonderful spawning hormone because it is in ready-to-use form, save money and time on out-dated technology. The recommended doses of ovaprim for carps are shown in Table 11.10.

The results of several trials conducted in many states of India clearly demonstrate the possibility of using ovaprim than pituitary extract. The overall response of carps (Table 11.11) also indicates that the use of ovaprim is more profitable than pituitary extract. However, though the cost of ovaprim is almost triple than the cost of pituitary extract, it has the advantage of known potency and assured breeding response. Spawing failure due to inferior quality of pituitary glands of varying potency involves high mortality of eggs and financial loss. When the loss is considered, the higher cost of ovaprim is offset by the lower risk of spawing failure. Ovaprim has many advantages over other spawning substances.

1. Reliable results are obtained.
2. It is easily available in readymade form.
3. It can be stored at room temperature.
4. It is highly stable with long life.
5. It is formulated to prevent over-dosing.

**Table 11.10 : Dose of Ovaprim Used For Breeding of Carps**

| Fish | Ovaprim (mg/kg of fish) | |
|---|---|---|
| | Female | Male |
| *Catla catla* | 0.4-0.5 | 0.1-0.2 |
| *Labeo rohita* | 0.3-0.4 | 0.1-0.2 |
| *Cirrhinus mrigala* | 0.2-0.3 | 0.1-0.2 |
| *Hypophthalmichthys molitrix* | 0.4-0.7 | 0.1-0.3 |
| *Ctenopharyngodon idella* | 0.4-0.8 | 0.1-0.3 |
| Fringe-lipped carp | 0.3-0.4 | 0.1-0.2 |
| Big-head carp | 0.4-0.5 | 0.1-0.2 |

**Table 11.11 : Comparison of Ovaprim and Pituitary Extract for the Production of Eggs and Hatching of Fish**

| Fish | Ovaprim | | Pituitary | | Percent difference | |
|---|---|---|---|---|---|---|
| | Eggs | Hatchlings | Eggs | Hatchlings | Eggs | Hatchlings |
| *Catla catla* | 1.17 | 0.71 | 0.73 | 0.39 | 60.27 | 82.05 |
| *Labeo rohita* | 1.41 | 0.93 | 1.15 | 0.53 | 22.60 | 75.47 |
| *Cirrhinus mrigala* | 0.87 | 0.45 | 0.47 | 0.22 | 85.10 | 104.55 |
| *Hypophthalmichthys molitrix* | 0.75 | 0.50 | 0.51 | 0.29 | 47.00 | 68.97 |
| *Ctenopharyngodon idella* | 0.42 | 0.28 | 0.31 | 0.23 | 35.48 | 21.74 |
| Fringe-lipped carp | 2.40 | 2.24 | 2.40 | 1.70 | – | 32.55 |
| Big-head carp | 0.65 | 0.43 | 0.60 | 0.20 | 6.60 | 104.76 |

Values are in lakh per kg of brood fish

6. It can be injected into the body of female fish at varying doses thus reducing post-breeding mortality.
7. Male and female fish can be injected simultaneously.
8. It is highly effective in cold and warm water species.
9. Ovaprim is completely free from pathogens.

### *Controlled Hatchery System*

In fish culture industry, adequate supply of fish seed has a greater influence on fish production. Any means of improving fish seed production strategies will aid in the return on fish farmer's investment. However, the main important considerations for the development of this type of system are (a) easy maintenance, (b) economic viability, and (c) rapid installation. Controlled hatchery system increases survival of eggs to more than 95 per cent as against the traditional method where the survival of eggs varies between 30 and 40 per cent. Several environmental factors of water such a dissolved oxygen, pH, temperature, and removal of metabolites as well as predators are easily controlled from maximum efficiency of the system.

During operation, the environment of the entire unit is controlled in such a way that the hatching rate of eggs is increased following the reduction of hatching time by 80 per cent. The circulated water is free from undesirable substances with high oxygen concentration (about 9 mg/l). For example, Indian major and exotic carp eggs take 24-28 and 18-20 hours respectively, for hatching at 30°C. But under controlled system, the time of spawning is reduced to 12-14 and 12-13 hours for Indian and exotic carps, respectively. Because of its controlled system, the unit is very effective in arid zones where unfavourable conditions prevail.

Hatchery series D-81 and D-85 and their models are made of earthen pools or galvanized iron or fibre glass or combination of portable plastic pools and fibre glass have been tested at different places in India by the Central Institute of Fisheries Education, Mumbai. The capacity of the D-81 hatchery model with 24 jars to produce hatchlings is 0.5 million and that of D-85 model with 10 jars is 10 millions. The latter models is, however, less expensive and is designed for large-scale production. Borehole water, after proper filtration, sedimentation and pH correction, is aerated and circulated slowly in the systems. This ensures clear filtered water having dissolved oxygen concentration 8-9 mg/l and temperature 27°C throughout the culture period. Slow current and aeration of water not only helps remove the metabolites but also keeps the eggs in floating condition.

## 11.29 Hatchery Units

To meet the fish protein requirement for ever-increasing populations, inland fishery has dramatically progressed through development of different types of hatcheries. Previously, fish seed collection was dependent on the riverine systems. But due to excessive application of chemical fertilizers and pesticides in agricultural sector, the potentiality of different river systems significantly decreased. For constant supply of fish seed during breeding season, however, different types of hatcheries such as Circular hatchery, Fibre glass jar hatchery, and Hapa hatchery are widely accepted. Large-scale production of hatchlings of fry from fertilized eggs with the aid of specially designed hatchery units is referred to as *hatchery*. The most effective fish seed production on commercial basis is the Circular hatchery and the Fibre glass jar hatchery. These types of hatcheries though cost-effective, could only be economical if they are constructed through fish farmers' cooperative society. Among different types of hatcheries, the 'Chinese Circular Hatchery' is most important because its efficiency of production is very high and for this reason this type of unit is widely used. Intensive fish culture recirculation system involves a number of step (Figure 11.10) which helps to produce fish seeds for development of high-tech aquaculture.

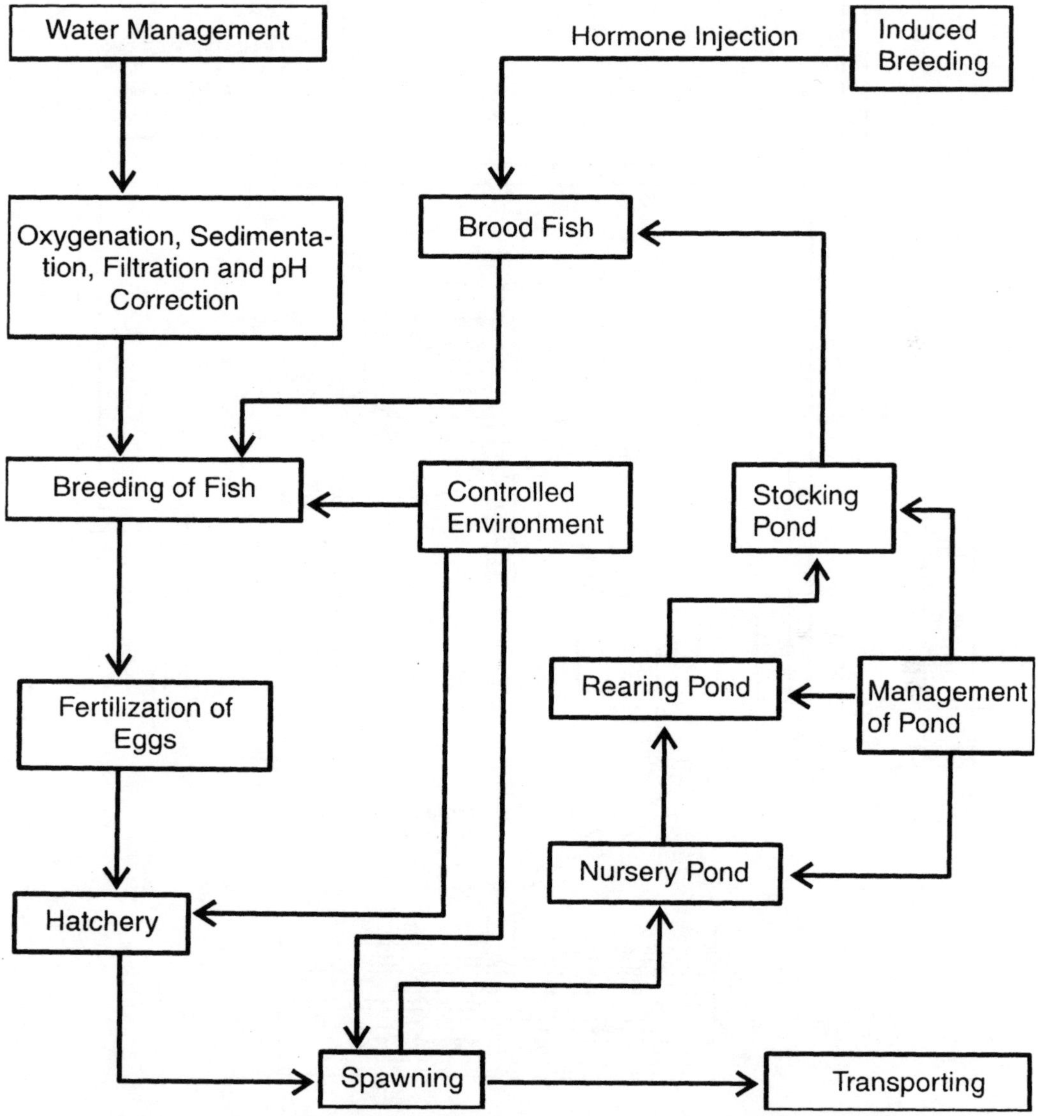


**Fig. 11.10 : A flow-sheet diagram showing different steps involved in fish seed production under controllec conditions.**

### *Chinese Circular Hatchery*

Each hatchery unit has four main parts such as one overhead storage tank, one spawning or breeding pond, two incubation or hatching ponds, and one hatchling receiving pond (Figure 11.11).

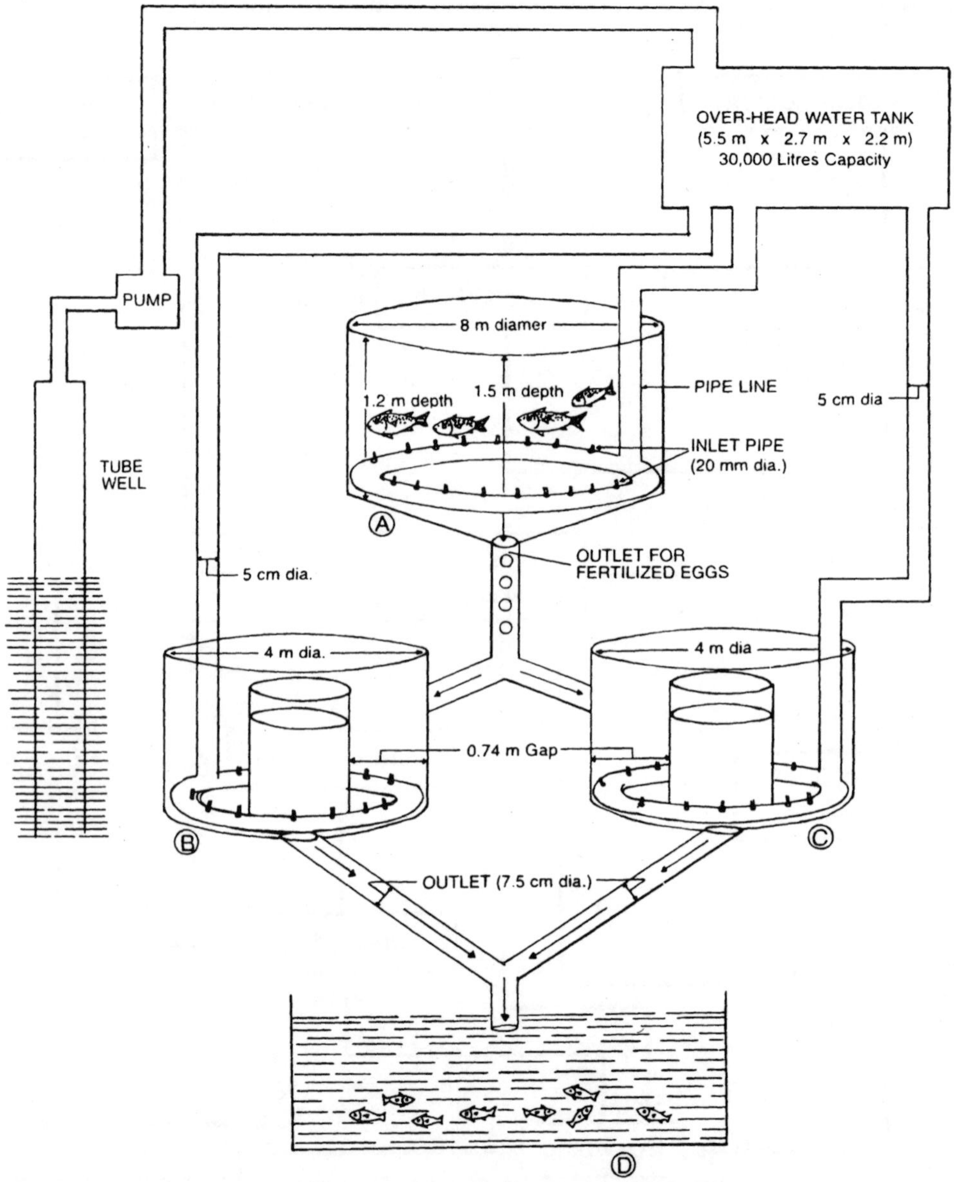


**Fig. 11.11 : Recirculation and rearing systems for intensive fish culture. A, Breeding pond; B and C, Hatchling pond; D, Hatchling receiving pond. Male and female brood fishes are injected with hormones and kept in the breeding pond. The fertilized eggs are drained into two hatchling ponds through outlet pipe. Within the hatching ponds, eggs are transformed into hatchlings. Hatchlings are collected in the hatchling pond through two outlet pipes. In both breeding and hatchling ponds, water is continuously circulated through main pipe lines. Hatchlings are transferred to nursery ponds for fry production.**

1. *Overhead Storage Tank* : It is placed at a height of 2.5 meter above the ground and it measures 5.5 meter X 2.7 meter X 2.2 meter with a capacity of about 30,000 litres of water. A deep tubewell with a pump-set is installed to assure continuous supply of water.
2. *Spawning Pond* : It is a concrete circular tank with a gradual slope from the periphery to the centre. The inner diameter of tank is 8 meter. Depth of the periphery and the centre is 1.2 and 1.5 meter, respectively. Water is supplied from the tank to the spawning pond through a pipe line. This pipe surrounds the tank from outside. From this circular pipe, 14-16 numbers of internal small outlets arise and each of which is fitted with a stop-cock. These outlets are placed at an angle of 45° along with the diameter of the pond. Centre to the pond, a pipe having a diameter of 10 cm with a stop-cock, is fitted that connects the incubation ponds.
3. *Incubation Ponds* : There are two concrete incubation ponds in each unit and both of them are circular in shape. Each pond has two chambers : outer and innner. The latter chamber is 0.74 meter away from the wall of the former. At the centre of each inner chamber, there are stop-cocked exit pipe holes through which excess water is drained out. From the overhead water tank, another pipe line with reduced diameter in succession as 7.5 cm and then 5 cm diameter is emerged from the spawning pond, divides into two halves each having 5 cm diameter.
4. *Hatchling Receiving Pond* : A rectangular concrete pond having the area of 4 m X 2.5 m X 1.2 m is situated at a lower level than the hatchling ponds for effective outflow of the water alongwith hatchlings. Each incubation pond is connected with hatchling receiving pond by an exit pipe.

Operation of the Chinese circular hatchery involves three consecutive steps. First, breeding in circular pond is carried out by releasing several male and female brood fishes after giving pituitary extract injection on appropriate doses (see Table 11.10). Four hours after injection, the valve of the jet pipe is opened so that the water of breeding pond is agitated. Second, the fertilized eggs are collected in the incubation ponds where hatchlings are formed. After four days, hatchlings are taken into the hatchling receiving pond. Third, hatchlings are kept within the hatchling receiving pond for at least two days from where they are transported to the nursery and rearing ponds for attaining fingerling stages.

### *Fibre Glass Jar Hatchery*

This hatchery is a specially designed apparatus which consists of several fibre glass jars each of 6.35 litre capacity. Each jar has an open mouth and the opposite end of the mouth is drawn into a narrow pipe. All jars are fitted on a wooden table having two rows of holes (Figure 11.12). Between the two rows runs an open pipe in such a way that the mouth of jars opens into the water pipe. A pipe line coming down from the water tank divides into two. While one line moves directly to a cistern and fitted with a shower, the other one passes under the table through the rows of the jar. The bottom end of each jar is fitted with this water line through rubber tubings.

About 50,000 fertilized eggs are placed in each jar and a continuous flow of water is maintained from the second pipe into the jar from bottom and then from open mouth to the open pipe which ultimately leads into the cistern. Eggs are hatched within 10-13 hours. The hatchling efficiency of this type of hatchery is almost 100 per cent.

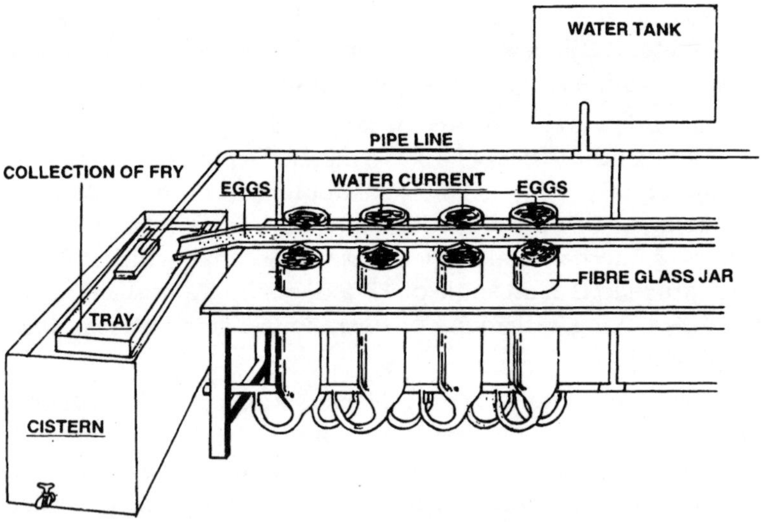


**Fig. 11.12 : Model of a fibre glass-jar hatchery. It is made by arranging different glass jars one after the other. At present, this type of hatchery is becoming very popular. Though its initial establishment cost is high, it has no cost of repairing or replacement. Consequently, it becomes more profitable. The hatching efficiency of this hatchery is very significant to fish culture and production generally.**

### *Hapa Hatchery*

An inverted rectangular-shaped mosquito net is termed a *hapa.* However, hapa hatchery is a very simple design and is composed of a series of hapa units. Each unit is formed by two hapas – one is placed within the other (Figure 11.13). The outer hapa – called the *hatching hapa,* is made

**Fig. 11.13 : Model of a hapa hatchery. In this type of hatchery, hatching of fertilized eggs is tried and successful results are obtained. In India and Bangladesh, most of the poor fish farmers use this type of hatchery for fish seed production. The percentage of hatching in the hapa varied between 40 and 90 per cent depending upon environmental factors such as temperature (26-30ºC), pH (7.0-8.0), dissolved oxygen (4-9 mg/l), ammonia-nitrogen (0.05-1.0 mg/l), nitrate-nitrogen (0.06-0.8 mg/l), and phosphate (0.05-0.5 mg/l).**

of the piece of cloth (2 meter X 1 meter). Four bamboo poles are fitted at four corners of the pond with which outer hapa is tied. The inner hapa is called the *breeding hapa,* is made a finely meshed nylon net (1.75 m X 0.75 m X 0.045 m). The breeding hapa is fitted with eight loops within the outer hapa.

About 75,000 fertilized eggs are released into the inner hapa. Hatchlings swim out through the mesh of the inner hapa. As soon as the outer hapa is filled with hatchlings, the inner one containing unfertilized and/or dead eggs are removed to prevent pollution of water.

### *Apprisal*

As soon as the success of induced breeding of carp was achieved, a variety of hatcheries have been designed in India for breeding and hatchling of carp eggs. These includes : hatchling pits, breeding hapas, hatchling hapas, floating hapas, earthen pot hatchery, tub hatchery, glass jar hatchey, bin hatchery, hanging dip net hatchery, and chinese hatchery. With the development of these hatchery units, environmental factors which are responsible for carp breeding have extensively been studied throughout the country. It has been found that dissolved oxygen, temperature, silt-free water, feeble water current, aeration, and removal of metabolites are the most critical factors for success of breeding of carps. However, the "D" series of carp hatchery systems (D80, D81, D84, D85 and D86 type) are operated throughout the country because these hatcheries are quite dependable and are economically viable. For this reason these hatcheries are widely used by the Government Organizations, enterpreneurs, and fish farmers in different states of the country such as Rajasthan, Uttar Pradesh, Madhya Pradesh, Andhra Pradesh, Tamil Nadu, Maharashtra and Haryana.

In India, a total number of 579 hatchery units (official record for the year 2000) have been set up for carp seed production. Most of the hatcheries are the original Chinese circular design, some are mini Chinese circular hatcheries. In some other cases, glass jar hatcheries are also common. Pituitary extracts/human chorionic gonadotrophic (HCG)/ovaprim/ovatide, depending on their availability in the local market, are used as inducing agents. However, hatchling and fry production potential in some states of the country is great (Table 11.12).

## 11.30 Spawn Mortality in Carp Hatcheries

Hatchery is an important primary source of good quality of fish seed. In hatchery systems, fishes are induced breed in confined waters. These water bodies are mostly subject to several factors such as low oxygen tension, hardness, pH, and microbial infection. These factors affect the spawning response of brood fishes and developing embroys as well as hatchlings. Besides these factors, improper management of hatchery farms also cause drastic reduction in seed prodution potential. Major factors of hatchery environment that are responsible for egg mortality are briefly discussed below.

**Table 11.12 : Production potential of Fish Seed in Different States of India (Data for the year 2000)**

| State | Total number of hatcheries | Number of Types | | | Production capacity/hectare (in lakh) | | |
|---|---|---|---|---|---|---|---|
| | | Jar | Chinese circular | Hapa | Hatchling | Fry | Fingerlings |
| Andhra Pradesh | 32 | 10 | 21 | 1 | 55,500 | 45,000 | 36,000 |
| Arunachal Pradesh | 4 | - | 4 | - | 20,000 | 15,500 | 9,500 |
| Assam | 62 | - | 62 | - | 200 | 1,700 | 1,200 |
| Bihar | 5 | - | 5 | - | 3,500 | 220 | 100 |
| Gujarat | 12 | - | 12 | - | 28,000 | 21,000 | 16,800 |
| Haryana | 21 | 5 | 10 | 6 | 5,975 | 1,375 | 1,000 |
| Karnataka | 28 | 2 | 24 | 2 | 6,343 | 1,269 | 3,172 |
| Kerala | 28 | - | 28 | - | 1,000 | 800 | 760 |
| Madhya Pradesh | 72 | - | 72 | - | 22,200 | 3,916 | 1,180 |
| Maharashtra | 28 | - | 27 | 1 | 9,960 | 2,320 | 760 |
| Manipur | 4 | - | 4 | - | 160 | 40 | 22 |
| Orissa | 37 | - | 37 | - | 19,672 | 17,705 | 4,965 |
| Punjab | 6 | - | 6 | - | 750 | 272 | 75 |
| Rajasthan | 19 | - | 19 | - | 6,300 | 1,300 | Not available |
| Tamil Nadu | 83 | 14 | 69 | - | 11,700 | 6,950 | 775 |
| Tripura | 5 | - | 5 | - | 14,210 | 478 | 360 |
| Uttar Pradesh | 45 | - | 45 | - | 17,445 | 8,805 | 1,050 |
| West Bengal | 88 | 43 | 45 | - | 33,300 | 24,850 | 17,450 |

### *Inefficient Hatchery Management*

Research on these aspects have demonstrated that proper hatchery management is a critical factor in a hatchery system, not only in producing good fish seed but also in sustaining long-term productivity. Some managment principles pertaining to hatchery manager are as follows :

1. Adoption of easy cleaning system for removal of debris and dead materials from the system.
2. Poor adjustment of duck mouth or obstruction of holes in the incubation unit cause low oxygen pockets in water. Therefore, these pockets should be removed.
3. Premature hatchling should also be prohibited.
4. Incubation unit should be stocked with recommended quantity of fertilized eggs (preferably at the rate of 70 lakhs/$m^3$ of incubation space).

### *Low Oxygen Tension and Hydrogen Ion Concentration*

Low oxygen tension provides serious limitations in the egg incubation system that causes mortality of developing eggs. At the same time, oxygen saturation may cause gas bubble disease to the hatchlings. To counteract this problem, oxygen level at 6-7 mg/l should be maintained.

Fluctuations of water pH is the most significant widespread constraint in hatchery units. However, maintaining water pH at 7.5-7.8 is safe for the system.

### *Total Alkalinity*

Total alkalinity of the water is a very important factor for the development of fertilized eggs and hatchling of the embryos. In general, total alkalinity should be kept below 150 mg/l. Experience has shown that total alkalinity of water more than 170 mg/l hamper the process of embryogenesis. Dilution of stock water with another freshwater of low alkalinity can help reduce mass mortality of fish seed.

In many carp hatcheries, groundwater is a matter of great concern, especially in areas where surface water is scarce. Groundwater sometimes contains low dissolved oxygen, methane, hydrogen sulphide, and carbon dioxide beyond the tolerance limits of the fertilized eggs. These factors have, however, been subject to severe damage to fish seed by whitening of hatchlings, and inhibiting the twitching movement of the embryo within the eggs, and declines in hatchery productivity and fish production. Groundwater should be exposed to sunlight in an earthen pond before use in hatchery system. Obviously, water threaten the capacity of hatchery to sustain quality fish seed production.

### *Application of Unstandardized Inducing Agents*

At present, a number of inducing agents (gonadotrohic hormones) under different trade names are available in the market. The recommended rates of these agents and their mode of application are not suitable for all the hatcheries because of variations in climatic conditions and maturity of brood fish. Unstandardized dose of inducing agents not only ovulates the non-viable eggs which leads to mass deformities but also exhibits partial or no spawning. In some cases, spawning continues for hours that results in differential development of embryos. The impact of standardized doses and their influence of brood fish under controlled conditions tend to overshadow the beneficial effects of hormones as inducing agents for quality seed production.

### *Other Factors*

Recent studies have recalled attention to signficant impact on the hatchery unit that do not exhibit harmful effects on seed production. Among the most significant are those involving sanitation. Adequte sanitation of the hatchery units and equipments with formaldehyde solution reduce the chances of mortality of embryos. Besides, fast food of formulated powdered feed or screened plankton are necessary for better survival of hatchlings.

## 11.31 Role of Cooperative in Fisheries Management

In general, fish farmers are characterized by low standard of living and hence belongs to the weaker section of the society. Fish farmers being mostly unorganized, illiterate, and heavily indebted having no proper institutional finance for the purpose. Besides these, fish farmers are always exploited by the middlemen and as result, they are bonded in various ways. Therefore, it is necessary to organize and to develop fishermen through formation of cooperatives among themselves. Cooperatives are usually considered as an ideal institution for self and mutual helps for a group of fish farmers. Keeping these values in mind, the Central and State Governments have been proposed to establish registered cooperatives for the fisheries sectors not only for the betterment of standard of living but also to improve their income.

To fulfil the need for fish farmers, the National Co-operative Development Corporation (NCDC) was established in March, 1963. After the act of NCDC was amended in 1974, the NCDC has formulated specific schemes for co-operative to take up various activities in marine, inland, and brackishwater sectors such as fishing inputs, marketing support, infrastructure, management, processing, training and extension work. However, the NCDC provides financial assistance to fisheries co-operatives for the following purposes :

1. For integrated fisheries development projects in different fisheries sectors.
2. Preparation of reports for fish farmers.
3. Purchase of boats, nets and engines.
4. Construction of shed and/or godown for storage of fish, fish products, nets, and equipments.
5. Development of inland fisheries (tanks, ponds, beels and reservoirs) and fish farming systems.
6. Establishment of fish seed farm and hatcheries.
7. Purchase of vehicle for fishing purpose.
8. Share capital to fish farmers' co-operative for marketing, supply of fish seed, fingerlings, feeds, and fertilizers.
9. Training of co-operative personnels and education to fish farmers.

## 11.32 Problems in Fish Culture Management

There is a general opinion among all the countries that fisheries development programs should be executed although the management techniques are not easy to implement and enforce. At present, both capture and culture fisheries are dominated by non-bonafide fishers and fish farmers. Bonafide fish farmers are not able to compete with the new fish farmers who have more capital or financial resources. This influx of additional 'new capital and labour' is because fish is much sought after due to its health attributes. Consequently, fisheries management is being dominated by these non-bonafide fish farmers.

For adequate fisheries management, fish farmers must appreciate the need for fish culture and usefulness of managing fisheries resources. Second, to accept fisheries management, it is important to describe and explain them what fisheries management is all about.

### *Levels of Social and Educational Conditions*

Because fish farmers belong to the lowest strata of the society and live an isolated life, management strategies will vary from one state to another or even from country to country. Different experimental results obtained by fisheries experts should be conveyed to fish farmers by personal contact, otherwise they will not accept the results. Unless primary education is given to fish farmers, no interest regarding fish culture techniques will be grown among them.

### *Lack of Opportunity in the Result Conveyed*

Results which are required to convey among fish farmers, should be brief and to the point, authentic and without any complexity. Complex results are sometimes difficult to accept by the fish farmers. Furthermore, uniform technical know-how regarding fish culture management protocols should always be adopted by the extension workers so that they may able to spread uniform knowledge to all fish farmers.

*Economic Conditions*

Fish farmers are very poor so far as their economic aspect is concerned. They are unable to save adequate money and hence the developmental strategies fail to continue in future. Therefore, if fish farmers are not able to adopt suitable management techniques due to lack of appropriate fund, the entire future fisheries/fish culture programs will be lost after several years of culture operation.

*Risk Factor*

Fishes are aquatic animals and are, therefore, succumbed to various infectious diseases and predators. This factor is predominant in semi-intensive and intensive fish culture systems and as a result, the entire crop may be lost within very short duration and thus fish farmers get disappointed and they are compelled to discontinue fish culture activities due to financial loss.

Different types of fishes, if cultured together, do not grow well in the same pond. Major carps, for example, may not grow well in ponds where soil and water are not fertile. In such ponds, culture of tilapia or air-breathing fishes are recommended. However, suitable species for culture, site selection, technical assistance, marketing facilities, and availability of inputs at cheaper rate are some of the risk factors for sustaining high production.

## 11.33 Risk Management in Fish Culture

Risk is an inevitable feature of fish culture sector. The growth of fisheries sector could not have taken place without some confidence. At the same time, the implications of losses which are associated with risks, could be vast in extent and could continue to represent constraints on the ability of the sector to develop. Hence, it is necessary to understand the extent of risks, which affect fish culture. It is also important to understand that in many cases it is not possible to control every area of risk but it may be possible to take into account some effects on management. Some of the important risks which are associated with fish culture are summarized below.

*Risk Category*

1. *Environmental* : Environmental factors play a decisive role in the development of the sector. Seasonal disruption, climatic conditions, diseases and parasites, and wide fluctuations of environmental conditions are the main constraints in determining the growth of fisheries sector.

   Other environmental disasters such as floods, drought and other weather changes can have serious effects. Though these events come by turns, their long-term effects are matter of grave concern. These natural calamities not only cause severe damage to fish culture industry but to farmers' community as well.

2. *Financial* : Most of the farmers live below the poverty line. Therefore financial factors, is in part, very important for marked expansion in fish culture industry. Fluctuations in cost of inputs, debt, repayment of loans, payment of interests on loan are some of the financial factors which must be considered.

   An important form of production-oriented fish farm credit is the short-term and/or long-term loans for fish culture which are issued to farmers to meet the outlay on inputs.

If loans are properly utilized by farmers for activities which are related to production and repayment of loans is satisfactory, fish culture sector will lead to an improvement in the production potential of the farms unless catastrophic events occur. At the same time, various activities for high production bring an increased income for the farmer.

The main cause of an increasing financial constraints is, obviously, the proverty of the farmer that has aggravated in recent years mainly due to growing pressure of population. Moreover, farmers are going beyond the limits of what is moderate of proper. Due to financial constraints farmers lose their productive efficiency.

3. *Biotechnical* : This category of risk involves (i) inappropriate failure of systems and components at different levels which are associated with physical systems and (ii) poor response to inputs. These factors undoubtedly cause severe loss of the entire culture system.
4. *Human and Social* : Breakdown of management systems, disagreement, theft, sabotage, human error, and negligency of staff are the most common risks of this category.
5. *Market* : The pattern of fish marketing systems in many Asian countries are traditional and not systematic. Since fish marketing systems are not well-organized, a drive is necessary to ameliorate the markets by modernizing the traditional fish marketing methods by adopting new management protocols. The strategy of fish marketing system can be formed by analyzing the objectives, developing fish demand, marketing operations and design of marketing. All the existing traditional markets, however, need improvement on the basis of modern management protocols. Different components of marketing systems also require betterment to such an extent that fish producers and fishermen may receive remunerative price for their products in wholesale fish markets where the exclusive control of a fish market by contractors always remain.

   The oscillation of fresh fish price is very conspicuous. The oscillations are so frequent to forecast any trend. Generally, the uncertaintity of supply and demand and gradual decaying propensity of fresh fish have a definite role in determining the price of fish in the market. Hence there is a great price risk in fish marketing systems. Unless marketing systems are developed, market risks will always exist and these risks are associated with the uncertainty of market conditions.

### *Strategies of Risk Management*

For successful technical management in fisheries sector, specific risks should be properly identified. The most important use of tricks in order to succeed in some risk management involve the followings : (i) improvement of security near farm site, (ii) improvement of training and motivation of staffs, (iii) supply of water to a fish farm, (iv) use of essential equipments on lease, (v) epidemic, (vi) algal problems in fish culture, and (vii) development of market facilities. Most of these strategies, however, involve heavy costs which must take these factors into account and assessed.

### *Structural Approaches*

Risk element in fish culture systems has been the long production cycle. It has been observed that in carp, trout, and catfish cultivation, the long time culture period is more critical than in prawn/shrimp culture. In such cases, however, investment of working capital is more and at the same time business risk is also high. Moreover, availability of fish seed, in some situations, becomes uncertainty. Consequently, farmers face some difficulties when restocking the ponds with fish is necessary.

To reduce the loss of fish stocks and to increase the revenue from the sale of produce, most fish farmers cultivate for short durations (3-6 months). This is due to the fact that longer production period for fish stocks involves greater management risk.

Inspite of significant increase in the technological control over fish culture, risks will not remove from this sector. If a fish farmer can able to understand risk, fish production potential will undoubtedly increase to light up the farmers' community.

### *Hazards and Insurance*

Insurance is a contract by which the insurer agrees to indemnify the insured. The main objectives of insurance business are : (1) To make it meaningful to the fish farmer, and (2) To help the growth of economy. Similar to other insurance schemes in agriculture and allied sectors, however, a comprehensive cover for inland fisheries has been devised since 1979 to protect fish farmers under Fish Farmers' Development Agency scheme.

The progress of insurance scheme in fishermens' community has only been limited. The benefits have so far been availed only by the farmer coming under the various progress such as Integrated Rural Development Program (IRDP), Small Farmers Development Agency (SFDA), and bank finance with great success. Therefore, it is necessary to extend this service on an individual farmer basis. This require a proper infrastructure of insurance that should be created in the country. So far as the present position is concerned, we have to proceed a long distance.

On the basis of insurance scheme, fish farmers may be well protected against risks arising out of fish crop failure on payment of a certain premium. The scheme covers the farmers availing of loans from commercial banks, regional rural banks, and co-operative credit institutions with the following objectives :

1. To restore the credit eligibility of fish farmers, after a crop failure, for the next season.
2. To stimulate fish production.
3. To provide financial support to farmers in the event of fish crop failure as a result of natural calamities such as floods and drought.
4. To protect farm income against fluctuations due to adverse weather conditions.

The insurance scheme envisages to cover only the borrowers of authorized financial institutions. Most of the farmers are reluctant to take loans for fisheries activity due to the fear of indebtedness. Therefore, they are not in a position to take advantage of this scheme. The purpose of this scheme is to alter the stage of the farming community which remains far from sight. On behalf of these farmers, the Government can insure, at least, for certain amount to help these farmers during adverse situations. The scheme may, however, pose an extra burden on limited budgetary resources.

Although the insurance scheme in fisheries sector is very important, it is best with a lot of problems in any developing economy. Some of the problems are mentioned below : (i) existing pond record and pond tenure system, (ii) ignorance and proverty of farmers, (iii) unavailability of trained personnel, (iv) limited financial resources, (v) wide variety of fish culture practices, and (vi) unavailability of data of fish production and losses.

Insurance scheme encompasses coverage of technical instability and hazard due to the factors beyond the control of farmers. Fluctuations of price for farm products have a deciding and dominant role in the rural economy and therefore need to be considered. Obviously, the scheme should focus attention not only fish production but also cover farm income and profitability of individual farmer.

Insurance scheme is a main step towards aversion of risk in fish farming and it is necessary to welcome from every corner. Although it is the first step towards fish farming activities, there are certain limitations mainly due to lack of experience on the part of implementing agencies. These limitations must be properly monitored, evaluated, and modified according to the need of farmers.

### 11.34 Research for Management

Aquacultural management is important for high fish production and fish culturists must play due attention to implement effective management strategies. Of course, strategies can be different from region to region, location to location, species to species, and country to country. In a fisheries circular (No. 886, 1995) published by the Food and Agricultural Organization (FAO), it has been reported that aquaculture is contributing over 14 per cent of the total fisheries production which is more than corresponding for the decline in capture fisheries. While aquaculture expansion has developed in the last several years, it is expected to continue to make an important and increasing contribution to total food fish supply and therefore, the potential of aquaculture is realised.

The need to introduce aquaculture to poor rural communities, where there is often a shortage of animal protein, is stronger than ever. Integration of aquacultural and agricultural development planning is essential so that culture-based fisheries can be implemented. However, to sustain high productivity of aquatic ecosystem, development and adoption of fish culture ecosystem technologies and management of agrochemicals in intensive aquaculture, promoting ecologically sustainable and economically viable techniques should be taken into consideration.

Various national and international fisheries research organizations which presented a diagnosis of existing development problems in both fisheries and aquaculture and sugggested a framework for organizations to follow is important. Expert consultant on fisheries research identifies a number of important constraints to further fish culture development. Some important research themes have emerged among regions. The followings are the items for developmental strategies :

1. Expansion of culture-based fisheries.
2. Improvement of product quality.
3. More efficient feed and increased use of natural feed in less intensive systems.
4. Improvement of pond design.
5. Economic, social, technical, and environmental criteria for fish culture planning.
6. Improvements in water quality and fish health management (for further detail, see Subasinghe *et. al.* (1995).
7. Application of unconventional feed ingradients and partial or complete replacement of fish meal.

8. Stocking of indigenous species for fish culture and species diversification.
9. Genetic selection for increased survival, growth, and maturity of fish.

## 11.35 Conclusion

Fish culture ecosystem is complex and, therefore, its quality has significant effects on fish biology. Environmental stress should be minimized by maintaining high standards of water quality with the aid of latest farming technology for fish production. Because environmental situations vary within a pond and among ponds in different places, different abiotic factors and their fluctuations are also responsible for high productivity of fish culture ponds. For successful adoption of various management techniques, these variations should not be ignored.

The introduction of improved culture systems, highly productive strains, highly improved feed formulations in different farming systems, development of co-operative societies, continuing research and development efforts, funding assistance to fish farmers, potential uses of computer vision, ozone, gas aerators, and expansion of production areas have all contributed to the exceptionally fast growth of fish culture in many countries of the world. This rapid expansion of fish culture shall be sustained in future as refined improved techniques, development of high quality feed, and higher yielding strains are genetically engineered.

The success of freshwater fish culture can only be obtained through proper management based on the knowledge of the culture environment and the biological processes involved in the culture operation as well as on the availability of quality fish feeds, fry, and the suitable data on the application rates of fertilizers as well as manures.

Similar to other countries, future prospects of Integrated Fish Farming (IFF) in India is no doubt bright. Of course, its viability depends upon the local agroclimatic conditions and topographical features. This system of farming has, however, been found to be highly profitable. For effective management of fish culture, use of poultry and cattle manures obtained from IFF is a common practice. Due to availability of plant and animal wastes and low investment capacities of fish farmers, organic recycling in IFF should be precisely considered.

### *References*

Agarwal, S.C. 1990. *Fisheries Management*. Ashish Publishing House, New Delhi, India.

Banerjee, R.K., P. Roy, G.S. Singh, and B.R. Dutta. 1979. Poultry droppings-- its manurial potentiality in aquaculture. *J. Inland Fish. Soc. India*. 2 : 94-108.

Beveridge, M.C.M. 1984. Cage and pen fish farming : Carrying capacity models and environmental impact. *FAO Fish Tech. Pap.* 255 : 131p.

Beveridge, M.C.M. 1996. *Cage Aquaculture*. Fishing News Book, Oxford, U.K.,352p.

Brady, N.C. 1991. *The Nature and Properties of Soils*. MacMillan Publishing Company, New York.

Chakraborty, P.K. 1990. Biofertilizers : An overview. *Everyman's Sci.* 25 : 79-82.

CIFRI. 1985. Rice-cum-fish farming system. Package of practices for increasing production. *Aquacultural Extension Manual*, New Series No. 4. CIFRI, Barrackpore, India. p. 1-14.

Cipriano, R.S. and G.L. Bullock. 1983. Furunculosis and other diseases in fishes caused by *Aeromonas salmonicida*. US Fish Wildlife Series Fish Disease Leaflet No. 66.

Cipriano, R.S. and G.L. Bullock. 1984. *Aeromonas hydrophila* and aeromonad septicemiasis of fish. US Fish Wildlife Series Fish Disease Leaflet No. 68.

Deepwater Rice-Fish Farming Systems Bulletin. 1991. Deepwater Rice Project. IRRI/ICAR Government of West Bengal. Rice Research Station, Chinsura, Issue No.1, February, 1991.

Desmukh, A.M. 1998. Biofertilizers and Biopesticides. Technoscience Publications, Jaipur, India.

Duby, K. 1993. Integrated Fish Farming in China. *Fish. Chim.* July, p. 8-13.

EPA (Environmental Protection Agency). 1973. Criteria for water quality. USEPA, Washington, DC.

Hoffman, G.L. 1984. Control and treatment of diseases of freshwater fishes. US Fish Wildlife Series Fish Disease Leaflet No. 28.

Jhingran, V.G. 1988. *Fish and Fisheries of India.* Hindustan Publishing Corporation, New Delhi, India.

Jhingran, V.G., S.H. Ahmad and A.K. Singh. 1998. Retrospect and prospect of integrated aquaculture-agriculture-animal husbandry farming in India. *In* : *Advances in Fisheries and Fish Production.* (*Ed* : S.H. Ahamad), p. 127-141. Hindustan Publishing Corporation, New Delhi, India.

NACA (Network of Aquaculture Centre in Asia and the Pacific). 1989. *Integratd Fish Farming in China.* NACA. Technical Manual No. 7. A World Food Day Publication of The Network of Aquaculture Centre in Asia and The Pacific, Bangkok, Thailand, pp. 1-278.

Proceedings, of The SEAFDEC/IDRC. 1979. *International Workshop on Pen and Cage Culture of Fish.* Tigbauan, Iloilo, Phillippines, 11-12 February, 1979.

Saha, C. 1998. Design of low-cost aeration device for use in aquaculture. *Fish. Chim.* 18 : 31-32.

Sarkar, S.K. 1998. Development of fish culture management techniques. *In* : *Advances in Fisheries and Fish Production* (*Ed* : S.H. Ahmad), pp. 76-90. Hindustan Publishing Corporation, New Delhi, India.

Sarkar, S.K. 1989. Growth rate and conversion efficiency of grass carp, *Ctenopharyngodon idella* in relation to temperature. *Geobios* 16 : 45-47.

Shukla, P.P., S.C. Shrivastava, and A.K. Garg. 1996. Case study of integrated fish farming in Bansagar Quarry, Rewa (M.P.). *Fish. Chim.* 16 : 39-40.

Sinhababu, D.P. 1993. Aquaculture – A component of sustainable rice-farming systems in rainfed lowlands of Eastern India. *Paper Presented During the Nat. Meet on Aqua-farming System, Practices, and Potentials.* Central Institute of Freshwater Aquaculture, Bhubaneswar, Orissa, February 10-11, 1993.

Sinhababu, D.P and B. Venkateshwarlu. 1998. Modern frontiers on rice-fish systems. *In* : *Advances in Fisheries and Fish Production* (*Ed* : S.H. Ahmad), pp. 206-228. Hindustan Publishing Corporation, New Delhi, India.

Sollows, J. and C.C. Dela. 1992. Rice Management in Rice-Fish Culture. Farmer Proven Integrated Agriculture–Aquaculture : A Technology Information Kit. International Institute of Rural Reconstruction, Phillippines.

Subasinghe, R.P., J.R. Arthur and M. Shariff. 1996. Health Management in Asian Aquaculture. Proceeding of The Regional Expert Consultation on Aquaculture Health Management in Asia and The Pacific. Serdang, Malaysia, 1995, pp. 1-142.

Talley, S.N. and D.W. Raims. 1982. Potential mechanization of *Azolla* cultivation in rice field. *In* : *Azolla as Green Manure : Use and Management in Crop Production.* (*Eds* : T.A. Lumpkin and D.L. Plucknett). West View Tropical Agriculture, Colorado, USA.

## Questions

1. What is the importance of management in fish culture?
2. How fertilization, manuring, and liming are related to fish culture management?
3. Suppose a fish culture pond has to be treated with lime at the rate 550 kg/hectare with calcium carbonate having a neutralizing value of 95 per cent and an efficiency rating of

90 per cent. How much amount of lime will be required to treat 0.5 hectare of water area? State the significance of liming.

4. Describe different methods of application of lime and fertilizer in fish ponds.
5. What is biofertilizers? How these fertilizers are used in fish culture?
6. Define the following terms : (a) pond conditioning, (b) gross conversion efficiency, (c) artificial fish habit, (d) hypophysation, (e) ovaprim, (f) gonadotropins.
7. How feeding of fish is related to water temperature?
8. What is turbidity? How it is determined? Why it is so important to fish culture?
9. Why treatment of pond ecosystem with chemicals is necessary? Name some chemicals and their application rates that are commonly used during fish culture operation.
10. State how computer is applied in fish culture management strategies.
11. What is ozone? How it can be used in fisheries management? Can you think of any problems?
12. What harm can result from exessive rates of fertilizer application?
13. Why aeration of pond water is necessary? How it is executed a fish culture pond?
14. What is integrated fish farming? Why this farming is so important in India?
15. Fish production in a sandy-loam pond has been declining. The fish farmer applied 15,000 kg/ha of organic material, 175 kg/ha of urea, and 250 kg/ha of single superphosphate. Unfortunately, the farmer did not receive a yield increase. Give possible explanations for this situation.
16. Why is the loss of nutrient more important today?
17. Which sub-section under the section integrated fish farming is most important for sustained production under Indian conditions? Explain.
18. What major changes have occurred in the last twenty years in the Integrated Fish Farming in India.
19. What is cage culture? How cage is operated in a lentic ecosystem? What are the advantages of this type of culture?
20. State some characteristic features of pond fish cage.
21. Why multiple stocking and harvesting should be practiced by fish farmers? What are the limitations of this method?
22. What is induced breeding? How fishes are induced for breeding under controlled system?
23. How fertilized fish eggs are hatched in different hatchery units?
24. Why mortality of carp hatchlings occur in hatcheries? Explain.
25. How co-operative societies are related to fisheries management?
26. What are the management that will be needed if fish culture strategies are improved in developing countries?

# 12

# Water, Soil and Chemical Pollution

Water is one of the most important natural resources for all living organisms. It has been estimated that about 75 per cent of the earth is covered with marine and freshwater. Moreover, out of the total water content in the earth, only 2 per cent of freshwater system is safe for human use. Human populations are entirely dependent on water and soil and they use for their own purposes. Soils and waters are the natural bodies in which biotic communities live and grow.

Soils and waters are also important in many other ways. They are used to absorb waste materials from animal sources, industries, sewage systems, municipal cities and agricultural lands and therefore, is considered as waste disposal sites. Besides this, they are important for construction of a suitable fish farm. Therefore, water and soil are significant to fish farmers.

Although good soil and water means bumper fish crop, farmers should keen aware of the value of a productive fish pond. Since water and soil undoubtedly helps to establish flourishing civilizations, continuous or frequent destruction of water and soil or mismanagement or unprotection is the principal factor in their downfall. Modern nations, though they are aware of harmful effects of waste materials, do not fully recognize the long-term significance of water and soil and they are ignoring about the consequence of future generations.

## 12.1 Pond Soil-Water Dynamics

Three important concepts about soil-water dynamics emphasize the significance in relation to productivity :

1. Soil holds a large number of nutrients with varying degrees depending on the amount of nutrient inputs added to a pond.
2. Soil micro-organisms take active part during the process of decomposition of inorganic/ organic waste materials.
3. Essential soluble nutrient elements (such as calcium, potassium, and phosphorus) are released from the soil to water column and ultimately form a solution and form a critical medium to sustain plant and animal life.

It is needless to say that water-holding capacity in a fish pond principally depends on the size of soil particles. Thus ponds having sandy soil, are not able to hold water which is lost through leaching. During leaching phenomenon, an appreciable quantities of nutrients are removed though water. In contrast, clay soils that contain high organic matter, are capable of holding both water and nutrients. For this reason, not all fish culture ponds are productive and hence cannot be used for the purpose.

Fish pond water contains small but significant quantities of essential nutrient elements (inorganic and organic) (Table 12.1). Some of these nutrient elements are very essential for the growth of natural food organisms. Besides this property, acidity or alkalinity of soil and water is also equally important. Many biological and chemical reactions take place in either acid or alkaline medium and the degree of reactions depends upon the levels of hydrogen ($H^+$) and hydroxyl ($OH^-$) ions in water and soil. These levels influence the solubility and availability of nutrients to aquatic life.

**Table 12.1 : Essential Nutrient Elements in Water and Their Sources**

| Used in large amounts | | Used in small amounts |
|---|---|---|
| From air | From fertilizers/Manures/Soil solids | From fertilizers/Salts/Soil soilds |
| Hydrogen (H) | Nitrogen (N) | Iron (Fe) |
| Carbon (C) | Phosphorus (P) | Manganese (Mn) |
| Oxygen (O) | Potassium (K) | Boron (B) |
| | Calcium (Ca) | Molybdenum (Mo) |
| | Magnesium (Mg) | Copper (Cu) |
| | Sulfur (S) | Zinc (Zn) |
| | | Chlorine (Cl) |
| | | Cobalt (Co) |

The activity of $H^+$ and $OH^-$ ions in water and soil is ascertained by determination of pH. The term pH can be defined as the negative logarithm of activity of $H^+$ ions ($-\log H^+$). Thus, each unit change in pH represents a tenfold change in the activity of the $H^+$ and $OH^-$ ions. However, it can be inferred that the pH of soil and water should be considered as a great significance in all aspects of aquatic ecosystem.

## 12.2 Aquatic Ecosystem – A Biological Laboratory

Aquatic ecosystems harbour a varied population of living organisms, both plants and animals. The shape, size, number, and weight of organisms vary greatly in different types of ecosystems. For examples, 1g of soil and 1cc of water may contain from a hundred thousand to several billion bacteria. The amount of micro-organisms present in any fish culture ecosystem may exhibit profound influence on the physical, chemical and biological properties so far as the productivity of fish ponds is considered.

Activities of pond soil organisms include the physical breakdown by oligochaete worms and some species of bottom-dwelling snails. These breakdown products are decomposed by bacteria, fungi and actinomycetes. Formation of humus also takes place due to soil micro-organisms.

After decomposition of organic materials, inorganic essential nutrients such as nitrogen, phosphorus, sulfur and potassium are released into the water. These nutrients are either absorbed by fish food organisms or oxidized, reduced and change the state of these elements. These changes have a significant impact on fish growth or affect the water and soil qualities.

## 12.3 Pollution

Man is responsible for releasing vast quantities of different chemical substances into the environment each year, these being waste products generated by industries and by consumers. For example, every one million inhabitants in a city may annually consume over 600,000 tonnes of water, 2,000 tonnes of food and 10,000 tonnes of fossil fuels. This leads to the production of 500,000 tonnes of sewage to be discharged into the aquatic environment and 2,000 tonnes of garbage to be buried. In addition to this, about 85,000 different chemical compounds are marketed and most of them are released into the environment.

The term pollution refers to contamination of soil and water. Pollution occurs when a harmful substance is released into the environment in sufficient quantities to damage living things. Water and soil pollution may be caused by waste materials from industries, agriculture or towns. In aquatic ecosystem, these may cause eutrophication or deoxigenation or simply kill by poisoning organisms.

The problem of pollution is global and pollution experts draw the careful attention when it becomes hazardous to human health and the environmental situation becomes more complicated. Running (lotic, such as rivers) and standing (lentic, such as lakes, ponds etc.) water bodies are frequently polluted by several kinds of pollutants.

## 12.4 Pollutants

Five kinds of pollutants commonly reach the aquatic systems. (1) *Pesticides* – Thousands of pesticides are widely used for agricultural purpose, (2) *Inorganic Pollutants* – Such as mercury, cadmium, lead, arsenic and others (3) *Organic Wastes* – Such as those from concentrated feedlots, industrial and municipal wastes, (4) *Nitrogen Fertilizers* – Such as urea, ammonium sulfate, diammonium phosphate, calcium ammonium nitrate, and nitrophosphate. These fertilizers undergo degradation by bacteria and produce toxic metabolites such as ammonia, nitrite and nitrate and (5) *Potassium Fertilizers* – Such as potassium nitrate, and potassium chloride. Although potassium fertilizers are very important nutrient for crop production, its usefulness in fish production is not known. Of course, field experiments have shown that potassium fertilizers suppress the growth and reproduction of some freshwater fishes.

## 12.5 Background of Pesticides

The term pesticide refers to all substances that are used to protect mankind against insect vectors or disease-causing pathogens, to protect crop plants from weeds and to protect crop, plant and live-stock from diseases. Thus pesticides have provided many benefits to society.

More than 10,000 insect species, 600 weed species, 1,500 plant diseases and 1,500 species of nematodes are known to be injurious to humans, animals and plants. Since the early Greek civilization, chemicals have been used to control these pests. In the nineteenth century, the use of chemicals has expanded but in the middle of twentieth century, synthetic pesticides came into existence.

The chemical revolution in agriculture actually began when the herbicidal effects of 2,4-D and insecticidal properties of DDT were discovered in 1941 and 1939, respectively. Scientists have observed that these chemicals kill pests and are very cheaper. After discovery of DDT and 2,4-D, thousands of such chemicals and multichemical formulations have been developed, examined their effects on different aspects of aquatic life and have been put to use. Although many chemicals have been restricted to use in developed countries, about 600 chemicals in 55,000 formulation are used worldwide (Figure 12.1).

When these chemicals are drained into the aquatic ecosystem from agricultural land, the entire ecosystem is degraded, ultimately become a complicated problem and has resulted in ecological imbalance with serious consequences. Some pesticides are not able to undergo biodegradation and hence persists in soil for many years. A large number of pesticides undergo biodegradation and are deterimental to both target and non-target species of aquatic ecosystems. Metabolites of pesticides are consumed by bacteria, which in turn, consume by earthworms, and snails. When these bacteria are consumed by fish, they are succumbed to various diseases and exhibits histopathological, biochemical, physiological as well as morphological abnormalities in fishes as chemicals reach the lethal levels. Therefore, damage to fish and fish food organisms can have devastating freshwater ecosystem complications.

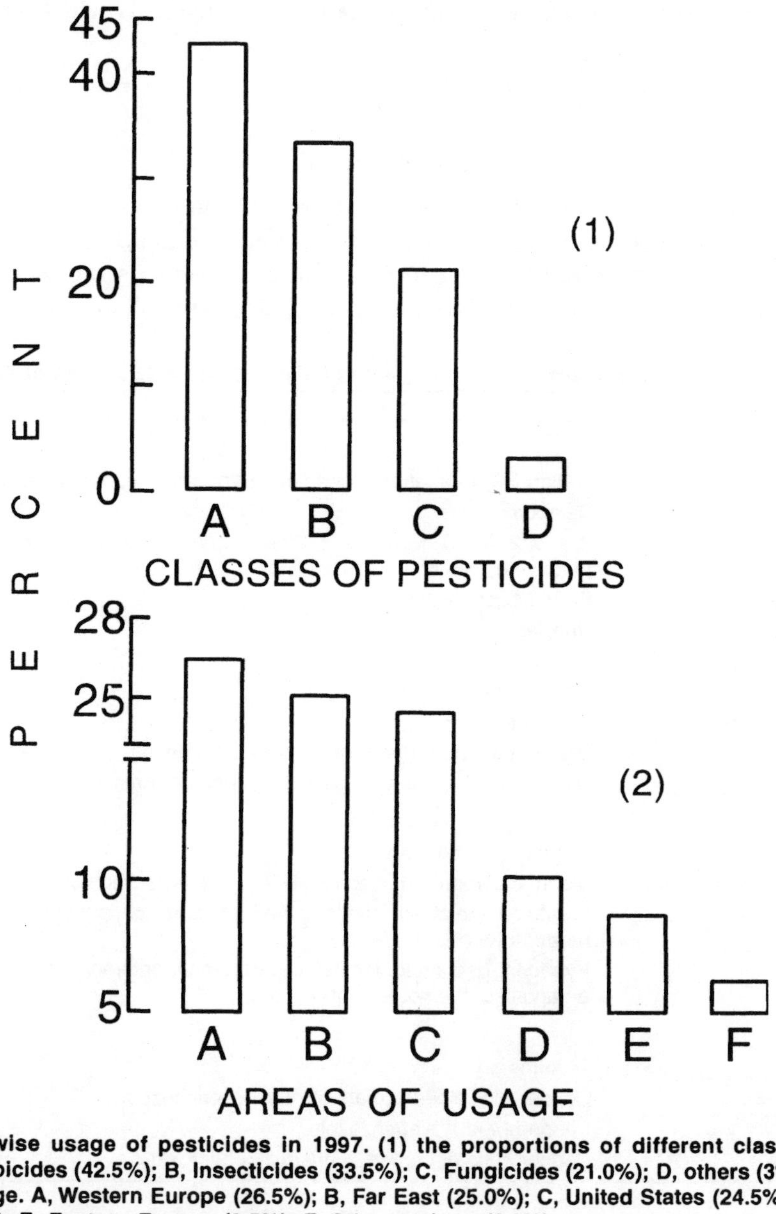


**Fig. 12.1 : Worldwise usage of pesticides in 1997. (1) the proportions of different classes of pesticides. A, Herbicides (42.5%); B, Insecticides (33.5%); C, Fungicides (21.0%); D, others (3%). (2) Major areas of usage. A, Western Europe (26.5%); B, Far East (25.0%); C, United States (24.5%); D, Latin America (10.0%); E, Eastern Europe (8.5%); F, Other regions (6.0%).**

## 12.6 Types of Pesticides

Pesticides are commonly classified according to the pest against which they are targetted :

1. *Fungicides :* These are used to protect crop plant nad fishes from fungal pathogens.
2. *Herbicides :* These are used to kill aquatic and terrestrial weeds.
3. *Insecticides :* These are used to kill insect pests and vectors of human diseases such as malaria, yellow fever etc.
4. *Molluscicides :* These chemicals are used against aquatic snails which are the vectors of certains diseases of fish.
5. *Nematocydes :* These are used to kill nematodes which is the most important parasite of the roots of crop plants.
6. *Rodenticides :* Which are used to control rats, mice etc.

In general, all finds their way into the aquatic systems through run-off from crop fields. As named above, the first three types of pesticides are extensively used and are therefore more likely to contaminate soil and water. Table 12.2 lists name of some commonly used insecticides, herbicides and fungicides and Figure 12.2 shows the variability in their chemical structures.

**Table 12.2 : Classes of Pesticides Commonly Used in India and Some Examples of Each Group**

| Group | Example |
|---|---|
| **Herbicides** | |
| *Triazines :* | Treflan EC, prometryn, dimethametryn, atrazine, simazine, metribuzin |
| *Aliphatic acids :* | Dalapon |
| *Phenoxyalkyl acids :* | 2, 4-D, 2, 4, 5-T, MCPA, 2, 6-Dichlorobenzonitrile, PGBE esters, dimethyl amine, butoxyethyl ester |
| *Carbamates :* | EPTC, thiobencarb |
| *Nitrophenols :* | NItrofen |
| *Dinitroanilines :* | Trifluralin |
| *Amides :* | Alachlor, metolachlor, propanil |
| *Benzoics, pathalates :* | Dicamba |
| *Bipyridinium :* | Diquat, paraquat, gramoxone, diquat-dibromide |
| *Phenylurea :* | Diuron, linuron, monuron, neburon, fenuron, bromacil |
| **Insecticides** | |
| *Carbamates :* | Carbaryl, carbofuran |
| *Chlorinated hydrocarbons :* | Aldrin, chlordane, heptachlor, DDT, lindane, dichlorvos, dimecron, toxapane |
| *Organophosphates :* | Parathion, malathion, methyl parathion, diazinon, phosphamidon, monocrotophos |
| *Phyrethrins :* | Pyrethrin, pyrethroid, dimilin, cypermethrin, entomopathogen, juvenile hormone antagonist, chemosterilant |
| **Fungicides** | |
| *Triazoles :* | Triadimefon |
| *Thiocarbamates :* | Copper oxychloride (maneb), zineb, mancozeb |
| *Benzimidazoles :* | Dithane-M-45, thiabenzazole |
| *Others :* | Copper sulfate, captan, triphamyltin hydroxide |

On the basis of similarities of chemical structure, pesticides can also be divided into two broad groups : (1) Inorganic pesticides and (2) Organic pesticides. However, they are further classified into several sub-groups as shown in the following chart :

- Pesticides
  - Inorganic (Arsenic, copper, Lead, mercury)
    - Bordeaux mixture
    - Arsenicals
  - Organic
    - Natural
      - Nicotine sulfate and Nicotine (Tobacco)
      - Pyrethroids (*Chrysanthemus* sp.)
      - Rotenone (*Derris elliptica*)
      - Red squill (*Scilla maritima*)
      - Strichnine (*Strychnos nuxvomica*)
      - Neem extract (*Azadiracta* sp.)
    - Synthetic
      - Phenols
      - Organomercurials
      - Pyrethroids (Cypermethrin, Fenvalerate, Permethrin etc.)
      - Triazine herbicides (Simazine, Atrazine etc.)
      - Carbamate Insecticides (Carbofuran etc.)
      - Organo-Chlorine Compounds (DDT, BHC, Endosulphan, Aldrin, Methoxychlor etc.)
      - Organo-Phosphorus Compounds (Malathion, Methyl Parathion, Dimethoate etc.)

## 12.7 Reaction of Pesticides in Soils

When pesticides are applied to crop field, they are drained into the ponds, lakes, rivers etc. through surface run-off. Pesticides are then moved in the following directions as the case may be depending on the type and structure of ecosystem :

1. They are absorbed by algae and aquatic plants and then detoxified within their body.
2. They are degraded by soil micro-organisms.
3. They may evaporate without chemical change.
4. They are adsorbed by clay particles and soil organic matters.
5. They are lost by leaching through soil in solution or liquid form.
6. They undergo chemical reactions.

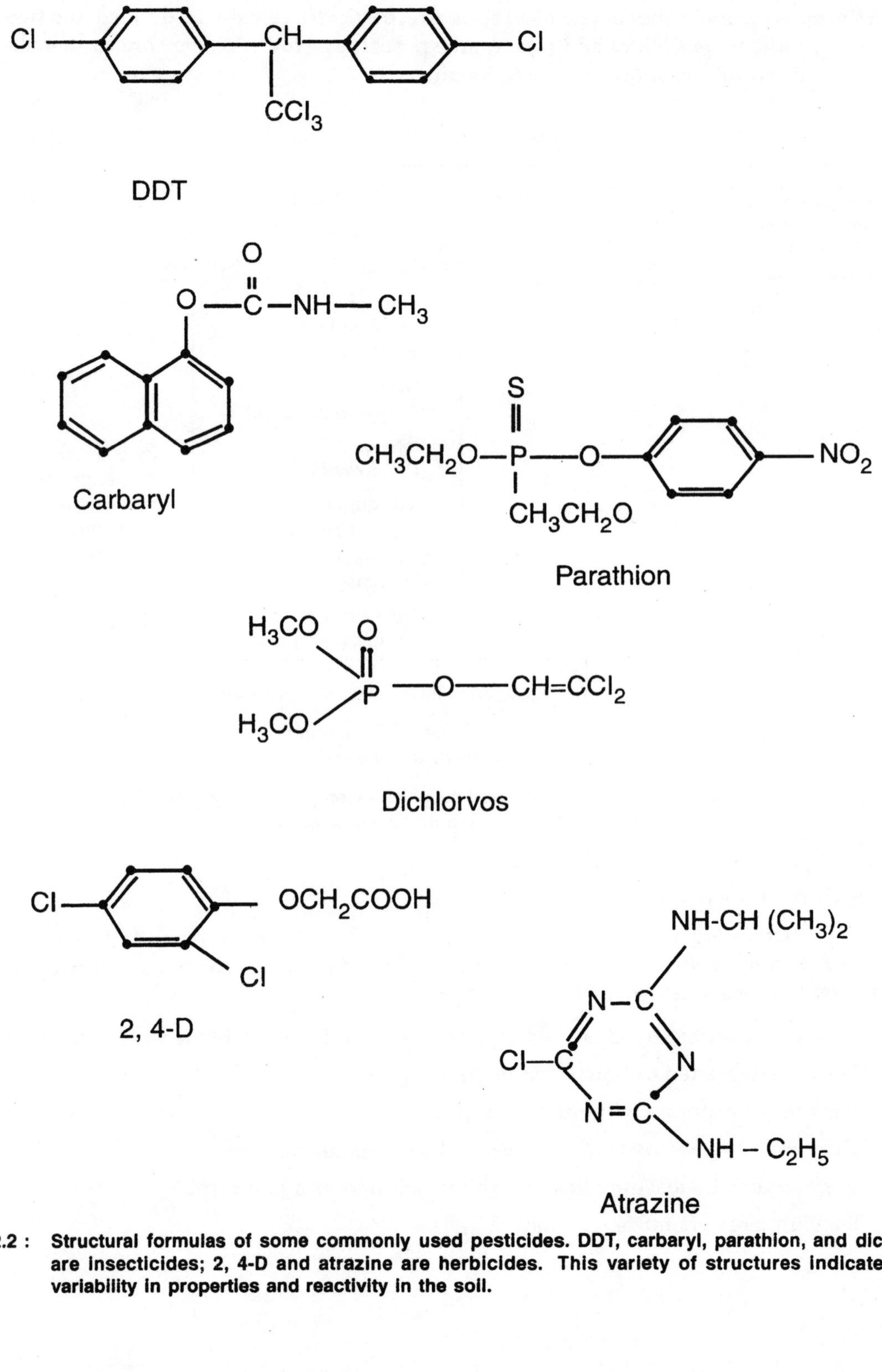


**Fig. 12.2 : Structural formulas of some commonly used pesticides. DDT, carbaryl, parathion, and dichlorvos are insecticides; 2, 4-D and atrazine are herbicides. This variety of structures indicates great variability in properties and reactivity in the soil.**

### *Adsorption*

The adsorption of pesticide molecules by soil depends largely on the characteristic features of soils and pesticides. In general, the larger the size of the pesticide molecule, the greater is its adsorption. Certain functional groups such as – $NH_2$, –NHR, –$CONH_2$, –OH, and –COOR are present in the structure of pesticides. These groups encourage adsorption in the humus. Some pesticides such as Paraquat and Diquat are also adsorbed by silicate clay because such pesticides have positively charged groups. Furthermore, the adsorption of pesticides depend on the concentration of $H^+$ ions present in silicate clay. Thus the lower the pH level, the greater is the adsorption of pesticides (Figure 12.3). This is because $H^+$ ions combine with the –$NH_2$, yields a positive charge on the pesticide, which ultimately binds with the negatively charged soil colloids.

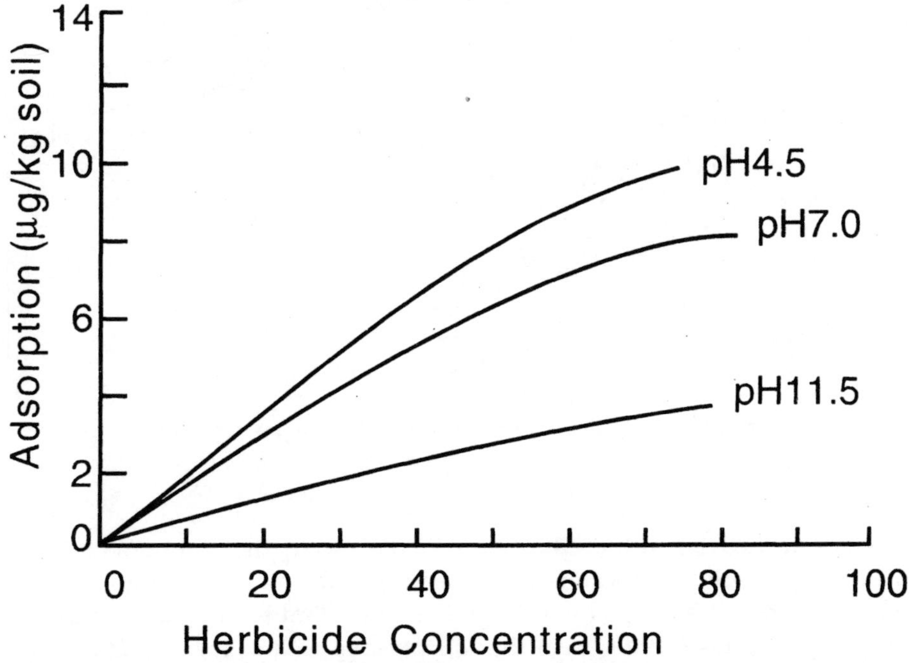


**Fig. 12.3 : The effect of pH of Kaolinite on the adsorption of glyphosate, a widely used herbicide. [Redrawn from McConnel and Hossner (1985)]**

### *Volatilization*

Previously it was assumed that the complete or partial disappearance of pesticides from soil and water was due to their breakdown by micro-organisms. But now it is known to all that pesticides that are lost in air, are again returned to the water during rainfall. Of course, a few pesticides such as PCNB, trifluralin, and methyl bromide are lost from the ecosystem through volatilization.

Some inorganic pesticides that contains mercury ions are highly toxic to biota because mercury gets biomagnified through different micro-organisms at different trophic levels in the aquatic ecosystem before its accumulation in fish and other aquatic organisms. Certain bacteria such as *Pseudomonas vibrio*, *Enterobactor* sp., and *Azotobactor lwoffi* have been found to volatilize various forms of mercury (Figure 12.4). Cold Vapor Atomic Absorption Spectrophotometric study reveals that bacteria can able to remove mercury as much as 86 per cent within 4 hours through volatilization. Therefore, the capacity to volatilize mercury would be of great importance in the abatement of mercury pollution of the aquatic ecosystem in short duration and at a very low cost.

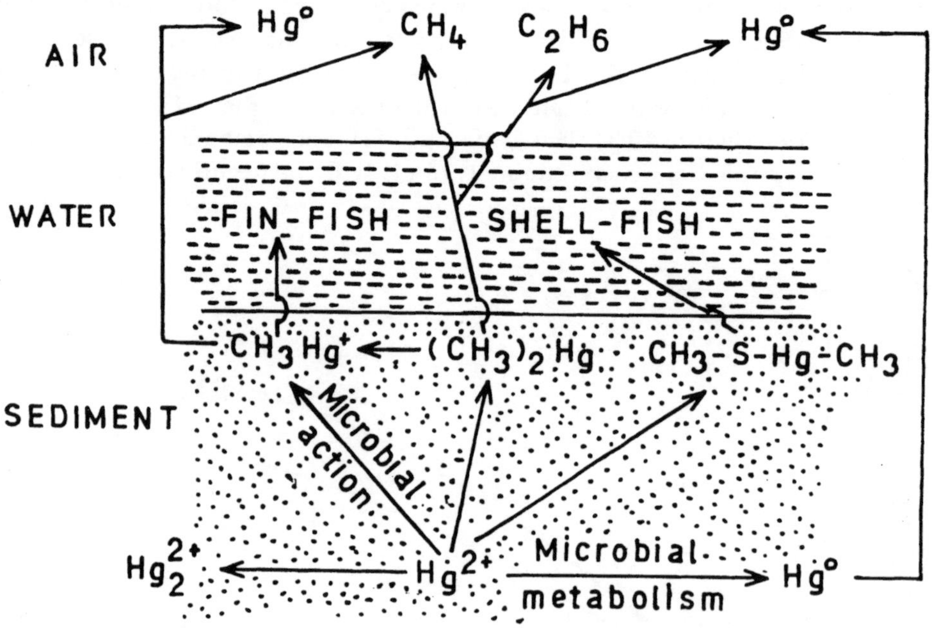


**Fig. 12.4 : Biological cycle for mercury in the environment.**

## *Leaching*

Leaching of pesticides depends on the degree of their solubility and potential for adsorption. Pesticides that are highly soluble in water are leached rapidly. Similarly, highly adsorbed pesticides on soil colloids discourage leaching (Table 12.3). It is important to note that sandy soils have the tendency to move pesticides rapidly than that of the soils having low organic matter and clay. Generally, herbicides seem to be more mobile than either insecticides or fungicides.

**Table 12.3 : Adsorption of Some Herbicides in Soil Colloids**

| Common Name | Trade Name | Degree of adsorption |
|---|---|---|
| Basagram | Bentazon | Weak |
| Amiben | Chloramben | Weak |
| 2, 4-D | 2, 4-D | Moderate |
| Ramrod | Propachlor | Moderate |
| Paraquat | Paraquat | Very strong |
| Treflan | Trifluralin | Very strong |
| Dacthal | DCPA | Very strong |
| Atrex | Atrazin | Strong |
| Lasso | Alachor | Strong |
| Eptan | EPTC | Strong |
| Karmex | Diuron | Strong |
| Dowpon | Dalapon | None |

## 12.8 Pesticide Metabolism by Micro-organisms

Removal of pesticides from water and soil principally depends on the biochemical degradation by micro-organisms. Pesticides interact differently on the micro-organisms. Micro-organisms attack the polar groups of pesticides such as $-NH_2$, $-OH$ etc. and facilitate degradation. Since some pesticides serve as nutrient source of micro-organisms, certain enzymes present in them trigger for degradation.

Several organophosphorus insecticides such as Parathion, are degraded rapidly by micro-organisms. On the other hand, some chlorinated hydrocarbons such as aldrin, DDT, dieldrin, heptachlor etc. undergo slow degradation, causing residue problems (Figure 12.5).

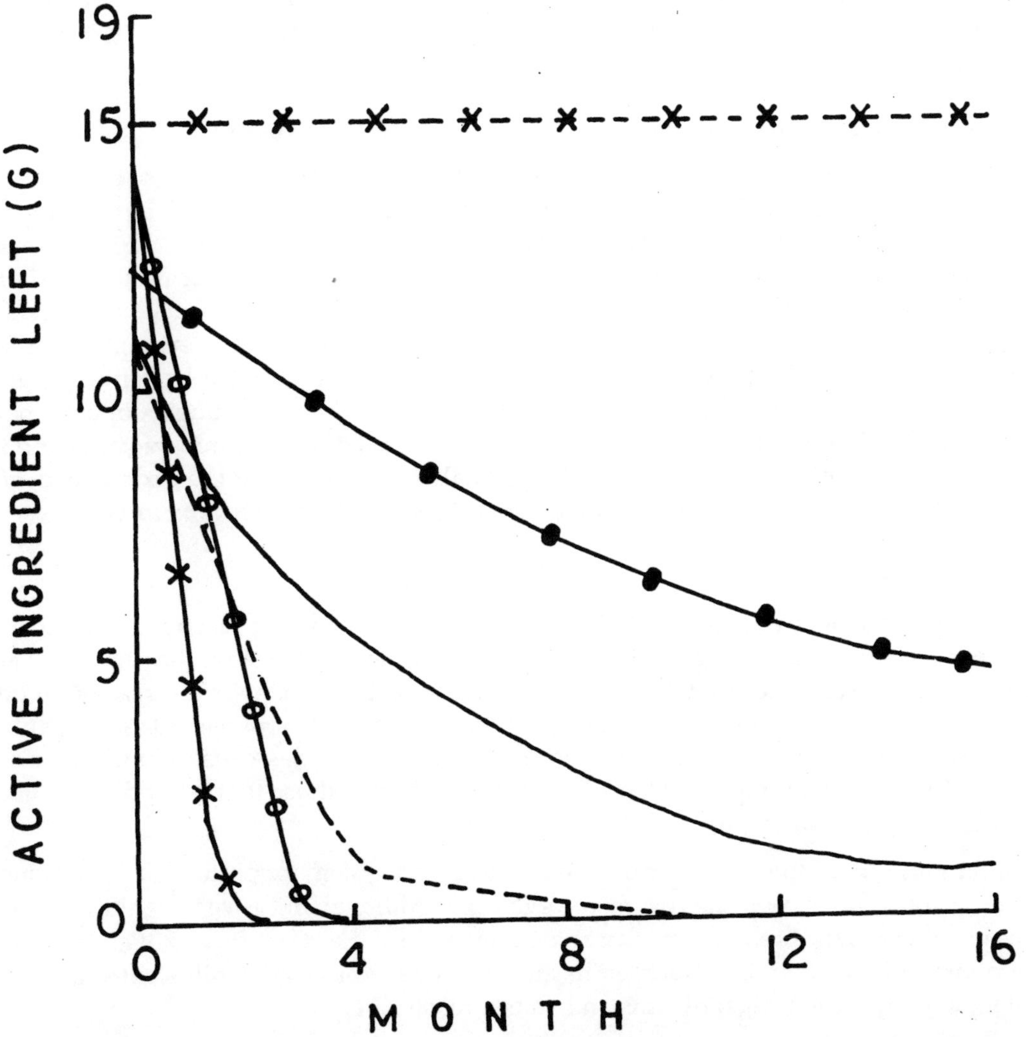


**Fig. 12.5 :** Degradation of some herbicides and insecticides. While atrazine is very slowly degraded, 2,4-D, parathion, and malathion are rapidly broken down. Carbaryl and heptachlor are moderately degraded. —×— 12, 4-D; —o— Parathion, ........ Malathion; ---×--- Atrazine; ——— Carbaryl; —●— Heptachlor.

While carbamates, aliphatic acids, 2, 4-D, and most organic fungicides are very susceptible to microbial degradation, some other pesticides such as atrazine and simazine are degraded by chemical action. Carbamates degrade in nature and forms alcohol and amines in presence of oxygen and water. Amines are converted into ammonia and then to nitrate.

## 12.9 Behavior and Distribution of Chemicals

The behavior and distribution of man-made chemicals in the environment depend on the physical and chemical features of the substances. Thus dissolved materials become components of the hydrosphere and the particulate matter may be transported by rivers. Persistent substances may remain in circulation and then they enter the sedimentary environment of the lakes, ponds and rivers. Many substances undergo transformation to forms that are potentially more hazardous. Thus certain bacteria in aquatic sediments are able to convert inorganic mercury to its more toxic methyl-mercury form.

## 12.10 Interactions of Herbicides and Insecticides in Ecosystem

In recent years, certain combinations of pesticides are advised to farmers for effective control of diseases. Pesticides in combination may interact differently on the soil microbes compared to individual ones. In fish culture ecosystem, a number of chemicals are deposited during rainy season and as a result the situation becomes more complicated. Therefore, it is necessary to know whether contamination of several chemicals have any antagonistic or synergistic effects on different biotic and abiotic factors of soil as well as water in fish survival and growth. For examples, application of fulchloralin (Basalin) alone decrease fungal populations but stimulated bacterial populations. Similarly, malathion and carbendazim (MBC) also increased bacterial and fungal populations (Table 12.4). But combination of MBC + fluchloralin and MBC + malathion favored bacterial and fungal proliferations. Obviously, combination of MBC + malathion neutralize the toxic effects of MBC whereas combination of MBC + fulchloralin exhibited synergistic effect on microbes.

### *Persistence in Soil*

The persistence of pesticides in pond soil is a summation of all the reactions, movements and degradations affecting these chemicals. Marked differences in persistence are the rule. For examples, organophosphate insecticides may last for a few days in soils. 2, 4-D persists in soils for only 2-4 weeks, but DDT and other chlorinated hydrocarbons may persist from 5 to 30 years (Table 12.5). However, rapidly degraded pesticides, herbicides and fungicides cause elimination of toxic metabolites from soils and improves the soil condition whereas those that resist degradation have potential for ecosystem damage.

Continuous run-off the same chemicals on the same ecosystem for prolonged periods increases the rate of degradation of chemicals by micro-organisms. Although this is very significant in relation to ecosystem quality, the breakdown may sometimes reduce the effectiveness of chemicals in fish culture ponds where chemicals such as herbicides, algaecides and mollusicides are used for controlling aquatic weeds, algal blooms and snails, respectively.

**Table 12.4 : Effects of Carbendazim on Soil Micro-organisms in Combination with Malathion and Fluchloralin**

| Treatment | Fungi (X $10^5$) | | | | | | Bacteria (X$10^5$) | | | | | |
|---|---|---|---|---|---|---|---|---|---|---|---|---|
| | Sampling interval (days) | | | | | | Sampling interval (days) | | | | | |
| | 0 | 5 | 15 | 30 | 60 | 120 | 0 | 5 | 15 | 30 | 60 | 120 |
| *Carbendazim* | | | | | | | | | | | | |
| Control-I | 2.1 | 4.0 | 2.4 | 2.8 | 2.3 | 1.4 | 26.0 | 43.0 | 40.5 | 41.0 | 39.4 | 33.0 |
| 50 ppm | 2.1 | 3.6 | 2.8 | 3.0 | 2.6 | 1.7 | 27.0 | 63.0 | 79.8 | 37.1 | 53.9 | 29.8 |
| 250 ppm | 2.0 | 3.3 | 2.1 | 3.0 | 2.6 | 1.6 | 26.0 | 37.7 | 74.3 | 30.0 | 44.2 | 26.2 |
| 500 ppm | 2.0 | 2.1 | 2.0 | 2.4 | 2.6 | 1.7 | 25.0 | 22.3 | 74.1 | 20.2 | 46.2 | 22.1 |
| *Carbendazim + Malathion* | | | | | | | | | | | | |
| Control-I | 2.1 | 2.4 | 2.1 | 2.3 | 2.2 | 2.0 | 26.0 | 44.0 | 44.1 | 34.8 | 31.0 | 28.2 |
| Control-II | 2.2 | 2.8 | 2.2 | 1.7 | 3.0 | 2.2 | 28.0 | 48.7 | 56.2 | 56.8 | 55.5 | 45.5 |
| 50 ppm | 2.0 | 3.0 | 2.8 | 2.1 | 3.6 | 2.4 | 27.0 | 69.1 | 41.1 | 96.1 | 74.5 | 55.8 |
| 250 ppm | 2.1 | 3.0 | 2.6 | 2.1 | 3.3 | 2.3 | 25.6 | 65.2 | 42.5 | 82.5 | 75.2 | 66.0 |
| 500 ppm | 2.0 | 2.7 | 2.4 | 2.2 | 3.1 | 2.3 | 26.0 | 81.2 | 40.0 | 71.0 | 67.3 | 48.0 |
| *Carbendazim + Fluchloralin* | | | | | | | | | | | | |
| Control-I | 2.2 | 2.3 | 2.0 | 2.3 | 2.4 | 2.0 | 25.1 | 27.6 | 73.8 | 110.7 | 90.1 | 84.6 |
| Control-II | 2.1 | 2.1 | 1.5 | 2.5 | 2.1 | 1.8 | 27.1 | 57.8 | 78.8 | 118.8 | 104.6 | 110.8 |
| 50 ppm | 2.0 | 2.0 | 2.1 | 2.0 | 2.0 | 1.8 | 26.0 | 71.4 | 47.1 | 125.0 | 111.2 | 118.8 |
| 250 ppm | 2.1 | 2.1 | 1.8 | 2.1 | 2.2 | 1.6 | 25.1 | 50.8 | 42.6 | 173.7 | 166.1 | 131.4 |
| 500 ppm | 2.0 | 2.0 | 1.6 | 2.1 | 2.2 | 1.6 | 25.0 | 37.7 | 64.6 | 200.0 | 169.4 | 125.4 |

*Source :* Ganesan and Lalithakumari (1990)
Control-I, Soil only; Control-II, Soil + Malathion (50 ppm)/Fluchloralin (100 ppm). Colonies are means of two replicates per g dry soil.

**Table 12.5 : Common Range of Persistence of a Number of Pesticides**

| Pesticide | Persistence |
|---|---|
| DDT, Chlordane, Dieldrin (Chlorinated hydrocarbon insecticides) | 2-5 years |
| Atrazine, Simazine (Triazin herbicides) | 1-2 years |
| Amiben, Dicamba (Benzoic acid herbicides) | 2-12 months |
| Monuron, Diuron, Linuron (Urea herbicides) | 2-10 months |
| 2, 4-D, 2, 4, 5-T (Phenoxy herbicides) | 1-5 months |
| Barban, CIPC (Carbamate herbicides) | 2-8 weeks |
| Carbamate insecticides | 1-8 weeks |
| Diazinon, Malathion, Methyl parathion (Organophosphate insecticides) | 1-12 weeks |
| Paraquat, Gramoxone, Diquat (Bipyridinium herbicides) | 1-5 weeks |

### *Reduction of Pesticide Level in Ecosystem*

In cases where certain chemicals were widely used during fish culture management strategies, adoption of large quantities of easily decomposable organic matter (such as animal manures) can easily reduce pesticide levels. Generally degradation of some non-resistant and resistant pesticides should be encouraged for favouring microbial action and availability of nutrients. For example, malathion is non-resistant and at the same time non-toxic to micro-organisms and their number is increased because malathion is rapidly degraded and phosphorus content in malathion is released and utilized by microbes as nutrient. Thus, pesticide level is reduced to a significant extent. Since aquatic animals are not able to metabolize pesticides in ecosystem, application of organic manures should be recommended to fish farmers. Moreover, microbial as well as photo degradation should also be allowed for favorable condition of ecosystem.

When pesticides are drained into the aquatic ecosystem through surface run-off, different steps such as decomposition, absorption, adsorption, and leaching are involved which affects the dissipation of pesticide molecules (Figure 12.6). It is important to note that the decomposition processes split the pesticide, wheras in transfer processes, the pesticide remains unaltered.

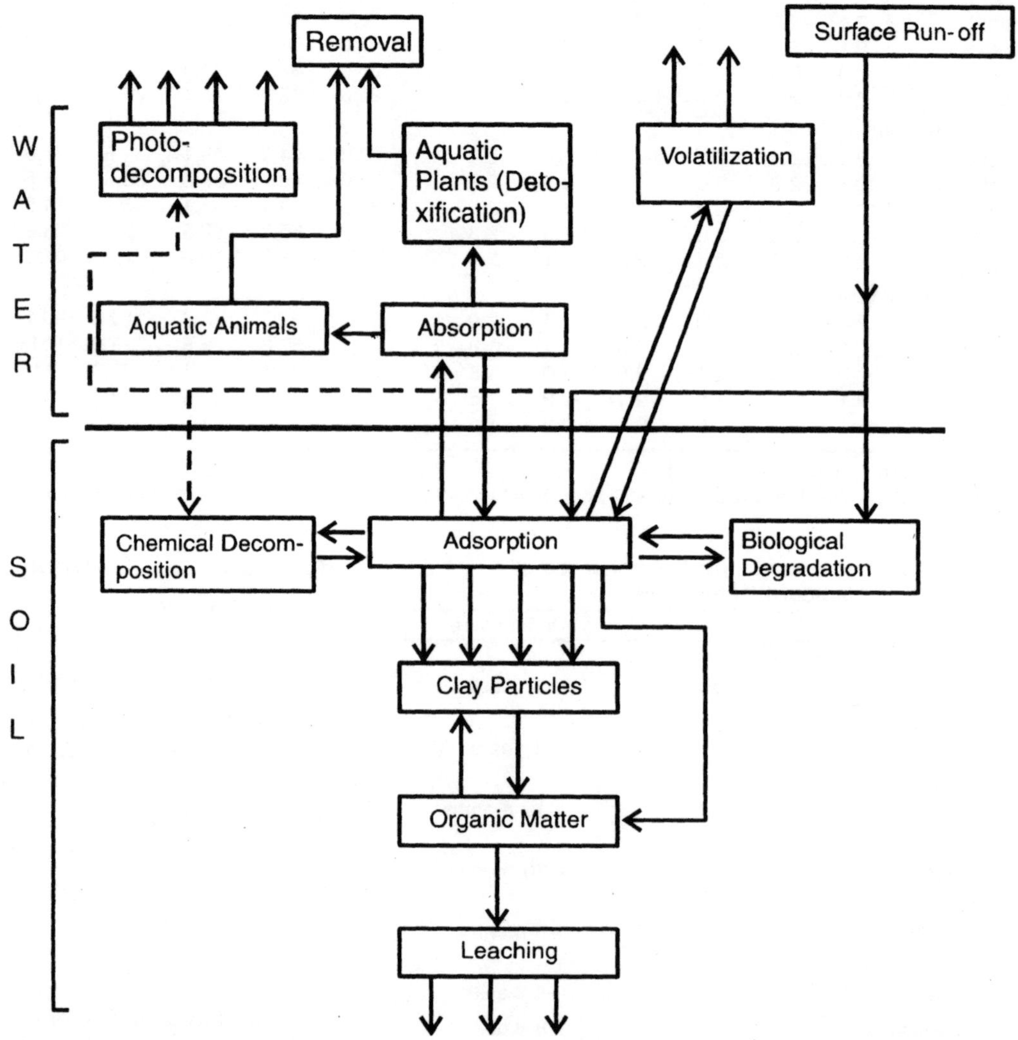


**Fig. 12.6 : Different steps affecting the dissipation of pesticides in an aquatic environment.**

Some floating and submerged plants are known to absorb several inorganic pesticides from water, thus reduce pesticide levels. However, since the uses of pesticides could never be ignored in near future, it is reasonable to conclude that the contamination of water and soil with pesticides would undoubtedly deteriorate much of their fish production potential by disrupting the food chains and nutrient dynamics. Therefore, the effects of chemicals on aquatic ecosystem must be evaluated carefully and precisely before their use is recommended.

### *Effects of Pesticides on Micro-organisms*

Effects of pesticides on micro-organisms are dramatic. Both insecticides and fungicides affect nitrogen fixation and nitrification than do most herbicides. Recent studies have shown that some pesticides can enhance biological nitrogen fixation by reducing the activity of protozoa and bacteria. Fungicides slow down the degradation of soil organic matter. But it is to note that the process of ammonification is benefitted by pesticide use.

Although negative effects of majority of pesticides on micro-organisms are temporary and varied between few days and a few weeks, the number of organisms generally recover. Inspite of this statement, it is necessary to dictate caution in the use of the chemicals and should apply at recommended levels.

### *Contamination of Ecosystem with Inorganic Pesticides*

Water and soil are frequently contaminated with inorganic compounds that are toxic to fish and aquatic organisms to a greater or lesser degree. Mercury, chromium, arsensic, and cadmium are highly toxic; lead, nickle, molibdenum are moderately so; and boron, copper, zinc, and manganese are very low in toxicity.

## 12.11 Source and Accumulation of Chemicals

There are many sources for inorganic compounds that can accumulate in the ecosystem and fish and produce effects on them (Table 12.6). During rainy season, pesticides are drained into the aquatic ecosystems. Lead, nickel and boron are gasoline additives that are released into the atmosphere and ultimately reached into the ecosystem through rain.

Chemical fertilizers are commonly reach the soil either through fertilization or surface run-off. Detergents contain borax that are daily released into the pond, lake and river ecosystems. Limestone and superphosphate contain small amounts of copper, cadmium, manganese, zinc and nickle. Chromium and cadmium are used in electro-plating of metals and the manufacture of batteries, respectively. Arsenic is used in the manufacture of insecticides. However, these elements are found in some organic pesticides and in industrial as well as demostic sewage sludge. Toxic metals are also accumulated into the aquatic environment as a result of weathering of rocks and soil.

Continuous discharge of these metals through various sources in pond water and soil can endanger public health after being incorporated in the food chain. Rapid increase in the use of these inorganic toxic compounds in recent years enhancing the opportunities for contamination.

## 12.12 Aeration in Pesticide-Treated Aquatic Ecosystem

Although metabolites of pesticides and fertilizers are gradually disappeared from soil and water through several processes, it does not indicate that the problems are solved. Question may arises whether disappearance of toxic chemicals indicates their complete or partial degradation or pesticides have been translocated to other parts of ecosystems.

**Table 12.6 : Sources of Some Inorganic Toxic Compounds in Soil and Water and Their Effects on Fish and Aquatic Ecosystem**

| Metal ions | Source | Effect | |
|---|---|---|---|
| | | Fish | Ecosystem |
| Cadmium | Industrial discharge; mining wastes; metal plating; water pipe; batteries | Kidney and brain damage; destruction of testicular tissues and red blood cells; interference with copper and zinc metabolism; metabolic disorders | Influence zooplankton production; decrease phytoplankton and bottom fauna concentration |
| Chromium | Metal plating; chrome-plated metals and refractory brick manufacture | Respiratory rate, feeding, growth, fecundity, and breeding are reduced | Deterioration of water quality variables; plankton concentration drastically reduced |
| Copper | Metal plating; industrial wastes; mining; fly ash; fertilizer | Liver, kidney and brain damage; growth and reproduction are hampered | Deterioration of water and soil quality variables; reduction of bottom fauna concentration |
| Lead | Industrial wastes; combustion of oil and coal | Anemia; kidney and brain damage; metabolic disorders | Plankton population and bottom fauna concentration greatly reduced |
| Mercury | Coal; pesticides; industrial wastes | Anaemia; kidney liver and brain damage; reduced growth, feeding rate and conversion efficiency | Destruction of food chain and food web |
| Zinc | Galvanized steel iron, alloys and batteries; rubber manufact-uring plants; industrial wastes | Anaemia; retarded growth and fecundity rate; kidney and liver damage | Minor changes of abiotic factors of soil and water |
| Arsenic | Pesticides, coal, oil, detergents, groundwaters | Damage of liver, kidney, intestine, and gills; reduce survival and growth | Reduction in the concentration of plankton and bottom fauna; destru-ction of food chain and food web |

The technique of artifical aeration, though expensive for poor fish farmers, is widely accepted to enhance the water quality criteria for survival of aquatic life. Aeration of a carbamate-treated water at the rate of 2.5 hour per day showed about 34 per cent rise in the concentration of dissolved oxygen with a corresponding decrease in the values of ammonia nitrogen (63.92 per cent), nitrite nitrogen (44.55 per cent) and nitrate nitorgen (51.56 per cent).

## 12.13 Effects of Chemicals on Freshwater Fish*

Use of pesticides for crop production cause deterioration of aquatic environment. Degradation of ecosystem is characterized by toxic impact on fish, fish food organisms, water-loving plants and disruption of food chain. Due to bio-accumulative (lipophilic) nature of pesticides, they are deposited in the fatty tissues of fishes and exhibits a number of symptoms such as alterations of behavior, histopathological damage to different organs (such as liver, kidney, gills etc.), disturbance of population dynamics of fish, change of feeding habit and reproductive potential, reduced growth and fecundity rates of fish, increased fish mortality and drastic reduction in the concentration of plankton and bottom fauna. While there exists a wide range of variations in the detrimental effects on fish exposed to various toxicants, it is very difficult to make a generalized conclusion on these aspects.

---

* For review, see Mc-Kim *et. al.* (1976), Murthy (1986), Metelev *et. al.* (1983), Motsumura (1985), Dhaliwal and Singh (1993).

## 12.14 Accumulation of Metals in Fish*

Metals are accumulated in aquatic environment, gradually deposited within the body of fish (Table 12.7) to a greater or lesser extent through successive stages (water-soil-bottom organisms-plankton-fish-human, Figure 12.7). Once metals are entered into the cycle, they are accumulated within the fish (particularly in flesh) to toxic levels. This is due to the fact that fish prey on organisms that contain greater concentrations of metals. This situation becomes more critical because contaminated fish when consumed by humans, obviously results in pathological manifestations.

**Table 12.7 : Concentrations of Heavy Metals (μg/g wet-weight) in Some Freshwater Fish**

| Fish | Metals | | | | | |
|---|---|---|---|---|---|---|
| | Arsenic | Cadmium | Copper | Mercury | Lead | Zinc |
| Common carp | 0.09 | 0.02 | 1.0 | 0.06 | 0.01 | 77.98 |
| Brown trout | 0.03 | 0.01 | 1.86 | 0.14 | 0.01 | 27.67 |
| Largemouth bass | 0.46 | 0.02 | 0.41 | 0.03 | 0.18 | 12.51 |
| Channel catfish | 0.08 | 0.01 | 0.37 | 0.02 | 0.05 | 16.60 |
| Lake trout | 0.56 | 0.01 | 1.39 | 0.15 | 0.02 | 12.21 |
| Tilapia | 0.14 | 0.06 | 23.1 | 0.03 | 0.92 | 27.33 |
| *Labeo rohita** | 0.17 | 0.09 | 13.00 | 0.08 | 1.20 | 16.70 |
| *Cirrhinus mirgala** | 0.20 | 0.13 | 10.40 | 1.00 | 1.13 | 15.00 |
| *Oreochromis mossambicus** | 0.10 | 0.07 | 2.60 | 0.07 | 0.75 | 10.25 |
| *Clarias batrachus** | 0.50 | 0.15 | 1.40 | 1.10 | 0.35 | 4.70 |

*Source* : Schmitt and Brumbaugh (1990)

* Compiled by the author from various published literature

Although metal concentrations in fish increased significantly at different freshwater ecosystems influenced by industrial discharges, regulatory measures adopted by several countries have successfully reduced the influx of different metals to the aquatic environment. Lower concentrations of metals in fish contradicts the usual trend of higher concentrations noted for waters. However, possible explanations for this apparent contradiction include : the dissolved component may contain metals that are not available for uptake by fish; concentrations of metals in fish may not reflect concentrations in water; or the dissolved species of metal elements may not include all available chemical forms.

## 12.15 Accumulation of Metals in Ecosystem

Accumulation of metals such as iron, copper, zinc and chromium in water, soil and aquatic plants are presented in Table 12.8. In general, the values of iron were found to be higher than the values of copper, zinc and chromium in all the samples studied. The concentrations were lower for chromium in all the samples.

* For review on the subject, see Schmitt and Brumbaugh (1990).

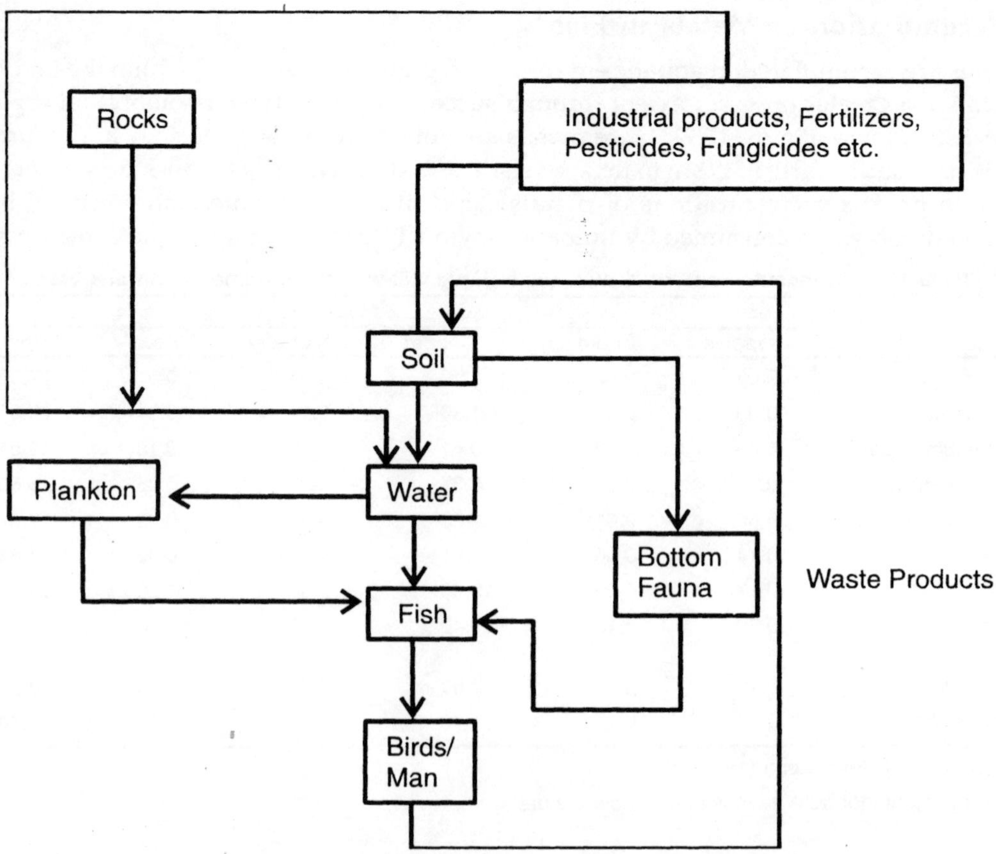


**Fig. 12.7 : Sources and cycle of heavy metals in water, soil, and aquatic ecosystem.**

**Table 12.8 : Estimation of Some Metals in Soil, Water and Aquatic Weeds in Sewage-Fed Ponds**

| Metals | Water (mg/l) | Soil (mg/g) | Aquatic weed (mg/g) (*Wolffia arrhiza*) |
|---|---|---|---|
| Copper | 0.034 | 0.049 | 0.035 |
| Iron | 0.124 | 12.72 | 0.376 |
| Zinc | 0.041 | 0.064 | 0.03 |
| Chromium | 0.016 | 0.031 | 0.018 |

*Source* : Pandey *et al* (1995)

Accumulation of metals in soil and water vary from one place to another and different geographical conditions of ecosystems fluctuate the concentration of metals. For example, the concentration of different metals in sewage-fed ponds located in West Bengal (Table 12.8) was lower than in the river water located in Bihar (Table 12.9). Accumulation of metals in water and soil also vary in their solubility and availability in water. The lower the solubility of metal, the greater is the availability in soil and is dependent upon the fluctuations of pH, temperature and alkalinity of water.

**Table 12.9 : Estimation of Some Metals in Soil and Water in Different Sites of the River Subarnarekha, Bihar**

| Metals | Site-A (Ratmuhan) | | Site-B (Dahigora) | | Site-C (Amainagar) | |
|---|---|---|---|---|---|---|
| | Water (mg/l) | Soil (mg/g) | Water (mg/l) | Soil (mg/g) | Water (mg/l) | Soil (mg/g) |
| Copper | 0.127 | 0.177 | 0.182 | 0.231 | 0.147 | 0.050 |
| Zinc | 0.205 | 0.080 | 0.292 | 0.194 | 0.242 | 0.274 |
| Iron | 0.475 | 21.000 | 2.025 | 32.500 | 0.950 | 15.000 |
| Chromium | 0.150 | 0.064 | 0.250 | 0.071 | 0.200 | 0.026 |

*Source* : Munshi and Singh (1989)

Some aquatic organisms are able to accumulate heavy metals (such as lead) is large quantities without apparant harm, while others are sensitive to metals and eliminated from the body of organisms. Among different soil fauna, earthworms are one of the most important animal which are responsible for mechanical mixing of the soil and thus maintain soil physical features such as aeration and mineral turnover. Different species of earthworms such as *Allobophora* sp., *Chlorotica* sp., *Lumbricus* sp., *Pheretima* spp. etc. were reported to accumulate heavy metals in their bodies.

### *Metallic Cyanides*

The cyanide compounds especially sodium and potassium cyanides (also called metallic cyanides) are used for the extraction of silver and gold as well as for electroplating. These compounds are, however, terrible poisonous, causing almost instantaneous death in tiny doses.

Cyanide compounds are drained and accumulated into the rivers and streams from contaminated dams at gold mines killed thousands of fish, and also fears of "a real graveyard" on the bottom of the river. These compounds persist in ground waters and soils, making it necessary to test crops grown nearby. Therefore, water samples should be collected at regular intervals to assess the extent of damage caused by cyanide spill.

On January 30, 2000, tonnes of dead and rotting fish were pulled out of the Tisza and the Danube rivers in Yugoslavia and Rumania and consequently, banned the sale of most freshwater fish. Cyanide spill has, in practical terms, eradicated most or all aquatic life from a stretch of upto 400 kilometers of these rivers. Of course, it is not known the real extent of the damage until the evaluation can be carried out.

The World Health Organization, however, expressed concern that heavy metals such as mercury, lead and cadmium might have escaped into the water along with the cyanide that could pose a far greater threat in the long run.

## 12.16 Seasonal Fluctuations of Inorganic Toxicants in Ecosystem

Generally toxic chemicals are applied to agricultural fields almost all the year round and their concentrations are extremely variable in different seasons of the year. For example, maximum and minimum values of different heavy metals in some freshwater ponds (area ranged from 517.5 to 88,093.87 square meters) have been recorded during summer and rainy seasons, respectively (Table 12.10). The values of all the heavy metals were below the limits prescribed by different agencies. This study has several practical significance during fish cultrue operations.

**Table 12.10 : Seasonal Fluctuations in the Concentrations of Heavy Metals (mg/l) in Motijheel, Surajkund and Ranital**

| Name of the water bodies and seasons | Heavy metals | | | | | | | |
|---|---|---|---|---|---|---|---|---|
| | Cu | Zn | Ni | Co | Pb | Mn | Cr | Cd |
| 1. **Motijheel** | | | | | | | | |
| (a) Summer | 0.024 | 0.065 | 0.001 | 0.003 | 0.003 | 0.009 | 0.020 | 0.010 |
| (b) Winter | 0.029 | 0.082 | 0.001 | 0.004 | 0.005 | 0.011 | 0.022 | 0.014 |
| (c) Rainy | 0.034 | 0.120 | 0.003 | 0.009 | 0.009 | 0.016 | 0.038 | 0.019 |
| 2. **Surajkund** | | | | | | | | |
| (a) Summer | 0.017 | 0.079 | 0.001 | 0.003 | 0.002 | 0.010 | 0.030 | 0.012 |
| (b) Winter | 0.019 | 0.090 | 0.002 | 0.005 | 0.003 | 0.012 | 0.039 | 0.014 |
| (c) Rainy | 0.026 | 0.116 | 0.004 | 0.009 | 0.009 | 0.016 | 0.048 | 0.018 |
| 3. **Ranital** | | | | | | | | |
| (a) Summer | 0.018 | 0.065 | 0.001 | 0.004 | 0.003 | 0.009 | 0.028 | 0.009 |
| (b) Winter | 0.022 | 0.070 | 0.001 | 0.007 | 0.004 | 0.012 | 0.032 | 0.011 |
| (c) Rainy | 0.028 | 0.116 | 0.003 | 0.009 | 0.009 | 0.016 | 0.042 | 0.015 |

*Source :* Kaushik *et.al.* (1997)

## 12.17 Relationship Between Heavy Metals and Ecosystem

Relationship between metals and pond ecosystem is also highly variable. Thus copper concentration in water was positively correlated with the concentration in soil and aquatic weeds and negatively correlated with fish. Iron concentration in water was found to be negatively correlated between metal concentration in soil but positively correlated with aquatic weeds. Zinc in water was negatively correlated with the concentration of soil but positively correlated with aquatic weeds and fish (Table 12.11).

**Table 12.11 : Correlation Coefficient Values Between Metal Concentration in Water and Accumulation in Soil, Weed and Fish**

| Factor | Metals | | | |
|---|---|---|---|---|
| | Copper | Iron | Zinc | Chromium |
| Water – Soil | 0.6109 | -0.2309 | -0.2472 | 0.1945 |
| Water – Weed | 0.2754 | 0.0061 | 0.2212 | -0.8122 |
| Water – Fish | -0.1083 | 0.2825 | 0.4584 | -0.0454 |

*Source :* Pandey *et.al.* (1995)

## 12.18 Mechanism of Toxicity of Metal Ions and their Detoxification

The toxicity of metal ions is due to one of the following mechanisms :

1. Displacement of the essential biological active sites of biomolecules.
2. Modification of the structure of biomolecules.
3. Blocking of the essential biological active sites of biomolecules.

The third mechanism is illustrated in the binding of mercury, arsenic and cadmium to – SH groups of cysteine residues in enzymes (Figure 12.8). The binding of metal ions to – SH groups inhibit the enzymatic activities.

The first mechanism may be explained by thermodynamic factor of the metal. Cadmium displaces zinc in zinc-activated enzymes since the thermodynamic factor (stability constant) of the cadmium system is higher that the zinc system. Cadmium forms more stronger bond with the enzymes than zinc and thus cadmium blocks the enzymatic activity. Enzymes are adenosine triphosphate, alcohol dehydrogenase, amylase and carbonic anhydrase.

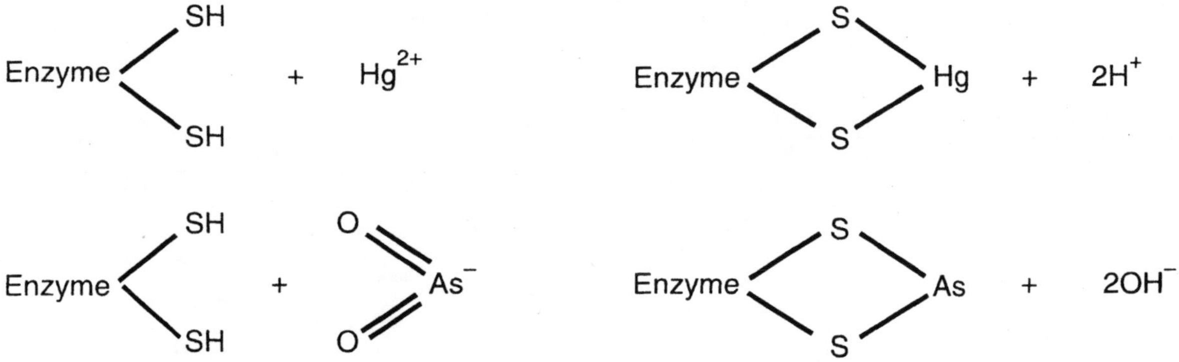


**Fig. 12.8 : Blocking of the essential biological active sites of molecules.**

The biomolecules assume certain specific confirmations for biological activation. The binding sites of mercury and cadmium, for example, modifies the confirmational structures of biomolecules such as proteins and polynucleotides. Thus disruption of structures of biomolecules can cause deformation and severe abnormalities in fishes.

Detoxification of lead is achieved by administration of calcium complex of ethylene diamine tetracetic acid (EDTA). Since the stability constant of lead complex of EDTA is higher than that of calcium complex of EDTA, the displacement of calcium by lead occures and the resulting lead chelate is excreted from the body of fish.

The toxic metal ions in pond and waste waters can be removed using cation exchange resins

$$2R - SO_3^{H+} + Ca^{+2} \longrightarrow (R - SO_3)_2\ Cd + 2H^+$$

$$2R - SO_3^{H+} + Hg^{+2} \longrightarrow (R - SO_3)_2Hg + 2H^+$$

$(R - SO_3H^+)$. The used resin can be regenerated by treatment with dilute hydrochloric acid and the regenerated resin can be further utilized.

$$(R - SO_3)_2Cd + 2HCI \longrightarrow 2R - SO_3^{H+} + CdCl_2$$

$$(R - SO_3)_2Hg + 2HCI \longrightarrow 2R - SO_3^{H+} + HgCl_2$$

The polymer-anchored chelating ligands have also been used for the removal of toxic metal ions from heavy metal polluted waters.

## *Methyl-Mercury Elimination from Fish*

It has been found that several anti-toxidants and vitamins play a key role in eliminating the toxic inorganic elements from fish body. Methyl-mercury decreases glutathione (anti-toxidant tripeptide) and vitamin B-complex levels in fish leading to histopathological and physiological changes. However, exogenous application of glutathione in suitable dose (20 mg/l) and vitamin

B-complex forte (88.4 mg/l) has been found to deplete methyl-mercury level in fish and consequently, mercury is significantly eliminated from the body (Table 12.12). It is also to note that combination of glutathione plus vitamin B-complex forte rapidly eliminated methyl-mercury than individual applications. Such therphy seems to be beneficial in detoxifying the methly-mercury poisoned fishes and making them suitable for human consumption. Of course, this technique has not been conducted at commercial level.

**Table 12.12 : Percentage of Mercury Alterations During Therapeutic Study as Compared to Withdrawal Group**

| Tissue | Group | Mercury content (μg/g) | Percentage of mercury alteration | |
|---|---|---|---|---|
| | | | Mercury remaining | Mercury removed |
| Brain | W | 45.45 | 100.00 | |
| | G | 17.85 | 39.27 | 60.73 |
| | B | 10.71 | 23.57 | 76.43 |
| | C | 13.64 | 30.00 | 70.00 |
| Intestine | W | 31.08 | 100.00 | |
| | G | 23.47 | 75.00 | 25 |
| | B | 18.25 | 56.72 | 41.28 |
| | C | 13.04 | 41.96 | 56.04 |
| Muscles | W | 6.15 | 100.00 | |
| | G | 4.50 | 73.18 | 26.81 |
| | B | 4.20 | 68.20 | 31.80 |
| | C | 3.62 | 51.50 | 48.50 |
| Gills | W | 39.65 | 100.00 | |
| | G | 4.28 | 10.80 | 89.20 |
| | B | 3.60 | 9.00 | 90.99 |
| | C | 0.61 | 2.52 | 97.47 |
| Eyes | W | 25.00 | 100.00 | |
| | G | 6.81 | 27.24 | 72.76 |
| | B | 5.36 | 21.14 | 78.58 |
| | C | 4.00 | 16.80 | 84.00 |

*Source :* Vaidehi et al (1997)
W, Withdrawal; G, Glutathione; B, Vitamin B-complex; C, Glutathione + Vitamin B-complex

Some heavy metals such as lead are found to be detoxified in freshwaters having higher alkalinity levels (Table 12.13). The increased availability of calcium in hard water may reduce the toxicity of metals and thus provide a protective effect of the aquatic animals including fish.

### *Detoxification of Insecticides*

Similar to detoxification of metals in fish, administration of vitamins in pesticide-intoxicated fishes cured anaemia, hyperglycemia, reduced serum urea level, and restored enzymatic activity and thus controls the toxicity of insecticides. For example, intra-muscular injection of vitamin $B_{12}$ in methyl parathion-intoxicated fish, *Channa punctatus*, at the dose of 0.25 ml on each alternate day for two weeks increased red blood cell and white blood cell counts, reduced serum urea and cholesterol levels, acid phosphatase and increased alkaline phosphatase (Figures 12.9, 12.10 and 12.11).

**Table 12.13 : Concentration of Lead (mg/g wet weight) in Gill and Muscle of *Macrobrachium malcolmsonii* Exposed to Sublethal Concentrations of Lead in Hard and Soft Waters**

| Hardness | Sublethal concentration of lead (mg/l) | Concentration of lead in tissue (μg/g) | |
|---|---|---|---|
| | | Gill | Muscle |
| 255 (Hard) | 9.15 | 4.0 | 2.6 |
| | 12.25 | 2.0 | 9.6 |
| | 22.87 | 165 | 12.0 |
| 100 (Soft) | 0.98 | 9.3 | 3.2 |
| | 1.60 | 93.1 | 29.0 |
| | 2.45 | 173.9 | 17.8 |

*Source* : Kabila *et.al.* (1996)

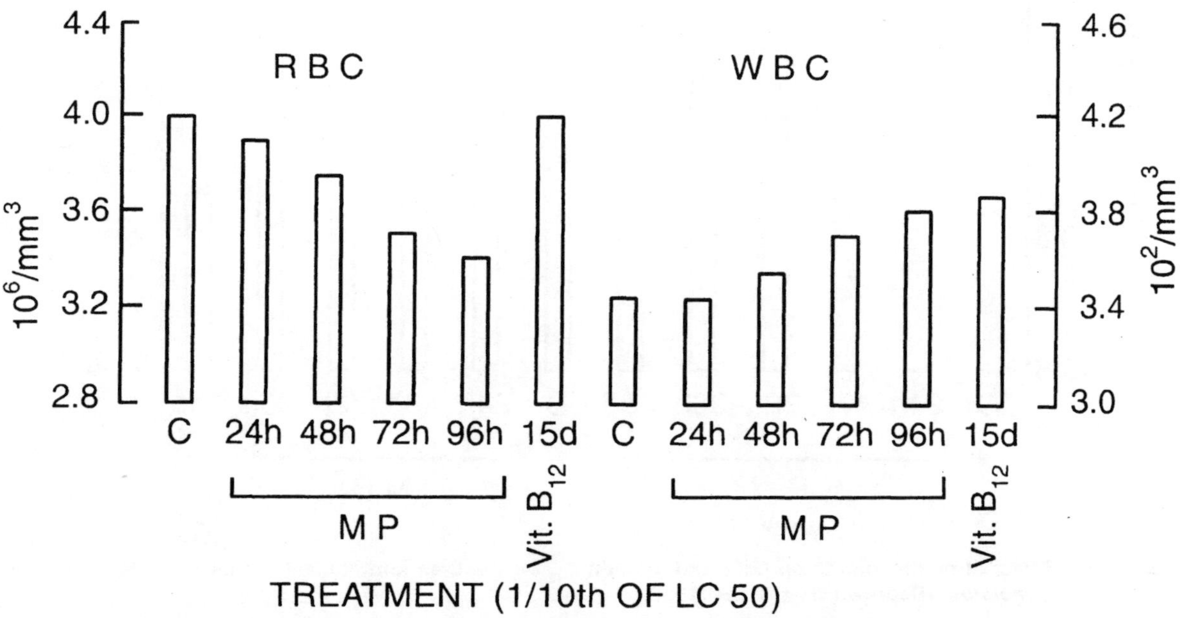


**Fig. 12.9 : Effect of methyl parathion (MP) and Vitamin $B_{12}$ on the Red Blood Corpuscles (RBC) and White Blood Corpuscles (WBC) in fish *Channa punctatus*. [Redrawn from Goel (1996)]**

## 12.19 Effects of Chemicals on Organisms

Since different types of chemicals are used to kill disease-producing organisms, it is not surprising that many of them are toxic to specific soil organisms in ponds. Some herbicides such as disodium salt of 2, 4-D, simazine, atrazine etc. when used in combination with chemical fertilizers and manures, have profound influence on the activities of soil organisms and the physico-chemical properties of water to some extent. For examples, many pesticides have only mild depressing effects on earthworm populations. Exceptions are most of the carbamates which are toxic to earthworms. Application of above-mentioned herbicides at sublethal levels increased the bacterial populations and bottom organisms such as *Chironomous* sp., *Lymnaea* sp. *Branchiura* sp., *Planorbis* sp., and larvae of Odonata.

The influence of combinations of lime, nitrogen, oil cake and herbicide (basalin) was found to be dramatic. Though the toxicity of basalin alone was found to be removed by use of lime yet liming increased the efficiency when fertilizers were mixed with basalin. Thus basalin exhibited synergistic influence on the total combinations of different nutrient inputs for the growth and existence of bottom fauna in fish ponds (Figure 12.12).

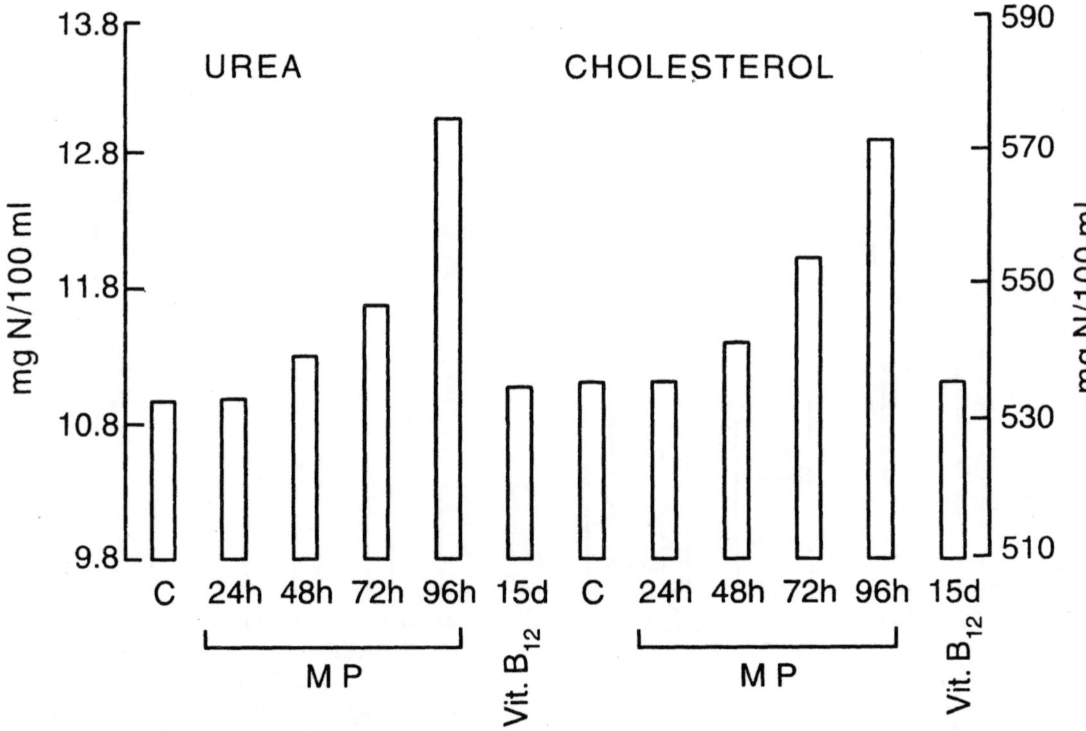


**Fig. 12.10 : Effect of methyl parathion (MP) and Vitamin $B_{12}$ on the urea and cholesterol levels in fish *Channa punctatus.* [Redrawn from Goel (1996)]**

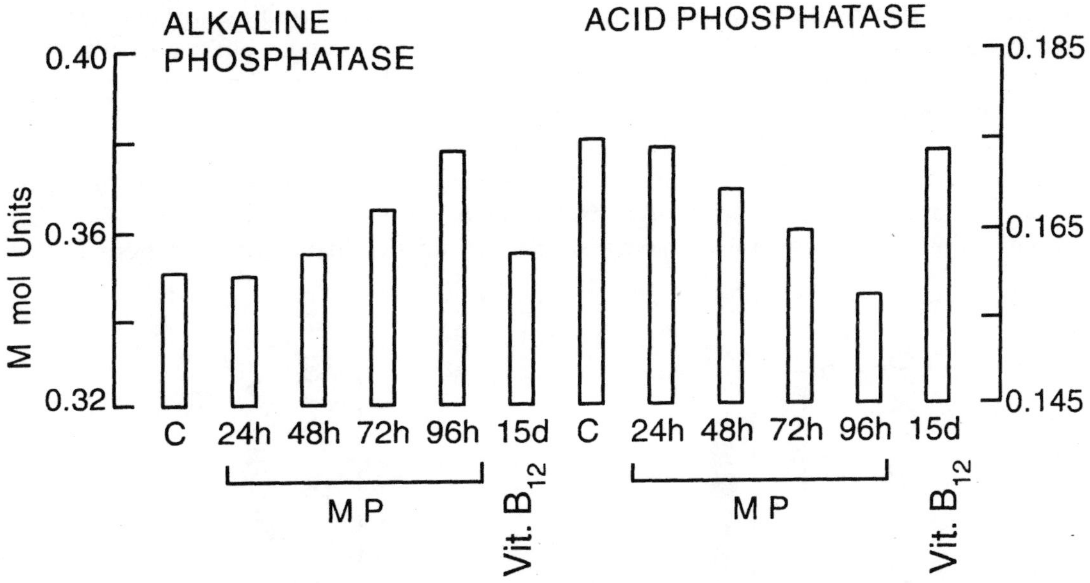


**Fig. 12.11 :** **Effect of methyl parathion (MP) and Vitamin $B_{12}$ on the alkaline and acid phosphatases in fish *Channa punctatus*. [Redrawn from Goel (1996)]**

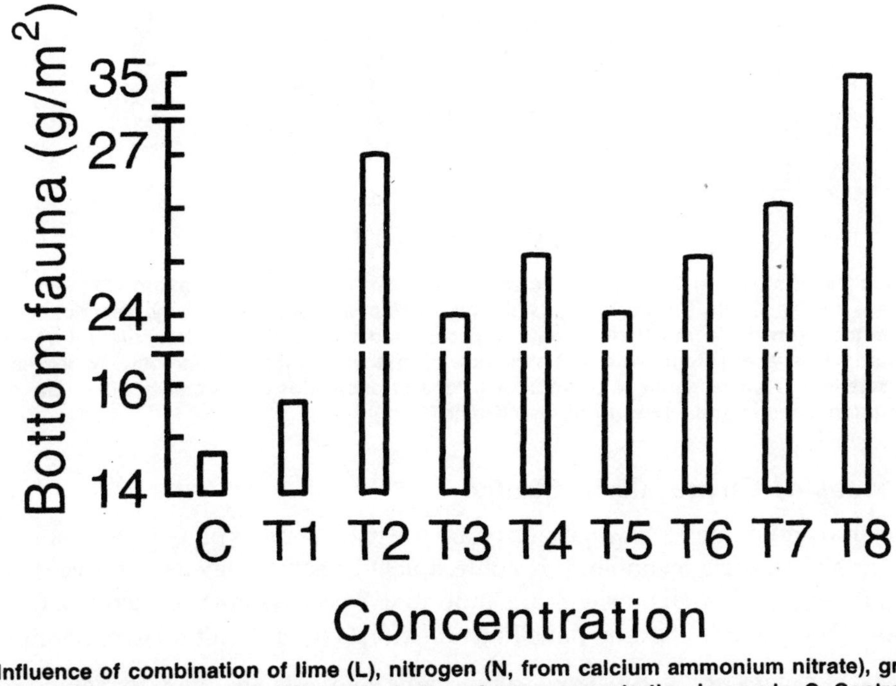


**Fig. 12.12 :** **Influence of combination of lime (L), nitrogen (N, from calcium ammonium nitrate), groundnut cake (G), and herbicide basalin (B) on the bottom fauna concentration in ponds. C, Control; T1, B; T2, B + L; T3, B + N; T4, B + G; T5, B + N + G; T6, B + N + L; T7, B + G + L; T8, B + N + G + L. Rates of basalin, groundnut cake, and lime (calcum oxide) are 1.0, 550, and 400 kg/ha/year, respectively. Rate of nitrogen is in terms of nutrient only. Experiments were conducted in earthen ponds over a period of one year. Bottom fauna was collected one month interval. Values are means of 13 observation. Species inlcuded *Tubifex, Chironomus* larva, dragon-fly nymph, *Planorbis,* and *Vipiparous.* Routine applications of these inputs to fish culture ponds are significant to the growth and development of natural food organisms. [Redrawn from Pramanik and Sarkar (1995)]**

The drainage of pesticides into the aquatic ecosystem has drastic and deleterious effects on certain trophic levels. These effects become more pronounced particularly when pesticide fluxes occur into the water intermittently. Since chemicals persist in soil for several months or years, it is enough to exhibit drastic effects on bottom organisms. For example, twelve exposures of methyl parathion at 7-day interval for 90 days decreased concentrations of plankton and bottom fauna (Figures 12.13 and 12.14). However, although rather extensive bibliographies give the impression that there is a vast amount of literature on the effect of chemicals on organisms, rield and laboratory studies indicate that it is not possible to make any scientifically justifiable generalizations on these aspects.

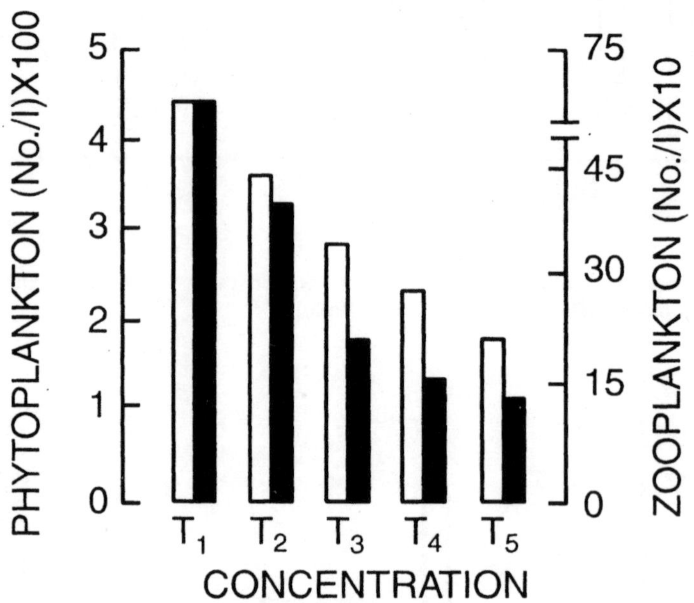


**Fig. 12.13 : Influence of methyl parathion on the phytoplankton (PP) and zooplankton (ZP). T1, Control; T2 to T5 are 0.1500, 0.3000, 0.4500, and 0.6100 mg/l of the pesticide, respectively. Test concentrations were selected from the acute toxicity data and exposed twelve times at an interval of seven days to experimental ecosystem. Note that exposure of methyl parathion progressively decreased plankton populations. Regular spillage of sublethal dose of pesticides into water is obviously detrimental to plankton populations. [From Pal and Konar (1990)]**

## 12.20 Hazardous of Chemicals in Sludge

The industrial and domestic sewage sludge are the main sources of toxic inorganic and organic chemicals when fish culture technology is contempleted because sewage sludge contains large quantities of nutrients. Although nutrients are important for ecosystem productivity, it also contains a variety of toxic compounds that can have harmful effects on fish culture ecosystem.

In Table 12.14 ranges of levels of several inorganic elements are given for sewage sludge and municipal wastes. Concentrations of heavy metals were much higher in these two types of waste materials than in cow manure. Table also shows that the concentrations of these elements in cities were more higher than from village.

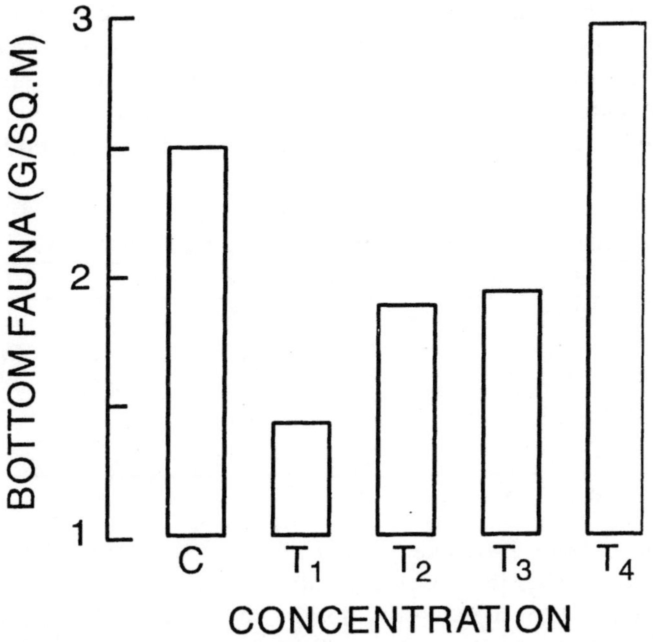


**Fig. 12.14 : Effect of methyl parathion on the bottom fauna concentration. Species includes Chironomid and Oligochaete. C, Control; T1, 0.15; T2, 0.30; T3, 0.45; T4, 0.61 mg/l. [Drawn from data as proposed by Pal and Konar (1990)]**

**Table 12.14 : Comparative Concentrations of Some Toxic Elements in Urban City (A), Village (B), Cow Manure (C), Sewage Sluge (D), and Solid Wastes (E)**

| Element | A (mg/kg) | B (mg/kg)* | | C (mg/kg) | D (mg/l)** | E (mg/l)*** (total) |
|---|---|---|---|---|---|---|
| | a | a | b | a | c | d |
| Iron | ND | ND | ND | ND | 2,360 | 5,526–9,253 |
| Cadmium | 7 | 1-2 | Trace | 1 | 11 | 1-2 |
| Chromium | 169 | 19-73 | 25.4-378.8 | 56 | 248 | 19-73 |
| Copper | 820 | 63-134 | 36.7-246.7 | 62 | 583 | 63-134 |
| Mercury | 11 | 4-18 | ND | 0.2 | ND | ND |
| Manganese | 128 | 250-680 | 23.7-444.0 | 286 | 296 | 250-680 |
| Nickle | 36 | 12-24 | 25.2-50.8 | 29 | 4 | 12-24 |
| Lead | 136 | 66-187 | 12.8-181.2 | 16 | 137 | 66-187 |
| Zinc | 560 | 152-359 | 102-760 | 70 | 1,326 | 152-359 |

*Source :* a, Selected data from Furr (1976)
b, Selected data from Mukherjee and Roy (1989)
c, Selected data from Jeyabaskaran and Sreeramulu (1998)
d, Selected data from Jeevan Rao and Shantaran (1996)
ND, Not detectable; *, Muncipal sewage; ** sludge from Coimbatore city (India);
***, Solid Wastes from Hyderabad city (India)

Presence of these elements in large quantities in sludge dictate caution in the application of sludge to soil and water. The effects of application of such metals on bottom organisms are severe. For example, the concentration of some heavy metals in the body of earthworm and in plankton (Table 12.15) clearly indicates that the concentration of different heavy metals is highly significant. Thus, it can be expected that further concentration to take place in the body of fish because fish consume these animals as food.

**Table 12.15 : Concentration of Some Heavy Metals in Phytoplankton, Zooplankton and Earthworm**

| Metal | Phytoplankton (mg/l) | Zooplankton (mg/l) | Earthworm (mg/kg) |
|---|---|---|---|
| Cadmium | 0.11 | 0.40 | 4.8 |
| Zinc | 4.70 | 10.80 | 228.0 |
| Copper | 0.61 | 1.18 | 13.0 |
| Lead | 0.36 | 0.33 | 17.0 |
| Mercury | 0.011 | 0.01 | 45.0 |

*Source* : Beyer *et al* (1982), Bhattacharya (1986)

Urban solid waste contains more heavy metals than either in cow manure or in village waste, hence addition of these wastes increases the metal load in soils. To avoid the heavy metal concentrations of wastes and to improve the quality of wastes as fertilizers, separate collection system for inorganic and organic wastes should be introduced.

Raw sewage is known to create health hazards and therefore, sewage farming is no longer considered as an efficient method for sewage disposal. Though problems of pollution can be alleviated by treatment processes (mechanical, chemical and biological), the cost of treatment is very high which obviously limits fish production to a great extent. It should be added as a note that, in many densely populated regions, sewage finds its way into waters, affecting productivity and fish production.

## 12.21 Fate of Inorganic Compounds in Soil

Owing to build-up of some important heavy metals in soil from application of sewage sludge in agricutural lands, has stimulated research on the fate of these metals in soil and water. Attention has been given to copper, zinc, nickel, cadmium, and lead because these elements are found in appreciable quantities. Studies have shown that these elements are bound in soil particles and hence are not leached from the soil. Although many acid soils where pH varies between 4.0 and 6.5 are able to percolate these metals along with water, application of lime can prevent leaching of these elements into ground waters.

### *Association of Metals with Soil*

It has been found that metals are associated with soil particles in the following four ways :

1. Association of metals takes place with carbonates and oxides of manganese and iron. These forms are less available to plants if the soils are not highly acidic.
2. Insoluble residual compounds (such as sulphides) that are less available to plants.
3. Either adsorbed or exchangeable forms that are available for plant uptake.

4. Elements that are bound with soil organic matter in the sludge. A high proportion of zinc and copper are found in this form, lead and chromium are not highly attracted.

### *Detoxification*

Some speçies of green algae such as *Scenedesmus* sp., *Nostoc* sp., *Anabaena* sp. etc. are capable of detoxifying pesticides such as dimethoate, thiometon, simzaine, and atrazine by adsorption and accumulation after biodegradation and/or photodegradation during cell metabolism thereby removing the pollutants from polluted waters.

Different strains of bacteria which have been isolated from industrial wastes are known to degrade toxic compounds. Reports have shown that *Pseudomonas pseudoalcaligenes,* for example, harbours a plasmid, an extra-chromosomal deoxyribonucleic acid which codes for the enzyme involved in detoxifying the pollutants.

Immobilization of the bacteria at the solid-liquid interface plays a key role in detoxifying the pollutants. Bacteria are immobilized on calcium alginate more efficiently and degraded aromatic substances such as benzoate and nitro-benzene. During this degradation, the nitro compounds are oxidized to nitrates. Aeration of the total suspended solids of effluents through agitation has been found to influence trans-conjugant formation of bacteria possibly by sharing the bacterial cell walls and permitting gene transfer. Further research *in vivo* genetic engineering studies on the bacteria to improve its efficiency is badly needed.

### *Lead*

Soil is contaminated with lead during pesticide application that is tied up in the soil as insoluble carbonates and sulphides and in combination with iron, aluminium, and manganese oxides.

The main sources of lead contamination of aquatic ecosystems are the industrial discharges, atmospheric fallout, and sewage effluents. Similar to other metals, the use of sewage and sludge as fertilizer might increase the level of lead in soil.

Accumulation of lead in algae exhibits more than 50 per cent reduction in $^{14}CO_2$ fixation that causes drastic decline in the rate of photosynthesis.

### *Boron*

Boron is accumulated in soil due to excessive application of chemical fertilizers. Though boron can be adsorbed by clays and organic matter, but it is available to water at low pH. It is highly soluble in water and soil. It is not highly toxic as in the case of other elements.

### *Arsenic*

Although the use of pesticides in agricultural fields are responsible for arsenic pollution in water bodies, it has recently been found that arsenic-contaminated groundwaters in several areas of India and Bangladesh are the main source of arsenic in waters. It is, however, present in soil as anionic form (such as $H_2^+ASO_4^-$) and absorbed by aluminium oxide that can lead to severe toxicity to fish, plankton and bottom fauna.

Arsenic compounds such as lead arsenates are used as insecticides. Some other compounds such as arsenic trioxide, sodium arsenate, calcium arsenate, nonosodium methane arsenate, and disodium methane arsenate are used as herbicides. Fertilizer plants and liquid effluents are reported to contain the element ranging from 0.25 to 3.5 mg/l. The element reacts with methane and forms $(CH_3)_3$ As in presence of soil micro-organisms. This is the most important form of arsenic identified in aquatic organisms. Different arsenic compounds are also bound with calcium and sulphides in combination with nutrients and minerals present in soil.

### *Mercury*

Among different metals, mercury is highly toxic and contamination of aquatic ecosystems has resulted in increasing toxic levels of mercury in many species of fish and other aquatic animals. Insoluble forms of mercury in soils are converted by micro-organisms to an organic form (methly-mercury). The organic form is soluble and available for absorption by aquatic animals including fish. Mercury is then deposited in fatty tissues of fish through food chain to levels that may be toxic to humans.

Some changes in mercury influxes could alter the mass of mercury in the water column. Two potential sources of mercury for the water column are atmospheric influxes and sedimentary remobilization (Figure 12.15). The estimated atmospheric influx of 1.5 g mercury to the water column

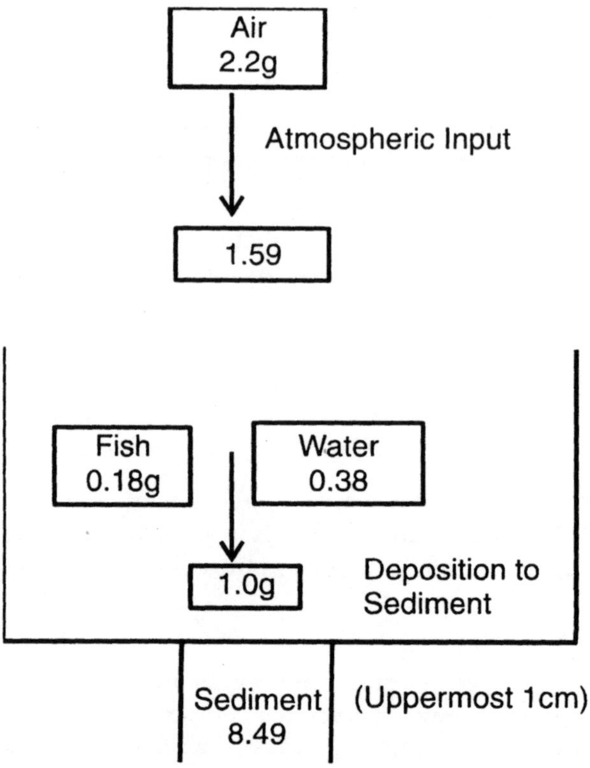


**Fig. 12.15 : Biogeochemical budget for mercury (Hg) in an acidified (pH 5.6) lake. [From Winner *et al* (1990)]**

is enough to supply the mass of mercury in fish and water and hence atmosphere is a potentially important source of methyl-mercury in waters. Moreover, the 8.4 g of mercury in the uppermost 1 cm of the sediment profile is an estimate of the amount of potentially biovailable mercury in the bottom sediment.

Because arsenic and mercury are the two important toxic elements to aquatic biotic communities to a significant extent, it is necessary to identify all arsenic and mercury-contaminated freshwater areas and mark them as hazardous to fish culture ecosystems through continuous/ frequent monitoring of the quality of water of all sources.

Arrangements should be made for training of fish farmers in diagnosis and treatment of affected fish at farm levels. Arsenic toxicity can also be reduced by application of sulfates of zinc, aluminium and iron. Heavy metal removal aquatic plants should be allowed to grow over water surfaces.

In general, inorganic compounds are gradually deposited in pond mud where they are completely or partially decomposed by soil microbes and forms insoluble or soluble compounds. While the insoluble compounds are adsorbed or absorbed on soil particles, the soluble forms are absorbed by phytoplankton and macrophytes and assimilated within the body of aquatic organisms. Seepage and volatilization of degraded molecules of inorganic compounds are also not uncommon in contaminated ecosystem. Insoluble compounds exhibit residual toxicity and causes adverse effects on biotic communities. It should be pointed out, however, that all these phenomena are greatly influenced by pH, temperature, and nutrient status in ecosystem. At the same time, knowledge of geochemical behavior such as precipitation and dissolution and physico-chemical processes that controls their persistence and transport are necessary for evaluation of the nature and extent of degradation of inorganic chemicals.

## 12.22 Prevention and Elimination of Chemical Contamination

Three methods may be considered so as to alleviate contamination of aquatic ecosystem by toxic inorganic compounds. These are : (1) management of ecosystem to prevent further recycling of the toxins, (2) elimination or reduction of the application of toxins and (3) prevention of surface run-off from agricultural fields.

### *Reduction of Recycle of Toxins*

As noted earlier, inorganic contaminants enter into the life cycle and ultimately exerts their toxic effects on different trophic levels. The cycle may be damaged by immobilizing the toxic materials in ecosystems. Toxic elements, for examples, are rendered less mobile and less available if the pH of soil remains either neutral or alkaline (Figure 12.16). Hence liming of soils having low pH should expedite for immobilization of toxic elements. Draining of water in large quantities from fish ponds may be beneficial because reduced forms of some toxic elements are removed from water to a greater degree.

Application of phosphorus carriers reduce the availability of toxic cations, but may have the opposite effects on toxic anionic forms. Leaching also effective for reducing the levels of toxicity of some elements such as boron.

### *Prevention of Surface Run-Off*

Run-off waters from agricultural fields are heavily loaded with large quantities of toxic elements that are detrimental to fish and ecosystem as a whole. Since prevention is the best possible control measure than cure, stringent action should be taken so that run-off water may not be entered into the systems or polluted water should be treated carefully prior to drainage.

### *Reduction of Application of Toxins*

Action should be taken to reduce unintentional contamination from industrial operations. Policy makers should be borne in mind that aquatic ecosystem must be considered as an important resources that can be damaged by accidental contamination with toxicants. Drastic reduction of application of fertilizers, waste materials, pesticides etc. should be avoided.

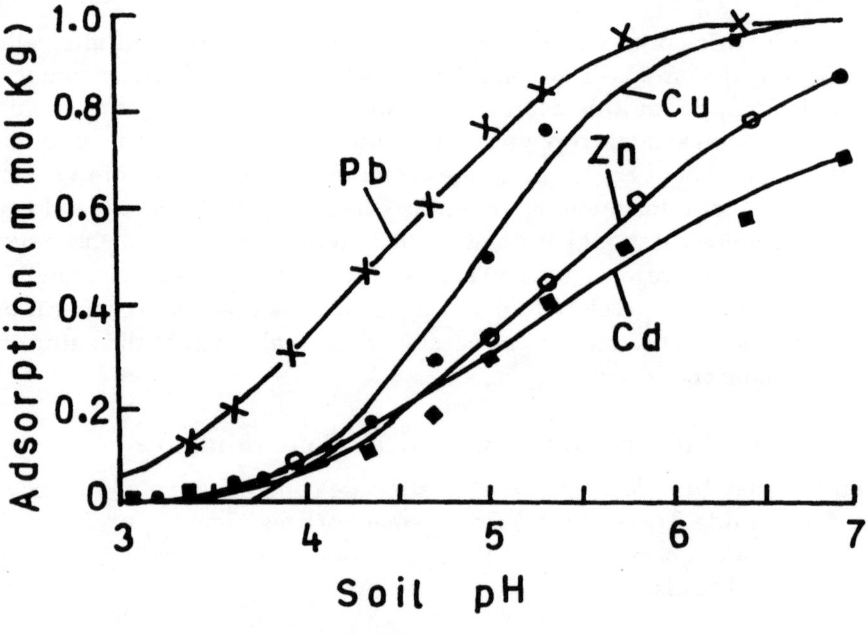


**Fig. 12.16 : Effect of soil pH on the adsorption of different heavy metals. Maintaining the soil near neutral provides the highest adsoprtion of each of these metal, particularly of copper and lead. The soil was silty clay loam. [From Elliot *et al* (1986)]**

## 12.23 Detergents

Results of several experiments on fish and aquatic ecosystem and their interpretations about the synthetic detergents clearly indicate that human society would be glad in future if they avoid the use of synthetic detergents and revert back to the practice of using organic soaps.

Use of synthetic detergents in excessive amounts is the one of the best example of man's interference with nature to the detriment of his own interest. There is an increasing concern amongst environmentalists about the propriety of using synthetic detergents. They are solely responsible for environmental degradations. At present, organic soaps and synthetic detergents are being used in almost all the countries although the latters are superior and less cost than the former.

### *Historical Background*

The era of synthetic detergents began with the synthesis of the first detergent from charcoal during 1930 in Germany. The detergent alkyl benzene sulphonate (ABS) was prepared in 1940 in USA from petroleum and coaltars. Since then detergents are being used as good substitutes for oil and fat soaps. Organic soaps have been discovered about 5,000 years ago. At that time, the primitive man baked animal meat with fat that mixed with ashes.

### *Classification and Chemistry*

Synthetic detergents lower the surface tension of water at the surface or at interphase between water and other molecules. Since most of them are surface active, they are also called as *surfactants.*

Surfactants are classified into (1) cationic, (2) ionic and (3) non-anionic forms depending upon whether they carry surplus positive or negative electric charges. The non-anionic forms are electrically neutral. Non-anionic surfactants are composed of a large hydrophobic (water-hating) alkyl group bonded to the highly polar hydrophilic (water-loving) group. Non-anionic detergents are represented by the formula $CH_2(CH_2)_{11}$ $(OCH_2{-}CH_2)_8OH$. Cationic surfactants are composed of water-hating alkyl group attached to the negatively charged water-loving group and are represented by the formula $(CH_3(CH_2)_{17})$ $2N^+(CH_3)_2Cl^-$. Anionic surfactants consists of sodium salts of alkyl benzene sulphonate. They usually contain $C_{10\text{-}15}$ chains of saturated hydrocarbons, attached indirectly or directly to sulfate or sulphonate groups. They are represented by the formula $C_{12}H_{25}C_6H_4SO_4Na^+$ or $CH_3(CH_2)_{110}SO^-{}_2Na^+$.

Among different anionic detergents, aklyl benzene sulphonate (ABS) and linear alkyl sulphonate (LAS) are the most important. The former is highly branched and non-biodegradable while the latter is unbranched and spraingly biodegradable. Owing to non-biodegradability, they persist in the aquatic environment for longer periods. The chemical structures of ABS, LAS and sodium sulphonate which are most widely used anionic surfactants are shown in Figure 12.17.

### *Permissible Limit*

American Public Health Association has earmarked safe and permissible concentration of anionic detergents at 0.5 mg/l in surface waters. Concentration of more than 0.1 mg/l detergents is accompanied by foaming, insufficient aeration leading to disturbances of biological processes in the ecosystem, imbalance of food chain and trophic levels.

### *Hazards to Aquatic Ecosystem*

Hazards of detergent pollutions are extensive. Synthetic detergents lower the surface tension of water and affect swimming ability of fishes. Fishes undergo lysis (destruction of tissues), particularly haemolysis (destruction of haemoglobin of red blood corpuscles). Experimental studies on fishes showed that 50 per cent of fertilized eggs maintained in water containing 20 mg/l detergent were killed within two weeks. Detergents exhibit profound toxic effects in fishes in concentrations between 0.4 and 40.0 mg/l. Toxicity of LAS is 2-4 times more than that of ABS. However, the toxic effects of detergents to fishes depends on the molecular structure of compounds, hardness of water, water temperature and dissolved oxygen concentration in water. Besides these, the effect also seems to vary from species to species of fish and their age. As detergents are known to remove oils from skin, fishes become prone to infections by parasites and pathogens.

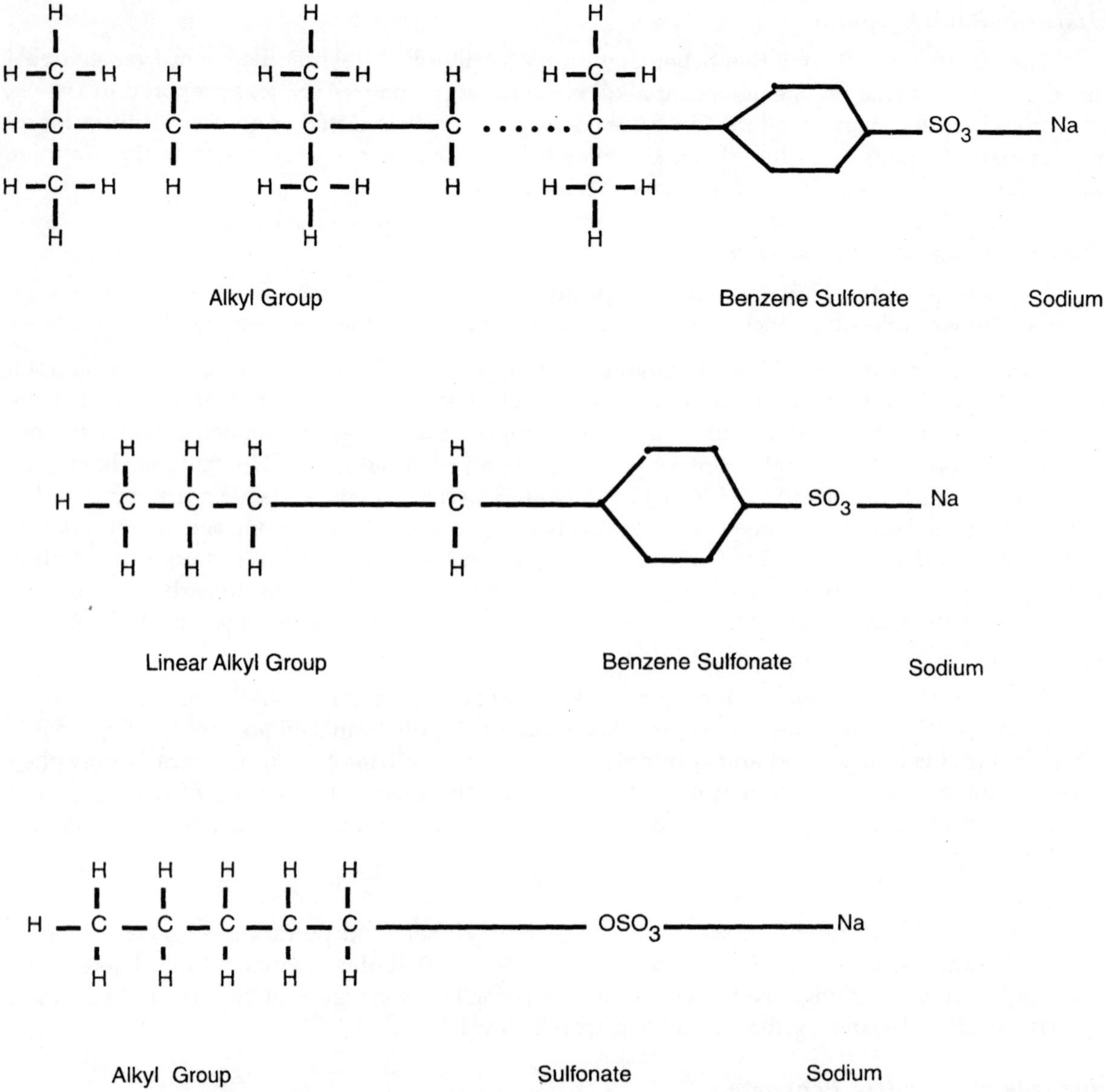


**Fig. 12.17 : Structural formulas of some commonly used detergents : Alkyl benzene sulphonate (ABS) (Top); Linear alkyl sulphonate (LAS) (Middle); Sodium alkyl sulphonate (Bottom).**

Surfactants also affect micro-organisms that form food of fishes. For example, surfactants affect growth and reproduction of *Daphnia magna* and *Cyclops viridis.* Detergents can reduce serum glucose level, kidney function and the activities of different digestive enzymes in fishes.

In detergent powder, soaps and liquids, sodium phosphate salts are used as additive which is more concern to environmentalists than the detergent ingradients. Phosphate compounds cause eutrophication in aquatic ecosystem. The consequences of eutrophication are significant. Enrichment of ponds or lakes by phosphates leads to rapid increase in algal population. Eutrophication manifests in algal blooms in freshwater and red tide in sea water. *Microsystis aeruginosa* (blue-green algae) and *Noctiluca* sp. (dinoflagellates) are the most common species and are widely distributed in freshwater and sea water in India. The algal bloom triggers rapid multiplication of zooplankton that leads to gradual depletion of dissolved oxygen and accumulation of carbon dioxide. This situation causes mass mortality of fishes and other aquatic organisms.

Since decomposition is an oxidative process, death and decay of animals aggregates further depletion of oxygen. The algal bloom, after its life span, dies in mass and undergo decomposition. If the situation become more severe, the organic matter having resulted from death and decay of plants and animals, may resort to partial decomposition an aerobically. Due to anaerobic decomposition, several obnoxious gases such as methane, carbon monoxide and sulfur oxide are released which contribute to the total extermination of biota from an aquatic ecosystem. Therefore, transient and rapid eutrophication of water is regarded as an indicator of the impending collapse of the ecosystem.

Non-degradable detergents in the ecosystem do not perish or break-down. Rather, they accumulate over years reaching the dangerous threshold. They get incorporated into the living systems and undergoes biological magnification as seen in case of DDT residue (DDE). The level of DDE in aquatic animals and plants goes on increasing from lower trophic levels to the higher ones. However, it is yet to know whether detergents get incorporated in the living bodies and undergo biological magnification.

### *Application of Detergents*

Detergents are substances used for washing off dirts, oils, oily stains from clothes, sanitary and kitchen wares and floors. They find wide use in houses, laundries, factories, laboratories, water supplies, and hospitals. At present, use of synthetic detergents are very popular because of their low cost, high washing efficiency and the ease with which dirt is removed.

### *Response of Detergents to Fish by Temperature*

Sublethal concentrations of synthetic detergents cause a thickening of the gill epithelium which probably accounts for lowered resistance to environmental hypoxia under different abiotic factors of water (such as temperature and pH).

Modification of the threshold response concentration of detergents by temperature is contradictory. Response of LAS to fish may or may not exhibit any alteration in threshold concentration with a 10° temperature change. This apparent contradiction may be a reflection of differences in species or type of detergent or both.

## 12.24 Influence of Detergents of Fish*

Decades of aquaculture research coupled with fish farmers' experience have clearly identified the effects of large number of detergents on various aspects of fish. These effects are principally included in the feeding, survival, growth and reproduction.

---

* For further information, see Kunchi *et. al.* (1980), Weinberg (1977), Pickering and Thatcher (1970), Holman and MaCek (1980), Umezawa *et. al.* (1979).

### *Feeding and Growth*

Feeding rates of fish exposed to some detergents such as alkyl benzene sulphonate and sandopan DTC decreased significantly (Figure 12.18). It is assumed that the detergent-treated fish generally take more time to consume food because they could not identify the palatable nature of food quickly. The response may be the damage of the olfactory epithelium in the nasal capsule of fish due to detergent exposure so that olfactory receptor response is blocked causing diminished sense of smell in fish.

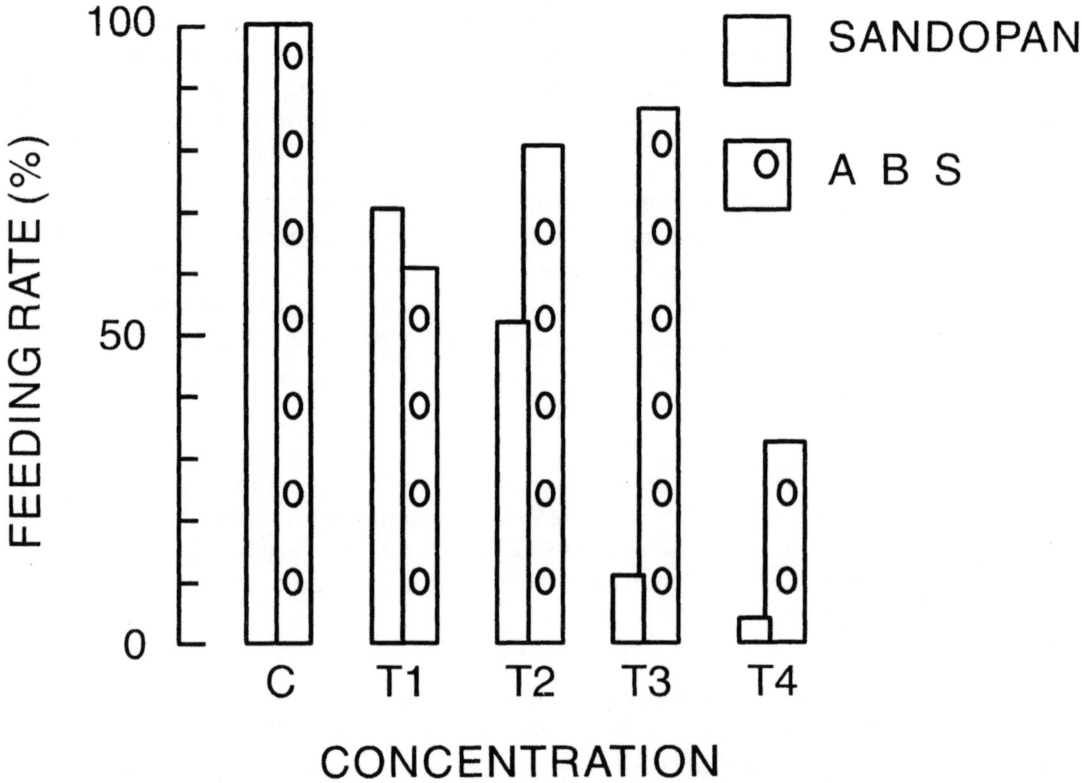


**Fig. 12.18 : Effects of some detergents on the feeding rate of the fish *Oreochromis mossambicus*. Experiments were conducted for 4 days in 15 l glass aquaria each holding 10 l of borehole water. Two fish (each having 7.13g in weight) were kept in each aquarium and live earthworm pieces of suitable size were added. Sublethal levels of detergents were added every 24h after renewal of test water. C, Control; T1, 2.23 and 0.25 mg/l; T2, 4.47 and 0.38 mg/l; T3, 10.95 and 0.51 mg/l, and T4, 20.37 and 1.1 mg/l for Sandopan DTC and alkyl benzene sulphonate (ABS), respectively. Note that feeding rate of fish decreased when compared to control values [Redrawn from data as proposed by Chattopadhyay and Konar (1984, 1985)]**

Reduced feeding rates of fish at different concentrations of detergents might explain the low yield of fish (Figure 12.19). Generally the fish in detergent-treated waters exhibit loss of appetite that causes drastic decline in fish production. Moreover, detergents accumulate in the gall bladder, intestine and liver that may produce functional disorders resulting in low digestive capability of fish.

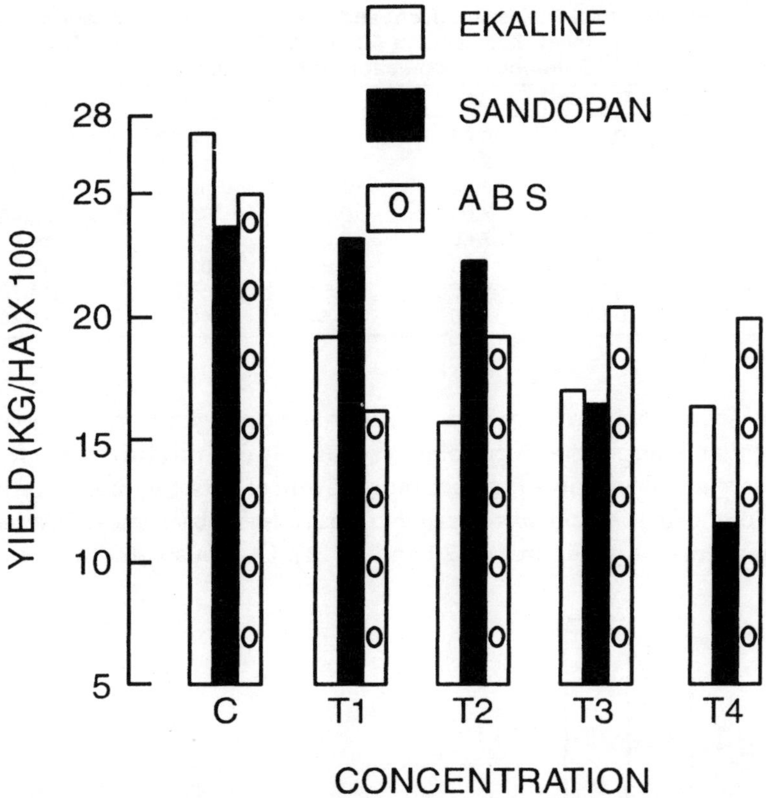


**Fig. 12.19 : Effects of some detergents on the yield of fish, *Oreochromis mossambicus*. Experiments were conducted for 90 days in earthen vessels, each holding 60 l of borehole water and 5 kg of uncontaminated soil. Fifteen fish (1g each) per vessel was stocked and exposed to detergents six times at 15 day intervals. C, Control; T1, 3.98, 2.23 and 0.25 mg/l; T2, 5.97, 4.47 and 0.38 mg/l; T3, 7.96, 10.95 and 0.51 mg/l; T4, 10.95, 20.37 and 1.1 mg/l for Ekaline, Sandopan DTC and Alkyl benzene sufonate (ABS), respectively. Note that fish production decreased when compared to control values. [Redrawn from data as proposed by Chattopadhyay and Konar (1986)]**

## 12.25 Interactions of Detergents and Effluents

Effluents and detergents which are discharged from various industries and oil refineries are also considered to be most important pollutants in aquatic ecosystem and has been recognized as a global problem. Generally, petroleum refinery effluent contains hydrocarbons which, in combination with detergents, may deteriorate the water qualities, productivity, fish growth and detritus food chain. For examples, dissolved oxygen and carbon dioxide levels, plankton and bottom fauna concentrations of water in some experimental vessels decreased significantly (Table 12.16). However, the biodegradation of detergents and effluents cause depletion of dissolved oxygen level and increase in bicarbonate ions resulting in the depletion of carbon dioxide.

**Table 12.16 : Effect of Mixture of Oil Refinery Effluent and Sandopan DTC on Dissolved oxygen (DO, mg/l), Carbon dioxide ($CO_2$, mg/l), Zooplankton (ZP, No./l), Phytoplankton (PP, No./l) and Chironomid Population (CP, No./sq.m)**

| Effluent (per cent) | DTC (mg/l) | DO | $CO_2$ | ZP | PP | CP |
|---|---|---|---|---|---|---|
| 0.00 | 0.00 | 10.08 | 1.80 | 382.20 | 1065.70 | 1541.80 |
| 0.16 | 10.95 | 8.45 | 1.01 | 114.71 | 868.50 | 1389.00 |
| 0.32 | 10.95 | 7.94 | 0.85 | 88.00 | 996.70 | 1266.00 |
| 0.64 | 10.95 | 7.60 | 0.72 | 76.53 | 935.40 | 1266.10 |
| 1.29 | 10.95 | 7.69 | 0.84 | 65.10 | 1110.80 | 942.40 |
| 2.58 | 10.95 | 7.12 | 0.82 | 53.70 | 1146.00 | 599.40 |
| 5.17 | 10.95 | 6.25 | 0.58 | 50.50 | 214.70 | 458.60 |
| 10.35 | 10.95 | 5.35 | 0.54 | 46.20 | 173.70 | 371.70 |

*Source* : Das and Konar (1987)

Effects of interactions of petroleum hydrocarbons and detergents on fish biology are dramatic. Evidences indicate that detergents and petroleum separately can produce adverse effect on fish but their mixtures in very small amounts produce more severe effects. Significant reduction in growth rate and fecundity of the fish *Oreochromis mossambicus* have been observed at different concentrations of effluent-detergent mixtures (Figures 12.20 and 12.21). Obviously, these ecotoxic effects of the

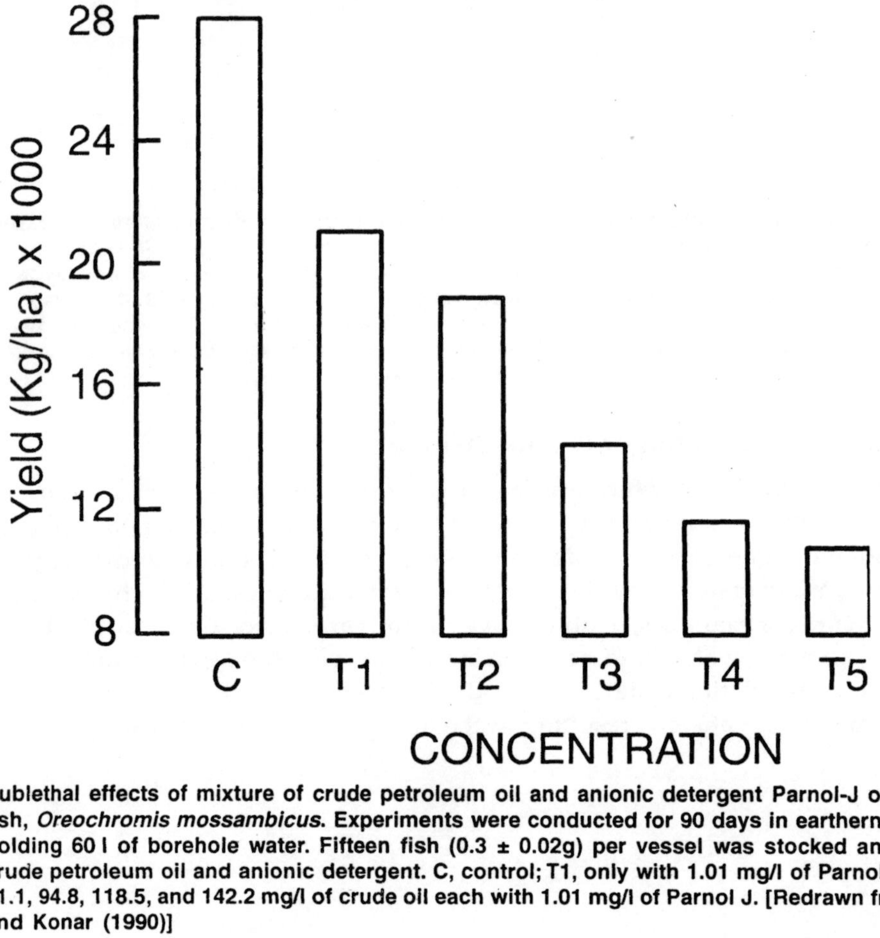


**Fig. 12.20 : Sublethal effects of mixture of crude petroleum oil and anionic detergent Parnol-J on the yield of fish, *Oreochromis mossambicus*. Experiments were conducted for 90 days in earthern vessels each holding 60 l of borehole water. Fifteen fish (0.3 ± 0.02g) per vessel was stocked and exposed to crude petroleum oil and anionic detergent. C, control; T1, only with 1.01 mg/l of Parnol J; T 2 to T5 = 71.1, 94.8, 118.5, and 142.2 mg/l of crude oil each with 1.01 mg/l of Parnol J. [Redrawn from Panigrahi and Konar (1990)]**

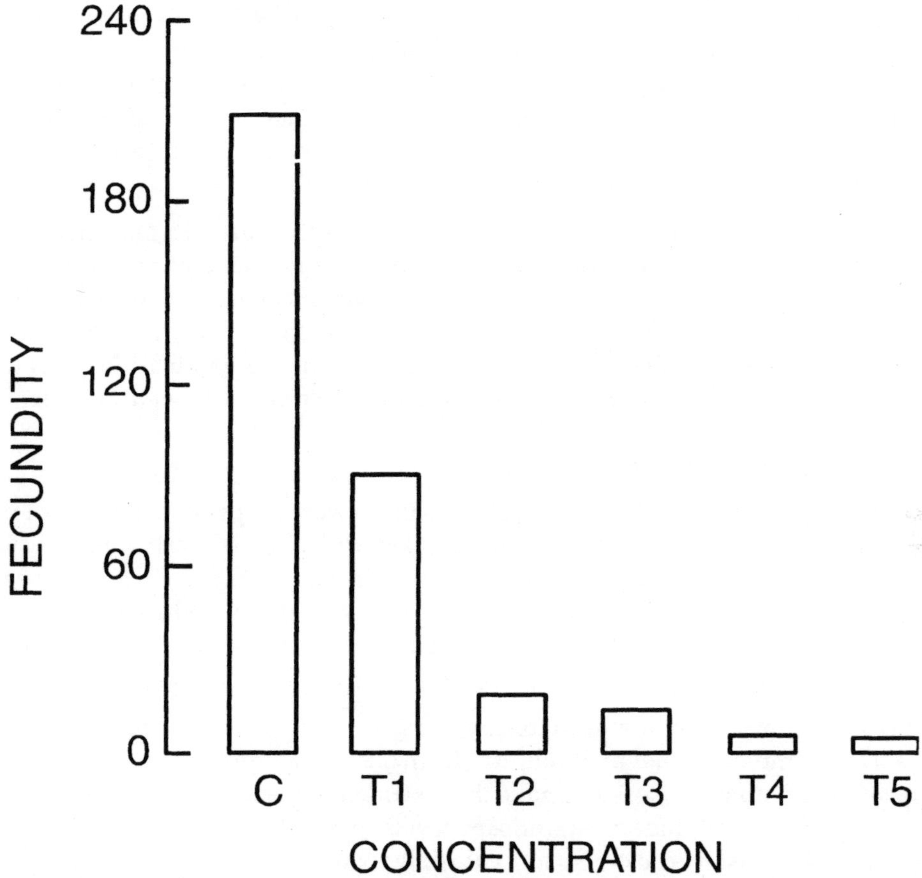


**Fig. 12.21 : Sublethal effects of mixture of crude petroleum oil and anionic detergent Parnol-J on the fecundity of the fish *Oreochromis mossambicus.* [Redrawn from Panigrahi and Konar (1990)]**

integrated pollutants must be considered and pragmetic steps should be taken to monitor the treatment and discharge the effluents from refineries and industries to avoid the chemical interactions in receiving aquatic ecosystem and to protect the fish stocks.

In natural situations where organic pollution may be severe, elevated temperature and pH will accentuate biological oxygen demand and thus impose environmentally induced hypoxia on animals. Detergents act primarily on the gills causing a restriction in gas exchange and blood chloride regulation and hence the combined effects of organic pollutants, detergents, and high temperature could be actually lethal.

## 12.26 Ecosystem as Waste Disposal Place

Aquatic ecosystems are able to utilize a lot of undesirable substances. Different types to toxic substances are drained into the aquatic system from industries, agricultural lands, domestic houses, animal sheds etc. Some waste materials at threshold levels not only improve physical and chemical properties of soil but also provide nutrients for ecosystem productivity. The beneficial effects must be encouraged, however, but when excess levels are drained, productivity of freshwater ecosystems may jeopardise resulting in water and soil pollution. Direct observations clearly indicate that when an aquatic ecosystem suffers from gradual disposal of waste materials day after day, it is completely or partially filled up, converted into barren ecosystem and ultimately becomes as a waste disposal place. These ecosystems are located in many urban areas and that built up by the dumping to create upland areas and uses as appartments and other facilities.

In many cases municipal refuse is covered with soils. In this case pits are constructed within which a variety of wastes are disposed of; although such pits are sometimes not so sanitary, because leaching and run-off from these areas are very intense which causes ground water and surface water contaminations due to accumulation of nitrate-nitrogen and heavy metals. It is difficult to control chemicals that are deposited into the ecosystems. While contamination of water and soil becomes dangerous, toxic materials in varied forms lead to serious fish health problems.

## 12.27 Salinity in Freshwater Ecosystem

The term salinity refers to the amount of soluble salts in water, expressed in terms of percentage, parts per thousand (ppt) or other convenient ratios. Contamination of freshwater ecosystem with different salts is one form of water pollution. Various inorganic salts are found in sweage sludge and domestic wastes in significant quantities. These waste materials when used in agricultural lands, release salts and drained into the ecosystem that has most extensive salinity problems. The salinity problem in small ecosystem is not so acute as in the case of rivers and lakes.

In most cases accumulation of salts in freshwater systems poses serious problems for existence of animal life. For example, the elevated salinity (from 0.29 to 38.0 ppt) of the lake water in Kerala (India) in pre and post-monsoon seasons make the lake ideal for retting. Decomposition of organic matter releases hydrogen sulphide in significant levels (6.8-11.6 mg/l) which dissolves into the water and reduces the dissolved oxygen content of the water, thus severely affects the flora and fauna. Hydrogen sulphide along with the saline conditions makes the ecosystem anoxic and untolerable to the fish. However, salinity problem can be alleviated by the application of gypsum or sulfur that help to eliminate sodium bicarbonate.

### *Problems of Salinity in Freshwater*

Gradual increase in population densities, townships and industrial complexes together have already put tremendous pressure on the limited freshwater resources particularly in coastal belt. The problem is further compounded by salinity intrusion from the sea into the river mouth as well as into underground aquifers. Periodic droughts and failure of monsoons have accentuated this phenomenon of salinity incursions into the river and underground aquifer.

### *Salinity Intrusion in Rivers*

The phenomenon of sea water intrusion in the coastal rivers is shown in Figure 12.22. Seawater tends to enter the river through the rivermouth but so resisted by freshwater discharge from the river. These two mutually opposing forces produce a dynamic equilibrium causing a salinity front. Water on the left side of the front within the river will be saline and on the right, it will be fresh.

The exact position of the salinity front ( distance of salinity front from rivermouth) depends on the magnitude of freshwater discharge from the river.

During agricultural activity, salt-leaden waters are lifted from different rivers where salinity problem is severe. In this situation salts gradually accumulate in irrigated soils.

### *Salinity Intrusion in Groundwater*

Salinity intrusion in the groundwater along the coastal belt is shown in Figure 12.23. Movement of sea water into aquifers is resisted by fresh groundwater discharged into the sea. Consequently, a dynamic equilibrium is established through the formation of a seawater - interfase. Since freshwater is lighter than sea water, the former floats over the latter. If the height of freshwater table at any point in the ground along the coast is 'h' above the mean sea level, freshwater will be available upto a depth of '40 h' before it meets the intruded seawater underground. Therefore, for every meter lowering of groundwater table in the coastal aquifers, there will be a 40 meter rise of the seawater - interfase from below, and consequently, the availability of fresh groundwater in the area. During summer and in drought years, groundwater table goes below and many of the coastal wells thus become saline. In many cases the increase in salinity was much beyond the permissible limit.

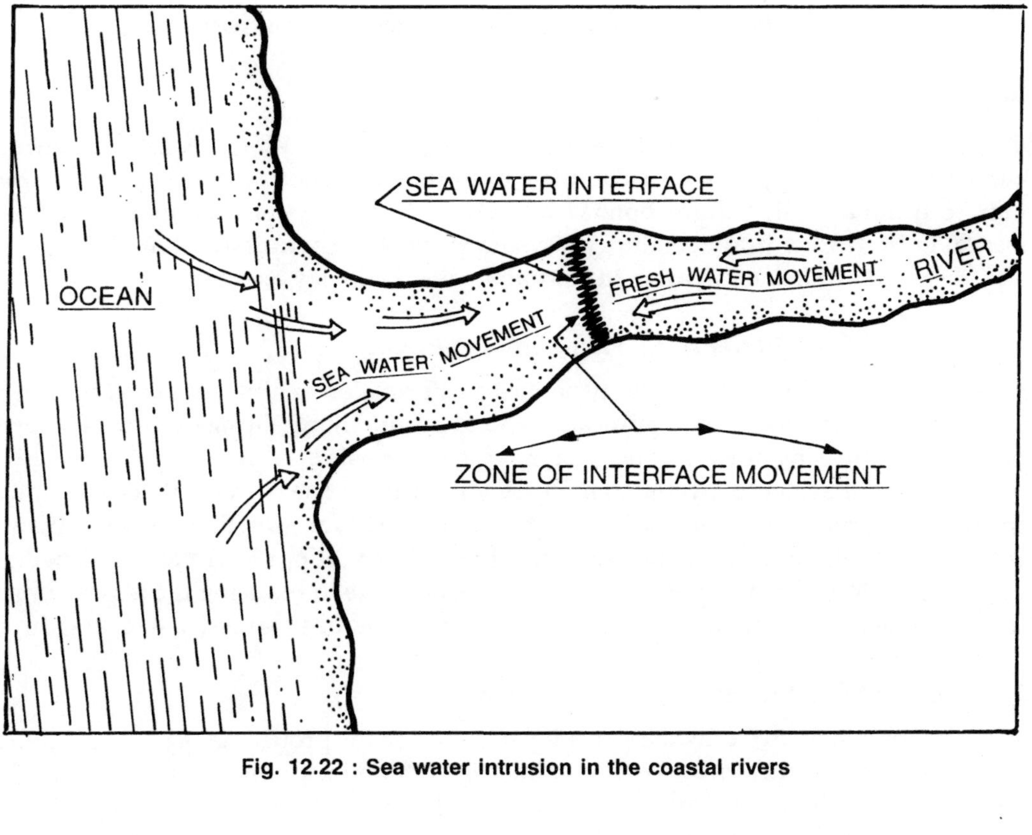


**Fig. 12.22 : Sea water intrusion in the coastal rivers**

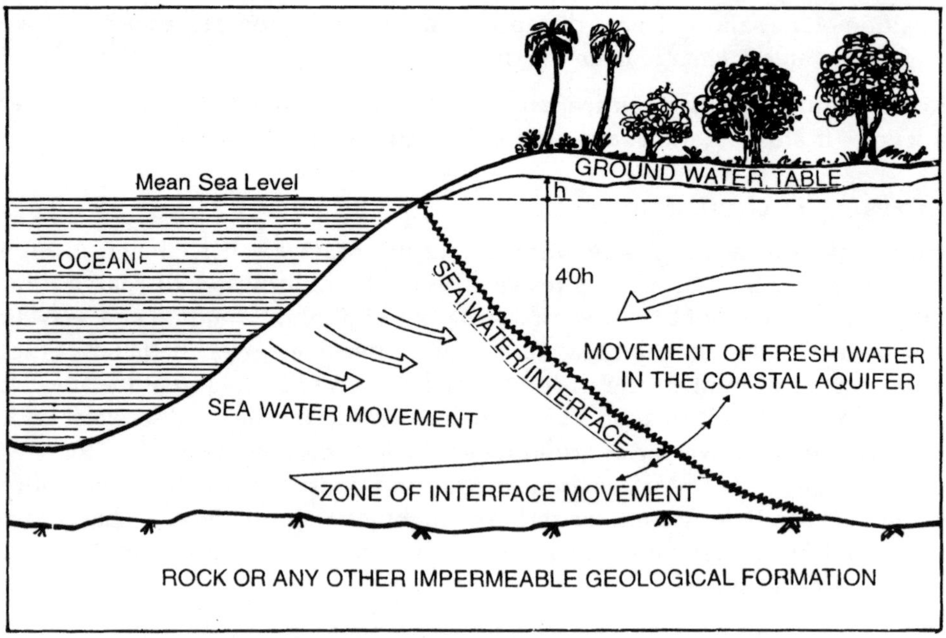


**Fig. 12.23 : Intrusion of salinity in the groundwater along the coastal belt.**

## *Contamination of Ponds and Lakes*

When salt-affected groundwaters are lifted through either deep-or shallow- tubewells for agricultural purposes, salts gradually accumulate in soils. In monsoon season, salts move from the crop lands to different fish culture ponds and lakes where it is concentrated. Although salt accumulation in freshwater prawn culture ecosystems below the permissible limit does not create problems, freshwater finfish cultivation in coastal areas is seriously affected.

## *Effects of Salinity on Nutrients in Fish Culture Ponds*

The effects of different levels of salinity on the transformation of nitrogen in fish culture ponds treated with nitrogen fertilizers are significant to pond productivity. Nitrogen content in fertilized soil exhibited a decreasing trend at the first stage of increasing salinity and, on reaching the minimum at 20 ppt, it began to exhibit an increase with further rise in salinity upto 40 ppt. It has been concluded that planktons must have a better environment for growth in the salinity range of 10 to 20 ppt and for those of soil algae, salinities higher than 20 ppt is preferred. However, the increased salinities from 10 to 40 ppt were found to be effective in reducing the losses of nitrogen from fertilizers (such as urea and ammonium sulfate) that had been added at 4, 6, and 10 week.

## *Control of Salinity in Freshwater Ecosystems*

The control of salinity in freshwaters entirely depends on the proper management. Application of gypsum or sulfur is highly effective in eliminating sodium bicarbonate and sodium chloride. Since fluctuations of salt concentration in prawn and shrimp culture ponds are detrimental that may affect their survival and growth, practical management protocols of aquacultural ecosystems is very important to prevent this happening.

## 12.28 Acid Water

Over recent decades, many research workers have called attention to significant increases in the acidity of precipitation. Acid precipitation, usually called *acid rain,* is due to the oxidation of nitrogen and sulfur dioxides that dissolve in the water vapour of the atmosphere to form nitric and sulphuric acids (Figure 12.24).

$$\underset{\text{Nitric oxide}}{2NO} + O_2 \longrightarrow \underset{\text{Nitrogen dioxide}}{2NO_2} \xrightarrow{H_2O} HNO_3 + \underset{\text{Nitrous acid}}{HNO_2}$$

$$2SO_2 + O_2 \longrightarrow \underset{\text{Sulfur trioxide}}{2SO_3} \xrightarrow{2H_2O} 2H_2SO_4$$

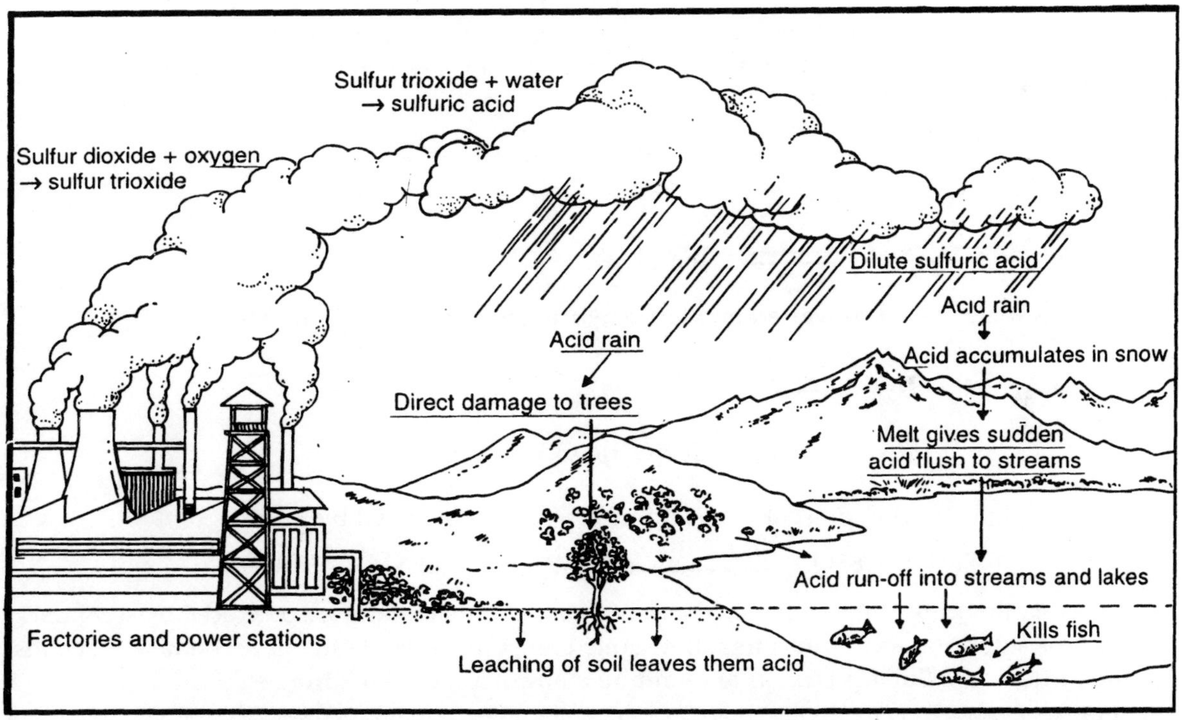


**Fig. 12.24 : Acid rain: A diagram illustrating its origins and effects. Note that acid run-off into streams, lakes, and ponds through successive stages may cause fish mortality. Adequate management is necessary to prevent this happening.**

Generally, sulfur dioxide and oxides of nitrogen, emitted from toller chimneys of factories, are widely dispersed and remain in the atmosphere for longer periods. In the atmosphere these compounds undergo various reactions involving oxidation and hydrolysis, to produce acids. Consequently, precipitation over extensive areas become acidic. A decrease of one unit on the pH scale represents a tenfold increase in acidity.

In ponds, lakes and reservoirs affected by acid deposition, freshwaters become more acidic, particularly those based on geological substrates with a limited capacity to neutralize acid inputs. As a result, the amount of dissolved aluminium is increased. These adverse conditions cause widespread reductions in biological productivity. For examples, in Southern Norway, about 2,650 lakes are seriously affected; in Eastern Canada, about 52,000 Sq Km of surface waters have undergone acidification. In Sweden, 2,500 lakes have suffered from fisheries damage.

## 12.29 Acid Drainage*

Different methods of surface and underground mining may create the problem of acid mine drainage pollution of aquatic ecosystem. However, one of the most important problems associated with mine operations is the disposal of refuse materials from mines and refuge piles are the major source of acid drainage in many areas.

### *Formation of Acid Drainage*

The following equations clearly illustrate the process of formation of acid drainage. Coal mines that contain iorn sulphide, are exposed to air and water, produces ferrous sulfate and sulphuric acid :

$$2FeS_2 + 7O_2 + 2H_2O \longrightarrow 2FeSO_4 + 2H_2SO_4$$

Ferrous sulfate is oxidized and form ferric sulfate :

$$4FeSO_4 + 2H_2SO_4 \rightleftharpoons 2Fe_2(SO_4)_3 + 2H_2O$$

Ferric sulfate is then converted into ferric hydroxide or basic ferric sulfate :

$$4Fe(SO_4)_3 + 6H_2O \rightleftharpoons 2Fe_2(OH)_3 + 3H_2SO_4$$

$$Fe_2(SO_4)_3 + 2H_2O \rightleftharpoons 2Fe(OH)(SO_4) + H_2SO_4$$

Iron sulphide (pyrite) present in coal mine can also be oxidized by ferric iron :

$$FeS_2 + 14Fe^{+3} + 8H_2O \longrightarrow 15Fe^{+2} + 2SO_4^{-2} + 16H^+$$

Several other constitutents are also present in mine drainage that are produced by secondary reactions of sulphuric acid with mineral organic compounds in the mine. Secondary reactions produce different pollutants such as aluminium, manganese, calcium, iron, etc.

Micro-organisms play a very important role in the formation of acid drainage pollution. Some species of bacteria are known to oxidize pyrite.

### *Pollution of Aquatic Ecosystem*

Pollution of aquatic ecosystem by acid drainage is considered as dangerous to aquatic life. Discharge of acid drainage to surface waters change the water quality by reducing the total alkalinities and pH, increasing total hardness, addition of inorganic elements and suspended materials. Such impact on the aquatic environment through acid drainage has changed the habitat in many ways.

* For review of this subject, see Muniz and Leivestadt (1980), Seip and Tollan (1985) and Tiwari (1996).

A number of factors also play a key role so far as the detrimental effects on aquatic life are taken into account. These factors are characterized by the presence of ions of iron and sulfate, mineral acids, zinc, copper, arsenic, and cadmium in lethal concentrations. Also, the toxicity of these factors when combined with any two of them exhibited synergistic influence between the following pairs : zinc and cadmium, zinc and copper, copper and cadmium, mercury and zinc, pH and toxic metals.

### *Effects on Fish Survival*

Survival of freshwater fish in acidified lakes and ponds drastically reduced. For examples, during the year 1975, fish populations in about 217 lakes in the New York (USA) dramatically declined that had pH below 5.0. In Southern parts of Norway, about 13,000 Sq Km water area is now completely devoid of fish. In many developed countries, acidification of lakes, ponds and rivers through acid drainage is a common phenomenon thus poses serious threats to ecosystem and are likely to become more critical in future. Where contamination of water bodies by acid drainage is very common, care must be taken to prevent the occurrence of fish mortality.

### *Effects on Aquatic Life*

Acidification of freshwater ecosystems due to acid drainage severaly affects the population of aquatic life. Producers, consumers, and decomposers are equally affected and consequently, the food chain is jeopardized. While fish populations dramatically declined in moderately acidified lakes (pH 5.0-5.5), some species of snails, midge larvae (*Tendipes* sp.), may fly, rooted and floating vascular plants, mosses, and filamentous algae were observed in streams heavily contaminated with acid mine drainage. This condition is prevailed for longer periods in many developed countries like USA, Norway and Sweden.

Some forms of algae become inactivate in acid drainage lakes and therefore, photosynthetic activity is reduced and hence the productivity of acidified lakes and ponds is lower. On the other hand, at pH 5.0 phosphorus is precipitated in acidified waters that contain appreciable amount of aluminium. Thus, phosphorus concentration in acid drainage ecosystem significantly reduced.

A large number of bacteria has been isolated from acid mine drainage waters. Microbial activity in acidified lakes leads to increase the concentration of organic matter and reduction in the availability of nutrients. In extreme acidified lakes and ponds where pH drops below about 4.0, some bottom fauna such as snails, insects and their larvae and mussels are completely exterminated.

## 12.30 Treatment of Acidified Ecosystem

There are four obvious ways to treat the acidified ecosystems :

1. Since pyrite oxidation is the main reaction during early stages of acid formation in acid mine drainage, 'convection transport mechanism' is adopted for the prevention of acid formation.
2. Anhydrous ammonia is also used for neutralization of acid mine drainage water. But this process has some disadvantages. It has high reagent and maintenance costs and ammonia-treated drainage may exhibit harmful effects on fish.
3. Sodium carbonate and sodium hydroxide solutions may also be used for the treatment of acid mine drainage but these reagents are very expensive.

4. Acidification of ecosystem can be alleviated by adding lime and limestone. Liming involves the addition of an alkaline agent, mixing, aerating and removal of the precipitate. Treatment with alkaline agent is the most common method. Limestone is cheaper than lime and the former produces less precipitate. However, for successful treatment, potential and economical solutions must receive attention.

## 12.31 Radionuclides in Ecosystem

For successful atomic weapons testing, nuclear fission reactions are carried out in many developed and developing countries. Besides this, naturally occurring radionuclides in soil (such as $^{14}C$, $^{40}K$ and $^{87}Rb$) are also added as a result of fission. Different types of terrestrial and aquatic ecosystems are victimized by a large number of radionuclides such as $Sr^{90}$, $Cs^{137}$, $Pu^{239}$, $Co^{60}$ and $C^{14}$. Because these nuclides have half life period ranging from few days (half life for $Sr^{90}$ = 50 days) to several hundred years (half life for $C^{14}$ = > 5,000 years), they obviously exhibit drastic effects on plants and animals. Although nuclear fission experiments are conducted in the deeper parts of the soils, nevertheless, accidentally they make their way to the water. Moreover, in the event of a catastrophic supply of fission products, accident of power plants and use of radioisotopes in biological researches significantly contribute to the contamination of radioactive substances to water. However, research in this field clearly indicates that seas, freshwaters and soils are not typically simple dilusion tanks and that animal life is significant in the transporting of radioactive substances.

### *Behavior of Radionuclides in Soils*

Appreciable research has been carried out regarding the behavior of some radionuclides. It has been known, for example, that the behavior of Strontium-90 is more or less similar to that of calcium. It enters soil from the atmosphere in soluble forms and is rapidly adsorbed by organic and inorganic colloidal fractions. It then undergoes cation exchange and is available to plants. Also, high concentration of calcium decreases the availability of Strontium-90.

In contrast to Strontium-90, Cesium-137 that is chemically related to potassium, are very less available in many soils. This is firmly fixed by clay soil minerals and oligochaetes. Thus the soil triggers to move Cesium-137 into the food chain of ecosystem.

The effects of radionuclides generally bring about the genetic changes that cause several deformities in both aquatic and terrestrial forms. $Sr^{90}$ and $Cs^{137}$ have been found to accumulate in bones and muscles resulting in damage the tissues and disrupt several physiological processes.

### *Half-Life of Radioisotopes*

It is the time period in which half the initial number of atoms of a radioactive element disintegrate into atoms of the element into which they change directly. The half-life of radioisotopes must be considered so far as their uptake and accumulation are concerned. Some products may be dangerous to use if consumed fresh but be harmless after storage of fish. For examples, phosphorus-32 has a radiological half-life of 13.3 days and strontium-90 has a half-life of 28 years. Cesium-137, with 27 years half-life, is one of the most dangerous fission elements.

## *Radioactivity in Aquatic Ecosystem*

Prior to 1945, radiation and X-rays mainly used for medicinal treatment, constituted the only important radiation hazard to man. With the gradual use of radioactive materials, the potential radiation danger has increased several times. Since oceans and rivers are generally considered as major recepients of radionuclides, it seems appropriate to conclude that radioactive materials enter the food chain and thus affect the productivity of the oceans and rivers.

Fishes may become contaminated in two ways : (1) through absorption of radionuclides via gills, and (2) through consuming contaminated food, unless the fish has a mechanism whereby radionuclide is screened and does not enter into the metabolic pathways. This situation may conversely be aggravated by radionuclides being selectively received and thereby risk exists that they become accumulated in the body of fish and other aquatic organisms if not rejected at a similar rate.

It is known that increased radioactivity has dramatically changed the world fisheries and aquaculture. It seems justified that the oceans and rivers are the recepients of the direct fallout. Moreover, the run-off from the continents carries additional loads to the ocean. Rivers also carry a substantial amounts of radioactive materials into the oceans. In this part of the chapter, a brief report has been given towards the effects of radioactive materials on freshwater organisms. The drastic effects of radioactive materials to marine organisms is highly significant and is beyond the scope of the present discussion.

Vast quantities of radioactive substances are accumulated in the atmosphere where radioactive cloud is formed. This cloud spreads gradually thousands of kilometers and during rainfall, different forms of water bodies are contaminated with radioactive substances.

## *Uptake and Accumulation by Aquatic Organisms*

Numerous studies have shown that aquatic organisms are capable to take up and accumulate radionuclides. For example, minnow (*Richardsonius* sp.) of Columbia River could accumulate $P^{32}$ 150,000 times that of the surrounding water. Gross radioactivity of plankton was about 2,000 times that of the surrounding water. However, estimation of the concentration factors for different organisms and radionuclides (Table 12.17) has shown that high values for $P^{32}$ are due to the fact that phosphate concentration in the Columbia River water is very low (0.02-0.10 mg/l).

**Table 12.17 : Estimated Concentration Factor (CF) for Various Radionuclides in Aquatic Organisms**

| Radionuclide | Place | Cf (1,000 times concentration) | | |
|---|---|---|---|---|
| | | Phytoplankton | Filamentous algae | Fish |
| $Na^{24}$ | 1 | 0.5 | 0.5 | 0.1 |
| $Cu^{64}$ | 1 | 2.0 | 0.5 | – |
| $Fe^{59}$ | 1 | 200.0 | 100.0 | 10.0 |
| $P^{32}$ | 1 | 200.0 | 100.0 | 100.0 |
| $Sr^{90}$ | 2 | 75.0 | 500.0 | 20-30 |
| $P^{32}$ | 2 | 150.0 | 850.0 | 30 |

*Source :* Krumholz and Foster (1957)

1, Columbia river; 2, White oak lake

During summer, some radiophosphorus was found to be taken up by organisms in the waters of the Canadian Ottawa River in which discharge of the fission products takes place. By the autumn, this acquired radioactivity disappeared owing to the short half-life of radiophosphorus.

During normal physiological activities in fishes, accumulation of radionuclides generally takes place. But is has been pointed out that uptake also governed by some physical factors such as ion exchange, pH and concentration differentials. It should be mentioned that most of the aquatic organisms are capable of absorbing nutrients and mineral ions directly from the water via the gills and skin. In other words, diet is not the main source of many compounds. Studies have shown that non-feeding tilapia fish took up no less than 60 per cent of the absorbed $Ca^{45}$ directly from the water.

It has been established that three species of freshwater algae such as *Scenedesmus* sp., *Cladophora* sp., and *Spirogyra* sp. showed weak affinity for $S^{35}$ and $Co^{60}$; lesser affinity for $Rb^{86}$; and strong affinity for $Zn^{65}$, $Fe^{59}$ and $Zr^{95}$. Of course, a large number of marine planktonic algae were found to be a very strong affinity for $Y^{90}$ and less affinity for $Sr^{90}$.

## 12.32 Radioactive Wastes

Besides radionuclides, low-level radioactive wastes substances either in the form of solid or dissolved state, are deposited in soils. In waste materials, nuclides of americium, neptunium, cesium, curium, plutonium, and uranium elements are found in appreciable quantities.

Transfer of radioactive elements usually depends on their rates of solubility in water. For examples, cesium is moderately soluble, uranium is highly soluble while americium and plutonium are insoluble. In any case, the long-term detrimental effects on aquatic life should not be ignored and care must be taken so that physico-chemical parameters of soil and water will discourage either leaching or movement of chemicals into the food chain. Therefore, monitoring of different places where nuclear fission experiments are conducted, is needed to ascertain minimum transfer of radioactive substances to different trophic levels.

## 12.33 Contamination of Water with Fertilizer Compounds

Application of various types of chemicals fertilizers in agricultural field in amounts far in excess can result in contamination of both ground and surface waters. Generally nitrates and phosphates are the most important that are responsible for water pollution. Concentration of nitrate is found in both surface run-off and ground waters whereas phosphate concentration is found in excessive levels in surface run-off.

When nitrogen and phsophorus is lost through various ways, soil fertility is drastically reduced. But the effect of water quality is serious. In extensive and intensive fish culture systems where feeding and fertilization programs are undertaken, the concentrations of nitrite and nitrate in water remain high which exceeds above about 4.5 and 6.7 mg/l, respectively and is considered fish health-hazard.

Over fertilization of ponds and lakes increases the levels of nitrogen and phosphorus that stimulate the growth of water-loving plants and algae. The effects of these flora on aquatic life and aquatic ecosystem as a whole are severe. However, the timing of fertilizer application should coincide with the productivity of ecosystem.

## 12.34 Human Waste

The influence of humankind on the ecology has become so marked that he has created extremely harsh environments of his own. The disposal of human waste is one of the most important factor affecting the quality of aquatic environments. Discharge of untreated sewage into freshwater bodies introduces a variety of proteins, fats and carbohydrates which provide an additional source of food for the organisms in the water. Oxygen is consumed as the organic matter is broken down and as a result the demand for oxygen exceeds the supply. Extensive areas become anaerobic and consequently uninhabitable for aerobic organisms.

It has been found that the quantities of faecal coliform (FC) and faecal streptococci (FS) bacteria – discharged by the human beings, are different from the animals. The ratio of FC to FS in human and cattle wastes clearly indicates that severe pollution of water might occur due to contamination with human waste (Table 12.18).

**Table 12.18 : Per Capita Contribution of Indicator Micro-organisms from Human Beings and Cattle**

| Source | Average indicator (density/g of faeces) | | Average contribution (per capita) | |
|---|---|---|---|---|
| | Faecal coliform ($\times 10^6$) | Faecal streptococi ($\times 10^6$) | Faecal coliform ($\times 10^6$) | Faecal streptococi ($\times 10^6$) |
| Human beings | 12.5 | 3.2 | 1,950 | 425 |
| Cattle | 17.23 | 116.3 | 29,300 | 3,26,000 |

*Source* : Kumar (1998)

Ratio (FC/FS) for human beings = 4.59 and for cattle = 0.089

## 12.35 Sewage and Sludge

Sewage is defined as a cloudy fluid produced from domestic wastes containing organic matter and mineral either in colloidal form in a dispersed state, or in suspension, or in solution or having particles of solid matter floating. However, sewage differs from sludge in that the latter includes liquid household wastes (not urine and faecal matter).

Various types of sewage/sludge are generally recognized such as municipal sewage, domestic sewage, industrial and human wastes. While municipal sewage consists of mixture of domestic sewage with trade wastes that are diluted with ground water, domestic sewage generally contains inorganic elements (such as zinc, cadmium, chromium, lead and nickle), high concentrations of organic carbon (200-400 mg/l) and total ammonia nitrogen (80-120 mg/l). On the other hand, industrial sewage contains excessive quantity of organic carbon (300-600 mg/l). Physico-chemcial qualities of different types of sewage (Table 12.19), however, clearly indicate how these pollutants create toxic conditions in aquatic environments. Pollutants which are deposited in aquatic system obviously contaminate fish. In humans, the clinical manifestations of poisoning by pollutants have been known for several centuries. There are instances where human populations have died or severely affected with various water-borne diseases from ingesting fish contaminated with, for example, dissolved solids, nitrate, phosphate, and chloride.

Composition and strength of sewage vary from place to place owing to significant differences in the dietary habits of people, water consumption and composition of trade waste materials. The strength of sewage is defined as the amount of oxygen necessary to oxidize the entire amount of organic matter and ammonia present in sweage.

**Table 12.19 : Physico-Chemical Characteristics of Industrial Wastes (A), Domestic Sewage (B), Municipal Sewage (C) and Agricultural Run-off (D)**

| Parameter | A | B | C | D |
|---|---|---|---|---|
| Alkalinity | 2,000-2,700 | 1,750-2,550 | 1,860-22,000 | 2,300-2,600 |
| BOD (soluble) | 1,670-2,600 | 425-440 | 260-325 | 17.8-30.0 |
| COD (soluble) | 3,000-4,800 | 1,075-1,135 | 667-835 | 42.8-72.5 |
| Total solids | 25,600-37,600 | 20,700-32,450 | 17,000-24,600 | NA |
| Dissolved solids | 22,170-31,100 | 70,00-80,000 | 625-1,000 | 144-325 |
| Suspended solids | 3,430-6,500 | NA | 3,000-5,000 | NA |
| Chloride | 770-14,900 | 1,000-1,100 | 175-250 | 9.4-46.5 |
| Sulphide (as $SO_4$) | 1,540-3,300 | 1,000-1,400 | 780-1,370 | NA |
| Sulphide (as S) | 55-130 | 2,500-3,000 | 60-86 | 50-85 |
| Nitrogen | 740-1,400 | 850-1,000 | 700-950 | 200-270 |
| Phosphate (as P) | 1.0-5.0 | 1,900-2,000 | 1.0-2.6 | 0.64-0.87 |
| Nitrate | 2.0-5.3 | 2.4-4.1 | 1.3-3.5 | 0.72-1.2 |
| pH | 8.2-9.2 | 4.6-6.5 | 6.5-7.5 | 7.5-8.5 |

*Source :* Sarin *et al* (1998), Kumar (1998)
Except pH, all values are expressed in mg/l;
NA, Data not available

The hazards of these waste substances are quite apparent when contaminated aquatic ecosystems drastically increase in mortality of fish. Raw sewage and sludge if they enter the environment of ponds, lakes, rivers, and oceans are generally associated with acute water pollution episodes. Unless they are properly treated through treatment systems, a number of notorious incident occurs particularly in lakes and rivers where high concentrations of carbon dioxide (30-100 mg/l), hydrogen sulphide (3-5 mg/l), total ammonia nitrogen (5.7-7.3 mg/l), organic carbon (150-300 mg/l) and suspended solids (150-400 mg/l) are observed. It is generally accepted that the combined effect of these ingradients is highly synergistic which is responsible for the severe deaths of fish and natural fish food and for exacerbating several diseases in fish populations.

### *Sewage and Sludge Treatment**

A number of 'waste stabilization ponds' are used for treatment of sewage and sludge. These ponds include both aerobic and anaerobic modes of stabilization. Generally sewage and sludge together constitute waste. Natural or artificial bodies of water are considered as stabilization ponds where organic waste-water or sewage are retained until the wastes are rendered stable through biological, physical and chemical processes. However, stabilization ponds are constructred with very low cost and are very easy to operate.

The sewage or waste is discharged into the waste stabilization ponds having suitable inlet and outlet arrangements. The treatment process depends on the effective use of bacteria for stabilization or breakdown of organic matter and on the presence of algal cells for supply of oxygen to hasten the reaction.

* For further detail, see Arceivala *et. al.* (1970), Sarin *et. al.* (1998) and Trivedy (1998).

Composite fish cultrue is carried out in both primary and secondary ponds. The main stabilization ponds should not be stocked with carps, but with air-breathing fishes such as *Clarias batrachus, Ophiocephalua punctatus, Heteropneustes fossilis*, and *Anabus testudineus.* It is also important to note that major carps can be cultured in secondary ponds more effectively where conditions can be controlled to ensure that depletion of oxygen does not occur. The water of secondary ponds is used for irrigation to agricultural lands. A layout for waste treatment and utilization for myriad purposes are shown in Figure 12.25.

It is needless to mention that most of the rivers becoming highly polluted since densely populated cities are situated in their banks. The rivers, ponds, lakes and beels serve as a source of water supply for various activities of human beings such as irrigation, domestic consumption and aquaculture. Large cities generate excessive quantities of sewage and sludge and it is beyond possible to treat these substances simply through waste stabilization ponds. To serve the purpose, various sanitary programs should be implemented. For example, the Government of India, in collaboration with the Government of Dutch, has established a project entitled "Environmental and Sanitary Engineering Project" in 1986, in Jaiman area of Kanpur. Under this project, a scheme was envisaged to collect different industrial wastes, solid waste, and domestic sewage for their treatment after diluting with an appropriate quantity of waste water in a UASS (Upflow Anaerobic Sludge System) treatment plant. However, different systems and units of this plant include industrial waste water feed system, domestic waste water feed system, mixing tank, pumping station, gas system, sludge system and inter-connecting piping. Objectives and aims of this project to remove pollutants from different freshwater ecosystems have been a subject of considerable research in recent years and elaborate assessment the treatment plant is beyond the purview of the present discussion.

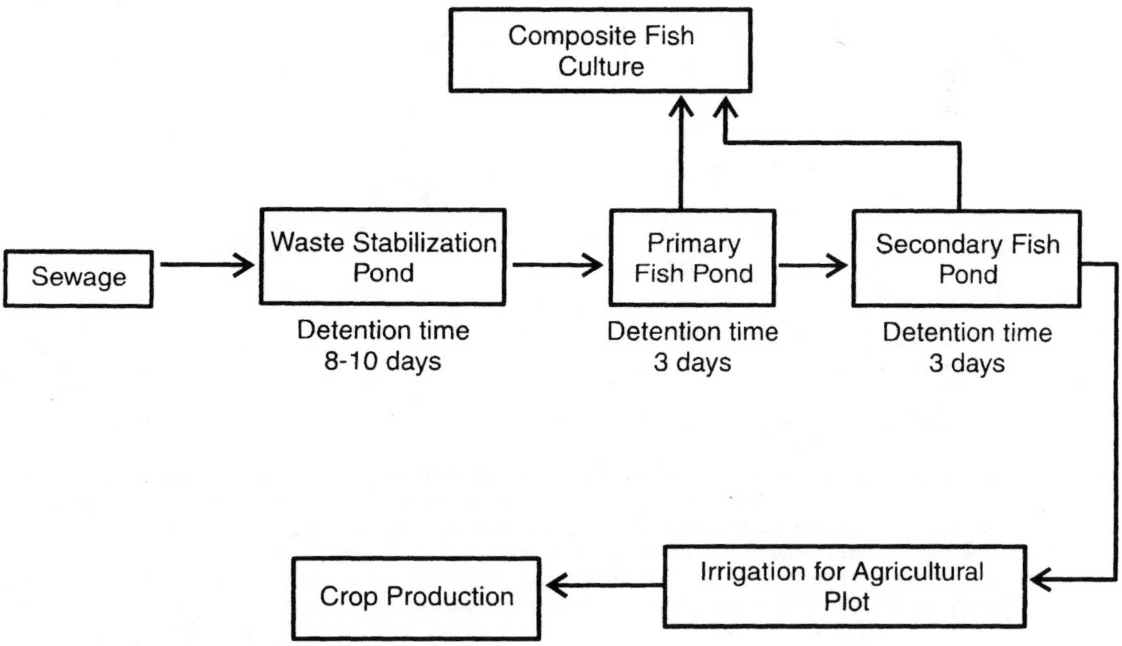


**Fig. 12.25 : Layout for wastewater treatment and utilization for fish culture and crop production.**

Removal of suspended solids from raw sludge is carried out through successive stages (Figure 12.26). While primary sludge treatment allows the separation of most of the solids from raw sludge, the secondary treatment triggers oxidation of organic substances and separation of more solids. Tertiary treatment generally involves the use of different compounds to remove inorganic elements from the sludge water. The water is then mixed with equal volume of freshwater and used for fish culture.

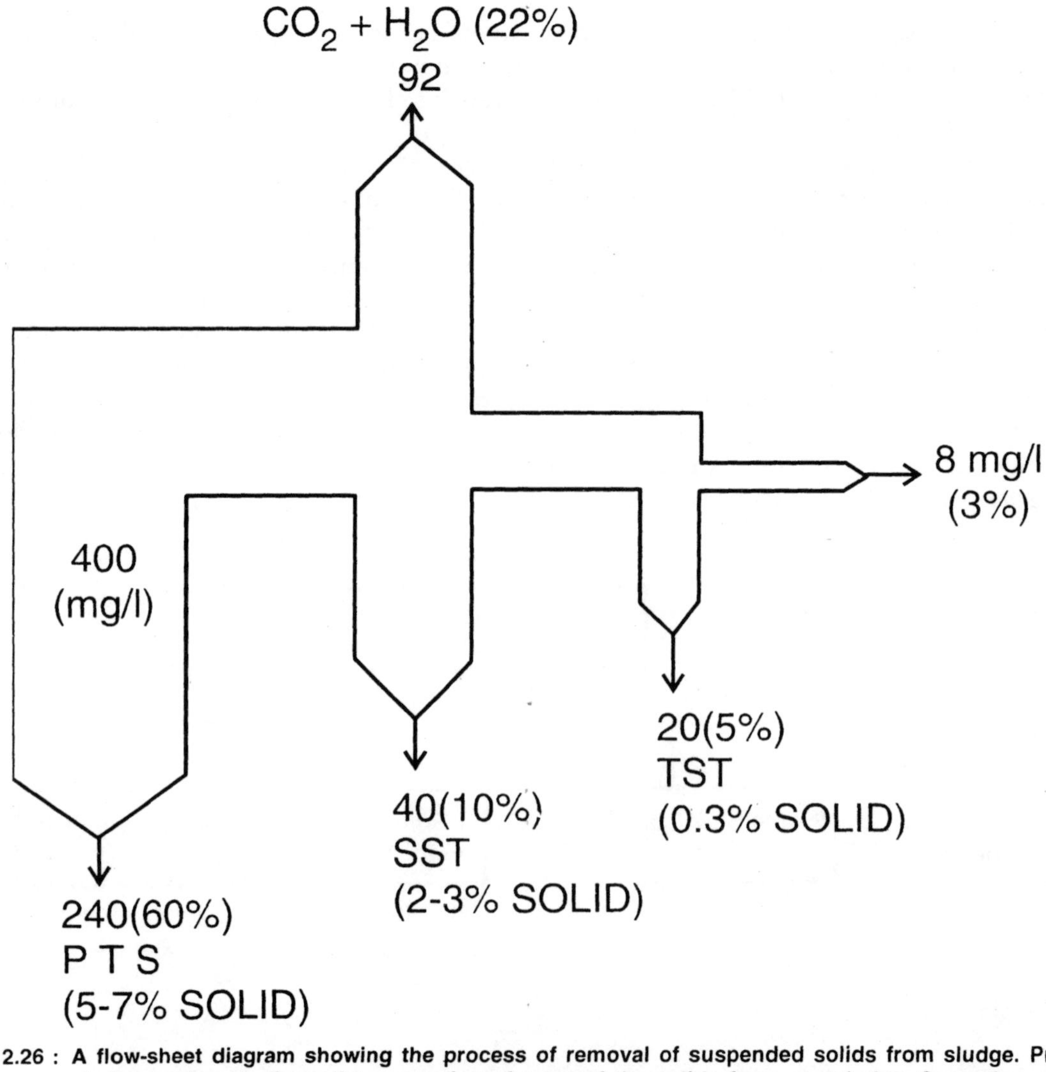


Fig. 12.26 : A flow-sheet diagram showing the process of removal of suspended solids from sludge. Primary sludge treatment allows the separation of most of the solids from raw sludge. Secondary sludge treatment triggers the oxidation of organic substances and separation of more solids. Tertiary sludge treatement involves the use of different compounds to remove inorganic elements from the sewage water. The water is then mixed with equal volume of freshwater for use in fish culture. PST, Primary sludge treatment; SST, Secondary sludge treatment; TST, Tertiary sludge treatment.

## 12.36 Garbage as Pollutant

When people crowd together in cities and towns, they replace natural ecosystems with man-made ecosystems. In these ecosystems man is the dominant species. Urbanization exhibits several serious problems when bulk-waste materials are dumped in many places. Generally bulk-waste materials contain a variety of ingradients such as cans, plastics, metals, papers, clothes, ceramics etc. Concentrations of waste in dumps harbour many of the scavengers together with a variety of other sorts of flies. Sometimes these materials are dumped at corners of fish culture ponds and consequently, the water is badly polluted and the fish population gradually diminished.

Garbage can be recycled very effectively. For example, Minamata, a small city in Japan's Southern Island of Kyushw, became infamous for Minamata disease. As a result of recycling, the amount of city waste to landfill has been reduced by half – from 2,760 ton in 1992 to 1,377 ton in 1996. The total amount of garbage was also reduced by 4.6 per cent from 10,000 ton in 1992 to 9,550 ton in 1996 which drove down the garbage disposal costs. Notwithstanding the fact, Minamata's accomplishments over a short time period persist highly extraordinary and advise future immanent. By implementing the exercise of recycling, the city has not only diminished its garbage, but fabricated closer links with citizens as well. However, garbage which is deposited at corners of most of the freshwater bodies, gradually results in permanent cessation of dynamic equilibrium of water and hence must be recycled at any cost so that water bodies may be altered from polluted to more productive aquatic environments.

## 12.37 Conclusion

Although chemical, biological and ecological methods for adequate management of water bodies have been tried in different countries, political and economical constraints delay to implement the programs. Whatever it may be the case, unless stringent rules and regulations are adopted and enforced, all management strategies will be damaged and it would never be possible to achieve successful results. The following conclusion may be drawn about water and soil pollution in relation to the quality of freshwater fish culture ecosystem :

1. Minimum use of toxic chemicals.
2. Anthropogenic interference damaging the water bodies should be checked.
3. Short - and long-term water management strategies should be adopted.
4. For washing purposes, low phosphate-containing detergents should be used because most of the detergents contain about 45 per cent phosphate compounds.
5. Recycling of different types of wastes at minimum costs.
6. Since aquatic ecosystems are valuable resources, they should be well protected from environmental contamination to prevent the ecosystem from permanent damage.
7. Aquatic ecosystems have the capacity to bind, absorb, adsorb and break-down of added substances and thus fish culture ecosystems offer promising mechanisms for utilization of many undesirable substances.
8. Acidification of freshwater ecosystems, release of pollutants from various sources and use of radioactive substances for myriad purposes should be restricted although these require high level political discussions and recommendations.

To gain a better understanding of how fish culture ecosystems might be used for high production, fishery scientists must devote a fair share of their research effort to environmental quality problems.

## *References*


Arceivala, S. J., J. S. S. Lakshminarayan, S. R. Algaraswamy and C. A. Sastry. 1970. Waste stabilization ponds, design, construction and operation in India. Central Public Health Engineering Research Institute, Nagpur, India.

Beyer, W. N., R. L. Chaney and B. M. Mulhem. 1982. Heavy metal concentration in earthworm from soil amended with sewage sludge. *J. Environ. Qual.* 11 : 381-385.

Bhattacharya, S. 1986. Live zooplankton in pollution research. *In : Environmental Biology : Coastal Ecosystem.* (*Eds* : R. C. Dalela, M. N. Madhyastha and M. M. Joseph), pp. 21-42, Academy of Environmental Biology, Muzaffarnagar, India.

Chattopadhyay, D. N. and S. K. Konar. 1984. Influence of an anionic detergent of fish. *Environ. Ecol.* 2 : 257-261.

Chattopadhyay, D. N. and S. K. Konar. 1985. Acute and chronic effects of linear alkyl benzene sulphonate on fish, plankton and worm. *Environ. Ecol.* 3 : 258-262.

Chattopadhyay, D. N. and S. K. Konar. 1986. Acute and chronic effects of a nonionic detergent on fish, plankton and worm. *Environ. Ecol.* 4 : 57-60.

Das, P. K. M. K. and S. K. Konar. 1987. Chronic effects of mixture of petroleum refinery effluent and detergent Sandopan DTC on aquatic ecosystem. *Environ. Ecol.* 5 : 769-772.

Dhaliwal, G. G. and B. Singh. 1993. *Pesticides, Their Ecological Impact in Developing Countries.* Commonwealth Publishers, New Delhi.

Elliot, H. A., M. R. Liberati and C. P. Huang. 1986. Competitive adsorption of heavy metals in soils. *J. Environ. Qual.* 15 : 214-219.

Furr, A. K. 1976. Multi-element and chlorinated hydrocarbon analysis of municipal sewage sludge. *Env. Sci. Tech.* 10 : 683-687.

Genesan, T. and D. Lalithakumari. 1990. Interaction of carbendazim with herbicide and insecticide and their influence on soil micro-organisms. *J. Ecobiol.* 2 : 319-324.

Goel, S. 1996. Role of vitamin $B_{12}$ against methyl parathion induced haematological indices in *Channa punctatus. Poll. Res.* 15 : 339-342.

Holman, W. F. adn K. J. MaCek. 1980. An acute safety assessment of linear alkyl benzene sulphonate : chronic effects on fathead minnow, *Pimephales promelas. Trans. Amer. Fish. Soc.* 109 : 122-131.

Jeevan Rao, K. and M. V. Shantaran. 1996. Micronutrient and heavy metal contents and their relative availability in stabilized urban wastes of Hyderabad. *Poll. Res.* 15 : 201-203.

Jeyabaskaran, K. J. and U. S. Sreeramulu. 1998. Effect of nursery application of sewage sludge on yield and heavy metal contents and uptake by rice in the field. *J. Environ. Biol.* 19 : 43-47.

Kabila, V., A. A. Yamuna and P. Geraldine. 1996. Water hardness as a determination on the potential toxicity of lead to the freshwater prawn. *Macrobrachium malcolmsonii. Poll. Res.* 15 : 39-42.

Kaushik, S., B. K. Sahu, R. Lawania and R. K. Tiwari. 1997. Occurrence of heavy metals in lentic waters of Gwalior region. *Poll. Res.* 16 : 237-239.

Kumar, A. 1998. Impact of industrial effluents on the ecology of river Ganga in Bihar, India. *In : Ecology of Polluted Waters and Toxicology.* (*Ed* : K. D. Mishra). Technoscience Publications, Jaipur, India. pp. 87-102.

Kunchi, M., M. Wakabayashi, H. Kojima and T. Yoshida. 1980. Bioaccumulation profiles of 35 - lebelled sodium alkylpoly (oxyethylene) sulphates in carp, *Cyprinus carpio. Wat. Res.* 14 : 1541-1548.

Krumholz, L. A. and R. F. Foster. 1957. Accumulation and retention of radioactivity from fission products and other radiomaterials, by freshwater organism. *Izvest. Akad. Nauk., USSR Ser.Biol.* 3 : 321-334.

McConnel, J. S. and L. R. Hossner. 1985. pH-dependent adsorption isotherm of glyphosate. *J. Agric. Food Chem.* 33 : 1075-1078.

McKim, J. M., R. L. Anderson, D. Benoit, R. L. Spehlar and C. N. Stoke. 1976. Effects of pollution on freshwater fish. *J. Wat. Poll. Cont. Fed.* 48 : 1544-1616.

Metelev, V. V., A. I. Kanaev and N. G. Dzasokhova. 1983. *Water Toxicology.* Amerind Publishing Co. Pvt. Ltd., New Delhi.

Motsumura. F. 1985. *Toxicity of Insecticides.* Plenum Press, New York.

Mukherjee, A. B. and P. Roy. 1989. Calcutta metropolitan waste, its characteristics, pollution load and disposal problem. *Environ. Ecol.* 7 : 1019-1022.

Munshi, J. S. D. and A. Singh. 1989. Heavy metal accumulation of fishes of the river Subarnarekha at Ghatsila. *Environ. Ecol.* 7 : 790-792.

Muniz, I. P. and H. Leivestadt. 1980. Acidification effects on freshwater fish. *In : Ecological Impact of Acid Precipitation* (*Eds :* D. Drablos and T. Token). Norwegian Institute for Water Research, Oslo, Norway.

Murthy, A. S. 1986. *Toxicity of Pesticides to Fish.* Vol. II, CRC Press, Florida.

Pal, A. K. and S. K. Konar. 1990. Pollution of aquatic ecosystem by pesticide methyl parathion. *Environ. Ecol.* 8 : 906-912.

Pandey, B. K., U. K. Sarkar, M. L. Bhowmik and S. D. Tripathi. 1995. Accumulation of heavy metals in soil, water, aquatic weeds and fish samples of sewage-fed ponds. *J. Environ. Biol.* 16 : 97-103.

Panigrahi, A. K. and S. K. Konar. 1990. Sublethal effects of mixture of crude petroleum oil and anionic detergent on fish. *Environ. Ecol.* 8 : 877-882.

Pickering, Q. H. and T. O. Thatcher. 1970. The chronic toxicity of LAS to *Pimephales promelas. J. Wat. Poll. Cont. Fed.* 42 : 243-254.

Pramanik, A. and S. K. Sarkar. 1995. Influence of basalin on the effectiveness of nitrogen and oil cakes on fish and aquatic ecosystem. *Proc. Acad. Environ. Biol.* 4 : 151-157.

Schmitt, C. J. and W. G. Brumbaugh. 1990. National contaminant biomonitoring program : Concentrations of arsenic, cadmium, lead, mercury, selenium and zinc is US freshwater fish, 1976-1984. *Arch. Environ. Contam. Toxicol.* 19 : 731-747.

Sarin, I., J. B. Dixit and A. Singh. 1998. Pollution management of river Ganga with special reference to Uttar Pradesh. *In : Ecology of Polluted Waters and Toxicology* (*Ed. :* K. D. Mishra), Technoscience Publications, Jaipur, India.

Seip, H. M. and A. Tollan. 1985. Acid precipitation. *In* : *Facets of Hydrology* (*Ed* : J. C. Rodda), John Wiley & Sons, New York.

Tiwari, T. N. 1996. Acid mine drainage in surface coal mining : Ecological hazards and control measures. *In: Assessment of Water Pollution* (*Ed* : S. R. Mishra), APH Publishing Corporation, New Delhi.

Trivedy, R. K. 1998. *Advances in Wastewater Treatment Technologies* Global Science Publications, Aligarh, India.

Umezawa, S. I., K. Ishida and M. Inoue. 1979. Effects of surfactants on fish. *Rep. USA Mar. Biol. Inst.* 1 : 65-78.

Vaidehi, J., A. P. Rao, N. Sinha, H. R. Dave and I. P. Sood. 1997. Elimination of methyl-mercury from fish tissues during glutathione and vitamin B complex therapy. *Poll. Res.* 16 : 183-188.

Weinberg, R. 1977. Studies on the effects of sublethal concentrations of ionogenic detergent on the activities of larvae and fry *Rivulus cylindraceus. Int. Rev. Gesamten. Hydrobiol* 62 : 71-79.

Winner, J. G., W. F. Fitzgerald, C. J. Watras and R. G. Rada. 1990. Partitioning and bioavailability of mercury in an acidified lakes. *Environ. Toxicol. Chem.* 9 : 909-918.

## Questions

1. What are the major types of pollutants?
2. What are the concepts regarding soil-water dynamics that are significant to productivity?
3. What is the fate of toxicants after they enter into the aquatic ecosystem?
4. What are the main types of pesticides?
5. What happens when toxicants are drained into a pond ecosystem?
6. How pesticides are metabolized in pond soils?
7. What are the effects of chemicals on fish?
8. State the mechanisms of toxicity of metal ions in an aquatic ecosystem.
9. What are the effects of chemicals on aquatic ecosystem?
10. Once products of thermonuclear explosions have reached the aquatic ecosystem, their uptake by fish and other organisms is variable. Why?
11. How radio-active substances are contaminated in fish and aquatic ecosystem?
12. Why use of chemical fertilizers can result in nitrate build up in groundwater?
13. What are the causes of acid rain?
14. Acid rain has, in some cases, brought about marked reduction in fish production and increased mortality. Why?
15. What are the hazardous effects of sewage sludge?
16. State the fate of inorganic compounds in soil.
17. How the contamination of aquatic ecosystem is prevented?
18. What management practices are important in preventing environmental degradation from the application of pollutants?
19. What management practices must be used to control salinity?
20. What is detergent? Classify it. Describe the hazardous effects of detergents on fish and aquatic ecosystem.
21. What is salinity? How freshwater ecosystems are contaminated with salinity on nutrients in ponds.
22. What are the causes of acid rain?
23. What are the causes of acid drainage? How aquatic ecosystems are affected by acid drainage?
24. What is sewage? How does it differs from sludge?
25. How sewage and sludge are treated in waste stabilization ponds?

# 13

# Acidity and Alkalinity in Fish Culture Ecosystem

The most important characteristic features of any fish culture ecosystem that should be kept in mind, is the acidity and alkalinity. Micro-organisms and aquatic macrophytes are responsible for controlling so much of the chemical environment in an aquatic ecosystem.

Acidity occurs where appreciable quantities of exchangeable base-forming cations ($Na^+$, $K^+$, $Ca^{2+}$, and $Mg^{2+}$) are leached from pond mud. In many situations, this condition is so widespread and the effects on aquatic life is so acute that acidity has become and important feature of any fish culture systems. Acidic ponds are not desireable for fish culture.

When a high degree of saturation with base-forming cations takes place, soil alkalinity occurs. Concentrations of calcium, magnesium, and sodium carbonates can result in a preponderance of hydroxy ions over hydrogen ions. Under this condition, the soil is alkaline. Ponds having alkaline soils, are recommended for fish culture. Because acidity and alkalinity significantly influences soil chemical properties and organisms, it is necessary to give a brief idea of how these conditions limit fish production.

## 13.1 Source of Acidity and Alkalinity

Two adsorbed cations such as hydrogen and aluminium are primarily responsible for soil acidity. The mechanisms by which hydrogen and aluminium exert their influence depends on the degree of soil acidity and on the nature of the soil colloids.

### *Strongly Acid Soils*

In cases where soil pH is less than 5.0, excessive aluminium becomes soluble and is either present in the form of aluminium hydrogen cations or is bound by organic matter. These exchangeable ions are adsorbed by the negative changes of soil colloids, which at low pH values are dominantly permanent charges associated with silicate clay soils.

The adosorbed aluminium is in equilibrium with aluminium ions in the soil. Aluminium ions have the tendency to hydrolyze and thus contribute to soil acidity.

$$\underset{\text{Adsorbed aluminium}}{Al^{3+}\,\boxed{\text{Micelle}}} \rightleftharpoons \underset{\text{Aluminium ion}}{Al^{3+}}$$

$$Al^{3+} + H_2O \longrightarrow Al(OH)^{2+} + H^+$$

The hydrogen ions thus releasd lower the pH of soil and are the major source of hydrogen ions. Most of the hydrogen, along with some iron and aluminium is bound by covalent bonds in the organic matter and contributes to the soil acidity. However, these $H^+$ ions are in equilibrium with the solution.

$$\underset{\text{Adsorbed hydrogen}}{H^+ \boxed{\text{Micelle}}} \rightleftharpoons \underset{\text{Hydrogen ion}}{H^+}$$

### *Moderately Acid Soils*

In cases where soil pH values vary between 5.0 and 6.5, hydrogen and aluminium compounds contribute to soil acidity but by different mechanisms. Moderately acid soils have higher percentage base saturations than the strongly acid soils. Aluminium ions are converted to aluminium hydrogen ions by the following reactions :

$$Al^{3+} + OH^- \longrightarrow Al\,(OH)^{2+}$$

$$Al(OH)^{2+} + OH^- \longrightarrow \underset{\text{Aluminium hydroxy ions}}{Al\,(OH_2)^+}$$

By the following hydrolysis reactions, aluminium hydroxy ions produce hydrogen ions using the simplest formulas for aluminium hydroxy ions :

$$Al(OH)^{2+} + H_2O \longrightarrow Al\,(OH)^+_2 + H^+$$

$$Al(OH)^{2+} + H_2O \longrightarrow Al\,(OH)_3 + H^+$$

The small amount of readily exchangeable hydrogen contirbutes to soil acidity. In addition, some hydrogen atoms were bound with the organic matter through covalent bonding, iron and aluminium oxides release $H^+$ ions.

$$\underset{\text{Bound hydrogen}}{{}^{H}_{H}\boxed{\text{Micelle}}} + \underset{\text{Calcium ion}}{Ca^{2+}} \longrightarrow \underset{\text{Calcium ion (exchangeable)}}{Ca^{2+}\boxed{\text{Micelle}}} + \underset{\text{Hydrogen ion}}{2H^+}$$

### *Alkaline Soils*

In this type of soils, the permanent charge exchange sites are occupied by base-forming cations such as $Ca^{2+}$, $Mg^{2+}$, $Na^+$ and $K^+$. Both hydrogen and aluminium hydroxy ions are replaced by these cations. Most of the aluminium hydroxy ions are converted to insoluble gibbsite by reactions such as

$$Al(OH)_2^+ + OH^- \longrightarrow Al\,(OH)_3$$

## 13.2 Association of Cations

The distribution of cations around soil colloids is changed along with the change in soil pH. Studies have shown that in some soils (such as silicate-clay and organic soils), the exchange capacity of base-forming cations declines as the pH is lowered.

Hydrogen and aluminium ions are strongly held by the covalent bonding is termed as *bound* hydrogen and aluminium. On the other hand, these ions are associated with permanent negative charges on the soil colloids and termed as *exchangeable.* Exchangeable ions have an immediate effect on soil pH; of course, bound and exchangeable forms are very important for determining how much lime is required to change soil pH.

Two dominant groups of element such as hydrogen and aluminium ions, are responsible for soil acidity. While these two ions generate acidity, base-forming cations resist it.

### *Source of Hydroxide Ions*

When liming substances such as calcium hydroxide are added to an acidic water bodies, $H^+$ and $Al^{3+}$ ions are replaced by $Ca^{2+}$ ions. Consequently, concentrations of $H^+$ and $Al^{3+}$ ions will decrease gradually following the increase in the concentration of $OH^-$ ions because there is an inverse relationship between $OH^-$ and $H^+$ ions in water. Therefore, cations can become indirect sources of $OH^-$ ions as they are adsorbed on soil colloids. Alkaline reaction takes place through the hydrolysis of soil colloids saturated with base-forming cations and the following reversible reaction would occur.

$$\boxed{\text{Micelle}} + Ca^{2+} + 2H_2O \rightleftharpoons {}^{H}_{H}\boxed{\text{Micelle}} + Ca^{2+} + 2OH^-$$

## 13.3 Types of Soil Acidity

Generally three types of acidity occur in the mud soil : (a) exchangeable acidity, (b) active acidity, and (c) residual acidity. These three types of acidity together constitute the *total acidity* of a mud.

### *Exchangeable Acidity*

This type of acidity is represented by the presence of $H^+$ and $Al^{3+}$ ions that are *excahngeable* (salt-replaceable) by other cations present in some salts such as potassium chloride.

$$\boxed{\text{Micelle}}^{H^+}_{Al^{3+}} + 4KCl \rightleftharpoons \boxed{\text{Micelle}}^{K^+\,K^+}_{K^+\,K^+} + AlClO_3 + HCl$$

It is important to note that in contrast to highly acid soils, $H^+$ and $Al^{3+}$ ion concentrations, in moderately acid soils is very low. But inspite of the presence of quite limited quantity of exchangeable hydrogen and aluminium, the limestone required to neutralize the acidity is about 130 times that required for very high acid soils.

### *Active Acidity*

This type of soil acidity is principally due to the $H^+$ ions in the soil mud. Though the concentration of $H^+$ ions in such soil is very small, it is important because in this environment microbes are exposed.

### *Residual Acidity*

This acidity is associated with alumimium hydroxy ions and with aluminium and hydrogen atoms that are bound in non-exchangeabe forms by silicate clays and organic matter. However, addition of lime to such soils increases the pH and the aluminium hydroxy ions are changed to uncharged gibbsite.

$$Al(OH)^{2+} \xrightarrow{OH^-} Al(OH)^+_2 \xrightarrow{OH^-} Al(OH)_3$$

$$\boxed{\text{Micelle}}\begin{matrix}Al\\H\end{matrix} + 2Ca(OH)_2 \longrightarrow \boxed{\text{Micelle}}\begin{matrix}Ca^{2+}\\Ca^{2+}\end{matrix} + Al(OH)_3 + H_2O$$

Bound Al and H (not exchangeable) — Exchangeable calcium

## 13.4 Changes of pH in Pond Soil

### *Acid-Forming Factors*

When fish culture ponds are fertilized with organic manures, they undergo decomposition and consequently, both organic and inorganic acids are formed. Perhaps the most widely acid found is carbonic acid. The persistent solvent action of this acid on the mineral constituents of the soil is responsible for the removal of base-forming cations by leaching and dissolution.

Generally, inorganic acids (such as sulphuric acid and nitric acid) along with organic acids encourage the development of acidic conditions (Figure 13.1) should be helpful in illustrating this explanation. Inorganic acids are formed by two processes : (a) from the microbial action on some nitrogen-containing materials such as ammonium sulfate and calcium ammonium nitrate, and (b) by the organic decay processes.

The precipitation of inorganic acids, oxides of sulfur and nitrogen from the atmosphere around industrial complexes is termed as *"acid rain"* since it has a pH value of 4.0-4.5 whereas in case of "normal ranifall", the pH value varies between 5.0 and 5.6. These substances also accumulate in aquatic ecosystem through acid drainage and consequently, $H^+$ ion concentration considerably increased. Although addition of $H^+$ ions are not adequate to bring about significant pH changes at once, over a long period of time their accumulation may have a significant acidifying effect.

### *Base-Forming Factors*

High values of the base-forming cations invariably contribute towards a reduction in acidity and an increase in alkalinity. The addition of agricultural limestone and other nutrient carriers make them available for adsorption by soil colloids and hence generate acidity. Therefore, liming and fertilization programs will permit the base-forming cations to remain in the mud that will encourage high pH values. This situation is favourable for ecosystem productivity.

### *Variability in Hydrogen Ions*

Considerable variation in the pH of the mud always exists in different areas of any aquatic ecosystem. However, this condition is not so prominent in small ecosystems, but in larger ones the

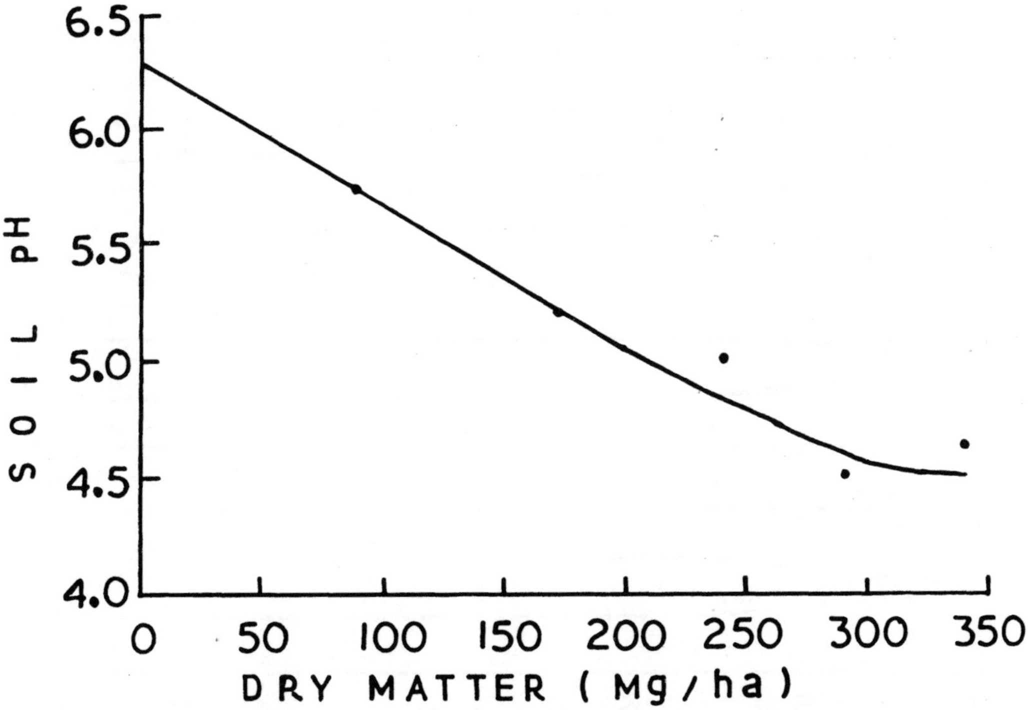


**Fig. 13.1 : Effect of large applications of sewage sludge over a period of 6 years of the pH of a fine sandy loam. The reduction was due to the organic and inorganic acids, formed during decomposition and oxidation of the orgnic matter. [from Robertson *et al* (1982)]**

variation is extreme. Such variation may result from microbial action due to the uneven distribution of organic matter. To avoid this condition, uniform distribution of organic manures is necessary.

Variability in hydrogen ions is very significant in some respects. For example, micro-organisms unfavorably influenced by a given $H^+$ ion concentration in one place may find a different environment to another place that is highly favourable. Thus the variety of environments may account for the great diversity in the population of micro-organisms present in the mud.

### 13.5 Role of pH in the Mud Soil

The pH of mud soil significantly influences soil micro-organisms and chemical properties. Figure 13.2 illustrates between soil pH and some chemical elements. The pH of soil dramatically affects/influence the availability of most of the chemical elements of importance to micro-organisms and fish food organisms. For examples, bacteria functions best at intermediate and high pH values. The availability of nitrogen is restricted at low pH value, whereas that of phosphorus is best at intermediate pH levels. The tendency for toxic elements (such as iron, cobalt, zinc etc.) is well pronounced at low pH values.

### 13.6 Measurement of pH of Soil and Water

For effective management of any fish culture ecosystem, the pH of soil and water should be tested. The test is very easy and can be made rapidly. The pH is measured directly in field, or the samples are brought to the laboratory for accurate determination of pH.

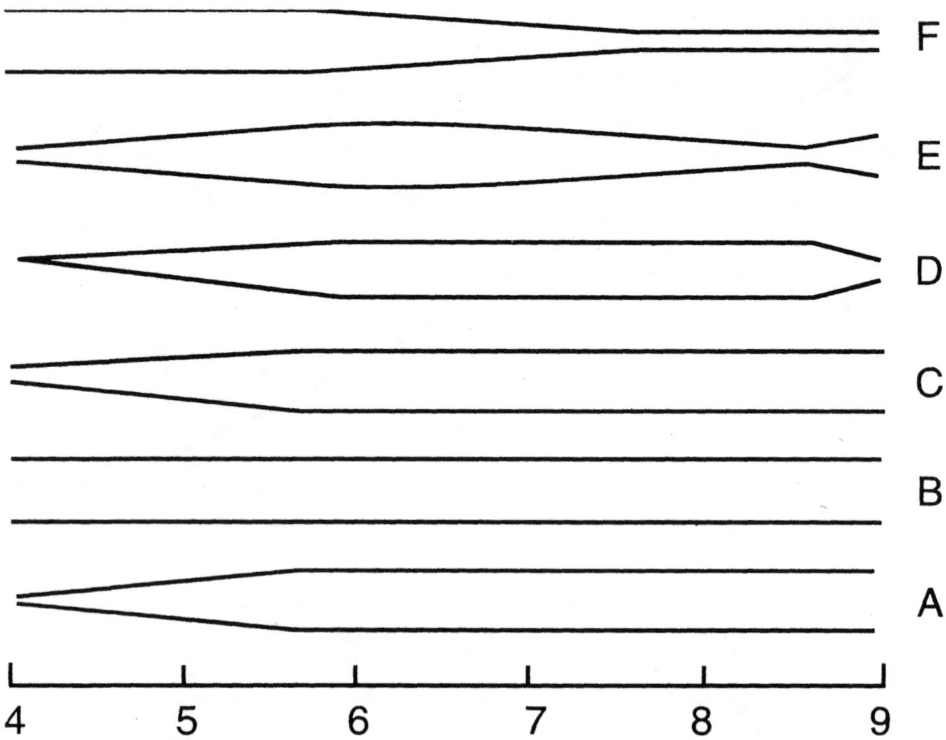


**Fig. 13.2 :** **Relationship between the activity of micro-organisms and the availability of nutrients on the one hand and the pH of soil on the other. The broad areas of the bands denote the zones of greatest activity of micro-organisms which depicts that the nutrients are readily available. Note that a pH range of about 5.6-8.5 trigger best the availability of nitrogen, phosphorus, calcium, and magnesium which are essential for the productivity of aquatic ecosystem. Furthermore, if soil pH is adjusted for phosphorus, the other nutrients will be available for the purpose. A, Bacteria and actinomycetes; B, Fungi; C, Nitrogen; D, Calcium and magnesium; E, Phosphorus; F, Manganese, cobalt, copper, zinc, and iron.**

### *Dye Methods*

These methods take advantage of the fact that some organic compounds change color as the pH is decreased or increased. Mixtures of dyes contribute to color changes over a wide pH range (4-9). Different types of dyes such as bromo-thymol blue and chloro-phenol red are generally used for effective determination of soil pH.

A few amounts (2-4 g) of dry powdered soil is kept on a white porcelain plate. Then a few drops of the dye solutions are placed in contact with the soil sample. After a few minutes, the color of the dye is compared to a color chart that indicates the approximate pH.

### *Litmus Papers*

For determination of soil pH, a suspension of soil in water (a ratio of 1:2 or 1:1) is made where a strip of litmus paper is inserted into the suspension (Figure 13.3). By changing the color of the paper indicates the pH.

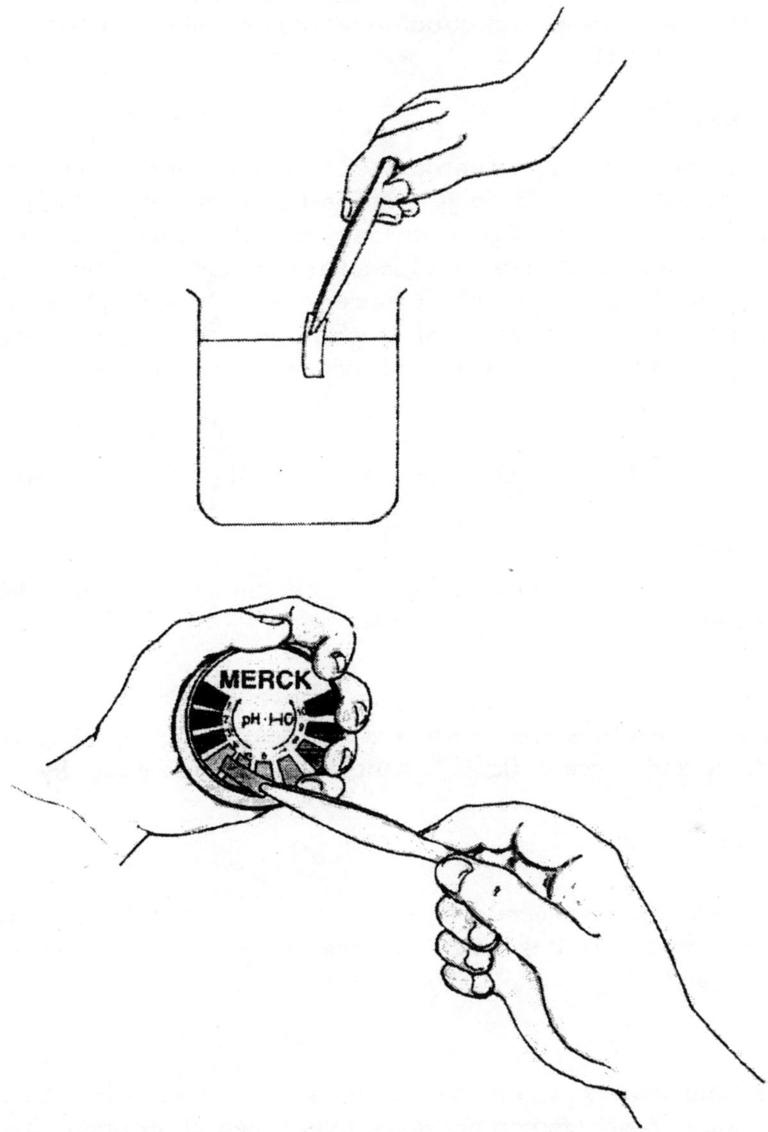


**Fig. 13.3 : For determining soil and water pH, the indicator method is widely used in the field. It is simple and is accurate enough for most purposes.**

For determination of water pH, pond water is taken in a beaker and a strip of litmus paper is immersed into the water. After standing a few seconds, the paper absorbs water and by compairing the color change of the paper to a color chart, the pH value is obtained.

### *Electrometric Method*

It is the most accurate and simple method of determining soil and water pH. In this method a sensing glass electrode is immersed into the sample (either water-soil mixture for testing soil pH or water only for testing water pH) that stimulate the solution. The difference between the $H^+$ ion activities in the sample and in the glass electrode gives rise to an electrometric potential difference that is related to the sample pH.

## 13.7 Increasing Acidity

Although acidic ecosystems are not encouraged to farmers for fish cultivation, some natural organisms which are used by fishes as food, grow best on water and soil pH of 6.0 and below. Therefore, it is sometimes necessary to generate the acidity of ecosystem. For this purpose, acid-forming organic and inorgnaic materials are added. Partial or complete decomposition of organic matter generate inorganic and organic acids that can reduce the soil pH. When the addition of organic matter is not possible, certain chemicals (such as ferrous sulfate) may be used. Hydrolysis of ferrous sulfate produces ferrous ion which enhances acidity by reactions such as.

$$Fe^{2+} + H_2O \rightleftharpoons Fe(OH)_2 + 2H^+$$

$$4Fe^{2+} + 6H_2O + O_2 \rightleftharpoons 4Fe(OH)^+_2 + 4H^+$$

## 13.8 Decreasing Acidity

Soil acidity is usually decreased by adding oxides, carbonates or hydroxides of calcium and magnesium compounds.

### *Oxide Forms*

Oxide of lime is referred to as *quicklime, burned lime,* or *oxide.* Oxide of lime is more costly and caustic than limestone and hence difficult to handle. It reacts more rapidly with the soil than limestone.

### *Hydroxide Form*

It is commonly referred to as *hydrated lime.* It appears on the market as a white powder and is more caustic than burned lime. It is quite expensive compared to limestone, and used where a rapid rate of reaction and a high soil pH is necessary.

### *Carbonate Forms*

Although the main sources of carbonate forms include oyster shells, marls, and basic slag, ground limestone is the most common and extensively used of all liming materials. The two important materials incuded in limestones are *dolomite,* which is calcium-magnesium carbonate $Ca.Mg(CO_3)_2$, and *calcite,* which is calcium carbonate ($CaCO_3$). If the ground limstone is entirely composed of calcium-magnesium carbonate and impurities, it is referred to as *dolomite.* When little or no dolomite is present, it is termed as *calcite.* The average total carbonate level of the crushed limestone is about 96 per cent.

## 13.9 Decreasing Alkalinity

In many oligotrophic ponds and lakes where pH of water drastically increases (more than 9.0), fish culture strategy has created several problems. High pH causes some of the calcium in the water to precipitate and thus, rises the sodium absorption ratio and the hazard of increased exchangeable sodium percentage. To counteract these problems, sulphuric acid is sometimes sprayed over water to reduce pH of the water. This practice, may, however, well recommend where there are economical sources of this acid and where the fish farmers using it have been trained and alerted to hazards of using the acid.

## 13.10 Reactions of Liming Materials in Aquatic Ecosystem

When liming substances are added to a fish pond, the magnesium and calcium compounds react with the acid colloidal complex and with carbon dioxide.

### *Reaction with Soil Colloids*

Different forms of lime react with acid soils, the calcium and magnesium replacing hydrogen and aluminium on the colloidal complex :

$$\boxed{\text{Micelle}}\begin{matrix}H^+\\H^+\end{matrix} + Ca(OH)_2 \rightleftharpoons \boxed{\text{Micelle}} \leftarrow [Ca^{2+} + 2H_2O$$

$$\boxed{\text{Micelle}}\begin{matrix}H^+\\H^+\end{matrix} + Mg(OH)_2 \rightleftharpoons \boxed{\text{Micelle}} \leftarrow [Mg^{2+} + 2H_2O$$

$$\boxed{\text{Micelle}}\begin{matrix}H^+\\H^+\end{matrix} + CaCO_3 \rightleftharpoons \boxed{\text{Micelle}} \leftarrow [Ca^{2+} + H_2O + CO_2$$

$$\boxed{\text{Micelle}}\begin{matrix}H^+\\H^+\end{matrix} + MgCO_3 \rightleftharpoons \boxed{\text{Micelle}} \leftarrow [Mg^{2+} + H_2O + CO_2$$

The adsorption of the calcium and magnesium ions increases the percentage base saturation of the colloidal complex, along with the pH of the soil.

### *Reaction with Carbon Dioxide*

When any forms of liming material is added to a pond, it reacts with carbon dioxide and water to produce bicarbonate form. The carbon dioxide partial pressure in the mud is several hundred times greater in atmospheric air which is high enough to trigger the following reactions.

$$CaCO_3 + H_2O + CO_2 \longrightarrow Ca(HCO_3)_2$$

$$Ca(OH)_2 + 2CO_2 \longrightarrow Ca(HCO_3)_2$$

$$CaO + H_2O + 2CO_2 \longrightarrow Ca(HCO_3)_2$$

### *Depletion of Calcium and Magnesium*

Since soluble calcium and magnesium compounds are moved from soil to water, algae and macrophytes are absorbed these elements from water. Consequently, the pH and percentage base saturation are reduced and therefore, application of lime is necessary. These elements, however, are depleted from acidic soils. Table 13.1 indicates losses of calcium and magnesium by leaching compared with those from the removal by aquatic plants and fish. Table indicates that the loss of calcium carbonate is higher (493 kg/hectare/year). It is significant to note that there is considerable variation in these losses among different fish culture ponds and in different geographical conditions. Hence the liming of soil must be repeated with regularity. To keep the nutrient of pond water and soil in balanced condition, application of these elements through liming should not be ignored.

**Table 13.1 : Magnesium and Calcium Losses from Water by Plant and Fish Removal and Leaching in an Acidic Pond. Values are in Kilograms/hectare/year**

| Type of removal | Magnesium expressed as | | Calcium expressed as | |
|---|---|---|---|---|
| | Calcium | Calcium carbonate | Magnesium | Magnesium carbonate |
| Leaching from a sandy-loam soil | 110 | 278 | 32 | 95 |
| Average fish (per hectare) of a standard production | 90 | 75 | 35 | 80 |
| Average plant (per hectare) of a standard growth | 75 | 140 | 30 | 92 |
| Total | | 493 | | 267 |

Compiled from different published literature

## 13.11 Chemical Equivalent of Calcium and Magnesium Compounds

Chemical equivalent of any liming substances indicates that one molecule of calcium carbonate ($CaCO_3$) will neutralize one molecule of either calcium hydroxide ($Ca(OH)_2$) or magnesium oxide (MgO) or calcium oxide (CaO). To determine the chemical equivalency of these materials, the ratio of the molecular masses is usually considered.

$$\frac{CaCO_3 \text{ (Calcium carbonate)}}{CaO \text{ (Calcium oxide)}} = \frac{100}{56} = 1.79$$

One quintal of pure calcium oxide has a calcium carbonate equivalent of 100 x 1.79. If calcium oxide is 90 per cent pure, then one quintal would supply 90 kg of calcium oxide so that 90 per cent calcium oxide would have a calcium carbonate equivalency of 90 x 1.79 = 161.1. Thus, by using molecular ratio, equivalency of following liming substances can easily be calculated :

$$\text{CaO equivalent of } CaCO_3 \quad \frac{CaO}{CaCO_3} = \frac{56}{100} = 0.56$$

$$CaCO_3 \text{ equivalent of } MgCO_3 \quad \frac{CaCO_3}{MgCO_3} = \frac{100}{84} = 1.19$$

$$\text{Mg equivalent of MgO} \quad \frac{Mg}{MgO} = \frac{24}{40} = 0.60$$

$$\text{Ca equivalent of } CaCO_3 \quad \frac{Ca}{CaCO_3} = \frac{40}{100} = 0.40$$

### *An Example*

If a fish farmer wants to know how much of limestone (calcium carbonate equivalent = 90) would be required to neutralize the same acidity as 1 metric tonne of a burned lime with a calcium oxide equivalent of 95 per cent, then the following steps should be worth remembering :

(i) The burned lime has the neutralizing ability of 1000 x 0.95 = 950 kg of pure CaO.

(ii) This amount of pure CaO is equivalent to 950 x 1.79 (calcium cabonate equivalent of calcium oxide) = 1700.50 kg of pure $CaCO_3$.

(iii) Since $CaCO_3$ equivalent is 90, the quantity of limestone required is

$$\frac{1700.50}{0.90} = 1{,}889 \text{ kg}$$

## 13.12 Fineness of Limestone and their Reaction

It is obvious that if the liming material is fine, it reacts more quickly with the soil particles. Generally calcium hydroxide and calcium carbonate are available in powder forms and therefore, there is no doubt about their fineness. Since different forms of limestones are used in fish culture, their particle size and hardness may vary considerably. To use these coarse and harder limestones, their fineness should be considered.

### *Measurement of Fineness*

The fineness of different forms of lime is measured by passing the liming substances through a series of screens openings of designated size. A number 200 mesh screen has 200 wires per inch and opening sizes of 0.075 mm. Similarly, a 100 mesh screen opening is 0.0375 mm and a 10 mesh screen opening sizes of 2 mm. Generally the size opening of a given screen indicates the maximum diameter of lime particles that can pass through the screen.

### *Reactions with Soil and Fish Production*

The effect of calcium oxide and calcium carbonate on the rate of reaction with soil was determined. It has been found that at the end of 2 months, about 80 per cent of calcium carbonate in the 100 mesh size had reacted with the soil, but less than 30 per cent of the 30 mesh size particles had reacted. On the other hand, at different size particles the reaction rate of calcium oxide with the soil was less than that of calcium carbonate (Figure 13.4). This is significant to ascertain liming rates.

Fish production in ponds is greatly influenced by different types of liming materials and their fineness. Figure 13.5 presents a data from several fish culture ponds to explain this statement. To obtain about 53 per cent of fish production, 2,000 kg/hectare/year of calcium carbonate would be required if 40 per cent of the particles passed through a 60 mesh screen. On the other hand, to achieve 90 per cent of the production about 500 kg/hectare/year of calcium carbonate would be needed if 95 per cent of calcium carbonate particles could pass through a 60 mesh screen. These data point to a conclusion that liming substances with about 95 per cent passing through a 60 mesh screen are satisfactory for higher production of fish in lateritic ponds. Although finer particles of liming substances may give higher fish production, the additional cost involved in grinding liming materials should not be overlooked.

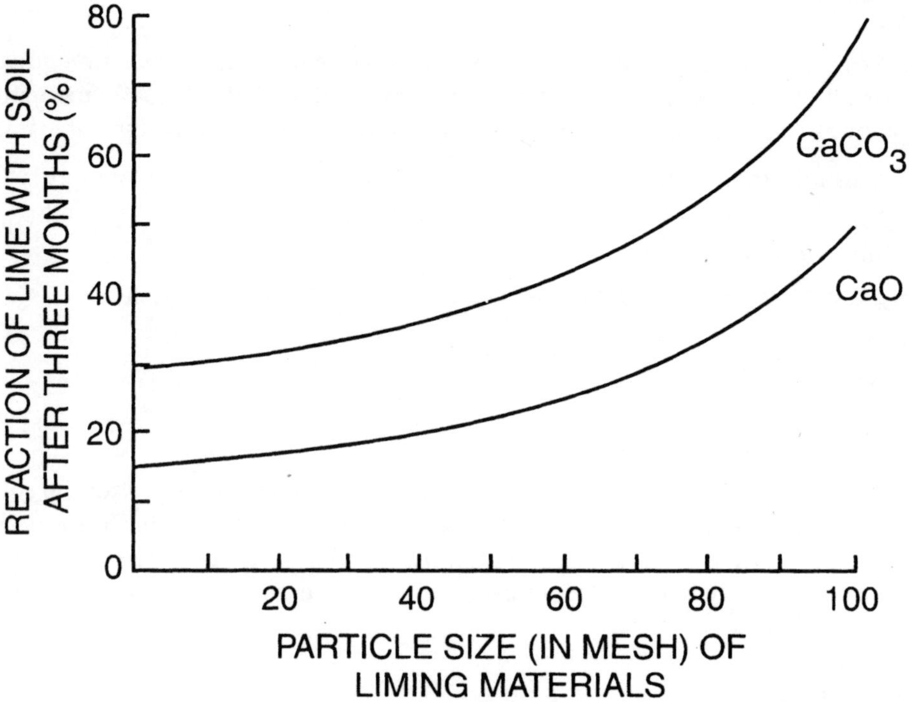


**Fig. 13.4 : Relationship between particle size of calcium oxide (CaO) and calcium carbonate ($CaCO_3$) and their reaction rates with the soil. Note that the rate of reaction of $CaCO_3$ particles with the soil at a given size is higher than that of CaO. The neutralizing capacity of the coarse fractions of these two materials is very poor. [Redrawn from Sarkar (1990)]**

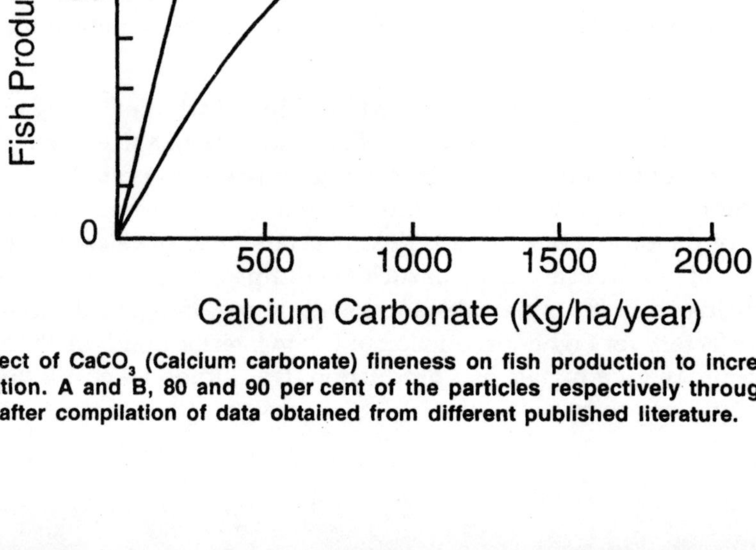


**Fig. 13.5 : The effect of $CaCO_3$ (Calcium carbonate) fineness on fish production to increase rates of $CaCO_3$ application. A and B, 80 and 90 per cent of the particles respectively through 60 mesh screen. Drawn after compilation of data obtained from different published literature.**

It should further be added as a note that treatment of lateritic ponds with calcium carbonate exhibited higher production of different species of major carps than that of calcium oxide (Figure 13.6). This indicates that for liming fish ponds, calcium carbonate is more superior than calcium oxide.

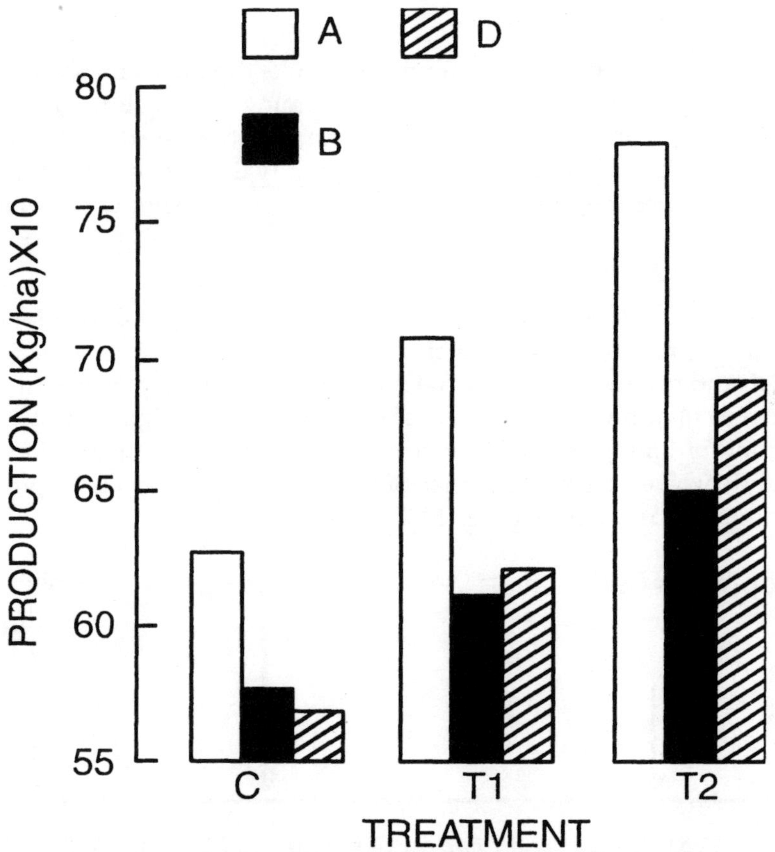


**Fig. 13.6 :** **Influence of calcium oxide (CaO, T1) and calcium carbonate ($CaCO_3$, T2) on the production of fish in acid lateritic ponds (Water area varied between 0.02 and 0.05 hectare). A, *Labeo rohita*; B, *Catla catla*; D, *Cirrhinus mrigala*; C control. The CaO and $CaCO_3$ were separately applied over the water surface in equal instalments at 20 days interval each and at the 2,000 kg/ha/year rate. Experiments were carried out for 3 months and fishes were daily fed with balanced fish feed at the rate of 6 per cent of body weight. [Redrawn from Sarkar (1990)]**

## 13.13 Practical Considerations

The characteristic features of soils and liming materials along with cost factors determine the type and quantity of lime to be used for pond productivity. Generally 1,000 kg/hectare/year of ground limestone is applied for pond management if the pH of soil is about 7.5. But even higher rate of ground limestone (about 2,000 kg/hectare/year) may be appropriate on acid lateritic pond (where pH drops below 6.0). Among different forms of lime, however, ground limestone (calcium carbonate) should be favoured to maintain adequate nutritional balance and to increase the activities of soil micro-organisms.

To achieve most satisfactory response of living substances to pond productivity as a whole, their uniform application over water surface through different methods (see Chapter 11) should be recommended to fish farmers. In ordinary practice, calcium oxide or calcium hydroxide is bulk applied for fish culture. But in many situations, sparse application of liming materials is recommended to fish farmers. This prevents uniform rise in soil pH, which might occur if liming substances were not uniformly applied.

Application of lime in excess than requirement results drastic increase in pH value which is very detrimental to any fish culture ecosystem. Over-liming is not common in fish ponds where soil pH varies between 7.5 and 8.3. It generally occurs in coarse-textured soils that are low in organic matter. The harmful effects of over-liming include availability of phosphorus and nitrogen. Under special circumstances, however, application of excess lime should be avoided.

## 13.14 Lime as Pond Fertility

Application of lime to fish ponds maintains the levels of exchangeable magnesium and calcium. Moreover, liming also provides a physico-chemical environment of an aquatic ecosystem that initiates the growth and production of fish and natural food. Liming materials also counteract the acid-forming tendency of nitrogen carriers whose use has increased significantly in the past several decades. Application of liming materials in fish culture ponds is a foundation for tropical and sub-tropical zones. However, the maintenance of fish pond fertility levels largely depends on the judicious use of liming substances (Figures 13.7).

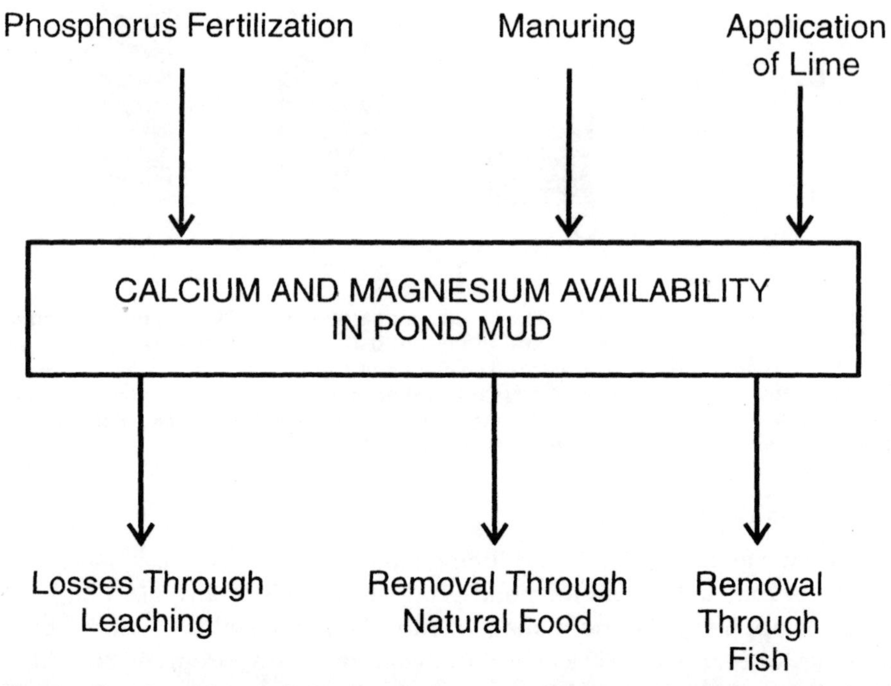


**Fig. 13.7 : Different ways by which available magnesium and calcium are added to and removed from pond mud. Calcium and magnesium are lost through leaching and fish harvest. These losses are replaced by the application of lime, manures, and fertilizers. Phosphorus fertilizers contain large amounts of calcium.**

## 13.15 Conclusion

Among different chemical characteristics of soil and water, pH is more important in determining the chemical environment of ponds, and its effect on growth of the fish and naural food. In contrast to acidic ponds, alkaline ones are usually considered as favourable for fish culture. This condition must be recognized in any fish culture management strategy.

The pH of pond mud is controlled by soil colloids and their associated exchangeable cations. Several base-forming cations encourage soil alkalinity, whereas hydrogen and aluminium increase soil acidity. The influence of pH on the availability of nutrients to plankton, its effect on fish survival and growth, and the control of acidity is exremely significant.

### *References*

Robertson, W.K., M.C. Lutrick, and T.L. Yuan. 1982. Heavy applications of liquid digested sludge on three Ultisols : I. Effects on soil chemistry. *J. Environ. Qual.* 11 : 278-283.

Sarkar, S.K. 1990. Lime treatment in acid lateritic ponds. *Environ. Ecol.* 8 : 1276-1280.

## Questions

1. What are the simplest means of determining soil and water pH and what are the principles involved?
2. What are the sources of acidity in a fish culture pond?
3. How the pH of a pond soil is changed?
4. State how liming materials react with water and soil.
5. A fish farmer wants to select between two types of limestones for his pond treatment. Limestone A has an analysis of 55 per cent calcium carbonate and 35 per cent magnesium carbonate and costs Rs. 500 per quintal. Limestone B has a neutralizing capacity (calcium carbonate equivalent) of 90 per cent and cost Rs. 450 per quintal. Which type of limestone he will select? Why?
6. Two ponds (Pond A and Pond B) have the same pH value (pH 6.5). To bring the soil pH of pond A and B to pH 7.5 about 2 and 1 tonnes of limestone respectivley are required. How do you account for this difference?

# 14

# Activities of Inland Fisheries Sector : Indian Scenario

Perhaps the most important feature of the inland fisheries sector is its development – that is, whether adequate manpower, extension, research, and developmenal strategies are available. Freshwater fisheries extension, research and developmental activities by scientific and technical personnels respond markedly to high production potential because they tend to control so much of fish culture environment.

In India, where population explosion is severe, manpower is easily available and may be engaged for various activities in fisheries sector. However, their influence is so pronounced that fisheries extension activities have become one of the most outstanding characteristic of fish culture under Indian conditions.

Research and development in fisheries sector requires comparatively high grade personnels and sophisticated laboratories with financial assistance. These personnels are able to develop fish culture resources for significant production. Though environmental conditions sometimes limit fish production, Indian aquaculture has witnessed an impressive development, but in some situations it is declining.

### 14.1 Production Potential

During the last two decades, India witnessed rapid growth in freshwater fisheries sector with an increase in fish production from 2.170 million metric tonnes in 1984-1985 to 980 million metric tonnes in 1997-1998 achieving an approximate annual growth rate of 6.2 per cent. About 6 million fishermen are actively engaged in this sector. However, in contrast to exploitation of marine fisheries resources, where it involves high capital investment, the inland fisheries sector has shown an impressive growth and immense scope for further development in the utilization of inland fisheries resources potential through adoption of improved technology. Moreover, Government of India has declared not to allow deep sea fishing through joint venture in collaboration with foreign companies. This strong decision has compelled to stimulate pressure on inland fish culture. High primary productivity and high growth rate of fish in tropical fish culture environments clearly indicate excellent production potential. However, a gap between the potential and its exploitation is seen to exist in different freshwater fisheries resources (Table 14.1).

### 14.2 Freshwater Fisheries Resources

Indian freshwater fisheries resources are very diversified. They vary from high altitude sub-temperate streams, rivers, lakes, reservoirs, and ponds to sub-tropical and tropical water bodies belonging to freshwater, brackish-water, and sea.

**Table 14.1 : Indian Freshwater Fisheries Resources, Total Area Available (TAA), Present Exploitation (PE), and Production Capacity (PC)**

| Resource | TAA | PE | PC | Technology adopted |
|---|---|---|---|---|
| Rivers | 28,720 km | 2,790 tonnes | 7,810-13,110 kg/km/yr | Traditional |
| Streams and canals | 1,39,580 Km | 1,920 tonnes | 25-60 kg/km/yr | Drag net |
| Lakes and reservoirs | 2 million ha | 47,500 tonnes | 65-90 kg/ha/yr | Traditional |
| Tanks | 9,35,500 ha | 1,99,850 tonnes | 670-1,150 kg/ha/yr | Extensive to semi-intensive |
| Ponds (under FFDA) | 3,90,515 ha | 2,100 kg/ha | 65,500-10,500 kg/ha/yr | Extensive to semi-intensive |
| Ordinary ponds | 2.75 million ha | 350 kg/ha/yr | 900-1,700 kg/ha/yr | Extensive to semi-intensive |
| Derelict waters and flood plains | 2,19,990 ha | 45 kg/ha/yr | 75-125 kg/ha/yr | Extensive |
| Ox-bow lakes and beels | 20,170 ha | 75 kg/ha/yr | 140-250 kg/ha/yr | Extensive |
| Bheries | 1,85,805 ha | 1,07,980 tonnes | - | - |
| Aquaculture | 1.5 million ha | 76,500 tonnes | 300-1,100 kg/ha/yr | Extensive to semi-intensive |

India has a wealth of rivers due to its unique geographical position and climate. Inland fisheries resources are supported by nine major river systems such as the Indus, the Ganga, and the Brahmaputra in the North, the Tapti and the Narmada in Central India, the Krishna, the Mahanadi, the Godavary, and the Cauvary in the South. With streams and canals of 3.10 million square kilometers and total length of 1.9 lakh kilometers. Reservoirs consturcted on these rivers have 2 million hectare with storage capacity of 16.250 cubic kilometers. Reservoir project on the Narmada will add 29 large, 135 medium, and 3,000 small reservoirs covering a water area of 2,67,500 hectare. However, the typical features of different freshwater fisheries sectors in India is shown in Table 14.2. The area under freshwater aquaculture in the form of ponds and tanks is estimated to be about 2.85 million hectare.

## 14.3 Manpower Engaged in Indian Fisheries Sector

It would not be exaggerated to say that development and management of freshwater fisheries resources in collaboration with fishermen and fish farmers require education, research, training, and technical as well as financial supports. Because of ample scope to upgrade techniques for re-exploration of resource potential in freshwater sector, no clear-cut formula by which adequate manpower required for fisheries activities could be decided. However, the inland fisheries sector requires experts in the fields of (1) fish biology, (2) limnology, (3) fish production technology, (4) fish breeding and hatchery technology, (5) fish pathology, (6) environment assessment, (7) pollution, (8) fish genetics, (9) fish physiology and nutrition, (10) aquaculture engineering, and (11) fisheries training, extension as well as sociology.

**Table 14.2 : Features of Freshwater Fisheries Sector in India**

| Feature | Specification |
|---|---|
| Total area of India | 3.7 million sq. km |
| Tanks and ponds | 22.42 lakh hectare |
| Derelict water area | 1,07,700 hectare |
| Ox-bow lakes and beels | 20,170 hectare |
| Rivers, streams, and canals | 1,68,300 kilometers |
| Paddy fields | 2.07 lakh hectare |
| Number of hatcheries | |
| Government sector | 339 |
| Private sector | 593 |
| Estimate of production potential | 3.2 million metric tonnes |
| Nursery ponds | 2,940 hectare |
| Fish Farmers Development Agencies | |
| Number | 419 |
| Water area under cultivation | 4.25 lakh hectare |
| Average production | 2,100 kg/hectare/year |
| Training to fish farmers | More than 4.6 lakh farmers |
| Per caput fish consumption per year | 5.9 kg |
| Average production/hectare/year | 2.5 tonnes |
| Number of co-operative societies | 7,098 |

*Source* : Modified after Dehadrai (1997) and Kohli (1998)

### *Scientific and Technical Manpower*

The principal scientific and technical manpower in different inland fisheries sectors in India under the financial assistance of Indian Council of Agricultural Research (ICAR) are given below :

1. The National Research Centre on Cold Water Fisheries at Haldwani has research and development activities for capture and culture of cold water fishes. This institute has a cadre strength of 20 scientists and 9 technical officers.
2. The Central Inland Capture Fisheries Research Institute located at Barrackpore, West Bengal, has a cadre strength of 100 scientists and 100 technical officers. This institute has been divided into three divisions at Bangalore, Allahabad, and Barrackpore for research and development programs of reservoirs, rivers, and estuaries, respectively.
3. The Central Inland Freshwater Aquacultrure at Bhubaneswar, Orissa, provides research support with 78 scientists and 63 technicians.
4. The National Bureau of Fish Genetic Research, Lucknow, U.P., has a cadre strength of 40 scientists and 12 technical officers.

## 14.4 Further Requirement of Manpower

It has been estimated that for successful development and monitoring of fisheries in about 2 lakh kilometers of rivers, a total number of about 1,700 development officers and 17 lakh fishermen/ fish farmers will be required. Most of the river systems are being gradually polluted and as a result of which fish population has been depleted considerably. Unless pollution of river water is controlled it would never be possible to rejuvenate the potentiality of fisheries resources in river systems; otherwise the utilization of manpower will be of no use.

For optimum exploitation of about 2.05 million hectare of reservoir area would need service of about 2,000 development officers and four lakh fishermen. But this effort requires heavy financial support.

## *Fish and Fish Seed Production*

At present, development of freshwater aquaculture has taken place at rapid rate and this sector contributes to about 75 per cent of the total inland fish production. Increasing trends in inland fish production in different states of the country during the three consecutive years are shown in Table 14.3. If this trend continues, however, the country will reach a plateau of inland fish production in future.

**Table 14.3 : Inland Fish Production (in tonnes) in Different States of India During Three Consecutive Years**

| State | Inland fish production | | |
|---|---|---|---|
| | 1994-95 | 1995-96 | 1996-97* |
| Andhra Pradesh | 1,95,130 | 2,03,970 | 3,15,700 |
| Andaman and Nicobar Islands | 50 | 50 | 85 |
| Arunachal Pradesh | 1,800 | 1,850 | 19,900 |
| Assam | 1,53,000 | 1,55,060 | 1,70,200 |
| Bihar | 1,95,370 | 2,37,580 | 2,75,850 |
| Chandigarh | 80 | 80 | 100 |
| Daman and Diu | 40 | - | 65 |
| Delhi | 3,900 | 4,000 | 4,300 |
| Goa | 3,440 | 3,610 | 3,900 |
| Gujarat | 70,100 | 60,000 | 75,500 |
| Haryana | 24,120 | 28,010 | 30,400 |
| Himachal Pradesh | 5,280 | 5,940 | 6,200 |
| West Bengal | 6,69,220 | 7,40,000 | 8,60,500 |
| Uttar Pradesh | 1,39,900 | 1,45,700 | 1,56,000 |
| Tripura | 25,100 | 25,710 | 26,100 |
| Tamil Nadu | 1,08,000 | 1,08,000 | 1,12,700 |
| Punjab | 24,000 | 26,000 | 27,500 |
| Sikkim | 100 | 150 | 210 |
| Rajasthan | 13,960 | 12,400 | 12,900 |
| Kerala | 48,190 | 49,590 | 50,700 |
| Karnataka | 70,290 | 87,350 | 95,000 |
| Jammu and Kashmir | 16,100 | 16,520 | 17,000 |
| Madhya Pradesh | 80,180 | 91,280 | 97,500 |
| Maharashtra | 89,880 | 77,000 | 85,700 |
| Manipur | 12,010 | 12,500 | 13,000 |
| Meghalaya | 3,950 | 3,580 | 4,200 |
| Mizoram | 2,000 | 2,500 | 3,150 |
| Nagaland | 2,500 | 3,000 | 3,700 |
| Orissa | 1,34,770 | 1,34,850 | 1,50,550 |
| Pondicherry | 4,190 | 4,000 | 4,500 |

*Source* : Fishing Chimes, Vol. 17(2), May, 1997, p. 64.

* Compiled by the author from various published literature.

Besides different institutions under the Indian Council of Agricultural Research, Government of India has implemented "Fish Farmers Development Agency" (FFDA) schemes in different states that has helped tremendous growth in fisheries sector. At present, 414 FFDAs are in active condition throughout the country covering the water area of about 4 lakh hectare. Average fish production in these areas has substantially increased form 900 kg/hectare/year during 1984 to about 2,610 kg/ hectare/year during 1997-1998. This production, though significant, is the overall productivity from ponds and tanks. But the average production is low mainly due to non-availability of adequate healthy fingerlings for stocking, and the lack of quality fish feed. Use of fertilizers has been restricted in many ponds due to multiple use of water that has resulted dramatic decline in fish production.

Freshwater fish seed production has been considerable increased in every states of India. More than 15,000 million fish seed is being produced every year. About 335 hectare have been earmarked for fish seed production in Government sector with a total capacity to produce about 3,040 million fish seed and about 58 hatcheries have been set-up in the Private sector with a capacity of about 12,290 million fish seed. To improve all freshwater areas, however, about 60 lakh farm workers and about 28,500 development officers will be required; of course, financial constraints delay to implement the program.

## 14.5 Creation of Inland Fisheries

In India, the inland fisheries extension unit was first established in 1952. Thereafter, a number of extension units were also gradually established and in 1962, the total number of unit come up to 10. In 1967, different Central Government extension units were closed down at the end of the third Five Year Plan because fisheries extension was declared as State Government affair. But during the Fifth Five Year Plan (1974-1979), attention has been given towards management, development, extension, research, and training on inland fisheries.

At present, five organizational set-up are actively engaged for enhancing inland fish and aquaculture production : (1) Ministry of Agriculture and State Fisheries Departments, (2) The Indian Council of Agricultural Research Unit, (3) Rural Development set-up (4) Non-Government set-up, and (5) Agricultural Universities and Fisheries Colleges. Moreover, these set-ups are also associated with two or more farming components which became part of the entire system – is referred to as *integrated farming.*

### *Present Status*

Although the country has achieved significant position in inland fish production during the past several years, the country is still laging far behind in mean levels of production when compared to other developed countries. On the other hand, various states of the country also exhibit wide fluctuations of environmental and agroclimatic conditions. For example, production in composite fish culture ponds in four zones of the country exhibited differential productivity (Table 14.4).

### *Status of Fish Farmers*

The glory for the fishermen came in 1980 when Government established the fishermen's co-operative society along with the financial assistance and subsequent implementation of different programs in fisheries sector that conferred the special status. With this status, however, fish farmers/ fishermen earned wide-spread recognition in the field of fisheries development across the country.

This status also spurred the fish farmers/fishermens society to improve their working results to a large extent. Consequently, the activities of the development and management continued to enhance greater levels of performance.

**Table 14.4 : Production in Composite Fish Culture Under Indian Condition**

| Zone | Production (kg/ha/year) | |
|---|---|---|
| | National demonstration level | Farmers level |
| Eastern states | 4,448 | 2,445 |
| Western states | 5,621 | 2,250 |
| Northern states | 3,754 | 2,520 |
| Southern states | 2,927 | 1,200 |

*Source* : Bhowmik (1997).

Since the demand regarding transfer of technology for ameliorating freshwater aquaculture production is fomidable, which is inter-disciplinary in approach, integrated activities in education, research, and training have already become imperative for further development of aquaculture strategies. At present, fisheries development and management sectors are well equipped with a large number of scientists, technicians, officers, planners, and administrators at National, State, District, and Block levels. But inspite of these advantages, there always exists a gap between fishermen and technology, thus creating antagonistic effect so far as the developmental activities are concerned.

### *Four Systems of Development*

1. The Indian Council of Agricultural Research is the top level agricultural research organization that assists research, education and training to the nation. Different fisheries research institutes under this organization execute research activities on various aspects of freshwater fish and fisheries.
2. Agricultural Universities and Fisheries Colleges have been set-up all states of the country for education and research on various aspects of fisheries science.
3. State Fisheries Departments in all the states regulate extension services in freshwater fisheries sector. The training centre of Central Institute for Fisheries Education at Kakinada, Andhra Pradesh, and the extension wing of Central Inland Capture Fisheries Research Institute, Barrackpore, West Bengal, provide in-depth current developmental strategies to extension personnels of the State Fisheries Departments. Moreover, Krishi Vigyan Kendras also provide adequate training to fish farmers. At the same time, several extension organizations whether Central or State Governments or Universities/Colleges or Non-Governmental, organize short-term training courses, Fish Farmers' Day, exhibition, group discussion and demonstration programs.
4. Although the main objective of various extension and training systems is to provide recent fish culture developmental informations to fish farmers/fishermen in due course of time, the extent of implementation of informations usually differ in the economic, cultural, and social features of farmers. Also, the geographical as well as climatic conditions in a particular fisheries sector play key role as to what extent the latest informations will be effective. However, about 60 per cent of fish farmers have been found to adopt recent technological know-how.

## 14.6 Benefits to Fish Farmers

In India, development and management of freshwater fisheries sector was not popular until recently when there was shortage in fish protein. To meet this short-fall, professional technical personnels advocated the importance of fisheries extension and training for increasing fish production. Using the conventional method of culture, fish cannot be increased beyond certain limits. For long time, demand for freshwater fish is high which has increased recently. Consequently, attention has already been given in recent years towards freshwater fisheries development and management using different methodologies as given below :

### *Lab to Land Program*

In this program, different technologies are transferred from the laboratories to experimental tanks and ponds so that marginal and small fish farmers may increase fish production.

### *Fish Farmers' Development Agency*

During Fifth Five Year Plan, Government of India has established an integrated program of freshwater fish culture as a pilot project in different states of the country under the scheme of Fish Farmers Development Agency. At present, the scheme is very successful.

## 14.7 Constraint in Fisheries Activities

While extension work, training, education, research, and development on various branches of fisheries and aquaculture are very complex, a number of constraints appear in fisheries sector are extremely grim and therefore, it is necessary to identify the constrants faced by fish farmers.

### *Finance*

During the last fifteen years far-reaching changes have taken place in the Indian fisheries scenario which have changed the shape of Indian farming policy. Actually, the changes started during the 1990s with the opening up of new fishery technology following the withdrawal of traditional farming practices. However, the financial support to fish farmers in recent years by Central and State Governments have given a new shape to fisheries industry in the country.

Economic growth has widely been accepted by the economists as a major goal of national policy. The economic development of fishermen's community depends upon the development of financial system of the country. The concept of financing to poor fish farmers through loan or subsidy as envisaged under planned economic development is a very important machine for the growth and development of fisheries activities by providing financial assistance to the needy farmers.

Financial constraint is the most important aspect that regulate fisheries development and planning as a whole. Since fish farmers live below the proverty line, they do not have adquate fund and consequently, they cannot cultivate their ponds scientifically. These aspects point to an obvious conclusion that there should be a balance in growing awareness among fish farmers through increasing fish production.

### *Fish Production Group*

During Sixth Five Year Plan (1981-1986), Different State Governments of the country have established fish production group with a view to provide technical know-how, financial assistance, and planning as well as developmental strategies among fish producers.

### *State Fisheries Departments*

The State Fisheries Departments have launched some important developmental programs to uplift the socio-economic status of fish farmers and creation of employment opportunities. Developmental programs have been undertaken by various agencies such as District Rural Development Agency, Industrial Refinance Development Program and National Co-operative Development Corporation. Some of the important programs on inland fisheries development launched by the State Governments are (i) preservation of brood fish, (ii) workshops on fish culture and fisheries, (iii) pollution control strategies, (iv) eco-friendly freshwater fish culture, (v) hatchery establishment, (vi) cage fish culture, (vii) induced breeding and fish seed production, and (viii) extension, training and other management protocols.

### *Krishi Vigyan Kendra*

This Government organization has been established in each state with a view to impart grass root level training programmes to fish farmers on the basis "learning by doing" and to solve the problems experienced by fish farmers during fish culture operations.

### *Demonstration Programs*

These programs are undertaken in several composite fish culture ponds with a view to implement and exhibit modern technologies stepwise among fish farmers. For example, demonstration programs were launched by the West Bengal State Fisheries Department in 500 composite fish culture ponds in different districts of West Bengal and average fish production was impressive — it is about 4,200 kg/hectare/year (range varied between 3,500 and 4,950 kg/hectare/year). These variations are, however, due to several reasons such as awareness among farmers and their standard of living different management protocols, and geographical as well as climatic conditions. Therefore, prior to the adoption of fish production technologies, these factors should be considered. The demonstration programs must be enforced among farmers and this will not only pave the way for augmenting fish production and creating additional employment in the rural sector but also will provide nutrition to rural people. Since lack of awareness is most prevalent among farmers, the progress of demonstration programs should be carefully monitored. At the same time, it is also necessary to identify how the cost of fish production can be reduced to a great extent.

### *Unavailability of Inputs*

It is very difficult for fish farmers to obtain quality fish seed from suitable locations for intensive fish cultivation. Consequently, farmers are compelled to collect fish seed from rivers. But riverine fish seed always contains a lot of undesirable and uneconomical fish seed. As a result, fish culture management strategies are jeopardized. Moreover, inspite of the recommendation for application of mahua oil cake as fish toxicant during pond preparation, it is not easily available in rural areas. Supplementary feed (such as mustard and groundnut cakes) and organic manures (such as cowdung and poultry litter) may not be available all the time because these ingradients are widely used in agricultural lands.

### *Ownership of Water Bodies*

Generally most of the ponds and tanks are under the control of small and rich farmers and they have least or no interest to undertake fish culture management. Furthermore, multi-ownership of large water bodies is a severe problem because they did neither pay any attention to develop such water bodies nor handed over to poor fish farmers on lease basis.

Much attention in aquaculture is focussed on the actions and the various programs of Sub-national Government. Farmers wake up when the outbreak of acute crises forcibly draws their attention. This phenomenon has been true of the reform process in aquaculture sector. Various technologies have been devoted to the changes introduced by research institutes. However, adoption of various new technologies will obviously foster significant increase in the overall level of production in the country and the production strategy will influence the decision regarding the construction of new fishery projects. Certainly, fish farmers/fishermen can reap the benefits from improved infrastructure.

### *Middleman*

The fishermen community means blackmail by middlemen and the urge to sale of their total production at throw-away prices would undoubtedly provide more income to midlemen that results in devastation of farmers community. This consideration is, however, very important because financial crisis of the poor fishermen togather with the activities of middlemen and the lack of suitable technology obviously fails to break the encourage of the farmers activity towards production strategies and it becomes more difficult for them to establish solid ground for their livelihood.

It should be added as a point that so far as the total production is concerned, the point of view of the fishermen and their efficiency have not taken into account. It should be remind us that the activities of middlemen that encompasses the fishermen class seems to be overwhelming among them. Additionally not taken into consideration is the exploration of other areas of resources with the aid of cost-effective technology to the benefit of the majority of the national population.

### *Social*

Due to non-cooperative atitude among fish farmers, most of the water bodies remain unutilized or if utilized, may exhibit several problems. For example, poisoning of fish culture ponds in rural areas is a common practice that drastically damage the entire program.

### *Lack of Extension Facility*

All inland fisheries development strategies should be supported by extension components to promote latest developmental activities and also to provide adequate data for further planning. It is obvious that most of the fish farmers are either fully engaged or partially engaged in fisheries activities and that they are ignorant to adopt recent developmental strategies. Therefore, if there is any gap between fish farmers and technological know-how, the rural economy will not be improved. In this case, timely advice to fish farmers should be given by extension personnels.

## 14.8 Development of Inland Fisheries

Since the beginning of Seventh Five Year Plan, (1987-88 to 1991-92), inland fish production increased by 6.6 per cent and the average growth of production also increased by 7.2 per cent. Out of the total yield of 49,49,380 tonnes during 1995-1996 eighth five year plan, the inland and marine sectors produced 22,42,320 and 27,07,060 tonnes, respectively. However, of the total present production of 3.07 million tonnes during 1997-1998, about 75 per cent of inland fish obtained from aquaculture.

Similarly, the various development programs launched by the State Governments in the Ninth Five Year Plan (1997-2002) exhibited encouraging results as evident from the fact that the inland fish production has further enhanced during 2000-2001 to the tune of 6.5 million tonnes. A target of 15 million tonnes by 2015 AD has been envisaged for the sector.

The inland fisheries sector has been gaining greater importance and fish production aimed at strengthening the economy of the country. This importance depicts that there has been the generations of significant interest among fish farmers and fishermen for commercial utilization of inland fisheries resources in the country. This sector has a great potential in India in view of the vast resources (about 58 lakh hectare) in the form of tanks, ponds, lakes, reservoirs, and other water bodies. But the national average inland fish production is considerable low (about 750 kg/hectare). By adopting fish culture management techniques, this production figure may at least be two-fold.

### *Supply Versus Demand*

In present day practice of fish culture in India, there always exists a gap between the demand of the increasing population and the production of the inland fisheries. This gap may be filled up to some extent by boosting up aquaculture production. An international conference was held in 1991 in *Puerto Rico* (A small island in North America) where scientists declared that world population would reach about 9 billion in 2025 when the production of capture fishery would be 100 million tonnes and the availability of fish and freshwater food would be to stay at 19.2 kg/caput/year. On the basis of these assumptions, aquaculture should fill up the gap of 19.6, 37.5, and 62.4 million tonnes by 2000, 2010, and 2025, respectively. But previous experience has shown that the production of capture fisheries reached at 90.1 million tonnes in 1994. Consequently, a wide gap between supply and demand will be observed. However, it has been well established that by introducing recent biotechnology in aquaculture, it would be possible to shrink the gap gradually.

### *Inland Fisheries Research*

Owing to rapid development in the field of inland fishery and subsequent implementation of technology, the national productivity from Fish Farmers Development Agency – supported ponds and tanks increased from about 50 kg/hectare/year (1974-1975) to about 2,610 kg/hectare/year (1996-1997). It has also been estimated that by the end of 2009-2010 fish production would increase to about 4,000 kg/hectare/year. This assumption has raised question how far the production rate may increase because several environmental factors (such as pollution, disease outbreak, political conflicts etc.), geographical and climatic conditions of ponds and tanks, and management protocols obviously limit fish production. These environmental factors are worth remembering.

Besides fish production through composite fish culture with biotechnology fish seed production from Indian carps[1] and Air-breathing fishes[2] are necessary for further development of aquaculture. Moreover, for better quality of fish seeds, upgradation of fish stocks through genetic engineering and biotechnology[3] are also necessary.

## 14.9 Fish Sanctuary – A Haven for Broodstock

Broodstock, particularly major carp species, is considerably overfished in India and consequently broodfish is being depleted because of the fact that they congregate together at the time of spawning and at the low water level during summer and winter months. The high mortality of the broodstock before they have a changes to spawn reduces their abundance in the succeeding years, thereby bringing down their population and production as a whole. In cases where induced breeding of carp species is executed, there is often unavailability of brood-fish or if available, it becomes much shorter than requirement. Furthermore, brood-stocks are also utilized for recreational purpose and festival programs. Therefore, it is necessary to protect the brood-stock of carps by establishing fish sanctuaries and maintaining breeding as well as spawning grounds.

In India, there are many lakes and ponds where local communities, on recreational and religious considerations, fishing have totally been prohibited. Bithoor (a religious place near Kanpur region of Uttar Pradesh) is one of the best example where fishing is strictly forbided in the mainstream of Ganga river. This has helped in the preservation of several endemic fish species such as *Pangasius pangasius* and *Rita rita*. These fishes are now luxuriantly available in sizes ranging from fry to adult.

### *Objectives of Fish Sanctuaries*

The main developmental objectives are to enhance and rehabilitate of aquatic environments and conservation as well as protection of fish stocks. Fish sanctuary programs would strength the capacity of the Department of Fisheries to undertake fisheries management. It is also necessary for the department to collaborate with the ministers of water and land associated with the environment for inclusion of fisheries aspects into development planning.

Immediate objectives for fish sanctuary project are (1) to identify suitable locations of major broodstocks, particularly their breeding and spawning grounds, and (2) to devise structural measures to reduce fishing efforts to an acceptable level.

### *Activities of Fish Sancuatries*

1. *Field Studies* : The Department of Fisheries would need to set up a planning and management unit to implement the fish sanctuaries project. This unit would carrry out field studies to identify breeding and spawning grounds.
2. *Awareness-Building* : A comprehensive education and awareness-building campaign should be carried out among stakeholders about the needs and benefits of fish sanctuaries and also the methods of sanctuary management and operations.
3. *Pilot Activities* : It is necessary to introduce a system of management to test the effectiveness of different structures in the sanctuary ground. The Department of Fisheries will assist fishermen time to time to undertake monitoring, control, surveillance program for breeding and spawning grounds of broodstocks.

[1,2,3] For review of the importance of these subjects, see Lakra (1998), Sinha (1998) and Thakur (1998).

## 14.10 General Consideration

Success in fish culture industry depends not only upon reducing production costs, but also improving product quality through adequate management protocol. A high quality product encourages consumers to purchase that product again, and in a new market where there is no tradition of fish consumption, it is fundamental if the market is to be successfully developed in the long term.

Whatever the focus of the recent biotechnological achievements in Inland fisheries and aquaculture, it is assumed that farmers' interest in fish culture industry will obviously transcends many obscures so that maximum production of fish may be obtained more efficiently. Recently, biotechnology has pursued its central mission of advancing research in different fields of agriculture, aquaculture, and broad area of technology, and of using that knowledge for the farmers. However, more importantly, it is assumed that the scientific management of fisheries sector will keep farmers at the forefront of fish culture developments in their own interest.

Development of technology indicates that there may be a good time for fish farmers to activate their attitudes towards aquaculture. As humam society move into the new century, there is hardly any area of fisheries science, that does not envisage significant break-through in solving problems of various aspects of aquaculture industry. Virtualy in every aspects of fisheries science, qualified technical personnels have so much to contribute, to using new knowledge for the upliftment of standard of living of fishermen and development of fisheries and aquaculture in general.

## 14.11 Conclusion

Three major conclusions may be drawn about different activities in relation to the development of inland fisheries resources. (1) Research and development in various fields of aquculture and fisheries point to an obvious conclusion that effective implementaion of new technologies will increase production both in terms of quality and value addition. It is expected that inland fish production can be increased several fold if technologies are transferred from research centres to fish farmers. (2) India has become the second member of the "billionaires club". Along with over-population come several ills of country – proverty, malnutrition, unemployment and so forth. Larger population means requirement of emergency food assistance which in turn means more manpower for the country. (3) Training and extension programs can be effective to fish farmers if they obtain financial assistance and if their awareness regarding fish culture strategies is developed.

To obtain better activities of how inland fisheries and aquaculture might be developed, scientists and technicians must devote their research and development efforts to significant fish production. The struggle to produce more fish is not yet lost. Of course, it requires extensive research effort not yet realized.

### *References*

Bhowmik, U. 1997. Status, Constraints and prospects of inland fisheries extension in India. *Fish. Chim.* 17 (9) : 21-25.

Dehadrai, P.V., 1997. Future priorities in fisheries research and development in India. *Fish. Chim.* 17 (1) : 11-16.

Kohli, M.P.S. 1998. Manpower requirement in Indian fisheries sector. *Fish Chim.* 18 (2) : 42-46.

Lakra, W.S. 1998. Advances in fish genetics and biotechnology. *In* : *Advances in Fisheries and Fish Production* (*Ed* : S.H. Ahmad), pp. 34-45, Hindustan Publishing Corporation, New Delhi, India.

Sinha, V.R.P. 1998. Breeding and culture of Indian major carp. *In* : *Advances in Fisheries and Fish Production* (*Ed* : S.H. Ahmand), pp. 63-68, Hindustan Publishing Corporation, New Delhi, India.

Thakur, N.K. 1998. Aquaculture of air-breathing fishes. *In* : *Advances in Fisheries and Fish Production* (*Ed* : S.H. Ahmad), pp. 99-112, Hindustan Publishing Corporation, New Delhi, India.

## Questions

1. Why manpower is considered as an important factor in the growth and development of inland fisheries sector?
2. Trace the genesis of inland fisheries in India.
3. What are the systems adopted in the development of inland fisheries sector?
4. Discuss the main constraints faced by fish farmers.
5. How inland fisheries sector can be developed?
6. What is a fish sanctuary? Why it is necessary?

# 15

# Principles of Management in Fisheries Sector

Management refers to all that a farmer does. Various functions in fisheries sector are carried out by farmers to make efficient use of available inputs so as to produce maximum fish at minimum cost and are generally termed as *management*. The functions such as planning, training budgeting, controlling, decision-making, production, maintenance, and marketing are the process of management. Fisheries management has some objectives such as planning for the future , utilization of different inputs, utilization of technical and trained personnels etc. If proper management principles are not adopted, different activities in fisheries sector will cease gradually. To make the fish culture a profitable business, the following principles of management are worth remembering.

## 15.1 Need for Management

It is needless to mention that fish farmers can do nothing without capital and neither can do anything without proper management. In fact, fish farm managers are essential in different types of fisheries activities to make things happen. The functions of management are essentially the same for any kind of fisheries sector. In all types of sector, however, farm managers operate by achieving goals through the coordinated efforts of fish farmers.

Principally, different fisheries sectors operate to achieve clearly stated and commonly held objectives. The objectives of a fisheries sector have to do with providing services to each fish farmer. In this sector it seems quite plausible that each farmer might do part of jobs which can thought important to fulfil the objectives. To ensure coordination of work and to accomplish the objective, farm managers are essential.

Management in fish culture strategies and in fisheries sector is necessary because without it, efforts of fish farmers would go waste. Consequently, management is necessary for two reasons : (1) to attain objectives and (2) to achieve efficiency and effectiveness.

For attaining various objectives in fisheries sector profitable venture should be offered, otherwise the entire activities of the sector will cease to function.

### *Efficiency Versus Effectiveness*

Two different terms of measuring the activities are usually considered such as *efficiency* and *effectiveness*. The term efficiency refers to the ability to get things done correctly. Efficiency is measured as the ratio of output to input. An efficient farm manager usually succeeds in achieving higher outputs (such as production of fish) relative to inputs (such as fertilizers, feeds, funds etc.). In other words, an efficient farm manager can reduce the cost of the resources used to yield a fixed level of output.

In contrast to efficiency, the term effectiveness refers to the ability to determine appropriate objectives and priorities for doing the correct things.

### *Definition of Management*

Though a number of definitions of management are in vogue, the following one is vrey common and may be considered for general purpose.

The term management refers to as guidance and supervision of the personnels to farmers/ fishermen to determine, interpret and achieve fish culture objectives by executing different activities of cultural aspects such as planning, fertilizing, liming, feeding, stocking, disease controlling, harvesting, and marketing. This definition, however, clearly indicates that farm managers and management make concreate decision to achieve fruitful results. Decision-making is a very important part of overall managment activities. It should be added as a note that farm managers must execute basic functions as referred to above (Figure 15.1).

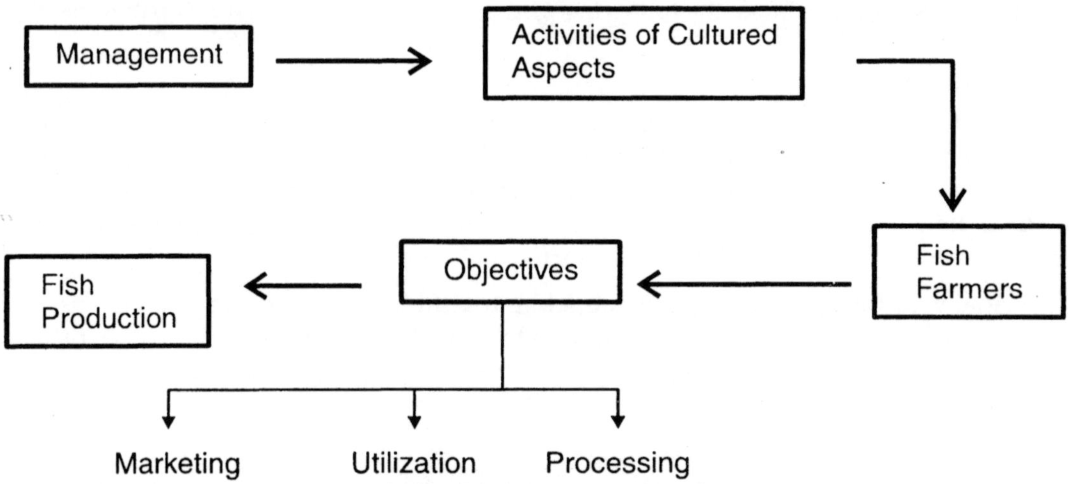


**Fig. 15.1 : Management as an art and a science**

### *Scientific Management*

It is generally agreed that management is the application of certain principles, technologies, and use of computers for decison-making – all these are examples of management as a science. Therefore, scientific management may be defined as the application of latest and systematic techniques during fish culture operation to significant production.

Scientific management which focuses on production efficiency is primarily attributed to the ideas and work. Management of fisheries sector on scientific basis is a very important aspect. At present, traditional management systems have been replaced by scientific one developing the most scientific and rational principles for handling equipments, inputs, and money and to secure maximum benefits for the fishermen. In general, three principles may be considered as basis for scientific management.

1. Management should cooperate with the fishermen to ensure that all activity is done in accordance with the scientific principles.
2. Fishermen should be selected on scientific basis with correct attitudes for the activity and then properly trained to perform the work.
3. The activity and responsibility are to be so divided between fishermen and management that such activity results in interdependence between management and the fishermen.

## 15.2 Planning

It is usually considered to be the basic function of fish farm managers because it has to be undertaken before performing other functions. Planning not only helps farm managers to avoid errors and waste but also aids the efforts of farm managers to become both efficient and effective. Planning involves the execution of several basic steps. Before planning, however, it is necessary for fish farmers to know why planning is essential.

Generally planning is prepared for future development of fish culture industry and the activity that allows farm managers to determine what they want and how it is possible to reach the ultimate goal.

Planning can be defined as selecting a course of action and deciding in advance what is to be done in what sequence, when, and how. Good planning attempts to consider the nature of the freshwater ecosystems in which planning actions is intended to operate.

### *Characteristics of Planning*

The planning function has three characteristics. First, planning is a system of decisions. It involves a decision-making process which will define as to what is to be achieved in the future. Second, it is focussed on desired future results. It means that the important fisheries objectives are accomplished as and when required. Third, a decision has to be made as to what to do and how before it is actually done.

### *Scope of Planning*

For high fish production at minimum cost, efficient farm managers always try to cultivate fish through proper planning. Planning in either small or large water bodies though basically the same, the kinds of plan they develop are different. Fish farmers' co-operative societies are directly concerned on commercial basis. Their planning includes the development of the standard of living of fish farmers.

Planning provides the basis of effective action resulting from management's ability to prepare for changes that might affect aquaculture objectives. Therefore, it forms the basis for integrating the management functions and is particularly essential for controlling fish culture operations.

### *Development of Planning Procedure*

The main reason why environmental aspects have not played an important role in the planning process is due to lack adequate information suited to planning applications. Suitable data must be provided on aquatic resources and presented in forms which can be directly related to the planning function before any significance can be derived. Coordinated programs of application of resource-base information in the planning process. It is necessary to convert technical knowledge into forms which are understandable by the planners and the decision-makers.

Because adequate resources are very important for economic and social development of fish farmers, sound and effective planning on regional or national basis must be accepted and implemented for the existence of aquatic resources. Freshwater fisheries resources, though limited, are subject to severe environmental problems through improper water use. For an efficient utilization of aquatic resources and improved environmental quality, the role of environmental factors in the planning process and establish effective means of communication among fishery scientists, the planners, and the farmers in order to facilitate functional interactions which will serve as a guide to develop policy decision-making. For most effective management, however, regional and national planning systems must be developed. This is accomplished by a coordinated program which interfaces the developmental and scientific planning of aquacultural resources.

### *Advantages of Planning*

Inspite of some disadvantages, the advantages of planning far outweight any problems involved. A few important advantages to fish culture industry include the following :

1. Planning helps minimize guesswork.
2. It tends to make objectives of fish farming system more specific.
3. It also saves time, money and effort.
4. Planning helps to reduce errors in decision-making.
5. Planning helps management to adapt changing fish culture environment.

## 15.3 Fish Culture Extension

### *Meaning and Objective*

Extension education and fish culture extension relate to the process of conveying the technology of scientific fish culture to the fish farmer in order to enable him to utilize the knowledge for better fish culture and economy.

Fishery extension services search (1) to give a share of the skills to the farmers for understanding improved fish culture operations, (2) to make available to them timely information, an improved practices in an easily understandable form suited to their level of awareness and literacy, and (3) to create a favourable attitude for introducing new things and change. Another important objective of extension is to change the outlook of the farming community. Fish farmers have to be educated about the impact of the use of inputs and the difference they make to yields and returns.

### *Extension Methods*

For fish culture extension programs, the following methods are followed : (1) Education and training, (2) Information and communication support, and (3) Demonstration program. These methods are briefly noted below :

1. *Education and Training* : The main objective of the farmer's education and training is to train farm managers who would guide fellow farmers. To spread the knowledge and techniques among farmers, the lab-to-land program has been organized. Under this

program the farming families are introduced to relevant technologies that would help in introduction of supplementary sources of income. Farmers are given regular training in the production methods. However, training makes available technical and financial helps from government and non-government organizations. Education and training have made significant contribution in relation to fish culture operations.

2. *Information and Communication Support* : Information and communcation support is considered as an essential component of farmers extension program for supporting the production activities of the farmers. Non-government and government organizations have been providing information and communicatin support by producing material for extension workers such as film show and audiovisual aid for educating farmers. Fisheries Universities and Colleges in different states of the country have established communication centres and extension departments.
3. *Demonstration* : Demostration is an important method of extension to upgrade the farmers and occupies a significant position in the extension program of the country. In general, there is need to develop for maximizing return with the new technology and to impress the farmers by the experts. The efficacy of new practices based on scientific knowledge and skill must be informed through demonstration programs. For this purpose a contact between the experts and the farmers is necessary in field level to discuss problems faced by farmers. For successful demonstration programs the aquaculture technology centre (TATC) has been set up by the Fish Farmers' Development Agency (FFDA). The TATC demonstrates the method of pond preparation, selective stocking with suitable combination of carp species, harvesting, and other management techniques.

### *Importance of Different Organizations*

The extension directorate of the Government of India and the Directorate of Fisheries of the state Governments are the main agencies looking after the extension activities and training programs. Moreover, extension and training programs are also undertaken by Fishery Universities. Extension works are carried out by the Fishery Extension Officers at block levels of each state. At the time of extension work, subject-matter experts afford technical advise in various areas.

### *Quality Estimation*

A quality estimation of the extension service indicates that the gap between the research station and the application of techniques in the field exists and that this should not be ignored by the extension services : (1) the extension workers tend to administer routine information and are not ready to respond to the particular problems of the farmers, (2) the extension service tends to be less concerned with needs of the farmers, and (3) the extension agency suffers from several internal problems. Consequently, evidence indicates that a large number of farmers demand other than the extension service as the main source of their information.

### *Fish Farmers' Development Agency*

The Fish Farmers' Development Agency has been launched in 1975 by the Government of India. It functions as an autonomous body under a centrally sponsored scheme. It brings together all categories of fish farmers/fishermen. This body takes full responsibility of financing, monitoring, training, and evaluation of the program. Under Fish Farmers' Development Agency (FFDA) scheme, about 0.50 million hectare of fallow ponds (data for 1999) have been brought under fish cultivation. Since FFDA's are based on district level, they are acting as good extension service agencies. However, the average fish production under the FFDA scheme has been reported as high as 3-5 tonnes/ hectare/year through improved stocking densities, aeration, artificial feeding, and other management schedules.

## 15.4 Training for Management in Fisheries Sector

Training is considered as an integral part for proper functioning of fisheries sector. It refers to improve a farmer's ability to do fisheries job and contribute to the development of fisheries and aquaculture. Generally technical personnels assess fish farmer's ability to undertake fish culture. Obviously, training supplies skill, knowledge, and attitude to individual farmer or group of farmers to improve their efficiency and to perform their activity.

Training is used to provide farmers' skill to the level necessary to adapt a new technology, as the technology changes from time to time. To fulfil the deficit of knowledge, more training to fish farmers is needed for additional skills. In fact, performance failures dictate that technical personnels assign farmers to training programs to improve their knowledge, skill, and future performance. The most important feature of the training is to prepare a task-force for fish culture industries for better production.

### *Training and Development*

Training and development is the process of developing knowledge and skills in fishermen that will enable them to better perform their activities. Training programs are primarily directed towards maintaining and improving aquaculture activity while development programs are intended to develop skills for future activities.

### *Need for Training*

It is important to note that the fishermen be inducted into training programs to improve their knowledge about different aspects of fisheries. The need for proper training is increased by the following considerations :

1. *Increased Productivity* : Adequate training improves performance skills which improves both the quantity of fish and the quality of fish products.
2. *Health Improvement* : Proper training helps prevent dramatic decline of fish production due to outbreak of various types of diseases. Knowledgeable and skilled fishermen are less prone to diseases.
3. *Stability of Fish Farm and Fisheries Activities* : Training and development programs foster the initiative and creativity of fishermen which increases a sense of belonging, thus preventing a manpower obsolescence.

### *Methods for Training*

Though there is no specific methods for training to farmers, training programs are matched on the basis scientific outlook, with the need of the sector, the fish farm owners, and the farmer being trained.

The most widely used training method is *'on-the-job training'*. In this case the farmer is trained during the time of fisheries activities. Generally one farmer learns about the activities by monitoring a fellow farmer. The trainee is taken to the fish farm and instructed about specific activity carried out by the technical personnel.

## 15.5 Decision Making

Decision making and solution of problems during fishery activities are the core function of adequate management because it is an integral part of overall activities in aquaculture. During fish cultivation, farmers always face a lot of problems that should be solved through right decisions. Right judgements must eliminate the root causes of the problems that have necessitated decisions. However, the decision may be a simple one such as selection of suitable combination of species to be stocked, use of balanced fish meal fortified with vitamins and minerals or it may be a major decision such as construction of a fish farm or hatchery in a suitable place. The decision, however, must be implemented carefully and accepted by all to get the best results.

### *Condition for Decision Making*

In some situations, it is not possible for farm owners to accurately predict the consequences of an implemented alternative. This is partly due to the dynamics of the environment. The more dynamic the environment, the higher the degree of uncertainty in predicting the result of a decision. However, decision under certainty is the most common and simplest form of decision making. The condition of certainty exists when decision makers know about the consequence of each alternative. The decision maker would select the alternative which has the best result. In case of small number of alternatives, the results can be compared with each other taking two alternatives at a time. The two alternatives can further be compared and the inferior one is discarded. The better one of the two is further compared with the next one until all results have been compared and the best one is selected. Suppose a fish farmer or a group of farmers wish to establish a fish farm or a hatchery. Once the decision to construct the farm has been made, there is a number of alternative ways of expenditure for the farm such as (i) farm may be constructed through loan, (ii) part loan and part cash, and (iii) hundred per cent cash. However, it is possible to calculate the total cost of each of these alternatives and choose the one which gives the minimum cost.

### *Group Decision Making*

In some situations, problems pertaining to aquaculture may come up which are not able to solve by a individual farmer. In such cases, the farm owner may assign the problem to a group of experienced farmers/experts for comments/decisions. It is often argued that cooperative societies can make higher quality decisions than individuals. There are assumptions that form the basis for this argument. First, groups can assess ideas better than individual. Second, groups are more vigilant than individual. Third, groups can generate ideas and develop more alternative solutions than individual. Bias may introduce into the decision if it is made by an individual farmer. Farmers' group can check for bias and evaluate ideas on an objective basis.

### *Advantages of Group Decision Making*

Group and individual decisions have their own weakness and strengths. A number of advantages have been identified. First, since group members have different experiences, they provide more reliable information and knowledge. Such information tends to be more comprehensive in nature. Second, the input from a large number of farmers eliminates the biases that are introduced due to individual decision making. Third, group decision making is more democratic in nature and consequently, democratic processes are generally accepted for the purpose.

### *Errors in Decision Making*

Because the importance of the right decision cannot be over-emphasized enough, it is imperative that some factors affecting the decision be properly investigated. This is due to the fact that the quality of the decision can make the difference between failure and success. Some important factors such as technical, financial, farmers' capabilities, perception of the environment etc. should be considered in the decision making process. Some investigators have pointed out that management in some areas of fisheries sector needs reassessment on the basis of advanced technology and where some mistakes (such as failure to isolate the causes of the problem and procrastination) are made that affect the decision making process as well as the efficiency of the decision and these mistakes should be avoided as far as possible.

## 15.6 Budget in Fisheries Sector

In different fisheries sectors, financial control is principally exercised through the budget. A budget is both a plan and a control as the budget preparation is an integral part of the planning process and the budget itself is the end point of the planning process.

A budget is an estimate of expenses or income for a specific time period and the estimates become the standards against which future activities will be measured and evaluated. The main objective of fish culture is not only optimum production but also ending with sound cost-benefit ratio for economic returns. Therefore, for making fish culture an economically viable industry and to keep a proper balance of cost-benefit ratio for sound economic returns, suitable management principles should be adopted. Fisheries sector, in general, prepare operating budget (expenditure versus income) that includes expense and revenue budgets for a fixed period of time.

### *Budget in Relation to Control and Planning*

Budgets are consolidated statements of planned revenue and expenditure of fund, personnel, equipments etc. Since planning is an integral part of any budget, a direct relationship among budget, control and planning always exists (Figure 15.2). Note that the budget provides the feedback for changing plans.

### *Benefits of Budget*

Budgets are used by any fish farmers' co-operative society to plan, monitor, evaluate, and diverse activities and operations at every level of fisheries sector. However, the proper use of budgets provide following benefits :

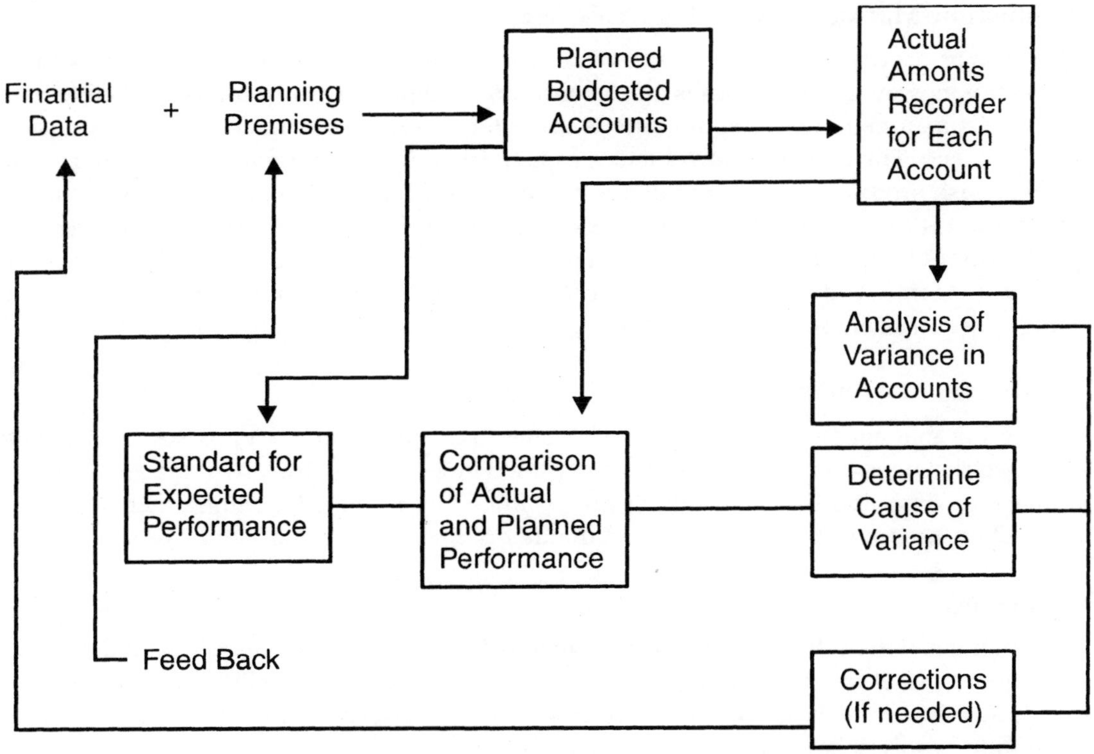


**Fig. 15.2 : How the budged relates to planning and control. [Modified after Haynes (1999)]**

1. Budget leads to better planning at all phases of fish culture operations.
2. Budgets enable to forecast future activities.
3. Budgets ensure a vivid statement and understanding entire fisheries sector in a particular area or state.

## 15.7 Importance of Manpower in Fisheries Sector

In every fisheries sector there are two distinct types of elements such *inputs* and *manpower*. Inputs consist of fertilizers, feeds fish seeds, etc. Manpower works with these inputs to make the fish culture industry viable. While introducing the principles of scientific management, management would mean dealing with adequate financial support, manpower, inputs,and technologies; of which manpower should be the predominant factor. Since man lives to learn throughout life, their improvement of efficiency or quality gradually increases and therefore, the importance of manpower is felt with the development of fisheries sector.

Manpower resource being the most significant one, manpower planning is neccesary. Manpower planning means determining the number of fish farmers to be employed in a particular fish farm to produce a target of fish production. The target must be fulfilled otherwise cost-benefit ratio will be upset.

## 15.8 Production Management in Fisheries Sector

The term *production* is used in the economic world to mean to yield fish and fish products. In fisheries sector, however, production is not a continuous affair. Generally fish is stocked to ponds, lakes or reservoirs for short or long duarations and after several weeks interval, fish are harvested and again stocked. For maximum production of fish, frequent harvesting and stocking system is undertaken. Fish production through this system requires an efficient management with the aid of skilled personnels throughout the culture period. Since a number of environmental factors limit fish production, efficient management strategies related to production should be considered. However, management should be made in such a manner that maximum production is possible according to requirements at minimum cost.

## 15.9 Marketing Management in Fisheries Sector

Marketing management is nothing but professionalization in carrying out the marketing functions. Marketing is a term which is very much used in the trade world and it practically replaces the term *selling*. Fish and fishery products are produced by farmers for selling them. Above all, fish production has to be made in a planned manner to meet the need of the buyers.

### *Fish Marketing*

A farmer generally sell his products through middlemen – termed as *merchant agent* who stock the entire quantities of fish after buying from the farmers and sell them to the retailers by breaking the bulk. Thus they relieve the producers from much of the risks by taking the responsibility of distribution. A whole-seller may sell a variety of fishes.

Usually two types of retail outlets are considered where the entire lots of fresh fish are sold. First, fish farmers bring the bulk into the market and the entire profit goes to farmers and second, farmers establish co-operative stores and they manage on co-operative basis and sell the fish at a reasonable price to the members (even to outsiders) by eliminating middlemen.

There are some other middlemen who neither buy or sell the bulk but bring into contacts a seller and buyer and earn brokerage on the total bulk of transaction. These types of middlemen, however, should be avoided.

### *Factors Affecting Fish Marketing*

A number of environmental factor affect fish marketing strategies : (1) varieties of fish with regard to its quality and price, (2) nature of the market which comprises the middlemen and the consumers, (3) socio-economic conditions which influence the demand, (4) transport facilities from the landing site to the market, and (5) the changes that take place in techniques of production and fish processing technologies.

## 15.10 Increasing the Knowledge of Management

There is general agreement about the importance of limiting environmental damage. It is also necessary to understand why actions of Government and Non-Government organizations are necessary to manage the fisheries sector. For example, pollution of water has adverse effects on others, but the pollution does not have to compensate them. When such spillover effects occur, the cost of pollution to fisheries sector is greater than the cost to the polluters. There is then too much pollution because individuals do not have the right to incentives to reduce it. An industry that discharges pollutants into a river has no incentive to consider the damage inflicted on those downstream.

Effective public politics provide incentives to reduce pollution and aquatic resource degradation by aligning social cost. In some cases, legal systems can provide such alignment without direct government action. For example, rules generally require polluters to compensate others for certain kinds of pollution damage. The assignment of property rights can also reduce the scope for degradation of aquacultural sector; for example, a lake that has one owner is not likely to be overfished. But the lake-owners of those right may incur large transactions costs in enforcing them, and the assignment of property rights is not feasible.

When information on discharges or the extent of their damage is not available, systems that monitor the actions of polluters, such as the required installation of pollution control devices, may be desirable. With the right information, pollution charges are superior. Unlike technology standards, they put firms under pressure to reduce pollution.

Information can also encourage pollution reduction and simultaneous dissemination of information on the implications of aquacultural degradation can offer opportunities for improvement, but the impact of better information depends on farmer's ability and willingness to use it.

## 15.11 Knowledge for Fisheries Management

The evaluation of aquacultural degradation centres on its relationship to economic developement. Some argue that degradation is the result of by-product of economic and social development of fishermen. Others state that economic and social development will not suffer if fisheries resources are adequately managed. Therefore, some consider management as a complement to development while others consider as conflicting. But several degradation can occur even without development, simply from population pressure. In general, good policies can support sustainable management strategies by improving and protecting the aquatic resources while promoting economic growth. Such strategies call for good institutions, good information, and better knowledge of the environmental impacts of alternative policies. The main aspects of integrating fisheries management with developement are as follows :

### *Using Aquacultural Information*

Management strategies argue that developing countries have plenty of opportunities to implement sustainable development policies. The information generated through monitoring the state of the aquaculture can be used in several ways. Some hazardous chemicals may have threshold concentrations, beyond which the risk of damage to farm fish jump from negligible to significant. It is also important to know whether the concentrations of such chemicals are nearing critical thresholds.

Better understanding of seasonal fluctuations throughout the culture period also have value. Consider a fish farmer's decision about which species and their combination to cultivate. The choice depends on the seasonal patterns expected over the coming months. Therefore, more reliable seasonal forecasts should provide significant benefits to fish farmers. Without reliable forecasts, fish farmers are forced to make farming decisions that are accurate for an average season, and take the risk of severe damage from an unforeseen extreme seasonal event.

Better knowledge about policy also an important contribution to aquacultural management. Questions may arise how do policies affect the ecosystem and how can cost-effective policies be best designed. However, with sophisticated knowledge management and decision support tools – and a better understanding of complex aquatic system, policy-makers are now implementing more integrated approaches to aquacultural management.

### *Managing Aquacultural Knowledge*

Managing aquacultural knowledge and building capacity for its use are as important as creating such knowledge. That is why more fisheries projects now include information systems. A number of organizations also take active part as a cost-effective means of disseminating aquacultural and development information. They have successfully built capacity and integrated aquacultural issues into reporting by print media.

Informations on the better fisheries management are voluminous. Decision-makers need tools that integrate and summarize informations on fisheries aspects. Application of latest technological innovations can exhibit cost-effective fisheries management to improve the technological know-how by simulating the aquacultural consequences of different courses of action.

### *Monitoring Ecosystem Quality*

It is also very important for fisheries management. But its effect on the benefit of fishermen depends on the framework in which this information is presented. To monitor ecosystem quality, a differnt information framework with additional indicators is necessary. Some indicators measure ecosystem bads such as deterioration of water and soil qualities. Others monitor the effects of ecosystem degradation, such as the incidence of fish diseases resulting in fish mortality and decline of fish production.

## 15.12 Integrated Fisheries Management

Farming systems that rely on polyculture and on the intensive use of feed, fertilizers, and manures are frequently held responsible for several negative environmental effects, direct public health risks, soil and water contamination. Integrated fisheries management encourages production of farmed-fish by using feeds and other nutrient inputs, stocking of healthy and disease-free fish, adopting cultural management, and as a last resort, judicious use of nutrients.

Integrated management is an information-intensive technology that needs continuous inputs from research source to maintain its dynamism at the farm level. At present, Government, Non-Government Organizations, donors, and the farmers groups take active part in the strategies of management. The challenge is to ensure that farmers remain focussed on the problem, through farmer to farmer extension, organizing farmer associations and development information media.

### *Productivity*

The term productivity is the ratio between the yield of fish and the inputs. Productivity and production are not identical terms. While the former is a relative term, the latter is an absolute one.

The most vital responsibility of fish culturist is to increase the productivity of fish culture ecosystems. This can be achieved by optimum utilization of the inputs. Maximum poductivity brings satisfaction to fish farmers. A number of factors play a decisive role for high productivity. These factors inlcude improved technology, supply of adequate funds, increasing the efficiency of farmers, improvement of socio-economic condition of farmers, efficient management, and research as well as development.

### *Research and Development*

Research and development is the latest innovation all over the world in the field of fisheries sector to stimulate high production of aquatic resources. Developed and developing nations spend lots of money for research to introduce changes in the techniques of production. The initial expenses are high but ultimately it becomes economic. At present, research activities are going on at such rapid rate that even it may not be possible to catch them up. Development is the ultimate goal of research. Seeking to improve fish production by bringing away from the constraints of conventional methods, several research organizations have successfully developed new technologies. Of them, biotechnology is very important. This is an advantage for the fish and a bonus for the farmer. Despite of financial constraints, however, aquaculture in many developing and under-developed nations has progressed to be able to contribute to the economy, supply of protein to their populations and become well established in overseas markets.

A sound research and development programs have been shown to directly increase the efficiency of biotechnology. With improved management systems application of biotechnology makes good sense. The technology can be delivered to fish farmers so it will have the greatest impact on fish production.

## 15.13 Evaluation of Aquatic Resources for Human Needs

The problems arising from massive exploitation of aquatic resources through technological advances, consumption and rapid population growth tend to threaten the existence of man. The degree to which future growth of fisheries sector can be expected will depend upon the geographical conditions of the local region and the resource needs of the nations.

To support the existing population, it is necessary to develop methods of intensive management implied by high levels of technological capacity. This persuit may even not produce a quality ecosystem. The more technology is involved in maintaining a quality ecosystem, the more aquatic resources are utilized for production. In developed and developing countries, an improved management technology to meet the nutritional deficiencies of rapidly expanding populations may lead to waste and expenditure of resources which cannot be tolerated. Therefore, some balance between resource utilization and technology must be established.

In consideration of the rapid growth of population and the increasing aquatic resource consumption, it is necessary to re-evaluate the human needs. If it is considered as a crisis solution, the human society can be more responsive to the environmental needs and implementation of comprehensive planning for aquatic resources.

## 15.14 Conclusion

The management of fisheries sector controls to a marked extent the ecosystem productivity and fish production. The type of management, research and developement of fish culture ecosystem have profound effects on fish growth and on all kinds of farm manipulation and use. These effects in turn influence overall management of fisheries sector thereby imparting better ecosystem for fish and high production at minimum costs.

Better management in fisheries sector is subject to farmers' control. Proper selection of fish seed, use of feed and fertilizers in suitable rates, and other management schedules help assure better management. Fish farm management, based on certain principles and practices, will obviously minimize the cost of fish production and water pollution. The need for management, therefore, should be realized.

Experience with ecology of fish culture ponds, fish culture technology, and management suggest that the three can be mixed, and even better, swallowed. It is assumed that the experience developed with the fish culture research program can serve as a model for the kinds of program that should be developed for fisheries management in the future.

### *Reference*

Haynes, W. 1999. *Principles of Management and Entrepreneurship Development.* New Central Book Agency Pvt. Ltd., Calcutta, India.

### Questions

1. What is management? Why management is so necessary?
2. Define the following terms : (a) Fish Farmers' Development Agency, (b) Production, (c) Extension, (d) Efficiency and Effectiveness.
3. What is scientific management? What is the principles of scientific management?
4. What is planning? Why it is so important in fisheries sector? How planning procedure can be developed?
5. How fish culture extension program can be considered in the development of fisheries sector?
6. State why training of fish farmers is necessary for proper functioning of fish culture sector.
7. State how decision making is related to fisheries management.
8. Describe how knowledge for fisheries management can be increased.
9. Why integrated fisheries management is necessary for sustained production?
10. State why aquatic resources should be evaluated for human society.

# 16

# Nutrient Recycling Through Organic Wastes and Wastewater in Fish Culture

Wastewater and solid waste recycling is becoming an increasing important strategy of environmentally sound sustainable aquaculture. Recycling involves return to the water and soil of essential elements that are taken up by plankton and bottom fauna and ultimately find their way into fish. Waste recycling not only provides organic matter but also reduces the need for additional fertilizer elements and artificial feed.

Development of wastewater treatment system based on ecological principles is perhaps the most important criteria that can drastically reduce environmental problems and facilitate the utilization of resources in wastewater. Recycling is often a logical consequence of ecological thinking. But it is necessary to recycle to obtain an ecologically sound solution. Several treatment systems have proved to be cost-effective, low-energy consuming, and high efficiency natural innovation/ecological alternative. In fish culture ecosystem, however, oxidation ponds, submerged and floating aquatic plants, and algal cells are used in wastewater treatement systems and are considered as a wise practice for nutrient conservations.

## 16.1 Quantity of Farm Manure

Huge quantities of farm manures are available each year for possible return to agriculture and aquaculture. For each 500 kg live weight of farm animals about 2 metric tonnes of manure is generated (Table 16.1). In India the total numbers of livestock and poultry have been estimated to be about 629, 892,000 and 146,528,000, respectively and the total quantity of manure voided by these animals is about 4.5 billion metric tonnes annually.

**Table 16.1 : Representative Annual Rates of Manure Production From Different Animals**

| Animal | Annual Production (Metric tonne/500 kg kg live weight) | |
|---|---|---|
| | Fresh excrement | Dry matter |
| Cow | 5.7 | 0.91 |
| Buffalow | 6.9 | 0.97 |
| Sheep | 5.9 | 2.00 |
| Poultry | 5.5 | 2.20 |
| Swine | 13.3 | 1.99 |
| Horse | 5.6 | 1.92 |

## 16.2 Significance of Farm Manure

For centuries the use of farm manure in fish culture industry has been synonymous. In addition to the supply of organic matter and nutrients to the water and soil, farm manure also closely associated with fish production, which stimulate the overall productivity of fish ponds. High proportion of solar energy captured by plants ulitmately is embodied in farm manure. Productivity of fish culture ecosystem and fish production is enhanced by the use of animal manure.

## 16.3 Chemical Composition of Animal Manure

Since animal manure as it is used in fish culture is a combination of urine and faeces along with feed waste, its chemical composition is highly variable (Table 16.2). Table indicates that the ratio of urine to faeces in farm manure (except poultry) varies from 1:2 to 1:4. Although more than one half of the nitrogen, all of the phosphorus and two fifths of the potassium are found in the faeces, this higher nutrient content of the faeces is offset by the availability of the constituents carried by the urine. Therefore, care should be taken in handling and storing the manure to minimize the loss of urine.

**Table 16.2 : Nutrient and Moisture Content of Farm Manure**

| Animal | Urine/Faeces ratio | Nutrient (kg/metric tonne) | | | Moisture (%) |
|---|---|---|---|---|---|
| | | N | $P_2O_5$ | $K_2O$ | |
| Poultry | 0:100 | 16.0 | 7.5 | 3.8 | 60 |
| Swine | 40:50 | 6.0 | 3.2 | 5.1 | 85 |
| Sheep | 30:70 | 10.0 | 3.1 | 8.5 | 65 |
| Horse | 20:80 | 7.0 | 2.0 | 6.2 | 70 |
| Dairy Cattle | 20:80 | 5.5 | 1.2 | 3.6 | 85 |

Table also indicates that the nutrient content of farm manure is relatively low in comparison with chemical fertilizer and a nutrient ratio that is considerably higher in nitrogen and potash than in phosphorus. Chemical fertilizers generally carry 20-46 times the nutrient content of manure and mixed fertilizers have higher ratios of phosphorus to nitrogen and potassium than that found in manures. The range of other nutrients commonly found in kilograms per metric tonne is shown in Table 16.3.

**Table 16.3 : Availability of Nutrients (other than N, $P_2O_5$, $K_2O$) in Animal Wastes**

| Nutrient | Quantity available (kg/metric tonne) |
|---|---|
| Magnesium (Mg) | 1.0-3.0 |
| Calcium (Ca) | 1.0-35.0 |
| Manganese (Mn) | 0.004-0.07 |
| Ferrous (Fe) | 0.04-0.88 |
| Zinc (Zn) | 0.010-0.08 |
| Sulphur (S) | 0.3-2.8 |
| Boron (B) | 0.01-0.05 |
| Copper (Cu) | 0.003-0.012 |

### *Organic and Moisture Constituents of Manure*

Manures are degraded plant materials since the animals utilize half of the organic matter they consume. Therefore, the bulk of the solid part of the manure is composed of organic compounds. Starches, sugars, and cellulose are readily decomposed, but lignin and hemicellulose have been modified as lignoprotein complexes. When manures are added to pond ecosystem, the soluble components of plant materials undergo degradation. Manures from cattle and sheep are teemed with micro-organisms and bacteria. Some of these organisms take active part in breaking down constituents in the voided faeces and participated in decomposition of the manure storage.

The moisture content of fresh manure is very high, generally varying between 60 and 85 per cent. This excess water is harmful if the fresh manure is added to the pond water. For this reason, the maure is allowed to keep on open air for some days before applying to ponds.

## 16.4 Storage and Management of Manures

Previously, storage and application of manure was very simple. Fish farmers allowed manure to pile up or spread in the vicinity of ponds until fish culture ponds permitted it to be used. At present, the concentrated animal management systems has drastically changed this situation. The possibilities of pollution of water and offensive odors from decaying manures have helped stimulate the storage and management strategies. However, to handle farm manures, four general management strategies are used as noted below :

### *Storing in Piles*

Manures from cattle feedlots and dairy barns are collected over a period of time and stored in piles where aerobic or anaerobic breakdown occurs. The type of breakdown depends on the degree of compaction of manure in piles. The products of breakdown are heat, water, and carbon dioxide although reactions involving nitrogen, phosphorus, and sulfur are of practical significance. Hydrolysis of urea, for example, yields ammonia and ultimately released to the atmosphere.

$$\underset{\text{Urea}}{CO(NH_2)_2} + 2H_2O \longrightarrow \underset{\text{Ammonium carbonate}}{(NH_4)_2\,CO_3}$$

$$(NH_4)_2\,CO_3 \longrightarrow 2NH_3 + CO_2 + H_2O$$

If conditions are favourable for nitrification, nitrate ions will appear in abundance. Inorganic nitrogen is lost and become a pollution hazard. While nitrate ions are subject to leaching or to movement in run-off water, ammonia is lost in gaseous from which may be captured by rain and returned to the ground.

### *Aerobic Treatment*

In this type of management system, animal waste is stored into an oxidation ditch where vigorous stirring is carried out to incorporate oxygen into the system. This brings about continuous oxidation on decomposition. On decomposition, water, carbon dioxide, and inorganic compounds are produced. The solid product is periodically applied to pond water.

### *Anaerobic Treatment*

This method is exactly similar to its aerobic counterpart except that no oxygen is added to encourage aerobiosis of liquid slurry. The reaction products contain 70 per cent methane and 30 per cent carbon dioxide. The methane gas is used as a fuel particularly in rural areas. Animal manures, human waste, agricultural waste, and wastewater is digested in a 400 $m^3$ continuous-flow stirred-septic tank high rate digester. After digestion the sludge is dried on sludge-drying beds. In this treatment method, about 260 tonnes of good quality sludge (dry matter) and 60,000 $m^3$ of methane annually are produced. The sludge is used for crop production.

### *Daily Use of Manure*

This system of management is encouraged where cattle-cum-fish cultivation is carried out. The manure is spread over ponds daily with the help of a shovel and sometimes, the manure is reinforced with superphosphate. Although daily spreading prevents nutrient loss, subsequent increase in probability of pollution of pond water must be considered.

## 16.5 Compost

Composting is the most common practice of encouraging partial rotting by soil micro-organisms of organic materials of animal or plant origin. In the process of composting the aerobic decomposition takes place in piles and kept sufficiently moist to stimulate the decay.

Leaves, weeds, lawn cutting, garden wastes etc. are considered as the primary sources of organic materials. These materials are sometimes supplemented with rice bran, wheat bran, saw dust, or furnace ashes. To expedite decay process, small quantities of nitrogen and phosphorus fertilizers may be added along with a little soil to ascertain the availability of micro-organisms for decay process. The materials can be kept either in a wooden bin or in a pile. The pile must be moist (60-70 per cent water) but not too wet because the breakdown must be aerobic. The materials are best packed to help keep the pile from drying out. Figure 16.1 explains how a compost pile is prepared.

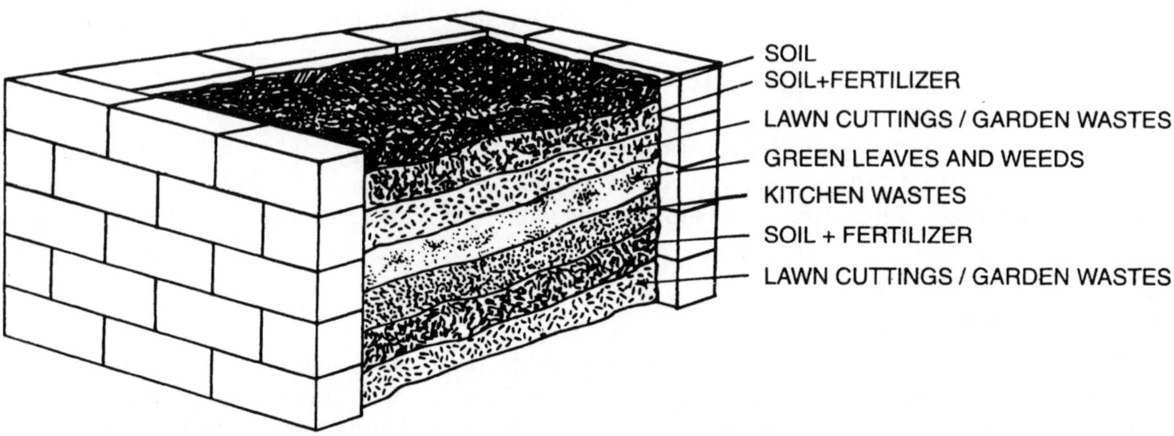


**Fig. 16.1 : Front view of a compost heap showing different layers of wastes, weeds, and fertilizer-soil. Various wastes from the farmyard, kitchen, and garden in combination with soil and a little fertilizer are placed in layers. These layers are kept most where partial decomposition by soil organisms takes place. After decomposition of the organic matter, it is added to ponds. Compost production is a blessing in disguise because composting ensures the conservation of some nutrients that are lost when wastes are burned.**

In the compost pile, temperatures of 50-65°C being reached when decay is occurring rapidly. At these temperatures most disease organisms are killed. During the composting period, visual observations with temperature and volume measurement are recorded to ascertain the extent of degradation. The pH of the pile is also an important parameter which greatly affects the composting process. The optimum ranges of pH for bacterial and fungal developments are 6.0-7.0 and 5.5-8.0, respectively.

### *Carbon-Nitrogen Ratio in Compost*

The C/N ratio is an important criteria to assess a successful composting process. To hasten the biodegradation process, nitrogen compounds are added. In a composting pile, there is an increase in nitrogen content and decrease in carbon content. Consequently, the C/N ratio gradually narrowed down.

### *Phosphorus in Compost*

Although phosphorus plays a vital role in fish production, its concentration in solid solution is only approximately 0.05 mg/l. For this reason, the possibility of practical use of rockphosphate for composting has received significant interest. The man approach for rockphosphate solubilization is the use of microbes able to excrete organic acids.

India, China, and some other Asian countries have probably made more extensive use of compost than any other country of the world. Animal manures and crop residues are well mixed with soil and sludge in compost piles. The piles are then plastered on the outside with wet soil to prevent rapid loss of moisture. Bamboo rods are used to make holes in the pile. These holes assure a supply of air throughout the pile. Different methods for the preparation of compost have been described in Chapter 2.

### *Fish Response to Compost*

Compost manures are widely used for fertilizing fish ponds in Java, Germany, Belgium Congo, and many other countries. In India, various combinations of cowdung, mustard oil cake and some species of aquatic plants (such as *Eichornia* sp., *Hydrilla* sp., *Pistia* sp., and *Najas* sp.) have been suggested for composting and manuring at the rate varying between 550 and 830 kg/hectare. In some situations, however, the application of compost at the rate of 5,000 kg/hectare is necessary because the nutrient content is low and the nutrients are readily available for the growth of fish and fish food organisms. If properly prepared compost is supplemented with lime and chemical fertilizers, it is as effective as animal manures in enhancing fish and plankton production. Application of compost material provides not only nutrients but improves soil physical properties as well.

## 16.6 Utilization of Manures

Farm manures supply a wide variety of nutrients along with organic matter that improves the physical characteristics of pond soils. Inspite of its handling and high labour costs and low amounts of nutrients, manures are considered as the most valuable soil organic resource and therefore, the cost is a secondary factor.

Organic manure stimulates the growth and development of zooplankton populations – an obvious source for fish growth. Generally the rate of application of manure depends on the productivity of fish ponds. Rates of 6,000-10,000 kg/hectare/year, however, are commonly employed.

Because micronutrients are essential for development of plankton populations, micronutrient deficiency in fish culture ponds can be ameliorated with manure application. Moreover, the water-holding capacity of sandy-loam and sandy soils is increased with heavy manure applications. Hence, the application of manure at high rates in such ponds is justified.

## 16.7 Quantity of Wastewater and Sewage and Their Nutrient Value

Recently, it has been estimated that the quantity of wastewater and sewage that are produced in India is about 35,000 and 15,000 million litres per day (MLD), respectively. In conventional treatment systems (such as primary and secondary settling tanks and biological filters), solar energy is not utilized. Consequently, a huge quantities of nutrients present in sewage and wastewater are not utilized for food production The total quantity and value of nutrients (nitrogen, phosphorus and potash) in waste materials are shown in Table 16.4. Table shows that for each 500 MLD of waste material about 3.6 tonnes of nitrogen, 0.6 tonne of phosphorus, and 1.8 tonnes of potash per day is obtained. At current prices, however, the equivalent fertilizer available from the sewage and wastewater is about Rs.32.27 crores.

**Table 16.4 : Fertilizer Value of Wastewater and Sewage**

| Volume of Waste materials (MLD) | Quantity of nutrients in waste materials | | | | | | Values of nutrients | | | Yearly return (Rs. in crore) |
|---|---|---|---|---|---|---|---|---|---|---|
| | Concentration | | | Tonne/day | | | N @ Rs. 2,500/- tonne | $P_2O_5$ @ Rs. 900/- tonne | $K_2OL$ @ Rs. 1,030/- tonne | |
| | N | $P_2O_5$ | $K_2O$ | N | $P_2O_5$ | $K_2O$ | | | | |
| 15,000 (Sewage) | 27.0 | 6.5 | 13.5 | 107.6 | 25.7 | 53.8 | 2,69,000 | 23,130 | 55,414 | 12.68 |
| 35,000 (Wastewater) | 62.7 | 15.0 | 31.4 | 251.0 | 60.0 | 125.6 | 6,27,500 | 54,000 | 1,29,368 | 29.59 |
| Total : 50,000 | 89.7 | 21.5 | 44.9 | 358.6 | 85.7 | 179.4 | 8,96,500 | 77,130 | 1,84,782 | 32.27 |

## 16.8 Waste Disposal

Much of the awareness has been brought about in recent years by active campaigns in the fields of air and sound pollution resulting today in third campaign, that of water pollution. In response to an arouse public, legislation has been passed on the state and national levels to aid the development of satisfactory disposal practices and to plan future waste management.

The most common repository for wastes used — the land — remains of paramount importance in almost every area of the world. Open dumps are still the predominant disposal method with severe pollution of natural waters.

People are aware of the interest in the concept of the recycling and reuse of waste materials. For decades, the secondary materials in industry have relieved a significant amount of the total waste disposal burden. The principle of recycling and reuse of waste materials, systematically developed and applied, is of extreme importance to the field of waste management. The technology of reuse, and the economic forces governing the efforts, do not admit recycling as an immediate answer to waste problem. However, when economic conditions become favourable for the recycling of waste materials, the damage for waste disposal site will decrease. The increasing waste load due to population increases of course, may serve to negate this reduction of disposal requirements.

### *Problems of Landfills*

Waste deposited in landfills degrade biologically and chemically to produce a number of contaminants. Decomposition of waste results in the release of gases and leachate. Water emanating from the disposal sites carries dissolved and suspended solid wastes and microbial waste products with it. Leachate that emerges at the ground surface may flow into the lakes, springs, streams, and rivers or may percolate into the ground.

Many landfill sites acceptable from an environmental control aspect are seriously opposed by nearby residents. Many objections are raised by opponents of landfill site location, a few of which are justifiable. Bad odor, smoke, water pollution, rat infestation, and disease-producing organisms are erroneously ascribed to the proposed landfill.

## 16.9 Wastewater Treatment Systems

For suitable wastewater treatment systems under a given set of conditions, it is necessary (a) to assess the problem, (b) to find out the most suitable system, and (c) to explore suitable solution on the basis of local premises and environment as well as cost comparisons. To date, however, the various cost of different wastewater treatment methods have not been fully evaluated — neither the environmental effects, nor direct economic costs.

### *Pond System*

For wastewater treatment and utilization, different types of pond systems have been developed in India, China, and many European countries. Aerated ponds are low-energy treatment solutions in cold climate. Removal efficiencies of 30 and 65 per cent for nitrogen and phosphorus respectively are obtainable in winter conditions at a system treating wastewater from 100 persons. Phosphorus is removed due to sedimentation. After removal of nitrogen and phosphorus, the effluent can be used for fish culture and irrigation, giving a high nutrient utilization.

In presence of sunlight, nutrients in the wastewater encourage the devleopment of healthy algal bloom, aerobic bacteria and other micro-organisms. Due to bacterial decomposition of organic matter, carbon dioxide is released which is then converted to algal cell material with the liberation of oxygen. The oxygen is utilized by bacteria for aerobic decomposition of organic matter.

The solar energy is trapped by the algal cells that results in the production of new algal cells. In waste stabilization ponds, the ratio of weight of oxygen released to weight of algae synthesized has been estimated to be about 1.25 to 1.75 for algae. However, activities of algal cells help remove huge quantities of nutrients from wastewater. Nutrients are concentrated in algal cells. Harvesting of algae from the pond effluent is a means of recovering nutrients from the wastewater. The nutrient-enriched algal cells are considered as a source of industrial raw material, poultry feed, and food.

1. *Problems of Nitrogen Removal* : Denitrification steps hasten the removal of nitrogen from wastewater. This increases energy consumption in the removal process, and hence removal costs. Nitrogen removal may also pruduce significant quantities of nitrous oxide, a very potent greenhouse gas. By removing nitrogen from wastewater, air pollution has

dramatically increased and hence increased the energy consumption. Thus, valuable nutrient is being lost to the atmosphere and increased the need for production of nitrogen fertilizers, which is also an energy-consuming process. The ecological soundness of this type of nitrogen removal can thus be questioned.

2. *Factors Affecting Nutrient Removal* : The removal of nitrogen and phosphorus depends upon the climatic conditions and discharge of wastewater. The lowest rate of the removal of nitrogen and phosphorus from the pond occurs during the period of low temperature and high discharge. Large varieties of aquatic plants are used during the process of removal. In most tropical countries, freshwater ponds and lakes are adapted to low levels of phosphorus and elevated levels of phosphorus can be toxic to many species of phytoplankton, thus it is encouraging that so many species can tolerate nutrient-enriched aquatic ecosystem.

### *Wetlands*

According to the International Union for Conservation of Nature and Natural Resources, 1971 wetlands are defined as *"area of marsh fen, peatland, whether natural or artificial, permanent or temporary, with water that is static or flowing fresh, brackish or salt, including areas of marine waters, and the depth of which at low tide does not exceed six meters"*.

Historically, wetlands are being looked upon as useless, unproductive, and disease-infested ecosystems. At present, this perspective has dramatically changed with specific ecological features, functions, and values. On a global basis, a total area of 62.77 million hectares of wetlands have been recognized However, India has 2,167 and 65,254 natural and artificial wetlands covering an extensive area of 1.75 million hectares. The inland wetlands include flood plains of rivers and littoral area of ponds and lakes, marshes and swamps.

Besides providing a source of drinking water, wetlands are utilized for agriculture, aquaculture, and for industrial applications. Wetlands are also considered as natural sewage and wastewater treatment plants. Hence they perform in chemical cycles and their function as reservoirs of wastes. Many wetlands are known to reduce nitrate and phosphate level upto 90 per cent and their role in the removal of heavy metals has been widely acknowledged. They reduce the quantity of suspended particles and consequently, the clarity of water is increased. The phosphorus removal is dependent on precipitation and absorption reactions. Although wetlands are, thus, converted into fish culture ecosystems for increasing the productivity, this has led to the degradation of the system generally.

### *Solar Drying of Sewage*

The mechanical dehydration of sewage with different pressing methods has its limits of about 40 per cent of solid mass. Most of the processes stop around 30 per cent. Therefore, a further dyhydration is only possible with thermal processes, at least for the time being.

In general, thermal drying process requires large quantities of energy, about 2,450 Kilo Jules/kg (0.68 khw/$m^2$) along with the efficiency of the drying process. This energy, however, can be provided either by *fossil fuel* or by *solar radiation*. The best solution is obviously the use of the second method which is being practiced in developed countries (such as Germany).

1. *Drying Process* : The sludge from the sewage pit is mechanically dehydrated by a press and then spread out at the end of the dryer. The machine transports the sludge to other end of the dryer. The aeration of the sludge starts as an additional composting effect which neutralizes the smell of fresh sludge. The radiation of the sun falls on the sludge and heats the surface up. Consequently, the warm surface heats up the air. The driving force for the drying process is the pressure difference between the water vapour in the air and in the sludge. The air is only used to evacuate the moistness rising out of the sludge. The evacuation of the air is increased with ventilators having a total electric power of about 2 Kilo-Watt. The sludge is formed into pellets of different sizes. This facilitates the handling of the product and reused for recultivated processes.

   It has been estimated that whether the disposing cost of sewage sludge is higher than the drying cost per cubic meter, the solar drying is cost-effective.

### *Aquatic Plants*

The recycling potential of floating and submerged plants is, to some extents, misunderstood and even ignored in intensive, high-tech modern trials to recycle wastewater. Progressive fish farmers should remind that aquatic plants have been used for hundreds of year in the economical recycling of wastewater on a commercial scale. However, progress in these area should encourage adoption of an efficient recycling agent – the aquatic plants. Different species of aquatic plants are used in different countries for wastewater treatment (Table 16.5).

**Table 16.5 : Use of Aquatic Plants by Different Countries in the Treatment of Wastewaters**

| Country | Species | Studies on wastewater management |
|---|---|---|
| India | *Eichornia crassipes*<br>*Pistia straitotes* | Removal of toxic metal ions and nutrients |
| USA | *E. crassipes*<br>*Lemna* sp. | Wastewater treatment, removal of ammonium and phosphorus-containing waste products |
| Israel | *Lemna gibba* | Clone selection for specific industrial products |
| Poland | *L. minor* | Nutrient removal |
| Switzerland | *Lemnaceae* | Physiological diversity among duckweed species and alones |
| Japan | *Imponoea aquatica*<br>*Nasturitium officinale* | Natural purification capability in polluted rivers |
| Latvia | *L. minor*<br>*P. straitotes* | Nutrient removal from dairy wastewater |

## 16.10 Potential of Aquatic Plants for Nutrient Recycling

Some common genera of aquatic plants used for treatment of domestic sewage are *Lemna, Wolffia, Spirodela* and *Azolla* and their estimated growth rates in the sewage-fed culture systems have been estimated as 270, 280, 350, and 160 g/m$^3$/day, respectively. The harvested weeds could be used for feeding of grass carp, for composting which is again utilized in ponds as manure.

It appears that a duckweed-fed fish pond provides a balanced diet for those fish that is directly consumed, while the faecal matter of duckweed fish is consumed by detritus feeders or through fertilization of natural food organisms. These organisms provide food for other species of fish.

Results on this aspect indicate that fish production through duckweed-fed fish polyculture varies between 10 and 15 metric tonnes/hectare/year and also increases the economic viability of the production system.

### *Duckweeds*

Duckweed species grow on or below the water surface are highly productive (130 kg dry weight/ha/day) and have a high protein content (45 per cent of the dry weight). They can tolerate and fix nitrogen, phosphorus and heavy metals. Hence, they are used for wastewater treatment and as an energy and protein sources. They are considered as regulator in aquatic ecosystems. The growth of different species of duckweed plant is temperature dependent. Some species grow at zero temperature whereas others need temperature above 35°C. Therefore, their utilization in wastewater treatment is effective in all regions of the world.

1. *Phosphorus Removal* : Removal of phosphorus in a duckweed controlled environment is achieved through a combination of plant uptake and chemical precipitation. The level of phosphorus uptake by the duckweed will depend on the concentration of phosphorus and the balance of the other nutrients.

   The removal of phosphorus from the system by the plant mat is dependent upon the total growth and subsequent harvest of the mat. A system can be designed and operated to remove the majority of phosphorus through this process.

   The density of the plant mat (measured in $kg/m^3$), represents the mass of fresh biomass per unit area. The density of biomass is maintained within a range which will provide the maximum growth rate than the older plant population of duckweed. A young plant population tends to have a higher phosphorus uptake rate than an older plant population due to its accelerated growth rate, specially designed aquatic mat at intervals of several days to several weeks. The frequency of harvesting depends upon the growth rate of the mat, the temperature, and the treatment objective.

   Some algal growth is inhibited by the shaded environment under the duckweed mat. Phosphorus tends to be available in solution where precipitation reactions occur. Some of the phosphorus, for example, will precipitate and settle as calcium phosphate. The extent of this natural precipitation will depend on the inorganic chemical make up of the water.

   During colder period of the year, when biological reactions slow down, the natural phosphorus removal capabilities of a lemna pond can be supplemented by alum injection. To minimize the amount of alum required, however, the process is carried out in the pond when the concentrations of biochemical oxygen demand and total suspended solids are lowest. The aluminium phosphate formed is minimal and will not create a significant deposit at the bottom of the pond. The deposited material is chemically stable and will not dissolve. The pond is so designed that this small amount of deposit does not require removal.

2. *Nitrogen Removal* : Plant uptake and microbial transformations are the two processes by which nitrogen can easily be removed. Nitrogen removal by the growth of the plant mat can be optimized by conducting frequent harvests during the growth season.

Nitrifying bacteria have been isolated from duckweed mats grown on wastewater. Denitrification of nitrate to gaseous forms can occur in anaerobic micro-sites within the mat or in the water column and sediments. The resulting loss of nitrates will reduce the total nitrogen in the wastewater. These nitrogen removal mechanisms are expected to be most significant in tropical areas and less significant in temperate areas.

### *Water Hyacinth*

The roots of water hyacinth plants provide a large surface area and colonize with bacteria. The plants translocate air to their roots. Oxygen flux to the roots is not sufficient for aeration of the treatment volume, but is sufficient to support aerobic bacteria. When roots are passing oxygen and carbon substrates to attached bacteria, physiologically significant gradients develop in the bioflim from the root surface to the wastewater interface. Along the roots, these gradients are likely to be based on redox potential and specific carbon substrates are controlled by plant physiology. While bacterial life cycles range from 20 minutes to two hours, water hyacinth live for months. Therefore, the physiological conditions induced by the roots tend to be a stabilizing factor in comparison to the physiological conditions established by bacteria alone in activated sludge.

## 16.11 Waste Recycling in Fish Culture

Waste substances such as animal manure, human excreta, rice bran and others are used as pond fertilizers or supplementary feeds for freshwater fish farming in most tropical and sub-tropical countries. The advantages include a partial elimination of waste disposal problem, a supply of materials at low cost and provision of cheap animal protein. In many Asian countries, traditional farming rely on polyculture (stocking of ponds with different species of fish), with manures as major nutrient input. The waste substance is either applied to fish ponds as fertilizer to enrich the water or incorporated in fish feeds. Fish grow rapidly in the tropics and if wastes replaced the need for expensive supplementary feed, the production cost is reduced. Therefore, it is necessary to extend this practice to countries where hunger and malnutrition are prevalent.

### *Action of Waste Substances in Fish Ponds*

When waste substances are added to fish ponds, the following three steps are involved :

1. The waste substances can serve as a base for production of bacteria and protozoa which are, in turn, consumed by fish.
2. The waste substances are directly consumed by fish.
3. Nutrients released from waste substances accelerate autotrophic production and ultimately trigger the growth of other organisms which are consumed by different species of fish.

In addition to the danger of direct consumption of sewage sludge containing various undesirable substances and high concentrations of trace metals by fish, metals can be incorporated into phytoplankton more efficiently and accumulate in different trophic levels.

### *Addition of Sludge to Fish Feeds*

Although the use of sludge as a feed for some species of fish (such as trout) has been proved to be promising and no evidence of metal accumulation in the tissue was reported, in many cases the trace metal contained in sludge might exert detrimental effects on fish. Studies on the trace metal concentration in different parts of tilapia fed with various diets (such as commercial feed alone, 95 per cent commercial feed + 5 per cent sludge cake and 70 per cent commercial feed + 30 per cent sludge cake) after 2 months have shown that values were highest in those fish fed the highest proportion of sludge, particularly in the viscera (concentration ranged between 3.62 and 387μg/g of dry weight basis). It is also important to note that among different trace metal incorporated, values were highest for zinc followed by copper, chromium, nickle, lead, and cadmium.

## 16.12 Use of Human Waste in Fish Culture : Pathological Considerations

Though use of human waste materials in fish culture is very important that produces fish at very low costs, the potential health risks and current epidemiological evindence for actual risks from pathogenic transimission through wastewater fish culture is worth remembering. Since wastewater treatment is a very important tissue to reduce pathogen transmission risks, several treatment processes are graded with respect to their pathogen elimination potential. Adequate treatment of human waste is essential for public health protection and for social acceptance of wastewater-based fish culture.

### *The Calcutta Wetlands*

The Calcutta wastewater-fed fish pond system is regarded as a model of wastewater-fed fish production. The Calcutta wetlands — the world's largest wastewater fisheries since about 1930, comprising an area of approximate 4,000 hectare. The fish ponds are batch-fed with raw sewage at low organic loading rates of 8-22 kg biochemical oxygen demand/hectare/day which lead to a safe hygienic quality of the pond water, with total coliform concentrations reportedly amounting to $10^2$-$10^3$/100 ml. If these values represent a long-term pond quality average, the potential health risk for fish consumers from pathogens contained in the waste-fed fisheries project could be judged minimal. Because all fish are cooked before consumption, the health risk is further reduced. The bulk of the pathogens are entered the ponds with the wastewater settles in the bottom mud.

It should be added as a note that from the end of last century until the 1950's, wastewater-based fish production was widespread in many countries of the world particularly in Germany. At present, however, this system of farming has likewise been abandoned due to increased land use and operational costs, in addition to the requirement of alternative treatment processes over the winter months when fish production has to be suspended.

### *Health Risks from Pathogens in Fish Culture*

Two risks to public health generally occur such as potential risk and actual risk. The potential risk occurs when (1) pathogens reach the pond, (2) pathogens multiply in an intermediate host residing in the pond, and (3) infective dose reaches a human host through consumption when infection causes further transmission of disease.

The actual public health risks occurring through waste use in fish culture are of three types such as (1) *Consumer risk* – those affecting consumers of the aquatic products grown in wastewater-fed waters, (2) *Workers' risk* – those handling and processing the fish and fish products, and (3) *Farmers' risk* – those affecting fish farmers who exposed to treat and/or diluted wastewaters.

Generally bacteria, virus, protozoa, and helminth eggs play a key role in fish culture. In most tropical countries, helminths are endemic, less in developed countries and have various potential transmission patterns through waste-fed fish culture.

### *Factors Regarding the Transmission of Infection and Their Importance in Human Waste Use in Fish Culture*

Survival of excreted pathogens is a very important factor when the public health dimension of waste-fed fish culture is considered. Survival of pathogen depends on the dryness, UV-light, and temperature. Survival rates increase in proportion to the level of intensity of these variables. Ammonia is known to be bactericidal. Hence, survival of bacteria is enhanced at pH more than 8.5 due to increase in the ammonia/ammonium ratio. Survival periods in faecal sludges and wastewater for tropical and temperate zones are shown in Table 16.6.

**Table 16.6 : Pathogen Survival in Wastewater (WW) and Faecal Sludge (FS)**

| Organisms | Tropical zone (20-30°C) | | Temperate zone (10-15°C) | |
|---|---|---|---|---|
| | FS | WW | FS | WW |
| **Bacteria** | | | | |
| Faecal coliform | 50 | 30 | 150 | 120 |
| *Salmonella* | 100 | 150 | 5 | 5 |
| **Viruses** | 100 | 50 | 20 | 50 |
| **Protozoa** | 30 | 50 | 15 | 15 |
| **Helminth eggs** | | | | |
| *Ascaris* | 2-3 years | | | 10-12 months |
| Tapeworm | 1 year | | | 6 months |

*Source* : Strauss (1996).

Large number of infections are endemic and the socio-economic status and level of hygiene are low. The risk of disease transmission through waste-fed fish culture in industrial societies might be of greater epidemiological significance than in developing countries, because the incidence of these infection have been drastically reduced.

### *Treatment of Human Waste*

The removal of excreted viruses, bacteria, protozoa, and helminth eggs is primarily by discrete settling, accumulation in settleable flocks or by adsorption onto settling particles (bacteria and virus). Most micro-organisms thus concentrate in the treatement plant sludge where they survive for varying periods of time depending on the sludge treatment applied.

1. *Rapid Sand Filtration** : It is very effective in removing excreted pathogens. While protozoan cysts and helminth eggs are likely to be completely eliminated due to their relatively large size (10-80 μm), the removal of bacteria and viruses is less reliable.

   Faecal colliform and coliphage elimination generally comprises activated sludge treatment (including nitrification) with simultaneous phosphorus precipitation, secondary sedimentation and rapid sand of filtration of coagulated secondary effluent (contact

* For further detail, see Heeb and Zust (1991).

filtration). Higher removal efficiencies (99 per cent for faecal coliform and 96 per cent for coliphages) were observed for bacteria than for viruses. In a system comprising ferric chloride coagulation (contact filtration), sedimentation, and filtration (so-called complete filtration) about 87 per cent of virus is removed.

In the sand filter, nitrification and denitrification takes place simultaneously. This means that the transfer of oxygen to reactor must be limiting. The average oxygen transfer is about 580 g per $m^3$ of wastewater treated. An important means of transferring oxygen in a sand filter is the forced ventilation due to intermittent loading. In a buried sand filter, oxygen is introduced through forced ventilation. By this process about 1.8 $m^3$ of air per $m^3$ of water is transferred to the filter (1 $m^3$ of air contains about 320 g of oxygen).

In the filter, however, preferential flow results in only little contact to the surfaces with microbial growth. Consequently, the water passing rapidly through the filter may contain relatively high concentrations of pollutants in the outlet. Water trickling slowly through the filter in a complicated way will have more contact to surfaces with micro-organisms, resulting in better degradation and filtration of pollutants.

In wastewater treatment, denitrification is normally limited by the amount of organic matter available. Chemical oxygen demand conservation determines how much organic matter is needed for denitrification (2.68 g chemical oxygen demand for denitrification of 1 g of nitrate-nitrogen). It is clear that the amount of organic matter or humus present in the soil is large enough to support denitrification for many years.

2. *Waste Stabilization Ponds* : These ponds are effective and important in tropical and sub-tropical countries for the production of quality effluent suitable for safe use in fish culture. Non-aerated ponds require a gross surface area of 3 $m^2$ per capita. But in temperate and cold zones, a gross area of 12-14 $m^2$ per capita is required. Wastewater used in fish culture ponds should have a maximum $10^3$-$10^4$ faecal coliforms/100 ml and an absence of eggs of flukes (trematodes) and tapworms (cestodes). In tropical and sub-tropical zones, the bacterial criterion can be satisfied with a total of 20-30 days retention. Bacteria and virus removal and their inactivation is achieved by accumulation in and adsorption onto settleable particles and flocs, by the algae-induced pH increase, by the effect of UV-light, time and temperature, pH level between 8.5 and 9.0, which reached in ponds in the afternoon, are detrimental to bacteria. Complete removal of helminth eggs is achieved within 10-15 days retention time.

### 16.13 Fish Production in Wastewater-Fed Ponds

In many countries of the world (such as India, Germany, Vietnam, and Indonesia) recycling of waste substances through current technological advances and reuse of treated effluents for myriad purposes centre around waste stabilization pond method. It has been observed that fish culture in a series of ponds fed with treated effluents could obviously improve the water quality by reducing biochemical oxygen demand, orthophosphate, ammonia, nitrite, and nitrate along with the increase in pH and dissoved oxygen concentration (Table 16.7). Treated effluents contain nitrogen, phosphorus, potash, carbon, and organic matter below the toxic threshold that permits the dramatic increase in fish production. Polyculture is usually recommended to upgrade sewage effluents by reducing biochemical oxygen demand, suspended solids and faecal coliforms.

**Table 16.7 : Ranges of Physico-Chemical Parameters of Raw Sewage (RW), Oxidation Pond Effluent (OPE), Primary Fish Pond Effluent (PFPE), and Secondary Fish Pond Effluent (SFPE). Except pH and Temperature, all the Other Values are in Terms of mg/l**

| Parameter | RS | OPE | PFPE | SFPE |
|---|---|---|---|---|
| pH | 6.0-7.0 | 6.0-7.5 | 7.0-7.8 | 7.0-8.0 |
| Temperature (°C) | 25-35 | 25-35 | 25-35 | 25-35 |
| Biochemical oxygen demand | 100-150 | 30-70 | 20-40 | 10-15 |
| Dissolved oxygen | Nil | 1-9 | 1-25 | 1-35 |
| Orthophosphate | 20-35 | 10-26 | 5-16 | 3-10 |
| Ammonia-nitrogen | 10-25 | 6-16 | 3-8 | 0.8-2.0 |
| Nitrite-nitrogen | 0.15-0.45 | 0.10-0.80 | 0.10-0.58 | 0.05-0.08 |
| Nitrate-nitrogen | Nil | 0.10-0.55 | 0.70-2.00 | 0.50-1.50 |

### *Achievement of Waste Treatement Technology in Fish Production*

Fish culture in effluent-treated ponds ensures to solve the problem of sewage disposal in an eco-friendly manner. Huge quantities of waste material are being used for the production of large quantities of fish. In this systems, however, one million litres per day is spread over two fish ponds (50 m x 20 m x 20 m), a set of 18 duckweed ponds (25 m x 8 m x 0.6 m), and two marketing/ depuration ponds (40 m x 20 m x 2 m) covering a total area of 0 .72 hectare. Various species of duckweed plants (such as *Azolla, Lemna, Spirodela,* and *Wolffia*) are used for sewage treatement. Several species of catfish, major and exotic carps, and freshwater prawns are stocked to bring down the biochemical oxygen demand of the sewage-fed waters from 100-150 mg/l to 10-15 mg/l. The sewage-fed fish farm could yield about 4 tonnes of fish/hectare/year thus giving return of 40 per cent over the working capital.

### *Polyculture*

The most important species cultured in wastewater-fed ponds in India is the Rohu (*Labeo rohita*), Mrigal (*Cirrhinus mrigala*), Catla (*Catla catla*), and Common carp (*Cyprinus carpio*). In a polyculture pond, 15 day-old fish weighing 5 g are stocked and fattened within 6 months to a market weight of around 650 g. The average weight of rohu, common carp and mrigal has been fond to be 420, 830, and 532 g/6 months, respectively. However, the total yield of fish obtained in different polyculture systems varied between 4,245 and 7,300 kg/hectre/year. This is considered to be extensive fish culture because no supplemental feed is provided and the production of fish biomass is primarily due to development of organic constituents, plankton, and benthic food organisms. In some cases additional food is provided to the fish, but this constitutes an additional cost. Several other fish species is also cultured but with limited success, mainly due to their unsuitability to the operational and environmental features of the wastewater-fed fish ponds. Freshwater fish production strategies in developing countries using animal manures and sewage sludge into fish ponds clearly indicate an increasing trend (Table 16.8).

## 16.14 Practical Significance of Oxidation Ponds

So far as the cost of wastewater sewage treatment through conventional system is considered, it has been calculated that a total of 331 mega watt (MW) of energy is required to treat about 50,000 million litre per day of waste material from about one billion persons. The capital cost of providing treatment by conventional methods for this quantity of waste material will be about Rs 1,017 crores and the operation, maintenance and recurring costs will be about Rs 530 crores (Table 16.9). Therefore, the application of conventional methods for waste treatement involves heavy costs along with the large scale use of energy.

**Table 16.8 : Fish Production from Ponds in Developing Countries**

| Type of Fish | Year | Asia (12 countries) | China | Sub-Saharan Africa (29 countries) | West Asia/ North Africa | Latin America/ Caribbean | Total |
|---|---|---|---|---|---|---|---|
| Carps and | 1985 | 5,45,916 | 28,38,786 | 389 | 40,293 | 1,147 | 29,71,531 |
| other | 1990 | 7,11,441 | 41,24,478 | 1,126 | 40,011 | 4,588 | 48,89,644 |
| Cyprinids | 1995* | 9,17,375 | 62,11,537 | 1,970 | 43,700 | 6,970 | 71,81,552 |
| | 1999* | 9,87,979 | 69,34,652 | 2,798 | 48,690 | 9,320 | 79,83,439 |
| Tilapia and | 1985 | 1,05,325 | 77,120 | 7,006 | 26,745 | 11,976 | 2,34,172 |
| other | 1990 | 1,54,716 | 1,60,369 | 8,799 | 30,555 | 26,622 | 3,81,061 |
| Cichlids | 1995* | 2,05,710 | 2,07,215 | 10,175 | 33,705 | 32,500 | 4,89,305 |
| | 1999* | 2,25,600 | 2,23,350 | 14,700 | 36,800 | 37,200 | 5,37,650 |

*Source :* Selected data from FAO (1992)

* Calculated after compilation of data from various sources

**Table 16.9 : Power Requirement and Costs of Treatment of Waste Materials**

| Volume of waste materials (MLD) | Power required at 6.6 KW/mld (in terms of MW) | Capital cost @ Rs. 2 lakhs/mld (Rs. in crore) | Operation, maintenance and recurring cost for power @ Rs. 1.6 crore/MW (R.s in crore) |
|---|---|---|---|
| 15,000 | 99.3 | 305.1 | 159 |
| 35,000 | 231.7 | 711.9 | 371 |
| 50,000 | 331 | 1,017 | 530 |

In constrast to conventional methods of treatements, however, the cost of treatment of waste material in waste stabilization ponds has been estimated to be about Rs. 241 crores and is always justified so far as the cost-benefit ration is considered.

## 16.15 Trace Metal Problems in Wastewater-Fed Fish Ponds

Besides supply of essential nutrients to ponds through suldge for raising plankton concentration and fish growth in sewage-fed ponds, comparatively higher levels of trace metals from sludge will be released into the pond water which, in turn, affects the fish biomass.

Presence of trace metals in sewage sludge cause a major problem when it is used as fish feed. Metals should be, therefore, either removed by employing chemical treatments or by aquatic plants.

When sludge is treated with mineral acids, the metallic ions are dissolved and present in the aqueous phase which could be separated from the sludge. The process involved is, however, complicated and excessive cost for the purpose may negate the benefits achieved.

Sewage sludge is generally considered as a substrate for growing algae which, in turn, used as food by fishes. It is commonly known that algae can absorb considerable amounts of trace metals.

Tilapia and carp fed with sludge-grown algae (*Chlorella* sp.) may exhibit higher concentration ratio of various metals (355-22, 182) and absolute values than those fed with algal-growth, shrimp and cladocerans, that indicates the elimination of trace metal.

## 16.16 Termination of Wastewater-Fed Fish Pond Operations

In many Western countries, high initial costs for the establishment of the treatment facilities and also high operational costs due to the maintenance of the extensive areas may negate the benefits achieved. Since such ponds are located on the urban fringes, land prices considerably increased due to expansion of cities. Problems of availability of required quantities of freshwater for dilution of raw sewage is becoming sparse.

Social acceptance of the fish from wastewater-fed ponds was cited as a problem, nevertheless the rural populations always consume fish from such ponds. Public awareness and information campaign should be implemented, stating the benefits for recycling wastes for nutrient recovery. If the fish are processed under accepted hygienic standards (such as cleaning, gutting and cooking), wastewater-fed fish culture will not be considered as a health risk to the consumers though health risk through contamination with pesticides and heavy metals have been reported.

Interest on wastewater-fed fish culture strategies has led to studies in some developing countries with success in respect to treatment, fish production, economics, social acceptance, and public health concerns.

## 16.17 Four Comments on Wastewater Fish Culture

The aquaculture workshop presentations for wastewater treatemt in India (1988), Sweden (1991), and Zurich (1995) comprised numerous scientific papers presented by researchers from different countries. The paper represented a variety of topics and it is difficult to put them into a common perspective. Regarding wastewater fish culture, however, the following comments are worth remembering.

### *Role of Wastewater Aquaculture in Wastewater Treatment System**

Wastewater aquaculture is a very important technique in tropical and sub-tropical areas of the world particularly in India and China. Historical aspects to population density and food availability have affected the social acceptance of fish produced through wastewater aquaculture. Today there is a difference in acceptance between ethnic groups. Therefore, the future role of wastewater fish culture is more of a public, social, and political matter than a technological one.

In temperate regions, wastewater fish culure has been an important treatment method rather than food production and wastewater treatment is executed through activated sludge methods. Short growing season, high cost of land, and social acceptance of fish produced in human excreta are the principal reasons for abandoning wastewater fish culture. At present, there is an increasing awareness among the public, and administrators about the importance for recycling of wastewater and for consuming fish produced in treated water. The health authorities and the people should understand that the fish is safe to consume. The general attitude towards fish production though critical, the recycling activity is positive.

---

* For excellent review on wastewater fish culture, see Edwards and Pullin (1990), and Elnier and Guterstan (1991)

### *Risks Associated with the Use of Wastewater in Fish Culture*

Wastewater fish culture is a management of a hypertrophic aquactic ecosystem. Though this system is extremely productive, it can be a disaster if not properly managed.

The main aspect for succcessful management is the recycling of nutrients. Key parameters such as pH, dissoved oxygen, ammonia, and nitrite which are toxic, should be monitored. The risk for the public is related to toxicity and hygiene. Disinfection of human pathogens is related to temperature and the pathogenic aspects of wastewater use should be considered. Transfer of toxic substances through the food chain in wastewater ecosystem should not be ignored. Moreover, pathogens must be measured in fish and fish products originating from wastewater environments.

### *Efficiency of Wastewater Fish Culture*

Wastewater fish culture is still practiced in Asia and the efficiency of fish culture system is on increase. In Europe, however, the simple technique has been replaced by activated sludge/sewage treatment technology. Hence the efficiency for fish culture in Asia is fairly advanced and in other countries it is at a very early stage of development.

### *Potential for Wastewater Treatment in Temperate and Tropical Regions*

Winter season in temperate zones has been a constraint for operation of fish culture. The main way to overcome the climatic constraint is to store the waste during the winter season. Though it is not possible to store all waste materials, a part of it may be considered for the purpose to replace the demand for nitrogen and phosphorus fertilizers. In tropical regions, the operation of wastewater fish culture all the year round is possible and the technique that are followed is simple and cost-effective. Therefore, the potential for the technique is enormus because it can save large amounts of energy. However, it has been estimated that the economic potential for wastewater treatment and fish production is very high.

## 16.18 Sewage and Wastewater Treatment Processes

Treatement of wastewater is necessary before disposal of wastewater without producing undesirable/harmful effects. Some municipalities and industries dispose of inadequately treated wastewater into natural bodies of water because they assume that the water body is so large that dilution will not encourage hazards. Moreover, they cannot rely on discharge of wastewater by dilution. However, disposal of untreated/inadequately treated wastewater leads to several consequences. First, increased danger in using natural water bodies for agriculture and aquaculture. Second, contamination of finfish and shellfish by the pollutants, making them unfit for human consumption. Third, dissemination of pathogens. Fourth, depletion of oxygen supply to the water by unstable organic matter that results in the increase in mortality of fish and fish food organisms. Fifth, toxic chemicals are desseminated and accumulated that endanger the ecosystem.

In the treatment of wastewater, a number of processes are involved. In this section, however, treatment processes of sewage and wastewater obtained from individual dwellings and from municipal areas have been briefly described below :

### *Individual Dwelling Processes*

Treatment of sewage and wastewater can be accomplished by anaerobic digestion and/or by aerobic metabolic processes. The most common technique used for the purpose is the septic tank.

1. *Septic Tank* : It is a sewage-settling tank specially designed to retain solids (sedimentation) of the sewage entering the tank that permit adequate decomposition (biological degradation) of sludge. Generally, sedimentation starts from the upper portion of the sewage that encourage a liquid with some suspended particles to discharge from the tank.

   Although septic tanks are the most satisfactory method for disposing of sewage from the tank, pathogenic micro-organisms which are loaded in the sewage, are not eliminated. As a result, the drainage from the tank should be prevented from contaminating the water used for fish culture.

### *Municipal Sewage Treatment Processes*

Treatment of sewage and wastewater from muncipalities involves a series of treatment processes as summarized below :

1. *Primary Treatment* : Wastewater is first treated to eliminate coarse solid materials by several methods such as grinding, screening, and grit chamber. After removal of soild materials, it is further treated to remove settleable solids.
2. *Secondary Treatment* : In this treatment process, oxidation of the organic material in the liquid wastewater is accomplished by microbial activity. The oxidation methods include the filtration by sand filter and trickling filter, activated sludge process, and oxidation ponds.

(i) *Trickling filter* : This filter consists of a bed of gravel, slag, or synthetic material with drains at the bottom of the tank. Through this filter, organic wastes or sewage trickle slowly. The liquid sewage is sprayed over the surface of the bed where the sewage is saturated with oxygen. Frequent application of the sewage allows maintenance of aerobic conditions in the bed. The filtering medium of the tank becomes coated with microbial flora, called *zoogloeal mass*. The mass is composed of protozoa, fungi, bacteria, and algae. These micro-organisms adsorb and metabolize the organic constituents to stable end products such as amino acids, nitrates, nitrites, organic acids, fatty acids etc. At the same time, micro-organisms are cultured on the filter bed due to continuous supply of nutrients through sewage. This operation is carried out over a period of a few weeks. During this operation, the growth of micro-organisms may become so extensive that impairs the operation of the filter. However, the interactions and the activities of micro-organisms in the filter are extremely complex.

(ii) *Activated sludge process* : Continuous aeration of sewage forms a *floc*. The colloidal and suspended matter of sewage forms aggregates, called the *floccules*. This floc is allowed to settle and then fresh sludge is added that is again aerated. This process is repeated until complete flocculation of the sewage occurs within a few hours. The floc, called the *"activated sludge"* contains large number of bacteria, protozoa, yeasts,and moulds. These micro-organisms are very effective in the oxidation of organic compounds. The growth of some filamentous micro-organisms (such as *Sphaerotilus, Streptothrix, Leucothrix, Microthrix* etc.) adversely affects the settlement of activated sludge flocs.

After an aeration period of 4 to 8 hours, the mixture is transferred to a sedimentation tank where considerable reduction of biochemical oxygen demand and suspended solids takes place.

(iii) *Oxidation ponds* : Oxidation ponds (also called stabilization ponds) having 2 to 4 feet in depth encourage algal growth on the wastewater effluents. Although oxygen for oxidation of nutrients is supplied from the air, the generation of oxygen during photosynthesis by the algae *Chlorella* provides an additional source of oxygen.

3. *Advanced Treatment* : This treatment process has been developed to remove additional objectionable substances, nutrients (such as nitrogen and phosphorus) and to further reduce biochemical oxygen demand. Although unit processes include filtration, nitrification-denitrification, carbon adsorption, reverse osmosis, and ion exchange, the cost of the advanced treatement processes is very high.

4. *Final Treatment* : After completion of different treatment processes, the liquid effluent is disinfected by different disinfectants (such as chlorine, ozone, ultraviolet light etc.) and then discharged to a body of water. Oxygen is also added to the treated wastewater by mechanical means prior to final discharge.

5. *Solid Processing* : From the primary, secondary, and the advanced stages of the treatment processes, solids are removed. Thickening is carried out to further concentrate the solids or sludge prior to stabilization. Although stabilization processes include anaerobic and aerobic digestion, chemical addition, heat treatment, and composting, the most common process is the anaerobic sludge digestion.

(i) *Anaerobic sludge digestion* : The solids (or sludge) are transferred to a sludge digestor where sludge is digested under controlled conditions. An anaerobic condition prevails in the digestor and the anaerobic bacteria are active. These micro-organisms degrade organic solids to soluble substances and gaseous products such as methane (60-70 per cent), carbon dioxide (20-30 per cent), hydrogen (5-10 per cent), and nitrogen (4-6 per cent). Rapid digestion of sludge takes place at temperature varied betwen 50 and 60°C and at pH 7.0. Digestion of sludge is completed within 3 weeks.

(ii) *Composting* : In this process, the dewatered sludge undergoes decomposition. Dewatered sludge is mixed with a bulking agent (such as wood chips). This mateial is added to increase air circulation through the sludge for effective stabilization process. The mixture of bulking material and sludge is kept in a pile where oxygen is furnished and the mixture is allowed to decompose for a period of 21 days. The sludge is then transferred to a material for use as a soil conditioner.

## 16.19 Efficiency of Different Treatment Methods

The efficiency of different wastewater treatement methods exhibits considerable variation (Table 16.10). This variation is mainly due to the design of the unit, operational methods, and the type of wastewater. However, all these methods are most effective and accomplish a high degree of wastewater purification.

## 16.20 Use of Biogas Slurry in Fish Culture

Biogas is produced after anaerobic decomposition of agricultural household wastes and cattle manure by the bacteria. These waste products are first broken down in a digester by the bacteria into simpler compounds. Then the acidifying bacteria act on the simpler compounds. Methanobacterium further reacts on these simpler compounds of the waste/manure and produces biogas. At the end of the reaction, biogas and biogas slurry are produced. While the former is used for lighting as it contains methane (50-65 per cent), the later is considered to be a suitable input for efficient use in fish ponds because the slurry contains the following nutrients : Nitrogen (1.4-1.8 per cent), Phosphorus (1.1-2.2 per cent), and Potassium (0.8-1.2 per cent).

**Table 16.10 : Efficiency of Different Sewage-Treatment Methods**

| Method | Percentage removal of suspended soilds | Gallons of sludge per million gallons of sewage | Percentage removal of | |
|---|---|---|---|---|
| | | | Bacteria | BOD |
| Sedimentation | 40-95 | 1,000-5,000 | 40-75 | 30-75 |
| Chemical precipitation | 75-95 | 5,000-10,000 | 80-90 | 60-80 |
| Septic tank | 40-75 | 500-1,500 | 40-75 | 25-65 |
| Sand filter | 95-98 | | 98-99 | 70-96 |
| Trickling filter | 0-80 | 250-750 | 70-85 | 60-90 |
| Activated sludge | 70-97 | 10,000-30,000 | 95-99 | 70-96 |

*Source :* Gainey and Lord (1952).

### *Fish Response ot Biogas Slurry*

Biogas slurry, if supplemented by small quantities of chemical fertilizers and lime when necessary, are as effective as organic manure in enhancing fish production. However, high production of tilapia (at the rate of 8 tonnes/hectare/year) and minor caprs (common carp and silver carp, at the rate of 10 tonnes/hectare/year) have been reported from ponds located in Philippines and in Israel, respectively. It has been estimated that 60 kg of fish could be produced per year from a 200 sq. m pond treated with the slurry from the 6.2 sq.m digester.

For most of the Asian countries, widespread use of slurry is a relatively recent phenomenon. In heavily populated areas of some Asian countries (such as India, Bangladesh, Japan, and China), use of slurry as food for fishes has long been practiced. Most importantly, the nutrients and organic matter are recycled and returned to the ecosystem for further utilization. Such conservation is likely to be prâcticed more widely by other countries.

## 16.21 Use of Sewage and Organic Wastes in Fish Culture*

Sewage-fed fish culture system uptil now restricted to some regions of several Asian countries is now being increasingly recognized as a major fish culture technique of importance as it would enable to utilize sewage as a nutrient carrier for increasing fish production after adequate treatment to adjust various parameters to acceptable levels. Treated sewage is first diluted with required quantities of freshwater and then it is discharged into the fish ponds. This type of farming technology should be implemented in countries where the increase in population is very high for years and in

* For further detail on the subject, see Chakrabarty (2001).

areas where it is possible to discharge treated sewage into the pond through special disposal systems.

Fish culture through organic wastes is being popularized in regions where fish production potential is very low or moderate. Also, this system of fish culture is very simple and cost-effective technology. In order to popularize this program, various demonstration activities with specific objectives should be implemented and enforced among fish farmers. Demonstration programs may include the utility of biogas slurry, *Azolla* as biofertilizer, combination of paddy straw and cattle manure, domestic wastes, decomposed aquatic plants, and other waste materials such as treated sewage water and industrial effluents to achieve sustainable fish production to the tune of about 4.5 tonnes/hectare/year for increasing per caput fish consumption.

It has been demonstrated that in many village ponds where sodic soils are more predominant and fish production generally varies between 300 and 500 kg/ha/yr, productivity of such ponds can be enhanced through application of this low-cost technology. In principle, however, the technology involves regular application of mixture of paddy straw and cattle manure (or other treated waste substances) with other management protocols which permits to neutralize the sodic effect of the soil.

This technology is an asset of poor farmers and it is presumed that it would be possible for them to cultivate fish in their own/lease ponds with minimum investment for good economic returns. To understand the usefulness and effectiveness of this technology, increasing awareness among fish farmers is the best solution, however.

## 16.22 Conclusion

Wastewater and animal waste are important sources of nutrients that can be recycled through various processes for use in fish culture. In India, more than 2.5 billion metric tonnes of farm manure per year and 15,000 million litres of wastewater per day is available for recycling. While in time these wastewaters may be more fully utilized for aquaculture, the cost of making them suitable for fish production and the environmental consequences along with the human health hazards make these choice of actions extremely questionable.

The use of treated waste to fish culture ponds not only provide physical and chemical effects on mud but also stimulate production of fish and natural food. These wastes help to maintain or increase soil organic matter that provide significant qunantities of nutrients.

The judicious recycling of wastes is a good practice that has characterized successful fish culture system through decades. At present, it is highly important in view of the high nutrient demand of aquatic resources required to provide food for ever-increasing human population.

It should be borne in mind that in many countries, the reuse of human wastes for aquaculture is prohibited either by social taboos or legally. But inspite of prohibition, wastewater-fed fish culture technologies still exist and the use is gradually expanding though little is known about the health risks associated with wastewater reuse.

In general, waste materials contain toxic levels of inorgnaic and organic substances that are harmful to fish and other aquatic organisms. These aspects are the subject of another chapter that has been discussed in Chapter 12.

### *References*

Chakrabarty, M.M. 2001. *Sewage – Fed Aquaculture*. Palani Paramount Publications, Palni, India.

Edwards, P. and R.S.V. Pullin 1990. Wastewater-fed aquaculture. Proceedings of the International Seminar on Wastewater Reclamation and Reuse for Aquaculture. Calcutta, India, December 6-9, 1988. Environmental Sanitation Information Centre, Asian Institute of Technology, Bangkok, Thailand.

Elnier, C. and B. Guterstan. 1991. Ecological engineering for wastewater treatment. Proceedings of the International Conference, March 20-24, 1991. Stensund Folk College, Sweden.

FAO (Food and Agricultural Organizations). 1992. FAO Fish Circular No. 815 (Rev. 4).

Gainey, P.L. and P.H. Lord. 1952. Microbiology of Water and Sewage. Prentice-Hall, Englewood Cliffs, New Jersy.

Heeb, J. and B. Zust. 1991. Sand and plant filter systems. *In* : *Ecological Engineering for Wastewater Treatment* (*Eds* : C. Etnier and B. Guterstan). *Proc. Int. Conf.* March 20-24, 1991. Stensund Folk College, Sweden.

Strauss, M. 1996. Health considerations regarding the use of human waste in aquaculture. *In* : *Recycling the Resources* (*Eds* : J. Staudenmann, A. Schonborn and C. Etnier). Transtec Publications, Switzerland.

## Questions

1. What are the advantage of organic manures compared to chemical fertilizers when they are added to ponds?
2. State the chemical composition of animal manure.
3. State how organic manures are stored?
4. What substances are generally used for compost?
5. Contrast the anerobic and aerobic systems of fermentation of animal wastes and state the advantages and disadvantages of each.
6. Why compost manures should be used in fish culture ponds?
7. How many tonnes of organic manures with 2.5 per cent nitrogen and 5 per cent phosphorus would be needed to provide the same total nitrogen and phosphorus as is supplied by 600 kg of a 10-10 chemical fertilizers?
8. Waste materials and animal manures are known as useful for aquaculture, but in many countries they are considered as a disposal problem. Why?
9. State how wastewater can be treated for further utilization in fish culture. Why treatment of wastewater is so important?
10. Why aquatic plants are used in the treatment of wastewaters?
11. The use of human waste substances in considered as fish production strategy, yet in many countries they are considered as the potential health risks. Explain.
12. State how human wastes are treated for utilization in fish culture.
13. Discuss the potentiality of treated wastewater in fish production.
14. What are the problems in wastewater-fed fish pond culture?
15. Why wastewater fish culture is important in many tropical countries?
16. Define the following terms : (a) oxidation ponds, (b) waste stabilization pond, (c) activated sludge, (d) primary and secondary treatments, (e) compost, (f) anaerobic sludge digestion, (g) biogas slurry.

# 17

# Role of Nitrogen Fertilizers in Fish Culture

Nitrogen is a very important component of many compounds including enzymes and chlorophyll, and essential for growth processes of phytoplankton. Phytoplankton is essential for pond productivity which is related to fish growth and production.

Phytoplankton responds rapidly to applications of nitrogen. This element encourages the growth of algae and aquatic macrophytes over pond surfaces and form a green color to the water. Deficiency of nitrogen is evident when the pond water is clear and transparent.

When too much nitrogen fertilizers are applied, excess growth of algae occurs, water becomes more toxic, and upset the overall productivity of water. Fish growth and reproduction is delayed or suppressed, and fishes are more susceptible to diseases. Detrimental effects on fish food organisms should not result unless excessive quantities of nitrogen are used. Although nitrogen is widely used for agricultural and aquacultural productivities, its potential adverse effects on water and soil qualities must be worth remembering.

## 17.1 Source and Distribution of Nitrogen

Although run-off water, atmosphere, plant debris, and dead animals are considered as the strategy of nitrogen supply to any water bodies, application of nitrogen fertilizers is a significant contribution to nitrogen input. The elemental nitrogen in pond water is derived mostly from the air, the other source being bacterial denitrification of nitrate and ammonia. Different species of blue-green algae such as *Anabaena flosaquae, Microsystis aeroginosa, Nodularia spumigena* etc. secrete extra cellular nitrogenous compounds.

The nitrogen contents of pond soil normally varied between 0.08 and 0.6 per cent. One hectare of such a soil would likely contain about 3.6 metric tonnes of nitrogen while one hectare of atmosphere contains 300,000 metric tonnes of the element. Therefore, the atmosphere is a limitless source of nitrogen although it is not readily usable by plants in elemental form.

Most of the pond soil nitrogen is in organic form. Nitrogen compounds are associated with clays and humus which protect them from microbial breakdown. Ammonium ions are fixed by clay may account for upto 10 per cent of the nitrogen in surface soil and about 40 per cent in subsoils. The quantity of nitrogen in the available ammonium and nitrate forms is seldom more than 2 per cent of the total soil nitrogen, except where nitrogen fertilizers have been applied. Of course, these available forms are lost from soil through leaching and volatilization.

## 17.2 Nitrogen Cycle

The dominant nitrogen compounds of an aquatic ecosystem are : nitrate-nitrogen, ammonia-nitrogen, dissolved molecular nitrogen, and large number of organic compounds from amino acids, amines to proteins and humic compounds of low nitrogen content. Combined nitrogen occurs as hydroxyl amine, ammonia, nitrate, nitrite, and pariculate as well as dissolved organic nitrogen. Organic nitrogen often accounts for more than one half of the total dissolved nitrogen. Nitrogen undergoes a number of transformations that involve volatile, inorganic, and organic compounds. These transformations, however, occur simultaneously, but individual steps often accomplish opposite goals. The reaction may be viewed in terms of a cycle in which the nitrogen is shuttled back and forth at the discretion of microbial flora.

A typical nitrogen cycle of aquatic ecosystems consists of nitrogen fixation, ammonification, nitrification, and denitrification. Different components of nitrogen cycle in terms of energy necessary for operation of the cycle are shown in Figure 17.1. The steps from protein to nitrate provide energy for organisms that accomplish the breakdown, whereas the return steps require energy from sunlight or organic matter. For example, ammonifying and nitriying bacteria obtain energy from the breakdown, whereas, nitrogen fixing and denitrifying bacteria require energy from sunlight or organic matter.

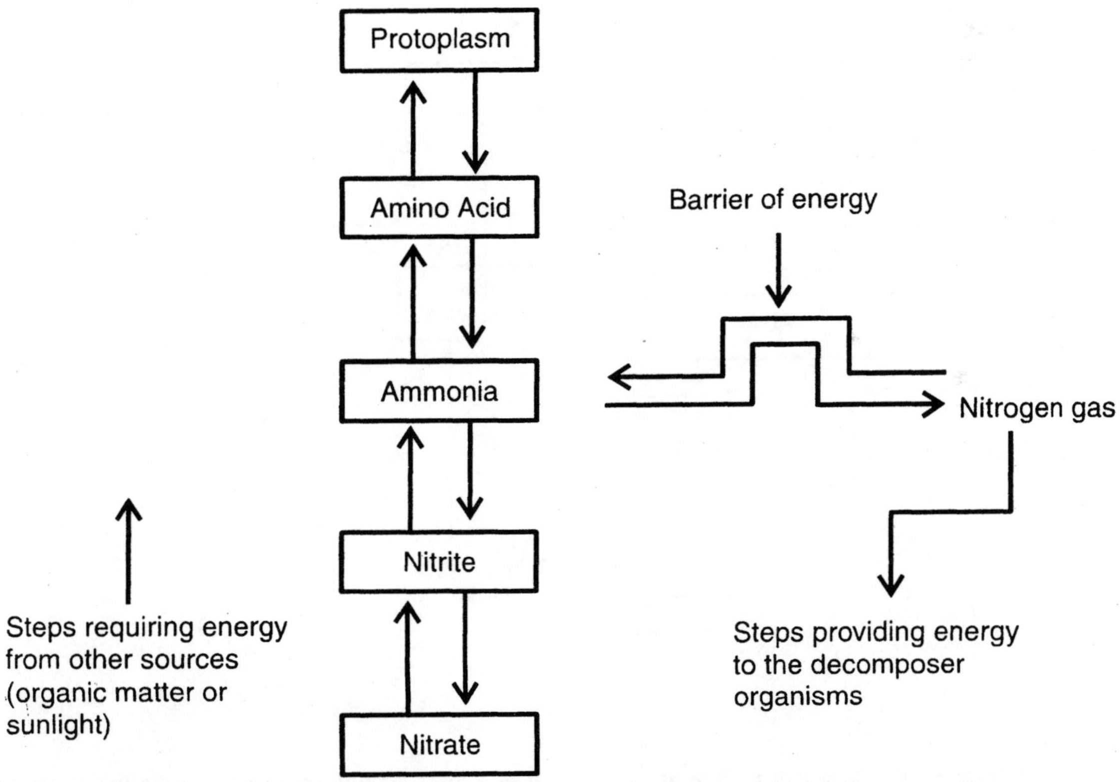


**Fig. 17.1 : Successive steps involved in the nitrogen cycle which are arranged in a ascending-descending series, with the high energy to distinguish steps that require energy from sunlight and organic matter. [From Odum (1983)]**

The term *nitrogen cycle* refers to the interaction among various forms of nitrogen in a pond soil, water, animals, and nitrogen in the atmosphere. A schematic representation of the processes of nitrogen transformations and removal from the soil and water of a fish culture pond is given in Figure 17.2. It has attracted scientific study for decades, and the practical significance of the cycle in a fish culture pond ecosystem for high production is beyond question.

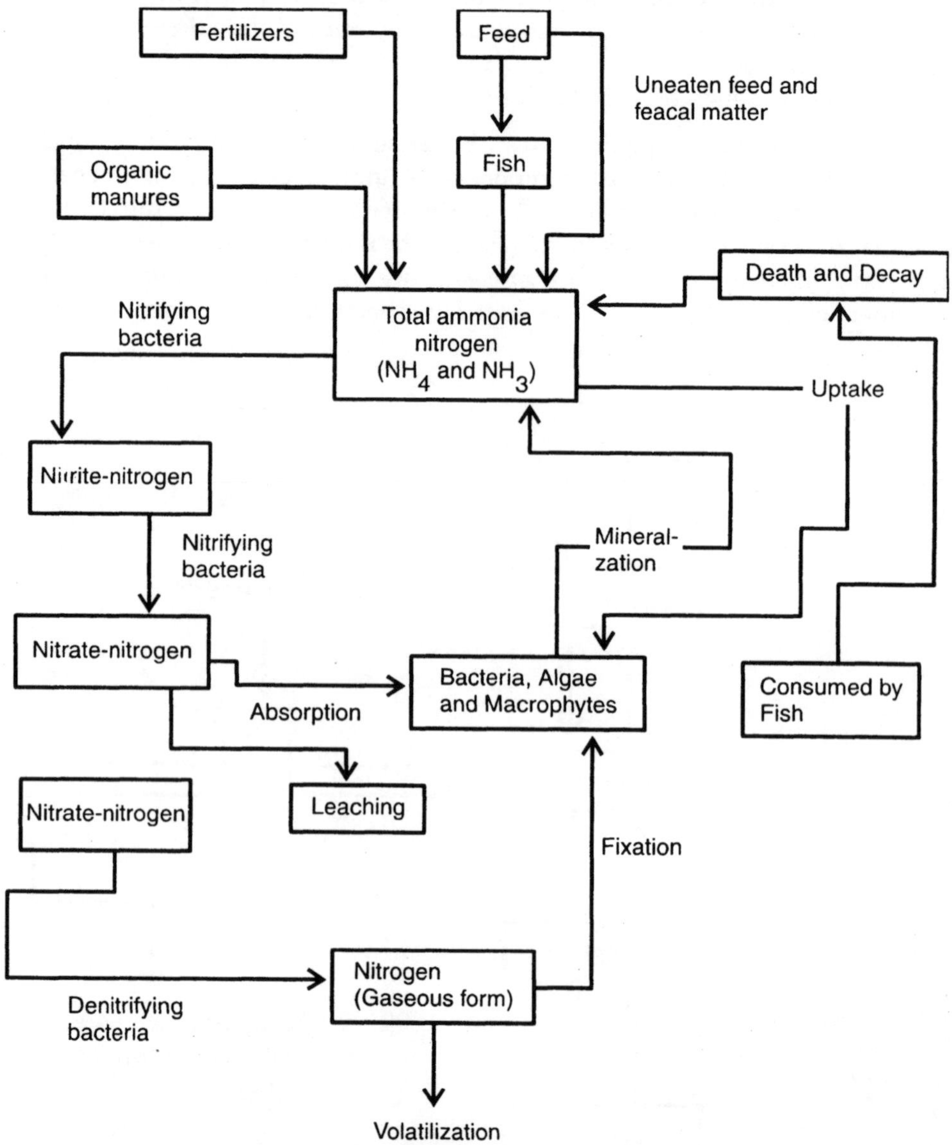


**Fig. 17.2 : Nitrogen cycle in a fish culture ecosystem. It is a term for biogeochemical cycle involving the element nitrogen. Nitrogen is fixed by the blue-green algae and certain types of bacteria. The nitrifying bacteria convert total ammonia-nitrogen into nitrate and returns to the soil. Under certain conditions, denitrifying bacteria obtain oxygen for respiration by breaking down nitrates and producing nitrogen gas to complete the cycle. Organic manures, fertilizers, fish feed, faecal matter, and dead bodies of all organisms are the chief sources of nitrogen input in fish culture ecosystems. Modern agriculture upsets the balance by applying chemical fertilizers to the soil, most of which is washed out into streams, rivers, lakes, and ponds.**

The nitrogen is derived from chemical fertilizers, organic manures, fish feed, and dead plants as well as animals. When total ammonia nitrogen is formed by the reduction of nitrogen, it is converted into nitrate and nitrite. These are consumed by algae and aquatic plants but after assimilation, enzymes help reduce nitrate to ammonia. Nitrogen depletion results from fish removal and absorption by plants.

Nitrogen fertilizers are readily soluble in water and soil solutions and are transformed into various forms (ammonia, nitrite and nitrate) through micro-organisms. Much of the nitrogen added to pond soil undergoes several transformations before it is remvoed. The nitrogen in organic combination is first transformed to simple amino compounds ($R - NH_2$), then to ammonium ion ($NH_4^+$), and finally to nitrate ion ($NO_3^-$). Even, nitrogen is either appropriated by micro-organisms or lost by denitrification and volatilization.

## *Classification of Nitrogen Cycle*

Though bulk quantities of nitrogen is present in soil as organic form, it is not largely available to plants. The organic nitrogen pool is several times higher than that of inorganic one (Figure 17.3). However, scientific effort has been devoted to the study of organic nitrogen regarding its release to forms usable by plants. The transformation of inorganic nitrogen to organic form of nitrogen is called *immobilization* whereas slow release of orgnic nitrogen to inorganic one is called *mineralization.*

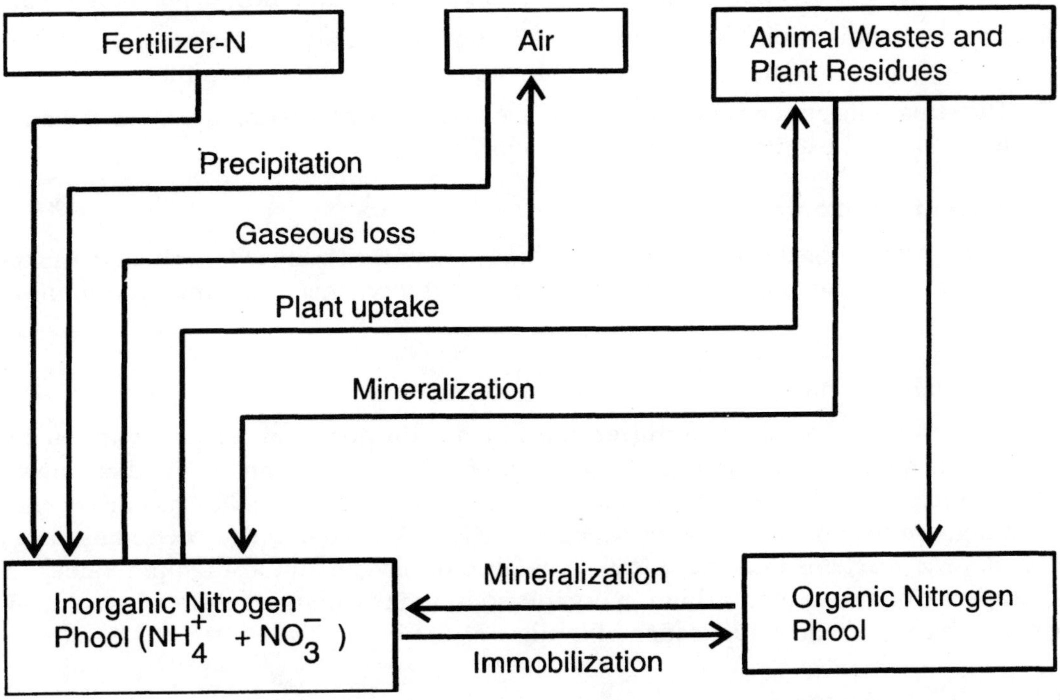


**Fig. 17.3 : Organic and inorganic are the two major pools of soil nitrogen. Note that atmosphere and nitrogen fertilizers are the basic sources of inorganic nitrogen. On the other hand, plant residues and animal wastes are the chief sources of orgnic nitrogen. Also note their processes of transfer.**

## 17.3 Mineralization and Immobilization

The release of organically bound nitrogen to inorganic forms (such as ammonium and nitrate) is called *mineralization*. Soil organisms simplify and hydrolyze the organic nitrogen compounds and produces $NH_4^+$ and $NO_3^-$ ions.

The reverse process of mineralization such as conversion of inorganic nitrogen ions ($NH_4^+$ and $NO_3^-$) into organic forms, is called *immobilization*. It occurs when animal and plant residues are added to pond soils. Residues are degraded by micro-organisms which, in turn, absorb the inorganic nitrogen ions and convert to organic tissue where the nitrogen is immobilized. After death of micro-organisms, some of the organic nitrogen in their bodies are converted into forms that make up the humus complex and some are released as ammonium and nitrate ions.

## 17.4 Fate of Ammonium Compounds

The ammonium nitrogen is moved in the following directions :

1. Considerable amounts of ammonium nitrogen are appropriated by soil micro-organisms. Consequently, nitrogen is incorporated in their bodies.
2. Phytoplanktons are about to use this form of nitrogen.
3. The ammonium ions are subject to binding by vermiculite and organic matter. In fixed forms, the nitrogen is not subject to oxidation, of course, in time it may become available.
4. Some ammonia gas is lost through volatilization particularly in alkaline soils. Such losses are significant when large quantities of ammonia are applied as fertilizers.
5. The remaining ammonium compounds are oxidized by bacteria, first to nitrites and then to nitrates. The entire process is called *nitrification*.

## 17.5 Fixation of Ammonia

In general, both inorganic and organic soil nitrogen have the ability to fix or bind the ammonia. It is also important to note that different compounds and mechanisms are involved in these two types of fixation.

### *Fixation by Organic Matter*

When ammonium-containing fertilizers are added to the pond soil, it reacts with soil organic matter and form compounds that resist decomposition. Therefore, it can be said that ammonia is chemically fixed or bound by the organic matter. Though it is not known the exact mechanisms by which ammonium fixation occurs, some specific reactions with components of humus have been established. The reactions take place in the presence of oxygen and at high pH values. Since fixation capacity of ammonium is high in organic soils, the reactions result in a loss of available nitrogen that recommend the application of fertilizers.

### *Fixation by Clay Minerals*

Several clay minerals are able to fix ammonium ions. The ions are fit between the crystal units of clay minerals and consequently, fixed or trapped. The ions are held in a non-exchangeable form, from which they are released to micro-organisms.

## 17.6 Nitrification

It is a process of enzymatic oxidation of ammonia to nitrates by micro-organisms in the pond soil. The oxidation process takes place in two steps. In the first step, nitrite ($NO_2^-$) ions are produced by the bacteria *Nitrosomonas* sp., apparently followed immediately by further oxidation to the nitrate ($NO_3^-$) form by the bacteria *Nitrobactor* sp.

***First Step :***

$$\underset{\text{Ammonium}}{NH_4^+} \xrightarrow{O} \underset{\text{Hydroxylamine}}{HONH_2} \xrightarrow{-2H} \underset{\text{Hyponitrite}}{\tfrac{1}{2}\,HONNOH} \xrightarrow{O} \underset{\text{Nitrite}}{NO_2^-} + H + \text{Energy}$$

***Second Step :*** $$NO_2^- \xrightarrow{O} NO_3^- + \text{Energy}$$

Of the two above-mentioned reactions, the second one is very significant because it prevents any great accumulation of the nitrite. It is highly signifiant because nitrites are generally toxic to fish and other aquatic animals.

### *Conditions Affecting Nitrification*

The nitrifying bacteria are more sensitive to their environment than most heterotrophic organisms. Soil conditions greatly influence the activity and population of nitrifiers (autotrophs) and therefore, the efficiency of nitrification will receive brief consideration.

1. *Level of Ammonia* : The process of nitrification occurs where there is a source of ammonia. High C/N ratio of residues prevent nitrification. Moreover, concentration of ammonia at high level also constraints nitrification. Application to alkaline pond soils of urea or anhydrous ammonia, appear to be toxic to nitrobactor, resulting in accumulation of toxic levels of nitrite ions.

2. *Temperature* : The temperature favourable for nitrification is 25-35°C. Therefore, tropical and sub-tropical fish ponds show rapid rate of nitrification and, consequently, the ability of ponds to provide nitrate for phytoplankton is enhanced. In temperate fish ponds, on the other hand, the accumulation of nitrate ions through nitrification is abridged. Nitrification rates gradually decline at temperature above 35°C and cease at temperature above 50°C.

3. *pH and Exchangeable Base-Forming Cations* : In acidic ponds, where pH value is below 6, there is remarkable accumulations of nitrate ions. Nitrifications rates decline at ponds where pH of soil is more than 7.5.

   The rate of nitrification proceeds rapidly where the levels of exchangeable base-forming cations (such as $Ca^{2+}$, $Mg^{2+}$, $K^+$, and $Na^+$) are high. The absence of these cations accounts in part for the slow nitrifications in acidic pond soils thus for the seeming sensitivity of the organisms to a low pH value.

4. *Pesticides* : Nitrifying bacteria are quite sensitive to some pesticides. Accumulation of pesticides to aquatic ecosystems through surface run-off almost completely inhibited the

process of nitrification; of course, there are some pesticides which slow the process of the nitrification down.

5. *Fertilizers* : Nitrifying organisms seem to have nutrient requirements. Therefore, application of nitrogen and/or phosphorus fertilizers is necessary to stimulate nitrification process. Use of large amounts of ammonium compounds to highly alkaline pond soils should be avoided to prevent loss of ammonia gas and to alleviate negative effects of the ammonium ion on nitrification process.

## 17.7 Fate of Nitrate Nitrogen

The nitrate nitrogen of the pond soil may go in three directions. It may (1) be leached from the soil, (2) be incorporated into micro-organisms, or (3) escape from the pond as a gas.

### *Leaching and Gaseous Loss*

Since the negatively charged ntirate ions are not adsorbed by the negatively charge soil colloids, they are subject to ready leaching from the soil and moved downward with the water. It has been calculated that about 15 per cent of the nitrogen is lost by leaching.

If excessive fertilization is made in sandy and sandy-loam pond soils, nitrate loss is increased by leaching but such losses can further be accentuated by increased use of organic manure of compost. Accumulation of nitrate in drainage water may rise to levels that could toxic to aquatic animals, and livestocks. Therefore, the need to take necessary steps to minimise nitrate leaching is obvious.

In pond mud, the process of *denitrification* is carried out by micro-organisms. This process involves chemical reactions and consequently gaseous losses of nitrogen take place. The mechanism for denitrification will be discussed later on.

### *Use of Nitrate-Nitrogen by Micro-organisms*

Both soil and plant micro-organisms readily assimilate nitrate-nitrogen. If soil micro-oganisms are provided with food, they use nitrates more rapidly than higher plants. On the other hand, if supplemental fertilizer nitrogen is not applied, they will suffer from nitrate-nitrogen.

## 17.8 Denitrification

### *Reduction by Micro-organisms*

Biochemical reduction of nitrate-nitrogen to gaseous form is a common phenomenon. Facultative anaerobic forms of micro-organisms take active part during the process of reduction. The five-valent nitrogen in nitrate is reduced stepwise to the zero-valent elemental nitrogen as noted below :

$$\underset{(+5)}{2NO_3^-} \xrightarrow{-2[O]} \underset{(+3)}{2NO_2^-} \xrightarrow{-2[O]} \underset{(+2)}{2NO} \xrightarrow{-[O]} \underset{(+1)}{N_2O} \xrightarrow{[O]} \underset{(0)}{N_2}$$

Each step in the reaction is accelerated by a specific reductase enzyme. It is important to note that the reaction can stop at any stage and the gaseous forms of NO, $N_2O$, and N can be released to the atmosphere. The oxygen atoms become incorporated into the bodies of the anaerobic bacteria.

### *Chemical Reduction*

In this process nitrite ions in a slightly acid soil will produce nitrogen gas when brought in contact with urea, phenol, and carbohydrates. The reaction is strictly chemical and does not require either the adverse soil conditions on the presence of micro-organisms and its practical significance is not great. The following reaction suggests what my happen to urea :

$$\underset{\text{Urea}}{CO(NH_2)_2} + 2HNO_2 \longrightarrow CO_2 + 3H_2O + 2N_2$$

### *Amount of Nitrogen Loss Through Denitrification*

In general, the exact magnitude of the losses of nitrogen through denitrification depends on the soil and cultural conditions (Table 17.1). Substantial losses of nitrogen through leaching and uptake by various species of plants and animals from different pond soil conditions, however, are not being uncommon. In rice-fish cultivation, loss of nitrogen by denitrification may be very high. About 65 per cent of the applied fertilizer nitrogen is volatilized as elemental nitrogen. By preventing the formation of nitrates by nitrification, losses can be reduced.

**Table 17.1 : Fate of Applied Nitrogen (as Urea) in Three Types of Pond Ecosystem**

| Fate of applied nitrogen @ 80 kg/ha | Sandy pond soil (%) | Clay pond soil (%) | Sandy-loam pond Soil (%) |
|---|---|---|---|
| Uptake by aquatic plants, algae and fish | 30 | 45 | 33 |
| Organic matter | 10 | 40 | 17 |
| Leached | 55 | 5 | 35 |
| Volatilization loss | 5 | 10 | 15 |

## 17.9 Ammonification*

A number of anaerobic or aerobic heterotrophic protein mineralizing bacteria, fungi, and actinomycetes in water and soil utilize organic nitrogen-rich substrate and convert it to ammonia. This process is called *ammonification.* The steps in ammonification are roughly the opposite of assimilation and amination.

### *Process of Ammonification*

There are three ways of producing ammonia from organically bound nitrogen : (1) extra-cellular organic nitrogen-containing compounds, biochemically or chemically, (2) from living bacterial cells during endogenous respiration, and (3) from death and lysed cells.

* For review on the subject, see Jana (1994).

1. *Breakdown of Extra-cellular Compounds* : Complex nitrogen-containing organic compounds are first deaminated to ammonia by enzymes on the cell wall and then transports the ammonia to the cell where it is used for synthesis. In organic wastes the amount of nitrogen-containing compounds is such that all the organic nitrogen is used in cell synthesis, but in cases where there is an excess of these compounds (low C/N ratio), ammonia will be formed in amount equal to the excess of the requirements for growth. The keto-acids and hydroxy-acids of the compounds are used for energy purposes.
2. *Endogenous Metabolism* : Although the process of endogenous metabolism is very complicated, it is known to take place in the presence of exogenous substrate (or endogenous substrate), so that even during growth there is the possibility of ammonia formation. The metabolism depends on the composition of the cells and environmental conditions.

   The rate of degradation of individual substances depends on their concentrations within the cells, which in turn depend on the nutritional status of the organism. Amino acids, proteins and ribonucleic acids give rise to ammonia. In fact, the presence of glycogen prevents the breakdown of nitrogen-containing substances so that ammonia is not released until the "shortage products" have been used. In cells which do not contain glycogen an imbalance between the essential components of the cell sometimes occurs when there is an amino acid deficiency. Thus, excess ribonucleic acid formed is broken down.
3. *Death and Lysis* : The cells which respire endogenously, lose their viability, that is, reproduction is hampered but continues to respire until lysis is reached when the cell membrane dissolves because of enzyme attack. Cell contents are exposed to the environment and production of ammonia occurs.

### *Proteolytic Activity of Micro-organisms*

Methylamines are produced by algae in freshwater and are likely to be decomposed by bacterial action to give ammonia. Protein-mineralizing bacteria can contribute to the liberation of ammonia which thus becomes available to the algae as a source of nitrogen either directly or after oxidation to nitrate. It has been suggested that the protein decomposition was initiated by proteolytic bacteria by hydrolyzing proteins, into peptides, and urea which in turn, are metabolized by ammonifying bacteria to liberate ammonia.

Organic nitrogen supplied by phytoplanktonic production or decomposition in water consists of about 85 per cent proteins and peptides. These organic materials with high molecular weight cannot be directly taken up by bacteria. Therefore, the hydrolysis of these materials is a first step of organic nitrogen utilization by micro-organisms.

### *Ammonifying Bacteria*

A variety of micro-organisms such as *Aerobactor cloacai*, *Pseudomons* sp., *Bacillus subtilis*, *Proteus vulgaris*, and *Escherichia coli* are capable of hydrolyzing proteins into simpler compounds which in turn, are metabolized by ammonifying bacteria to liberate ammonia to ammonium sulfate.

Ammonia is associated with a waste product overflow in microbial metabolism, the accumulated ammonium representing the amount of substrate nitrogen in excess of the microbial demands. Some of the ammonifying bacteria are substrate specific. They use only peptone but not amino acids, or use urea but not uric acid. Some other ammonifying bacteria are able to use a wide variety of organic nitrogen sources. Bacteria removes amino group and utilize the resulting ammonia as a source of nitrogen. Organic matter in the pond mud is transformed into ammonia by ammonifying bacteria.

## 17.10 Sources and Estimates of Ammonia in Fish Culture Ponds

It is known to all that ammonia play a key role in limiting fish growth in intensive fish culture ponds where fish production is dependent upon the application of protein-rice feed which serve as an important source of ammonia. The diets which are used are often far from balanced and about 80 per cent of the feed nitrogen is excreted by the fish as ammonia. In addition, decomposition of unconsumed feed and the use of chemical fertilizer and organic manure further add ammonia to fish ponds. Excretion of ammonium by zooplankton and autolysis after cell death are chief sources of ammonia in fish ponds.

The trapping and binding of ammonia in the pond sediments seems to be an another factor determining ammonia concentrations in the water. The sediments serve both as source from which ammonia may diffuse into the pond water and a sink and trap for nitrogenous matter from the water.

### *Ammonia Pool*

High concentrations of ammonia are found adsorbed on to the soil particles. The size of the ammonia pool in the growth cycle of an intensive fish pond in Israel has shown that the concentration of ammonia is increased along with the increase in sediment depth (Figure 17.4) while the size of the ammonia pool did not change during the growth cycle. Generally, ammonia pool is reduced when the ponds are dried.

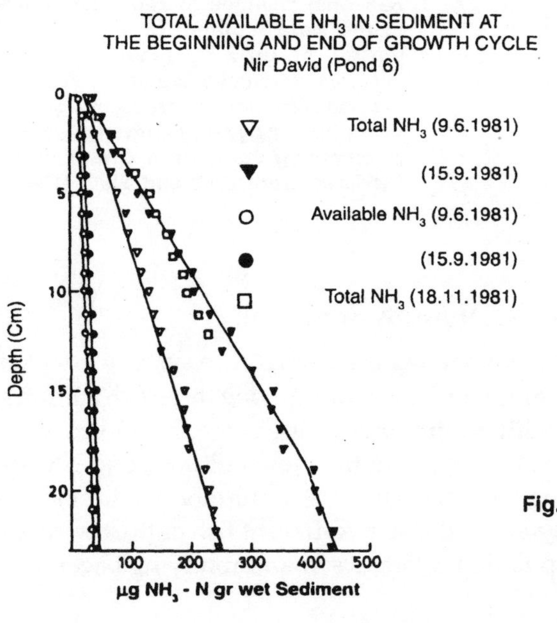


**Fig. 17.4 : Distribution of total and available ammonia at the beginning and end of the growth cycle in a fish pond. [Redrawn from Shilo and Rimon (1982)]**

Most of the intensively cultured fish ponds in Israel are characterized by a low ammonia content (below 0.1 mg $NH_3$-N/l) for most of the growth period. Experiments have, however, shown relatively high concentration of ammonia only early in the season but rapidly declined to trace amounts (Figure 17.5). This clearly indicates effect of drying and refilling of the pond. Immediately after refilling the ponds accumulation of ammonia in the sediments resumed, and in the water column a transient increase in ammonia (Figure 17.6) and nitrite was observed. In some cases, this temporary increase in nitrite upon refilling could be related to fish mortality, since loss of nitrite from the water causes disappearance of nitrite toxicity. At the time of drying the accumulated nitrate in the sediment is lost by denitrification upon refiling of the ponds. In drained fish ponds, ammonia concentrations in the sediment progressively increased, reaching a peak at the end of the growth period. Most of the ammonia is bound by soil particles, and only a small fraction remains free in the core water.

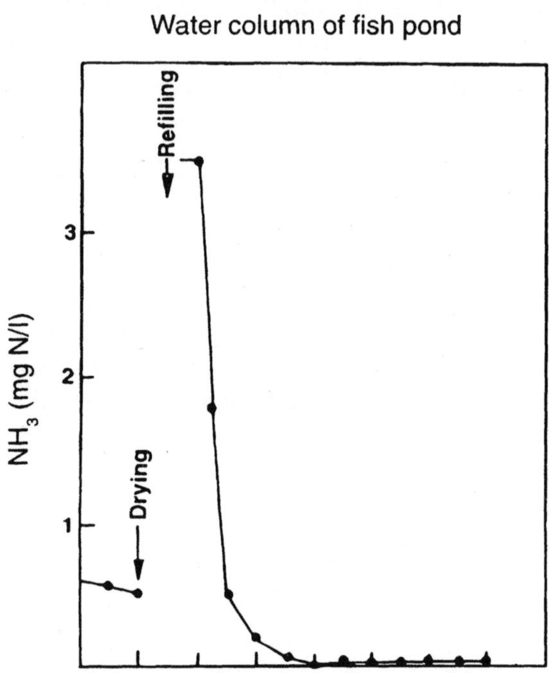


**Fig. 17.5 : Seasonal changes of ammonia concentration in the water column of a typical fish pond during the growing season of 1985. Water samples were collected at the depth of 20 cm, the samples were filtered through fibre glass filters and analyzed. Arrows indicate the times of emptying the pond and refilling the pond. [Redrawn from Diab and Shilo (1986)]**

## 17.11 Ammonification in Relation to Fish Culture System

The rates of ammonification varied markedly among the fish culture system. For example, a significant difference has been observed in the ammonification rate among three fish culture systems (such as traditional, mono and polyculture) with minimum activity in the traditional system and maximum in monoculture. The maximum nitrogen input in the monoculture pond is responsible for the massive development of ammonifying bacteria. The rate of ammonification is, however, extremely variable in different seasons of the year and also in different fish culture systems. Such changes are directly dependent upon the population dynamics of ammonifying bacteria.

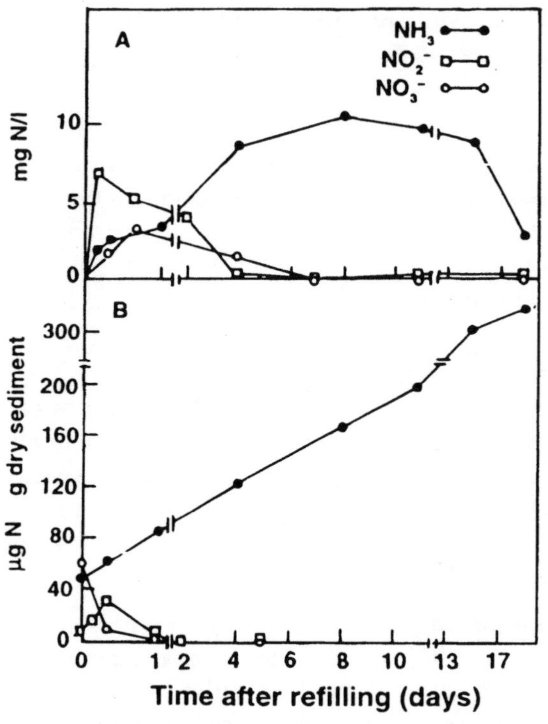


**Fig. 17.6 : The effect of refilling the drying fish pond sediment on ammonia, nitrite, and nitrate concentrations in water body and sediment. A, water body; B, fish pond sediment. [Redrawn from Diab and Shilo (1986)]**

## 17.12 Nitrogen Fixation

The fixation of nitrogen is the biochemical process by which elemental nitrogen is combined into organic forms. It is carried out by a number of organisms including a few actinomycetes, blue-green algae, and several species of bacteria. The quantity of nitrogen fixed globally each year through different systems is enormous, having been estimated at about 170 million metric tonnes. This exceeds the amount used in chemical fertilizers. Nitrogen fixation is said to be the most important biological process on earth.

### *System of Fixation*

Biological nitrogen fixation occurs through a number of systems such as symbiotic fixation with legumes, symbiotic fixation with nodule-forming non-legumes, symbiotic nitrogen fixation without nodules, and non-symbiotic nitrogen fixation. Of all the systems mentioned, the non-symbiotic nitrogen fixation is common in fish culture pond ecosystem. Pond soil and water certain free-living micro-organisms that are able to fix nitrogen. Since these organisms are not directly associated with plants, the transformation is called as *non-symbiotic* or *free-living*.

### *Fixation by Autotrophs*

The light and the bacteria/algae constitute the system for this type of nitrogen fixation. In the presence of light, certain photosynthetic bacteria and blue-green algae are able to fix both nitrogen and carbon dioxide simultaneously. Though the contribution of the photosynthetic bacteria is not certain, blue-green algae is thought to be some significance. Under normal conditions of an aquatic ecosystem, about 25 mg N/hectare is fixed each year. This estimate is no doubt to much variation depending upon the conditions peculiar to a particular ecosystem.

### *Fixation by Heterotrophs*

Some heterotrophic aerobic bacteria such as *Azotobactor* sp., *Clostridium* sp., and *Beijerinckia* sp. are able to fix nitrogen. The amount of nitrogen fixed by these organisms depends on the pH, soil nitrogen level, and the source of organic matter to the organisms for energy. Under normal conditions the rate of nitrogen fixation by these organisms are somewhat lower than that of autotrophic organisms, being the range of 5-10 kg N/hectare/year.

## 17.13 Addition of Nitrogen to Aquatic Ecosystem Through Precipitation

The atmosphere contains nitrogen compounds released from the plants, soils, automobiles, truck engines, and the combustion of petroleum products and coal. Nitrates exist, too, in small quantities, resulting from the oxidation of atmospheric nitrogen oxides and from lightening in the atmosphere. These atmospheric-borne nitrogen compounds are added to the soil and water through rain. Although the rates of addition of nitrogen per hectare are small, the total quantity of nitrogen added is significant and may be 10 kg/hectare particularly in highly polluted areas.

The quantity of ammonia and nitrates in precipitation varies with location and with season. The quantities are greater in comparable areas in the tropics than in humid temperate regions and greater in the latter than under semi-arid temperate climates. Rainfall additions of nitrogen compounds are highest and near huge animal feedlots, cities, and near industrial areas. There is special concern for the deposition of oxides of nitrogen from these areas of concentration because they are associated with increased soil acidity. The environmental impact of "acid rain" and its effect on aquatic ecosystems tend to overshadow the beneficial effects of precipitation-supplied nitrates as nutrients for growing phytoplankton.

## 17.14 Reactions of Nitrogen Fertilizers

The use of nitrogen-containing fertilizers has expanded in recent years. Consequently, nitrogen in the soil solution often is dominated by fertilizer-applied materials. While the nitrate and ammoniun ions coming from fertilizers react in a similar way to comparable ions released by microbial breakdown of organic materials, thier tendency to acidify the soil deserve special attention.

### *Acidity of Soil*

Ammonium-containing fertilizer and those that form ammonia upon reacting in the soil can increase soil acidity. Nitrification releases hydrogen ions that become adsorbed on the soil colloids. For good fish production, substantial and continued use of acid-forming fertilizers must be accompanied by application of lime.

The formation of nitrate-containing materials does not increase soil acidity. This is due to the fact that metallic cations (such as sodium or calcium nitrate) present in fertilizer materials have a slight alkaline effect.

### *High Concentration*

Application of urea, ammonium sulfate, ammonium sulfate nitrate, and anhydrous ammonia stimulates several reactions. The fixation of ammonium ions by organic matter and clays is increased. High concentration of ammonia inhibits the second stage of nitrification, resulting in the accumulation of nitrite ions. In alkaline pond soils, a high concentration of ammonium ion can result in the release of ammonia gas.

The microbial processes of fixation and gaseous nitrogen loss are affected by the application of nitrogen-containing materials in high amounts. Generally, fixation by free-living organisms is depressed by high levels of nitrogen fertilizers. Gaseous losses are often encouraged by abundant nitrates. Therefore, application of nitrogen fertilizers at high rates followed by nitrification tends to increase losses of nitrogen from the soil.

## 17.15 Management of Soil Nitrogen

Usually, deficit of nitrogen is met from three sources such as chemical fertilizers, green manures, and farm manures. Fish production levels, however, can be maintained through the use of these nutrient materials in required amounts.

### *Balance Sheet of Nitrogen*

Significant losses and gains of available soil nitrogen are shown in Figure 17.7. While the relative losses and additions by several mechanisms will vary greatly from soil to soil, the principles of soil nitrogen management are justified.

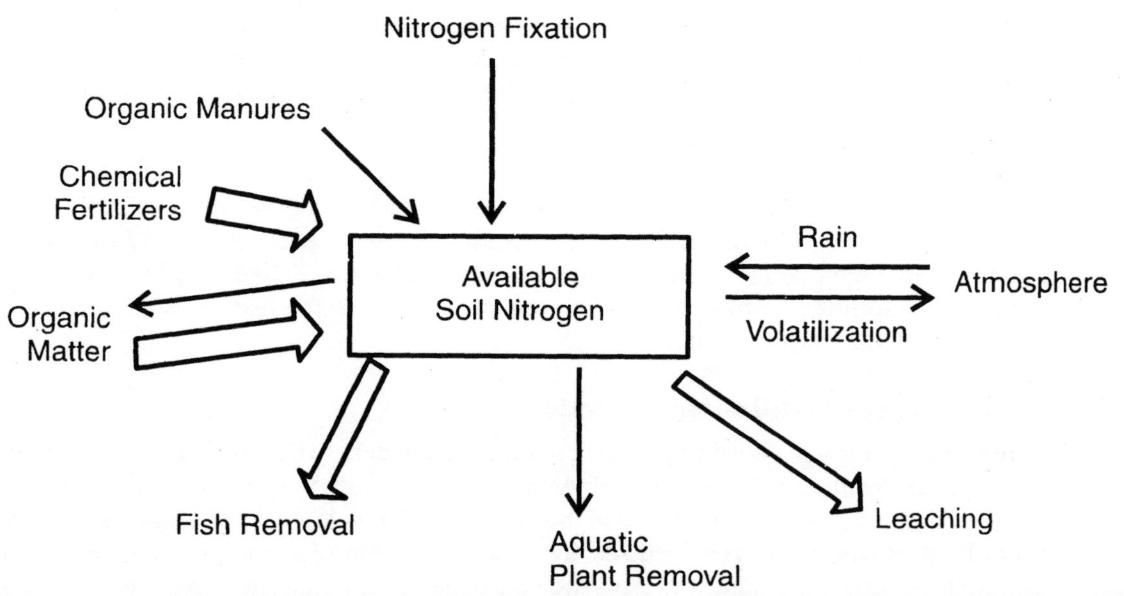


**Fig. 17.7 : Significant losses and gains of available soil nitrogen. Note that the widths of the arrows indicate the magnitude of the losses and the additions often encountered. The figure represents mean conditions and there is much variability in the relative and actual amount of nitrogen involved.**

Maximum quantities of nitrogen are removed by aquatic plants, leaching, and fish. In intensive and semi-intensive fish cultivations, substantial amount of nitrogen (about 135 kg N/hectare) is removed from fish biomass and about 40 per cent of the total nitrogen removed is returned to the soil through faecal matter. In some situations, well-fertilized and/or well-managed pond ecosystems may remove more than 150 mg N/hectare.

Volatilization and leaching losses are determined by pond management practices. Their magnitudes are so dependent upon specific situations that it is difficult to make a justifiable generalization on this aspect. Pond management practices that gives optimum fish production and in turn, organic manuring will perhaps reduce nitrogen loss to a satisfactory minimum.

## 17.16 Role of Nitrogen Fertilizers on Plankton

Plankton constitutes the natural food of fishes and is very important for ecosystem productivity. Their abundance and concentrations significantly increased by the application of nitrogen fertilizers. Generally phytoplankton concentration is increased when urea, ammonium sulfate, and calcium ammonium nitrate are used at high rates (160-1,280 kg N/hectare). Excessive development of phytoplankton is not desirable because they form green scum over the surface of water and bottom mud although some species of phytoplankton absorb and store solar energy and for good source of nutrient to aquatic life and also eliminate metabolites of nitrogen fertilizers from fish ponds. However, it has been found that application of nitrogen fertilizers at low rates (5-80 kg N/hectare) are beneficial to fish ponds.

In many tropical fish ponds where water temperature varied between 20 and 36°C, algal blooms grow well when there is high concentration of nitrogen. Some algal cells liberate some toxic substances into the water only when they die. However, zooplankton depends upon the production of phytoplankton. Zooplankton do not consume toxic algae; but under certain conditions of pond ecosystem, zooplankton is compelled to consume such algal forms and consequently zooplankton concentration completely disappeared from the water.

In an intensive fish culture pond, phytoplankton and zooplankton concentrations in water with nitrogen at 40-320 kg/hectare increased for most of the growth period respectively by 80.7-149.3 and 41-175 per cent over ponds that had recevied no nitrogen carrier (Figure 17.8). It is also evident that relatively high plankton concentrations were found within 7-10 days after fertilization, but these gradually declined to trace concentrations at the end of the test showing a clear cut evidence of the utilization of plankton by fish.

## 17.17 Role of Nitrogen Fertilization on Bottom Fauna

The efficiency of nitrogen caused by adding nitrogen fertilizers to different fish culture ponds has shown significant results. Addition of nitrogen fertilizers at rates varying between 40 and 160 kg/N hectare exhibited the highest efficiency in the growth and development of different species of bottom fauna. In cases where nitrogen fertilizers increase the number of mico-organisms, the bottom fauna consumes these micro-organisms as food and consequently, the concentration of bottom fauna considerably increased. Although bottom fauna concentration is increased by use of nitrogen fertilizers, excessive development of phytoplankton due to high nitrogen fertilization decrease their numbers by shadding.

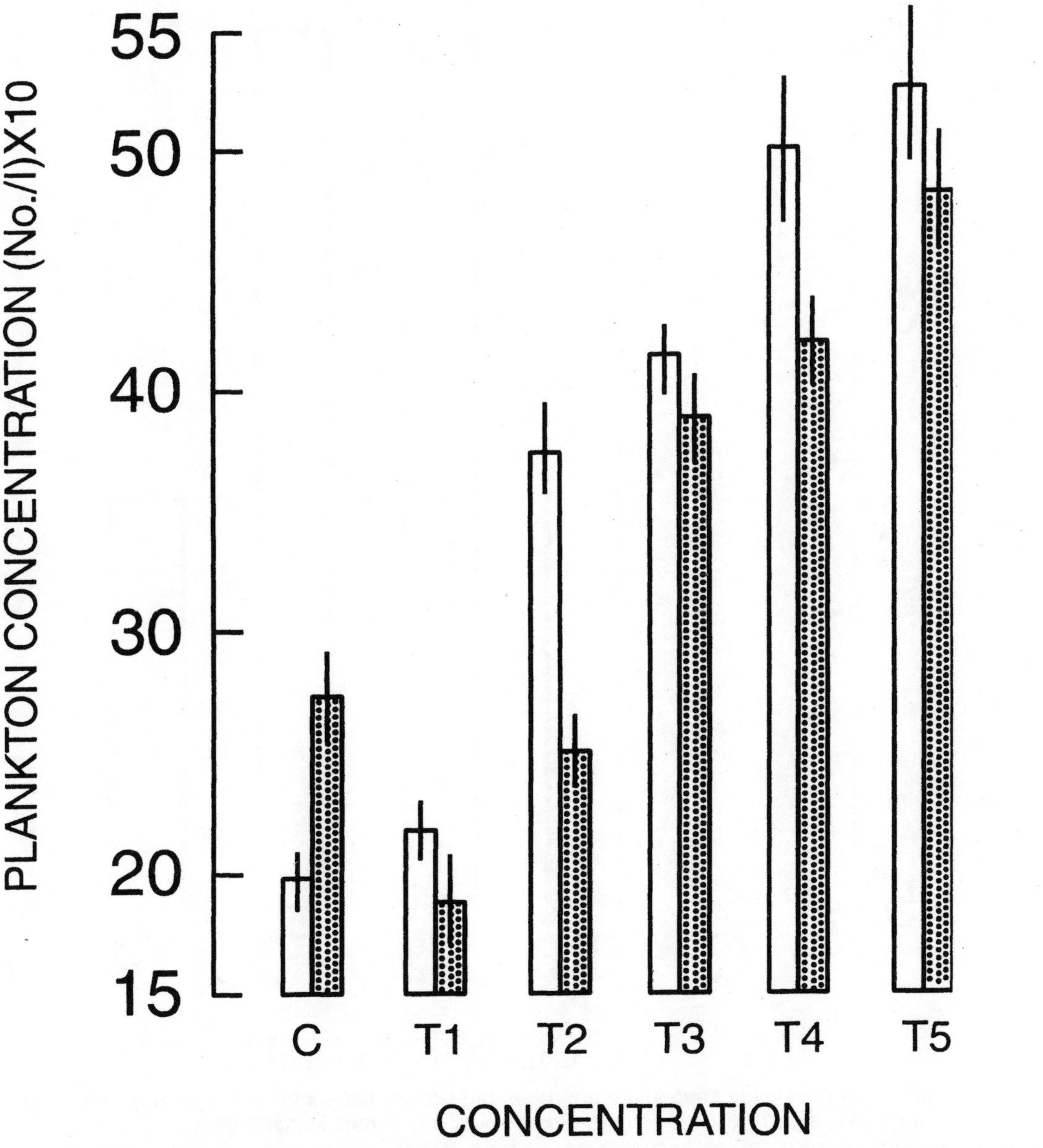


**Fig. 17.8 : Influence of calcium ammomium nitrate on the plankton concentration (Number/litre of water) in ponds. C, Control; T1, 20 kg; T2, 40 kg; T3, 80 kg; T4, 160 kg, and T5, 320 kg/ha N. Values are means of 13 observations. [From Sarkar (1996)]**

The population of different species of bottom fauna significantly increased in pond soils with nitrogen at 40 and 80 kg/hectare (Figure 17.9). This increase is attributed to the effect of greater accumulation of nutrient in soils. In most fish culture ponds where nutrients are applied for productivity, the density of bottom fauna exceeded than that of the untreated ponds. This suggests that application of nutrient at certain intervals and at optimum rates favour to maintain constant growth and development of bottom fauna in ponds.

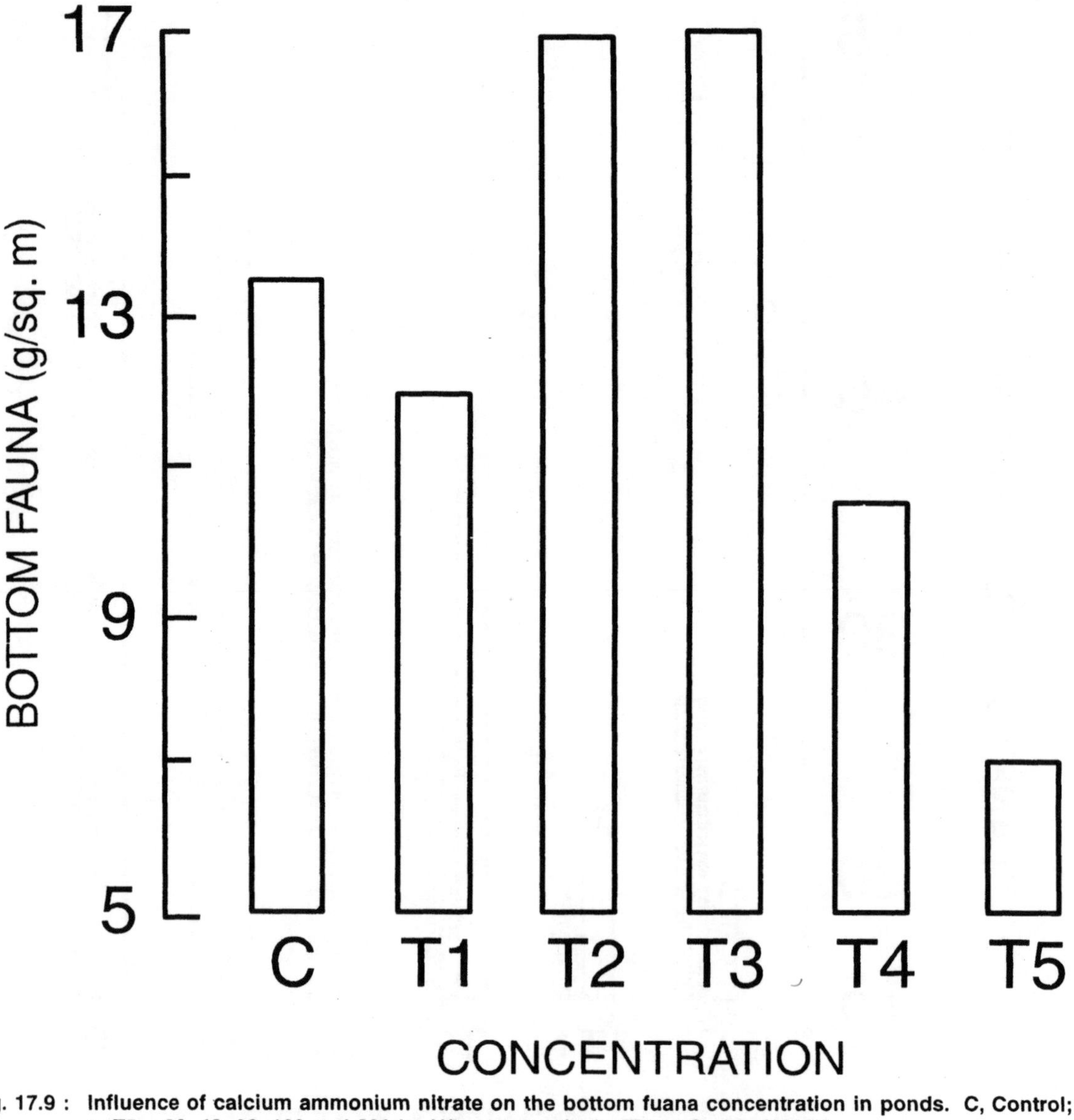


**Fig. 17.9 : Influence of calcium ammonium nitrate on the bottom fuana concentration in ponds. C, Control; T1 to T5 = 20, 40, 80, 160 and 320 kg N/ha, respectively. [From Sarkar (1996)]**

### 17.18 Role of Nitrogen Fertilizers on Fish Production

The most critical time to provide nutrients for fish ponds is during the early development of fish. Shortage of fertilizer use during fry and early fingerling stages can cut fish production by 30-50 per cent. Therefore, fertilization of fish ponds is essential for significant production. Fertilization increases fish yield and thus allows for a reliable supply of fish to the market.

Evaluation on the use of nitrogen fertilizers in freshwater fish production from well-managed ponds did not exhibit uniform results. Nitrogen fertilizers alone do not produce significant effect in fish production unless it is mixed with phosphatic fertilizers. In some situations, however, nitrogen fertilizers may cause mortality of fish particularly in their early stages and consequently, fish production is diminished.

Fish production in freshwater ponds is extremely variable from country to country and even from one region to another and it is directly related to the nutrient concentration in water. Variations in topographical features make fish production potential more significant. Sandy, sodic, and acid-sulfate soils are generally less productive than other types of soils because loss of nutrients is very intense in these soils. Also, interactions between environmental factors and fish biomass are not unimportant so far as fish growth is considered. It has been found, for example, that fish growth is either positively correlated with organic nitrogen or negatively correlated with ammonia-nitrogen/ nitrate-nitrogen/nitrite nitrogen. Metabolites of nitrogen fertilizers are toxic to fish and fish food organisms and suppress fish growth and therefore, caution must be taken to avoid nitrogen fertilizer toxicity if fish culturists feel it is necessary for ponds. Since it is very difficult to make a justifiable generalization on fish production, it is generally well recognized that nitrogen fertilization is effective when nitrogen deficiency in ponds occurs but during fish cultivation, fertilizer application rates should either be greatly reduced or discontinued for short/long periods having moderate or high organic matter content in soils.

### *Variations in Fish Production*

The need for fertilization and the optimum application rate largely depend not only upon water and soil characteristics of any given region of a country but also its climatic conditions and cultural as well as management techniques. For examples, in fertilized polyculture ponds in Egypt stocked with common carp, tilapia, silver carp, and grass carp obtained a total yield of 1,878 to 2,654 kg/hectare. A carrying capacity of about 2,000-3,000 kg/hectare in ponds stocked with tilapia in Thailand, about 2,500-3,500 kg/hectare in polyculture ponds in India, and about 8,000-10,000 kg/hectare in China have been observed. Production of fish as high as 12,000-16,000 kg/hectare has been obtained in earthen ponds in USA. Therefore, no definite application rates of nutrient carriers can be recommeded for fish culture. Studies conducted in different countries (such as USA) reveal that application rates of nitrogen and phosphorus for fish culture may be greatly reduced or even omitted without reducing fish production and that several monoculture ponds may not need any fertilization. Consequently, the cost of fish production and pollution will be lowered and savings of fertilizers on this account could be used in agricultural sectors.

### *Efficiency of Fertilizers in Fish Production*

Although nitrogen fertilizers are important in fish culture, their excessive applications can cause hazardous effects on fish biomass. During the last two decades, extensive studies have been made in aquaculture sector using some nitrogen fertilizers such as urea and ammonium sulfate to demonstrate their usefulness in freshwater fish culture. Recent investigations on the use of other nitrogen fertilizers in fish culture ponds have revealed the need for reassessment of efficiency of nitrogen fertilizers in fish production. From investigations, it is possible to suggest that the use of calcium ammonium nitrate, ammonium sulfate nitrate, diammonium phosphate, and urea at the 60, 60, 22.5, and 60 kg N/hectare rates respectively, are suitable for fish culture. Urea, diammonium phosphate and calcium ammonium nitrate are better nitrogen fertilizers than that of ammonium sulfate and ammonium sulfate nitrate because the first three types are less toxic to fish than the last two nitrogen carriers.

## 17.19 Nitrogen Toxicity to Fish Physiology

In water, ammonia is present primarily as $NH_4^+$ and as undissociated $NH_4OH$, the latter being highly toxic to many organisms particularly fish. The proportion of $NH_4^+$ and $NH_4OH$ is dependent on the dissociation dynamics as governed by pH and temperature. The toxicity of ammonia, nitrite, and nitrate to fish have been extensively studied. It has been demonstrated, however, that the extent of toxicity of ammonia depends primarily upon the application rates of nitrogen fertilizers. The toxicity of ammonia has been considered to be relatively independent of pH, while ammonium is regarded as having little or no toxicity and its toxicity has also been ascribed to the fact that the unionized form of ammonia can diffuse across gill membranes whereas, the ionized form occurs pass through the gill membranes.

The toxicity of nitrogen to fish physiology is very important. Metabolites of nitrogen fertilizers (such as ammonia, nitrate, and nitrite) that affect the physiology and toxicity of fishes and, as a result, the abnormalities of fish will receive brief consideration.

1. *"Gas bubble" Disease* : Supersaturation of nitrogen can occur at the water-air interface which frequently causes "gas bubble" disease in fishes, a physiological condition also caused by super saturated oxygen.

2. *Histological Effects* : Short-term levels of ammonia in water which is capable of killing fish over few days, start at about 0.6 mg/l. Chronic exposure to ammonia levels as low as 0.06 mg/l can cause gill and kidney damage, reduction in growth and oxygen-carrying capacity of fish.

   In fish culture ponds, eutrophication of metabolites of nitrogen fertilizers cause several abnormalities in fishes such as extensive hyperplasis of the gill epithelium, liver, kidney and intestine; morphological erythrocytic abnormalities and changes in histochemical activities of glycogen, golgi, mitochondria, and chromatophores. Therefore, before application of nitrogen fertilizers caution should be taken so that their adverse effects may be avoided during fisheries management.

3. *Brown Blood Disease* : This disease occurs in fish when nitrite level in water is increased. Nitrite enters the blood through gills and consequently, the color of blood becomes chocolate brown. Generally haemoglobin binds with nitrite to form methemoglobin, which is not capable of oxygen transport.

   Use of sewage for fish culture is now being advocated due to its nutrient content. Sewage not only contain nitrates but a variety of sewage bacteria such as *Pseudomonas, Bacillus, Micrococcus, Alcaligenes, Aerobactor* and *Proteus.* While the first four types are considered as nitrate respirers, the other two converts nitrate to nitrite. However, the transitory production of nitrite may result in nitrite toxicity to fish. At the same time, the transient concentration of nitrite formed due to nitrate reduction, may account for more than 40 per cent of methemoglobin in fish.

4. *Blood Chloride* : In nitrogen-enriched fish culture ponds, the concentration of blood chloride in fish is altered which affect fish physiology and gradual accumulation of metabolities is associated with the depression of blood chloride level in fish.

5. *Behavioural Symptoms* : Application of nitrogen fertilizers at high rates (160-1,280 kg N/ hectare) may exhibit several abnormal behavior of fishes such as irregular swimming movement, erratic and jerks, rapid opercular movement, mucus secretion over body, respiratory distress, enlargement of liver and abdomen, and cessation of feeding activities.

## 17.20 Toxicity of Nitrogen Fertilizers to Fish : Temperature Effect

Sensitivity of fish to nitrogen fertilizers at different temperature of water is significant. In cold water, toxicity of nitrogen fertilizers to fishes is low. As temperature increases, dissociation of nitrogen fertilizers occurs that results in accumulation of metabolits. Also, toxic metabolites are adversely affect not only the normal behavior survival, and growth of fish but the quality of fish as well and all these phenomena are directly related to temperature fluctuations.

In general, total ammonia nitrogen toxicity to fishes increases with an increase in temperature and pH of water. It has been established, however, that in order to evaluate the amount of un-ionized ammonia present, it is necessary to obtain the fraction of ammonia for a particular pH and temperature as presented in Table 17.2. This fraction should be multiplied by the total ammonia nitrogen in water to obtain the level of un-ionized ammonia in terms of mg/l.

**Table 17.2 : Fraction of Un-ionized Ammonia in Aqueous Solutions at Various pH and Temperatures**

| | Temperture (°C) | | | | | | | | |
|---|---|---|---|---|---|---|---|---|---|
| pH | 14 | 16 | 18 | 20 | 22 | 24 | 26 | 28 | 30 |
| 7.0 | 0.0025 | 0.0029 | 0.0034 | 0.0039 | 0.0046 | 0.0052 | 0.0060 | 0.0069 | 0.0080 |
| 7.4 | 0.0063 | 0.0073 | 0.0085 | 0.0098 | 0.0114 | 0.0131 | 0.0150 | 0.0173 | 0.0198 |
| 7.8 | 0.0157 | 0.0182 | 0.0211 | 0.0244 | 0.0281 | 0.0322 | 0.0370 | 0.0423 | 0.0482 |
| 8.2 | 0.0385 | 0.0445 | 0.514 | 0.0590 | 0.676 | 0.0772 | 0.0880 | 0.0998 | 0.1129 |
| 8.6 | 0.0914 | 0.1048 | 0.1197 | 0.1361 | 0.1541 | 0.1737 | 0.1950 | 0.2178 | 0.2422 |
| 9.4 | 0.3884 | 0.4249 | 0.4618 | 0.4985 | 0.5348 | 0.5702 | 0.6045 | 0.6373 | 0.6685 |
| 9.8 | 0.6147 | 0.6499 | 0.6831 | 0.7140 | 0.7428 | 0.7692 | 0.7933 | 0.8153 | 0.8351 |
| 10.2 | 0.8003 | 0.8234 | 0.8441 | 0.8625 | 0.8788 | 0.8933 | 0.9060 | 0.9173 | 0.9271 |

*Source :* Emerson et al (1975)

It has been observed that the development of fish eggs is inhibited following exposure to ammonium sulfate at 32 and 36°C. At different water temperatures (23.5, 29.5, and 34.5°C), fingerlings of *Cyprinus carpio* were 7.1-7.5 and 3.8-5.7 times more sensitive to ammonium sulfate and calcium ammonium nitrate respectively than urea. The toxicity of different nitrogen fertilizers to fish at various water temperatures are shown in Table 17.3

## 17.21 Conclusion

Nitrogen is held by soil colloids in forms that become slowly available. It is found in different organic substances as part of the soil organic matter. Formation of $NH_4^+$ and $NO_3^-$ ions is accomplished by micro-organisms. Anaerobic organisms are able to convert ammonium and nitrate into gaseous forms which are then released to the atmosphere to be combined by similar gases released from urban and industrial areas. The gases are deposited on aquatic environments in forms that are popularly called "acid rain" which has serious consequences to fish culture.

**Table 17.3 : Toxicity of Nitrogen Fertilizers to Fish at Different Water Temperatures**

| Fertilizer | Species | Temperature (°C) | | | | | |
|---|---|---|---|---|---|---|---|
| | | 20.0 | 23.5 | 25.0 | 29.5 | 34.5 | 36.0 |
| Urea | *Cyprinus carpio* | | 1,340 | | 1,000 | 560 | |
| | *Oreochromis mossambicus* | 660 | | 1,085 | 1,240 | 1,321 | |
| | *Labeo rohita* | 128 | 139.4 | | | | 93 |
| Ammonium sulfate | *C. carpio* | | 180 | | 141 | 75 | |
| | *O. mossambicus* | 330 | 175 | | 89.3 | 76 | |
| Calcium ammonium nitrate | *C. carpio* | | 255 | | 175 | 125 | |
| Ammonium sulfate nitrate | *Catla catla* | 247 | 235.7 | | | | 80 |
| Diammonium phosphate | *L. rohita* | 1,700 | | 1,610 | | | 115 |
| | *C. catla* | 2,310 | | 1,986 | | | 171 |
| | *C. carpio* | 2,785 | | 2,350 | | | 208 |
| | *O. mossambicus* | 2,799 | | 2,426 | | | 477 |

Nitrogen, which is removed from fish ponds through several processes, must be replenished regularly by chemical fertilizers or organic manures.

In water, ammonia is found in the form of $NH_3$ and $NH^+{}_4$. Ammonium hydroxide generated in alkaline conditions is extremely toxic to fish. Sources of ammonia are diverse, ranging from surface run-off to management strategies of fish culture ponds. Ammonia is sorbed to particulate and colloidal particles particularly in alkaline water containing high concentrations of humic dissolved matter. Transformation of nitrogen is often encountered in tropical water. Some problems always exist in fish culture which is characterized by the turn-over and removal or large amounts of nitrogenous compounds resulting from use of nitrogen-rich feed. Moreover, use of fertilizer and manure and decomposition of unconsumed feed lead to the accumulation of toxic intermediate formed in the transfer of the organic matter. These problems should be worth remembering.

Nitrogen dynamics is very important so far as the productivity of fish culture ponds is concerned. Micro-organisms and their activities control the ammonification, nitrification, and denitrification and hence the nitrogen status in soil and water. Periodic applications of nitrogen carriers increase the concentrations of plankton and bottom fauna in ponds. Transformation of nitrogen from sediment to water and its accumulation within the fish and macrophytes in the form of protein is utilized by higher trophic levels. Judicious applications of nitrogen fertilizers obviously help reduce the loss of nutrient thus surface and groundwater pollutions can be prevented. Frequent assessment of fish culture ecosystems with regard to nitrogen demand must be undertaken to solve the environmental quality problems.

## *References*

Diab, S. and M. Shilo. 1986. Transformation of nitrogen in sediments of fish ponds in Israel. *Bamidgeh.* 38 : 67-88.

Emerson, K., R.C. Russon, R.E. Lund and R.V. Thurston. 1975. Aqueous ammonia equilibrium calculations : effects of pH and temperature. *J. Fish. Res. Bd. Can.* 32 : 2379-2383.

Jana, B.B. 1994. Ammonification in aquatic environments : A brief review. *Limnologica*. 24 : 389-413.

Odum, E.P. 1983. *Basic Ecology*. Saunders College Publications, Japan.

Sarkar, S.K. 1996. Influence of calcium ammonium nitrate on fish and aquatic ecosystems. *Indian J. Fish.* 43 : 87-95.

Shilo, M. and A. Rimon. 1982. Factors which affect the intensification of fish breeding in Israel. 2. Ammonia transformations in intensive fish ponds. *Bamidgeh*. 34 : 101-114.

## Questions

1. How fixation of ammonia takes place when nitrogen fertilizers are added to pond soils?
2. Ammonia is converted into nitrite and nitrate by the bacteria *Nitrosomonas* sp. and *Nitrobactor* sp., respectively. Which one is more significant and why?
3. What are the conditions affecting nitrification?
4. The contamination of water by the use of nitrogen fertilizers is very common. What relationship do the processes of nitrification and denitrification have to these concerns? Discuss.
5. How ammonification occurs in an aquatic ecosystem?
6. What are the sources of ammonia in a fish culture pond?
7. State how nitrogen is fixed. Why nitrogen fixation is considered as the most important biological processes in an aquatic environment?
8. You have come to know about "acid rain". What are the advantages and disadvantage to fish culture?
9. State how nitrogen fertilizers are related to fish production.
10. State how nitrogen is balanced in an aquatic environment?
11. Nitrogen is an important nutrient to fish culture ponds, yet it exhibits toxic effects and abnormalities in fishes. Explain.
12. Explain how temperature is related to the toxicity of nitrogen fertilizers to fishes.
13. Describe the role of microbes in determining the availability of nitrogen.
14. Define the following terms : (a) ammonification, (b) denitrification, (c) nitrogen cycle, (d) mineralization, (e) immobilization, (f) nitrification, (g) ammonia pool.

# 18

# Role of Phosphorus Fertilizers in Fish Culture

Next to nitrogen, phosphorus is most critical essential element in influencing fish growth and production. Unlike nitrogen, phosphorus is not supplied through fixation but must come through fertilizers, animal manures, green manures, and domestic as well as human wastes.

## 18.1 Importance of Phosphorus

Phosphorus is an abundant element and occurs in a wide range of materials in combination with other elements. It is the most likely element to be limiting the development of biosphere and is widely distributed in plants, nucleic acids, adenosine triphosphate (a high energy compound), animals, water, soil and other minerals. Phosphorus though required in small amounts, has been known for many years to be very imporant single element regulating biological productivity in water bodies. Because fish production increases almost linearly with increase in application rates of phosphorus as long as phosphorus is the limiting factor in management system [For further comments, see Hickling (1962)], phosphorus fertilizers are used to enhance plankton production which, in turn, triggers the growth and production of fish.

Fish growth requires phosphorus for the synthesis of protoplasm, bone formation in fish and phytoplankton and therefore, phosphorus plays an important role in the biological productivity of aquatic systems [For further detail, see Jhingran (1988)]. Dominant phosphorus compounds of water are phosphate, organic and inorganic phosphorus. Transformation of phosphorus occurs with a number of steps and at the discretion of some abiotic as well as biotic factors.

## 18.2 Phosphorus Eutrophication

Eutrophication occurs where there is an accumulation of phosphates in ponds. This nutrient accelerates the growth and development of plants. Accumulation of phosphates is often encountered in tropical water where algal blooms are developed and create a dense green mat over the water. Rapid growth of algae and mass decomposition of plant materials eventually use up all of the oxygen in water. This leads to the death of the algae and proliferation of bacteria and consequently, hydrogen sulphide is generated which is extremely toxic to fish.

The amount of dissolved phosphorus present in pond water depends upon the nature of phosphorus fertilizers and the amount to be used. In general, inorganic and organic materials contain high and low amounts of phosphorus respectively, providing a basis for differences in the productivity of ponds.

## 18.3 Phosphorus as Pollutant

In contrast to nitrogen, leaching of phosphate is not significant and consequently threat to groundwater pollution from phosphorus fertilization is not intense. Generally phosphorus is accumulated in ponds and lakes through surface run-off and fertilization, thus resulting in deterioration of water quality. It has been reported that about 80 per cent of the phosphorus gets into the surface waters, directly through effluents. Moreover, cultivation of agricultural crop contributed only 17 per cent; of which, two third being contributed by the erosin of phosphorus-containing soil and one third by intensive live-stock farming.

Presence of heavy metals in different types of phosphorus fertilizers are considered as the most important pollutants. Among them, cadmium is given much concern because it can easily be transformed from soil to plant. Through the application of phosphatic fertilizers, an appreciable amount of cadmium is incorporated into the food chain. It has been estimated that super phosphate, rock phosphate, and diammonium phosphate contain 187, 303, and 109 mg/l cadmium/kg of $P_2O_5$ respectively. However, by using calcination techniques (trapping of cadmium in flue gases) and by using ion-exchange resins, heavy metals from phosphatic fertilizers can be removed to some extent.

## 18.4 Phosphorus Cycle

The phosphorus cycle in a pond environment [For review on the subject, see Sarkar (1989)] is illustrated in Figure 18.1. Note that the pond ecosystem receives, to a lesser or greater extent,

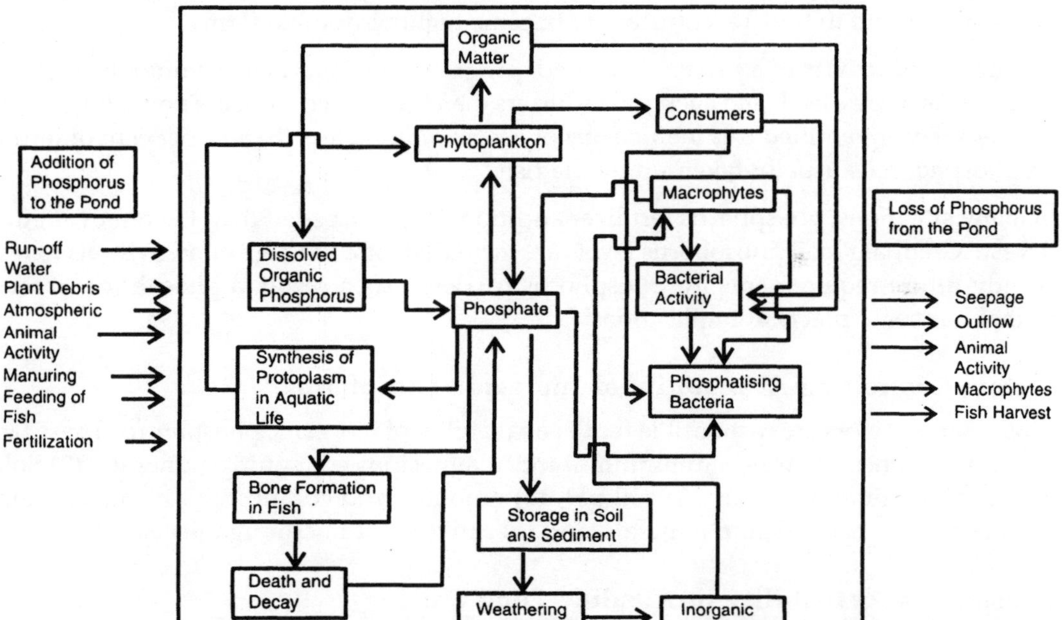


**Fig. 18.1 : A model of the phosphorus cycle in a pond environment. Phosphorus is found chiefly as inorganic phosphate, as soluble organic phosphate, or as a part of the mineral phosphate in sediments. This phosphate then moves through the food chain from the producers to the consumers. Some phosphates are released through excretory products by aquatic animals. After the death and decay of aquatic animals, the dead bodies are decomposed by the phosphatizing bacteria and consequently, soluble phosphates are released in the pond soil. This soluble phosphate is again absorbed by phytoplankton, followed by zooplankton and then fish. Since this cycle has no atmospheric phase and hence, it is also termed as the *sedimentary cycle*. Note that the addition of phosphorus to fish culture pond ecosystem takes place through a series of steps whereas losses of phosphorus from the pond occur mainly through fish harvest and seepage.**

phosphorus from various sources and translocated to fish, plankton, macrophytes, and other aquatic animals. After the death and decay of animals, the phosphorus is returned to the soil. Micro-organisms decompose the residues. Some of it becomes associated with the soil organic matter. Some is converted to the soluble forms that aquatic plants can absorb, thereby starting a repeat of the cycle.

In most fish culture ponds, the amount of phosphorus in the available form is very low, seldom exceeding about 0.03 per cent of the total phosphorus in the soil. Therefore, available phosphorus levels must be supplemented on soils by adding phosphorus fertilizers. It is unfortunate that much of the added phosphorus is converted to the less available secondary mineral forms (iron, aluminium, calcium, and clays), from which it is also released and becomes useful to plankton over a period of months. Hence, the problem of maintaining phosphorus in an available form should be considered first.

## 18.5 Phosphorus Problem

In a pond ecosystem, the phosphorus problem is two-fold. First, the phosphorus compounds are mostly unavailable, some being highly insoluble. Second, when fertilizers and manures are added to ponds, they are settled at the pond bottom where they are either fixed or are changed to unavailable forms and in time react further to become highly insoluble forms.

For high productivity of a fish culture pond, progressive fish farmers commonly apply more phosphorus as manures, feed, and chemical fertilizers than is removed from the pond. Investigations on this aspect have quantified this inefficiency of use, exhibiting less than 12 per cent of fertilizer-applied phosphorus is usually taken up by the fish.

In many situations, phosphorus fertilizer additions have exceeded than the removal through fish harvest. Gradually, soil phosphorus levels are increased, often to high enough levels to reduce significantly future requirements for phosphorus carriers. As a result, a phosphorus reserve is formed that has some practical implications.

## 18.6 Factors Controlling Availability of Inorganic Phosphorus

A number of factors are responsible for the availability of inorganic phosphorus. These factors include : (1) presence of iron-, aluminium-, and magnesium-containing minerals, (2) soluble manganese, aluminium and iron, (3) soil pH, (4) amount and composition of organic matter, (5) available calcium and calcium minerals, and (6) activities of micro-organisms.

## 18.7 Phosphorus Availability in Alkaline Pond Soils

The phosphorus availability in alkaline pond soils is determined by the solubility of calcium compounds. If $H_2PO_4^-$-containing fertilizers are added to an alkaline pond (pH 7.5-8.0), the $H_2PO_4^-$ ion quickly reacts to form less soluble compounds. Although a number of intermediate compounds are formed, tricalcium phosphate is the most important one. The reaction involving calcium carbonate and $H_2PO_4^-$- containing superphosphate in the soil can be represented as follows :

$$Ca(H_2PO_4)_2 \cdot H_2O \quad + \quad CaCO_3 \longrightarrow Ca(PO_4)_2 \; + \; 2CO_2 \; + \; 2H_2O$$

The solubility of the compounds and the availability to plankton and fish, decrease as the phosphorus changes from $H_2PO_4^-$ ion to tricalcium phosphate. This insoluble compound is converted further to even more insoluble compounds. Oxy-, carbonate-, and hydroxy compounds may be formed (Table 18.1). These compounds are, however, thousand times more insoluble than freshly formed tricalcium phosphate.

**Table 18.1 : Some Calcium Compounds of Phosphorus Found in Soils Listed in Order of Decreasing Solubility**

| Phosphorus compounds | Chemical formula |
|---|---|
| Monocalcium phosphate | $Ca(H_2PO_4)_2$ |
| Dicalcium phosphate | $CaHPO_4.2H_2O$ |
| Tricalcium phosphate | $Ca_3(PO_4)_2$ |
| Oxyapatite | $3Ca_3(PO_4)_2.CaO$ |
| Hydroxy apatite | $3Ca_3(PO_4)_2.Ca(OH)_2$ |
| Carbonate apatite | $3Ca_3(PO_4)_2.CaCO_3$ |

The reversion to insoluble calcium phophate may also found in some pond soils and the problem is more serious where excess calcium carbonate is present.

## 18.8 Phosphorus Availability in Acidic Pond Soils

### *Precipitation by Iron, Aluminium and Manganese Ions*

Some soluble iron, aluminium, and manganese are generally found in highly acidic soils. These ions immediately react with $H_2PO_4^-$ ions resulting in the formation of insoluble hydroxy phosphates. This chemical precipitation may be represented as follows, using the aluminium cation as an example.

$$Al^{3+} + H_2PO_4^- + 2H_2O \rightleftharpoons Al(OH)_2H_2PO_4 + 2H_2O$$

### *Reaction with Hydrous Oxides*

The $H_2PO_4^-$ ion also reacts with insoluble hydrous oxides such as aluminium oxide ($Al_2O_3.3H_2O$) and ferrous oxide ($Fe_2O_3.3H_2O$). For example, when aluminium hydroxide reacts with $H_2PO_4^-$ ions insoluble hydroxy phosphate is formed which can be shown as follows :

$$(HO)_2{>}Al - OH + H_2PO_4^- \rightleftharpoons (HO)_2{>}Al - H_2PO_4 + OH^-$$

By this reaction the formation of several phosphate minerals containing either aluminium or iron both occurs. Consequently, phosphorus fixation by this mechanism possibly takes place over wide pH range. However, the presence of the readily available $H_2PO_4^-$ ions in soils also results in conditions conducive to the vigorous fixation or precipitation of the phosphorus by manganese, iron, and aluminium compounds.

### *Fixation by Silicate Clays*

The fixation of phosphorus by silicate minerals are due to the surface reaction between the $H_2PO_4^-$ ions and the -OH groups on the mineral crystal. Iron and aluminium ions are removed from the edges of the silicate crystals and then form hydroxy phosphates of the same formula as those already discussed.

## 18.9 Availability of Phosphate

When soluble phosphates are added to soils, insoluble phosphates are formed with aluminium, calcium or iron. The total surface area of these phosphate containing particles are high and as a result, phosphorus availability is appreciable. Thus, even though the water-soluble phosphorus may be precipitated in the soil, the freshly-precipitated compounds will release much of their phosphorus to phytoplankton.

### *Relation to Soil Texture*

Phosphorus reacts with the finer soil fractions. Consequently, phosphorus fixation tends to be more pronounced in clay soils than in the coarse textured ones. Thus clay soils have the tendency to reduce phosphorus availability.

## 18.10 Organic Matter, Micro-organisms and Available Phosphorus

Organic form of phosphorus can be immobilized and mineralized by the same process pertinent for nitrogen. The following reaction clearly explains this statement :

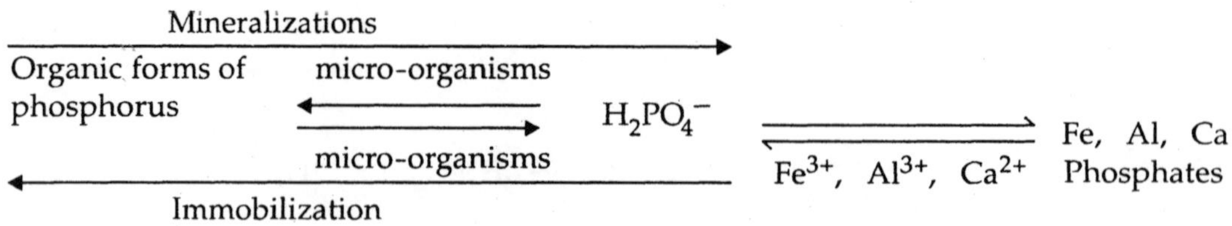


Decomposition of organic residues and humus results in the formation of soluble phosphorus compounds. The resulting soluble phosphate ions ($H_2PO_4^-$) is subject to uptake by plants or fixation into insoluble forms.

Organic matter influences the availability of phosphorus in two other ways. First, organic phosphorus (such as the nucleic acids) are adsorbed by humic compounds and by silicate clays. Such adsorption possibly protects the organic phosphorus from microbial attack. Second, organic compounds form complexes with aluminium and iron ions and hydrous oxides, thereby preventing these materials from reacting with phosphates. Organic manure is known to influence the availability of inorganic phosphorus compounds (Figure 18.2).

## 18.11 Control of Phosphorus Availability

Continued application of phosphatic fertilizers tends in time to increase the level of this nutrient in the soil and its level in the labile forms that can release phosphorus to the soil solution. Hence, even though much of the phosphorus added in fertilizer is not used during the year of application and it may provide a source in future.

The small amount of control that can be exerted over phosphorus availability seems to be organic matter and liming. The effective utilization of phosphorus along with organic manures is evidence of the importance of organic matter in increasing the availability of phosphorus. Moreover, by holding the pH of soil between 6.0 and 7.5, the phosphate fixation can be kept at a minimum. Inspite of these precautions, a major portion of the added phosphorus fertilizers still reverts to less available forms (Figure 18.3).

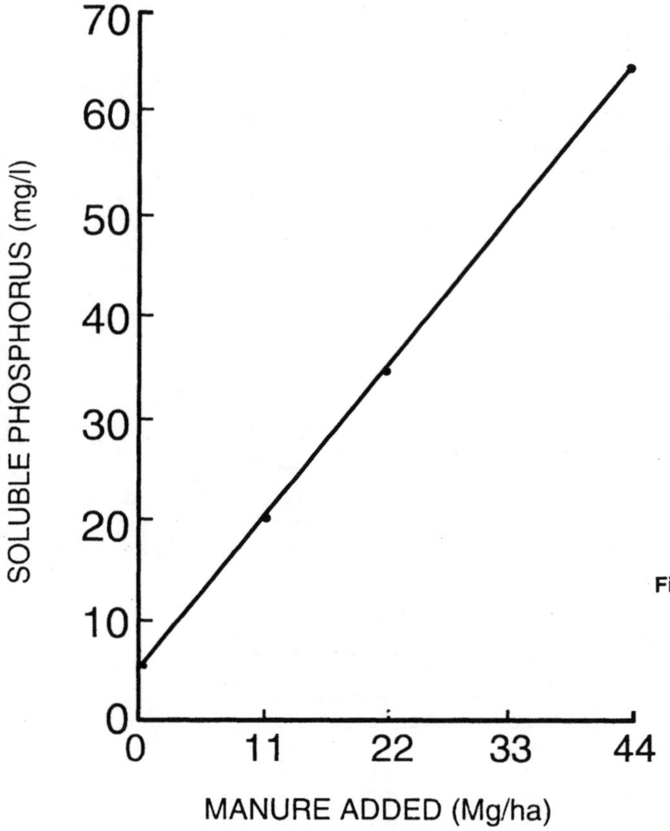


**Fig. 18.2 : The effect of organic manures on the soluble phosphorus level of soil at pH 7.2. On decomposition, organic manure produces organic acids that forms stable complexes with aluminium and iron compounds and also affects the solubility of calcium phosphate. [Drawn from data as proposed by El-Baruni and Olsen (1979)]**

### 18.12 Phosphorus in Relation to Phytoplankton and Bottom Fauna

Phosphorus limits phytoplankton production in fish culture ponds. Fish production is increased by applying phosphatic fertilizers that stimulates the development of phytoplankton. Phytoplankton is the basis of the food web that culminates in fish flesh. At high rates of phosphorus fertilizers, phytoplankton concentration is increased that limits fish production due to depletion of dissolved oxygen.

Increasing phosphorus levels did not affect benthic algal population in two types of clay soils having lower (Type - I) and higher (Type - II) values of pH (5.9 and 7.9), available phosphorus (9.28 and 11.44 mg/l), exchangeable ferrous (58.48 and 89.56 mg/l), total exchangeable bases (30.40 and 3.45 mg/100 g), and exchangeable $Ca^{+}$-$Mg^{+}$ (31.50 and 39.45 me/100 g) which may have been due to the harmful effect of phosphorus rates greater than 30 mg/l (Table 18.2). This suggests that there is an optimum level of phosphorus above which it becomes antagonistic with iron. Excessive phosphorus can inactivate iron in the plant by forming insoluble iron phosphate, which may in turn, decrease chlorophyll content and finally the rate of photosynthesis.

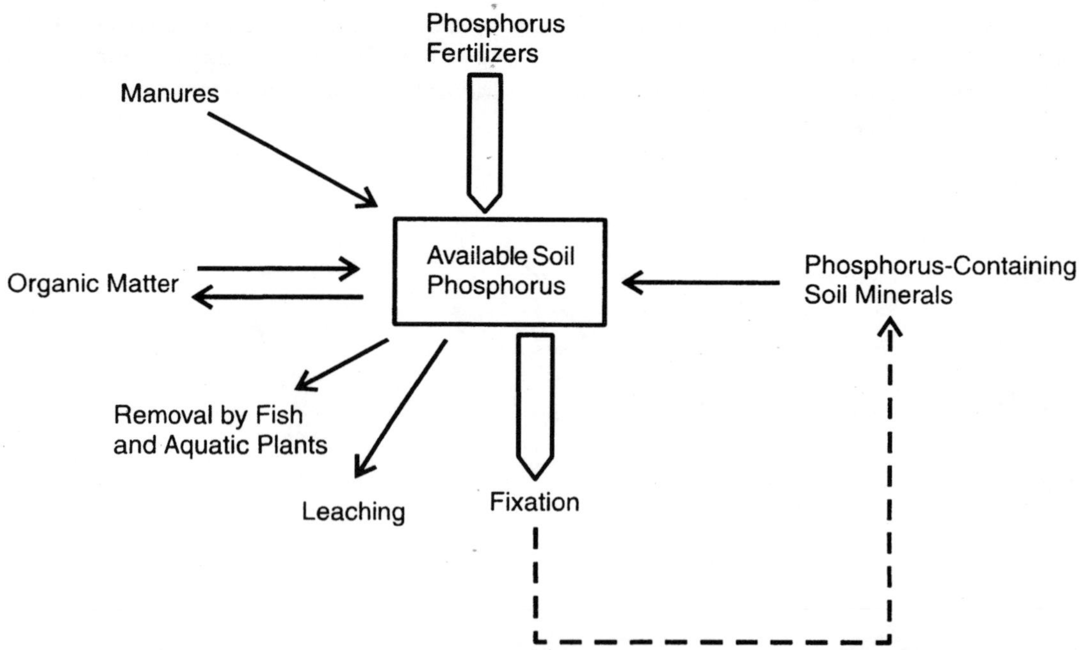


**Fig. 18.3 : Replenishment and depletion of the available phosphorus in soil. The addition of phosphorus fertilizers and the fixation of phosphorus in insoluble forms are the two important features. In general, the amount of available phosphorus in the soil at anyone time is small when compared to the amounts of magnesium, potassium, and calcium.**

**Table 18.2 : Benthic Algal Population in Ponds During the 8-Week Period**

| Treatment (mg/l P) | Weeks 2 | 4 | 6 | 8 | Mean |
|---|---|---|---|---|---|
| Soil Type - I | | | | | |
| 0 | 41.45 | 88.16 | 73.90 | 0.0 | 50.90 |
| 30 | 218.64 | 85.80 | 72.52 | 0.0 | 94.24 |
| 60 | 41.32 | 86.91 | 72.60 | 69.12 | 62.99 |
| 90 | 147.70 | 93.02 | 79.54 | 83.93 | 101.04 |
| Mean | 112.30 | 88.47 | 74.69 | 33.26 | |
| Soil Type - II | | | | | |
| 0 | 312.2 | 120.44 | 186.35 | 133.89 | 188.48 |
| 30 | 192.21 | 215.97 | 186.26 | 195.26 | 197.42 |
| 60 | 251.44 | 170.01 | 103.28 | 139.38 | 166.02 |
| 90 | 160.00 | 165.68 | 120.08 | 85.80 | 132.89 |
| Mean | 229.23 | 168.02 | 148.99 | 138.08 | |

*Source* : Pahila (1990).

Frequent low and moderate application of phosphorus along with some organic manures have some advantages over nitrogen fertilizers. Such conclusion is based on the growth and development of bottom fauna. Experiments conducted in fish culture ponds using single superphosphate, triple superphosphate, rock phosphate, and diammonium phosphate at different rates exhibited significant development of different species of bottom fauna such as *Chironomus* sp., *Tubifex* sp., *Viviparus bengalensis*, larvae of Odonata etc. (Table 18.3).

**Table 18.3 : Concentration of Bottom Fauna in Ponds Fertilized with Phosphorus Fertilizers**

| Fertilizer | Application rate (kg/ha) | Bottom fauna ($g/m^2$) |
|---|---|---|
| Single superphosphate | 250 | 17.5 |
| Triple superphosphate | 170 | 15.5 |
| Rockphosphate | 250 | 9.5 |
| | 350 | 14.9 |
| Diammonium phosphate | 45 | 10.0 |

*Source* : Sarkar and Pramanik (1990), Sarkar (1991)

## 18.13 Phosphorus in Relation to Fish Production

In general, the type and amount of phosphorus fertilizers, their effectiveness, physico-chemical as well as biological properties of pond environment have profound influence on fish production. Although different combinations of nitrogen, phosphorus, and oil cakes are important for fish culture, only phosphorus fertilizers along with organic manures at medium or low rates obviously exhibit synergistic influence on the potentiality of the combination of fertilizers and manures in fish production. However, experiments conducted in ponds using different types of phsophorus fertilizers have shown different patterns of fish production (Table 18.4). Note that high fish production is possible within 4-6 months by using rock phosphate and diammonium phosphate although geographical conditions of ponds, climatic and edaphic factors determine the application rates of phosphorus fertilizers. Moreover, fish production from freshwater fish culture ponds is extremely variable regionally and seasonally in relation to soil types and the level of phosphorus in soil and water.

**Table 18.4 : Fish Production in Ponds Treated with Phosphorus Fertilizers**

| Fertilizer | Application rate (kg/ha) | Production (kg/ha) | Species | Duration of experiment (month) |
|---|---|---|---|---|
| Single superphosphate | 250 | 470 | CM | 3 |
| Triple super-phosphate | 85 | 668 | IMC | 12 |
| | 170 | 670 | IMC | 12 |
| Rockphosphate | 250 | 3,515 | IMC | 6 |
| Diammonium phosphate | 45 | 3,310 | IMC | 6 |
| | 22.5 | 1,450 | IMC | 4 |

CM, Common carp; IMC, Indian Major Carps (Rohu, Catla and Mrigal)

*Source* : Sarkar and Pramanik (1990), Sarkar (1991), Sarkar and Das (1996), Sarkar and Konar (1983)

Generally the exact amount of phosphorus varies according to soil texture and structure. Consequently, it is difficult to correlate the data on the rates of phosphorus fertilizers and fish growth for general consideration. An appraisal of fish production using different types of phosphorus carriers as indicated by a vast amount of literature reveals that optimum fish production is possible by using phosphorus carriers at the rates varying between 25 and 150 kg $P_2O_5$/hectare; of course, application of phosphorus carriers at high rates did not exhibit any residual toxicity to fish.

The high capacity of soil for fixing phosphorus explains why much fertilizer- supplied phosphorus is quickly rendered unavailable to overlaying water. Although the fixation has definite conserving features, these tend to overweighed by the disadvantages of luxury consumption by aquatic plants and leaching. In some fish ponds, the release of phosphorus from mineral form is very slow to support pond productivity. Consequently, increased usage of commercial phosphorus fertilizers must be expected if fish production is to be increased or even maintained.

## 18.14 Conclusion

Phosphorus availability to pond water has a double constraints : (1) the low total level in soils, and (2) the small percentage of this level that is present in available forms. Even when soluble phosphates are added to ponds, they are rapidly fixed into insoluble forms that in time become unavailable. In alkaline soils, the phophorus is fixed by magnesium and calcium; in acids soils by magnesium, iron, and aluminium. This fixation reduces the efficiency of phosphate carriers so that little of the added phosphorus can be utilized by benthic organisms. In time, this fixed phosphorus can built up and can serve as a reserve pool.

Fish production from freshwater ponds treated with phosphorus fertilizers at suitable rates is highly variable since the production plateau depends on various factors of the pond ecosystem such as structure and texture of soil profile, physico-chemical parameters of soil and water, and biotic factors. A problem associated with fish culture is the removal of an appreciable amount of phosphorus by aquatic plants, leaching, and formation of insoluble phosphorus compounds.

### *References*

El-Baruni, B, and S.R. Olsen. 1979. Effect of manure on solubility of phosphorus in calcareous soils. *Soil Sci.*, 112 : 219-225.

Hickling, C.F. 1962. *Fish Culture.* Faber and Faber, London, England.

Jhingran, V.G. 1988. *Fish and Fisheries of India.* Hindustan Publishing Corporation, New Delhi, India.

Pahila, I.G. 1990. Soluble phosphorus in relation to phosphorus fertilization of lab-lab (benthic algae) in brackish water pond. *In* : The Second Asian Fisheries Forum (*Eds*: R.Hirano and I. Hanyu), Phillipines, pp. 189-192.

Sarkar, S.K. 1989. Phosphorus and aquatic ecosystem. *Ind. Rev. Life Sci.*, 9 : 227-251.

Sarkar, S.K. 1991. Role of phosphorus fertilizers on the effectiveness of urea and lime in fish survival and yield and their influence on lateritic pond ecosystem. *J. Environ. Biol.*, 12 : 287-298.

Sarkar, S.K. and S.K. Konar. 1983. Influence of NPK fertilizer combinations on fish. *Environ., Ecol.*, 1 : 145-150.

Sarkar, S.K. and A. Pramanik. 1990. Role of combined use of calcium ammonium nitrate, triple superphosphate and groundnut cake on fish .nd aquatic ecosystem. *Symp. Acad. Environ. Biol.*, 10 : 307-313.

Sarkar, S.K. and R.N. Das. 1996. Efficiency of fertilizers, lime and oil cakes in fish culture. *Fish. Chim.* 16 : 21-26.

## Questions

1. State the importance of phosphorus in fish culture.
2. Why a phosphorus reserve has practical significance?
3. What is the fate of phosphorus fertilizers added to fish ponds?

4. The same amount of a phosphorus fertilizer was applied to two fish pond : (a) pond A with a sandy soil and (b) pond B with a clay soil. Fish was stocked in each pond. In which case more phosphorus will be removed? Why and how?
5. State how alkaline and acidic conditions of ponds are related to phosphorus availability.
6. How phosphorus availability in an alkaline pond can be controlled?
7. Explain how phosphorus fertilizers are related to fish production and fish food organisms generally.

# 19

# Global Inland Capture Fisheries and Freshwater Fish Culture

The increasing globalization of the world economy affect fisheries and those who are responsible for formulating and implementing the national policies in the fishery sector find that the nature and scope of their task is changing. Today one important aspect of this task is the control, monitoring, surveillance and analysis of fisheries and aquaculture developments at international level in a systematic manner. The policy-makers for fisheries and aquaculture find that an understanding of only the national conditions affecting the sector is not sufficient. The international context must be understood and, while this has already been the case for policy-makers and administrators in the fishing nations for the last two decades, it is becoming essential to all fishing nations.

The explosion of large number of information clearly indicates that there is continuous supply of information concerning fisheries and aquaculture. But it is significant to note that most of the information available so far is heterogeneous and therefore, needs to be evaluated, monitored and shaped into scenarios for future development.

Inland capture fisheries and fish culture are the basis for development of freshwater fisheries sector. Consequently, no understanding of fisheries sector is complete without at least a brief knowledge of the inland fisheries and fish culture by geographical region of the world. Trends in freshwater fish production, utilization, management issues and outlook will be covered first and then regional production, management and prospects will receive attention.

## 19.1 Trends in World Fish Production and Utilization

World fish production clearly indicates that production from marine fish species has increased from about 14 million tonnes in 1950 to about 81 million tonnes in 1998. This figure also indicates that the trend in marine fish production is by far the best and far exceed than that of inland capture fisheries and fish culture.

### *Fish Production*

In 1994, the cultivation of freshwater fish under controlled conditions reached a record level of 25 million tonnes. The total global production of finfish and shellfish contributing over 18.5 million tonnes (17 per cent of world fisheries production). However, of the total aquaculture production of 18.5 million tonnes during 1994, the contributions from marine, brackish and freshwaters were 5.3, 1.5 and 11.7 million tonnes, respectively. This figure clearly indicates how freshwater fish culture has developed through growing of possible technological developments.

About 2 million tonnes came from inland aquaculture production mainly from Asia. Table 19.1 indicate a new peak of total inland fishery production of 26.75 million tonnes in 1998. If this trend of production continues, it is expected that through cost-effective techniques, strategies and managemenι, fish production will further increase in the near and far future.

**Table 19.1 : World Inland Fishery Production (Million Tonnes)**

| Resource | Year | | | | | | | |
|---|---|---|---|---|---|---|---|---|
| | 1980 | 1984 | 1988 | 1990 | 1992 | 1994 | 1996* | 1998* |
| Aquaculture | 2.0 | 6.0 | 7.7 | 8.26 | 9.55 | 12.46 | 16.05 | 18.00 |
| Capture | 1.0 | 3.4 | 5.0 | 6.54 | 6.21 | 6.71 | 8.05 | 8.75 |
| Total | 3.0 | 9.4 | 12.7 | 14.80 | 15.76 | 19.17 | 24.10 | 26.75 |

*Source* : Selected data from FAO (1990, 1997)

* Compiled from different published data

The rapid growth in aquaculture production is due to the increased predominances of carp species. In 1994 carps accounted for almost half of the total volume of aquatic products. As a result of slow geographical spread of aquaculture and increase in the number of species under culture, the predominance of traditional species increased.

### *Utilization*

The net increment in 1996 of fish available for human consumption was due to fish culture production. Despite record levels, food fish production from inland water resources was 2.5 million tonnes higher than 1995. The quantity of fish available for human consumption totalled about 17.7 million tonnes in 1996. Globally, the increase in the total quantity available for human consumption resulted in a very small increase in average per caput availability of fish in 1994 to 13.6 kg (Figure 19.1).

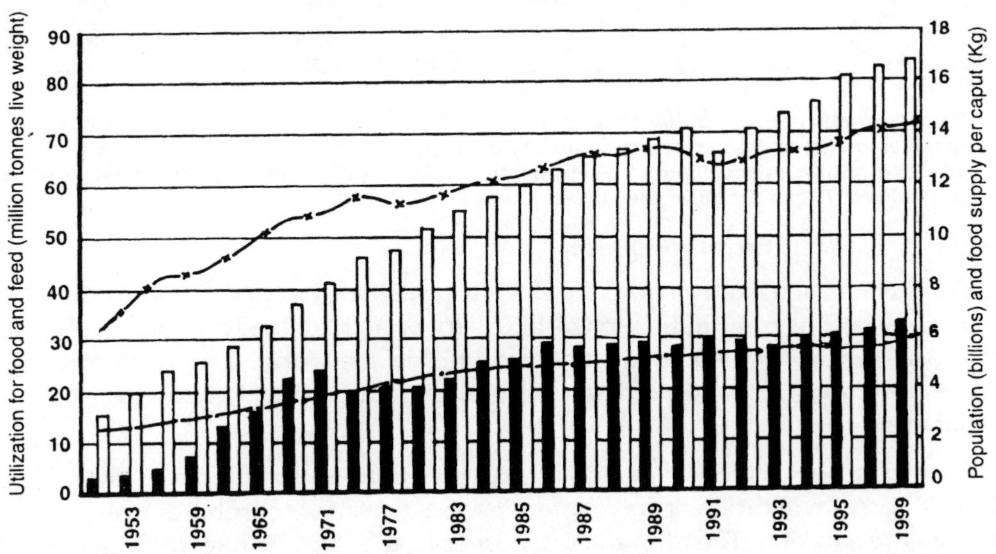


**Fig. 19.1 : World fish production. Solid and hollow columns indicate feed and food, respectively. Broken crossed line indicates the rate of consumption in terms of kilogram per caput. Broken dotted line shows population growth. [Redrawn from FAO (1997); Used with permission of Food and Agricultural Organization of the United Nations]**

The availability of freshwater fish for human consumption in 1996 was estimated to be 17.7 million tonnes, 2.3 million tonnes more than in 1995, representing a greater increase than the estimated population growth rate in the same year. The average annual per caput availability of food fish increased to 14 kg.

### *Fish Production in Low-Income Food-Deficit Countries*

Gross production of freshwater fish in the low-income food-deficit countries (LIFDCs) continued the pattern of high growth that has characterized in recent years. However, in 1994, LIFDCs accounted for 35 per cent of total production, compared with 26 per cent in 1988 (Figure 19.2). This increase has taken place in LIFDCs that are large fish-producing countries such as India (5 per cent), Bangladesh (6 per cent), Morocco (7 per cent), Indonesia (7 per cent), Philippines (14 per cent) and China (14 per cent). These countries also account for about 73 per cent of the global population of LIFDCs. At present, production of fish has changed over recent years and in some cases it has dropped significantly.

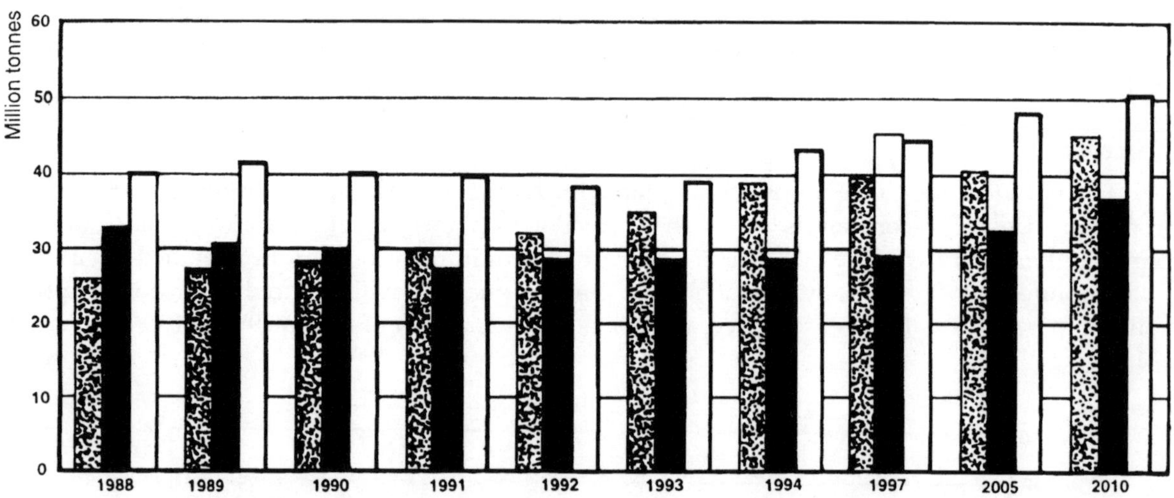


**Fig. 19.2 : Fishery production for low-income food-deficit countries (dottes columns) in relation to industrial (solid columns) and other countries (empty columns). Anticipated data for the years 2005 and 2010 are also shown in the figure. [Redrawn from FAO (1997); Used with permission of Food and Agricultural Organization of the United Nations]**

## 19.2 Trends in Aquaculture Production

The expansion of aquaculture since the 1980s was sustained in 1998 (Figure 19.3). Obviously, aquaculture increased its contribution to world fishery prodcution and maintained its position as one of the fastest-growing food production activities in the world. If environmentally sound technologies are adopted, it is hoped that aquaculture production will gradually increased in an unhampered manner.

The Asian continent increased its dominance as an aquacultrue producer (over 85 per cent) of finfish in 1993. Among different freshwater species, cyprinids contributes over 75 per cent of total freshwater production. The most important species which constitute the bulk of production (77.7 per cent of total freshwater aquaculture production) are represented by 13 species (Table 19.2).

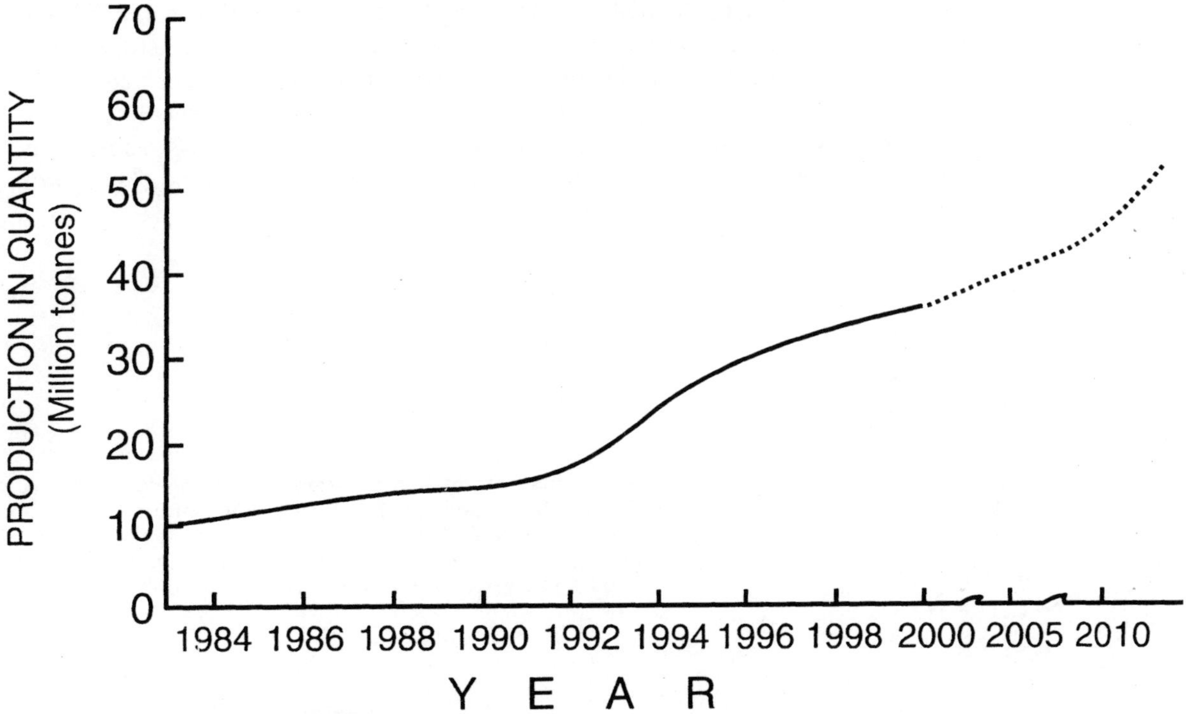


**Fig. 19.3 : Global trend in aquaculture production. [Redrawn from FAO (1997); Used with permission of Food and Agricultural Organization of the United Nations]**

**Table 19.2 : Major Species Contributing to the Global Freshwater Aquaculture Produce**

| Species position | Production (in Tonnes) | |
|---|---|---|
| | 1993 | 1998 |
| *Hypophthalmichthys molitrix* (Silver carp) | 1,889,021 | 2,143,500 |
| *Ctenopharyngodon idella* (Grass carp) | 1,482,192 | 1,986,620 |
| *Cyprinus carpio* (Common carp) | 1,240,551 | 1,600,630 |
| *Aristichthys nobilis* (Big head carp) | 919,253 | NA |
| *Oreochromis niloticus* (Nile tilapia) | 346,615 | 750,435 |
| *Labeo rohita* (Rohu) | 317,676 | 705,375 |
| *Catla catla* (Catla) | 307,964 | 790,354 |
| *Cirrhinus mrigala* (Mrigal) | 301,548 | 590,421 |
| *Carassius carassius* (Crusian carp) | 294,518 | 298,250 |
| *Oncorhynchus mykiss* (Rainbow trout) | 269,488 | 632,760 |
| *Ictalurus punctatus* (Channel catfish) | 228,550 | 537,800 |
| *Parabramis pekinensis* (Bream) | 218,921 | NA |
| *Cirrhinus mullitorela* (Mud carp) | 100,136 | NA |

*Source :* Modified after Ayyappan and Jena (1997)

NA, Data not available

In 1994, total production of finfish, shellfish and aquatic plants were found to be 25.5 million tonnes (Figure 19.4). Asia increased its dominance as an aquaculture producer of finfish, shellfish and aquatic plants. In 1984 and 1994, India and China produced 42 and 66 per cent of total world production respectively, while other five Asian countries accounted for about 24 and 15 per cent respectively (Fig. 19.5). Sharp decline in freshwater aquaculture production in 1994 was possibly due to sudden/rapid oscillations of environmental conditions including pollution. There are, however, ten leading countries that provides bulk of the aquaculture produce and are shown in Table 19.3.

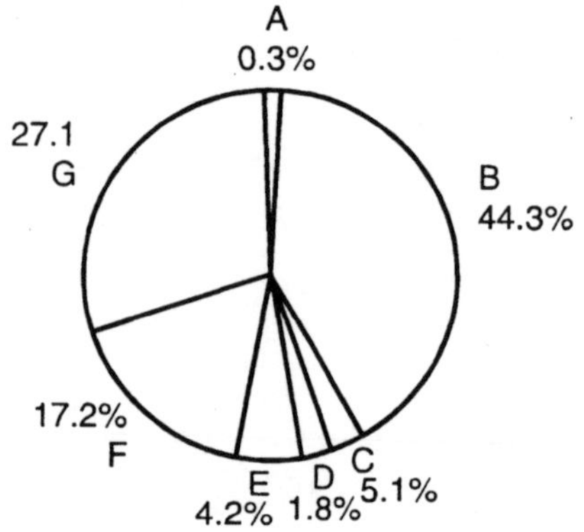


**Fig. 19.4 : Aquaculture production by categories of specie in 1994. A, Others; B, Freshwater fishes; C, Diadromous fishes; D, Marine fishes;E, Crustanceans; F, Molluscs; G, Aquatic plants. [Redrawn from FAO (1997); Used with permission of Food and Agricultural Organization of the United Nations]**

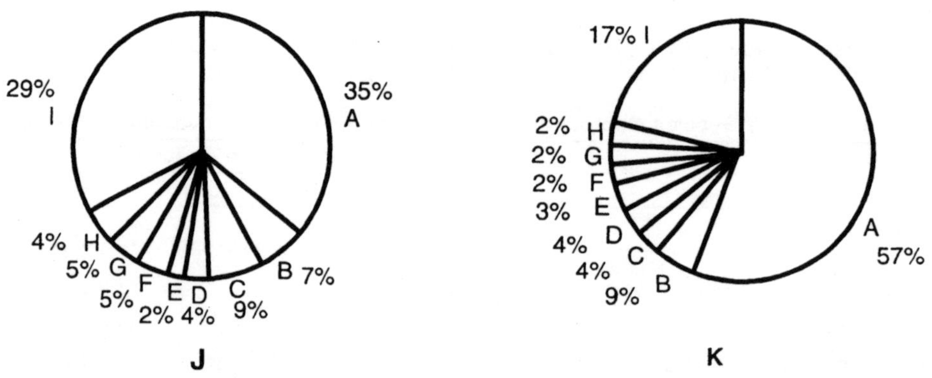


**Fig. 19.5 : Contribution of principal countries to global aquaculture production of finfish and shellfish in the years 1984 (J) and 1994 (K). A, China; B, India; C, Japan; D, Indonesia; E, Thailand; F, United States; G. Philippines; H, Republic of Korea; I, Others. [Redrawn from FAO (1997); Used with permission of Food and Agricultural Organization of the United Nations]**

Although cultured fish and shellfish contribute significantly to total fishery production, farming activities in most countries are dominated by a few species such as carps in China and India, oysters and mussels in Japan, France and the Republic of Korea. Most of the world's production of milkfish is reported from the Philippines and Indonesia. Cultured milkfish account for 42 and 27 per cent of total production in the Philippines and Indonesia, respectively.

The culture of Cyprinids, in particular freshwater herbivorous Chinese carps produced largely under extensive and semi-intensive culture systems, dominated finfish production and 9.2 million tonnes were farmed in China. The four Chinese carps – the grass, silver, bighead and common carps – represented the top four culture species by weight and made up half of total finfish production.

**Table 19.3 : Finfish and Shellfish Production Through Aquaculture by Leading Countries**

| Country | Freshwater finfish and shellfish production through aquaculture | | Country | Total finfish production through aquaculture | |
|---|---|---|---|---|---|
| | 1993 | 1998 | | 1993 | 1998 |
| China | 6,464,948 | 8,453,700 | China | 8,880,167 | 10,247,300 |
| India | 1,384,680 | 3,650,445 | India | 1,438,915 | 3,654,700 |
| Indonesia | 261,460 | 860,550 | Japan | 833,032 | 1,760,542 |
| USA | 242,161 | 680,350 | Indonesia | 592,081 | 960,525 |
| Bangladesh | 224,691 | 580,960 | USA | 433,698 | 780,540 |
| Vietnam | 135,000 | 370,650 | Thailand | 414,269 | 675,890 |
| Thailand | 134,579 | 450,550 | Philippines | 991,703 | 570,876 |
| Philippines | 97,110 | 175,800 | Korea Rep. | 391,424 | NA |
| Russian Fed. | 89,614 | 100,760 | France | 270,880 | 560,430 |
| Ukraine | 65,327 | NA | Bangladesh | 247,816 | 650,880 |

*Source :* Modified after Ayyappan and Jena (1997, NA, Data not available

## 19.3 Opinion of Initiatives in Management Issues

Recently, the international community addressed several of the management issues connected with sustainable fisheries; how to (1) reduce overfishing and control fishing capacity and (2) reduce environmental degradation. These issues are briefly discussed.

### *Overfishing and Fishing Capacity*

Inland capture fishery resources are always succumbed to overfishing that affects capture fisheries both in developed and developing countries, often severe in densely populated and productive areas. Unless effective action is taken, overfishing will move from bad to worse condition. In many countries, shortage of alternative employment opportunities and population pressure, together with the lack of effective management and conservation strategies, will increase the attraction of fisheries as a last resort of employment.

While the problems differ from one situation to another, followings are the most important factors contributing to excessive fishing effort :

1. Many Governments are reluctant to take necessary conservation and management decisions.
2. There is lack of technical and financial resources to formulate and implement the management actions in many developing countries.
3. Slow or no growth in employment in many developing countries.
4. Lack of commitment to international cooperation towards joint management.

### *Government Initiatives*

Inspite of difficulties, actions are being taken to reduce excess fleet capacity and to improve management by adopting community-based management. The existing overfishing and fishing capacity generally contributes to reduce broodfish and causes major economic losses to fishermen communities. These consequences contribute serious problems that are difficult to overcome by developing countries, funds are frequently not available to pay for the operational and cost-effective implementation and enforcement of conservation and management measures. However, in recent years a number of countries such as Namibia, Chile, Malaysia, Argentina, Austrialia, Canada, Japan, New Zealand, Norway and the United States, have introduced management and conservation measures that limit fisheries inputs and outputs.

Fisheries conservation and management are high-cost activities. Regarding better management, however, industry is assuming financial responsibility and increased conservation and management decisions. In Australia and Japan, the industry is involved in setting research priorities in support of fisheries conservation and management.

## 19.4 Environmental Degradation

In many inland water resources, fish habitats are being degraded to greater degree by industrial and agricultural pollution, damming of rivers, sedimentation, mining and oil exploration etc. Many flood plain areas are drained for agricultural and other uses. Fisheries departments face difficulties to preserve flood plains inspite of the fact that these are the most productive areas.

In general, fisheries sector suffers from environmental damage mainly due to local pollution, and irresponsible fishing practices. A severe problem has been found to create environmental degradation due to intensive aquacultural practices. For example, tropical intensive shrimp farming has been associated in some countries like India, China, Thailand, and some other Asian countries with severe environmental degradation. However, integrated management policies of inland fisheries need to cover the complete extent of the basin in order to be effective.

### *Government Initiatives*

Regarding environmental damage from fishermen, many countries are aware of the problems and several have tried to solve the problem. Although the adverse environmental impacts of intensive aquaculture are perhaps, not being properly addressed to fishermen/fish farmers by several Government/Non-Government organizations, yet a number of countries that produce prawns, shrimps and carps have instituted stringent controls on production to ensure that pollution is kept within acceptable limits. Many South and South-East Asian countries have instituted a temporary ban on new technological development of tropical shirmp farming until an acceptable envorinmental policy is adopted. Furthermore, a number of other countries also have set up the required legislative frameworks. As in other areas of management, however, enforcement is often difficult.

The integrated management of inland catchment areas has received less attention. In catchment areas the management of large inland water bodies together with their riparian areas has generally been more actively pursued than riverine management. For example, in Latin America, Peru and Bolivia have cooperated for a number of years in managing Lake Titicaca and in Africa, the agreement of the riparian countries to establish the Lake Victoria Commission points to the introduction in the not too distant future of cross-sectoral management of the lake and its surroundings.

With few exceptions, river management, paticularly where rivers cross national/regional boundaries, has very often been restricted to such topics as establishing water use and navigation, rights, rather than basin environmental management. Nevertheless, cases of such catchment area environmental management are emerging, for example in Australia where an earlier emphasis on land and water-use management in catchment areas is being widened to take account of such sectors as fisheries.

## 19.5 Outlook for Fisheries

### *Demand and Supply : Possible Scenarios in 2010*

The demand of food fish is determined by three factors such as population growth, change in per caput income and the pace of urbanization. Interactions of these factors was considered in a review prepared by the Food and Agricultural Organizations for the Kyoto Conference, Japan, in 1995. The review declared that the estimated demand for food fish will remain within the range of 110 to 120 million tonnes (live weight) for the year 2010, compared with 75 to 80 million tonnes in 1994-1995.

Projections indicate that Oceania, North America and Europe will have the highest per caput demand, at more than 20 kg per year (live weight equivalent), but the large population in Asia means that region could account for about two-thirds of total demand.

Fish meal is the main product derived from the fish used for non-food purposes. As fishmeal is one of the most expensive ingredients in animal feeds, livestock, poultry and fish/shrimp producers have a clear incentive to reduce the amounts used. In response to possible price increases in the future, fishmeal usage in poultry finisher and layer diets may disappear completely and less fishmeal may go into pig grower diet. This may make it possible for more fishmeal to be used in aquaculture without causing its price to increase. Thus it is expected that the demand, and the supply of fish for reduction will remain stable at between 30 million and 33 million tonnes until further oscillation occurs in this regard.

### *Supply*

Per caput supplies of food fish increased in 1994, 1995 and 1996. But it is not clear that growth in aquaculture production can compensate for the possible stagnation in food fish production from capture fisheries.

In Asia and Europe, growth of fish cultrue is faster than elsewhere and has become very popular in these countries for two reasons : (1) it can be incorporated into local agricultural systems to diversify the production base and (2) it provides a source of income. Consequently, flexible integrated fish culture systems are being adopted in many regions. Global freshwater fish culture scenarios clearly indicate that there is considerable potential for further expansion and under controlled and favourable conditions, production could be about 40 million tonnes by 2010 (Figure 19.6).

In low-income countries, the growth of commercial freshwater fish culture will be stimulated by wealthy consumers in high-income countries and by the adoption of macro-economic policies aimed at providing a better environment to small-scale entrepreneurs. During the last decade or so, the rate of growth of commercial carp fish in cultured water areas was very fast but at present, this rapid growth of cultured carps considerably decreased as fish farmers have switch over to higher-priced species.

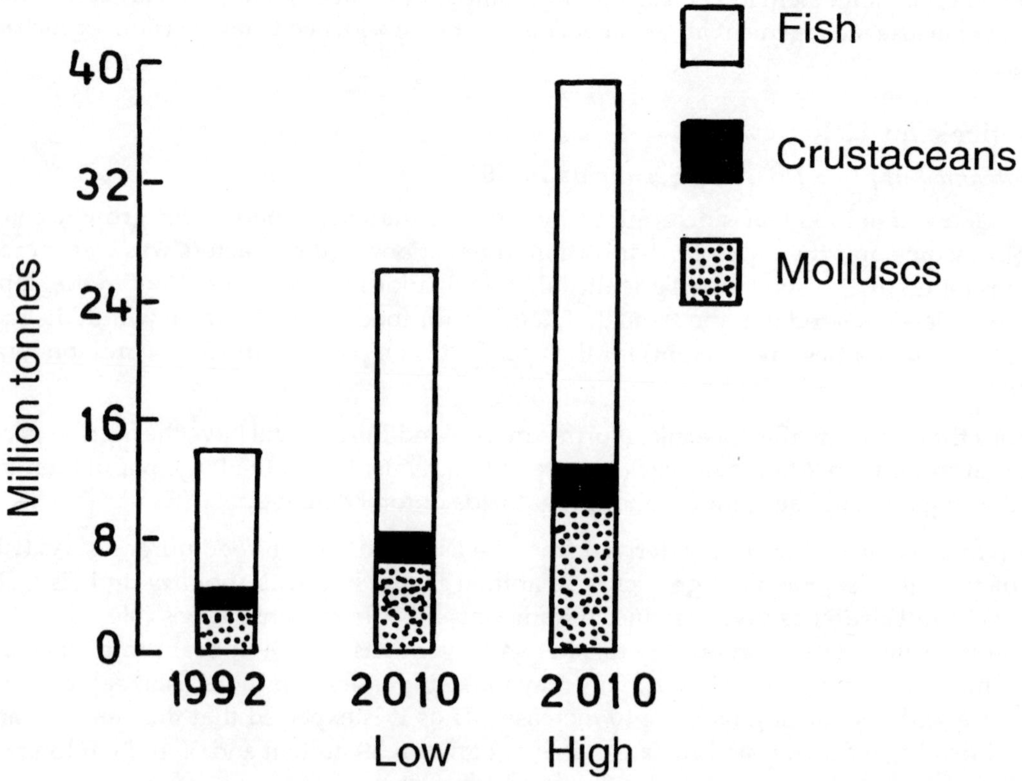


**Fig. 19.6 : Possible aquaculture production in 2010 (excluding aquatic plants). [Redrawn from FAO (1997); Used with permission of Food and Agricultural Organization of the United Nations]**

The three main constraints on aquacultrue are the environmental degradation, lack of quality fish seeds and the availability of land and water. The first two constraints may results from the mismanagement of aquaculture facilities. Third constraint originates from competition with agriculture. However, these factors will undoubtedly limit the growth of aquaculture.

It has been estimated that fish supplies from capture fisheries and aquaculture production for human consumption are less than the fish used for reduction and other purposes. As indicated in Table 19.4, supplies for reduction will continue at the level of 30 to 33 million tonnes for the future and it is assumed that aquaculture production will be used for food.

**Table 19.4 : Projected Supplies of Fish for Human Consumption, 2010**

| Source of production | Peessimistic scenario (million tonne) | Optimistic scenario (million tonne) |
|---|---|---|
| Aquaculture production | 27 | 39 |
| Capture fisheries | 80 | 105 |
| Subtotal | 107 | 144 |
| Less (for reduction) | 33 | 30 |
| Available for human comsumption | 74 | 114 |

*Source* : FAO (1997)

## 19.6 Production and Consumption

It has been ascertained that a number of factors come to play in which production and consumption develop in the future. These factors include the fish trade regime and trade rules. While trade regime helps influence consumption patterns, trade rules also help to provide insentives for increased production and better management of production. In this case, Governments have a role to play in assuming that capture fisheries achieve a sustainable yield of 105 million tonnes. Of course sustainable production levels is generally influenced by how Non-Government organizations and Governments deal with the protection of aquatic ecosystem management of fish stocks, financial support to fisheries sector and fish trade.

### *Protection of Aquatic Ecosystems*

Although protection of aquatic ecosystems for the long-term sustainability of capture fisheries is an important issue, fish farmers/fishermen can do very little about on its own. Densely inhabited LIFDCs have fewer resources to take steps to protect aquatic ecosystems and in many situations, political pressure fails to carry out the plan of work. This problem does not arise in case of high-income food-enriched countries. However, it is presumed that LIFDCs will be characterized by further damage to aquatic ecosystems, a decline in capture fisheries production and increasing conflicts in fisheries management.

### *Overfishing*

Adequate management of existing fishes produces higher incomes and greater catch volumes. Gradual increase of fish production as shown in Table 19.1 indicates that if effective management strategies of resources are adopted in those fisheries that are depleted, production would grow only gradually. Effective management of fish stocks that are at present overexploited will ensure high production, although it will take time for stocks to recover, however, particularly those stocks that require both a reduction of fishing effort and better environmental conditions to recover.

Without effective action, there is a considerable danger that overfishing will continue to get worse. In many developing countries, population pressure and the shortage of alternative employment opportunities, together with the lack of effective management policies, will make fisheries more attractive to poor people as an employment of last resort. If no management action is taken, annual production from capture fisheries for human consumption could fall significantly.

### *Fish Trade*

Interest in the trade of commercially important species of fish will be stimulated by the various agreements concluded at the establishment of the World Trade Organization. Discussions among the members of this organization aimed at liberalizing fish trade policy, have paved the way for facilitating the flow of fish to markets with strong purchasing power.

Some environmental organizations are moving to enlist the support of consumers, to use their purchasing power to force producers of fishery products and certify so as to gurarantee that their fish and fish products are the result of ecologically correct production methods. However, this may have a major impact on the activity of fisheries and aquaculture.

## 19.7 Impact on Fish Consumption and Production

It seems plausible that average world per caput fish consumption by the end of 2005 will be approximately 17 kg. By the year 2010, per caput consumption may have grown in south-east Asia and the Near East and North Africa and declined in Sub-Saharan Africa and South Asia. In Sub-Saharan Africa, per caput consumption will probably continue to decline owing to continued low imports and the inability of local production to keep up with population growth. Significant growth in production in inland capture fisheries are at a stage of advanced exploitation. Aquaculture starts from a group of producers to be able to achieve production increases that will make a contribution to total supplies. Besides this, it seems that in the future African aquaculture entrepreneurs will follow the culture of high-value species for overseas markets.

It therefore appears that fish prices will increase as supplies will not be sufficient to satisfy demand. Consequently, it will be economically possible for fishermen to continue fishing, inspite of declining catches. The conservation and management of stocks will become increasingly urgent and difficult.

### *South Asia*

While low per caput fish consumption in South Asia will not have declined, per caput fish consumption will possibly have increased to some extent in south-east Asia. However, freshwater fish consumption is significant in South Asia and supply is maintained through the culture of various species of carps. Massive fish imports are not an economically viable solution for South Asia and therefore, fish culturists should apply latest technologies to their farms for better production. This is very important so far as fish supplies and the food security of marginal populations are concerned.

### *North America*

No significant changes are observed in the production and consumption patterns of North America. This is due to the fact that higher per caput income is going to be reflected in an increasing proportion of high-priced products in the fish consumption basket of the average consumer. The economy will be able to import fish and fish products to make up any shortfalls in local production. Aquaculture will expand as fast as fish consumption in response to local market opportunities and fisheries as well as aquaculture should become more sustainable.

### *Latin America*

In Latin America, per caput fish consumption will possibly decline gradually, only small volumes are required that will come from local production. Freshwater capture fisheries and aquaculture will continue to be a marginal activity essentially for local markets and also continue to focus on high-priced products for export markets.

### *South-East Asia*

In South-east Asia, economic development will possibly guarantee increasing or constant food fish supplies. Different trade patterns will obviously cause per caput supplies to decrease in some countries and increase in others. Fishermen has little influence over the state of aquatic ecosystems that is very essential in determining the future role of inland fisheries and freshwater fish culture.

### *China*

In the most pessimistic scenario, per caput fish consumption in China will stabilize over the next few years because China is unlikely to become a substantial importer of fish. However, the economic and social environment in China has changed and has contributed to very rapid expansion in fish culture and inland capture fisheries production. As a consequence, aquatic ecosystems have been damaged. It seems that the observed rate of increase in fish production cannot be sustained over the next several years.

### *North Africa and Near East*

In the North Afirca and Near East, per caput consumption will increase to some degree. This increase is probably due to greater imports from outside the region. Production of fish from inland waters and fish culture within the region will be continued in immediate future and consumed locally, with two major exceptions such as fish production in Oman and in Morocco.

### *Eastern Europe*

In Eastern Europe, per caput fish consumption will stabilize over the next several years as the adaptation to market economics continues. In the past, fish consumption in these regions was high. But recently drastic increases in prices and low incomes have turned fish into a food item that is replaced by other low-priced items. As bulk amounts of fish will not be imported, consumption will have to depend on local supplies and therefore, the culture of freshwater fish will be enhanced to greater degree in Eastern Europe.

### *Western Europe*

The increase in demand for freshwater fish will be low due to very slow population growth and increase in income levels. Demand, if required, will be met through increased imports and higher price levels from the share of ready-to-eat items.

## 19.8 Development of Inland Fisheries and Aquaculture by Geographical Regions

In the following sections, discussions on various aspects of inland fisheries and fish culture by geographical regions have been presented concerning the geographical boundaries, resources

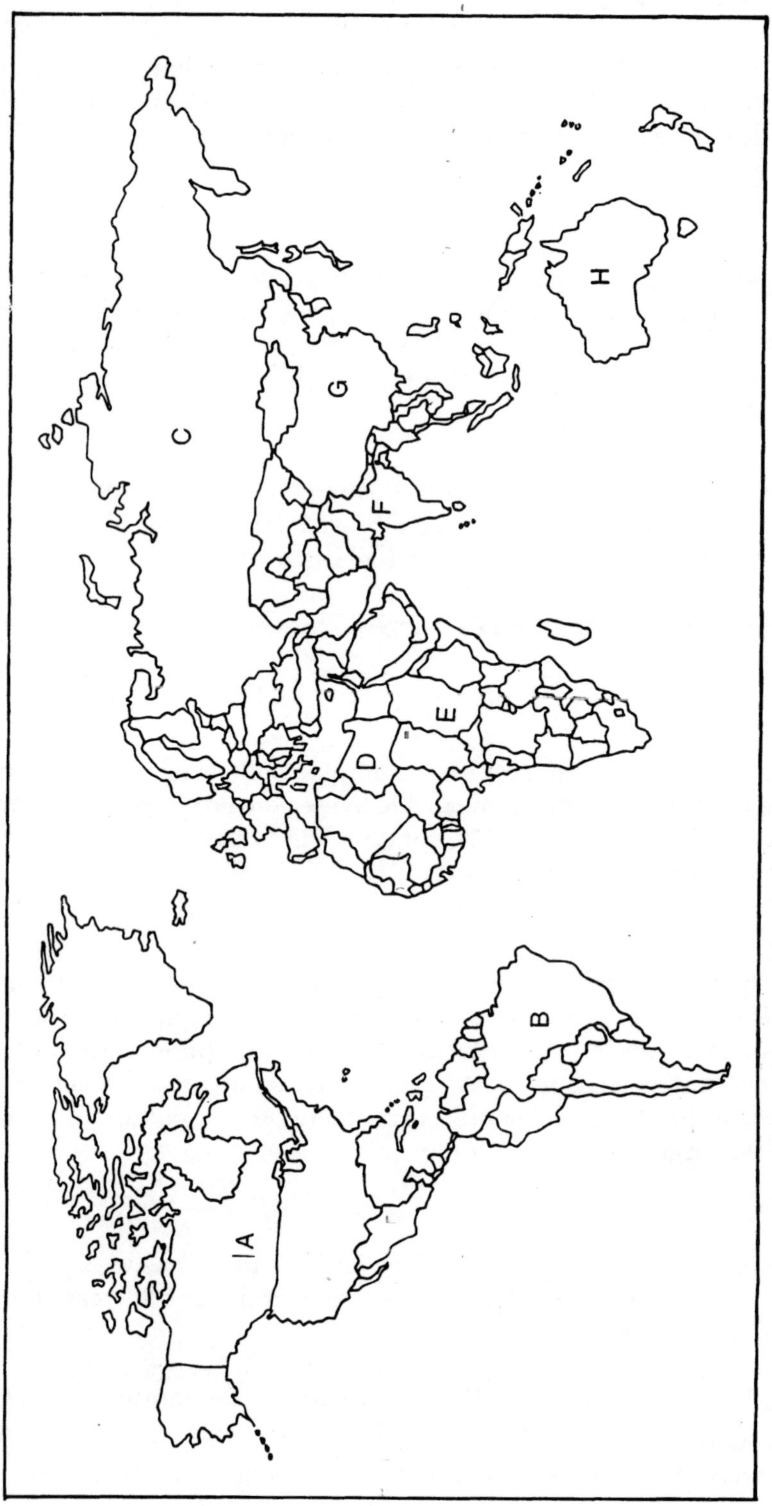


Fig. 19.7 : Map showing the geographical boundaries of the world concerning inland fisheries and fish culture. A, North America; B, Latin America and the Caribbean; C, Europe; D, Near East and North Africa; E, Sub-Saharan Africa; F, South and South-east Asia; G, East Asia; H, South Pacific.

and production and management as well as prospects. For this purpose the world has been divided into eight regions : South-east and South Asia, Near East and North Africa, Latin America and the Caribbean, North America, Europe, East Asia, Sub-Saharan Africa and South Pacific (Figure 19.7).

## 19.9 South Pacific

The region encompasses the western and central Pacific Ocean, stretching from Australia in the west to Pitcarian Island in the east. There are 16 independent states and dependent territories of France, the United Kingdom and the United States. New Zealand and Australia are well developed states and the remaining states and territories are developing states. The seas around the region include central and south-east Pacific Ocean and the eastern Indian Ocean.

The region provides only 2 per cent of total world fishery production, but the fisheries sector plays a key role in the economies of the region. While food fish consumption in Australia and New Zealand is high in the region at an average of 20 kg per caput per year (live weight equivalent), the average for some small island developing states would be twice as high. Fish production in this region and share of world production is shown in Figure 19.8.

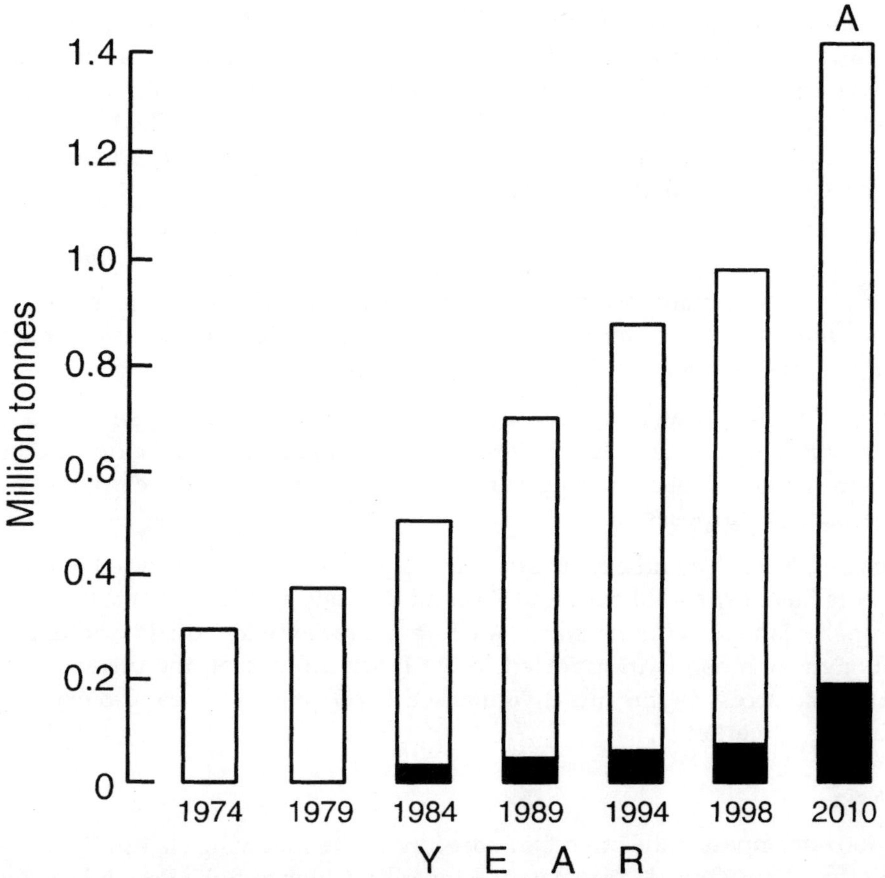


**Fig. 19.8 :** **Regional fish production (South Pacific) in inland waters (solid zones) and marine waters (hollow zones). A, possible production. [Redrawn from FAO (1997); Used with permission of Food and Agricultural Organization of the United Nations]**

### *Resources and Production*

1. *Inland Water Fisheries :* Due to good supply of freshwater, inland fisheries are restricted to the land masses of Australia, New Zealand and Papua New Guinea. In the first two regions, inland fisheries are valued as a recreational resource. In some areas of Papua New Guinea, inland fisheries are important due to limited production of other sources of animal protein. Some small inland developing states have inland water resources. The total inland capture fishery harvest in this region in 1994 and 1998 has been reported as 25,102 and 28,300 tonnes, respectively. River eels, tilapia, rainbow trout and brown trout are common in this region.
2. *Aquaculture :* In 1994 and 1998, aquaculture production was 75,000 and 78,500 tonnes, respectively due to increase in development and management of aquaculture. Rapid increases have been observed in two developed states in the culture of salmon (*Salmo salar*). Although most aquaculture production is derived from coastal aquaculture, the physical potential for freshwater aquaculture development could be considerable in larger countries.

### *Management*

Many inland freshwater resources have declined as a result of habitat degradation due to population growth and overfishing. Efforts to reserve the situation include the construction of artificial reefs, stocking, habitat improvement and the introduction of exotic species. Nearly 30, 20 and 30 types of exotic species have been introduced to Australia, New Zealand and Papua New Guinea, respectively. Smaller island states have introduced tilapia (*Oreochromis* spp.), mollisca (*Trochus niloticus*) and green snail (*Turbo* spp.).

### *Prospects*

Fisheries play a social and economic role in this region. Fish for human consumption will have to be considered as the most important source of animal protein. The fisheries sector will one of the main trigger of economic development.

The potential for freshwater aquaculture development differs among sub-regions. Economic, physical, biotechnical, and institutional issues disturb the aquaculture development in this region. Of course, the freshwater aquaculture potential does exist for careful development in some areas for food and economic purposes.

Australia and New Zealand export canned tuna fish to Japan, the United States and Europe. Recently, exports have expanded for New Zealand mussels and it is hoped that the export will continue. In small island developing states, the forecast for excellent quality of tuna fish export is positive. Although tuna fish is transported to the Japanese market, the distances involved will probably hamper to access to this lucrative market. Papua New Guinea also exports high grade tuna to the European market.

## 19.10 East Asia

This region encompasses the inland jurisdictions of the following states : Democratic People's Republic of Korea, Hong Kong, Macao, Japan, Mongolia, China and the Republic of Korea, the east coast of the Russian Federation and Taiwan (province of China). The seas within this region consists of the north-west Pacific Ocean including the sea of Japan, the East China and the Yellow Sea.

While fish production is an important economic activity in this region, aquaculture production contributed more than 70 per cent of the total global volume. Per caput food fish consumption is generally high – about 21 kg on average. This region is an active trading partner on the international market. Fish production in this region and share of world production is shown in Figure 19.9.

### *Resources and Production*

1. *Inland Water Fisheries* : Production from freshwater capture fisheries continues to be dominated by China. Environmental degradation combined with overfishing have affected capture fisheries in all major rivers due to pollution. This damage has resulted in the loss of many commercially important species and drastic reduction in gross yield. Inspite of declining contribution of river fisheries, significant increases in yield have been obtained through extensive exploitation of lakes and reservoirs such as fertilization, improved stocking, environmental engineering, and habitat modification.

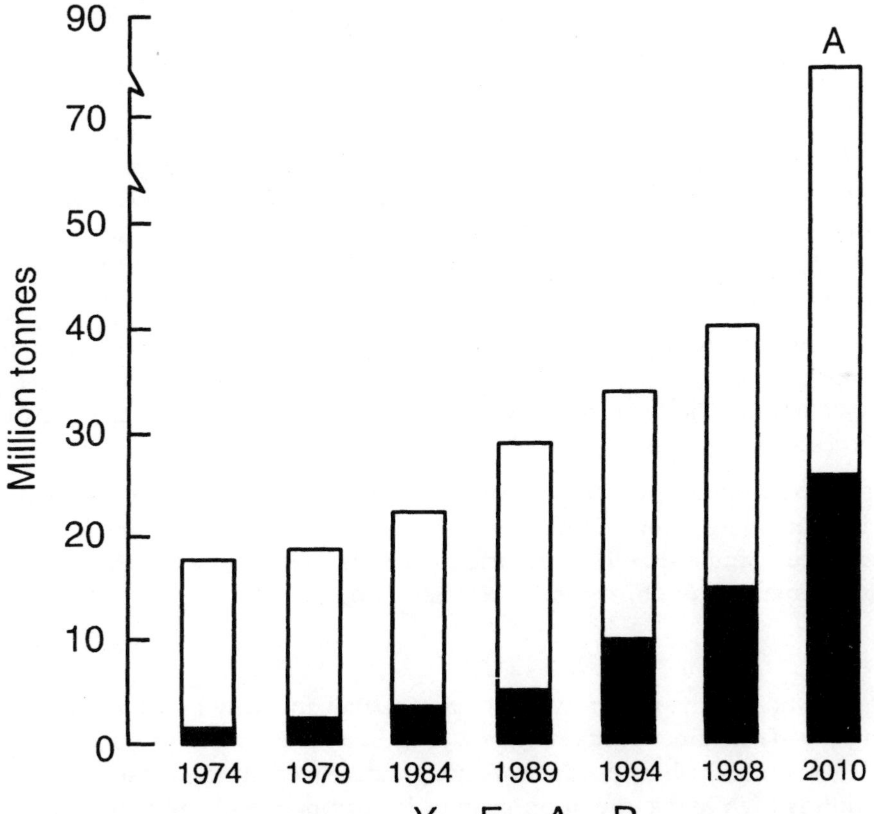


**Fig. 19.9 : Regional fish production (East Asia) in inland waters (solid zones) and marine waters (hollow zones). A, possible production. [Redrawn from FAO (1997); Used with permission of Food and Agricultural Organization of the United Nations]**

2. *Aquaculture* : Total aquaculture production has been reported to be about 18.4 million and 20.7 million tonnes in 1994 and 1998, respectively, representing 73 and 76 per cent of the total world production. However, China is the important freshwater fish and shellfish

producer, representing over 60 per cent of world aquaculture – 15.4 million tonnes in 1994. Finfish is cultivated at low stocking densities within polyculture, semi-intensive and pond-based farming systems. The principal species cultured are grass carp, silver carp, common carp, bighead carp, and tilapia. Molluscs and crustaceans are also extensively cultured.

Environmental factors play an important role in aquaculture development throughout the East Asia. For example, shrimp farmers in China suffered losses in production due to disease outbreaks from poor soil and water conditions in ponds. High densities of shrimp in many ponds resulted in introduction of pathogens. Intensification trends in resource use and yield in freshwater aquaculture are very significant so far as the release of waste matters are considered.

In the Republic of Korea, shellfish and finfish farmers have to face harmful algal blooms, resulting in anoxic bottom waters and toxin contamination in fish. The increasing pollution in rivers, reservoirs, and lakes in China causing dramatic economic losses to fish farmers. At present, considerable attention is being given towards sustainable aquaculture development in East Asian countries.

### *Management*

International fisheries issues in the region have been dealt with through bilateral fisheries agreements. However, regional cooperation is also carried out through Asia-Pacific Fishery Commission. In the Republic of Korea, the Fishery Act was amended in 1995 to accommodate new international management schemes. Moreover, to tackle environmental degradation, the Government has designated a number of "fisheries resource conservation areas" and increased research on pollution-related aspects.

Japan has a comprehensive system of fisheries management. Inland fisheries are controlled by fisheries cooperatives through a system of fishing rights. Fisheries Resource Protection Law and Fisheries Law constitute the legal basis for fisheries management with the aid of Ministry of Agriculture, Forestry and Fisheries.

China has also adopted a pragmatic approach towards developing its fisheries administration and managerial efficiencies through adequate Fisheries Laws and Regulations for the conservation of fisheries resources. Similar to other countries, China is facing same fishery management issues such as environmental pollution, and habitat destruction.

### *Prospects*

Fish consumption in the region will increase further in many areas along with population growth and improved consumer purchasing power. In East Asian countries, aquaculture has become a more important source of fishery products. Among different countries, Japan is the top most where fish consumption is high and population growth is very negligible and the composition of the Japanese fish consumption is expected to continue change from lower-value to higher-value products. The sophisticated nature of fish consumption attitude has bearings on domestic production strategies which need to focus on those types of products where Japanese producers have competitive advantage over foreign suppliers.

In China, the freshwater culture fisheries have practically reached a plateau of production and rapid growth of economy will undoubtedly increase per caput consumption. Significant growth potential exists for aquaculture, through the rehabilitation of existing ponds, and utilization of paddy fields and water-logged areas. However, it is expected that economic growth will trigger purchasing power to satisfy domestic demand-supply gap with imports.

## 19.11 Europe

This region covers two groups of countries : (1) the Eastern and Central Europe including the Russian Federation and the other European Republics of the former USSR and (2) the European Union (EU), the European Economic Area (EEA), Malta, Andorva and Monaco. The main seas neighbouring the countries of the region are the North-east Atlantic (the North and Baltic Seas, the Norwegian sea, the Black Sea, the Mediterranean Sea and North-west Pacific).

Marine fish production in many countries of Europe is very important than inland and aquaculture sectors. However, per caput fish consumption varies from 10 kg per year (in some inland countries) to 30 kg per year (in Nordic and Mediterranean countries). Fish production in this region and share of world production is shown in Figure 19.10.

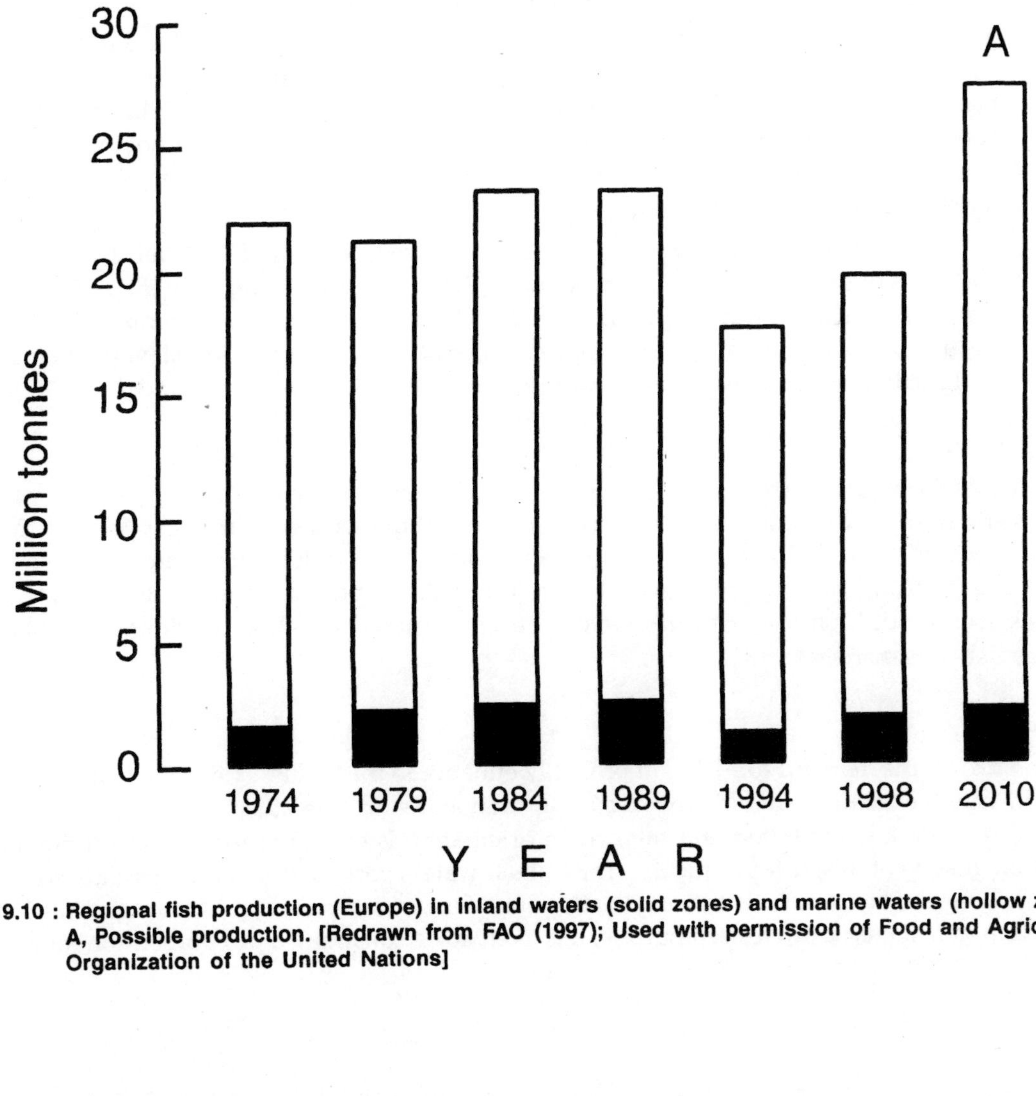


**Fig. 19.10 : Regional fish production (Europe) in inland waters (solid zones) and marine waters (hollow zones). A, Possible production. [Redrawn from FAO (1997); Used with permission of Food and Agricultural Organization of the United Nations]**

### *Resource and Production*

1. *Inland Water Fisheries :* During the last one and half decade, capture fishery production has decreased by about 50 per cent (from 810,000 tonnes in 1982 to 416,000 tonnes in 1997). Inspite of having an important role for food supplies in many countries, commercial catches have declined owing to potential changes and collapse of previous distribution patterns and infrastructure projects. Negative impact on fisheries production was probably due to pollution, competition for water resources from other sectors and infrastructure failure. All countries utilize inland fisheries resources through broodstock management, intensive stocking and fertilization to improve inland fisheries sectors. In the Russian Federation, inland water fisheries activity is being continued but in other countries, it has been discontinued due to economic difficulties. In the industrialized countries, inland commercial fishing activities are considered as secondary and activities are carried out for sport fisheries. The stocking of selected species and stringent anti-pollution laws are the primary components of inland fisheries management programs. The main species in inland water systems include rainbow trout and common carp.
2. *Aquaculture :* In 1994, regional aquaculture production accounted for 6 per cent of total global production in quantity (1.3 million tonnes and 195,000 tonnes from industrialized countries and transition states, respectively). In the industrialized countries, about 45 per cent of production is molluscs (only 10 per cent of the gross weight of molluscs is consumed and 90 per cent of the production volume is shell) and the rest of the production comes from salmon and rainbow trout. Salmon fish farming in this region has undergone a revolution in aquaculture sector.

   While the total aquaculture production between 1991 and 1993 fell by half in the countries of the former USSR, population in Poland, Hungary and former Czechoslovakia remained stable. Farming of common and Chinese carps is most common but if the production of higher-value species is being introduced, the production levels will increase and further switch towards aquaculture development could be observed.

### *Management*

During the years of transition, fisheries administrative structure has undergone several changes and need modifications such as a strengthening of the socio-economic capacity. Although natural scientists are available in all countries, social scientists are lacking and therefore, the concept of planning the fisheries sector needs to be reintroduced. Attempts have been made to privatise the fisheries sector, but high interest rates have been encountered which is considered as a barrier towards fisheries management.

### *Prospects*

In Europe, the demand for fish will probably enhance in the future. However, fish production in the Russian Federation should stop declining soon and start to recover slowly. Inland fishery resources are already exploited and their main orientation is changing from food production to recreation. Inspite of this, fish production from inland waters particularly in the Russian Federation has been increased to some extent owing to adoption of better management. Although aquaculture production may increase in Western Europe, it is likely to be constrained by competition for aquatic resources, limited site availability and cheaper imports from other regions. Given the potential market in industrialized countries, aquaculture production in the Russian Federation will shift to the culture of high-value species such as eels and salmon.

## 19.12 Latin America and the Caribbean

The region covers the South American continent, Mexico, Central America and the island states as well as territories of the Caribbean. The region is surrounded by southern parts of Pacific and Atlantic Oceans and includes the semi-closed seas of the Caribbean seas and Gulf of Mexico.

Similar to European countries, this region is entirely surrounded by large bodies of water such as Atlantic and pacific Oceans, Gulf of Mexico and the Caribbean Sea and therefore, marine fish production reached a record level in 1996 of 26.5 million tonnes. Countries of this region are the major exporting countries of fish and account for 12 per cent of world exports. Average food fish consumption is about 10 kg per year. Caribbean islands do not have any inland water resources of importance. Consequently, inland fisheries are non-existent there. Regional fish production and share of world production is shown in Figure 19.11.

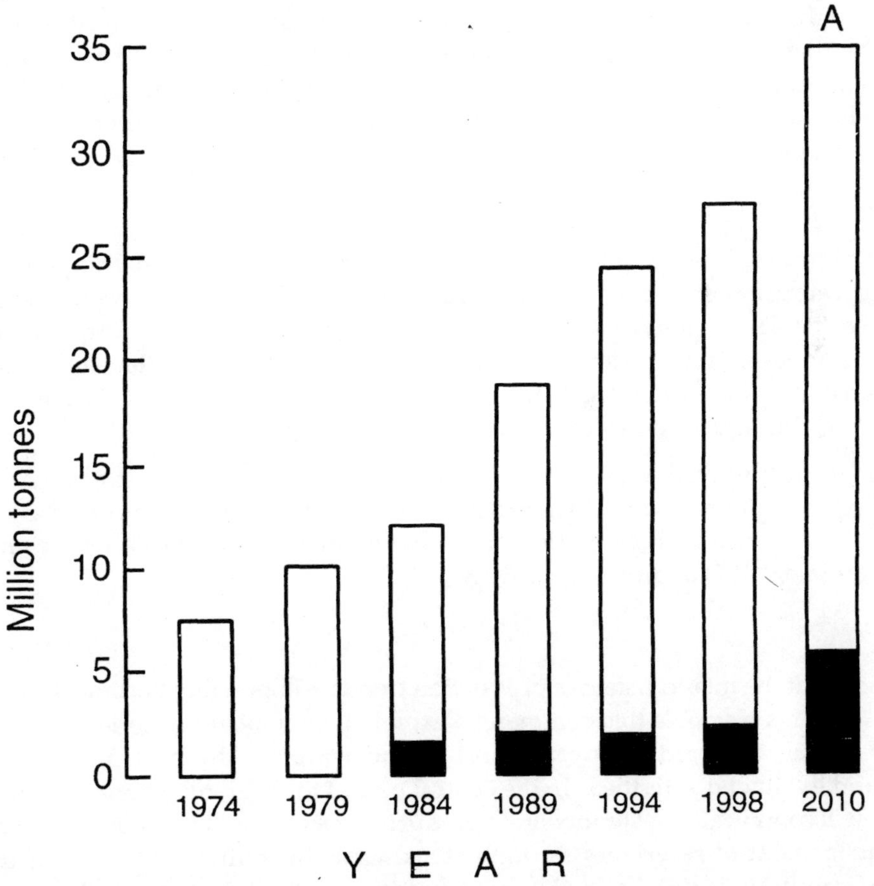


**Fig. 19.11 :** Regional fish production (Latin America and the Caribbean) in inland waters (solid zones) and marine waters (hollow zones). A, Possible production. [Redrawn from FAO (1997); Used with permission of Food and Agricultural Organization of the United Nations]

### *Resources and Production*

1. *Inland Water Fisheries* : During early 1980s, the growth of inland water fisheries was very rapid. But in 1991; the inland fish catch declined to about 450,000 tonnes as against 580,000 tonnes in the late 1980s. In 1994, total reported catches attained some 500,000 tonnes, but this yield is much lower than reported production from similar areas in Africa and Asia. This low production is possibly due to low productivity of inland water systems.

   Inland fisheries are restricted in areas near the main water courses where overfishing is very intensive and the effects of overfishing have been exacerbated by environmental degradation. In the Southern part of Argentina and parts of Brazil, commercial fisheries have been closed and water resources are reserved for recreational and livelihood activities. In the central part of Chile, Argentina, and Brazil, fisheries on rivers and reservoirs are less intensive whereas in the Northern part of Brazil, Cuba and Mexico, intensive management of reservoirs through stocking has resulted in highest growth of inland fisheries in recent years.

2. *Aquaculture* : In 1994 and 1998, total aquaculture production reached 472,000 tonnes and 500,200 tonnes, respectively, representing about 2 and 4.5 per cent of the world production. Although shrimp and salmon cultures have increased rapidly, profit margins foı both species have been declining. Freshwater mollusc, tilapia, trout and carp cultures are also carried out. However, aquaculture oriented towards producing low-cost products has developed and industrial export-oriented aquaculture has expanded having moderate growth potential.

### *Management*

In many countries, fisheries policy has been influenced by macro-economic policies. Measures include the privatization of production units, reduction or elimination of economic incentives in the fisheries sector. These policies have aimed at streamlining the administrative and technical structures of fisheries administration. The research, management and development capacity of fisheries adminstration has been affected in terms of budget allocation, administrative and technical staffs. Of course, this situation has improved in some countries.

The regional institutional framework for fisheries cooperation, management and development is formed by some regional bodies. Among different regional bodies, the commission for inland fisheries of Latin America and the 'Latin American Organization for Fisheries Development' are the two FAO and non-FAO bodies, respectively.

### *Prospects*

It is hoped that the inland fisheries production trends will possibly become more pronounced in the future. It seems plausible that commercial exploitation of inland fisheries in Argentina, Chile and parts of Brazil will gradually cease and inland waters will be utilized particularly for recreational and livelihood activities. In the central part of the region, fisheries activities on rivers are reservoirs will continue at low or moderate productivity levels. In the Northern part of the region, extensive management of reservoirs through stocking will continue. However, if these trends continue, the productivity of Latin American reservoirs and ponds will be increased with concomittent increase in fish production, of course, it is possible subject to careful management strategies.

Export-oriented industrial aquaculture has expanded in the region and has moderate growth potential. Pond-based fisheries in reservoirs and freshwater fish culture have developed less than expected. In general, regional aquaculture potential depends not only from land, water, temperature and agriculture but also from the existing institutional set-up and research capacity. These factors, however, create several problems so far as the slow growth of aquaculture in this region is taken into consideration. As a result, very little developement has taken place in socially-oriented aquaculture and aquaculture for low-income social sectors. Therefore, adequate measure should be given top priority towards the existing potential for aquaculture production.

### 19.13 North America

The region inlcudes Greenland, Canada and the United States (including Bermuda and Saint-Pierre-et-Miquelon and Alasca but excluding Caribbean and the South Pacific Islands) and the adjacent fishing areas of the North-west and Central Atlantic and the North-east Pacific.

In 1994 and 1998, the North American region contributes about 6 and 8.7 per cent global fish catch, respectively. Inland fisheries and aquaculture show a more steady growth. Food fish consumption averages a per caput supply of 22 kg (live weight equivalent) annually. Fish production in this region and share of world production is shown in Figure 19.12.

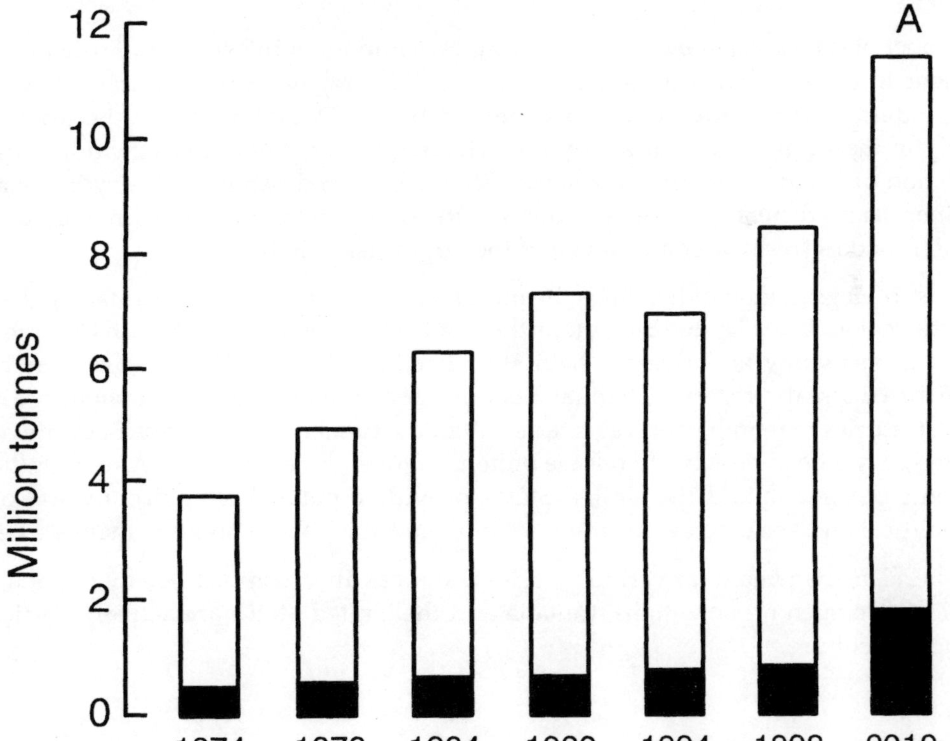


**Fig. 19.12 : Regional fish production (North America) in inland water (solid zones) and marine waters (hollow zones). A, Possible production. [Redrawn from FAO (1997); Used with permission of Food and Agricultural Organization of the United Nations]**

### *Resouces and Production*

1. *Inland Water Fisheries* : The catch from commercial small - scale fisheries sectors is steadily decreasing owing to vast expansion of recreational fisheries. At present, freshwater recreational catch level of United States exceeds the commercial catch for all of North America. In 1995, however, seven federal agencies have been established to develop recreational fishery programs. Beside this, in the Arctic and central regions of Canada, recreational catch exceeds the commercial production from inland water capture fisheries. In 1994 and 1997, the total reported commercial catch was 71,000 and 78,300 tonnes respectively.
2. *Aquaculture* : In this region, aquaculture is a diversified industry which includes freshwater fishes, crustaceans, molluscs, and plants. Canadian aquaculture industry has been established on the basis of coldwater species such as salmon, trout, and molluscs. In the United States, the main species include rainbow trout, golden shiner, salmon, catfish, crawfish, tilapia, and shrimp. While aquaculture industry in the United States grew at about 2 per cent a year over the decade to 1994, Canadian aquaculture exhibited 20 per cent growth. In 1994 and 1998, total aquaculture production attained almost 0.5 and 2.3 million tonnes, respectively.

### *Management*

A number of state and federal agencies play a very important role for inland fisheries resources management in the respective jurisdictional areas. With regard to freshwater inland fisheries resources, federal fisheries jurisdiction is exercised by the Department of Fisheries and oceans particularly in the Arctic and Central regions. The management objectives include conservation and protection of resources to ensure a sustainable fishery and fishing industry in collaboration with commeriaı and recreational users. However, the Arctic and Central Regions contain about 67 per cent of Canada's freshwater and seven of the largest lakes in the world.

The loss of degradation of habitat is the most serious environmental issue facing the fisheries sectors in the region. In the United States (North-west Pacific) and Canada (British Columbia), about 80 per cent of spawning and riverine habitat of Pacific Salmon and Steel-head has been lost. Development of aquatic reserves, habitat restoration and hatchery management are the most important strategies to rebuilt capture fisheries. With the tremendous increase in the development of hatcheries, it has been possible to release billions of juveniles every year. Although the Federal Governments of Canada and the United States provide a parasol of environmental protection policies, political and socio-economic interests influence the policies to implement and enforce.

In general, management for various species and areas are recommended by regional fisheries organizations. In such organizations, Canada and the United States are actively participated in developing fish culture.

### *Prospects*

Although it is difficult to quantify the exact impact of different issues upon consumption patterns, it is expected that the demand for fish will increase in the future. Some marine fish species have not immediate market and therefore, increased production will obviously come from aquaculture through technology-oriented development and for that due attention should be paid to environmental issues.

Fish supply could be boosted by imports, increasing the potential in South-east Asia where fish culture industries are expanding to significant extent, to increase exports to North America. The United States will therefore continue to be a major importer of fish although environmental aspects might influence imports to the region.

### 19.14 South-east and South Asia

The region includes the countries of South and South-east Asia, from Pakistan in the West to Indonesia in the East. The main fishing areas of the region includes the northern part of the Indian Ocean, the Bay of Bengal, the Arabian Sea, the South China Sea and the Western Central part of the Pacific Ocean.

This region is characterized by the most important productive fishing water in the world. Total regional fish production reached record level in 1994 and 1998 of 19.5 and 21.7 million tonnes respectively, representing about 27 and 30.3 per cent of the global catch. It has been estimated that about 12 million people are engaged in fisheries. Consumption of food fish is extremely variable among the different sub-regions and countries. However, in 1994 and 1998, average per caput fish consumption were 9 and 12 kg, respectively (live weight equivalent). Over the last decade, fish trade in the region has expanded significantly. Among different countries, Thailand is the world's leading exporter of fish and fishery products. Regional fish production and share of world production is shown in Figure 19.13.

#### *Resources and Production*

1. *Inland Water Fisheries* : Regional production from inland water fisheries exhibited slight increase from 2.3 million tonnes in 1984 to 2.6 million tonnes in 1996. About 25 per cent of the total catch is taken from the extensive inland water fisheries of Bangladesh where production reached 570,000 tonnes and 596,600 tonnes in 1994 and 1998, respectively. India and Indonesia have considerable inland water fisheries resources and together contribute about 35 per cent of total regional output. Some other countries have recently undertaken large-scale stocking programs and the increase in freshwater fish production in Bangladesh is partly owing to enhancement of fisheries.

   According to one estimate, the inland fisheries in the region has stabilized at about 1,00,000 tonnes annually over the period 1994 to 1999. This stability probably reflects a balance between increasing exploitational pressure and declining production due to increasing environmental disturbance of the water bodies throughout the region, removal of forest cover, and the pressures of agricultural developments on the catchments have dramatically changed the flow patterns of the rivers and their flood plains and increased the sediment flows, usually to the detriment of their fish production.

2. *Aquaculture* : In the region, total aquaculture production increased from 1.8 million tonnes in 1984 to 6.7 million tonnes in 1998. However, the main producers are India, Phillippines, Indonesia, and Thailand. Prawns, shrimps, carp and carp-based fish are the main species produced in these countries. Prawn and shrimp represented more than 55 per cent of the total volume in 1998.

In 1998, total production of farmed finfish was about 5 million tonnes. The bulk of freshwater production is based on polyculture within traditional and semi-intensive pond-based farming systems. This farming system contributes a low-priced source of food fish for mass domestic consumption, particularly in India. The main species cultivated in the region belong to the Cyprinid family, including rohu, catla, mrigal, common carp, grass carp, and silver carp. Other species are also cultured in pens and cages for tilapia, *Amabas* sp., *clarias* sp., and *Heteropneustes* sp.

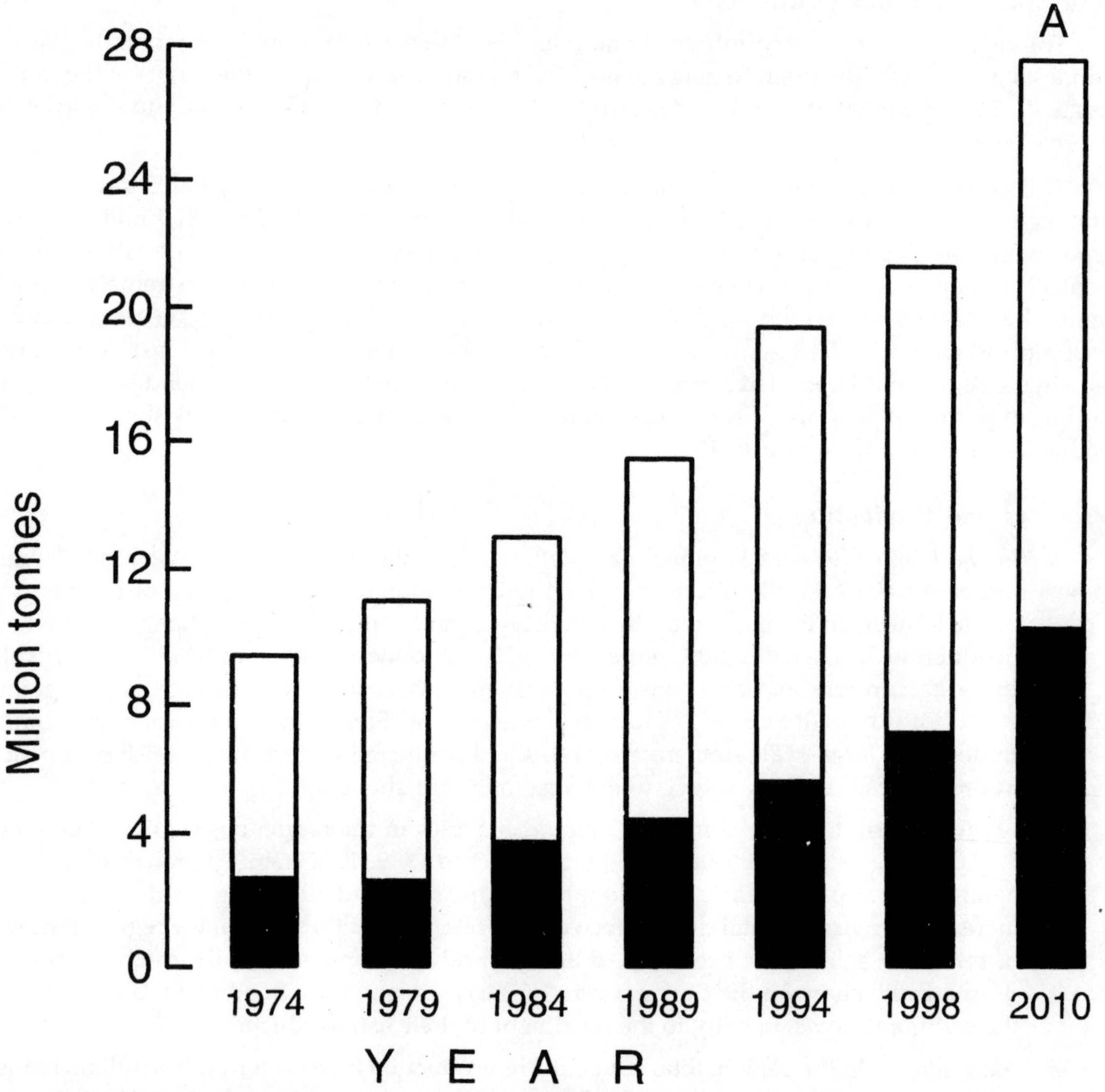


**Fig. 19.13 : Regional fish production (South-east and South Asia) in inland waters (solid zones) and marine waters (hollow zones). A, Possible production. [Redrawn from FAO (1997); Used with permission of Food and Agricultural Organization of the United Nations]**

Dramatic expansion of prawn and shrimp culture industries has taken place over the last decade and production in 1997 was 705,000 tonnes. The region contributes 75 per cent of total world production of cultured shrimp, the giant tigar prawn is the most popular species cultured. Thailand and Indonesia are the major shrimp-producing countries.

The rapid growth and expansion of prawn and shrimp farming has provided the developing countries in the region with considerable foreign currency earnings. The expansion of shrimp pond areas has significantly contributed to the loss of mangroves. Inspite of this, many shrimp farms have been set up in more suitable non-mangrove areas. Other environmental impacts include the salination of ground-waters and agricultural land and pollution from the effluents of shrimp areas. Despite the recent availability of reports on environmental impacts, the severity and extent of the effects have not always been studied with the necessary objectives. However, it has been recognized that public perception of shrimp farming might be damaged by the irresponsible practices of some entrepreneurs who have caused serious environmental degradation and social disruption.

### *Management*

The sustainability of fishery resources is a central issue in many countries. In some countries, a partnership between local fishermen communities and Central/State Goverments is evolving to develop a community-based fisheries management system for local resources, with the result that conflicts among different types of fishermen/fish farmers decreased signficantly. Environmental degradation caused by humans through aquaculture and fish farming is a serious problem. The long-term sustainability of shrimp farming is being addressed at national level. Many private-sector representatives, Governments and Non-Governmental Organizations now advocate shrimp farming practices that are environmental and socially acceptable. Regional activities include environmental management in the various farming systems, augmenting the sustainable intensification of the traditionally extensive farming systems, integrating shrimp production and silviculture.

### *Prospects*

The population in the region is growing rapidly and fish is generally considered as a main source of animal protein for most people. In high-income society, domestic consumption of high-value fish are expected to grow rapidly and higher prices on international markets will undoubtedly help to expand exports of high-value farmed fish. However, it has been estimated that by 2015, fish supplies will need to increase by 7.8 million tonnes to maintain current per caput fish consumption levels.

Fish culture and inland fisheries will provide considerable opportunities for further development to enhance fish production in many countries such as Bangladesh, India, Cambodia, Loas, Malaysia, Indonesia, Myanmar, Phillippines, Thailand, and Vietnam. Inspite of significant increase in fish production from freshwater resources (also from marine water resources), the region will probably have to depend more on imports of fishery products.

## 19.15 Sub-Saharan Africa

The region covers the African continent except for the North African States bordering the Mediterranean. Marine fishing is concentrated the Eastern Central Atlantic, the South-east Atlantic and the Western Indian Ocean.

Many countries of the region depend on the fishery resources and play an important rọle for supplying animal protein. About 10 million people are directly or indirectly employed in fishery sector. Total regional fish production in 1994 and 1997 was 4 million and 6.2 million tonnes, respectively. Food fish consumption has declined from 9 kg in 1990 to less than 7 kg in 1994. However, regional fish production and share of world production is shown in Figure 19.14.

***Resources and Production***

1. *Inland Water Fisheries* : Inland water capture fisheries production in the region has increased over the past decade at an annual rate of about 3.5 per cent. In 1994 and 1998, regional production were 1.6 million and 2.3 million tonnes, respectively. The main species include

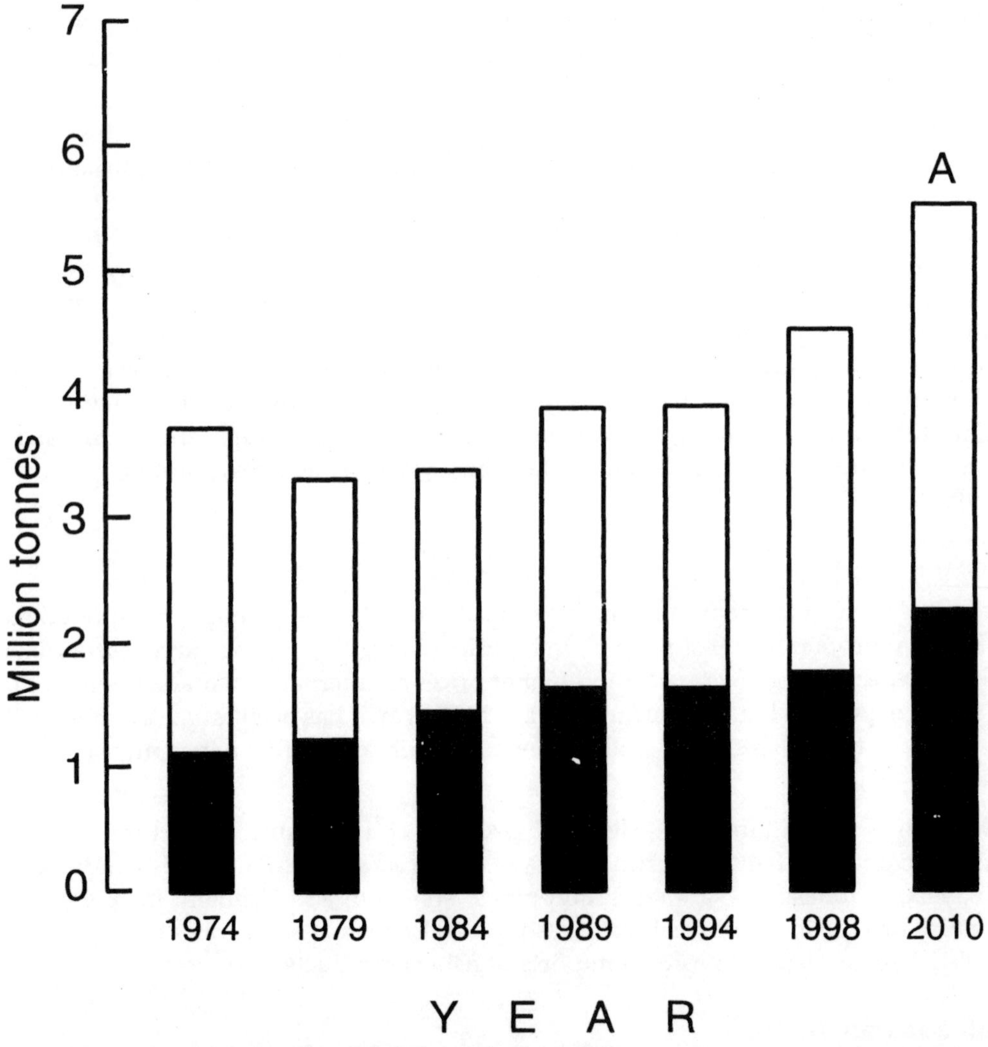


**Fig. 19.14 : Regional fish production (Sub-Saharan Africa) in inland waters (solid zones) and marine waters (hollow zones). A, Possible production. [Redrawn from FAO (1997); Used with permission of Food and Agricultural Organization of the United Nations]**

catfish, Nile perch and tilapia. A few tonnes of Nile perch are exported and the entire inland production is supplied to local market. Among different countries of the region, Uganda, Zaire, the United Republic of Tanzania, Kenya, and Nigeria are considered as the top freshwater fish producers, contributing 70 per cent of the total harvest. Lake Victoria contributing one-quarter of the total fish production.

2. *Aquaculture* : Aquaculture in the region is starting to expand. Inspite of recent development on aquaculture, the region contributes 0.2 per cent of the total world production. In 1994 and 1998, about 33,000 tonnes and 37,300 tonnes of freshwater fish were produced respectively by the region. Kenya, Nigeria, Madagaskar, South Africa, and Zambia produced about 1200 tonnes each. Over the 12 years these countries have doubled their annual production several times.

   The most important species cultured include tilapia, catfish, carp, molluscs, and shrimp. Freshwater fish make up about 80 per cent of the total aquaculture harvest and almost all fish farming in the region is carried out by rural fish farmers in small freshwater ponds. Although shrimp culture at commercial basis is developing in Madagaskar, Guinea, Kenya and Mozambique, aquaculture development is jeopardised by several reasons such as climatic change, dependence on external assistance for aquaculture development project and weak financial support. These suggest that efforts on aquaculture development strategies are not sustainable.

### *Management*

As fish is a popular food item in the region and provides 18 per cent of total annual protein intake, suitable managment strategy is urgently needed as most fishing grounds now show signs of intensive exploitation. During the last 16 years, several countries have aimed at developing medium-term sectoral plants to support Government policies, but these plans have failed to sustain good fisheries management. The causes of failure are probably due to low budgets and lack of political will to implement management measures and policies.

Annual variability of freshwater fisheries potential is the main feature of the region. Systems that fluctuate seasonally and from year to year in reservoirs, swamps and river flood plains, account for almost 60 per cent of the total water surface area. Consequently, the biological and social management of such fisheries is difficult.

Recently efforts have been made to strengthen the mandate of fisheries orgnization in aspects related to management or to establish new bodies such as the Ministerial Conference of African States bordering the Atlantic Ocean and the Lake Victoria Fisheries Commission. However, it is expected that these bodies will function in fisheries matter but which, with the exception of the Southern African Development Community, have not built up the necessary capacity to implement effective fisheries management strategies.

### *Prospects*

Population growth in the region indicates a population of 700 million by the year 2010 and 915 million by 2030. Therefore, an increase of total supplies in the order of about 1 million tonnes would be needed to meet demand in 2030. Productivity enhancement programs in small water areas, fish culture development, increased imports, implementing sound fisheries management regimes, and distribution networks are the main future possibilities for enhancing food fish supplies in the region. Growth forecasts on the gross domestic product over the next 15 years clearly indicates that future prospects seem to be poor. Trends include constraints on imports and continued demand for low-value species. Simultaneously, lower public subsidies will increase production costs and weaken competitiveness on export markets.

## 19.16 Near East and North Africa

The region extends form the Atlantic coast of Morocco along the Mediterranean coast of North Africa (including Cyprus) to the coast of Eastern Mediterranean and Turkey and also include Jordon, Egypt and the countries of the Arabian peninsula (bordering the Gulf of Aden, the Red Sea, the Gulf of Oman and the Arabian sea) and the countires of central Asia. Other main water bodies include the Caspian Sea, the Aral Sea and the Black Sea.

No country of this region depends substantially on fish as a mainstay of its economy. Fisheries are diversified, ranging from abundant resources of the Atlantic Coast of Morocco to inland water fisheries with poor resources. In 1994 and 1998, total production reached about 2.8 million and 4.2 million tonnes, respectively. Food fish cunsumption varies widely throughout the region. Per caput fish consumption varies between 0.1 kg ( in Afghanistan) and 40 kg (in Yemen). However, the region is not a substantial contributor to international trade in fisheries. Regional fish production and share of world production is shown in Figure 19.15.

### *Resources and Production*

1. *Inland Water Fisheries* : In 1994 and 1997, the total inland water production of the region amounted to about 452,000 tonnes and 478,600 tonnes, respectively. Egypt accounts one-third of the total volume due to Nile River and Nasser Lake. Israel, Turkey, Iraq and Iran also have inland capture fishery resources.

   In the Caspian Sea, which dominates Iran's inland water resources, silver carp, Caspian shad and sturgeon are the main species exploited. The Caspian Sea is suffering from environmental and severe pollution problems over the past decades. Fish resources have been affected by hevy fishing and sturgeon stocks are damaged.

   In Central Asia, the creation of reservoirs, the damming of rivers and gradual environmental degradation of the Aral Sea have obstructed the healthy development of fisheries. The Aral Sea is fed by the Amu-Darya and Sur-Darya rivers. Since 1960, the sea has been subject to desiccation. During 1970s, a series of dry years, the sea has been drained by the diversion of water by about 80 per cent for agriculture purpose. Consequently, the depth of the sea has been significantly reduced, fish stock have been affected and there has been a substantial reduction in spawning and nursery habitats. As a result, the sea has ceased to be of importance to fisheries and all commercial fishing came to an end in 1982.

2. *Aquaculture* : In 1994 and 1997, the total aquaculture production amounted to about 148,000 tonnes and 157,300 tonnes, respectively. The main important species cultured in six countries consists of common carp, Nile tilapia and silver carp. About 90 per cent of the total aquaculture production come from the culture of above-mentioned species. Crustaceans make up the small balance (10 per cent of the total finfish production).

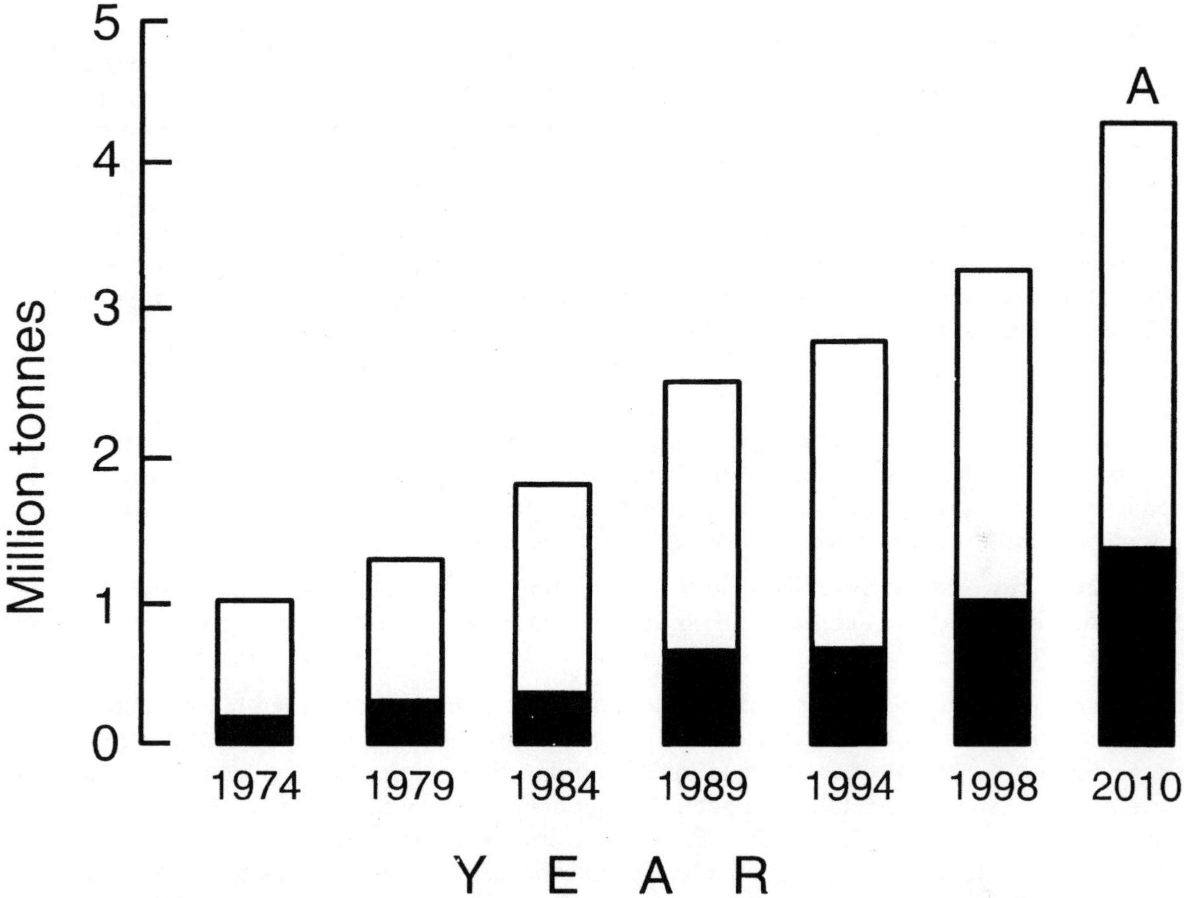


**Fig. 19.15 : Regional fish production (Near East and North Africa) in inland waters (solid zones) and marine waters (hollow zones). A, Possible production. [Redrawn from FAO (1997); Used with permission of Food and Agricultural Organization of the United Nations]**

Culture of combinations of omnivorous and herbivorous finfish species are based on extensive and semi-intensive systems using local sources. In Egypt, the culture of common carp is undertaken in rice fields. However, it seems quite plausible that aquaculture production has continued to grow well above global rates and there is good potential for further expansion. In some countries, aquaculture makes a useful contribution to fish production : 17 per cent in Egypt, 50 per cent in the Syrian Arab Republic and 70 per cent Israel.

### *Management*

As fisheries play a very significant role in most regional countries, main functions of national fisheries administrations are gradually shrinking with virtually no funds available to provide support to the sector and management decisions are not enforced. Many Governments in this region have taken few conservation and management decisions, only a few countries have undertaken fishing efforts of different species.

*Prospects*

So far as regional aquaculture development is concerned, future prospects are characterized by the competition for freshwater, suitable sites and feed ingredients. There is a lack of development planning between aquaculture and agriculture. Countries facing these constraints are turning to fish culture in inland waters to avoid the use of arable land. Since fish consumption in the region is assumed to be relatively low, it would reasonable to expect that at least until 2015, a slight increase in demand could be met from higher regional landings of fish if fish are not exported to the European market. European markets are high-value fish led to many small-scale fishing enterprises being set up in countries (such as Morocco) while finfish, shellfish and fish products are exported to Japan, Israel, Saudi Arabia and the United Arab Emirates.

## 19.17 Conclusion

Global fish production through inland fisheries and freshwater fish culture has increased dramatically. Freshwater fish culture holds the key for increasing fish production in the years to come and for this purpose different organizations are playing their role in increasing productivity not only within the country but also throughout the world.

Trends in world fish production and consumption clearly indicate that inland fisheries and fish culture increased and maintained their positions as one of the fastest-growing food production activities in the world.

So far as development of inland fisheries and fish culture are concerned it is concluded that management, resources, and production in different geographical regions of the world vary widely depending on the socio-economic conditions, population pressure, financial constraints, lack of development planning, technology-oriented development, and environmental issues. While culture-based fisheries in some regions such as East Africa, South-east Asia and South Asia have reached a plateau of production, fish production in Europe and Latin America significantly declined due to recreational and livelihood activities.

### *References*

Ayyappan, S. and J.K. Jena. 1997. Freshwater aquaculture – An upcoming sector in global fisheries. *Fish. Chim.* 17 (1) : 17-21

FAO (Food and Agricultural Organization). 1997. *The State of the World Fisheries and Aquaculture.* FAO Fisheries Department, Rome.

FAO (Food and Agricultural Organization). 1990. *Review of The State of World Fishery Resources.* FAO Fisheries Circular N. 710.

## Questions

1. What are the current trends in world fish and aquaculture production?
2. For sustainable fisheries, environmental degradation should be reduced. Explain.
3. State how production and consumption of fish can be increased?
4. Describe the production and consumption patterns of fish in the following countries : (a) China, (b) South Asia, (c) North America, (d) India.

5. In which regions of the world fish production potential is great and why?
6. Is there any possibility to increase fish production in developing regions? How does the developing regions differ from those of developed ones?
7. Why inland water fisheries and aquaculture production in South-east and South Asia is higher than in other regions?

# 20

# Carp Culture in India

Carps are widely distributed in different lakes, rivers, and ponds in the world. Although the native place of carp is believed to be in the Asia, they have introduced to other parts of the world for cultivation. Carps are valuable food fishes. India, China, and other South-east Asian countries produce considerable amounts of carp.

Rice, vegetables, and fish have for ages been drawn up by the people of India as their major food sources. This dependence is not likely to be changed in the light of rapidly growing populations.

In India, the freshwater resources are among the most productive to be found anywhere. Fast growing and fertile carps of different types make up the major portion of the freshwater species found in rivers, lakes, ponds, and reservoirs of the country. Scientific techniques are now followed to enhance the productivity of the waters by using more efficient cultivation methods.

## 20.1 Types of Fish Culture in Freshwaters*

In freshwaters, different types of culture are undertaken on commercial basis such as (1) monoculture (when single species of fish is cultured in a pond), (2) composite culture (when more than one species of fish is cutured together in a pond), (3) paddy-cum-fish culture (when different species of fish such as *Oreochromis mossambicus, Ophiocephalus punctatus, Cyprinus carpio, Labeo rohita,* and *Labeo calbasu* are cultured in the paddy fields), (4) sewage-fed fishery (treated sewage waters are drained into freshwater systems where several species of carps are stocked), and (5) air-breathing fish culture (different types of air-breathing fish such as *Anabas testudineus, Heteropneustes fossilis, Clarias batrachus, Ophiocephalus punctatus, O. striatus,* and *O. marulius* are cultured in shallow and derelict waters ).

## 20.2 Characters of Culturable Freshwater Carp

The success of fish culture principally depends upon the selection of right type of carp species. Generally improved variety of fish is selected for freshwater fish culture. On the basis of the following characteristic features, different species of carp is selected.

1. Capability of consuming supplementary feeds.
2. Capacity to grow rapidly.
3. Abilitv of carp to adapt with other species.
4. Capacity to live in a very limited space.
5. Adapting capacity of carps along with the change in physico-chemical conditions of water.

* For further detail, see Huet (1986) and Horvath *et. al.* (1992).

6. Ability to breed in confind water through hormone injections.
7. High demand and market price.

## 20.3 Species of Carp Cultured in Freshwaters

Most of the cultivable species of fish belong to the Order Cypriniformes. Besides above mentioned features, there are some other body characters by which carp is considered as a very suitable fish for culture in freshwaters. They can easily consume both natural food (phytoplankton, zooplankton, and bottom fauna) and supplementary feeds. Moreover, the pharyngeal bone is provided with teeth and consequently different minute food particles are effectively utilized by the fish.

Carp is a group of edible and culturable herbivorous fishes which do not have teeth in mouth. Except hard region, the entire body is covered with cycloid scales and they have an air bladder. Carps are important from the economic point of view.

In India, good species of carp are being cultivated such as *Labeo rohita* (Rohu), *Catla catla* (Catla), *Cirrhinus mrigala* (Mrigal), *Labeo calbasu* (Kalbasu), *Hypophthalmichthys molitrix* (Silver carp), *Ctenopharyngodon idella* (Grass carp), and *Cyprinus carpio* var. *communis* (Common carp). The first four types are called *Indian major carps* and the last three ones are minor or *exotic carps* (Figure 20.1)

### *Characteristic Features of Some Carps**

Identification and segregation of different types of carp fry and fingerlings from a mixture of fish seed is the most important task for successful rearing of carps in ponds. Some specific characteristic features are evident in each species of carp fry and fingerlings by which progressive fish culturists can easily recognize the species for stocking in ponds. The characteristic features of some commonly used carps in ponds are briefly noted below :

**1.** ***Major Carp Fry***

(i) *Catla catla* : Head large, dorsal profile is convex and ventral profile is concave; no distinct spot on the caudal fin; margins of dorsal and caudal fins are dark; opercular region is reddish; no maxillary barbels; lip thick but not fringed.

(ii) *Labeo rohita* : A dark diffused band at the caudal peduncle; a pair of light greyish maxillary barbels are present; lips fringed.

(iii) *Labeo calbasu* : Body dark in color and has alternative black and yellow bands; the basal parts of the fins are black; four black barbels are present; narrow mouth.

(iv) *Cirrhinus mrigala* : Head small and slender body; a triangular dark spot at the caudal peduncle; no barbels; lips are thin and not fringed; the lower lip of the caudal fin has a reddish tinge.

**2.** ***Major Carp Fingerlings***

(i) *C. catla* : Head large; no distinct spot over the body; no barbles; lips thick and not fringed; dorsal, anal and caudal fins are dark greyish in color.

(ii) *L. rohita* : Dark band at the caudal peduncle; reddish tinge in dorsal, caudal, pelvic and anal fins; lips are fringed; prominent barbels.

---

* For further detail, see Talwar and Jhingran (1993).

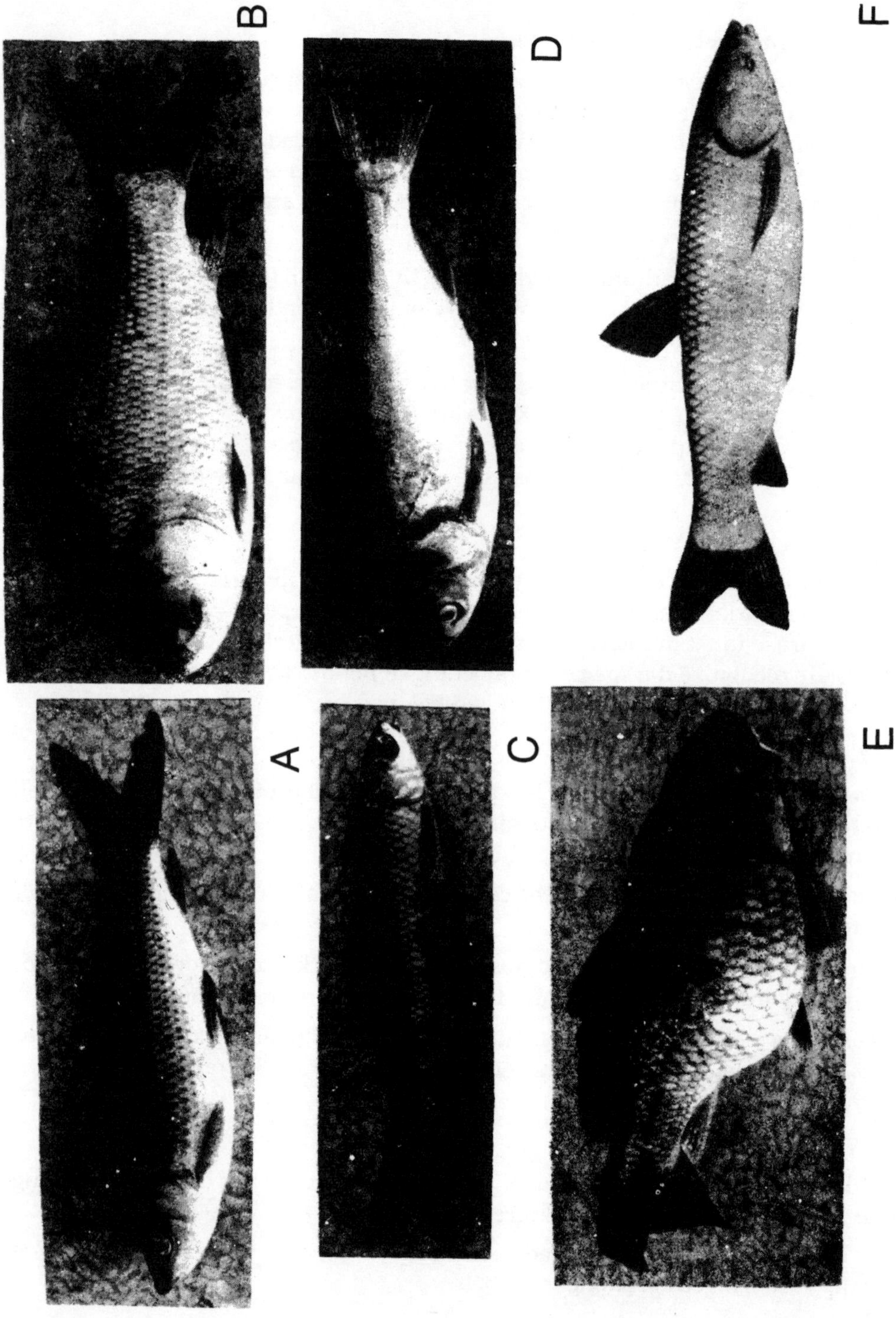


Fig. 20.1 : Carps are commercially important species generally cultured in freshwater pond/tanks. They are planktophagus (except grass carp), compatible, and exhibit fast growth rate. For freshwater fish culture under Indian conditions, a few indigenous and exotic species of carps are extensively cultured together with high profit. Three Indian major carps and three exotic carps are shown in this figure. A *Labeo rohita*; B, *Catla catla*; C, *Cirrhinus mrigala*; D, *Hypophthalmichthys molitrix*; E, *Cyprinus carpio*; F, *Ctenopharyngodon idella*.

(iii) *L. calbasu* : Body color gradually turns black; basal parts of dorsal and caudal fins are black; lips are fringed; presence of four prominent black barbels.

(iv) *C. mrigala* : The spot of caudal peduncle becomes diamond-shaped; barbels are faintly visible; a few longitudinal lines appear on the body; the tips of the lower lobe of caudal fin is vermilion red in color.

3. *Exotic Carps*

(i) *Hypophthalmichthys molitrix* : Body oblong and slightly compressed; pointed head, round snout and lower jaw slightly protruding; lower jaw with a tubercle and upper one notched; scale small; keeled abdomen; fins dark; gill rakers are large.

(ii) *Ctenopharyngodon idella* : Head broad with short round snout; upper jaw slightly longer than the lower; no barbel; mouth subterminal; anal fin rounded; gill raker small; dark grey color above the body and silvery on the belly; all fins are dark in color.

(iii) *Cyprinus carpio* var. *communis* : Scantily clothed in a few patches of large, irregular small scales; head comparatively small and tapering at the anterior end; abdomen round.

## *Biology of Carp*

Although carp are distributed widely between tropical and temperate region, in general, they prefer a warm climate. When the temperature of water is between 15° and 35°C, the appetite of the carp increases; but when the water temperature is lower than 10°C, the apetite declined.

The spawning of Indian major carps takes place in the flood regions of the rivers soon after the beginning of the monsoon. Spawning grounds are found in the middle reaches of most rivers where flood water spreads over fertile flats well above tidal reaches. The advancing flood water is warm, well oxygenated and the bottom vegetation supply the necessary stimulation that causes sexual display. In South India, the spawning of carps has been observed in all rivers during the time of high flood both during the day and night.

Normally, the major and minor (except comon carp) carps do not breed in ponds but they breed in the lowlands adjacent to the pond when the pond water is overflooded in rainy season. The collection of hatchlings from these places has some disadvantages such as limitation of hatchling collection and mixture of unwanted species of fish. To overcome these difficulties, major carps are induced to breed in the pond with the aid of gonadotrophic hormones.

Indian major carps are known to breed near confluences of rivers and streams. The male and female fish gather in shoals. They discharge milt and eggs while rubbing against each other's bodies. At this moment, the eggs are fertilized and settled at the bottom of the spawning ground. In case of common carp, fertilized eggs are firmly attached to the surface of adjacent water plants. However, the number of eggs that a carp deposits their eggs depends on the age of the carp. The length of time needed for hatching differs, depending on the temperature and oxygen of the water and the type of species to be reared. In case of common carp, for example, if the water is 20°C, fry emerge after 5 days. In case of Indian major carps, fry emerge after 16 hrs when water temperature varied between 26° and 29°C. These youngs just hatched, are about 5 mm long and have a yolk sac in the abdominal area. Carp grow the first three to four days by utilizing the yolk sac, but afterward devour small zooplankton and phytoplankton. Some species of zooplankton such as *Daphnia magna*, *Cyclops viridis*, *Mysis* sp., *Diaptomus* sp., and *Nauplii* are suitable as carp food. During the early period of growth, carp consume phytoplankton, but when they grow up they eat zooplankton and become polyphagous. Bottom-dwelling carps (such as common carp and mrigal) eat benthic organisms (such as Chironomous larvae),vegetable matter, and organic debris.

## 20.4 Preliminary Idea About Freshwater Fish Culture

In India, carp production can be increased several times through adoption of scientific management procedures. The following important points should be kept in mind when fish culture operation is contemplated.

1. The depth of the pond should vary between 1.5 and 2.0 metres.
2. The pH of water should vary between 7.5 and 8.5.
3. Depending upon the environment of pond ecosystem, fish species should be selected.
4. The number of fish to be stocked should be determined on the basis of the abundance of natural food species and the size of the pond.
5. Application of inorganic and organic fertilizers should be made to increase the fertility of ponds.
6. Growth of aquatic plants should be checked.
7. Reproductive capacity of culturable species should be high so that eggs and fry can be obtained for fish cultivation.
8. Construction of fish ponds and their conditions should be good.

## 20.5 Site Selection of a Freshwater Fish Farm

Site selection is the most important critical for the establishment of a fish farm. In freshwater fish culture, ponds may either be rain-fed or supplied with freshwater from different sources such as surface run-off, canals, reservoirs, deep wells, and springs.

In considering each of the four major factors – water supply, drainage system, soil characteristic, and topography – emphasis should be placed on their impact on pond construction. Note, however, that the impact of one factor is dependent of those of the others. Thus soil which supplies nutrients to water, stimulates to develop natural food organisms for the growth of fish. Likewise, the accurate and systematic description of a region guides the planning of drainage and water supply systems as well as construction costs. A typical lay-out of a freshwater fish farm is shown in Figure 20.2.

### *Criteria for Site Selection*

There are five main criteria that a progressive fish farmer may follow to construct a good fish farm for better carp production : (1) cost and availability of manpower, construction material, equipments, fertilizers, manures and feeds, (2) source and availabilty of stocking materials, (3) system of culture to be adopted (whether intesive, semi-intensive, and extensive), (4) operational method (whether monoculture or polyculture), and (5) location of market for the produce and transport facility. However, each criteria should be studied separately with a view to producing more food fish. Progressive fish farmers generally have choice these criteria. Fish farmers can built a good fish farm, they can grow as many fish annually as feasible. By increasing annual yields per hectare can they produce more fish.

## *Soil Characteristics*

Although nation has been made regarding soil in relation to fish culture, attention must be given to fish farm construction and to the maintenance of soil structure throughout the fish-growing season. Soil characteristic feature is central to such a discussion.

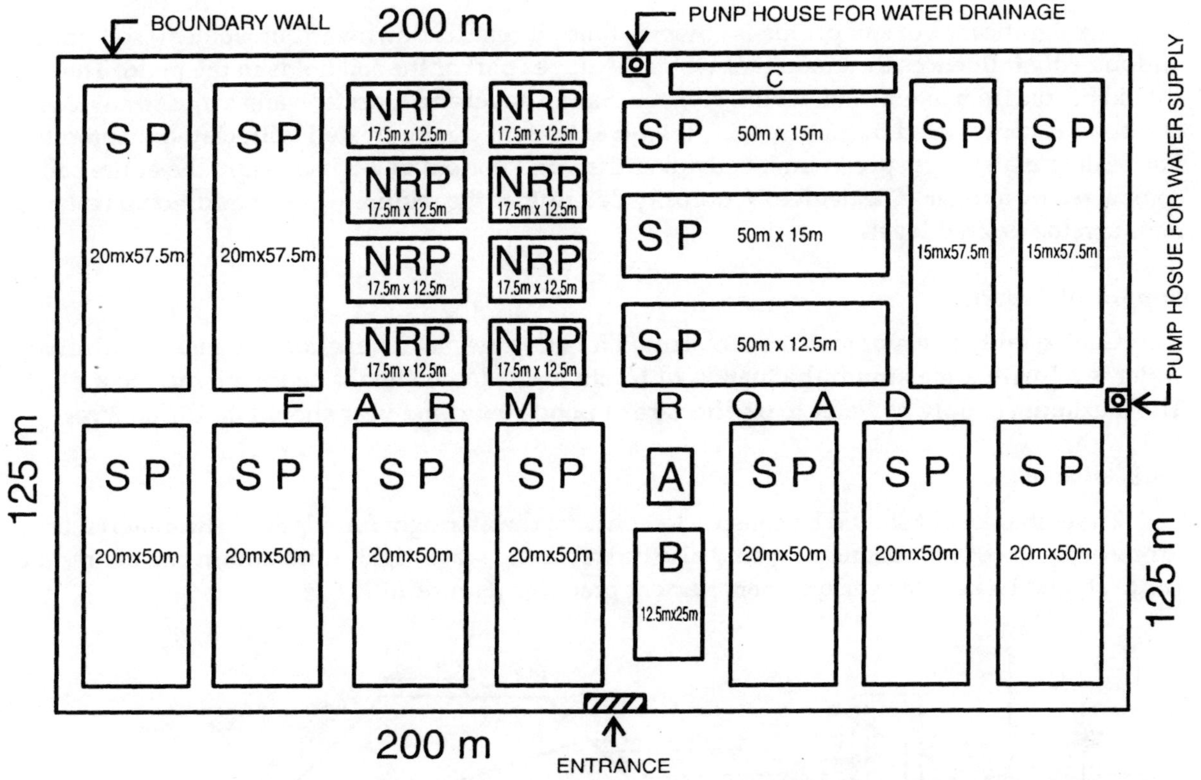


**Fig. 20.2 : Lay-out of a 5 hectare freshwater fish production farm (Plan at GL). SP, Stocking ponds; NRP, Nursery-cum – rearing ponds; A, Laboratory-cum-office; B, Handling-cum-marketing pond; C, Area for housing complex. A total number of twenty three ponds may be constructed in a 5 hectare farm. The breakup of different types of ponds is as follows : Stocking ponds = 14 (Total area 2.635 hectare), Nursery-cum-rearing ponds = 8 (Total area 0.204 hectare), Handling-cum-marketing pond = 1 (Total area 0.063 hectare).**

The most suitable types of soils for fish pond construction are usually impervious and are ordinarily quite heavy. Impervious soil areas remain water for prolonged period – the loss of water being from evaporation. Soil characteristic features have profound effects on the use of soils for fish farm construction.

The most important feature of a soil for pond construction is *soil strength.* Soil strength is determined by a number of related soil characteristics, among which are *soil compactability* and *soil compressibility.* Generally heavy clays, clays and clay loams are considered. In soils where quartz and feldspars are present along with the sand fraction, are conducive to high soil strength. Soil compaction controls the seepage of water from the pond bottom. Sandy and peaty soils must be avoided as these soils will not permit soil compaction and consequent increase downward movement of water. In case of impervious silt or clay soils, the downward movement is impeded. However, if the soil profile do not contain a layer of substance that is impervious and thick to check seepage, expensive sealing procedures will be carried out. The sub-soil should be checked by taking soil samples to ascertain that there is presence of at least 1.0 metre layer of good soil under the bottom of the pond. Layers differing in physical make up from the horizons are common. These layers have profound influence on water movement and deserve specific attention.

The significance of soil profile is obvious. Since it retards downward movement, soil profile undoubtedly influences the amount of water the upper part of the soil holds in the pond. The soil should be free from lateral and ventral seepage. Sand, gravel and limestone areas must be avoided. If a pond is constructed in porous soils, the pond bottom can be treated with clay or bentonite.* The sealing material is spread uniformly over the pond bottom in 2-3 layers and the entire soil is compacted by a roller. The degree of porosity determines the requirements of additional water to maintain the desired levels.

### *Supply of Water*

Good quality of water must be ascertained for effective fish cultivation. In general, adequate water level with a maximum fluctuation of 60 cm in its depth should be maintained in a pond. The maximum supply of water to one hectare of pond round-the-year should be 18,000 litres.

### *Embankment*

The embankment should be constructed on a solid water tight foundation. The material used for construction of embankment, should also be stable and water tight. A minimum of 2 metres top width should be kept for various management practices (Figure 20.3).

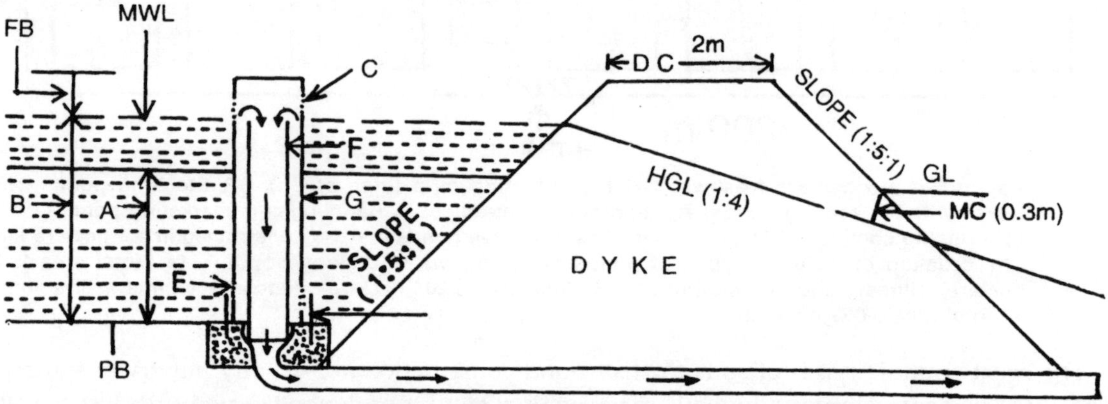


**Fig. 20.3 : Cross section of a pond dyke along with the outlet structure. When a dyke is constructed, its stability is verified by drawing a hydraulic gradient line (HGL). This line must be passed through the base of the dyke with a cover of 0.3 m between the HGL and the foot of the dyke. The draining of water is necessary, the socket which is embeded within the concrete, is opened and the water is drained both from the bottom and the surface of the pond. A, earth work in cutting (1.5 m); B, water depth (2 m); C, openings for draining water from the surface; D, PVC socket inserted within the concrete; E, openings for draining water from the pond bottom; F, 150 mm PVC pipe G, 200 mm PVC pipe; DC, dyke crest; FB, foot board; MWL, maximum water level; PB, pond bottom; MC, minimal clearance.**

* It is a soft plastic light-colored clay formed by chemical alteration of volcanic ash. It is composed essentially of montomorillionite (a dioctahedral clay mineral of the smectite group) and related minerals. In general, sodium, calcium and potassium bentonites are important where $Na^+$, $Ca^+$, and $K^+$ respectively are the dominant exchangeable ion. Sodium bentonite absorbs large quantities of water, increasing in volume as much as 8 times. It is used in bonding foundry sands, in pelletizing pulverized iron ore, and in oil well drilling muds.

The slope of the embankment depends upon the type of the soil . The slopes define as the distance in horizontal axis for each foot of height. A 3:1 slope means 90 cm of base for 30 cm of height. This ratio is suitable for loamy, silty, or sandy soils. However, the recommended site slopes of embankments for different soil types are given in Table 20.1

**Table 20.1 : Side Slopes of Embankment for Different Types of Soils**

| Soil type | Ratio of vertical (V) and horizontal (H) slopes |
|---|---|
| Clay and silty clay | IV : 1.5 H |
| Sandy clay | IV : 2 H |
| Sandy loam | IV : 2.5 H |
| Losse and wet earth | IV : 3 H |

*Source* : Saha (2000)

For estimation of the volume of earth required as fill, the cross section of embankment should be worked out. All embankments should, however, have extra height – called as *freeboard* above water level to prevent water washing out and from overflowing. The height of freeboard varied in different regions. In heavy rainfall areas, for example, a 60 cm freeboard is required for nursery and rearing ponds. For every 2 cm of average rainfall, a freeboard of 1 metre is constructed. During construction, the embankement should be 15 per cent high than required. All embankments are constructed by soil in layers not exceeding 20 cm in thickness.

## *Arrangement of Inlet and Outlet*

For effective pond management, fish culture ponds should be provided with inlet and outlet arrangements (Figure 20.3 and 20.4). These arrangements should be made in such a way that the entry of undesirable aquatic animals into the pond is prevented.

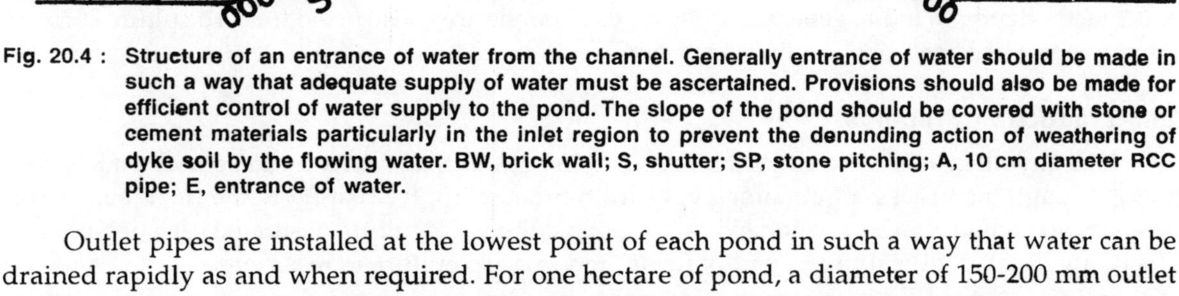


**Fig. 20.4 :** Structure of an entrance of water from the channel. Generally entrance of water should be made in such a way that adequate supply of water must be ascertained. Provisions should also be made for efficient control of water supply to the pond. The slope of the pond should be covered with stone or cement materials particularly in the inlet region to prevent the denunding action of weathering of dyke soil by the flowing water. BW, brick wall; S, shutter; SP, stone pitching; A, 10 cm diameter RCC pipe; E, entrance of water.

Outlet pipes are installed at the lowest point of each pond in such a way that water can be drained rapidly as and when required. For one hectare of pond, a diameter of 150-200 mm outlet pipe is required. Cast iron, PVC pipe, fibre glass, corrugated metal, concrete tiles, and asbestos cement are generally considered as materials for outlet pipes. At present, PVC pipe is widely used. In small ponds, however, galvanized iron is preferred. Inlet pipes should be at least 150 mm above the water level of the pond. Both inlet and outlet pipes should be provided with a bag type screen to prevent the entry of unwanted fish into the pond as well as the escape of stocked fish from the pond. By installing outlet and inlet pipes, continuous exchange of water is made simultaneously. Outlet pipe helps remove excess faecal matter from the pond bottom. Although inlet and outlet systems are very important in fish culture management, the cost of construction is very high.

Other management schedules include : (1) removal of trees, bushes and other foliage, (2) removal of woody material from pond bottom, (3) turfing of embankment to prevent erosion, (4) fencing around the fish farm particularly with multiflora rose – the bush of which will provide a guard fench, and (5) initial filling the pond with great care to avoid slumping of sliding of embankments. However, before any construction work is started, the economics should be worked out.

## 20.6 Pond Structure

For construction of a fish culture pond, agricultural structure of the region must be considered. However, it should be pointed out that the size of the pond is an important factor in fish production. In the northern (such as West Bengal, Uttar Pradesh, Bihar, and Orissa) and North-Eastern (such as Assam and Tripura) parts of India, the average size of a pond is about 0.4 hectare which is common, while in Southern parts of India such as Andhra Pradesh, Karnataka, and Maharashtra large enterprises of above 10 hectares are common. The size of ponds at all India level in terms of percentage is shown in Table 20.2.

**Table 20.2 : Comparative Data of Size of Freshwater Fish Culture Ponds**

| Range of pond area (hectare) | Contribution at all India level (per cent) |
|---|---|
| Upto 0.4 | 30.5 |
| 0.4 - 1.0 | 24.5 |
| 1.0 - 2.0 | 21.5 |
| 2.0 - 3.0 | 21.0 |
| 3.0 - 4.0 | - |
| 4.0 - 10.0 | - |
| Above 10 | 2.0 |

The auxilliary pond culture is observed to significant extent in many states of India where it is closely tied to agriculture and dairy. Farmers rear some species of carps for short duration of the year to food-fish size.

While the maximum size of a carp-rearing unit in most of the states of India is around 1-3 hectares, those for seed production unit are small, generally 0.04 hectare and 1 metre depth. In India, generally three types of ponds are constructed for carp culture such as nursery pond, rearing pond, and stocking pond.

## 20.7 Production Methods

Like any other livestock, carps are raised from eggs to the size ready to be used as food by going through the process of obtaining eggs from brood carp, fry rearing to the final table carp. Production methods can be categorized as : (1) pond culture, (2) culture of stunted carp fingerlings, (3) culture in lakes, (4) culture in paddy fields, and (5) carp culture in reservoirs.

## 20.8 Pond Culture

Ponds are prepared particularly for the cultivation of carp. In India, generally three types of pond are constructed for carps culture such as nursery, rearing, and stocking ponds.

## *Nursery Ponds*

Effective carp culture requires special preparation of ponds to obtain hatchlings. Usually small, shallow and seasonal ponds are considered as they facilitate effective control of the conditions of ponds. During summer season, shallow ponds are allowed to dry that helps to remove excess organic matter. In deeper perennial and undrained ponds, special steps are undertaken for preparation of nursery ponds such as control of predatory and weed fishes, fertilization, control of algal blooms and control of predatory insects. The size of nursery pond is shown in Table 20.3.

**Table 20.3 : Size of Nursery Ponds**

| Type | Area |
|---|---|
| Shallow and Seasonal | 90-150 cm depth |
| | 1,520 cm X 1,520 cm |
| | 1,828 cm X 914 cm |
| Seasonal | 1,535-1,828 cm X 914-1,220 cm X 90-122 cm |
| | 1.0-1.5 m depth |
| Perennial | 1.8-3.5 m depth |

1. *Weed* and *Predatory Fishes* : The term weed fish is used to include all species of small-sized and uneconomical fish that naturally occur or accidentally introduced in ponds. Predatory fishes are harmful to carps as they directly prey on carps and compete with them. Some common weed and predatory fishes occurred in nursery ponds are shown in Table 20.4.

   Repeated drag netting is the common method of removing these fishes from nursery ponds. However, some bottom-dwelling fishes (such as murrels, *Anabus* sp., *Heteropneustes* sp., and *Clarias* sp.) are difficult to remove. In such cases, dewatering of nursery ponds is the most effective means of removing these fishes. If dewatering is not feasible and not economical, the pond should be poisoned. To serve the purpose, certain fish toxicants are used in nursery ponds.

(a) *Fish toxicants* : A fish toxicant must be the following characteristic features : (i) it is effective in killing the target organisms at low application rates, (ii) it does not render the affected fish unsuitable for consumption, (iii) it should be economical and available in the local market, and (iv) it will be neutralized in water and leaves no adverse effects on ponds.

Several types of chemical substances are used in nurseries for controlling undesirable fishes. These substances can be divided as (i) Organophosphates, (ii) Chlorinated hydrocarbons, and (iii) Plant derivatives. Although organophosphorus compounds are less toxic to fish, they adversely affect other aquatic life. On the other hand, chlorinated hydrocarbons are the most toxic to fishes.

(i) *Organophosphates* : A number of organophosphates (such as Thiometon, Nuvan 100 EC (DDVP), and Phosphamidon) have been found successful for killing fish under laboratory conditions. Laboratory studies with DDVP, for example, reveal that most of the insects and unwanted fishes are killed at a concentration ranging from 0.0032 to 0.5 mg/l and 3 to 30 mg/l, respectively. Different species of zooplankton tolerate 0.2 mg/l DDVP for 168 hours.

**Table 20.4 : Weed and Predatory Fishes in Nursery Ponds**

| Weed fish | Predatory fish |
|---|---|
| *Ambassis ranga* | *Anabas testudineus* |
| *A. nama* | *Anguilla bengalensis* |
| *Amblypharyngodon mola* | *Amphipnous cuchia* |
| *Barilius barila* | *Ailia coila* |
| *B. vagra* | *Channa striatus* |
| *B. bola* | *C. gachua* |
| *Chela laubuca* | *C. marulius* |
| *C. gora* | *C. punctatus* |
| *Esomus dendricus* | *Clarias batrachus* |
| *Oxygaster bacaila* | *Clupisoma garua* |
| *Puntius conchonius* | *Eutropichthys vacha* |
| *P. sophore* | *Glyptothorax sp.* |
| *P. ticto* | *Glossogobius giuris* |
| *Gadusia chpra* | *Mystus seenghala* |
| *Setipinna phasa* | *M. aor* |
| *Colisa fasciata* | *M. cavasius* |
| *Botia dayi* | *M. vittatus* |
| *Changunius chagunio* | *Mastocembelus armatus* |
| | *M. pancalus* |
| | *Ompak pabda* |
| | *O. pabo* |
| | *Nandus nandus* |
| | *Notopterus notopterus* |
| | *N. chitala* |
| | *Heteropneustes fossilis* |
| | *Wallago attu* |

(ii) *Chlorinated hydrocarbons* : Several chlorinated hydrocarbons such as dieldrin, aldrin, and endrin are sometimes used for controlling unwanted fishes from nurseries. These chemicals are effectively killed all types of unwanted fishes, prawns, and insects (Table 20.5). The calculated quantity of toxicant is mixed with water and spread uniformly over the water surface of ponds. However, toxicants in nursery ponds should be applied six weeks before stocking of ponds with hatchlings.

It should be added as a note that the toxic effects of chlorinated hydrocarbons last for several weeks. Suitable agents such as potassium permanganate, sodium hydroxide, calcium oxide, sulphuric acid, and cowdung are sometimes used for detoxification of treated waters. The application of raw cowdung at the rate of 18,000 kg/hectare/month, for example, is found to be very effective for detoxifying the effects of endrin.

(iii) *Plant derivatives* : A large number of indigenous plants have been successfully used to control unwanted fishes. Some of the common plant toxicants used in nurseries include : *Dioscorea* sp. (banalu), *Juglans regia* (akhrot), *Croton tighum* (seed kernel), *Strychmos thopsus* (kuchla), *Berberis arisata* (dar-hald), *Artemisia vulgaris* (nogdona), *Albizzia procera* (siris), Tea seed cake, *Tamarindus indica* (tamarind seed), *Millettia pachycarpa, Walsura piscidia, Phyllanthus wrinaris* (hazarmani), *Derris trifoliata* (derris powder), and *Bassia latifolia* (mahua oil cake).

Derris powder of roots and twigs of the indigenous plant *Derris trifoliata* is found to contain 2.2 per cent rotenone. The effective dose of this powder at water temperature above 25°C in shallow water (1 metre depth) varies from 11 to 39 mg/l to kill unwanted fishes from nurseries. Application of derris powder should be made one month before the release of hatchlings in nurseries. Among different types of plant toxicants, tea-seed cake and mahua oil cake are widely used in nursery ponds. However, for complete removal of unwanted fishes, mahua oil cake at the rate of 250 mg/l at water temperature above 29°C is recommended. Tea-seed and mahua oil cakes serve not only as a toxicant but also as a fertilizer. While derris powder contains 2.2 per cent rotenone, tea-seed cake and mahua oil cake contain 10 and 6 per cent saponine (mowrin), respectively. (Table 20.6) These substances are highly soluble in water, enter into the blood streams of fish and exhibits strong haemolytic properties.

**Table 20.5 : Some Commonly Used Chemicals in Nursery Ponds for Controlling Unwanted Fishes**

| Chemical | Recommended rate (mg/l) | Target organism |
|---|---|---|
| Aldrin | 0.001 | Prawns and *Anisops* spp. |
| | 0.2 | Unwanted fishes |
| Dieldrin | 0.01 | Unwanted fishes |
| | 0.5 | Pawns and Insects |
| Endrin | 0.1 | All types of unwanted fishes |
| Rogor | 4.9 | *Heteropneustes fossilis* |
| Endosulphan | 1.3 | *H. fossilis* |
| | 4.0 | *Channa striatus* |
| Quin alphos + Phenthoate (1:1) | 5.51 | *Oreochromis mossambicus* |
| Metasystox | 29.5 | *H. fossilis* |
| Glyphosate | 14.6 | *H. fossilis* |

**Table 20.6 : Application Rate of Different Toxicants Derived from Plants and Their Content of Active Ingredients**

| Toxicant | Application rate (mg/l) | Content of active ingredient (per cent) | |
|---|---|---|---|
| | | Saponine | Rotenone |
| Mahua cake | 200-250 | 4-6 | - |
| Tea-seed cake | 100-150 | 10 | - |
| Tamarind seed cake | 5-10 | 4 | - |
| *Millettia* (root power) | 2-6 | 3 | - |
| Seed kernel | 3-5 | 3.5 | - |
| *Barringtonia* sp. (seed powder) | 20 | 2-4 | - |
| *Walsura piscidia* (bark powder) | 10-20 | 4 | - |
| Derris powder | 4-20 | - | 2-5 |

2. *Fertilization* : Fertilization of ponds is the most important means of intensifying production. The main aim of fertilization is to improve the natural productivity of the pond. After removal of unwanted fishes, nursery ponds are first treated fish agricultural lime (rate depends upon the pH of soil). Different types of fertilizers, manures and their role in fish culture have been described in chapters 2, 17 and 18. However, the quantity of fertilizers and manures are determined empirically depending upon the fish toxicant used in nurseries. If mahua oil cake is used as toxicant, the pond is fertilized with cowdung at the rate of 5,000 kg/hectare. But in case of other toxicants having no manurial effect, cowdung or pig manure (if available) at the rate of 10,000 or 5,000 kg/hectare respectively, is applied. Since major carps prefer a large variety of zooplankton, initial rate of 10,000

kg/hectare of cowdung 15 days before stocking with hatchlings followed by a second application at the rate of 5,000 kg/hectare after stocking is essential for maintaining sustained production of zooplankton. In India, cowdung or poultry litter is preferred in nursery ponds over inorganic fertilizers as the former triggers rapid production of zooplankton population. Application of cowdung is made when one crop of fry is harvested from nursery ponds. If more crops are to be harvested, nursery ponds may be treated with cowdung at the rate of 5,000 kg/hectare immediately after the first crop of fry is harvested.

For high survival of hatchlings in nursery ponds, adequate supply of zooplankton of desired species must be ascertained. For this reason, mass culture of different species of zooplankton is very important and has been delth with in Chapter 8 (See sections 8.3, 8.4, and 8.5).

Manuring of nursery ponds may sometimes result in the development of phytoplankton blooms and macrophytes. Algal blooms and macrophytes must be controlled. Weed control has been discussed in Chapter 4 (See section 4.9).

3. *Aquatic Insects and Their Control* : Nursery ponds are heavily infested with a large number of aquatic insects almost all the year round particularly during rainy season. They severely damage the carp nurseries either in their larval and/or adult stages. In general, eleven orders of the class Insecta inhabit the aquatic ecosystem. Of them, members of the orders Coleoptera, Odonata, and Hemiptera are very important. Some common predatory insects which inhabit in nursery ponds are shown in Table 20.7.

   Most of the predatory aquatic insects multiply rapidly in nursery ponds and fly from one pond to other. Consequently, a nursery pond which has been cleared of its insect population, is again reinfested with insects. For the survival of carp hatchlings, repeated cleaning of ponds with a fine-meshed net is carried out. This method, of course, is not quite effective because it does not completely remove the predatory insects. Therefore, their control by insecticides is necessary.

   The most cheapest and common method for controlling predatory insects consists of spraying an emulsion of mustard oil and cheap soap or mustard oil and Teepol B-300 in the ratio of 56:18 kg/hectare or 56 kg/hectare : 500 ml/hectare, respectively. The use of diesel oil at 501/hectare mixed with one-third to one-fourth of its weight of cheap soap is also recommended.

4. *Fish Food Organisms and Their Production* : Carp hatchlings require adequate supply of natural food and only feed upon plankton of restricted sizes. Nursery ponds that contain appreciable concentration of zooplankton (particularly Cladocerans and Rotifers), should be selected. Concentration of plankton in a nursery pond can be easily determined by a simple field method. In this method, 55 litres of water taken from different sections of a pond are filtered through a muslin ring net having a 2.5 cm diameter glass tube fitted to the lower narrow end of the net. The glass tube is detached from the net and a pinch of sodium chloride (common salt) is added to the tube to kill plankton. Dead plankton organisms will settle to the bottom of tube. If the column of the plankton sediment is 6-9 cm high from the bottom of tube, the pond is considered as sufficiently rich in plankton

and accordingly hatchlings may be stocked in nursery ponds. For qualitative study, plankton in the tube is preserved by adding formaldehyde and the total plankton volume is seggregated species-wise and identified. A minimum of 1-2 ml of zooplankton in 55 litres of water is necessary for stocking of ponds with hatchlings.

**Table 20.7 : Some Aquatic Insects Commonly Found in Nursery Ponds**

| Order : Coleoptera | Order : Hemiptera | Order : Odonata |
|---|---|---|
| **Family : Dytiscidae** | **Family : Notonectidae** | **Family : Gomphidae** |
| *Cybister confusus* | *Anisops bouvieri* | **Family : Aeschnidae** |
| *Eretes sticticus* | *A. barbata* | *Anax* sp. |
| *Hydaticus vittatus* | *A. breddini* | **Family : Libellulidae** |
| *Canthydrus laetabilis* | *A. sardea* | *Bradinophygia* sp. |
| *Agabus* sp. | Family : Pleidae | *Libellula* sp. |
| *Bidessus* sp. | *Plea* spp. | *Macaromia* sp. |
| *Laccophilus parvulus* | **Family : Nepidae** | **Family : Agrionidae** |
| *Nepostermus* sp. | *Ranatra filiformis* | |
| **Family : Gyrinidae** | *R. digitata* | |
| *Dineutus indicus* | *R. elongata* | |
| *Gyrimus* sp. | *Laccotrephes* sp. | |
| **Family : Hydrophilidae** | **Family : Belostomidae** | |
| *Hydrous indicus* | *Belostoma indicum* | |
| *Hydrous* sp. | *Diplonychus rusticum* | |
| *Berosus* sp. | **Family : Corixidae** | |
| *Enochrus* sp. | *Corixa distorta* | |
| *Laccobius* sp. | *Micronecta haliploides* | |
| | *M. thyesta* | |
| | *M. albifoons* | |

5. *Stocking Rate of Hatchlings* : The carp hatchlings for stocking must be the offsprings of carp that are superior in quality. A superior strain implies that the growth is fast, physically healthy, and the survival rate will be high. However, the stocking of hatchlings in nurseries vary between 10 and 20 lakhs/hectare. Sometimes much higher stocking densities (80 lakhs/hectare) are recommended. In a well-manured nursery pond, 10-20 lakhs hatchlings/hectare generally recommended. It has been estimated that from one hectare of water area, eight million fry can be produced within 15 days. Two to three crops can be produced from the same nursery pond. The stocking of a nursey pond is best done late in the evening and in different areas of the pond.
6. *Feeds* : Carp hatchlings are voracious eater and they consume zooplankton and thus depend on animal diet. It has been observed that a carp hatchling (6-7 mm in length) consumes 3 to 30 water fleas per hour, the exact number depends upon the size of hatchlings. For the first four days, the hatchlings entirely depend upon natural food. After this period, application of artificial feed is necessary. Certain general qualifications of a good carp feed are evident. It should be such that (1) hatchlings like to eat it, (2) it is inexpensive

and plentiful, (3) it is easy to accept and digest, (4) it has no detrimental effect on the hatchlings, and (5) it has high conversion value.

Concerning artificial feeding, it is important to pay great attention to the effect on the health and growth of hatchlings. At the same time, consideration should be given to cost, favouring less expensive feeds. The commonly used feed for carp hatchlings are rice bran and oil cake (either mustard oil cake or groundnut cake). They are mixed together in the ratio of 1:1 and daily application of artificial feed should be related to the weight of the hatchlings stocked. The feeding schedule in terms of weight (g) for 15 days of rearing of hatchlings is shown in Table 20.8.

**Table 20.8 : Feeding Schedule of Hatchlings in a Nursery Pond**

| Period after stocking (day) | Feed per day in terms of weight (g) |
|---|---|
| 1-5 | Equal or double |
| 6-10 | Two or three times |
| 11-15 | Three or four times |

To raise carp fry on one single type of feed often tends to create nutritional imbalance, hence it is more common to have mixed feeds. Vitamins, proteins and minerals are essential nutritive elements for the survival and growth of carp hatchlings. An artificial feed consisting a mixture of dried, powdered and sieved aquatic insects, small shrimps and prawns, and cheap pulses (cowpea) in the ratio of 5:3:3 gives better results in increasing the survival and growth of hatchlings than the rice bran and mustard cake mixture (Table 20.9). A very high survival of over 80 per cent can be obtained by using a diet consisting a mixture of groundnut cake, rice bran and fish meal in the ratio of 1:1:1. However, to ensure high survival and fast growth, artificial feed must be fortified with micronutrients, yeast and other ingredients such as powdered algae and aquatic weeds. Yeast is a rich source of amino acids, vitamins and hence act as a growth promoting substance.

**Table 20.9 : Survival and Growth of Carp Hatchlings After 15 Days Rearing in Nursery Ponds**

| Species | Rice bran + Mustard cake (1:1) | | Mixture of feed | |
|---|---|---|---|---|
| | Survival (per cent) | Growth (per cent) | Survival (per cent) | Growth (per cent) |
| *Labeo rohita* | 47.67 | 0.0715 | 53.8 | 0.1655 |
| *Catla catla* | 51.50 | 0.0635 | 74.37 | 0.1615 |
| *Hypophthalmichthys molitrix* | 44.01 | 0.1582 | 45.52 | 0.2825 |

*Source* : Lakshmanan *et al* (1967)

Protein is very important in the growth of carp hatchlings. When protein is short in the feed, hatchlings will suffer and will lose weight. The body protein is provided by the diet. The maintenance protein of carp hatchling is 0.3 mg per gram of carp per day when the temperture of the water is 22-24°C.

It is also necessary to know the weight increase of carp as related to the amount of feed (also called food conversion rate). This depends upon the density of hatchlings, various environmental factors, fertility of water, age of the fish season, and feeding methods. The food conversion rate or growth coefficient is defined as the amount of feed necessary to bring about a weight increase of one unit and is computed as follows :

$$\text{Growth coefficient or Food conversion rate} = \frac{\text{Quantity of feed}}{\text{Weight increase (flesh)}}$$

The food conversion rate of some important carp feeds is given in Table 20.10. Since the assessment of weight increase is complex and it depends upon several conditions mentioned above, no reliable data have been obtained on the rate of growth coefficient of feeds into fish flesh.

**Table 20.10 : Conversion Rate of Some Carp Feed**

| Feed | Conversion ratio |
|---|---|
| Meat powder | 2.0 |
| Fish Meal | 1.5-3.0 |
| Soyabean cake | 2.22 |
| Oats | 2.60 |
| Dried silkworm pupae | 1.3-2.1 |
| *Mysis* sp. | 2.0 |
| Chironomids | 2.3-4.4 |
| Wheat flour | 7.2 |
| Rice bran | 5.08 |
| Wheat bran | 4.22 |
| Cotton seed cake | 3.0 |
| Maize | 4.0-6.0 |
| Peanut cake | 2.13-2.70 |

7. *Hydrobiological Conditions* : Productivity of a nursery pond is determined by some abiotic factors of water which is governed by the management strategies. Hatchlings are frequently affected by environmental fluctuations. However, Table 20.11 presents guidelines for water quality variables which are conducive to high survival of hatchlings in nurseries.

8. *Harvesting and Transportation* : The fry arrive at nursery ponds within 15 days after they are caught. Most of the little fish are sold within 15 or 30 days being caught. Whenever it is desirable to retain the fry in the nursery ponds for later sale or delivery, the fish growth can be almost suppressed. Thus the size of fish is kept small, to facilitate transportation, by means of crowding and supplying less feed.

**Table 20.11 : Hydrological Conditions in a Nursery Pond Required for Optimum Survival and Growth of Carp Hatchlings**

| Water quality variable | Range |
|---|---|
| pH | 7.5-8.8 |
| Total alkalinity (mg/l) | 70-110 |
| Dissolved oxygen (mg/l) | 4.5-10.0 |
| Total ammonia nitrogen (mg/l) | 0.01-0.09 |
| Orthophosphate (mg/l) | 0.1-0.2 |
| Plankton concentration (mg/451 of water) | 1.0-1.5 |
| Water temperature (°c) | 24-29 |

In order to transfer fry from nursery ponds, different methods of transportation are employed. For great distances, truck or train are used. For short distance transportation, boxes are used. For transporting fry to remote places, they are carried by sealed container charged with oxygen. It has been estimated that the hatchlings and fry of Indian major carps can live for 24 hours or more in container water containing 0.5 and 1.0 mg/l oxygen, respectively. For two to three days prior to transporting, no feeding should take place. Sterilized containers and speedy handling safeguard the delicate young.

For transporting fish seedlings, improved open metal containers are used in the Indo-Pacific regions. It is a round vessel with wide mouth closed with a perforated lid. The larger type is about 53 cm in diameter and 38 cm in height; the mouth is about 20 cm in diameter and permits the introduction of a dip-net for moving moribund fish. The lid permits change of water, but does not allow the removal of faecal matters. The vessel is insulated by framing it with wood, which is kept wet during the journey. If clean water is not available for replacement, a semi-rotary pump is used. In this vessel, the fry can be transported by road vans upto a distance of about 480 kilometres with 5 per cent mortality only.

## Rearing Ponds

This is a pond to nurse the fry that are 15 days old. Fry are collected with the aid of a drag net of desired mesh size and released in rearing ponds. The methods of preparation of rearing ponds are more or less similar to that of nursery ponds.

The main objective of fish culture in rearing ponds is to culture of carp fry to fingerlings within short time (usually three months). The management of rearing ponds involves the removal of weeds, and predatory fishes, manuring and fertilization, stocking of carp fry in suitable combinations, supplementary feeding and harvesting. Removal of aquatic weeds and unwanted fishes are similar to those adopted for nursery ponds. However, the size of rearing ponds is shown in Table 20.12.

**Table 20.12 : Size of Rearing Pond**

| Type | Area |
|---|---|
| Perennial or Seasonal | 122-183 cm depth, 0.60-1.2 hectare and paddy fields with a depth of 45.5 cm or more |
| Seasonal (water is retained for long time) | 183 cm depth, long and narrow for easy fishing operation and paddy fields with 45-60 cm depth |

As in other animal production, higher returns are obtained through selective management schedules. These include : (1) pond preparation, (2) stocking of fry at the rate varying between 60,000 and 1,25,000 number/hectare, (3) species composition and raito (Catla : Rohu : Mrigal : Common carp = 3:4:1:2 or Silver carp : Grass carp : Common carp=4:3:3), (4) stocking density (at the rate of 6,000-10,000 per hectare), (5) fertilization (cowdung @ 10,000 kg/hectare 10 days before stocking; Ammonium sulfate + Single superphosphate+Calcium ammonium nitrate (11:5:1) at 700 kg/hectare, 2 months after stocking), (6) artificial feeding (mustard cake and rice bran in equal ratio fortified with vitamins, yeast, and minerals), and (7) netting. However, the degree of adaptation to local soil and climatic conditions, as well as disease resistance, is an additional feature essential for raising carp fry to fingerlings. Through these management schedules, average production of carp fingerlings, may reach upto 2,205 kg/hectare/3-months.

For raising carp fry to fingerlings in rearing ponds, artificial feeding is recommended. The individual components of fish food should be evaluated on the basis of their importance in fish metabolism. Artificial food should in its composition be similar, as far as possible, to that of the natural food. In artificial food, protein and fat are the main constraints for successful rearing of carp hatchlings. It is evident from the Table 20.13 that the composition of protein and fat is highly variable between vegetable and animal foodstuffs, viz., the relatively low protein content in artificial food. In the natural feeding of the fish, the relation between protein and fat is around 1:1, but in artificial food it is lower. This difference is neutralized by mixing the artificial food with natural one. This should be considered particularly in the rearing of carp fry.

**Table 20.13 : Some Common Important Feed Used in Nursery and Rearing Ponds and their Protein and Fat Contents. Values are in Terms of Per cent**

| Ingredient | Vegetable | | | | | | |
|---|---|---|---|---|---|---|---|
| | Rice bran | Wheat bran | Corn | Rye | Potato | Oats | |
| Protein | 14.8 | 14.6 | 6.2 | 11.6 | 2.1 | 11.4 | |
| Fat | 18.2 | 4.8 | 5.5 | 1.7 | 0.3 | 9.9 | |
| | Animal | | | | | | |
| | Silk worm pupae (dry) | Dry fish | *Mysis* | *Tubifex* | *Chironomus* | *Viviparus* | *Daphnia* |
| | 57.5 | 65 | 16.3 | 8.1 | 8.2 | 14.0 | 3.5 |
| | 29.7 | 9 | 3.3 | 2.0 | 2.0 | 0.5 | 0.6 |

The amount of feed to be added is adjusted to the appetite of the carp. This is influenced by the amount of oxygen dissolved in the water, the pH, water temperature and other abiotic factors. Therefore, some caution is necessary, as these factors should favour a good appetite. It is necessary to take adequate measure that has to be taken to the risk of an increase in the level of excretory substances. The degree of water renewal and oxygen balance have to be followed closely. In rearing ponds, environmental stress and disease problems, as well as parasitic infections can be reduced by maintaining high water quality. Some abiotic factors of water which are very important for survival and growth of fry are shown in Table 20.14.

**Table 20.14 : Hydrological Conditions in a Rearing Pond Required for Optimum Survival and Growth of Carp Fry**

| Water Quality Variable | Range |
|---|---|
| pH | 7.5-8.5 |
| Total alkalinity (mg/l) | 40-80 |
| Dissolved oxygen (mg/l) | 4.0-10.0 |
| Total ammonia nitrogen (mg/l) | 0.02-0.08 |
| Nitrate nitrogen (mg/l) | 0.05-0.10 |
| Orthophosphate (mg/l) | 0.1-0.3 |
| Plankton concentration (ml/451 of water) | 1.0-1.5 |
| Hardness (mg/l) | More than 15 |
| Water temperature (°C) | 26-32 |

In rearing ponds, frequent harvesting is practiced by many, rendering a continuous income and providing more space for the remaining fish. The netting of fish is done at dawn when the water temerature and dissolved oxygen concentration in water decrease. This situation makes it easier to handle the catch than during the heat of the day.

### *Stocking Ponds*

It is a special form of fish culture pond constructed in freshwater areas. Stocking ponds are large perennial water areas, varying between 0.2 and 2.0 hectare, preferably long and narrow in shape for easy and inexpensive netting operations. The depth of water should be 1.5-2.0 m with a deeper spot of an additional 0.5 m in the central part of the pond. The main aim of the management of stocking ponds is to attain high production of fish for the table. The management of stocking pond involves several steps :

1. *Pond Preparation* : First, aquatic weeds and unwanted fishes should be removed and the method is similar to that of nursery ponds. In case of larger and deeper ponds, it is not possible to remove unwanted fishes and other aquatic animals completely unless ponds

are treated with poisons. Old ponds having excessive silt should be desilted, if possible, during summer season.

Most of the rearing and stocking ponds in Tropical countries are heavily infested with some marginal, emergent, and floating plants such as *Ipomoea* sp., *Nymphea* sp., *Pistia* sp., *Eichornia* sp., and *Marsilea* sp., (Figure 20.5). Growth of these aquatic plants is one of the most important problems faced by fish culturists. Aquatic plants not only curtail the productivity of pond water but also poses serious menace to fish culture in several ways. Prior to initiation of fish culture, however, these plants must be removed either by mechanically or by the application of suitable chemicals (For further detail, see Chapter 4).

2. *Fertilization* : Many fish culturists recognize the favourable effect of the fertilizers and manures. In India, it was known for more than forty years ago that the development of natural organisms as carp food was enhanced by using organic and inorganic fertilizers to the culture ponds. Different types of fertilizers such as urea, ammonium sulfate, calcium ammonium nitrate, single superphosphate and rockphosphate are widely used in fish culture. At the same time, cattle manure, poultry litter, and pig manure are also used to stimulate the production of plankton populations.

   Contrary to conditions in agriculture, nitrogen fertilizers added to ponds are of little effect, as they are consumed by bacteria. Phosphatic fertilizers are more effective particularly in calcium-rich silted ponds, but less effective in sandy soil. For further discussion on fertilizers and manures for fish culture, see section 11.2 and chapters 17 and 18.

Naturally, climatic and geographical conditions are essential for the maximum effect of fertilizers and manures in stocking ponds. Under Indian conditions, however, cow-manure (at the rate varying between 20,000 and 25,000 kg/hectare/year) or poultry litter (at the rate varying between 2,000 and 8,000 kg/hectare/year) and chemical fertilizers (mixture of ammonium sulfate, single superphosphate and calcium ammonium nitrate in 11:5:1 ratio at the rate varying between 1,000 and 1,500 kg/hectare/year or mixture of urea and single superphosphate at the 175 kg and 250 kg/hectare rates, respectively) are used in fish culture along with the application of lime (calcium oxide). These quantities of nutrients are belived to be the most effective for high fish production. Common practice is to apply the above-mentioned nutrient carriers annually, divided into several lots.

3. *Species Combination and Stocking Rate* : The success of fish culture in stocking ponds not only depends upon the species combination but also their stocking rate. In general, 4-6 species of fish fingerlings of major and exotic carps are considered. Among them, three are indigenous and the other 2 or 3 are exotic carps. The stocking density of carp fingerlings has, however, been recommended as 5,000 and 10,000 number per hectare to yield 3,000 and 8,000 kg of fish, respectively. This point will further be discussed in the next chapter.

The stocking rate of carp fingerlings in a stocking pond can be computed by the following formula :

$$\text{Number of fishes to be stocked/hectare} = \frac{\text{Total expected increase in weight}}{\text{Expected increase in weight of individual fish}} + \text{15 per cent mortality}$$

Fig. 20.5 : Growth of different types of aquatic plants has serious implications to fish culture ecosystems. They severely disturb the fish culture management strategies in various ways. For high productivity of fish culture ponds, control of aquatic plants is urgently needed. Some common aquatic plants grown in most tropical countries are shown in this figure. A, *Ipomoea* sp.; B, *Nymphea* sp.; C, *Eichornia* sp.; D, *Marsilea* sp.; E, *Pistia* sp.

Suppose one hectare pond can yield 2,000 kg of fish if the pond is stocked with rohu, catla, and mrigal and under normal condition their growth is 900, 780, and 550g respectively. Then the number of fingerlings to be stocked is given by

$$\text{For rohu} = \frac{2{,}000 \times 1{,}000}{900} = 2{,}222$$

$$\text{For Catla} = \frac{2{,}000 \times 1{,}000}{780} = 2{,}564$$

$$\text{For mrigal} = \frac{2{,}000 \times 1{,}000}{550} = 3{,}636$$

Therefore, the ratio of rohu, catla, and mrigal will be 2,222 : 2,564 : 3,636. If mortality is taken into account, the resultant number to be increased by 15 per cent will be 2,555, 2,948, and 4,181 for rohu, catla, and mrigal respectively.

4. *Supplementary Feeding* : In ponds where plankton population is rich, artificial feeds are applied. This combination provides fish to grow rapidly, reaching a marketable size in 8 to 10 months. Discussions on this aspect have, however, been made in nursery and rearing pond management schedules and in chapter 9.
5. *Carrying Capacity* : The term carrying capacity of stock in ponds refers to whether the way by which they are managed, can support a fish biomass only upto a certain weight limit. This is also called as the *maximum standing crop.* The carrying capacity of stocking depends upon the natural productivity and different management schedules such as fertilization, feeding, stocking, type of species, food conversion rate and pond depth. For examples, (a) the fast growing species which attain maximum size producing more than slow growing ones, since the latter types exhibit a check in their growth rate under natural conditions of the stocking pond. This natural check is ceased completely before the carrying capacity is reached; of course, the production to some extent, can be increased by repeated harvesting and stocking. In general, the carrying capacity is considerably lower in ponds stocked with either carnivores or zooplankton feeders than herbivorous fish since the latter subsist on the base of the food chain, (b) A shallow pond will support less fish than a deeper one because of the more living area in the latter case, and (c) In areas where the growth period is short (as in case of temperate water), the carrying capacity is not attained in the time available.

Tropical freshwater ponds are biologically more productive than temperate zone counterparts. This is due to the fact that the temperature of water in tropical regions is high and fluctuates between 20 and 40°C. At the same time, prolonged growing seasons (10-12 months) exhibit high fish production. Even in different zones of the same country, the productivity of water varies considerably due to variations in soil fertility, soil types, mineral constituents and abiotic factors of pond ecosystem. Artificial feeding and fertilization are the most important means of increasing carrying capacity of stocking ponds. The carrying capacity in control, fertilized, artificial fed and fertilized + artificial fed ponds have, however, been established as 160, 350, 600, and 2,680 kg/hectare, respectively.

For good production and economic returns, carp fingerlings are stocked in stocking ponds at a rate well below the carrying capacity level and fish is allowed to grow up to or near the carrying capacity in the shortest time. To produce large-sized fish, stocking the ponds with fingerlings is made at very low rate, although this may cause a decrease in fish production per unit area. A sustained fish production is possible by maintaining a gap between the standing crop and the carrying capacity at a point of time by periodical harvesting. A portion of the fish crop is removed from the pond which decrease the biomass below the carrying capacity level and triggers the growth of the remaining fish crop till the carrying capacity is approached.

6. *Harvesting* : Although the harvesting size of fish varies considerably, they are, in general, immature and small. The major and exotic carps reach the market weighing between 0.5 and 1.5 kg. Fishes are frequenlty harvested when they weigh slightly more than 1 kilogram.

## 20.9 Culture of Stunted Carp Fingerlings

In India, a large number of seasonal ponds (also called as *derelicts*) exist that comprise about 38 per cent of the available 1.6 million hectare of lentic water bodies. These water bodies can be utilized for fish production within short span of time (for 4-5 months) by stocking with stunted carp fingerlings.

The technology of production of stunted carp fingerlings in stocking ponds has been developed by fish farmers of Andhra Pradesh. Farmers produce fish in ponds by using stunted carp fingerlings. First, they stock early fingerlings of carps at high densities varying between 1 and 1.5 lakh number/hectare and rear them fo 4-5 months. Due to high stocking density with feeding at very low rates, unhealthy and weak fingerlings die and the existing ones exhibit stunted growth. Second, these stunted fingerlings, are again stocked in grow-out ponds at the normal stocking rate with regular feeding at optimum levels that results in high fish production and at the same time, is more economical than the stocking of grow-out pond with normal fingerlings. The estimated production of stunted carp fingerlings during five months of culture period was observed to be about 65.5 per cent higher (2,679 kg/hectare) than that of normal fingerlings (1,755 kg/hectare) having survival rate of 90 per cent in both the cases.

## 20.10 Culture in Lakes

In India, natural lakes are estimated to have an area of about 0.72 million hectare. Natural lakes are formed by depressions on the surface of the earth and gradually filled up with water. However, a lake can be defined as a large and shallow body of standing water which is enough to stratify thermally, completely isolated from the sea and infested with aquatic plants.

Among different types of lakes in India, three upland lakes of Tamil Nadu (such as Yercand lake in Shevaroy Hills, Kodaikanal lake in Palni, and Ooty lake in Nilgiri Hills) and Logtak lake in Manipur are widely known. The average depth of these lakes is 2 metres. Among all these lakes, Logtak lake has a water area of 4,480-27,300 hectare. The annual fish production in Kodaikanal, Yercand, Ooty, and Logtak lakes was 65.5 kg, 132 kg, 175 kg/hectare, and 482 tonnes, respectively (Published data for the year 1994). However, dominat species of fish in these lakes include *Anabus testudineus, Clarias batrachus, Ophiocephalus* spp., *Puntius* spp., and *Cyprinus carpio.* Besides these, *Labeo rohita,* and *Oreochromis mossambicus* are also stocked.

Fish culture in lakes generally involves the stocking of fingerlings. They are stocked every year and recover them after a lapse of some months or years. In these periods, measures are taken to prohibit fishing. Therefore, the fish can grow naturally and propagate.

## 20.11 Culture in Paddy Fields

The method of carp production in paddy field was started in India about 1,600 years ago. In the month of February 1994, a total number of 4,670 common carp fry were released in one hectare of paddy fields located in the Konkan region of Maharashtra. The final crop amounted to 145 kilogram without providing any additional feed and the paddy production went up by 38 per cent in the paddy plot with fish culture. The duration of experiment was 80 days.

Since there are vast areas of paddy fields in India, it is extremely beneficial to utilize these water for fish culture. Experiments have confirmed that fish culture in the rice fields never results in a reduction of the rice crop.

When the rearing is accomplished by natural food organisms available in the paddy fields, the number of carp fry released to the fields generally should be lower. If the size of the fry is about 3 cm (range 2.5-4.0 cm), the suitable number to release is about 3,000 per hectare. In many cases where supplementary feed is regularly applied to culture fields, 3,000-4,000 numbers carp fry can be released in each hectare. Thus, fry will grow in size to 8-20 cm, the yield would be 40-80 kg, when no supplementary feed is given, and 105-130 kg when supplemented.

A variety of natural food organisms especially zooplankton that grow in pady fields are very good for the growth of carp fry, manuring to promote the increase of natural food is also practice. In India, a total of 10,000 kg of compost or cow-manure and 100 kg mixture of ammonium sulfate and single superphosphate (1:1 ratio) are applied. It is known to be effective for the development of zooplankton as well as for the growth of rice plants. The rice-fish cultivation has been described in detail in Chapter 11 (see section 11.20).

A number of precautions are considered for rice-fish cultivation : (1) there should be no risk of any flooding, (2) the ditch for the incoming water supply and the drainage from the field should be covered by a mesh so that fish cannot escape from the fields, and (3) the fields should be so located as to allow the supplying of water.

## 20.12 Carp Culture in Reservoirs

A reservoir can be defined as a large impoundment water body artificially constructed by putting an earthen, stone masonry or concrete dam or bundh across the river water. Reservoirs for flood control, irrigation, power generation, recreation, navigation, sport fishing, and fishery developement are found in many places of the world. Reservoirs are utilized for fish culture.

Generally fish production from reservoirs can never be as high as production from ponds because of differences in ecological conditions for the production of fish food organisms. Reservoirs are unique biotopes representing both fluviatile and lentic characters. It is needless to mention that through scientific management, reservoir fisheries offer great opportunity for high fish production. There are 3.1 million hectare of reservoir water area in India and the area is expected to increase by 6.5 million hectare by 2020. Although no exact figures for fish production from

Indian reservoirs are available, a total of about 60,000 metric tonnes of carps have been found to produce from different types of reservoir waters. Well planned development of reservoir fisheries (For further detail, see Sugunan (1997)) may result in increased fish production. Reservoirs in tropical and sub-tropical regions exhibit high primary productivity than the temperate zone counterparts and therefore considered as an excellent source of fish for the table.

### *Types of Reservoir and Production Potential*

Reservoirs are generally classified into three types such as large (more than 5,000 hectares), medium (1,000-5000 hectares), and small (less than 1,000 hectares) from which 15.5, 16.0 and 54 kg of carp/hectare/year respectively, are produced (Published data for 1999). These production values are, of course very low in contrast to other countries like Russia and Sri Lanka where 90 and 100 kg of carp/hectare/year respectively are produced. Lower production of fish from Indian reservoirs is possible due to improper management. There is good production potential from different reservoirs in India if several management strategies are adopted. Management strategies include pollution control of reservoir waters, mesh size regulation, optimum fishing effort, selection of suitable species for periodical stocking, and biological as well as environmental factors. For sustaining production from reservoirs round-the-year, however, periodical stocking of quality fish fingerlings is essential. For this purpose, breeding farms have been established near the reservoirs to ensure the supply of fish seed for stocking.

In India, reservoir management largely depends upon the development of carp fishery, particularly Indian major carps due to their fast growth rate. At the same time, major carps are not well equipped to utilize phytoplankton –the most dominant component in reservoirs. Therefore, some suitable species of fish such as *Hypophthalmicthys molitrix* (silver carp) and *Thynmichthys sandhkhol* (sandhkol carp) which are excellent phytoplanton feeders, should be stocked particularly in North Indian reservoirs such as Rana Pratapsagar of Rajasthan, Rihand of Uttar Pradesh, and Govindasagar of Punjab.

### *Development of Reservoirs*

During the last four decades, a number of large freshwater impoundments have been constructed in India as a result of the completion of different river-valley projects. These impoundments have practical significance for the development of fishery resources in the country. Fish production in Indian reservoirs varies considerably from one type of water to another depending upon the development strategies. However, various method adopted in the development of the Indian reservoirs are :

1. Survey of the fish fauna of the river.
2. Clearance of submerged obstruction to permit easy exploitation.
3. Construction of fish farms.
4. Stocking of the reservoir with fingerlings
5. Construction of fish seed farm.
6. Rehabilitation of fishermen communities near the reservoir.
7. Formation of co-operative societies for proper management.

8. Conservation and management.
9. Transport and marketing.

Development of reservoir fisheries involves heavy cost and therefore, must take this factor into consideration. Some of the methods which are pertinent to the development of reservoirs, are briefly described below:

In a few of the river valley projects such as Pipiri and Rihand, Damodar Valley Corporation, Hirakund, Gandhisagar, Tungabhadra, and Bhabanisagar, fish and fishery survey were conducted. Some recommendations have been made on stocking, conservation and management with varying degrees of success at different reservoirs. However, a detail survey of the fish and fisheries of the reservoir along with the fish farmers/fishermen, fishing craft and gear and other management schedules are very essential for successful development of the reservoir fisheries.

For easy fishing and the use of effective methods of exploitation, the reservoir bottom should be free from obstructions. These obstructions include rocks, boulders and tree trunks. Removal of obstruction is taken up during summer season when the water level decreases and shore areas are exposed.

Aquatic weeds should be removed from the reservoir because most of the reservoirs are heavily infested with a variety of aquatic plants. Aquatic plants hamper fishing and suppress the growth of bottom fauna. Removal of aquatic plants with mechanical winches are most effective.

For effective stocking of reservoirs with fingerlings, fish seed farms should be constructed near the reservoirs. In cases where reservoirs are connected with the river system have a natural stock of carps. In case of large reservoirs, however, natural stocks is supplemented with carp fingerlings at regular intervals. Some fish farms have been constructed in close connection with many reservoirs such as Mettur and Poondi of Tamil Nadu, Krishnasagar of Karnataka, and Govindasagar of Punjab. In cases where loss of water through soil is a serious problem (such as Govindasagar, Krishnasagar, and Mettur reservoirs), fish farms may either be manured with raw cow-manure or concrete materials.

For sustained production on an economically sound basis, heavy stocking of commercially important species of fish is made. Some of the carp fishes stocked and cultured in most of the Indian reservoirs are : *Labeo dero, L. pangusis, L. calbasu, L. gonius, L. bata, L. boga, L. boggut, L. fimbriatus, L. rohita, L. porcellus, L. potail, L. nigrescens, L. kontius, Cyprinus carpio, Cirrhinus mrigala, C. reba, C. cirrhosa, C. fulungee, C. horai,* and *Catla catla.*

The recommended rate of stocking of carps in Indian reservoirs varies between 200 and 450 (without catfish) and 400 and 600 (with catfish) fingerlings/hectare. In some small and highly productive reservoirs (such as Krishnagiri reservoir in Tamil Nadu, Meenkar and Culliyar reservoirs in Kerala, and Bachhra reservoir in Uttar Pradesh, the stocking rate should be at higher levels that varies between 760 and 1,225 fingerlings/hectare. Since there is wide variations in the stocking rate in different types of reservoirs, their carrying capacity levels are extremely variable and therefore, fish production potential cannot be explained in a general way. Stocking should be maintained at levels always higher than carrying capacity of reservoirs. Production potential and present production from different types of reservoirs, however, indicate that the existing fishery resources in the country has the potential to yield 2.4 lakh tonnes of fish with targets of average production (Table 20.15).

**Table 20.15 : Production Potential and Present Yield from Different Types of Indian Reservoir**

| Type | Total area of reservoir (ha) | Production potential | | Present yield | |
|---|---|---|---|---|---|
| | | Total fish production (lakh tonne) | Average produ-ction (kg/ha) | Total production (lakh tonne) | Average production (kg/ha) |
| Large | 11,40,347 | 0.56 | 50 | 0.164 | 15.5 |
| Medium | 5,27,485 | 0.46 | 75 | 0.067 | 16.0 |
| Small | 14,85,613 | 1.57 | 100 | 0.775 | 54.0 |
| Total | 31,53,445 | 2.59 | 225 | 1.006 | 85.5 |

Source : Modified after Sinha (1998)

The population of many predatory fishes such as *Channa* sp., *Wallago attu, Mystus* spp., and *Oreochromis mossambicus* is abundant in most reservoirs. The catch composition of some resrvoirs of Uttar Pradesh and Karnataka has shown that catfishes constitute more than 60 per cent of the total fish population. However, effective control of these predatory fishes is necessary so that commercially important fishes of short food chain can be established in reservoirs.

### *Stocking Density of Carp Seed Versus Fish Production*

Fish production is greatly influenced by adequate feeding and especially by stocking density or size of reservoirs. Figure 20.6 shows that the stocking density is definitely an important criteria for high fish production. In case of small reservoirs, fish production potential is great than in larger ones. This is due to the fact that small reservoirs are easy to govern effective management.

The influence of stocking density on fish yield is illustrated by data shown in the Figure. Although the carp production increases gradually with the increase in stocking density, after a definite point, however, the carp production declined. This situation is, perhaps due to the competition for food and space in the reservoir. These factors obviously retard the growth rate of carps followed by mortality. Hence, it seems plausible that the sustainable stocking is necessary for better production. The significant consideration of stocking densities was limited mostly to small reservoirs. Although agro-climatic conditions are very important, the effects of stocking densities have been extended to fish culture in waters that have adopted mixed fish farming practices.

### *Transition in Reservoir Productivity*

When a reservoir is constructed, the soil which contains terrestrial vegetation, start decaying soon after submergence. Decomposition of submerged vegetation causes initial fertilization of the water and consequently benthic region is teemed with flora and fauna as well as plankton in water. Thus, reservoir water becomes enriched with nutrients which stimulate photosynthetic activity. This rise and fall of biota with a great rolling motion is observed in the early stages of reservoir impoundment and may last for 3 to 4 years.

After the initial high fertility, trophic depression begins. This is caused by the increase in the volume of water, in the rate of nutrient release due to continuous sedimentation and absorption of

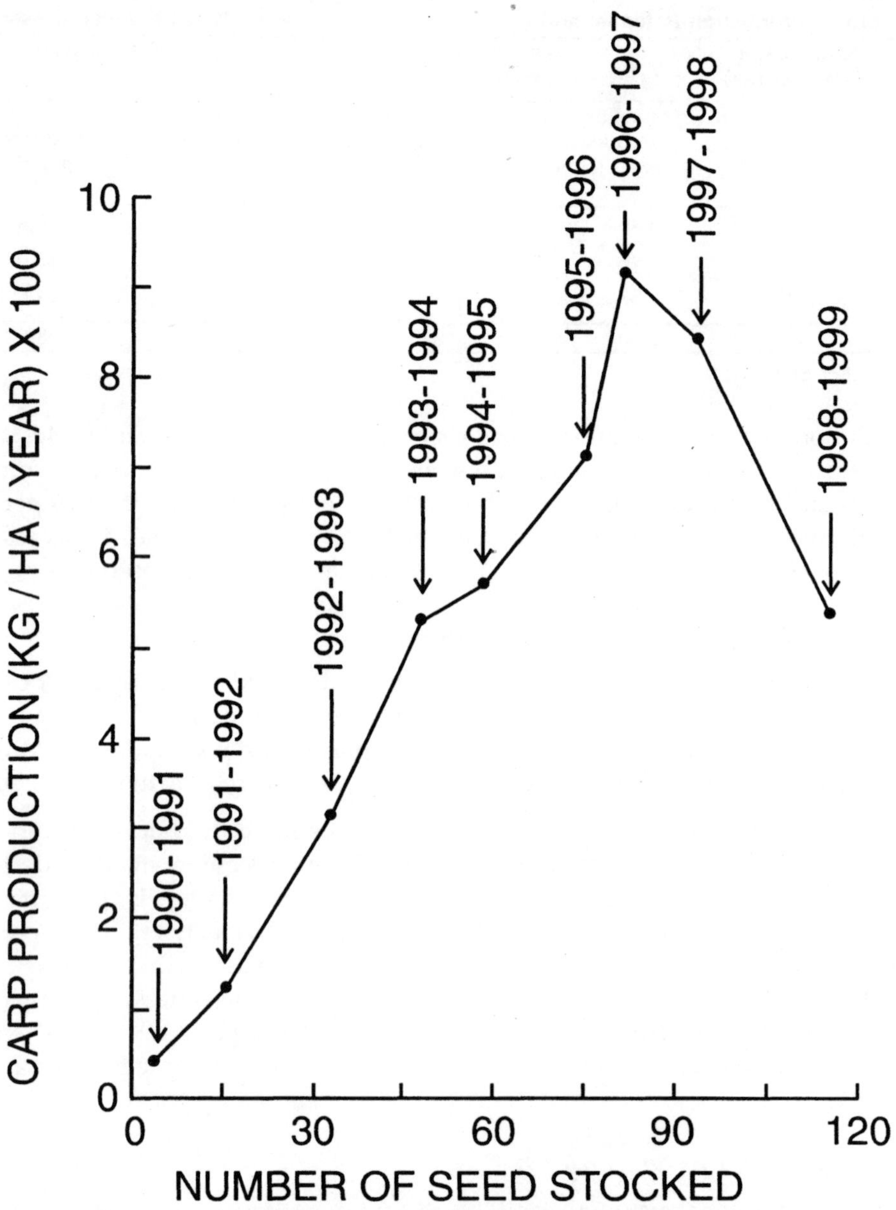


**Fig. 20.6 :** **Relationship between carp production and stocking density in a small reservoir (water area is 25 hectare) located in Shathamraj (near Hyderabad). Fry of *Labeo rohita, Catla catla, Cirrhinus mrigala, Ctenopharyngodon idella,* and *Cyprinus carpio* were stocked during 1990-1991 to 1998-1999 at the rate ranged from 500 to 11,500 numbers/hectare/year. Carp production increased from 13.5 to 98 per cent during the nine consecutive financial year. Among different carps, *C. catla* exhibited high production (33.8 per cent of the total) followed by *L. rohita* (22.5 per cent), *C. carpio* (21 per cent), *C. mrigala* (17.6 per cent), and *C. idella* (5.1 per cent). For optimum carp production, however, 850 numbers of fry/hectare/year should be recommended for small reservoirs. [Drawn from data as proposed by Piska (2000)]**

nutrients by aquatic plants. Moreover, regular withdrawal of water from reservoirs causes decrease in the level of nutrients brought in by the influx of water. The duration of trophic depression varies and may last for about 25 years.

As soon as the trophic depression stage is over, a productivity level is attained due to gradual accumulation of organic matter in the bottom soil leading to eutrophication of the reservoir.

The slope and size of reservoirs greatly influence their productivity and no uniform patterns of productivity exhibit in resrvoir ecosystems. Generally, deeper reservoirs such as Mettur, Bhabanisagar, Aliyer, Amravathy, and Trimoorthy of Tamil Nadu developed algal bloom round the year. On the other hand, the shallow reservoirs such as Poondi, Krishnagiri, and Vidur of Tamil Nadu exhibit poor production of plankton.

### *Trophic Status of Reservoirs*

The trophic status of a reservoir is determined by edaphic factors. For determining the trophic status of reservoir ecosystem, adequate knowledge about the soil where a series of chemical, physical, biochemical and microbial reactions are going ceaselessly on, is essential. The main physical and chemical features of soil are primarily related to particle size distribution, types of clay colloids, cation and anion exchange capacity, oxidation-reduction potential, and nutrient dynamics. Extreme fluctuations of water level due to influx of water during rainy season and release of water through sluice gates are the factors affecting particle size distribution. Bottom soil should not be too adsorptive as to eliminate nutrients from water phase and at the same time it must not be too porous to allow loss of nutrients. Therefore, soil types should consist of an organic layer followed by a clay loam layer for effective utilization of nutrients. The productivity ranges of different nutrients in Indian reservoirs for high production are shown in Table 20.16.

**Table 20.16 : Production Range of Different Nutrients in Indian Reservoirs for Fish Production**

| Parameter | Production range |
|---|---|
| *A. Soil*: | |
| pH | 6.5-8.5 |
| Available nitrogen (mg/100g) | 25-75 |
| Available phosphorus (mg/100g) | 4.7-6.2 |
| C/N ratio (%) | 1.5-2.5 |
| Specific conductivity (m mhos/cm) | 0.40-0.80 |
| *B. Water*: | |
| pH | 6.0-8.0 |
| Nitrate nitrogen (mg/l) | 0.10-1.0 |
| Orthophosphate (mg/l) | 0.50-2.00 |
| Dissolved oxygen (mg/l) | 6-11 |
| Total alkalinity (mg/l) | 40-150 |
| Specific conductivity | 225-662 X $10^6$ |

The bottom sediment profile has two layers such as anaerobic (reduced) layer – below the aerobic one and aerobic (oxidized) layer – at the water-soild interface. The co-existence of these two layers in the reservoir soil has practical significance. Highly aerobic conditions favour solubility and availability of nutrients for maintaining nutrient supply to fish food organisms.

Soil organic matter is also an important critical factor that dictate the establishment of trophic status of reservoir ecosystems. This is due to the fact that organic matter serves three main purposes: (1) the dissolved organic matter acts as exogenous growth substances and chelating agents, (2) it is a substrate for bacterial growth, and (3) it is a food source of bottom-dwelling fishes. Generally, reservoir ecosystems accumulate excessive organic matter through loading of allochthonous and autochthonous substances. However, most of the reservoirs are often fortified with facultative and obligate anaerobes. These microbes help release organically-bound nutrients and at the same time, some obnoxious gases such as methane, ammonia, and hydrogen sulphide are developed which adversely affect the productivity of reservoirs to a considerable extent.

Although there is a vast scope for increasing fish production from reservoirs, due attention has not been given till recently. Furthermore, it is impracticable to control physico-chemical factors for providing a healthy environment particularly in large reservoirs. Greater attention should be given so that a better trophic status can be established in reservoirs.

## 20.13 Procurement of Fish Seed for Carp Cultivation

The main problem towards the development of carp cultivation in India is the non-availability of pure carp seed in due time. In the early 20th century, carp culture was mainly dependent on the procurement of fish seed from the riverine system in the form of hatchlings and fry. Major carps are known to breed in rivers and certain reservoirs and in artificially constructed bundh-type ponds or tanks where the riverine conditions are established at the time of spawning season. Mass spawning of carps takes place in flooded zones of rivers during south-west monsoon months. In some rivers of peninsular India, carps breed during north-east monsoon months. Spawning grounds are located in close connection with the river banks that are inundated with water during monsoon season. Carp seeds are collected with the aid of a specially-designed seed collecting net.

With the advent of induced breeding technique through hypophysation, the problem of procurement of hundred per cent pure seed of desired variety has been solved to a large extent. Carp seeds are collected from three sources such as River, induced breeding, and bundh breeding. The technique of induced breeding has been summarized in Chapter 11.

### *Riverine Seed Collection*

Prior to innovation of induced breeding method, river systems were the most important sources of fish seed collection. At present, however, negligible quantity of seed is collected from the riverine sources. Funnel-shaped net with leno-weave or round meshed mosquito net made of cotton/nylon filament or twine are used for seed collection. There are various types of net specially developed by experience of commercial hatchling collectors.

The climatic and diverse geographical conditions of India are reflected in the riverine resources of the country. Different river systems exhibit variations with regard to the distribution and abundance of fish seed, fry and fingerlings. There are nine major river systems in India : the Ganga, the Indus, and the Brahmaputra in the north, the Tapti and the Normada in central India, and the Krishna, the Mahanandi, the Godavary, and the Cauvery in the south. Among them, the Ganga system is the largest and perhaps contains the rich source of carp eggs, hatchlings, and fry.

Large scale seed collection is possible where suitable breeding grounds are identified. Eggs are, however, collected from 1 to 2 feet deep water. At the time of collection of eggs, bottom sediment is disturbed and then scooped the egg with a rectangular open piece of cloth which is refered to as "*gamcha*" but has the shape of a hood. In some cases, drifting eggs are collected by fixing a shooting net. The net has a funnel-shaped structure of finely woven netting and is operated in shallow margins of flooded rivers. The mouth of the net is directed against the current of water. The hatchlings move along with the current of water are collected in *gamcha* from where hatchlings are periodically scooped, sieved through a strainer and kept in *hapas*. Some breeding grounds have been located in Bihar (such as one on the river Mahananda at Dingrahaghat and the other on river Badua near Badua reservoir) and these breeding grounds are most productive : more than 90 per cent of carp eggs (particularly Rohu and Catla) are collected.

Hatchling collection from the Narmada and the Tapti river systems reveal that the Indian major carps constitute about 25-30 per cent of the total collection. In the lower stretch of the Narmada, the percentage of major carps is high and the upper stretch is unsuitable for seed collection.

Hatchling collection from the Indus river system in the erstwhile Punjab state was not known till 1960. At present, this river source produces negligible quantity of hatchlings. Eggs and hatchlings of major carps are also collected from the Cauvery river system in large numbers. However, no egg and/or hatchling collection centre has so far been reported from the Godavari in Maharashtra and the Mahanandi in Orissa.

Pollution of river water through indiscriminate discharge of the domestic sewage and industrial effluents causes complete or partial destruction of breeding grounds. Consequently natural seed production potential significantly declined and deserves specific attention. This situation, however, forces increased emphasis on the collection of pure seed through other sources such as bundh-type breeding and induced breeding.

### *Bundh Breeding*

Bundhs are special type of seasonal and perennial ponds or tanks or impoundments where riverine conditions are created during monsoon months. Hence, they are considered as natural and environment-friendly hatcheries. Where riverine fish seed collection is not common, seed production through bundh systems is widely practiced. The contribution of carp hatchling production from bundh systems are significant because bundhs produce 100 per cent pure hatchlings.

1. *Origin and Development of Bundh Breeding* : Although the origin of bundh breeding in India as a source of carp seed production is not known, it is assumed that bundh systems have been originated in the year 1980 in the laterite zone of West Bengal (particularly in the districts of Purulia, Bankura, and Midnapore). On the basis of experience gained by farmers, certain technological modifications have been made in the system. Most of the Bundh-type tanks are located in some districts of West Bengal (established in 1926), in Chhattarpur District in Madhya Pradesh (established in 1958), at Ajmer in Rajasthan (established in 1964), and in Bhandara district in Maharashtra (established in 1973). However, reports on bundh breeding in other states are fragmentary and not much attention has been given in this regard.

   At present, 10-15 per cent of brood fishes in the bundh are injected with pituitary hormone. This injection triggers the sympathetic breeding of non-injected fishes and results in complete spawning of the total brood stock. Sympathetic breeding method is now widely practiced. A total number of 1,752 bundhs (data for the year 2000) have been set up in four states of India and the breakup is as follows : West Bengal -1400, Rajasthan - 257, Maharashtra - 37, and Madhya Pradesh - 58. From these units, about 6,440 million hatchlings are obtained.

2. *Types of Bundhs* : Bundhs are of two types such as (a) Wet bundh and (b) Dry bundh. A brief account of these bundhs are given below:

(a) *Wet bundh* : A typical wet bundh is perennial tank/pond located in the slope of vast catchment area of an undulating stretch of land with adequate dyke and having elevated stretch of land for inflow of water with an outlet on the opposite side. During summer months, the vast shallow area is dried up but the water is retained round the year in the deeper part of the tank/pond where brood fishes are stocked.

During monsoon months, rain water from the upland areas rushes into the bundh (locally called *Dhals* in West Bengal) and most of the region of the pond bed gets submerged. Excess water is simultaneously drained out through the outlet – the mouth of which is guarded by a net/wire mesh to prevent the loss of eggs from the tank. The shallow sloping area of the bundh is called *"Moans"*. Moans are the main spawning ground of the bundh. Run-off water in the bundh stimulates the brood fishes, they migrate from the deeper zone to the shallow marginal areas and starts breeding. Shallow breeding grounds with gradual slope encourage the brood fish to lay eggs more efficiently. It has been found that Rohu and Mrigal breed in the shallow catchment area as compared to Catla which breeds in the deeper area of the bundh. In general, the depth of water where breeding takes place, varied between 8 cm and 1.2 metre.

(i) *Characteristic features of wet bundh* : The most important characteristic features of wet bundh are as follows : Perennial tank/pond, small or large or irrigation reservoir having suitable topography for breeding; it is very difficult to manage; mixed type of hatchling is produced; breeding is not controlled and can be operated once a year; egg collection is difficult; and less economical as compared to dry bundhs.

(b) *Dry bundh* ; A dry bundh is considered as a shallow depression circumscribed by an earthen wall, called as *bundh* on three side which restrain rain water within limits from the catchment area. In most part of the year, such impoundments remain more or less dry and hence called the *dry bundh*. In West Bengal, the catchment area consists of agricultural land or barren and is usually covered with bushy plants such as *Bassia latefolia* (Mahua), *Magnifera indica* (Mango), *Tamarindus indica* (Tamarind), *Shorea robusta* (Sal) etc.

The topographical features of the soil of dry bundh play an important role in the construction of bundhs. The soil is, however, mostly red laterite – it is sticky during rainy season and become hard in the dry season. Moreover, the wavy land provides a large catchment area and rapid filling of the bundh with even an intermittent rain. The water can be easily drained from the bundh that creates suitable environment.

(i) *Characteristic features of dry bundhs* : The main characteristic features of dry bundh are as follows : Seasonal small tank or pond; it can be controlled and managed without any difficulty; egg collection is very easy; desired quality of hatchlings can be produced; it is more economical as compared to a wet bundh; it can be operated 4 to 5 times in the season.

3. *Design and Construction of a Dry Bundh* : Little improvement of seasonal bundhs can be used as dry bundh by providing breeding grounds at different levels on both the sides of the incoming water current. A dry bundh can be constructed at any place – particularly in sandy, clayey or laterite or rocky soil type in close connection with the source of water. The bundh should be guarded with fine mesh iron netting through which excess water can flow over. The most latest constructions are generally stone masonry structures with arrangement of sluice gate in the deepest zone of the bundh so that the water can be easily drained. In some cases, a dry bundh unit consists of stocking ponds for rearing brood fishes.

   The slope gradient of the catchment area is highly variable. The ratio of the bundh proper and the catchment area is generally 1:5 as in the case of West Bengal. In the case of Nowgong bundh in Madhya Pradesh, the ratio is 1:25.

4. *Breeding Techniques* : Spawning usually occurs after continuous heavy rains for days when large quantity of rain water rushes into the bundh. However, a selected number of carp breeders in the ratio of 2 males to 1 female are introduced. Within a few hours of releasing the brood fish into the bundh, courtship behavior occurs. The coiling of the two partners exerts pressure on the abdomen of the pair, resulting in the release of ova and milt. All eggs are not laid at one place but at different places and at intervals. Eggs are deposited in shallow water. When breeding is over, the eggs are collected from the bundh with the aid of mosquito netting cloth, called *"gamcha"*. Eggs are then stocked for hatching either in double-walled hatching hapas or in cement hatcheries. The degree of release of eggs and milt depends upon the rush of water into or out of the bundh : the greater the rush, the more complete the release. However, the level of water in the bundh is reduced by opening the outlet and the spent fishes are netted out for the market or they move, if suitable, to the deeper areas. It has been reported that 4-5 successful breeding operations can be made in one breeding season.

5. *Factors Necessary for Spawning in Bundh* :

(a) Sudden heavy rains that results in abrupt rise of water level in the bundh, permit spawning of carps.

(b) Flood in the early phase of south-west monsoon is necessary for spawning. If monsoon is delayed, carp do not spawn.

(c) There is controversy in the opinion on the patern of water current. However, spawning generally occurs in strong or moderate current in the spawning ground.

(d) Cloudy days and thunder storm seen to have some influence on spawning of carps in the bundh.

(e) At the time of spawning, water temperature should be varied between 22° and 32°C in different environments.

(f) High oxygen content (8-9 mg/l) in flood water, fluctuation in pH (6.0-7.8), alkalinity, chloride and presence of minerals do not seem to play any significant role on spawning of carp in the bundh. However, highly turbid water, low alkalinity and temperature (varying between 27 and 29°C) are the favourable conditions for spawning in bundh.

(g) 'Gonadal hydration' is considered as the factor that stimulates fish for spawning.

(h) Presence of a hormone-like substance – called the *'Repressive factor'*, which retards spawning. This substance is secreted by the fish and released into the water. This factor is nullified as soon as the fresh rain water is added into the water.

Although a correlation of increasing the efficiency of spawning with rainfall and water level exists, it must be pointed out that the degree of efficiency of brooders in any bundh is influeced by some abiotic factors of water. Climatic factors always work together.

6. *Brood Stock Management* : The productivity of brood fish depends as much on proper management of brood fish as that of fish farms. Prior to breeding season, brood fish require a minimum seasonal care (for 3-4 months). The brood fish must be reared in a healthy pond environment. The brood should be stocked in a pond at the rate of 25,000 kg fish/ hectare and the pond must be free from any aquatic plants and pollutants. Feeding with balanced diet containing 30 per cent protein which often must be used in brood ponds, is necessary. On the contrary, application of lime, fertilizers, and manures is less necessary unless ponds have developed in regions low in nutrients. In acid soils, the situation is different. The acid conditions result in the dissolution of aluminium, iron and manganese in toxic quantities. Under these conditions large amount of lime may be necessary but overliming should be avoided.

   One month before the onset of breeding operation, it is advisable to segregate the female and male brood fishes for their management in separate ponds. If brood fishes, even after care, are not properly developed, hormone injections with pituitary extract/ ovatide/ovaprim/human chorionic gonadotrophin are necessary for the fish. Hormone injection helps restore the vitality of brood fishes to release eggs.

### *Evolution of Spawning Techniques*

During the course of evolution, the technique adopted for bundh breeding has changed to a large extent and can be classified into following four steps :

1. Previously, no importance was given to maturity of brood fishes and sex ratios. With the onset of monsoon, brood fishes were stocked to the bundh and the breeding was dependent on favourable conditions.
2. In course of time, techniques were further improved by considering the maturity of brood fishes and the sex ratio (male and female ratio is 2:1 and 1:1 ratio in weight). By maintaining this situation, it was paved the way for successive spawning operations as many as 5 times in one season in the bundh.
3. The third stage was the initiation of an advanced stage through the introduction of modern techniques. In this technique, a few pairs of brood fish in the bundh were injected and kept them with other non-injected pairs and generally referred to a *sympathetic breeding* in bundhs. This type of breeding counteracts several factors which inhibits spawning of brood fishes.
4. The last and recent bundh system of breeding is generally termed as the *Bangla Bundh* and has become very popular among farmers. This type of bundh was first set up by the

farmers of Mogra Village in West Bengal and then in the District of Rajpur in Madhya Pradesh. The bangla bundh is composed of a cemented tank/pond of about 75 feet x 25 feet with side walls. The ratio of slope of the pond is 1:1. The pond bottom is divided into two zones : one zone has the gradual slope from the top level and the slope extends upto 50 feet length. The other zone has a continuous slope upto 25 feet length. In the bundh, fine river sand is spread to make 5 inches thick layer at the bottom. During monsoon season, the pond is filled with water. While the first zone has about 1 metre depth, the other one has 2 metre depth. Male and female brood carps in suitable ratio are placed in the bundh. Carps are injected with a lower dose of inducing agents. About 80 lakhs of hatchlings are produced at a time. At the same time, successive breeding of carps has also been reported in a single breeding season.

The dry and wet bundhs have been constructed in different states of the country such as Maharashtra, West Bengal, Rajasthan, Madhya Pradesh and Andhra Pradesh and adequate quantity of hatchlings is being produced in bundhs. However, modern breeding practices in bundh help increase carp seed production for further management in rearing and stocking ponds. This technique not only reduces the cost of seed production but also encourages the farmers.

## 20.14 Conclusion

Ponds, paddy fields and reservoirs are ideally suited for fish culture. Ponds and paddy fields should be constructed in such a way that management practices become effective. However, a major breakthrough is being achieved by the progressive fish culturists and experts in the field of fish culture technology. This will open up the possibilities of a manifold increase in fish production and employment opportunities for the rural people. Moreover, the development has take place in the technology of breeding of carps under controlled conditions. This will enable a round-the-year rearing of carps through continuous supply of seeds.

India has huge potential of resources in freshwater fish culture sector particularly in the form of ponds, reservoirs and paddy fields. Although carps are cultured in ponds and rice fields on scientific basis, reservoirs are also important source of commercially important species of fish where fishes are stocked for several months or years. Development of these resources will undoubtedly increase carp production to a considerable extent.

The nature and efficiency of fish farming differ considerably in different types of water areas and soil conditions. The technology of fish farming should, therefore, be modified keeping in view of the level of culture conditions that can be provided. For the villages where culture conditions are in a still primitive stage, a program of culturing of indigenous carp with exotic one should be followed. Such technique would help in increasing fish production several folds.

Fish culture strategies have been realized in most fish production countries of the tropics. But in many rural areas of India, the extent of these strategies is still less well-known. The management principles, however, should be helpful in this regard. Since fish culture strategies involve heavy costs, this factor must be seriously considered.

### *References*

Horvath, L., G. Tames, and C. Seagravec. 1992. *Carp and Pond Fish Culture*. Fishing News Books Ltd., England.

Huet, M. 1986. *Text Book of Fish Culture : Breeding and Cultivation of Fish*. 2nd Ed. Fishing News Books Ltd., England.

Lakshmanan, M.A.V., D.S. Murthy, K.K. Pillai, and S.C. Banerjee 1967. On a new artificail feed for carp fry. *FAO Fish Rep.* 44 : 373-387.

Piska, R.S. 2000. Impact of stocking densities of major carp seed on fish prodcution in a minor reservoir. *Fishing Chimes*. 20 : 38-41.

Saha, C. 2000. Site selection, design and construction of fish ponds. *Fishing Chimes*. 20 : 9-13.

Sinha, M. 1998. Policy options for integrated development of reservoir fisheries from production to marketing. *Fishing Chimes*. 18 : 54-59.

Sugunan, V.V. 1997. *Reservoir Fisheries of India*. Daya Publishing House, Delhi, India.

Talwar, P.K. and A.G. Jingran. 1993. *Inland Fishes*. Oxford and IBH Publishing Co. Pvt. Ltd., New Delhi, India.

## Questions

1. What are the types of fish culture generally carried out in freshwaters? What are the features of cuturable carps?
2. Mention some characteristic features of carps.
3. What are the prerequisites for carp culture in ponds?
4. What management schedules should be considered before construction of freshwater fish farm? Discuss.
5. What are the methods necessary for carp cultivation? Describe any one method which is generally carried out in rural areas.
6. Why fish toxicants are used during pond preparation? What are the advantages and disadvantages of using the toxicants? Name some commonly used fish toxicants.
7. For pond preparation, either chemical substances or plant derivatives are used. Which one you will prefer? Give reasons.
8. Explain how carps are cultured on commercial basis in ponds.
9. What are the characteristic features of nursery, rearing, and stocking ponds?
10. What is the role of feeds and fertilizers in carp culture to achieve fish production goals.
11. Why species combination and the rate of stocking are important for sustained production of carp?
12. A fish farmer wants to cultivate carp fish in his own pond. What will be your recommendation?
13. What is carrying capacity? How it is related to carp production?
14. Explain how carp culture can be executed in rice fields. State the potentiality of carp culture in rice fields under Indian conditions.

15. What is reservoir? Trace the development of reservoirs in relation to fish production.
16. What are the characteristic features of reservoirs?
17. How carp seed is collected from different freshwater sources?
18. What is bundh breeding? What are the types? Discuss.
19. What are the factors necessary for carp spawning in bundh?
20. Define the following terms : (a) bundh, (b) bungla bundh, (c) dry bundh, (d) sympathetic breeding, (e) repressive factor, (f) carrying capacity.

# 21

# Composite Fish Culture

Composite fish culture or polyculture with different Asian and African fishes have been used. The leading fishes in polyculture systems are mainly carnivores and herbivores. Due to the climate of their habitat, they could develop rapidly because plankton and macrophytes grow together throughout the year.

Cultivation of fast-growing compatible fish species and management of fish culture ponds are the most critical factor in schemes to increase food supplies. At the same time, employment generation in rural areas and economical stability of fish farmers have expanded and have created strong competition for culture techniques. This competition forces increased emphasis on efficiency of the technology of polyculture for fish production. The patterns and techniques of polyculture have been discussed in this chapter.

## 21.1 Definition and Explanation

Polyculture or mixed fish farming or composite fish culture is defined as the culture of fast-growing, compatible species of fish of different feeding habits and habitats in the same pond to occupy the same ecological niches and where phytoplankton, zooplankton, and aquatic weeds are effectively utilized to maximize fish production.

This expression indicates that polyculture or mixed fish farming comprises an association of various types of fish in the same pond. If polyculture is considered in another way, it will be seen that it is an association of certain definite fish population other than carps and it is termed as *balanced fish population.* The characteristic features of a balanced fish population is that it should have (1) a definite range in the ratios of the weights of forage and piscivorous groups, (2) a narrow range in the ratios of weights of small forage fishes to the weight of piscivorous groups, and (3) more than 33 per cent of the total population weight of harvestable sized fishes.

The objective of polyculture is that when compatible fishes of different feeding habits are cultured together, they live in well condition and in the most efficient manner, and factors responsible for the growth of fish are available. Further, there is no severe competition between different species and at the same time, each species may have a beneficial influence on growth of the other.

For successful polyculture, technological progress must be considered. Technological progress can be defined as those changes in the production processes which reduce the cost of output. The changes can occur by introducing new factor of production either by replacing old ones or simply as additional inputs.

A number of factors determine the technological progress. Among different features, the following ones are important : (1) The pattern of ownership and size of water bodies, (2) Institutional factors which act as incentives to fish culture practices, and (3) The attitude and awareness of farmers to modern methods of cultivation.

As noted earlier, pond productivity and production can be increased substantially through the inputs of quality fish seeds, fertilizers, manures, and feeds. These inputs become more effective and their potential will better be utilized if appropriate technologies are available to the farmer. Most of the fish farmers still confide in the traditional practices. The dependence on these practices does not go in harmony with the new modern practices. A stage has been reached in polyculture sector where it may be essential to shift from traditional culture to more advanced method of cultivation.

A number of field experiments have been conducted to determine the efficiency of polyculture. All the studies have indicated that polyculture promotes high fish production. The difference between monoculture and polyculture may range upto as high as between 60 and 65 per cent.

Traditional techniques are evolved over generations. They are continuously adjusted within a restricted frame to changing circumstances. The pursuit of traditional techniques involves less uncertainty. Since traditional techniques are passed on from one generation to other, there is practically no input cost and relatively less uncertainty of output. These are the two main reasons for the reluctance of the farming community to change to modern techniques.

Experiments have shown that significant production of fish cannot be accomplished with the aid of traditional techniques and practices. A change is almost a necessary condition for the growth of fish culture. Two techniques and practices have been evolved over the years : (1) Integrated fish culture and (2) Polyculture.

## 21.2 Expansion of Polyculture

Technology is in part essential for the marked expansion of fish culture in population growth. Previously, monoculture (culture of any one carp fish) was widely practices in India. By this method, only 300-500 kg/hectare/year of fish was obtained. During the past 2-3 decades, advances in technology and their application throughout the country have changed this figure of production. Using carp species of different feeding habitats in a single water body, fish production has been dramatically increases. Transfer of technology has taken place in areas where production potential is great. Improved quality feeds have helped in increasing food fish. The result is unprecedented yield of fish biomass. Average yield of fish through polyculture with three Indian major carps has been reported as 3,000 kg/hectare. Experts have advocated that by adopting management schedules, high fish production through polyculture is possible. At present, polyculture with three species of indigenous fish has been modified by introducing some exotic carps (particularly common carp, silver carp, and grass carp). This has resulted significant fish production from 6,000 to 10,000 kg/hectare/year.

## 21.3 Scientific Outlook of Polyculture

The main objective of polyculture is to produce maximum fish per hectare water area in minimum period of time. The following scientific principles are generally encountered for significant production :

1. In polyculture, fast-growing and non-predatory fish as well as food fish species are cultured together. These fishes are able to utilize food organisms and supplementary feeds.
2. Stocking of different compatible fish species which has much shorter food chain, is made in suitable combinations and numbers.

3. A proportionate mixture of naturally compatible fish species is introduced which a pond can support and due attention should be given so that natural food organisms and supplementary feeds can be fully utilized by the fish in relation to their stocking densities.
4. Application of right type of fertilizers and manures stimualate natural food organisms for the growth of fish in ponds. At the same time, use of supplementary feeds is also essential. Consequently, the carrying capacity level significantly increased.
5. By changing the stocking density of fish, intermediate harvesting is done during the entire growing season (12 months).

## 21.4 Factors Affecting Fish Production in Polyculture

The capacity of a pond under polyculture schemes is determined by social and economic factors that affect the farmer's incentives to produce. Fish production is also affected by some other factors such as the following :

1. The resources of freshwater available for culture.
2. Available technology and the knowledge of proper management of pond ecosystems.
3. Fish seed production efficiency and size specification of fish seed that respond to proper management.
4. Composition of fish species and their stocking rates.
5. Supply of production inputs such as fertilizers and supplementary feeds.

How these factors apply to fish production depends on the regions and quality of soil as well as water – their natural productivity and response to management. It is very difficult to say which factor should place high on the list of requisites for adequate fish production because each factor is linked with the other to make the production process into success.

## 21.5 Species Selected for Polyculture

The success of All India Co-ordinated Research Project for polyculture started in 1971 by the Central Inland Fisheries Research Institute made possible to popularize the scientific fish culture in the country. This technique of culture is being followed by many Asian countries. In polyculture system, the following carps are stocked in well-managed freshwater ponds : *Labeo rohita, Catla catla, Cirrhinus mrigala, Hypophthalmichthys molitrix, Cyprinus carpio* and *Ctenopharyngodon idella*. These species belong to the family Cyprinidae. Moreover with a view to increase fish production or catering to local demand, certain other species (such as *L. calbasu, Clarias batrachus, L. fimbriatus, Mystus seenghala, M. aor, Puntius javanicus, Mugil cephalus, Macrobrachium rosenbergii, and M. malcolmsonii*) are also stocked in polyculture system (Figure 21.1)

Quality fish seeds with suitable combinations are essential for increasing fish production. Unless the fish farmer has capacity to select fish of suitable varieties, he cannot get the best out of their inputs such as fertilizers and feeds. Following suitable species and their combinations, it becomes possible for him to take polyculture systems because of the resultant high production and good economic returns. The evolution of selection of different species in a single pond not only helped to change the entire scene of fish culture economy, but also brought into focus the importance of significant factor of polyculture.

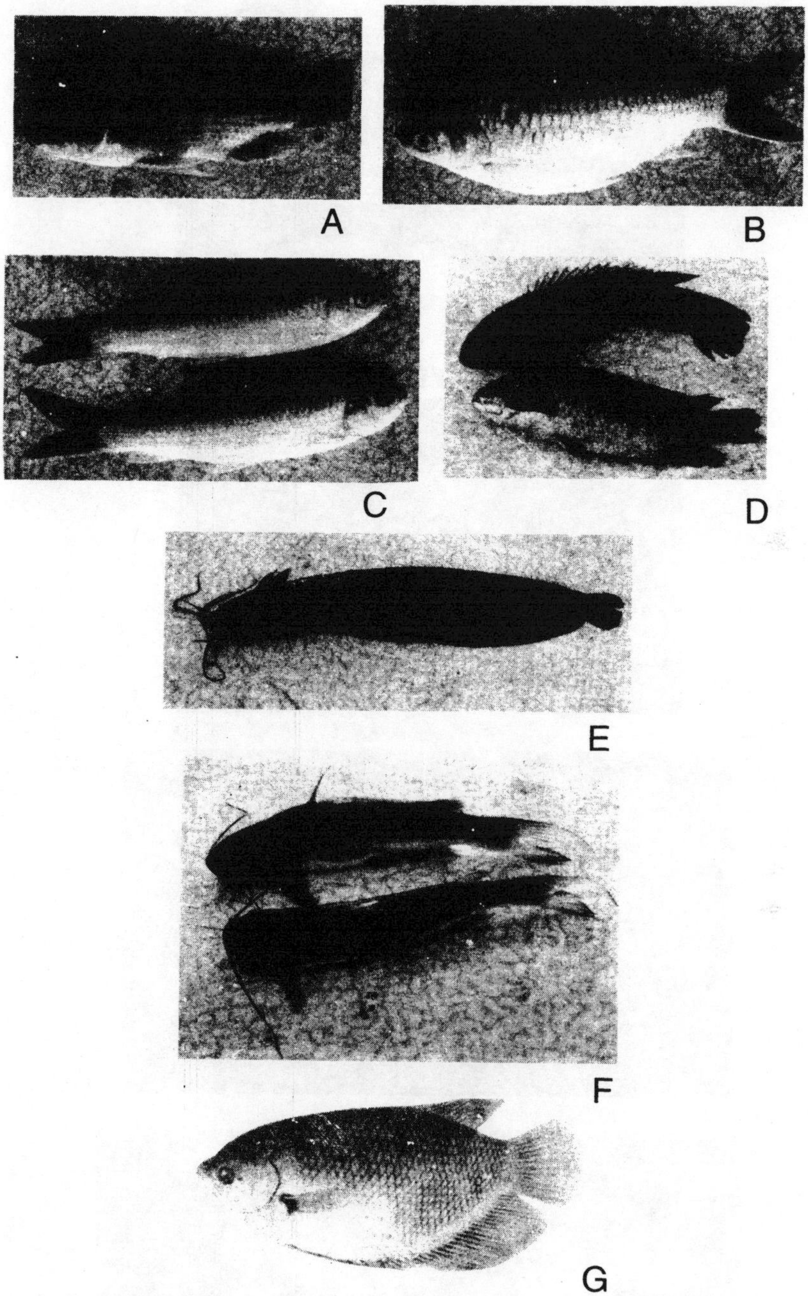

**Fig. 21.1 :** **Some freshwater fish species are important due to their high prices and good taste. Although they are cultured through special techniques, it is advisable to farmers, depending on local demand, to cultivate other species of fish along with the carps. Some important species of such fish (and also prawn) are shown in this figure. A, *Oreochromis mossambicus;* B, *Puntius javanicus;* C, *Labeo bata;* D, *Anabus testudineous;* E, *Clarias batrachus;* F, *Mystus seenghala;* G, *Osphronemus goramy;***

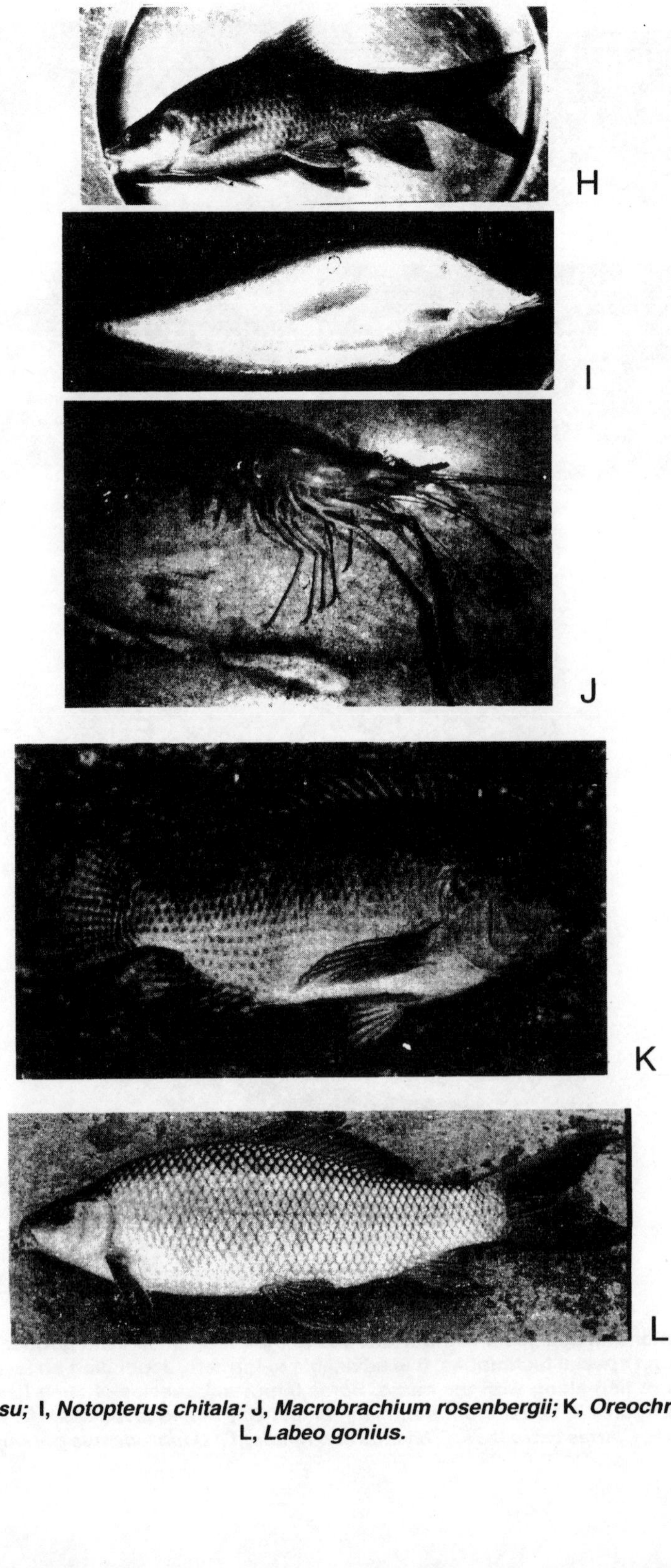

**H, *Labeo calbasu*; I, *Notopterus chitala*; J, *Macrobrachium rosenbergii*; K, *Oreochromis nilotica*; L, *Labeo gonius*.**

## 21.6 Food and Feeding Habit of Selected Fish Species

For fast growth of above-mentioned species in a polyculture pond, a proper balance of phytoplankton, zooplankton and water weeds must be ascertained. Fish with different feeding habits are cultured together to achieve a balance where these natural food organisms are effectively utilized for increasing fish production.

Silver carp and catla are planktophagous surface feeders. Although the latter species consume predominantly zooplankton, the former prefers phytoplankton and some decaying aquatic plant materials. Due to their similar feeding habit and feeding zone, it is assumed that there exists competition between catla and silver carp in the same ecological niche. In India, various combinations of silver carp and catla have been stocked in a composite fish culture pond. In general, ponds are stocked with 20 and 15 per cent of silver carp and catla, respectively.

Rohu feeds on plankton in the underwater zone and is a column feeder. Since the fish grows well in deeper ponds, ponds having more than three metres depth of water should be stocked with 20 per cent of total fish population. In shallow ponds, however, about 10 per cent of rohu is stocked.

Grass carp is both surface and column feeder. Although the fish consume zooplankton and unicellular algae, they prefer aquatic macrophytes. Common carp and mrigal are both bottom-feeders. They consume decaying plant and animal materials, algae, and plankton. Grass carp is, however, also able to utilize supplementary feeds (such as oil cakes and rice bran) most efficiently.

The association of common carp and mrigal with grass carp has an important indirect benefit. The faecal matter of grass carp is utilized as food by bottom-dwelling fish. To avoid competition among mrigal, common carp, rohu and grass carp for supplementary feeds, it is advisable to supply aquatic weeds in suitable quantity for the growth of grass carp. The stocking density of grass carp generally depends upon the availability of aquatic weeds.

Indian major carps accept different types of supplementary feeds which principally consist of brans (rice and wheat), oil cakes (mustard and groundnut), fish meal, blood meal and silk worm pupae.

## 21.7 Different Steps Followed in Composite Fish Culture Technique

In India, a number of management schedules are generally practiced for high fish production. In a polyculture pond, the following steps are adopted : (1) Preparation of pond, (2) Liming, fertilization and manuring, (3) Selection of fish species and stocking, (4) Supplementary feeding, and (5) Harvesting.

### *Pond Preparation*

If possible, the pond water should be drained to remove undesirable gases and harmful micro-organisms. Unwanted aquatic plants and animals are also removed by physical and chemical means (see chapter 4). For removal of unwanted aquatic animals, different plant products are also used (see section 20.8).

### *Fertilization, Manuring and Liming*

Application of chemical fertilizers (such as urea or ammonium sulfate, calcium ammonium nitrate, and single superphosphate, organic manures (such as cow-manure or poultry litter), and agricultural lime are most common in a polyculture pond. In general, nitrogen and phosphorus are applied in the ratio of 2:1. The NP fertilizers are applied in such a way that the total available nitrogen and phosphate contents of water do not exceed 1.0 and 0.3 mg/l, respectively. Raw cow-manure or compost manure is applied at the rate varying between 10,000 and 25,000 kg/hectare/ year depending on the level of organic carbon in soil. However, organic carbon content in soil should not be less than 0.5 per cent.

The level of calcium is also important. This is due to the fact that if calcium level is not adequate, the carbon dioxide in water forms carbonic acid and the water becomes acidic (pH 5 or less) particularly in night. Similarly during day time, the carbon dioxide is utilized by aquatic plants during photosynthesis and the water becomes alkaline. Such changes in the pH of water are detrimental to fishes. To counteract this problem, liming of ponds is essential. The application of lime however, depends upon the pH value of soil and water.

The rate and type of fertilizers and manures principally depend on the characteristic features of soil, concentration of nutrient substances, plankton and stocking density of fishes. In general, the break-up of various types of fertilizers is estimated to be as follows : 25,000 kg of cow-manure, 825 kg of ammonium sulfate, 175 kg of urea and 550 kg of single superphosphate. These ingredients should be applied in 10 equal instalments. If mahua oil cake is used during pond preparation, there is no need to apply cow-manure in first instalment. In cases where excessive floating algae are developed, the oxygen content of water drastically reduced and mortality of fishes occur. Therefore, the application of fertilizers and manures should be stopped for short period of time. The application methods of fertilizers and lime have been described in Chapter 11.

### *Species Combination*

Research on polyculture suggests that stocking of six species of fish in a pond is the most scientific to intensify fish production. The break-up of different species of fish stocked has, however, been calculated as follows :

Silver carp:Catla:Rohu:Grass carp:Mrigal:Common carp =
2.5:1:2.5:1:1:2

The contribution to the yield from exotic carp and Indian major carp are 55 and 45 per cent, respectively. At all India level, however, the Indian major carp contributes 86.6 per cent while exotic carp contributes 14.4 per cent to the total fish production for the polyculture pond (Table 21.1). Note that the Indian major carps are accounted for the major portion of stocking in polyculture ponds. The contribution of exotic carps is negligible (only 15.8 per cent of the total species combination). Since very small percentage of grass carp and silver carp is stocked at national level, it can be said that the polyculture program has remained four species culture. In many states, four species are used in polyculture as against the recommended six species. In cases where four species are cultured, the ratio between Rohu, Catla, Mrigal, and Common carp should be followed as 2.5:2.5:1:1.

### *Stocking Rate*

In composite fish culture schemes, the stocking rate is considered in terms of fingerling stage. Different rates of stocking in different water areas and their production figures are extremely variable in different states of India. (Table 21.2).

**Table 21.1 : State-wise Composition of Fish Fingerlings Used in Polyculture. Values are in Terms of Per cent**

| State | Rohu | Catla | Mrigal | Silver carp | Grass carp | Common carp | Minor carp | Others |
|---|---|---|---|---|---|---|---|---|
| West Bengal | 24.60 | 31.83 | 22.20 | 4.50 | 2.22 | 4.23 | 2.17 | 10.50 |
| Andhra Pradesh | 24.60 | 23.85 | 16.25 | 0.25 | - | 23.33 | 1.83 | 9.89 |
| Assam | 31.57 | 35.80 | 17.83 | 0.53 | 4.34 | 7.10 | 2.13 | 0.70 |
| Bihar | 26.40 | 37.35 | 24.28 | - | - | 7.70 | 3.20 | 12.50 |
| Gujarat | 21.83 | 32.17 | 23.20 | - | 1.70 | 4.43 | - | 16.67 |
| Karnataka | 15.85 | 15.93 | 20.48 | 0.53 | 0.47 | 23.70 | 8.62 | 15.05 |
| Madhya Pradesh | 23.00 | 37.95 | 18.50 | - | - | 4.80 | - | 10.75 |
| Maharashtra | 18.30 | 23.16 | 16.64 | 0.14 | 2.14 | 9.44 | - | 30.18 |
| Orissa | 17.40 | 21.00 | 14.23 | 5.33 | 5.87 | 10.30 | 3.33 | 22.81 |
| Rajasthan | 30.59 | 16.73 | 27.46 | - | - | 6.67 | - | 18.57 |
| Tamil Nadu | 19.37 | 21.25 | 14.82 | 0.35 | 5.30 | 18.10 | 4.98 | 15.83 |
| Uttar Pradesh | 32.83 | 26.18 | 25.00 | 2.00 | 1.14 | 3.93 | 2.22 | 6.70 |
| Himachal Pradesh | 16.87 | 14.45 | 20.30 | - | - | 20.00 | 2.00 | 26.38 |
| Kerala | 27.50 | 25.00 | 18.90 | 1.50 | 0.80 | 15.00 | - | 11.30 |
| Tripura | 26.40 | 28.90 | 20.31 | 3.00 | 1.55 | 3.60 | 1.80 | 14.44 |
| Manipur | 20.00 | 18.60 | 19.70 | 3.00 | 1.00 | 2.50 | - | 35.20 |
| Nagaland | 25.40 | 23.80 | 20.00 | 2.40 | 1.35 | 2.20 | - | 24.85 |
| Haryana | 22.70 | 26.86 | 18.90 | 2.10 | - | 2.20 | - | 27.24 |
| Average | 23.62 | 25.60 | 19.94 | 1.40 | 1.55 | 9.40 | 1.79 | 16.70 |
| | | 69.16 | | | | 14.14 | | 16.70 |

*Source :* Modified after Agarwal (1990)

**Table 21.2 : State-wise Stocking Rate and Fish Production in India**

| State | Stocking Rate of fingerlings (Number/hectare) | Fish Production (kg/hectare) |
|---|---|---|
| West Bengal | 4,840 | 1,340 |
| Andhra Pradeh | 1,586 | 490 |
| Assam | 2,910 | 871 |
| Bihar | 2,799 | 1,366 |
| Gujarat | 3,158 | 550 |
| Karnataka | 3,070 | 535 |
| Madhya Pradesh | 2,660 | 942 |
| Maharashtra | 4,055 | 456 |
| Orissa | 5,800 | 1,580 |
| Rajasthan | 4,310 | 810 |
| Tamil Nadu | 5,490 | 1,110 |
| Uttar Pradesh | 4,000 | 1,060 |
| Himachal Pradesh | 1,500 | 745 |
| Kerala | 1,750 | 1,070 |
| Tripura | 2,700 | 1,115 |
| Manipur | 1,200 | 650 |
| Nagaland | 1,000 | 680 |
| Haryana | 9,530 | 1,902 |
| Average | 3,464 | 910 |

*Source :* Modified after Agarwal (1990)

The stocking rate in polyculture has been recommended as 6,000 fingerlings per hectare water area. National Commission on Agriculture has, however, recommended 5,000 and 10,000 fingerlings/hectare for water area below 10 and above 10 hectares, respectively. The average stocking rates for intensive, semi-intesive, and extensive culture units have been recommended by the Fish Seed Committee as 6,000, 3,000, and 1,000 fingerlings/hectare, respectively. If artificial feeding is done along with the aeration of pond water, high stocking rate (10,250 fingerlings/hectare) with Rohu, Catla, and Mrigal in the ratio of 3:3:4 may be recommended for sustained production.

### *Supplementary Feeds*

Various types of supplementary feeds are used in polyculture. Excessive feed should, however, be avoided as it not only causes water pollution but also wastage of feed. The quantity of feed required principally depends upon several factors such as condition of ponds, stocking rate, and the number to be stocked as well as individual fish weight. Supplementary feeds and feeding have been discussed in Chapter 9.

Grass carp consumes large-sized aquatic weeds of the genus *Wolffia, Lemna, Azolla, Sprirudela, Hydrilla, Vallisneria, and Najas.* However, the stocking rate of grass carp depends predominantly on the availability of suitable weeds.

### *Growth and Production*

On an average, fishes normally attain table-size within 10-12 months. Under suitable management, silver carp, catla, grass carp and common carp can attain 1 kilogram within 8 months. Periodic harvesting of these fishes and restocking the pond with fingerlings not only helps growth of other stocked fish but also increase more than one crop within a year. Adoption of advanced technology in polyculture with an average survival of 85 per cent will undoubtedly increase fish production to the tune of 9,000 kg/hectare/year.

### *Harvesting of Fish*

Harvesting is done in summer months when the water level falls or when the market demand is high. In some situations, however, harvesting is also executed in winter months when water temperature decreases. At the time of harvesting, farmers face a serious problem. About 80 per cent of surface and column feeders are caught by drag net but the bottom-feeders are slip out. This problem is most common in large and deeper ponds. In this case, a trap net is used.

## 21.8 Models of Polyculture Used in Different Countries

Different models of polyculture or mixed fish farming have long been recognized in many countries of the world as essential for high fish production and hence, practiced for centuries. It is known to be indispensable for efficient use of different ecological niches of the pond ecosystem. Promotion of this practice to yield more fish would alleviate malnutrition and produce animal protein. A variety of carp species which adapt to feed on various organisms within the fish culture pond to fully utilize all the resources derived from various management schedules must be taken into consideration. Depending on the geographical and climatic conditions of any given region, different countries follow the specific models for fish production.

### *China*

In China, fish culture is seemingly older than in any othe region of the world. However, at present, 4-6 species of fish are cultured together on a large scale in the ponds of China and they are all native to the natural waters of the country. The following species of fish are cultured : *Mylopharyngodon piceus* (black carp) – the fish consumes some molluscs at the pond bottom as food, *Ctenopharyngodon idella* (grass carp) – feeds on coarse vegetable matter, *Hypophthalmichthys molitrix* (silver carp) – subsists on plankton, *Aristichthys mobilis* (big head) – consume macroplankton, *Cyprinus carpio* (common carp) – an omnivorous scavenger, *Cirrihinus molitorella* ( mud carp) – a bottom feeder. In general, the depth of pond, nature of water supply, type of feeds available, climatic condition, and pond fertility are the factors which determine the stocking rate and combination of fishes. For example, in deeper ponds (4-7 metres in depth), fingerlings for five species (such as big head, silver carp, grass carp, common carp, and mud carp) weighing 200 to 500g are stocked at the rate of 12,000/hectare. In shallow ponds (2 metres in depth), the same species of fingerlings ranging from 120 to 350 mm are stocked at the rate of 9,5000/hectare. In cases where the climate is warmer (such as in Hong Kong), mud and black carps are not widely used. Ponds having 1-2 metre depth, are stocked with fingerlings of other varieties at high rates (varied between 26,000-34,000/hectare), the length of fry being 2 to 7 cm and periodical harvesting is practiced.

1. *Culture Methods* : To prepare the ponds to stock fingerlings, it is customary to drain the water so that the bottoms may be exposed to the sun for some days. Further, the bottom is treated with tea cake and lime to kill predatory organisms and encourage the growth of the minute animal and plant life upon which the fry normally feed.

   Replenishing the ponds has become a highly intricate procedure. Different species of fish are stocked together and each fish farmer has his own particular formula for governing the association of species under his control, derived from both market needs and the feeding habits of each species.

   With the exception of the grass carp, which depends on any type of vegetation, fishes are fed great quantities of silk worm pupae, soya plant wastes, brewery, animal excrement and oil cakes (such as soyabean and peanut). The feeding of fish is reduced in quantity when the water temperature rises above 25°C or drops below 15°C.

2. *Prospects* : Polyculture is expanding at a remarkable rate – a matter of great significance to the future of the Chinese people. Since there are limits to the productivity from natural waters, the production of fish through polyculture (and paddy cultivation) constitute important aspects for mass consumption.

### *Singapore, Malaysia and Thailand*

Polyculture in these countries has been dominated by Chinese tradition because the conditions of these regions are ideal for more intensive fish cultivation. However, polyculture includes species of grass carp, big head, silver carp and common carp in varying combinations. Fishes weighing between 300 and 600 g, are stocked at the rate of 650 to 1,750 number/hectare. While Malaysia and Thailand extend the culture of carp for eight months, the duration of cultivation of fish in Singapore is 6 months only.

### *Taiwan*

The Chinese is credited with introducing fish cultivation prior to the eighteenth century. In Taiwan, fish culture is concerned with the rearing of the milkfish and the grey mullet along with the carps such as silver carp, grass carp, big head and common carp. In polyculture, seven species of fish such as silver carp, big head, grass carp, common carp, mullet, milkfish, and tilapia are stocked in the number of 2,000-3,000, 100-300, 100-300, 1,000-1,500, 3,000-4000, 1,000-2,000, and 3,000-5,000/hectare, respectively. Yields of fish in well-managed ponds reported to attain an average level of 6,000 kg/hectare/year. Tilapia constitute an element of real significance to the fish culture of Taiwan.

### *Japan*

Polyculture is extensively carried out in Japan where the following species is used : *Cyprinus carpio speculario* (mirror carp), *C. carpio nudus* (leather carp), *C. carpio* (scale carp), *Carassius* spp. (crucian carp), and *Carassius auratus* (gold fish). Grass carp, common carp, and big head have been introduced in Japan. For cultivation, however, superior strains of carp are selected. Carps are also cultured in association with the common freshwater eel, *Anguilla japonica* and the pond mullet, *Hypomesus olidus*. Fishes weighing 20-50 g, are stocked at the rate of 390-550 numbers/hectare.

The pond mullet and the carp are bottom-feeders and the eel is distributed throughout the pond water. Therefore, these three species together make a very efficient team to consume food. Although a large number of species of fish reared in ponds at present, the eel and the common carp are intensively cultivated in Japan and produce substantial quantity of fish that enter the commercial food markets.

### *Indonesia*

The potentialities of polycultue in Indonesia are great. This country has well-developed fish culture industry that makes an important contribution to the diet of the people. However, the species cultured in the country include : *Helostoma temminaki* (tambakan or kissing gourami) – a plankton feeder, *Puntius gonionotus* and *P. javanicus* (tawes) – subsists on coarse vegetable matter, *Osphronemus gorami* (gourami) – a herbivorous fish, and *Osteochilus hasselti* (nilem) – feeds on periphyton and soft decaying vegetation, *Cyprinus carpio* (common carp), and Tilapia. In Indonesia, 2 or 4 types of species combinations are practiced such as : common carp : tilapia (80 : 20), tilapia : common carp : nilem : gourami (35 : 30 : 20 : 15) or tambakan : tawes : common carp : nilem (50 : 10 : 20 : 20).

### *Bangladesh and Sri Lanka*

These countries have considerable culture-based fisheries. However, the development of culture-based fisheries on the polyculture systems of ponds in Bangladesh and Sri Lanka has raised yields from about 300 kg/hectare to about 800 kg/hectare and from 280 kg/hectare to about 740 kg/hectare, respectively. In these countries, the polyculure includes species of rohu, catla, mrigal, and common carp.

### *Burma*

In Burma, polyculture of catla, rohu, mrigal in association with common carp and gourami is undertaken on commercial basis. Fish production in a pond stocked with rohu, catla, mrigal, and gourami reported in suitable combination to reach a level of about 10,000 kg/hectare/year. However, the development of culture-based fisheries suggest that there is a large opportunity to increase fish production. Indian major carps and tilapias represent real possibilities for pond cultivation.

### *Pakistan*

Despite the prevalence of arid conditions, this country has, to some extent, fish production potential in the form of polyculture. Different stocking combinations of catla, rohu, and mrigal in the ratio of 50 : 30 : 20 or 30 : 35 : 35 or 40 : 30 : 30 or 30 : 40 : 30 have been found suitable for moderate fish production. Fish production through polyculture usually varies between 550 and 1,655 kg/hectare/year. Most of the freshwater ponds are generally of a low basic productivity.

### *India*

In India, the total production of freshwater from a composite fish culture pond amount to 10,150 kg/hectare/year. It should not be overlooked, however, that the culture-based freshwater fish include varieties of high value which is of great preference. There is evidence that the natural water fisheries are in some decline brought about by environmental degradation and high fishing pressures. This shortfall is being made up by an increasingly effective management of composite fish culture ponds.

During 1960's, a number of experiments on polyculture at different places of the country were carried out. Table 21.3 provides an example of production of polyculture system. Note that experimental ponds stocked with indigenous and exotic carps in different combinations were permitted to produce good results. At present, fish production from a polyculture pond can be enhanced 1-2 folds by adopting some management practices such as superior quality of artificial feed (feed fortified with several industrial products such as Agrimin and Fishmin), periodical harvesting and stocking, and aeration. However experiments on polyculture at different states of the country have shown the possibilities of increasing fish production from 3,448 to 7,500 kg/ hectare/year under various establishments (Table 21.4). It is optimistic about the future of mixed fish farming in India. Warm to hot climates provide production potential of ponds. Harvesting of fish several times a year can be made from the same pond. The potential for food production is enormous.

### *Israel*

In Israel, two systems of polyculture are undertaken. First, two sizes of common carp are stocked together. Second, either tilapia or grey mullet or both are cultured with tilapia. Ponds are fertilized with ammonium sulfate, superphosphate, and poultry manure at 15 days interval. In the ponds where tilapia (*Tilapia aura)*, mullet (*Mugil cephalus*), and common carp (*Cyprinus carpio*) are stocked together, the tilapia grow rapidly, increasing fish production by 30 per cent within 120-150 days. In general, mullet causes a decrease of 15 per cent in carp production while trying of tilapia to bring about friendly relations with the mullet depressed the carp and mullet production only by additional 2.8 and 0.5 per cent respectively. Tilapia constitutes a new element or real significance to the fish culture of Israel.

At present, the culture of silver carp along with other fishes is being carried out with great success. During growing season, ponds are systematically fertilized with chemical fertilizers and

**Table 21.3 : Experimental Details of Polyculture in Some Ponds of India**

| Details of experiment | Number of ponds | | | |
|---|---|---|---|---|
| | 1 | 2 | 3 | 4 |
| Area (hectare) | 0.12 | 0.12 | 0.138 | 0.15 |
| Location | Cuttack (Orissa) | Cuttack (Orissa) | Cuttack (Orissa) | Kalyani (West Bengal) |
| Depth (metre) | 2.5-3.5 | 1.2 | 1-2 | 1-2 |
| Duration of experiment | 1 year | 1 year | 1 year | 6 months |
| Species combination : | | | | |
| (a) C : R : M | 3 : 3 : 2 | - | - | - |
| (b) SC : CC : GC | - | 4 : 3 : 2 | - | - |
| (c) C : R : M : SC : GC : CC : Gourami | - | - | 2 : 6 : 2.5 : 5 : 2 : 2.5 : 3.0 | - |
| (d) C : R : M : SC : GC : CC | - | - | - | 1.76 : 2.74 : 1.15 : 2.29 : 0.90 : 1.15 |
| Stocking density (Fingerlings/hectare) | 3,750 | 5,000 | 5,075 | 5,473 |
| Initial total weight of fingerlings (kg/ha) | 54.6 | 75.8 | 240.5 | 315 |
| Survival (per cent) | 91 | 70 | 87 | Data not available |
| Production (kg/hectare) | 2,535 | 2,909 | 2,210 | 3,232 |
| Average feed (kg/hectare / year) | : | | | |
| (a) Oil cake + rice bran (1 : 1 ratio) | 1,500 | 3,390 | 2,320 | 480 |
| (b) Weeds for grass carp (*Lemna, Hydrilla, Azolla, Spirodella*) | - | 10,067 | 13,875 | 17,700 |
| Cow-manure (kg/hectare/year) | 15,000 | 3,833 | 25,000 | 5,000 |
| Fertilizers : | | | | |
| (a) AS + SSP + CAN (11 : 5 : 1) | 1,816 | 1,383 | 1,725 | AS (450 kg) + SSP (250 kg) + MOP (40 kg) + Urea (8 kg) |
| (b) Frequency of fertilizers : | | | | |
| (i) Inorganic | Monthly | Monthly | Monthly | Monthly |
| (ii) Organic | - | At irregular interval | - | - |
| Net profit (kg/hectare) (At price index of 1972) | Rs. 5,954.26 | Rs. 7290 | Rs. 12,160 | Not calculated |

*Source :* Selected data from Anon (1969), Lakshmanan *et al* (1971), and Sinha *et al* (1973). C, catla; R, rohu; M, mrigal; SC, silver carp; GC, grass carp; CC, common carp; AS, ammonium sulfate; SSP, single superphosphate; MOP, muriate of potash.

organic manures in suitable combinations. Artificial feeding is also done regularly. Mullet and carp fingerlings are stocked at the rate of 1,000 fish/hectare and marketed after 6 months.

A method has been developed for evaluation of the efficiency of composite fish culture over that of monoculture in terms of increase in fish yield in ponds. An index of competition has been suggested and can be estimated as follows :

$$C = (a - b) / a$$

Where C = Index of competition, a= Carp production in monoculture per day and b = Carp production in mixed culture per day.

Table 21.4 : Production of Fish in Some States of India

| State/place | Duration of Experiment (month) | Production/Production range (kg/hectare) |
|---|---|---|
| Assam (Gauhati) | 12 | 4,084 |
| Andhra Pradesh (Badampuri) | 12 | 3,535-4,648 |
| Bihar (Ranchi) | 12 | 3,526 |
| Haryana (Karnal) | 6 | 3,448-5,894 |
| | 8 | 6,191-7,332 |
| Maharashtra (Poona) | 8 | 2,200 |
| Orissa (Cuttack) | 12 | 7,500 |
| Tamil Nadu (Bhavanisagar) | 12 | 3,000 |
| Uttar Pradesh (Gujartal) | 12 | 7,037-7,370 |
| West Bengal | | |
| District Nadia | 12 | 3,393-6,053 |
| District 24 Parganas* | 12 | 4,947 |
| | 4 | 2,025 |

*Source :* Compiled from the reports of the All India coordinated research project on composite fish culture and fish seed production (1976), CIFRI, Barrackpore.

*Sarkar (1991,1996)

### *Europe*

In Europe, the common carp is the main fish cultured in ponds. Besides the common carp, pike-perch (*Lucioperca lucioperca*), sheat fish (*Silurus glanis*), black bass (*Micropterus salmoides*), tench (*Tinca tinca*), crussian carp (*Carassius auratus*), grass carp (*Ctenopharyngodon idella*), and silver carp (*Hypophthalmichthys molitrix*) are also stocked in different combinations. Due to low temperature, culture period varies between 2 and 3 years to produce table size fish because fish growth is directly related to temperature of water. Fish cultivation is still rather intensive in Central Europe. Artificial feeding, fertilization and liming are the measures which allow increase in production. Fish culture is, however, favoured in the development schemes. Several cross-breeds (such as white fish), due to their rapid growth rates, are considered for culture in ponds. These breeds are also cultured along with the traditional fish species.

### *United States of America*

Warm-water fish culture in the USA is principally based on the establishment and management of balanced fish populations. Several species of fish such as the large mouth bass (*Micropterus salmoides*), the bluegill (*Lepomis macrochirus*), crappies (*Pomoxis* sp.), bullhead (*Amevius* sp.), red-ear (*Lepomis microlophus*), and channel catfish (*Ictalurus punctatus*) are cultured. If satisfactory amount of fish is harvested, the interrelationship in fish population is also satisfactory when the fertility of pond water is taken into consideration. Such population are considered to be *balanced populations* and the species present within the population are said to be in *balance.*

Grass carp, common carp, and tilapia are also cultured though common carp has a very limited market. The range of production of different types of fish is shown in Table 21.5.

Fertilization of pond usually consists of 10-12 periodic applications of inorganic fertilizers. Ponds develop communities of algae and aquatic plants before the onset of fertilization. The abundance of phytoplankton required for optimum yield of different species of fish generally varies. For example, experience suggests that greater abundance of phytoplankton is necessary in the culture of *Tilapia* than in the culture of *Lepomis*.

**Table 21.5 : Production Range of Different Types of Fish**

| Species | Range of production (kg/hectare/season) |
|---|---|
| Channel catfish | 2,243 - 2,647 |
| Java tilapia | 2,010 - 4,916 |
| Nile tilapia | 2,666 - 4,486 |
| Bullhead | 1,008 - 1,344 |
| Common carp fry* | 1,300 - 1,900 |
| Common carp fingerlings | 1,456 - 2,120 |
| Major carp fingerlings* | 1,730 - 5,020 |

*Source :* Selected data from Swingle (1960)

* Sarkar (1993,1994)

## 21.9 Problems in Polyculture

Polyculture is generally undertaken in an organized commercial scale. As a result, a number of problems continue to play an important role in polyculture development. For example, fish farmers suffer drastic losses in production owing to disease outbreaks from high stocking densities in their ponds. Moreover, addition of freshwater to polyculture ponds, if required, that may contains toxic substances and pathogens, causes decrease in production. Availability of quality fish seeds, trained personnels, and financial constraints are also equally responsible for decline in fish yield. Accumulation of metabolites due to application of fertilizers, feeds and addition of excretory product of fishes cause pollution of water that results in considerable economic losses to fish farmers. In heavily fertilized/manured composite fish culture ponds or in cases where feeding is done during the culture period, algal blooms are excessively developed (Figure 21.2). This situation sometimes causes high mortality of carp fry due to gradual depletion of oxygen concentration in water that results in poor production. If these problems are however mitigated, the polyculture systems are bound to yield high dividends. Several other problems in polyculture development are receiving considerable attention in areas concerned, and efforts are under way to ensure more sustainable development of the sector.

## 21.10 Control of Nitrogen in Polyculture

Nitrogen in fish culture ponds is contributed to high-energy supplemented feeds, organic manures and fertilizers in different forms such as nitrate-nitrogen and ammonia-nitrogen. Although these factors influence the degree of phytoplankton bloom in ponds, over-production of phytoplankton changes the hydrology.

### *Nitrogen Elevation in Polyculture*

In carp culture ponds where intensive or semi-intensive culture systems are adopted, nitrogen is released into pond water through artificial feeds. This occurs in three ways : (1) When fishes are

**Fig. 21.2:** In many controlled freshwater habitats algae often undergo spectacular outbursts of growth. These phases, known as the algal blooms, occur in response to an increase in nutrient levels beyond the threshold limits. The foul smell associated with algal blooms is caused by algal cells consuming up all the oxygen in the water and releasing toxic gases such as hydrogen sulphide. Excessive growth of algal blooms due to heavy fertilization and manuring in a composite fish culture pond is shown in this Figure. The blooms were dominated by two species such as *Spirogyra* sp. and *Oscillatoria* sp. Effective control of the bloom is badly needed.

fed with water-soluble feeds, (2) If there is poor absorption of feed, nitrogen is retained within the body of fish. This nitrogen is excreted either through faecal matter or gills, and as a result, nitrogen concentration in water is increased, and (3) Feeding of fish at a time when hunger is not occur or overfeeding. Nitrogen is, however, lost through various ways such as (1) uneaten feed, (2) excess use of organic manures and chemical fertilizers, and (3) an increase in temperature of water. In general, excretion of nitrogen by fishes is related to the amount of nitrogen ingested through feeds. Greater the ingestion of nitrogen, higher rate is the excretion of nitrogen by fish. It has been observed that the range of nitrogen output at food conversion rate of 1.5 is increased along with the increase in the production of carps (Table 21.6).

Although protein in fish feeds stimulates better growth of carps, regeneration of ammonia by heterotrophic bacteria is increased if carbon-nitrogen ratio is not taken into consideration. This situation is more important in ponds where stocking densities are high. Feeds and organic matter sink into the bottom of pond, decomposed by heterotrophic bacteria and consequently, nitrogen concentration ratio in water is increased. A substantial amount of nitrogen is lost in the sediment and alters the soil nutrient status and the structure of bottom organisms.

**Table 21.6 : Estimated Range of Nitrogen Output in a Carp Culture Pond**

| Fish production (kg/hectare) | Nitrogen (kg/hectare) |
|---|---|
| 250 | 3.0 - 6.0 |
| 500 | 6.0 - 10.5 |
| 1,000 | 12.5 - 21.0 |
| 1,500 | 20.0 - 32.5 |
| 2,000 | 25.2 - 40.0 |
| 2,500 | 33.5 - 46.8 |
| 3,000 | 38.5 - 58.5 |
| 4,000 | 44.5 - 66.0 |
| 5,000 | 50.0 - 75.0 |

## *Management of Nitrogen*

The following management systems are adopted to control nitrogen in polyculture :

1. *Activated Microbial Suspension* : Nitrogen in pond water can be controlled by activated microbial suspension. It is a system by which a dense microbial population is developed. These microbes degrade organic waste substances into simpler forms thus reduces the level of nitrogen from the pond water. Recently it has been established that the application of some carbohydrates such as starch and cellulose to pond water triggers to uptake of inorganic nitrogen which, in turn, produces protein for the growth of microbes. This protein is utilized by fish.
2. *Shifting of the Pond Environment* : To improve the quality of pond water, it is necessary to shift the pond environment from an autotrophic to a heterotrophic microbial population because heterotrophic micro- organisms rapidly degrade organic wastes and ultimately nutrients are released into the pond water. This system maintains a healthy pond environment for sustained production.
3. *Water-Exchange System* : This system helps reduce the concentration of nitrogen in pond water although this system is a cost factor. Retention of pond water for longer periods is beneficial to poor farmers to remove nitrogen by natural processes. If draining of pond water is necessary, drainage velocity must be minimized so that the sediment remains in undisturbed conditions since high drainage velocity would reduce the amount of organic nitrogen.
4. *Nitrogen Retention Efficiency* : If the amount of nitrogen in fish is estimated, the nitrogen retention efficiency and the evaluation of diet can be computed. The nitrogen retention efficiency is determined by the following formulae :

   (i) Nitrogen retention = Final body nitrogen (absorbed into the blood from the diet) – Initial body nitrogen × 100 – Total dietary nitrogen supplied

   (ii) Nitrogen balance = Nitrogen consumed – Nitrogen retained – Nitrogen loss – Faecal nitrogen
5. Water-stable and quality feeds should be provided. This results in good consumption by fish and less faeces as well as metabolic waste.
6. Over-stocking of ponds with fry/fingerlings should be avoided.
7. Excessive use of inorganic and organic fertilizers should be avoided although farmers have already reduced their use because of environmental and economic concerns. It is expected that this trend will continue in the future.

## 21.11 Conclusion

Polyculture or mixed fish farming is a fish rearing activity which generally involves growing, feeding, and other management schedules of different species of fish together in a well-managed pond under controlled conditions. It is highly dependent upon the inputs provided by the farmer.

Fish production may be optimized by the application of management techniques to control the loss of revenue. Inspite of the difficulty of predicting the potential level of production of polyculture ponds, a pattern can be seen where the technologies are now reaching their practical limits. Actions for future management strategies in this regard must take this condition into account.

The application of technology in polyculture has expanded greatly in recent years. Consequently, success of mixed fish farming, particularly in the tropical and sub-tropical zones often is dominated by fish production potential. While excess use of nutrients and feed as well as high stocking densities cause pollution of pond ecosystem, high concentration of toxicants and tendency to reduce fish growth and increased mortality deserve special attention.

Although polyculture is considered as a high risk bio-industry as there is every possibility of loss of fish biomass at any stage of management, it is found to be more profitable when compared with other livestock enterprises.

### *References*

Agarwal, S.C. 1990. *Fishery Management.* Ashish Publishing House, New Delhi.

Anon. 1969. High fish yields from mixed culture in small freshwater ponds. Central Inland Fisheries Research Institute, Barrackpore.

Lakshmanan, M.A.V., K.K. Sukumaran, D.S. Murthy, D.P. Chakraborty, and B.T. Philipose. 1971. Observations on intensive fish farming in freshwater ponds by the composite culture of Indian and exotic species. *J. Inland Fish. Soc. India.* 2 : 1-21.

Sarkar, S.K. 1991. Role of phosphorus fertilizers on the effectiveness of urea and lime in fish survival and yield and their influence on lateritic pond ecosystem. J. Environ. Biol., 12 : 287-299.

Sarkar, S.K. 1993. Combined effects of atrazine, nitrogen, phosphorus and lime on fish and aquatic ecosystem. *Ind. Fert. Scene Ann.* 6 : 43-51.

Sarkar, S.K. 1994. Role of diammonium phosphate in pisciculture. *Fisheries World.* 1 : 29-34.

Sarkar, S.K. 1996. Influence of calcium ammonium nitrate of fish and aquatic ecosystem. *Ind. J. Fish.*, 43 : 87-95.

Sinha, V.R.P., M.K. Banerjee, and D. Kumar. 1973. Composite fish culture at Kalyani, West Bengal. *J. Inland Fish. Soc. India.* 3 : 201-207.

Swingle, H.S. 1960. Comparative evaluation of two tilapias as pond fishes in Alabama. *Trans. Amer. Fish. Soc.*, 89 : 142-148.

## Questions

1. What do you mean by composite fish culture and what are its advantages over the monoculture system?
2. What are the objectives of polyculture? State some factors that affect fish production in mixed fish farming.

3. Which species are considered in polyculture system and why?
4. Although polyculture is undertaken on commercial basis, this system of farming causes, in many cases, serious problems. Why?
5. What major changes have taken place in polyculture industry during the last twenty five years?
6. What are the techniques adopted in polyculture system in India?
7. How would you explain the carp production potential of India and China?

# 22

# Water and the World's Fish Supply as Food

Water is one of the most important and precious commodity of life. Although it can be found anywhere it is seldom available in the qualities and quantities necessary for the purpose.

Sustainable freshwater fish culture management strategies require a uniform supply of fresh and unpolluted water throughout the culture period. Therefore, requirement of unpolluted water is essential to successful fish production and consequently, water increases yield and allows for a reliable supply of healthy fish for market. However, reduction of water level in any fish culture ecosystem can cut production by 30-40 per cent. Water offers the richest resources available to rural commodities.

Fish is a very important dietary animal protein source for the human society. Malnutrition is an old to the world and it has already been a threat to human survival. In any rural and urban places of the world, protein deficiency has brought human disease, illness and death. This threat is due to inadequate production of fish in several countries, excessive pollution of freshwater and marine ecosystems, financial constraints, political conficts, etc. The main problem of fish protein supply to human society particularly in Low-income Food-deficit Countries depends on their increasing population in a gallopin rate. Production of aquatic species through freshwater fisheries and aquaculture for protein supply is being or has been carried out rapidly in developed and developing countries but in under-developed countries, it is declining. Since fish is considered as one of the most important source of nutrients, its importance and significance should receive attention first.

## 22.1 World Population Expansion

With increasing population requiring increased production of fish and fishery products, the world has been forced to lead the way with sensible fisheries management strategies for sustainable fish production.

Advancement of science and technology is very important for the greater expansion in world population growth which has occurred in the developing countries. Until about the midpoint of the 20th century, high birth rates in Africa, South America, and Asia were negated by equally high death rates. Poor health facilities suffering from malnutrition, high infant mortality, and disease-spreading insects had drastically checked the population growth.

During the past several decades, advances in medical sciences and application strategy of different technologies have triggered the expansion of global population that has changed the entire situation (Figure 22.1). Since medical assistance is now available in remote areas of almost all the nations, it has paved the way for reducing the death rates among the child and young. Of course in low-income food-deficit countries, this scenario is still not in existence owing to inadequate medical personnel. Thus population growth has taken place in an unhampered manner. Statistical evaluation has shown that the population is doubling every 20-30 year in developing countries where about two thirds of the world's population now survive. As per recommendation of expert committee, it is evident that the world population will be between 8 and 8.5 billion by the year 2020 – about 34 per cent higher than it is today. However, about more than 70 per cent of this increase will occur in those countries where supply of aquaculture products is not adequate. This increase in the number of mouths to be fed is so large that it is necessary to exploit freshwater fish culture resources (and also marine water resources) and the capacity of different types of aquatic ecosystems could soon be reached a plateau using current farming technology. Hence the world's fish supply as food continues to be the most serious problems which is very difficult to overcome unless specific measures are adopted.

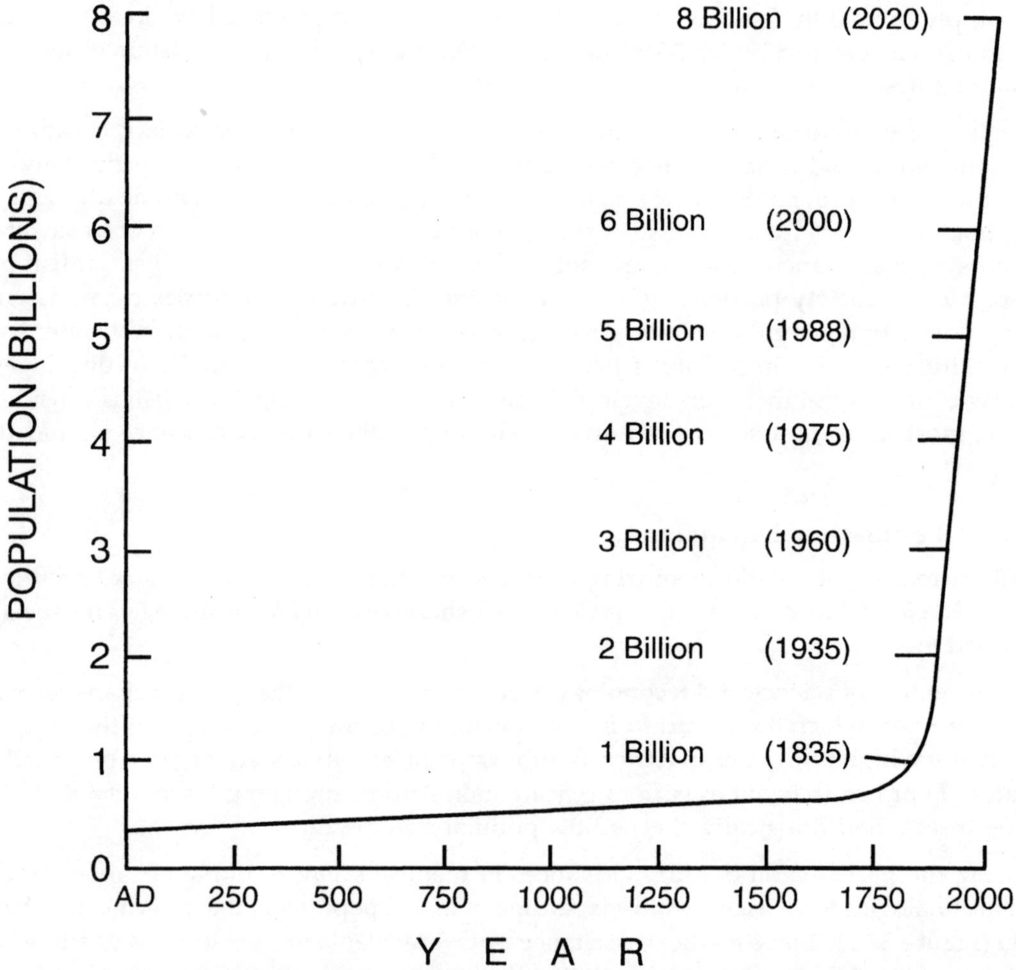


**Fig. 22.1 : From the beginning of the human race until 1935, the population of the world increased to 2 billions. Note that every 16 years, the world's population increased by 1 billion. It is also important to note that most of the population growth will be observed in countries where there is severe malnutrition, inadequate technology, and food supplies are critical.**

Although the economic systems of some countries have, to some extent, managed the demographic pressure of the population bomb, the time has come to re-examine the relationship between population and local ecosystem including environmental resources. If the nation is to reach a balance between population and resources, policies that encourage the farmers to develop aquaculture activities will undoubtedly help to increase finfish and shellfish production and thus contribute to further development of the human society.

## 22.2 Parameters Affecting World Food Supplies

It seems quite plausible that the capacity of a country to produce fish and fishery products is determined by several factors such as economic, social, and political. These factors combined or individual factors dramatically affect the farmers' activity to yield aquaculture products at commercial levels. Aquaculture production is also affected by the following factors :

1. Distribution of production materials such as manures, fertilizers and chemicals that check aquatic weeds and diseases of fish.
2. Balanced artificial feeds having quality control certified by appropriate authorities.
3. Improved fish breeds that respond to adequate management.
4. Available technology and knowledge of suitable management strategies of aquatic ecosystem.
5. Availability of good quality of water for culture.
6. Availability of pure seeds of commercial fish; fry and fingerlings of fishes and juveniles of shellfish.

Application of all these factors to aquaculture production depends on the area and quality of freshwater systems, their overall productivity and response to management on scientific basis. In order to sustain adequate supply of fish, fish products, shellfish and aquatic plants for human consumption, three main priorities must be emphasized towards production such as financial assistance, management strategies and implementation of new technologies.

## 22.3 Fish as a Source of World's Nutrition

To feed an ever increasing population, a substantial part of the arable lands of the world has been brought under cultivation through man's quest for food. Concurrently, the fishing frontiers have been pushed to freshwater and marine of all countries. For several centuries, farming and fishing advanced at more or less comparable rates. Since the middle of the nineteenth century, agriculture has far out distanced the fishing industry through the world. As long as more/and could be brought into production, less consideration was given to the vast water areas in the world as a source of food. During the last two decades, greater attention has been given to explore fishery resources and tapping their vast stores of animal protein food.

### Catch Potential

In many countries of the world such as Japan, Russia and European countries chief emphasis was put on fishing from marine and inland resources as a means of rapidly acquiring needed high quality protein. It has been estimated that the global fish and shellfish production from capture and culture fisheries has reached an annual level of 109.6 million metric tonnes (Table 22.1) with annual growth rates of over 6-7 per cent. Considering marine fish catch in 1995 (Figure 22.2), it is evident that Asia is the leading community (41 per cent of the total world catch) followed by south America (25 per cent), Europe (15 per cent), and North America (8 per cent). It is important to note that not all landed fish is for direct human consumption. Out of 109.6 million metric tonnes, about one-fifth of the catch is used for the production of oil and meat. Moreover, three-tenth of the catch is comprised by shellfish. This leaves an amount of 55 million metric tonnes of food fish, about 40 million metric tonnes of which is considered as marine and the remainder as freshwater areas (lakes, ponds, river etc.).

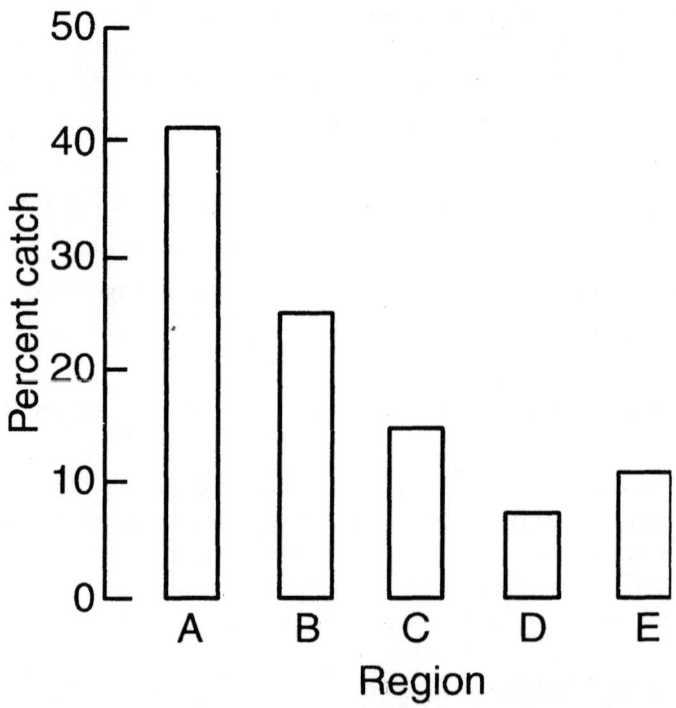


**Fig. 22.2 : Marine fish catch in different regions of the world during the year 1995. Note that Asia (A) is the leading community (41 per cent of the total catch) followed by South America (B, 25 per cent), Europe (C, 15 per cent), Others (D, 11 per cent), and North America (E, 8 per cent).**

**Table 22.1 : World Catch Potential of Finfish and Shellfish**

| Year | Million metric tonne |
|---|---|
| 1950 | 23.0 |
| 1960 | 37.8 |
| 1970 | 51.5 |
| 1980 | 77.0 |
| 1990 | 97.3 |
| 1998 | 109.6 |
| 2010 | 135.0* |

*Anticipated

### *Freshwater Fisheries*

In many countries, freshwater fisheries supply a substantial part of the nation's protein food. Almost half of the world's freshwater catch (approximately 4 million metric tonnes) refers to China, where fish production has been predominant. Fish is captured from an area no less than 16.5 million hectare, of which 6,000-500,000 hectare constitute 6.3 million hectares.

Russia presumably ranks second and its inland water account for 6 million metric tonnes. But with the effects of rapid industrialization has drasitcally affected freshwater fishing. Adequate maintanence of ponds and fish-rising area constitute major endeavour to hold the freshwater trout. Fishing from natural fishing water is supported by stocking with fingerlings and feed organisms and other production-increasing strategies.

The freshwater fisheries of African continent are concentrated in Lake Nyasa, Lake Victoria, Congo, and Nile basins. Tanzania and Uganda together account for 150,000 metric tonnes. The Victoria Lake is rich in fish.

The Great Lakes of Canada and the United States provide about 70,000 tonnes of fish a year. In Northern and Eastern Europe, Japan and Asia, freshwater fisheries are of considerble importance, although hampered by population growth and industrialization. Inland fisheries are important as food resources in Brazil, Argentina and Venezuela.

## 22.4 Productive Capacity of a Sea and Freshwater Areas

Fish production obtainable from catches give an indication of the productive capacity of various waters, as influenced by climate, latitude, availability of nutrients etc. However, the Baltic and the Mediterranean show low productivity for different reasons. Temperate latitudes deliver more fish and Arctic regions are sparse. Freshwaters are more readily productive than marine areas, and peaks in production are higher in controlled ponds (Table 22.2).

**Table 22.2 : Productive Capacity of Sea and Freshwater Areas**

| Fish catch from some sea areas | | Fish catch from some freshwater areas | |
|---|---|---|---|
| Seas | kg/ha | Freshwater | kg/ha |
| East China sea | 75 | Oligotrophic lakes | |
| Japanese sea | 50 | (temperate) | 8.0-16.5 |
| North sea | 40-50 | Eutrophic lakes | |
| Caspian sea | 12 | (temperate) | 30-250 |
| Baltic sea | 10-15 | Fish ponds (unfertilized) | 50-150 |
| Mediterranean sea | 10 | Fish ponds (tropics) | 350-1,250 |
| Bay of Bengal | 40-75 | Fish ponds (Fertilized) | 250-9,500 |
| | | Tropical lakes | 2,200 |

*Source* : Modified after Mortimer and Hickling (1954)

### *Production by Principal Producers and Region*

Twenty countries account for about 80 per cent of total world production, while ten countries account for almost 70 per cent (Figure 22.3). China has been the largest producer since 1988, when it overtook Japan, having started to increase production dramatically in the early 1980s.

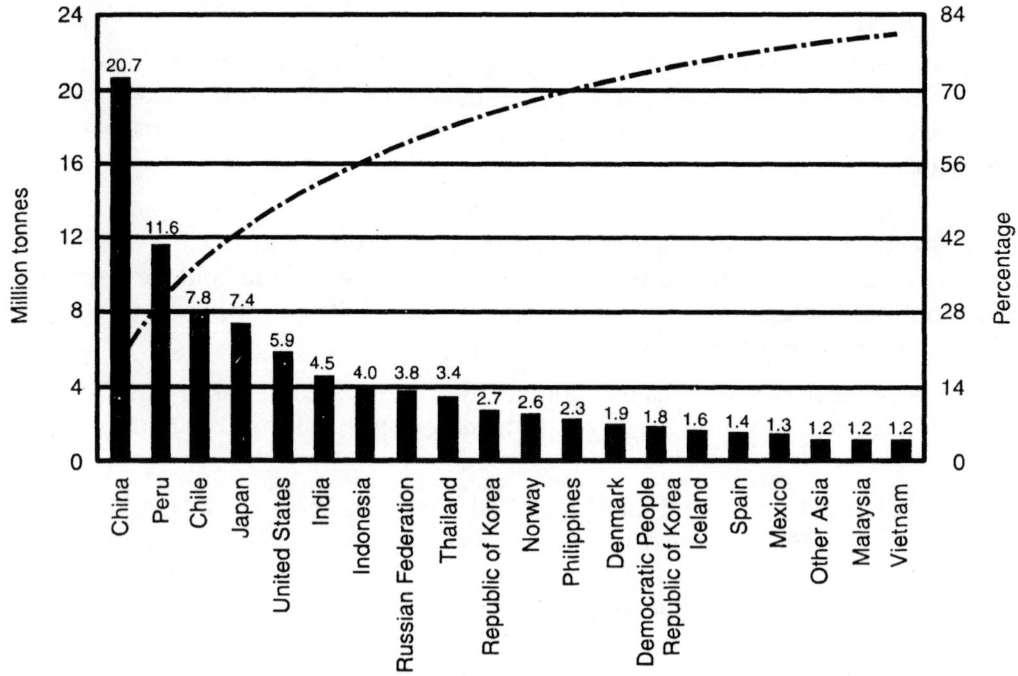


**Fig. 22.3 : Production by principal producers (solid column) in 1994 and cummulative production as percentage of world total (dotted line). Countries listed are those with a production above 1.1 million tonnes. [Redrawn from FAO (1997); Used with permission of the Food and Agricultural Organization of the United Nations]**

In examining the changes in fish production by the comparative performance of each major region, the region showing a marked decline in production in Europe, which here includes the European republics of the former USSR (Figure 22.4).

The decline in European production is confined to the former centrally controlled economics in Eastern Europe and the former USSR. The expansion of production in China accounts for most of the growth achieved by the East Asia group, while the increase in anchoveta catches in the south-east Pacific has driven the expanded production of Latin America and the Caribbean. Production in South Asia and South-east Asia has contracted slightly, while a greater decrease in production in North America is owing to the contraction of Canadian east coast fishing activities. Other regions have shown little change in production. Unless environmental quality deteriorate/fluctuate widely, fish production rate will increase continuously in future through adequate management strategies with small or large oscillations reflecting natural variations of resources productivity.

So far as the regional fish production is concerned, it is evident that among different regions, production from different fisheries sectors have been estimated to be highest in East Asia, Latin America, Europe and South-east and South Asia (Table 22.3). It is however, expected that with the application of latest technologies, fish production will undoubtedly increase in different fishery sectors to meet the demand of ever-increasing populations.

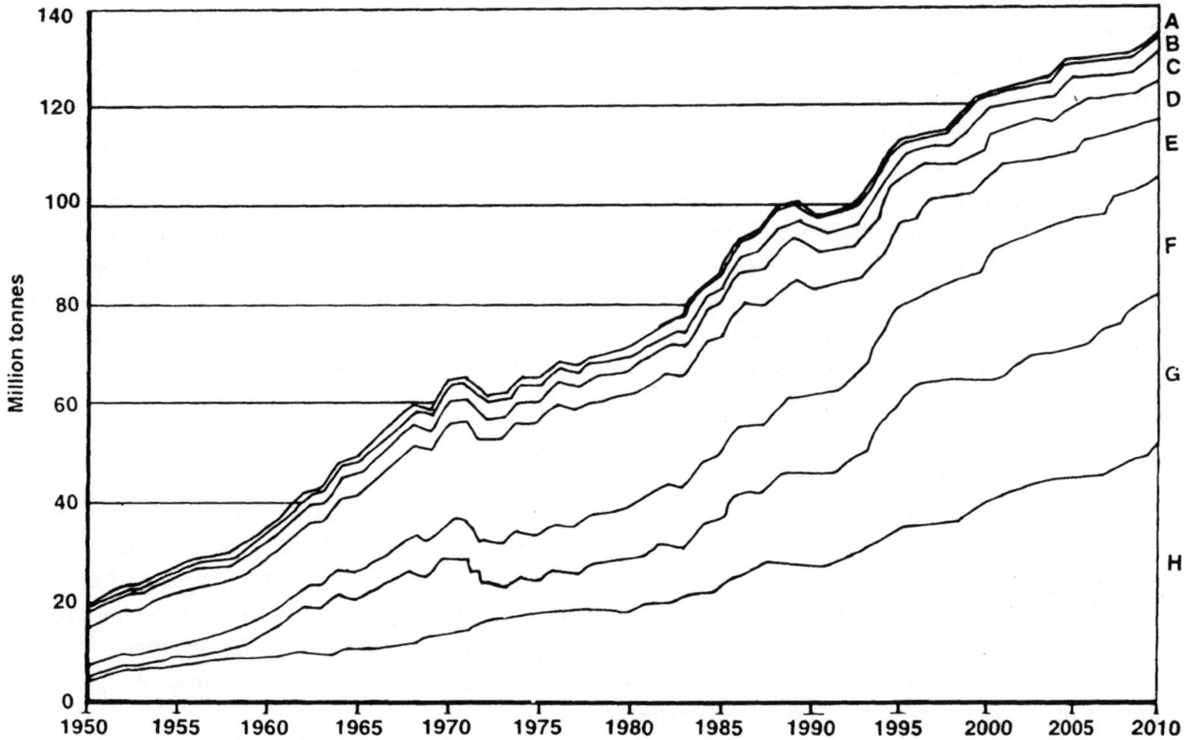


**Fig. 22.4 : Total fishery production by region. A, South Pacific; B, Near East and North Africa; C, Sub-Saharan Africa; D, North America; E, Europe; F, South and South-east Asia; G, Latin America and the Caribbean; H, East Asia. Anticipated data are shown in the year 2005 and 2010. [Redrawn from FAO (1997); Used with permission of the Food and Agricultural Organization of the United Nations]**

**Table 22.3 : Regional Finfish Production (in Million Tonne)**

| Region | Marine | | Inland | | Aquaculture | |
|---|---|---|---|---|---|---|
| | 1994 | 2010* | 1994 | 2010* | 1994 | 2010* |
| South Pacific | 0.77 | 2.80 | 0.025 | 0.90 | 0.075 | 1.0 |
| East Asia | 22.60 | 30.70 | 1.4 | 3.5 | 18.4 | 27.5 |
| Europe | 15.3 | 27.40 | 0.42 | 1.5 | 1.5 | 3.6 |
| Latin America and the Caribbean | 23.1 | 37.00 | 0.5 | 2.0 | 0.47 | 3.6 |
| North America | 6.5 | 13.50 | 0.71 | 3.4 | 0.50 | 3.5 |
| Near East and North Africa | 2.2 | 8.00 | 0.45 | 4.0 | 0.15 | 3.5 |
| South-east and South Asia | 13.4 | 20.00 | 2.3 | 8.0 | 4.4 | 14.5 |
| Sub-Saharan Africa | 4.2 | 15.00 | 1.6 | 10.0 | 0.033 | 0.70 |

*Source* : FAO (1997) * Anticipated

## 22.5 Consumption

Despite the growing world population, the per caput fish consumption is increasing more rapidly. Fish is unique when compared to other food items. This indicates the dependence on fish in the feeding of the mankind. Consequently, the fish culture industry is getting better organized. But there are some exceptions to this general rule. Declines per-caput consumption are reported in South and South-east Asian and Sub-Saharan African countries (less than 7 kg in 1994) where the extreme shortage of food mainly due to excessive population growth. The population pressure is inducing scientific methods of freshwater aquaculture. Therefore, the greater degree of dependency of the human society on fish as food constitute triple challenge : (1) improved methods of culture and capture of fish, (2) the need for better protection of freshwater resources from pollutants, and (3) the need for better protection of fish from diseases. Partial or complete spoilage of fish raises the question of the consumption of fish which has an influence on the fish in nutrition.

## 22.6 Global Water Resources

There is a total of about 362, 149, 700 Square Kilometers area of marine and freshwater in the world. Out of this total water area, the average freshwater area has been estimated to be about 34, 114, 970 Square Kilometers. But most of the freshwaters are not suitable for capture and culture fisheries. In many places water is highly saline (more than 25 ppt.), polluted, unproductive or moderately productive, too cold, and therefore, not fit for fish life. While a major amount of water bodies are not fertile, only a very limited amount of water areas are actually productive so far as fisheries and aquaculture aspects are considered.

Most of the inland water areas are suitable for fisheries and aquaculture, but for so many reasons cannot be properly utilized. However, the types of waters and their response to management of aquaculture and inland water resources may hold the key to adequate finfish and shellfish production in many countries of the world.

### *Regional Differences*

As noted earlier, waters may play a key role for meeting the world's food requirements to large extent. While the total potential water areas is more than double that being utilized today, there is great variation from region to region. In Europe and Asia, where population pressure is high for years, most of the potential freshwater ecosystems are under cultivation. In Oceania and South America, the physico-chemical potential for greater utilization of cultivated waters is high.

It is not fortunate that cultivable waters are not better distributed in relation to population densities. Per caput area of culturable water is high in Russian Federation and North America. It is low in Africa, Asia and Europe. This does not create a serious problem in Europe and economically developed parts of Asia. Although they are able to purchase fish from the countries with excess supply, trade policy and marketing problems are the main constraints.

In South and South-east Asia, Latin America and Africa, the situation is very critical. This is due to the fact that their populations are increasing at gallopin rate and as a result they are compelled to cultivate fish at high levels. Countries that previously exported fish products now import them. The national economic growth rate of these countries is very slow to provide resources to pay for the needed fish as food. They must either increase their own capacity to produce fish or must be provided with fish by their fortunate neighbours.

### *Action Taken for Fish Production*

In order to augment fish production, nations may follow the following three actions such as : (1) intensify fish production already under cultivation, (2) clear fallow water areas that have heretofore not been carried out, and (3) increase fish culture intensive (number of fish culture period in ponds, lakes etc. per year). Asian and European countries try to intensify fish production with the aid of recent technologies. They increase their annual fish yield per hectare of water area which ultimately produce more fish. South America and Africa exhibits physical potential for increasing water areas under cultivation is great. However, large water areas in these continents have tropical climates and for this purpose sustainable management of inland waters and aquaculture is yet to be considered.

### *Environmental Quality Constraints*

There are other reasons for caution in expanding water under cultivation. For this purpose, unproductive and fallow waters should be brought under cultivation and replaced with different modern farming systems. This invariably leads to increase in pollution and consequently, productivity of water drastically decreased. In densely populated areas, most of the freshwater systems have resulted in serious damage due to pollution. These water bodies should not be allowed to contaminate with pollutants, a stringent management strategy that has been followed in developed countries.

The degradation of freshwater fisheries resources results in a loss of productivity and consequently, commercial species of plants and animals will be lost. Moreover, intensive fish culture and integrated fish culture systems in most of the tropical countries accumulate appreciable amounts of metabolites of toxic compounds in aquatic environments which can markedly influence fish culture industries. Since there are no alternatives to aquaculture extension as means of providing future protein needs, these systems of fish cultivation must be adopted through careful management.

## 22.7 Potential of Different Waters

Many humid tropical freshwater areas are dominated by acids soils through which water is easily leached. As a consequence, productivity of such water areas is low. With adequate management such as application of organic and inorganic materials, these waters can be made quite productive.

In contrast to this, water is not leached from semiarid tropical fish ponds. With proper management, these water bodies in this tropics can also be made productive. But it is unfortunate that the semiarid tropical ponds are subject to considerable variability in annual levels of water. This means that periods of low level of water in fish culture ponds and lakes are common. Depletion of water level in fish ponds indicates the difficulty of maintaining sustainable fish production in this semiarid areas.

Some tropical and sub-tropical countries such as India, Bangladesh, Thailand, Indonesia, Queit, Yemen, Oman, UAE, Katar, Southern parts of Saudi Arabia, Iran, Iraq, Jordon and Northern parts of Saudi Arabia are very important so far as the freshwater aquaculture is concerned. These countries produce appreciable quantities of finfish and shellfish from freshwater areas when these waters are treated with nutrient carriers. High solar radiation in these countries indicates that water bodies are highly productive which stimulate high fish production.

Japan, China, Peru, Chile, Latin America, Korea, Taiwan, and Russian Federation are countries of high shellfish and finfish production. Fish production in Europe, North America, Near East and North Africa are moderate while in Sub-Saharan Africa and South Pacific regions, fish production is comparatively less than other parts of the world. Therefore, attention is focused on the regions where fish production is moderate or less and where water resources are not being effectively utilized.

## 22.8 Constraints for Production

There are some sorts of limitations of waters for fish culture in any geographical areas of the world. However, a deficiency of nutrients in water and soil, wide fluctuations of temperature and inadequte supply of nutritionally balanced diets are the most significant world-wide constraints. These constraints are most notable in South Asia, South-east Asia, and Africa. Moreover, excess water due to heavy monsoon rains is a serious problem in several countries.

### *Disease and Marketing Constraints*

Spread of fish/shrimp diseases are considered as a major threat to aquaculture. In the early 1990s, for example, the epizootic ulcerative disease of freshwater fish reached the Philippines, Sri Lanka, Bangladesh, India, Bhutan and Nepal, causing severe mortalities not only in fish ponds but in natural waters too. The disease raised serious public concern in the affected countires disrupting traditional fish distribution and consumption patterns. It is unfortunate that there was little progress in understanding the transmission of the diesease and the arresting of its further spread.

Marketing constraints in Asian and South-east Asian countries are emerging increasingly as major problems in the further developement of aquaculture. Domestic marketing difficulties for freshwater fish have been observed in several regions of the world such as in Malaysia, Indonesia, and the Philippines. Marketing of more value-added products on the export markets, improvement of the internal marketing systems and promotion of domestic fish counsumption have been proposed to overcome marketing problems.

Despite the emerging constraints, aquaculture proved to be the most rapidly expanding food production branch in these regions. However, unless marketing and disease constraints are realized, food production through aquaculture and fisheries may not be sustainable.

## 22.9 Problems and Prospects in Tropical and Sub-Tropical Countries

Under special situations, question may arise why water of tropics and sub-tropics have not been more effectively utilized for fish culture. Because they have some advantages over their temperate zone counterparts. In tropical and sub-tropical regions, a large number of fishermen and fish farmers are available and they are either temporarily or permanently engaged in this sector. Moreover, the fish crop growing season is usually all the year round and waters have physical and chemical features for better to the waters of the temperate zones.

In acid and moderately fertile soils and water in some regions such as Asia and Africa helps illustrate why waters of these areas have not been effectively utilized. This is due to the fact that waters in these regions are low in nutrients and often it is difficult to improve the productivity of waters and soils.

### *Limiting Factors*

Although tropical and sub-tropical water bodies are exposed to high solar radiation almost all the year round which indicates high productivity, deficient in nitrogen, calcium, and phophorus and high toxic levels of different agrochemicals in waters and bottom sediments are the primary factors limiting more effective utilization of tropical and sub-tropical waters for fish culture. In acid-sulfate soils, aluminium toxicity is more common. Several essential elements in tropical and sub-tropical soils of freshwater ecosystems are depleted gradually.

Special problems relate to tropical and sub-tropical water areas are characterized by dense forest. Since the compounds that make the humus are synthesized in the soil from the plant material, the quantities of major chemical nutrients such as potassium, nitrogen, phosphorus, calcium, magnesium and organic matter in soil and water are greater than in temperate zone. However, so far as the comparative distribution of nutrients and organic matter in soil in a tropical lake and pond (Table 22.4) are considered, it is evident that the soil is rich in phosphorus – limits the productivity of freshwater ecosystems. Therefore, phosphorus is the dominant nutrient element of a nutrient cycle that is primarily responsible for productivity in this area.

**Table 22.4 : Concentration of Nutrient Elements and Organic Matter in a Tropical Pond and Lake Soils**

| Element | Allochthonous Pond | Lake |
|---|---|---|
| Nitrogen (mg/100 g) | 15.57 | 4.70 |
| Phosphorus (mg/100 g) | 49.10 | 18.00 |
| Potassium (mg/100 g) | 328.00 | 27.47 |
| Organic matter (per cent) | 1.22 | 0.73 |

In many states of tropical and sub-tropical countries, periodic depletion of water level in cultured ponds and lakes during summer season is often a serious limiting factor for survival and growth of fish. Management strategies to reduce the evaporative loss of water must be paramount important for fish production. Moreover, industrially developed states cause water pollution that results in mass mortality and reduced growth of fish, thus fish production is drastically hampered. Environmentalists and aquaculturists are challenged to devise systems that will undoubtedly change this situation.

### *Research on Tropical and Sub-Tropical Waters*

It would not be exaggerated that little but fragmentary reports are known about water and soil in these regions inspite of adoption of new management strategies. However, research on these ecosystems and their potential for fish produciton is not significant when compared to temperate region waters. Of course, some informations on the water and soil in some tropical countries such as Bangladesh, India, Thailand, Oman, Indonesia and Jordon received so far is encouraging. For examples, characteristic features of ecosystems of Bangladesh, China, Japan and Indonesia are better for shellfish production. They respond well to modern management. In contrast, inspite of the adoption of modern management in shellfish and finfish farmings, there was dramatic failure for the shrimp culture scheme in several countries. Discharge of effluent, poor water quality, high stocking density, lack of extension service and break out of viral and bacterial diseases resulted in catastrophic decline in production. For example, in July 1994, two diseases of suspected viral etiology have had disastrous effects on shrimp farming which has upset the industry. It has been estimated that diseases have caused a loss of nearly Rs 600 crores to the industry. Therefore, the rise, fall and recovery of shellfish culture are common in these regions.

The chemical features of soil in the regions differ drastically from those of soil in temperate regions. The high humus and hydrous oxide content dictates phosphate-fixing capacities. The low cation exchange capacities result in removal of nutrients. Hence the level of technology required to manage the soils in the tropics is as high as that needed for temperate zones soils.

### *Management Strategies in Tropical and Sub-Tropical Waters*

Management systems of aquatic systems have been successful in increasing production capacity of finfish and shellfish. Shrimps are usually cultured to earn foreign exchange and therefore, require modern mechanized farming and financial inputs for high production. Best available technology for management strategies are imported from developed countries. Although it has already been successful in producing fish and shrimp to significant degree, it provides little benefit for indigenous small-scale fish farmers.

## 22.10 Integrated Farming System

At the opposite extreme from the management strategies are indigenous systems that have evolved by trial and error by the native fish farmers. The bulk of freshwater finfish production is based on different types of pond-based farming systems. These farming systems contribute a source of food fish for mass domestic consumption.

The aquaculture systems comprise an array of systems such as cage culture, pond-based composite fish culture, pen culture, sewage-fed fish culture, paddy-cum-fish culture etc. to suit the prevalent ecological conditions and local demands of fish. However, different types of culture systems such as composite, integrated, brackish water, and air-breathing culture systems trigger fish production to significant degree. Note in Table 22.5 the comparative production level in different farming systems in India.

### *Advantages of the System*

Integrated fish farming is a system of nutrient accumulation, utilization and recycling. This practice utilizes the waste from various components of the system such as poultry, duck, livestock and agriculture by-products for fish production. It has been estimated that about 45 kg of organic wastes are converted into 1 kg of fish. In fish-cum-sericulture system, pupae are used as fish feed, while worm faeces and wastewater from silk processing industries are used as pond fertilizer. Ducks and geese are reared in ponds, embankment of ponds is used for horticulture and agricultural crop production, and rearing of cattle, pig, poultry, goat, and mushroom. In addition to fish, the system provides eggs, fruits, fodder, meat, milk, and mushroom. The water areas and it adjacent land are used for increased food production for human consumption. It also holds a potential for increasing optimum production of high class protein at very low cost, employment generation and improvement of socio-economic status of rural communities. Although integrated fish farming is very promising, it deserves careful study. In some tropical and sub-tropical countries such as China, Taiwan, Thailand, Hong Kong, Malaysia, Bangladesh, Indonesia and some European countries like Hungary, Poland, Germany and others integrated farming system is more successful than that of more efficient temperate zone systems.

**Table 22.5 : Fish Production Level from Different Farming Systems in India**

| Farming system | Production level (Tonne/ha/year) |
|---|---|
| Composite fish farming | 4-15 |
| Integrated fish farming | 4-7 |
| Paddy-cum-fish culture | 0.5-1.0 |
| Biogas slurry-fed fish culture | 4-5 |
| Sewage-fed fish farming | |
| Strong sewage-fed fisheries | 10-14 |
| Moderate sewage-fed fisheries | 4-8 |
| Weak sewage-fed fisheries | 2-3 |
| Air-breathing fish farming | 4-5 |
| Prawn farming | 1-2 |
| Brackishwater fish farming | |
| Extensive | 1-2 |
| Semi-intensive | 5 |

*Source* : Shyam (1998)

At present, sewage-and rice-fish integrations are being extensively carried out in many tropical countries except India where, in addition to these integrations, poultry-fish, duck-fish, goat-fish, mushroom-fish, horticulture-fish, seri-fish, cattle-fish, and pig-fish integrations, (see section 11.20) are being conducted although these practices still remain very limited. Considering the availability of large quantities of plant and animal residues and low investment capacities of rural communities, organic recycling is the basis of integrated fish farming that holds great promise in the coming years.

Integrated fish farming systems not only maintain yields of crops, fruits, eggs, meat and proteins but also maintain soil organic matter and nutrient element levels. Waste materials of different animals are incoporated into the pond soil, degraded by soil micro-organisms, thereby enhancing organic matter and mineral element contents while simultaneously maintaining fish yields.

### *Loss of Productivity from Fish Culture Ecosystem*

To sustain high fish production from freshwater resources, fish farmers have adopted intensive and/or semi-intensive culture techniques to reduce the period of fallow – to recultivate a given water body after only few month fallow compared to several years in the past. Moreover, periodical harvesting and stocking the ponds with fry/fingerlings are also carried out. Thus, inadequate time is allowed for the rejuvenation of fish culture ecosystem between culture periods. In areas where shorter fallow periods are common, insufficient regeneration time is available and the fertility level declines rapidly. At the same time, disease out-breaks in fish populations due to intensive fish farming systems and undraining the culture ponds (either partial or complete drainage) are also the most important factors for reduction of fish production potential in many South-east and South Asian countries where production potential is obviously great.

### *Water Quality Deterioration*

Tremendous increases in human population have placed great strain on the integrated fish farming systems in some areas of the tropics and sub-tropics. More than 280 million people who depend upon integrated fish farming systems for most of their food represent a dramatic increase in numbers during the last several years. To meet the food demand for growing populations, fish farmers have been forced to adopt this system of farming and produced different types of food items. As a result, environmental and water qualities have drastically been reduced. Unless a few centimeters of pond soil is removed after few years of continuous fish cultivation, accumulation of toxic substances will continue throughout the culture with increasingly toxic effects on fish. Possibly, it is one of the reasons for a decline per caput fish protein production for large population in these regions.

### *Potential of Tropical and Sub-Tropical Fish Culture*

One should be optimistic about the future prospect of fish culture through various culture systems in these regions. Warm to hot climate and high solar radiation undoubtedly provide photosynthetic potential. Integrated aquaculture system encompasses the process of trapping solar energy for organic matter production by the producers which is then utilized by the consumers. After the death and decay of consumers by the saprotrophs, nutrients are released for producers. However, several species of fish are cultured in suitable combinations. Scientists have been successful to culativate varieties of freshwater fish in ponds and lakes under different geographical and climatic conditions. Usage of chemical fertilizers and organic manures becoming more common in three regions. Therefore, there is tremendous scope for producing more food in the form of meat, eggs, vegetables, crops, fruits in addition to fish.

## 22.11 Closed Versus Open Circulation Systems in Fish Production

Fishing and fish culture are the two oldest sources of food for mankind. In many places, however, catches from traditional fishing in rivers and in seas have come down because of pollution by industrial wastes and effluents. Even small ponds and lakes far away from industrial centres have upset the equilibrium by acid rain. To overcome these problems it is necessary to apply suitable methods which will control variable aspects of fish culture activity. Almost all intensive fish production farm must be developed for maximum fish production.

In general, two systems of farming are in vogue such as (1) open system and (2) closed system. In the first case, water is collected from rivers, seas or other sources and polluted waters are rejected. Open system consists of cage system in the open sea or ponds or lakes. Gradual accumulation of toxic compounds, high rate of mortality, loss of fish stocks, and pollution of waters are the main disadvantages of open systems.

In closed system of farming, with optimum controlled environment throughout the culture period, high quantities of good quality fish can be produced from a small or medium fish farm. The net weight of individual fish cultured in a closed system is several times higher than that of open one. However, the closed circulation system is widely practiced in many European countries.

## 22.12 Sewage Utilization

The sewage utilization for fish farms means the conversion of effluents through sewage-treatment plants into fish food. Diluted sewage (sewage : water = 1 : 1) induces production of major carps at the rates varied between 5,400 ad 8,000 kg/hectare/year. However, one hectare of water purifies the mechanically classified sewage from 2,000 persons, yielding 27.5 kg of protein. Raw sewage is extenisvely converted into food via fish in the Far East.

## 22.13 Future Requirements

The ability of the world to supply fish as source of protein depends upon several factors. Of them, improved technology in the developing nations is very important. Improved technology primarily depends upon education and research which have direct relevance to the countries. However, the technology of the United States or of Europe should be transferred to the underdeveloped countries.

There are two main requirements for increased aquacultural production in the tropics and sub-tropics. First, the economic, social and political climate in the developing countries must be such as to make it profitable for the farmers to adopt the new technologies. Second, improved technologies must be adopted for future generations in the tropical and sub-tropical regions.

### *Fertilizers and Manures*

To increase food production from agriculture sector, dramatic increases in supplies of fertilizers and manures have occurred. The total $N$-$P_2O_5$-$K_2O$ consumption in the developing countries was about 45 million Megaton (MMg) in 1997. But so far as aquaculture production is considered, it is obvious that application of nitrogen fertilizers should be reduced as far as possible as metabolites of nitrogen fertilizers are highly toxic to fish and consequently, limits fish production. Phosphatic fertilizers are very important for productivity of fish culture ecosystems. Use of organic manures is by far the best as they are least toxic than chemical fertilizers. Tropical and sub-tropical countries have tremendous vegetation and consequently the soils in these regions are highly productive. Therefore, organic matter is continuously accumulated in soil. Vegetative portion is converted into organic manures through several processes such as compost. Moreover, organic manures improve the soil physical properties than that of chemical fertilizers. In highly leached pond soils, adequate evaluations should be made of nutrient deficiencies such as nitrogen, phosphorus, potash and calcium. Considerable attention is being given to various combinations of fertilizers and manures for synergistic impact on fish production.

While the use of commerical fertilizers is being encouraged, practical considerations dictate the search for alternative sources, particularly nitrogen, to check water pollution. Different types of compost materials which are produced from plant residues, should be used. Manures of domestic animal origin, will help supplement the commercial fertilizers as animal manures produce an appreciable quantities of fish, eggs, and meat (Table 22.6).

### *Species Combination*

Combination of 4, 5 or 6 species in a pond or tank with suitable management practices has increased fish production. Although the yield of fish depends on feed, water and soil characteristic

**Table 22.6 : Production of Fish, Eggs and Meat Through Use of Different Animal Excreta Under Indian Condition**

| Item | Duck | Cow | Pig | Poultry |
|---|---|---|---|---|
| Amount of cattle manure/ha water/year (tonne) | 10-15 | 15-20 | 15-20 | 10-15 |
| Amount of cattle manure/animal/year/kg | 35-45 | 5,000-10,000 | 500,-600 | 20-25 |
| Number of animals to be reared for providing cattle manure/ha/year | 200-300 | 3-4 | 30-40 | 500-600 |
| Stocking of fish (Number/ha) | 5,000-6,000 | 5,000-6,000 | 5,000-6,000 | 5,000-6,000 |
| Fish production (tonne/ha/year) | 3-4 | 3-4 | 4-7 | 4-5 |
| Animal production (kg) | 500-600 kg meat and 4,000-6,000 eggs | - | 4,000-4,500 kg meat | 1,250 kg meat and 70,000 eggs |

*Source :* Ayyappan et al (1998)

features in a particular region and adequate combination of fish species, per hectare production of fish is highly variable from country to country and even region to region. For examples, combination of Rohu, Catla, Mrigal, and Prawn increased production potential than that of combination of Rohu, Catla and Mrigal. Similarly, combination of exotic (silver, common and grass carps) and Indian carps (rohu, catla and mrigal) significantly increased fish production than that of the combination of either exotic or Indian carps. However, the production of finfish and shellfish in India, China, Bangladesh, and Thailand has doubled in the past 10 years as a result of introducing suitable stocking densities and species combinations. It is quite fortunate that minor carp varieties are adapted to food-deficit countries. Disease problems have been alleviated through the use of quality fish feeds.

### *Water Management*

It is needless to mention that adequate management of water (see Chapter 10) is a critial factor in tropical and sub-tropical aquaculture for sustaining long-term productivity. Though waters of these regions differ from those of temperate regions, some temperate-zone management practices are pertinent for the tropics and sub-tropics. High solar radiation in the tropics and sub-tropics is also important for ecosystem productivity. Note that Table 22.7 shows high production of fish food organisms (bottom fauna) and primary productivity of water in a well-managed fish culture ponds than in unmanaged one that reflects high fish production from managed ponds.

**Table 22.7 : Primary Productivity and Bottom Fauna Levels in Managed and Unmanaged Fish Ponds in the Tropics**

| Pond type | Primary productivity (mg C/m$^3$/h) | Bottom fauna (G/M$^2$) |
|---|---|---|
| Managed | 1,471 | 3,029 |
| Unmanaged | 176 | 424 |

*Source :* Ghosh and Banerjee (1996)

It is always necessary to maintain the water quality criteria where fish are cultivated in ponds under intensive and semi-intensive systems. Different parameters of water such as pH, dissolved oxygen, alkalinity, phosphate and calcium are very important for production of high quality and healthy fish and also fish food organisms. These parameter should be kept at optimum levels. If their levels deteriorate in any way (such as pollution), fish production level will substantially decrease.

### *Human Resources*

Since production of finfish and shellfish through inland fisheries and aquaculture is essential for food supply and the market, a number of factors are responsible that contributes towards the success of fish culture industries. Among different factors (such as manpower, capital, and inputs), human factor is the most significant one, because it is the people who have to use the other resources. Without the productive efforts of the skilled personnels, the utilization of aquaculture resources would not be optimal.

The human resource activities have become increasingly important in today's fast changing world. Several trends such as global competition, educated work force, and technological dynamics are all forcing aquaculture sectors to organize more responsively.

> *Human Resource Planning* : It is a process which involves objective and systematic assessment of skilled personnels of the aquaculture sector, identifying the available personnels to meet the requirement of the current needs, forecasting the production potential and supply fish and fish products and continuous monitoring and evaluating the source of food supply from aquatic resources. Human resource planning is very important in providing some benefits to the fishery and aquaculture sectors. First, it improves the utilization of human resources for significant development of the sector. Second, it makes provisions for replacement of traditional methodologies. Third, it helpes identify the possible sources of fish supply. Fourth, it also helps provide production-oriented improved technologies to fish farmers and fishermen generally.

Trained personnel is the main requisite for increased production. The range required goes from scientists to the fish farmers/fishermen. It is also necessary for researchers/experts whose interest is directly related to the solution of the world food problem. In today's aquacultural and fishery sectors, dynamic economic, political, and technical environment, and the most suitable personnels are becoming more crucial and indispensible assest for effective performance. Field services personnel, technicians, and individuals trained in different aspects of fish culture (such as management, culture methods, production, processing, and marketing) are also necessary. To feed the world population, it is urgently needed to utilize scientific and technological inputs that are related to water science, fisheries, and aquaculture.

## 22.14 Nutritional Role in Fish

The nutritional quality of the protein contained in various species and types of fish and shellfish compares very favourably with that muscle meat of beef, pork, poultry, and mutton. Generally fish protein provides six essential amino acids such as lysine, methionine, arginine, tryptophan, valine, and leusine. This statement clearly points towards the potentialities for increased use of fish.

Fish maintains a unique position as a protein source, being the most potent in this respect whether the protein content is calculated on the basis of 100 calories (Table 22.8). Note that lean fish differ from fatty fish with regard to the content of protein.

**Table 22.8 : Grams of Protein in 100 Calories of Some Foods**

| Food | g/100 Cal |
|---|---|
| Banana | 1.0 |
| Sweet potato | 1.1 |
| Wheat flour | 3.3 |
| Maize | 2.6 |
| Groundnut | 4.7 |
| Soyabean | 11.3 |
| Eggs | 7.9 |
| Cow's milk | 7.9 |
| Fatty fish | 10.2-13.2 |
| Lean fish | 15.0-21.0 |

*Source :* Borgstrom (1962)

Fish flesh also contains an appreciable quantity of lipids. Lipids provide maximum energy, vitamins (such as A, D, E, and K), essential fatty acids (such as mono-and polyunsaturated fatty acids particularly of n-3 type). This gives fish an uncomparable nutritional merit over all other animal meat which are rich in saturated fatty acid. The two n-3 polyunsaturated fatty acids that confer the greatest health benefit to humans such as eicosapentaenoic acid (20 : 5 n-3) and docosahexaenoic acid (22 : 6 n-3) are present in fish lipids. Besides lipids, fish flesh also contains vitamins (such as A, D, E, K and $B_{12}$) and other essential minerals (Table 22.9). Evidences have shown that risk of ischemic heart disease, blood pressure in diabetic persons, and blood viscosity can be reduced by n-3 fatty acids. Furthermore, a diet having high fish oil such as containing n-3 polyunsaturated fatty acids + highly unsaturated fatty acids was found to increase insulin-dependent glucose transport across cellular membranes and lower blood sugar level. Experiments have, however, shown that cultured fishes which are fed with formulated feed containing n-3 and n-6 fatty acids, accumulate these acids in muscle fillet thereby influencing n-3/n-6 balance in a manner which will benefit the consumer in regard to health and nutrition.

## 22.15 Fish Production and World Feeding

Most of the countries long ago realized the importance of fish to human nutrition. The population groups subsist almost entirely on fish and plant products. The nutritive attributes of fish have, consequently, long been well establisehd as empirical knowledge is endorsed by experiments and analyses through sophisticated instruments and laboratories.

The total aquatic production (finfish and shellfish) of inland, brackish and marine water resources (Table 22.10) has been increasing slowly but steadily from 1984 to 1996. In different waters, dominant groups of finfish include Cichlids, Miscellaneous freshwater fish, Eels, Salmonoids, Miscellaneous diadromus fish, Redfishes and Bass, Mullets, Herrings, Flatfishes, Cods, Tunas, Miscellaneous marine fish, Sturgeons and that of shellfish include Shrimps, Prawns,

Freshwater and marine Crustaceans, Crabs and Lobsters. However, freshwater aquaculture has recorded modest increases for the years listed in the table and is approaching 55 per cent of the total production which is mainly due to the rapid expansion of shrimp/fish culture particularly in South and North America, Asia and the Pacific.

**Table 22.9 : Contents of Some Minerals and Vitamin $B_{12}$ in some Common Freshwater Fishes Values are in Terms of mg/100 g Flesh**

| Species | Calcium | Phosphorus | Iron | Vit.$B_{12}$ |
|---|---|---|---|---|
| *Labeo rohita* | 650 | 175 | 1.0 | 3.0 |
| *L. bata* | 790 | 280 | 2.0 | 3.0 |
| *Catla catla* | 530 | 235 | 4.0 | 5.0 |
| *Cirrhinus mrigala* | 350 | 280 | 1.0 | 3.0 |
| *Clarias batrachus* | 430 | 305 | 4.0 | 5.0 |
| *Heteropneustes fossilis* | 670 | 650 | 4.0 | 5.0 |
| *Anabus testudenius* | 410 | 390 | 1.0 | 2.0 |
| *L.calbasu* | 320 | 380 | 1.0 | 3.0 |
| *Notopterous chitala* | 180 | 250 | 3.0 | 4.0 |

*Source* : Gopalan *et al* (1987)

**Table 22.10 : World Production (Million Tonne) of Finfish and Shellfish by Different Environments**

| Source | 1984[1] | 1988[2] | 1992[3] | 1996[4] | 2010* |
|---|---|---|---|---|---|
| Inland Waters | | | | | |
| Total finfish | 3.8 | 6.2 | 7.7 | 8.9 | 15.5 |
| Total shellfish | 0.03 | 0.05 | 0.9 | 1.7 | 5.0 |
| Brackish waters | | | | | |
| Total finfish | 0.3 | 0.4 | 0.7 | 1.3 | 5.5 |
| Total shellfish | 0.2 | 0.5 | 0.9 | 1.7 | 5.0 |
| Marine | | | | | |
| Total finfish | 0.3 | 0.5 | 0.6 | 1.2 | 4.8 |
| Total shellfish | 0.02 | 0.04 | 0.08 | 0.7 | 3.5 |
| Freshwater aquaculture | 4.1 | 6.6 | 11.6 | 16.8 | 25.0 |

*Sources* : 1,2 FAO (1990); 3,4 Compiled from various published data.
* Anticipated

Considerable improvement was registered in Asia and the Pacific in seed supply of freshwater fish species and also of cultured freshwater prawn and marine shrimp species. Large-, medium- and small-scale fish and crustacean hatcheries emerged in high numbers in several countries of Asia, contributing significantly to the availability of fish, shrimp and prawn seed at reasonable prices.

In Asian and South-east Asian countries, increased production has been recorded for lakes and reservoirs practising culture-based fisheries. Fish production predictive models have been/are being evolved which will allow better approximations to be made about the carrying capacities of lakes and reservoirs in these regions. However, the daming of rivers and uncontrolled pollution remain the major constraints for increasing fish production from capture fisheries. Projects aimed at rehabilitation of fish stocks and fisheries in natural freshwater bodies are on the increase throughout the region and more attention should be given to maintain the fishery sustainability.

So far as the world population is concerned, it is evident that the mean yearly consumption of fish amounts to 16 kg per caput. Moreover, more than 6 kg per caput are utilized as food via meal or as fertilizers. In better-fed countries where there is an enlargement of fish consumption (30 kg per caput in Europe), no nutritional deficiencies have been detected. But in low income food deficit countries, lack of high quality protein constitutes one of the major nutritional problems. Therefore, fish constitute an easily available source of cheap protein and consequently, can be placed in both the national and the global diets.

## 22.16 Conclusion

The world population explosion presents mankind with one of its most complex challenges. This challenge is greatest in the countries of the world where population explosion is severe. In developing countries, the challenge is most severe. Adequate management and improvement of freshwaters is one of the key to meeting the challenge. Fish culture management systems must be developed and adapted to increase fish production and at the same time prevent water quality deterioration. Fishery scientists will play an important role in developing different systems.

The struggle to produce finfish and shellfish is not yet lost; of course, it requires extensive research and development efforts not yet realized. Fishery scientists should acquire knowledge of the potential and features of uncultivated water areas. Such knowledge is very essential because it will assure new freshwater areas to bring into aquaculture production to meet the world's fish demands.

Global interest in the production of fish and fishery products as food through aquaculture has already been stimulated by many countries by adopting new technology aimed at rising of rural living standards, constant supply of edible fish and fish trade at national/global levels. For this purpose fishermen/fish farmers will have to produce as more fish as possible, not only to satisfy the need of animal protein for large population but also to satisfy the demands.

The capacity of freshwater resources throughout the world to produce adequate fish for current and future generations is very high. This high capacity will principally depends on the knowledge that is used in managing waters for fish cultivation. This knowledge will be based on the active involvement of both fishery scientists and fish farmers around the world.

### *References*

Ayyappan, S., K. Kumar, and J.K. Jena. 1998. Integrated fish farming -- Practices and potentails. *Fish. Chim.* 18 (1) : 15-18.

Borgstrom, G. 1962. *Fish As Food*. Vol. II. Academic Press, New York.

FAO (Food and Agriculture Organization). 1990. Review of the State of World Fishery Resources. FAO Fisheries Circular No. 710.

FAO (Food and Agriculture Organization). 1997. The State of World Fisheries and Aquaculture. Food and Agriculture Organization of the United Nations.

Ghosh, M.K. and S. Banerjee. 1996. Macrobenthos of two tropical freshwater carp culture ponds under variable management techniques. *In* : *Assessment of Water Pollution* (*Ed* : S.R. Mishra), APH Publishing Corporation, New Delhi, India, pp. 253-269.

Gopalan, C., B.V. Rama Sastri, and N. Balsubramanian. 1987. *Nutritive value of Indian foods.* National Institute of Nutrition, Hyderabad.

Mortimer, C.H. and C.F. Hickling. 1954. *Fertilizers in Fish Pond.* Fishery Publications, London, 5 : 156 pp.

Shyam, R. 1998. Status of fisheries in India. *In* : *Advances in Fisheries and Fish Production* (*Ed* : S.H. Ahmad), Hindustan Publishing Corporation, New Delhi, pp. 6-21.

## Questions

1. Why fish production problems are severe in many tropical countries?
2. What are the methods for augmenting fish production?
3. Vast areas of potential but unculturable water bodies exist in many tropical countries. What are the reasons against and for the conversion of these water areas to use for fish culture?
4. What is the role of fishery experts in helping the world to achieve its fish production goals?
5. What are the requisites for increased production of fish in the tropics and sub-tropics?
6. Why fishes are considered as a unique source of protein than muscles? Explain.

# GLOSSARY

Abiotic. Pertaining to the absence of living organisms.

Abyssal. Characteristic feature of a deep water lakes and oceans. It begins at about 2,000 meters from the surface.

Acidity. The quantity and kinds of compounds present in a lake/pond that collectively shift the pH to the acidic side of neutrality.

Acid rain. Atmospheric precipitation with pH values less than about 5.6, the acidity being due to inorganic acids such as sulphuric and nitric that are formed when oxides of nitrogen and sulfur are emitted into the atmosphere.

Acid soil. A soil with a pH value less than 7.

Acid-sulfate soil. Pyrite-containing soil that oxidizes and form sulphuric acid.

Actinomycetes. A group of mycelial, gram-positive organisms intermediate between the fungi and the bacteria. They are mostly saprophytes, some are parasitic.

Activated-sludge process. The use of biologically active sewage sludge to stimulate the breakdown of organic matter in raw sewage during secondary treatment.

Adsorption. The attraction of compounds or ions to the surface of a solid. Soil colloids adsorb large amounts of ions and water.

Aerobe: An organism that requires oxygen for growth and can grow under an air atmosphere (20 per cent oxygen).

Aflatoxin. The toxin produced by some strains of the fungus *Aspergillus*.

Agar. A dried polysaccharide extract of red algae used as a solidifying agent in microbiological media.

Agglutination. The clumping together of bacteria (or Red Blood Corpuscles) as a result of the action of antibodies.

Aggregate (soil). Soil particles held in a single mass or cluster such as clog, scrumb, or block.

Algae. Any member of heterogeneous group of eucaryotic, photosynthetic, an unicellular or multicellular organisms.

Alkalinity. The quantity and kinds of compounds present in a lake/pond that collectively shift the pH to the alkaline side of neutrality.

Alkaline soil. A soil with a pH value more than 7.

Alum. A double salt, generally of alkali metal and aluminium sulfate and have the empirical formula, $M.Al\ (SO_4)_2.12\ H_2O$, where M is the alkali metal.

Amendment (soil). Any substance other than fertilizers such as lime gypsum, etc., used to alter the physical or chemical properties of a soil, to make it more productive.

Amino acid. An organic compound containing both amino (-$NH_2$) and carboxyl (-COOH) groups. Amino acids are the building stones of proteins.

Ammonification. The decomposition of organic nitrogen compounds. For example, proteins, by micro-oganisms with the release of ammonia.

Ammonium phosphate. A mineral compound containing ammonium and phosphate ions. It is used as a good fertilizer.

Ammonium sulfate. A nitrogenous chemical compound containing two -$NH_4$ groups attached with sulfate radical. It is used as an important fertilizer.

Anaerobe. An organism that does not use oxygen to obtain energy. Cannot grow under an air atmosphere because oxygen is toxic.

Anion. A negatively charged ion, during electrolysis it is attacted to the positively charged cathode.

Anion exchange capacity. The sum total of exchangeable anions that a soil can adsorb.

Antagonism. The injury, inhibition, or killing of growth of one species by another when one animal adversely affects the environment of other.

Antibiotic. A substance of microbial origin that has antimicrobial activity in very small amounts.

Anti-nutritional factor. A number of feed ingredients which are responsible for the deleterious effects to fish health.

Antitoxidants. Chemical substances capable of neutralizing other chemicals which are toxic.

Apatite. A group of hexagonal materials consisting of calcium phosphate together with chlorine, hydroxyl, flourine or carbonate in varying amounts. The apatite mineral occurs as accessory minerals in igneous and metamorphic rocks. It is the chief constituent of phosphate rock.

Aquaculture. A population of aquatic animals and plants of commercial importance cultivated in aquatic and productive environments on scientific basis.

Aquazyn. A blended concentrations of selected and cultured bacterial formulations grown on a cereal and mineral substrate along with enzymes and buffers.

Assimilation. Changing of dissolved food and other essential materials from water to aquatic plant body.

Autotroph. A plant/micro-organism that use inorganic materials as a source of nutrients; carbon dioxide is the chief source of carbon.

Bacteria. Unicellular plants which are usually colorless, a few have chlorophyll and have great economic and environmental importance.

Bactericide. An agent that destroy bacteria.

Balance diet. A diet contains all the items of food such as carbohydrates, fats, proteins, vitamins and minerals in such quantities that can provide necessary energy and maintain normal growth as well as proper functioning of the body of animals.

Benthos. A collective term for the organisms living along the bottom of any aquatic ecosystem.

Biochemical oxygen demand (BOD). A measure of the amount of oxygen consumed in biological processes that breakdown organic matter in water. A measure of the organic pollutant load.

Biodegradation. Subject to decomposition of any compound by micro-organisms.

Biofertilizers. Fertilizers of biological origin which serve as manure such as Azolla.

Biogeochemical agents. Organisms that mineralize organic carbon, nitrogen, sulfur, phosphorus etc.

Biogeochemical cycle. The cycle of chemical constituents through a biological system.

Biological nitrogen fixation. The symbiotic bacteria, *Rhizobium* sp. ar associated with roots of leguminous plants. The bacteria convert atmospheric nitrogen to ammonia while the legume supplies the bacteria with carbohydrates. Free-living bacteria that can fix nitrogen includes members of the genera *Azotobactor, Clostridium, Anabaena* etc.

Biomass. The mass of living matter present in a specified area.

Biota. The animal, plant, and microbial life–a characteristic feature in a given area.

Biotechnology. A developmental strategy by which genes are manipulated to have novel products/ varieties for use in fish culture technology for high production.

Biosphere. The part of the earth in which life exists.

Bloom. A colored area on the surface of a body of water caused by heavy growth of phytoplankton.

Budget. An estimate of expence and income for a specific period of time to have maximum production from any farming unit.

Bundh breeding. Bundhs are special type of seasonal and perennial ponds or impoundments where riverine conditions are created for breeding of carps. Bundhs are of two types : Dry bundh and Wet bundh.

Calcium carbonate. A solid occurring in nature chiefly as mineral calcite and argonites.

Calori. A unit of heat; the amount of heat required to raise the temperature of 1g of water by 1°C.

Carbohydrates. An organic compound composed of a chain of carbon atoms to which H and O are attached in the ratio of 2 : 1. They are essential part of living matter. Upon breakdown, carbohydrate substances generate energy.

Carbon cycle. The sequence of transformations whereby carbon dioxide is fixed in living organisms by photosynthesis or by chemosynthesis, liberates by respiration and by the death and decomposition of the fixing organism, used by heterotrophic species, and ultimately returned to its original state.

Carbon-nitrogen ratio. The ratio of the weight of organic carbon (C) to the weight of total nitrogen (N) in a soil or in organic material.

Carotenoid. A water-insoluble pigment, usually yellow, orange, or red, which consists of a long aliphatic polyene chain composed of isoprene units.

Cation. A positively charged ion, during electrolysis it is attracted to the negatively charged anode.

Cation exchange capacity. The sum total of exachangeable cations that a soil can adsorb. Sometimes called base-exchange capacity, total-exchange capacity, or cation-adsorption capacity.

Cell. The microscopic, functionally and structurally basic unit of all living organisms.

Chelate. A chemical compound in which a metallic ion is firmly combined with an organic molecue by means of multiple chemical bonds.

Chelating agent. An organic compound in which atoms form more than one coordinate bond with metals, keeping them in solution.

Chemosynthesis. The process of conversion of inorganic molecules into food by use of energy released through different chemical reactions.

Chlorophyll. A light-trapping green pigment essential as an electron donor in photosynthesis.

Chlortech. A very powerful liquid safe versatile disinfectant.

Classification. The systematic arrangement of units (such as organisms, soils etc.) into groups and often further arrangement of those groups into larger groups.

Clay. A detrital mineral particle of any composition having a diameter less than 1/256 mm (4 microns). The mineral particle shows colloidal properties.

Commensalism. An association between two species which require the same food and which may benefit other species and is harmful to neither.

Commercial fertilizer. Any inorganic salt which is used for acceleration of high production in ecosystem by supplying nutrient.

Competition. The interaction of two or more species or between individuals of a single species in which a required resource is in limited supply and as a result one or both of the competitors suffer in their growth or survival.

Compost. An organic residue or a mixture of organic residue that have been piled, moistened and allowed to undergo biological decomposition.

Computer. A sensitive and information processing machine, a tool for storing, manipulating, and correlating data.

Consumer. In an ecosystem those which obtain energy from producer (primary consumers) or from other consumers (secondary consumers).

Culture. A population of organisms cultivated in a medium.

Decomposer. Organism that utilizes dead plant or animal materials and food and releases the component element to the environment, thus contributing to circulation of these elements in the ecosystem.

Decomposition. The process of bacterial breakdown of plant and animal materials by which chemical reaction (hydrolysis, oxidation, carbonation, ion exchange etc.) transform minerals into new chemical combinations that are further utilized by plants.

Demineralization. The process by which acid produced by bacteria dissolved the salts.

Denitrification. The reduction of nitrates to nitrogen gas.

De-odorase. A product extraced from the plant of the genus *Yucca* that contains certain glycocomponents (such as gitigenin, sapogenin, chlorogenin etc.) and has the capability of binding with ammonia and hydrogen sulphide – developed in fish culture ponds. Consequently, these obnoxious gases are removed from pond water.

Detergent. A synthetic cleansing material containing surface-active agents which do not precipitate in hard water.

Anionic. Detergents which ionize with the detergent property resident in the anion.

Cationic. Detergents which ionize with the detergent property resident in the cation.

Non-ionic. Detergents which do not ionize.

Detoxification. A process by which toxic action (or toxicity) of any compound in the environment or ecosystem is partially or completely eliminated.

Detritus. Dead organic material in particulate or solid form.

Detritus food chain. A food chain that begins with dead organic matter into mico-organisms and then to organisms feeding on detritus.

Diatoms. The microscopic unicellular or colonial algae of the class Bacillariophyceae having siliceous cell walls that persist as a skeleton after death. They found abundantly in salt and fresh waters.

Diet. The habitual food consumption of an individual. A fish diet should contain the several varieties of nutrients essential for growth and survival. Nutrients must be supplied in the correct relative proportions, otherwise various nutritional disorders may develop.

Diffusion. The spreading-out of molecules or ions into a vacuum, fluid, or porous medium, in a direction tending to equalize concentrations in all parts of the system.

Disinfectant. An agent that are free from infection by destroying the vegetative cells of micro-organisms.

Dissimilation. Chemical reactions that relaease energy by the breakdown of nutrients.

Dolomite. A common rock-forming mineral, Ca.Mg $(CO_3)_2$. Part of the magnesium may be replaced by ferrous iron. It is white to light-colored and has perfect rohombohedral cleavage.

Drainage. The frequency and duration of periods when the water (with or without hazardous substances) is removed by surface flow.

Earthworms. Animals of the Annelida phylum that burrow into and live in the soil. They improve soil fertility and several bottom-dwelling fishes consume earthworms as food.

Ecological pyramid. A graphical representation of trophic structure and function at successive trophic level in the form of pyramid.

Ecology. The study of the inter-relationships that exist between organisms and their environment.

Ecosystem. An organizational unit of biosphere which inlcudes living organisms and non-living substances interacting to produce an exchange of materials between living and non-living part.

Effectiveness. Ability to determine correct objectives for doing the correct things.

Efficiency. Refers to the ability to get things done correctly.

Effluent. The liquid waste of sewage and industrial processing.

Endogenous metabolism. Metabolism that takes place within the cell.

Environment. The surroundings of an organism, including both the non-living world and the other organisms inhabiting the area.

Environmental stress. An environmental factor, such as temperature, water availability of acidity, which attains a level close to the tolerance limit of a species and hence causes problems for its continue survival.

Enzyme. An organic catalyst produced by an organism.

Equilibrium. A static condition of a chemical reaction in which nothing is changed.

Estuary. A semi-enclosed coastal body of water which opens to the sea.

Eutrophication. An aging process in ponds or lakes, during which the water becomes overly rich in dissolved nutrients. This results in the excessive development of algae and other microscopic plants, causing a decline in the levels of dissolved oxygen.

Evaporation. Gradual loss of water from soil surface, pond, lakes, streams, and oceans.

Exchangeable ions. Ions which are associated with permanent negative charges on the soil colloids.

Farm Yard Manure (FYM). A material which is produced from the decomposition of the mixture of animal excreta and urine.

Fecundity (fish). Potential capability to reproduce reproductive units such as sperms and eggs.

Feldspar. A group of abundant rock-forming minerals which consitute 60 per cent of the earth's crust. These are white and gray to pink and have a hardness of 6.

Filter. A special device through which micro-organisms, helminth eggs, cysts, etc.cannot pass.

Fishery. A vast domain which includes the exploitation of natural resources of water (such as fishes, prawns, crabs, oysters, snails etc.) for the consumption and benefit of human beings.

Floc. An aggregate of the finely suspended and colloidal matter of sewage.

Floccule. An adherent aggregate of micro-organisms or other materials floating in or on a liquid.

Floodplain. The land bordering a stream, built up of sediments from overflow of the stream and subject to inundation when the stream is a flood stage.

Food chain. Series of organisms through which food energy moves before it is completely spent; the pathway of chemical energy within a community from producers to consumers and then to decomposers.

Food web. The complex feeding interactions between species in a community.

Fungi. Simple plants that lack a photosynthetic pigment. The individual cells may be linked together in long filaments called *hyphae*, which may grow together to form a visible body.

Fungicide. An agent that kills fungi.

Genus. A group of very closely related species.

Germicide. An agent capable of killing germs, usually pathogenic micro-organisms.

Glucan. A polymer of glucose.

Gonadotrophin. Hormones are collectively glycoproteins, secreted by the pituitary gland of vertebrates and mammalian placenta which control activity of gonads responsibe for the onset of sexual maturity and for breeding season rhythms in fishes.

Grazing food chain. A food chain characterized by autotrophs (phytoplankton and algae), grazing animal (zooplankton and fish), and their predators.

Green manure. Plant materials incorporated with the soil while green, or soon after maturity, for improving the soil.

Groundwater. All sub-surface waters, particularly that occurring in the zone of saturation.

Gypsum. A widely distributed mineral consisting of hydrous calcium sulfate and is frequently associated with halite and an hydrite forming thick, extensive beds, especially in rocks of Permian and Triassic ages.

Habitat. The natural environment of an organism.

Half-life. The time period in which half the initial number of atoms of a radioactive element disintegrate into atoms of the element into which they change directly.

Hatchery. A large-scale production plant through which hatchlings/fry are produced from the fertilized eggs.

Heavy metals. Metals with particle densities more than 5 $mg/m^3$.

Herbicide. A chemical that kills plants or inhibits their growth; intended for weed control.

Herbivore. A plant-eating animal.

Heterotroph. An organism capable of driving energy for life processes only from decomposition of organic compounds and not capable of using inorganic compounds as sole sources of energy or for organic synthesis.

Hormones. Organic substances produced in minute quantity in one part of an organism and transported to other parts where it exerts a profound effect.

Humic acid. Black acidic organic matter extracted from soil. It is insouble in acids and organic solvents.

Humus. More or less stable fraction of the soil organic matter remaining after the major portions of plant and animal residues have decomposed. It is dark in color.

Hydrogen ion concentration (pH). Th negative logarithm of the hydrogen ion concentration of a soil or water. The degree of acidity (or alkalinity) of a soil or water as determined by means of a glass or other suitable electrode or indicator at a specific moisture content of soil/water ratio, and expressed in terms of the pH scale.

Hydrologic cycle. The complete cycle through which water passes from oceans, through the atmosphere, to the land, and back to the oceans.

Hypersalinity. Extensively saline, with a salinity substantially greater than that of normal aquatic ecosystem.

Hypophysation. A technique by which sex hormones are administered into the body of fish through injection for successful spawning in confined waters.

Hyposalinity. Having a salinity below the lowest level.

Igneous rock. A group of the same general type of occurrence (plutonic or volcanic) having in common certain mineralogical features and exhibiting a continuous variation from one extremity to the other.

Immobilization. Transformation of inorganic nitrogen to organic form of nitrogen.

Impoundment. The process of forming a lake or pond by a dam, dike, or other barrier.

Indicator. A substance that alters color as conditions changes such as pH indicator (litmus paper) reflect changes in acidity or alkalinity.

Infection. A pathological condition due to the growth of micro-organisms in a host (such as fish).

Infrastructure. Different forms of strategies involved in the development of fish culture activities.

Inland fishery. Impounding water areas such as lakes, reservoirs, ponds, etc. of both fresh and brackish water where aquaculture is practiced.

Insecticide. A chemical that kills insects; intended for insect control.

Integrated fish farming. Association of two or more farming components which become part of the entire system is referred to as *integrated farming* and when fish is included as a component, it is regarded as integrated fish farming.

Iodophores. Organic compounds of iodine.

Laterite soil A iron-rich subsoil layer found in some weathered tropical soils that, when exposed and allowed to dry, becomes hard and will not soften when rewetted.

Leaching. The downward movement of soluble minerals or nutrients from the soil through percolating water.

Lentic ecosystem. An ecosystem in which water always remains in closed area and it means "still water".

Lime. In chemical term, calcium oxide. In practical term, a material containing the carbonates, oxides and/or hydroxides of calcium and/or magnesium used to netralized soil acidity.

Lime requirement. The quantity of agricultural limestone, or the equivalent of other specified liming material, required to increase the pH of the soil to a desired value under field condition.

Limestone. A sedimentary rock composed primarily of calcium carbonate. If dolomite is present in adequate amounts, it is called as *dolomitic limestone.*

Limnetic zone. Zone away from the littoral zone and extends upto the depth where the rates of respiration and photosynthesis are equal.

Limnology. The study of the chemical, physical, geological, and biological aspects of lakes, ponds, and streams.

Lipids. Lipids are fatty acids, neutral fats, waxes, steroids, and phospholipids. They are greasy and hydrophobic.

Litmus. A lichen extract used as an indicator for pH and oxidation or reduction.

Littoral zone. A shallow water zone of a large lake or sea with light penetration to the bottom and often occupied with rooted aquatic plants.

Loam. A textural class name for soil having a moderate amount of sand, silt, and clay.

Lysis. The disruption or disintegration of bacteria or red blood corpuscles by the action of specific substances.

Macronutrients. Minerals that are required in relatively large amounts for the healthy growth of plants and animals such as Ca, N, P, Mg, H, O, C, K, and S.

Management (fishery). The sum total of all fish culture operations, liming, manuring, fertilization, pond preparation, netting, disease control, etc. conducted in a fish culture ecosystem for the production of fish.

Metabolism. The total of chemical process occurring within an organism or biological system.

Metamorphic rock. Any rock derived from pre-existing rocks by chemical, mineralogical, or structural changes, esentially in the solid state, in response to marked change in temperature at depth in the earth's crust.

Microbe. Any microscopic organism.

Micro-organisms. Any organisms of microscopic dimensions and are associated with the health of animals.

Mineralization. The process by which a mineral or minerals are introduced into a rock, resulting in a potential ore deposit. A process in which the organic components are replaced by inorganic mineral.

Micronutrient. A chemical element necessary in only extremely small amounts for the growth of plants and animals such as copper, iron, zinc, manganese etc.

Mixed fertilizer. Fertilizer that has been prepared by combining two or three single fertilizers of different classes such as single superphosphate and ammonium sulfate may be mixed to get a mixed fertilizer.

Mortality. Refers to death of living forms.

Mutualism. A symbiosis in which two or more organisms living together benefit each other.

Nectoplankton. Active small planktons that moves great distance by water currents.

Nekton. Animals such as fish, which can control their position by swimming.

Nutrient. A chemical substance that contributes to the growth of an organism.

Neuston. Organisms live on water surface.

Neutralization value. Refers to the relative power of live substances to neutralize acidity.

Nitrate reduction. The reduction of nitrates to nitrites and ammonia.

Nitrification. The transformation of ammonia nitrogen to nitrates.

Nitrogen cycle. A term for a biogeochemical cycle involving the element nitrogen. Nitrogen gas is used/fixed by some species of bacteria and blue-green algae. These organisms convert nitrogen gas into ammonia, which they use to make nitrogenous compounds.

Nitrogen fixation. The formation of ammonia from free atmospheric nitrogen.

Nitrogen solution. A nitrogenous solution – prepared by dissolving ammonium nirate and uiea in water that contain about 30 per cent nitrogen.

Oil cake. The residue left after extraction of oil from oil-seeds (such as groundnut, neem, mustard, etc.).

Organic manure. A nutrient carrier obtained from plant or animal sources and contains very small amounts of nitrogen, phosphorus, and potash.

Osmoregulation. Maintenance of internal osmolarity with respect to environment by an animal. It is termed as the balance of water and salts within the body of an animal.

Oxidation. The process of combining with oxygen.

Ozone (gas). An allotrophic forms of oxygen. It is unstable, highly toxic, and blue oxidizing gas with a pungent odor.

Parasite. An organism that derives its nourishment from a living plant or animal hosts. A parasite does not necessarily cause disease.

Parasitism. The relationship of a parasite to its host.

Pathogen. An organism capable of procuding disease.

Pesticide. A chemical by virtue of its toxicity is used to kill pest organisms. A term of wide application which includes insecticides, fungicides, herbicides, etc.

Phosphate rock. Any rock that contains one or more phosphatic materials, especially apatite, of sufficient quantity and purity to permit its commercial use as a source of phosphatic compounds or elemental phosphorus.

Phytoplankton. Microscopic plant community in marine and freshwater ecosystems which floats free in the water and contains many species of diatoms and algae.

Plankton. A collective term for the passively floating or drifting flora and fauna of a body of water, consisting largely of microscopic organisms.

Planning. Advance selection of a course of action.

Pollution. The disruption of a natural ecosystem as a result of human contamination.

Polyculture. A biotic community dominated by several species such as cultivation of 4 or 6 species of commercially important species of fish.

Polysaccharide. A carbohydrate formed by the combination of many molecules of monosaccharides; examples are starch, glycogen, and cellulose.

Population. A collection of individuals of the same species.

Primary treatment. The first stage in wastewater treatment in which settleable or floating solids are mechanically removed by sedimentation and screening.

Producer. The photosynthetic plants in any ecosystem.

Productivity. The capacity of an aquatic ecosystem for producing fish and other aquatic species to a large extent for human use under a specified system of management. In fish culture system, productivity emphasizes the efficiency of water to produce especially finfish and shellfish and expressed in terms of yields.

Profundal zone. Zone that extends beyond the depth of effective light penetration.

Protein. One of a class of complex organic nitrogenous compound composed of an extremely large number of amino acids joined by peptide bonds. Proteins are essential parts of living matter.

Protozoa. An unicellular, parasitic or free-living, heterotrophic and often motile eucaryote.

Radioactivity. The emission of energetic particles, and/or radiation during radioactive decay.

Radioisotope. An isotope of a radioactive substance that is unstable and disintegrates, emitting radiation energy.

Reduction. A chemical process involving the removal of oxygen.

Repressive factor. A hormone-like substance that retards spawning of fish.

Resource. The basic requirements of an organism, including water, food, energy, and minerals.

Resource management. The informed manipulation of an ecosystem by man in order to gain some products from the ecosystem.

Salinity. The total salt content of a volume of sea or freshwater, defined in a complex and precise way.

Sedimentary rock. A layered rock resulting from the consolidation of sediment.

Septic tank. A unit using an anaerobic system for the treatment of a limited volume of sewage.

Sewage. Liquid or solid refuse (industrial and domestic) carried off in sewers.

Sewage sludge. Settled sewage solids combined with varying amounts of water and dissolved materials, removed from sewage by screening, sedimentation, chemical precipitation, or bacterial digestion.

Silt. A soil separate consisting of particles between 0.05 and 0.002 mm in equivalent diameter.

Single cell protein. Micro-organisms cultivated on industrial wastes or by-products as nutrients to yield a large cell crop rich in protein.

Sludge. The semisolid part of sewge that has been sedimented or acted upon by bacteria.

Slurry. A mixture that contains adequate amounts of organic manurs, fertilizer, and oil cake used to fertilize the soil and water for productivity.

Soil. A dynamic natural body composed of mineral and organic materials, weathered substrate, and living forms in which plants grow.

Soil conditioner. Organic manures which help to maintain and improve physical properties of pond/ lake soil.

Soil fertility. The amount of a soil that enables it to provide essential chemical elements in quantities and proportions for the production of plants.

Soil organic matter. The organic fraction of the soil that includes animal and plant residues at various stages of decomposition, cells and tissues of soil organisms, and substances synthesized by the soil population.

Soil structure. The arrangement of primary soil particles into secondary particles or units. These secondary units may be arranged in the profile in such a way as to give a distinctive characteristic pattern. The secondary units are characterized and classified on the basis of shape, size, and degree of distinctness into classes, types, and grades, respectively.

Soil texture. The relative proportions of the various soil separates in a soil.

Spawning. The production of eggs and sperms in those animals, such as most fish species, where fertilization is external and eggs and sperms are just released into the water.

Stakeholders. Different co-operative societies, private sector, fishermen, fish culturists, technicians, etc. who mark the water bodies for fish culture with stakes to prove ownership.

Steroid. A complex substance containing the tetracyclic carbon ring system of the sterols; used as therapeutic agents.

Strategy. Mode of behavior adopted by an individual when it could, either actually or theoretically, behave in a dfferent way.

Suphala. The trade name of nitrophosphate that contains 18-20 per cent N, 15-20 per cent $P_2O_5$, and 2-15 per cent $K_2O$.

Sympathetic breeding. A technique where a few pairs of brood fish are injected and kept them with other non-injected pairs. Consequently, the latter type of fish starts breeding along with the injected one.

Synergism. The ability of two more organisms to bring about changes (generaly chemical) that neither can accomplish alone.

Synergistic. The action of different agencies that together have a greater influence than the sum of their independent effect.

Taxonomy. The classification, nomenclature, and identification of plants, animals, chemicals, and soils.

Threshold. The level of a stimulus required to produce an effect.

Tillage. The mechanical manipulation of soil for myriad purpose; but in fish culture, it is restricted to the modifying of soil physical condition for overall productivity of ecosystem.

Total acidity. Active, residual, and exchangeabe acidity together called the total acidity of a mud.

Toxicity. Ability to poison, or to interfere adversely with vital processes of the organism by physical and chemical means.

Trace elements. Chemical elements used by organisms in very small amounts and essential to physiology.

Trickling filter. A secondary treatment process in which sewage is trickled over a bed of rocks so that bacteria can breakdown organic wastes.

Trophic interactions. Organisms interact with one another through feeding relationship.

Trophic level. The feeding status of an organism, such as plant, herbivore, carnivore, and top carnivore.

Trophogenic zone. Includes littoral and sub-littoral zone and is distinguished by abundant growth of plants.

Tropholytic zone. Upper part of profundal zone and is characterized by the absence of aquatic plants.

Turbidity. A condition in water or that reduces the transparency.

Ultraviolet rays. Radiations from about 150 to 2000 Å.

Upwelling. The rise of water from a deeper to a shallower depth in an ocean/river/lake.

Urea. A water soluble nitrogenous compound, $H_2N\text{-}CO\text{-}NH_2$, and used as an important fertilizer.

Vermi-compost. A solid organic waste material prepared from agricultural washes with the aid of some species of earthworms.

Virus. An obligate intra-cellular parasitic micro-organisms that is smaller than bacteria.

Vitamins. Organic chemical compounds that are essential in small quantities for metabolism. Most of them seem to act as catalyst for essential chemical changes in the body, each one influencing a number of vital processes.

Volatilization. Gaseous loss of compounds to atmosphere.

Wetlands. The flood plains of different river systems, oxbow lakes, meander scroll depressions, and tectonic depressions which are dominated by aquatic flora and fauna.

Zonation. The distribution of organisms in zones particularly a stratification of certain kinds of algae at certain depths and locations in the pond/lake/ocean.

Zoogloeal masses. Masses composed of micro-organisms which are embedded in a common matrix of slime.

Zooplankton. Microscopic animal community in marine and freshwater ecosystems which floats free in the water and moving passively with the currents.

# INDEX

## C



## D